Birkhäuser

Birkhäuser Advanced Texts

For further volumes:
http://www.springer.com/series/4842

Pavel Drábek • Jaroslav Milota

Methods of Nonlinear Analysis

Applications to Differential Equations

Second Edition

Birkhäuser

Pavel Drábek
Faculty of Applied Sciences
Department of Mathematics
University of West Bohemia
Plzeň
Czech Republic

Jaroslav Milota
Faculty of Information Technology
Department of Applied Mathematics
Czech Technical University
Praha 6
Czech Republic

ISBN 978-3-0348-0386-1 ISBN 978-3-0348-0387-8 (eBook)
DOI 10.1007/978-3-0348-0387-8
Springer Basel Heidelberg New York Dordrecht London

Library of Congress Control Number: 2012955049

Printed on acid-free paper

Springer Basel is part of Springer Science+Business Media (www.birkhauser-science.com)

Contents

Preface

In this second edition we have made minor revisions and added new material to Chapter 8 and partly also to Chapters 2, 3 and 6. We quote from the Preface to the first edition in order to clarify the original motivation behind publication of this book:

"There are many books and monographs devoted to the methods of nonlinear analysis and their applications. Typically, such a book is either dedicated to a particular topic and treats details which are difficult to understand for a student, or it deals with an application to complicated nonlinear partial differential equations in which a lot of technicalities are involved. In both cases it is very difficult for a student to get oriented in this kind of material and to pick up the ideas underlying the main tools for treating the problems in question. The purpose of this book is to describe the basic methods of nonlinear analysis and to illustrate them on simple examples. Our aim is to motivate each method considered, to explain it in a general form but in the simplest possible abstract framework, and finally, to show its application (typically to boundary value problems for elementary ordinary or partial differential equations). To keep the text free of technical details and make it accessible also to beginners we did not formulate some key assertions and illustrative examples in the most general form.

In order to make the text self-contained, we decided to comment on several notions and statements in footnotes. To place the material from the footnotes in the text could disturb a more advanced reader and make the exposition more complicated. In order to emphasize the role of the statements in our exposition we identify them as Theorem, Proposition, Lemma and Corollary. However, the reader should be aware of the fact that this by no means expresses the importance of the statement within the whole of mathematics. So, several times, we call important theorems Propositions, Lemmas or Corollaries.

Although the book should primarily serve as a textbook for students on the graduate level, it can be a suitable source for scientists and engineers who have need of modern methods of nonlinear analysis.

At this point we would like to include a few words about our good friend, colleague and mentor Svatopluk Fučík to whom we dedicate this book. His work in the field of nonlinear analysis is well recognized and although he died in 1979 at the age of 34, he ranks among the most important and gifted Czech mathematicians of the 20th century.

We would like to thank Marie Benediktová and Jiří Benedikt for an excellent typesetting of this book in LaTeX 2_{ε}, excellent figures and illustrations as well as for their valuable comments which improved the quality of the text. Our thanks belong also to Eva Fašangová, Gabriela Holubová, Eva Kaspříková and Petr Stehlík for their careful reading of the manuscript and useful comments which have decreased the number of our mistakes and made this text more readable. Our special thanks belong to Jiří Jarník for correction of our English, Ralph Chill and Herbert Leinfelder for their improvements of the text and methodological advice."

The exposition of the material in the second edition is at two levels, visually differentiated by different font sizes. The basic material is contained in the body of the first seven chapters. The more advanced material is contained in appendices to a number of sections and is presented in a smaller font size. The last chapter contains selected special material from the theory of partial differential equations. The basic material is independent of the more advanced material, is self-contained, and can be read by students new to the subject. It should prepare an undergraduate student in mathematics to read scientific papers in nonlinear analysis and to understand applications of the methods presented to more complex problems. Each chapter contains a number of exercises which should provoke reader's creativity and help develop his or her own style of approaching problems. However, the exercises play an additional role. They carry some of the technical material that was omitted in simplifying some of the basic proofs. They are thus an organic part of the exposition for graduate students who already have experience with the methods of nonlinear analysis and are interested in generalizations.

In the second edition we organize the material differently than in the first edition. We dedicate an extra chapter to monotonicity methods, where we newly include the elements of maximum principles both for ordinary differential equations and partial differential equations. Most of the material about boundary value problems for partial differential equations is now subsumed into the body of Chapters 5–7 as an immediate application of the methods presented there. The last Chapter 8 contains technically more involved material which was not included in the first edition. It deals with semigroups, semilinear evolution partial differential equations, application of fixed point methods to solve some quasilinear partial differential equations as well as some partial differential equations on manifolds. In particular, to prepare the reader for subsections on evolution partial differential equations we extend some parts of Chapters 2 and 3.

We would like to thank Soňa Königsmarková for the typing the new text in the second edition and to Jiří Benedikt for creating new illustrations. Our special thanks belong to the editorial staff of Birkhäuser, in particular to Sylvia Lotrovsky and Thomas Hempfling, for the agreeable collaboration.

Plzeň–Praha, *Pavel Drábek*
September 2012 *Jaroslav Milota*

Chapter 1

Preliminaries

1.1 Elements of Linear Algebra

This section is rather brief since we suppose that the reader has got some knowledge of linear algebra. Therefore, it should be viewed mainly as a source of concepts and notation which will be frequently used in the sequel. There are plenty of textbooks on this subject. As we are interested in applications to analysis we recommend to the reader the classical book Halmos [76] to find more detailed information.

A decisive part of analysis concerns the study of various sets of numbers like $\mathbb{R}$, $\mathbb{C}$, $\mathbb{R}^M$, …, sets of functions (continuous, integrable, differentiable), and mappings between them. These sets usually allow some algebraic operations, mainly addition and multiplication by scalars. We will denote the set of scalars by $\mathbb{A}$ and have usually in mind either the set of real numbers $\mathbb{R}$ or that of complex numbers $\mathbb{C}$.

Definition 1.1.1. A set X on which two operations – addition and multiplication by scalars – are defined, is called a *linear space over a field* $\mathbb{A}$ if the following conditions are fulfilled:

(1) X with respect to the operation $x, y \in X \mapsto x + y \in X$ forms a commutative group with a unit element denoted by o and the inverse element to $x \in X$ denoted by $-x$.

(2) The operation $a \in \mathbb{A}, x \in X \mapsto ax \in X$ satisfies
   (a) $a(bx) = (ab)x$, $a, b \in \mathbb{A}$, $x \in X$,
   (b) $1x = x$, $x \in X$, where 1 is the multiplicative unit of the field $\mathbb{A}$.

(3) For the two operations the distributive laws hold, i.e., for $a, b \in \mathbb{A}$, $x, y \in X$, we have
   (a) $(a + b)x = ax + bx$,
   (b) $a(x + y) = ax + ay$.

If $\mathbb{A} = \mathbb{R}$ or $\mathbb{C}$, then X is called a real or complex linear space, respectively. If a subset $Y \subset X$ itself is a linear space with respect to the operations induced by those defined on X, then Y is said to be a (linear) *subspace* of X.

In the rest of this section the character X always denotes a linear space over $\mathbb{A}$. If $\mathbb{A}$ is not specified, then it always means that a definition or a statement holds for an arbitrary field $\mathbb{A}$. For $x_1, \dots, x_n \in X$, $a_1, \dots, a_n \in \mathbb{A}$ the sum $\sum_{i=1}^{n} a_i x_i$ is well defined and determines an element $x \in X$ which is called a *linear combination* of $x_1, \dots, x_n$ (with coefficients $a_1, \dots, a_n$). Notice that only finite linear combinations are defined since infinite sums cannot be defined without any topology on X.

If $\mathcal{A}$ is a subset of X, then the set of all linear combinations of elements of $\mathcal{A}$ is denoted by $\operatorname{Lin}\mathcal{A}$ and is called the *span* of $\mathcal{A}$. A span is always a subspace of X. We can ask whether $x \in \operatorname{Lin}\{x_1, \dots, x_n\}$ can be expressed in a unique way as a linear combination of $x_1, \dots, x_n$. This uniqueness holds if and only if $x_1, \dots, x_n$ are *linearly independent*, i.e., the condition

$$\sum_{i=1}^{n} a_i x_i = o \qquad \Longleftrightarrow \qquad a_1 = \cdots = a_n = 0$$

is satisfied. More generally, we have the following definition.

Definition 1.1.2. A set $\mathcal{A} \subset X$ is said to be *linearly independent* if every finite subset of $\mathcal{A}$ is linearly independent. A set $\mathcal{A} \subset X$ is called a *basis*[1] of X provided $\mathcal{A}$ is linearly independent and $\operatorname{Lin}\mathcal{A} = X$.

Theorem 1.1.3. *Every linear space $X \neq \{o\}$ has a basis. If $\mathcal{A}$, $\mathcal{B}$ are two bases of X, then there is a bijective (i.e., injective and surjective) mapping from $\mathcal{A}$ onto $\mathcal{B}$.*

We will give the proof of the existence part since it contains a very important method which is frequently used. To see the idea of the proof notice that a basis is a linearly independent set which is maximal in the sense that by adding an element it will cease to be linearly independent. The question why such a maximal set has to exist concerns generally mathematical philosophy. There are several equivalent statements of the set theory which guarantee this existence result. As the most useful we have found the following one.[2]

Theorem 1.1.4 (Zorn's Lemma). *Let $(\mathcal{A}, \prec)$ be an ordered set in which every chain has the lowest upper bound. Then for any $a \in \mathcal{A}$ there is a maximal $m \in \mathcal{A}$ such that $a \prec m$.*[3]

---

[1]It is sometimes called a *Hamel basis* in order to emphasize the distinction from a Schauder or orthonormal basis in a Banach space or a Hilbert space, respectively (see Section 1.2).

[2]It can be viewed also as an axiom of the set theory.

[3]A binary relation $\prec$ on $\mathcal{A} \times \mathcal{A}$ is called an *ordering* if

(1) $x \prec x$ for all $x \in \mathcal{A}$,

(2) if $x \prec y$ and $y \prec x$, then $x = y$,

We now return to

Proof of Theorem 1.1.3. Let $\mathscr{A}$ be a collection of all linearly independent subsets of X and define $\mathcal{A} \prec \mathcal{B}$ for $\mathcal{A}, \mathcal{B} \in \mathscr{A}$ if $\mathcal{A}$ is a subset of $\mathcal{B}$. Choose $\mathcal{A} \in \mathscr{A}$ ($\mathscr{A} \neq \emptyset$ since $X \neq \{o\}$) and let $\mathcal{M}$ be a maximal element of $\mathscr{A}$, $\mathcal{A} \subset \mathcal{M}$, whose existence is guaranteed by Zorn's Lemma (if $\mathscr{B}$ is a chain in $\mathscr{A}$, then $\sup \mathscr{B} = \bigcup\limits_{\mathcal{B} \in \mathscr{B}} \mathcal{B}$). Then

$$\operatorname{Lin} \mathcal{M} = X.$$

The proof of the latter part of Theorem 1.1.3 is more involved (the construction of an injection of $\mathcal{A}$ into $\mathcal{B}$ is also based on the application of Zorn's Lemma) and it is omitted. □

Definition 1.1.5. Let X be a linear space and let $\mathcal{A}$ be a basis of X. Then the cardinality of $\mathcal{A}$ is called the *dimension* of X.

Example 1.1.6.

(i) Assume that $\mathcal{A}$ is a basis of a linear space X. Then there is a set (the so-called index set) Γ and a bijection $\gamma \in \Gamma \mapsto e_\gamma \in \mathcal{A}$ onto $\mathcal{A}$. We will also say that $\{e_\gamma\}_{\gamma \in \Gamma}$ is a basis of X. For any $x \in X$ there is a finite subset $\mathcal{K} \subset \Gamma$ and scalars $\{\alpha_\gamma\}_{\gamma \in \mathcal{K}}$ such that

$$x = \sum_{\gamma \in \mathcal{K}} \alpha_\gamma e_\gamma.$$

These scalars are uniquely determined and will be called the *coordinates* of x with respect to the basis $\{e_\gamma\}$.

(ii) The space $\mathbb{R}^M$ of real M-tuples with the usual operations is a real linear space and the elements

$$e_k = (0, \dots, 0, 1, 0, \dots, 0), \qquad k = 1, \dots, M$$

(1 is at the kth position), form a basis of $\mathbb{R}^M$. It will be called the *standard basis* of $\mathbb{R}^M$. If $x = (x_1, \dots, x_M) \in \mathbb{R}^M$, then $x_1, \dots, x_M$ are the coordinates of x with respect to the standard basis.

---

(3) if $x \prec y$ and $y \prec z$, then $x \prec z$.

If $(\mathcal{A}, \prec)$ is an ordered set, then $\mathcal{B} \subset \mathcal{A}$ is called a *chain* if for every $x, y \in \mathcal{B}$ we have either $x \prec y$ or $y \prec x$.

An element $b \in \mathcal{A}$ is called the *lowest upper bound* of a subset $\mathcal{B} \subset \mathcal{A}$ ($b = \sup \mathcal{B}$) if

(1) $a \in \mathcal{B} \implies a \prec b$;

(2) if $a \prec c$ for all $a \in \mathcal{B}$, then $b \prec c$.

Similarly, we call $d \in \mathcal{A}$ the *greatest lower bound* of a subset $\mathcal{B} \subset \mathcal{A}$ ($d = \inf \mathcal{B}$) if

(1) $a \in \mathcal{B} \implies d \prec a$;

(2) if $c \prec a$ for all $a \in \mathcal{B}$, then $c \prec d$.

An element $m \in \mathcal{A}$ is called a *maximal element* of $\mathcal{A}$ if $m \prec x$ for an $x \in \mathcal{A}$ implies $m = x$.

(iii) Similarly, $\mathbb{C}^M$ is the space of complex M-tuples and the set of elements $e_1, \dots, e_M$ defined as above is the standard basis of $\mathbb{C}^M$. More generally, if X is a real linear space and $\mathrm{i}X$ is defined by $\mathrm{i}X := \{\mathrm{i}x : x \in X\}$ where $\mathrm{i}^2 = -1$, then

$$X_{\mathbb{C}} := X + \mathrm{i}X \qquad (= \{x + \mathrm{i}y : x, y \in X\})$$

is the *complexification* of X. The equality $x + \mathrm{i}y = o$ holds if and only if $x = y = o$. The operations in $X_{\mathbb{C}}$ are defined as follows:

$$(x_1 + \mathrm{i}y_1) + (x_2 + \mathrm{i}y_2) := (x_1 + x_2) + \mathrm{i}(y_1 + y_2), \qquad x_1, x_2, y_1, y_2 \in X,$$
$$(a + ib)(x + \mathrm{i}y) := (ax - by) + \mathrm{i}(bx + ay), \qquad a, b \in \mathbb{R},\ x, y \in X.$$

It is easy to verify that $X_{\mathbb{C}}$ is a complex linear space.

(iv) Let $\mathcal{P}$ be the family of all polynomials of one variable with real or complex coefficients. Then $\mathcal{P}$ is respectively a real or complex linear space and the polynomials

$$P_k(z) = z^k, \qquad k = 0, 1, \dots,$$

form a basis of $\mathcal{P}$.

(v) The space $C[0,1]$ of all real (complex) continuous functions on the interval $[0,1]$ is a real (complex) linear space. According to Theorem 1.1.3, $C[0,1]$ has a basis but it is uncountable (this is not so easy to prove). We will not distinguish among different infinite cardinals and refer to spaces like $\mathcal{P}$ and $C[0,1]$ as infinite-dimensional spaces and use (incorrectly) the symbol $\dim = \infty$.

(vi) We can consider $\mathbb{R}$ as a linear space over the field $\mathbb{Q}$ of rational numbers. For example, the elements 1 and $\sqrt{2}$ are linearly independent in this case. In this case a basis is uncountable, and serves as a tool for the constructions of "pathological" examples in analysis, like a noncontinuous (or, equivalently, non-measurable) solution f of the functional equation

$$f(x + y) = f(x) + f(y), \qquad x, y \in \mathbb{R}. \qquad \bigcirc$$

Remark 1.1.7. In the sequel we will use the symbol

to warn the reader that the statement in question is true only in linear spaces of finite dimension.

Next we state a corollary of Theorem 1.1.3.

Corollary 1.1.8. *Let X be a linear space and let Y be a subspace of X. Then there exists a subspace Z of X with the following properties:*

(i) *for every $x \in X$ there are unique $y \in Y$, $z \in Z$ such that $x = y + z$;*

(ii) $Y \cap Z = \{o\}$.

Notation. $X = Y \oplus Z$ and X is called the *direct sum* of Y, Z, and Z a *direct complement* of Y in X.

Proof. Let $\mathcal{A}$ be a basis of Y and

$$\mathscr{A} = \{\mathcal{B} \text{ linearly independent subset of } X : \mathcal{A} \subset \mathcal{B}\}.$$

By Zorn's Lemma, $\mathscr{A}$ has a maximal element $\tilde{\mathcal{A}}$. Put $\mathcal{C} = \tilde{\mathcal{A}} \setminus \mathcal{A}$ (the set complement). It is easy to see that $Z := \operatorname{Lin} \mathcal{C}$ satisfies both (i) and (ii). □

Notice that the elements $y \in Y$, $z \in Z$ are uniquely determined by x in (i). If $\{o\} \neq Y$ and $Y \neq X$, then Z is not uniquely determined. A simple example can be given in $\mathbb{R}^2$ and the reader is invited to provide one!

Definition 1.1.9. Let X and Y be linear spaces over the same field $\mathbb{\Lambda}$. A mapping $A\colon X \to Y$ is said to be a *linear operator* if it possesses the following properties:

(1) $A(x + y) = Ax + Ay$ for all $x, y \in X$;

(2) $A(\alpha x) = \alpha Ax$ for every $\alpha \in \mathbb{\Lambda}$, $x \in X$.

The collection of all linear operators from X into Y is denoted by $L(X, Y)$. We will use the simpler notation $L(X)$ if $X = Y$.

Remark 1.1.10.

(i) A linear operator $A \in L(X, Y)$ is uniquely determined by its values on the elements of a basis $\mathcal{A} = \{e_\gamma\}_{\gamma\in\Gamma}$. Indeed, let $f_\gamma := Ae_\gamma$, $\gamma \in \Gamma$, and

$$x = \sum_{\substack{\gamma\in\mathcal{K} \\ \mathcal{K} \text{ finite } \subset\Gamma}} \alpha_\gamma e_\gamma.$$

If A is linear, then Ax has to be equal to $\sum\limits_{\gamma\in\mathcal{K}} \alpha_\gamma f_\gamma$. On the other hand, if $\{f_\gamma\}_{\gamma\in\Gamma}$ are given, then

$$Ax := \sum_{\gamma\in\mathcal{K}} \alpha_\gamma f_\gamma \qquad \text{for} \quad x = \sum_{\gamma\in\mathcal{K}} \alpha_\gamma e_\gamma$$

satisfies (1) and (2) from Definition 1.1.9.

(ii) Assume that both X and Y are finite-dimensional spaces and

$$\{e_1, \dots, e_M\} \qquad \text{and} \qquad \{f_1, \dots, f_N\}$$

are bases of X and Y, respectively. If $A \in L(X, Y)$, then

$$Ae_j = \sum_{i=1}^{N} a_{ij} f_i, \qquad j = 1, \dots, M, \quad \text{for some scalars} \quad a_{ij}. \tag{1.1.1}$$

These scalars form a matrix

$$\boldsymbol{A} = (a_{ij})_{\substack{i=1,\dots,N\\ j=1,\dots,M}}$$

(N rows and M columns; the jth column consists of the coordinates of Ae_j). This matrix $\boldsymbol{A}$ is called the *matrix representation* of the linear operator A with respect to the bases $\{e_1, \dots, e_M\}$ and $\{f_1, \dots, f_N\}$.

On the other hand, if $\{e_1, \dots, e_M\}$ and $\{f_1, \dots, f_N\}$ are bases of X and Y, respectively, and $\boldsymbol{A}$ is an $N \times M$ matrix, then the formula (1.1.1) determines a linear operator $A \in L(X, Y)$.

(iii) If $A, B \in L(X, Y)$ have matrix representations $\boldsymbol{A}$ and $\boldsymbol{B}$ (with respect to the same bases), then

$$\boldsymbol{A} + \boldsymbol{B} := (a_{ij} + b_{ij})_{\substack{i=1,\dots,N\\ j=1,\dots,M}}$$

is the matrix representation of

$$A + B \colon x \mapsto Ax + Bx.$$

Similarly,

$$\alpha\boldsymbol{A} := (\alpha a_{ij})_{\substack{i=1,\dots,N\\ j=1,\dots,M}}$$

is the matrix representation of

$$\alpha A \colon x \mapsto \alpha Ax.$$

It is obvious that $L(X, Y)$ is a linear space (over the same scalar field Λ) under these definitions of $A + B$, αA. This is true without any restrictions on the dimensions of X and Y.

(iv) If X, Y, Z are linear spaces over the same scalar field Λ and $A \in L(X, Y)$, $B \in L(Y, Z)$, then

$$BA \colon x \mapsto B(Ax), \qquad x \in X,$$

is a linear operator from X into Z. Moreover, if X, Y, Z are finite-dimensional spaces and $\boldsymbol{A} = (a_{ij})_{\substack{i=1,\dots,N\\ j=1,\dots,M}}$, $\boldsymbol{B} = (b_{ki})_{\substack{k=1,\dots,P\\ i=1,\dots,N}}$ are matrix representations of A and B, respectively, then

$$\boldsymbol{BA} := \left(\sum_{i=1}^{N} b_{ki} a_{ij}\right)_{\substack{k=1,\dots,P\\ j=1,\dots,M}}$$

is the matrix representation of BA. This product of operators is non-commutative in general, even in the case $X = Y = Z$.

For $A \in L(X,Y)$ we denote by

$$\operatorname{Ker} A := \{x \in X : Ax = o\}$$

the *kernel* of A, and by

$$\operatorname{Im} A := \{Ax \in Y : x \in X\}$$

the *image* of A. Evidently, $\operatorname{Ker} A$ and $\operatorname{Im} A$ are linear subspaces of X and Y, respectively.

Definition 1.1.11. A linear operator $A \in L(X,Y)$ is said to be

(1) *injective* if $\operatorname{Ker} A = \{o\}$,

(2) *surjective* if $\operatorname{Im} A = Y$,

(3) an *isomorphism* if A is both injective and surjective.

Remark 1.1.12.

(i) If $A \in L(X,Y)$ is injective and $e_1, \dots, e_n$ are linearly independent elements of X, then $Ae_1, \dots, Ae_n$ are linearly independent elements of Y. Further, $A \in L(X,Y)$ is an isomorphism if and only if $\{Ae_\gamma\}_{\gamma\in\Gamma}$ is a basis of Y whenever $\{e_\gamma\}_{\gamma\in\Gamma}$ is a basis of X. In other words:

linear spaces X, Y (over the same scalar field) have the same dimension if and only if there is an isomorphism $A \in L(X,Y)$.

(ii) Assume that $A \in L(X,Y)$ is an isomorphism and put $A^{-1}y = x$ where $y = Ax$. Then $A^{-1} \in L(Y,X)$ and

$$AA^{-1} = I_Y, \qquad A^{-1}A = I_X$$

where I_X and I_Y denote the identity maps on X and Y, respectively. A^{-1} is called the *inverse* of A. If $X = Y$ and $\boldsymbol{A}$ is a matrix representation of A, then A^{-1} has the inverse matrix $\boldsymbol{A}^{-1}$ as the representation in the same bases.

(iii) (Transformation of coordinates in a finite-dimensional space)
Let $\mathcal{E} = \{e_1, \dots, e_M\}$ and $\mathcal{F} = \{f_1, \dots, f_M\}$ be two bases of a linear space X. There are two questions:

(a) What is the relation between the coordinates of a given $x \in X$ with respect to these bases?

(b) Let $A \in L(X)$ have matrix representations $\boldsymbol{A}^{\mathcal{E}}$ and $\boldsymbol{A}^{\mathcal{F}}$ with respect to these bases. What is the relation between $\boldsymbol{A}^{\mathcal{E}}$ and $\boldsymbol{A}^{\mathcal{F}}$?

The answer to the first question is easy: Put

$$Te_j = f_j, \qquad j = 1, \dots, M,$$

and extend T to a linear operator on X. Then T is an isomorphism. Denote by $\boldsymbol{T} = (t_{ij})_{\substack{i=1,\dots,M\\ j=1,\dots,M}}$ its matrix representation with respect to the basis $\mathcal{E}$,

i.e.,

$$Te_j = \sum_{i=1}^{M} t_{ij} e_i, \qquad j = 1, \dots, M.$$

For $x = \sum\limits_{j=1}^{M} \eta_j f_j$ we have

$$x = \sum_{j=1}^{M} \eta_j \sum_{i=1}^{M} t_{ij} e_i = \sum_{i=1}^{M} \left(\sum_{j=1}^{M} t_{ij} \eta_j \right) e_i.$$

This means that the column vector $\xi = \begin{pmatrix} \xi_1 \\ \vdots \\ \xi_M \end{pmatrix}$ of the coordinates of x in the basis $\mathcal{E}$ is given by $\xi = \boldsymbol{T}\eta$ where

$$\xi_i = \sum_{j=1}^{M} t_{ij} \eta_j.$$

The second question can be answered by the same method but a certain caution in computation is desirable. Write

$$Af_j = A \left(\sum_{k=1}^{M} t_{kj} e_k \right) = \sum_{k,i=1}^{M} t_{kj} a_{ik}^{(\mathcal{E})} e_i = \sum_{k=1}^{M} a_{kj}^{(\mathcal{F})} Te_k = \sum_{k,i=1}^{M} a_{kj}^{(\mathcal{F})} t_{ik} e_i.$$

This equality can be expressed in matrix notation as

$$\boldsymbol{A}^{\mathcal{E}} \boldsymbol{T} = \boldsymbol{T} \boldsymbol{A}^{\mathcal{F}}.$$

Since the matrix $\boldsymbol{T}$ has an inverse, we get

$$\boldsymbol{A}^{\mathcal{F}} = \boldsymbol{T}^{-1} \boldsymbol{A}^{\mathcal{E}} \boldsymbol{T}. \tag{1.1.2}$$

Example 1.1.13.

(i) Let $X = Y \oplus Z$. Define

$$Px = y \qquad \text{where} \quad x = y + z, \qquad y \in Y, \quad z \in Z.$$

Then P is the so-called *projection* of X onto Y and has the following properties:

(a) $P^2 := PP = P$,

(b) $\operatorname{Ker} P = Z$.

It is easy to see that the properties (a), (b) determine uniquely the projection P and hence also the decomposition $X = Y \oplus Z$ ($Y = \operatorname{Im} P$).

(ii) Let Y be a subspace of X. For $x \in X$ put

$$[x] := x + Y = \{x + y : y \in Y\}.$$

If $x, y \in X$, then either $[x] = [y]$ ($\Leftrightarrow x - y \in Y$) or $[x] \cap [y] = \emptyset$. Define

$$[x] + [y] := [x + y], \quad \alpha[x] := [\alpha x] \qquad \text{for} \quad x, y \in X, \ \alpha \in \mathbb{\Lambda}.$$

These operations are well defined and endow the set

$$X|_Y := \{[x] : x \in X\}$$

with the structure of a linear space. The space $X|_Y$ is called a *factor space* or simply a Y-factor. Put

$$\kappa \colon x \mapsto [x], \qquad x \in X.$$

Then κ (the so-called *canonical embedding* of X onto $X|_Y$) is a linear, surjective operator from X onto $X|_Y$, and $\operatorname{Ker} \kappa = Y$. If $x = y + z$ where $y \in Y$, $z \in Z$ and $X = Y \oplus Z$, then the mapping $j \colon [x] \mapsto z$ is an isomorphism of $X|_Y$ onto Z. In particular, $X|_Y$ and Z have the same dimension. The dimension of $X|_Y$ is sometimes called the *codimension* of Y ($\operatorname{codim} Y$) and

$$\dim X = \dim Y + \operatorname{codim} Y. \tag{1.1.3}$$

Warning. *If X is an infinite-dimensional space, then the sum on the right-hand side is the sum of infinite cardinal numbers!* ○

Proposition 1.1.14. *Let $A \in L(X, Y)$ and let κ be the canonical embedding of X onto $X|_{\operatorname{Ker} A}$. If $\hat{A}[x] := Ax$, then $\hat{A}$ is injective and the diagram in Figure* 1.1.1 *is commutative, i.e., $A = \hat{A}\kappa$.*

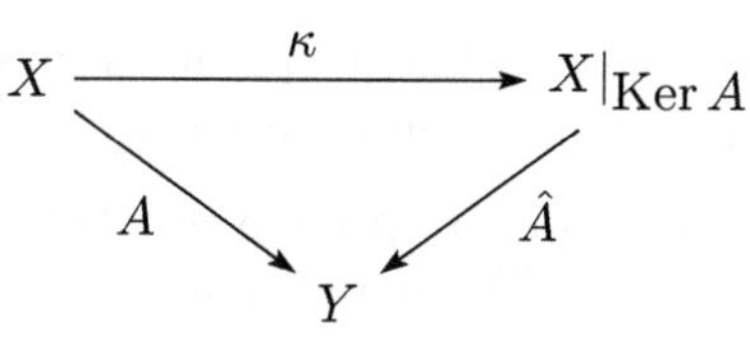

Figure 1.1.1.

Proof. The assertion is obvious but do not forget to prove that $\hat{A}$ is well defined. □

Corollary 1.1.15. *Let $A \in L(X,Y)$. Then*

$$\dim X = \dim \operatorname{Ker} A + \dim \operatorname{Im} A. \tag{1.1.4}$$

In particular, if $X = Y$ and $\dim X < \infty$, then $A \in L(X,Y)$ is injective if and only if it is surjective.

Proof. We have

$$\operatorname{codim} \operatorname{Ker} A = \dim X|_{\operatorname{Ker} A} = \dim \operatorname{Im} \hat{A} = \dim \operatorname{Im} A$$

since $\hat{A}$ is an isomorphism of $X|_{\operatorname{Ker} A}$ onto its image. Equality (1.1.4) follows immediately from (1.1.3). If A is injective, then

$$\dim X = \dim \operatorname{Im} A,$$

and this implies (only in the case of X and Y having the same finite dimension) that $Y = \operatorname{Im} A$. If $\operatorname{Im} A = Y$, then (finite dimensions!)

$$\dim \operatorname{Ker} A = 0,$$

i.e., A is injective. □

Example 1.1.16. Let X be the space of bounded (real) sequences $l^\infty(\mathbb{N})$ and define the *right-shift*

$$S_R\colon x = (x_1, \dots) \mapsto (0, x_1, x_2, \dots)$$

and the *left-shift*

$$S_L\colon x = (x_1, \dots) \mapsto (x_2, x_3, \dots).$$

Then S_R is injective but not surjective and S_L is surjective but not injective. Moreover,

$$S_L S_R x = x \qquad \text{for every} \quad x \in X.$$

What is $S_R S_L$? ○

The following special case of linear operators plays an important role both in the theory of linear spaces and in applications.

Definition 1.1.17. Let X be a linear space over a field $\mathbb{A}$. A linear operator from X into $\mathbb{A}$ is called a *linear form*. The linear space of all linear forms on X is called the (algebraic) *dual space* of X and is denoted by $X^{\#}$.

Example 1.1.18.

(i) Let $\{e_1, \dots, e_M\}$ be a basis of X, i.e., for every x there is a unique M-tuple $(\xi_1, \dots, \xi_M) \in \mathbb{A}^M$ (coordinates of x) such that $x = \sum_{i=1}^{M} \xi_i e_i$. The mapping

$e^i : x \mapsto \xi_i$ is a linear form (the ith coordinate form). It is straightforward to show that $e^1, \dots, e^M$ are linearly independent and

$$\operatorname{Lin}\{e^1, \dots, e^M\} = X^{\#},$$

i.e., $\{e^1, \dots, e^M\}$ is a basis of $X^{\#}$ (the so-called *dual basis* of $X^{\#}$, dual to $\{e_1, \dots, e_M\}$).

(ii) If $f \in X^{\#} \setminus \{o\}$, then

$$\operatorname{codim} \operatorname{Ker} f = 1.$$

To see this choose $x_0 \in X$ such that $f(x_0) = 1$. Then

$$x = (x - f(x)x_0) + f(x)x_0 \in \operatorname{Ker} f \oplus \operatorname{Lin}\{x_0\}.$$

On the other hand, if Y is a subspace of X of codimension one,[4] then

$$X = Y \oplus \operatorname{Lin}\{x_0\} \qquad \text{for an} \quad x_0 \in X.$$

For $x = y + \alpha x_0$, $y \in Y$, we put $f(x) = \alpha$. Then

$$f \in X^{\#} \setminus \{o\} \qquad \text{and} \qquad \operatorname{Ker} f = Y.$$

Moreover, if $f, g \in X^{\#}$ are such that $\operatorname{Ker} f = \operatorname{Ker} g$, then there is an $\alpha \in \mathbb{K}$ for which $f = \alpha g$. ○

This fact has the following generalization, which will be used in Section 7.8, more precisely in the proof of Theorem 7.8.2.

Proposition 1.1.19. *Let $f_1, \dots, f_n$, g be linear forms on X. Then*

$$g \in \operatorname{Lin}\{f_1, \dots, f_n\} \qquad \textit{if and only if} \qquad \bigcap_{i=1}^{n} \operatorname{Ker} f_i \subset \operatorname{Ker} g.$$

Proof. The "only if" part is obvious. For the "if" part notice that the assertion $g \in \operatorname{Lin}\{f_1, \dots, f_n\}$ can be interpreted as the existence of a linear form $\lambda \in (\mathbb{K}^n)^{\#}$ such that

$$g = \lambda \circ F \qquad \text{where} \quad F(x) := (f_1(x), \dots, f_n(x)). \tag{1.1.5}$$

Let $\mathbb{K}^n = \operatorname{Im} F(X) \oplus Y$ (Corollary 1.1.8). If $\alpha = \beta + \gamma$, $\beta = F(x)$, $\gamma \in Y$, then the mapping $\lambda(\alpha) = g(x)$ is a well defined linear form (by assumption). This means that (1.1.5) holds. □

Definition 1.1.20. Let $A \in L(X, Y)$ and $g \in Y^{\#}$. Then the linear form $f(x) := g(Ax)$ is denoted by $f = A^{\#} g$ and $A^{\#}$ is called the *adjoint operator* to A.

[4]Such a subspace is often called a *hyperplane*.

Remark 1.1.21.

(i) $A^{\#} \in L(Y^{\#}, X^{\#})$.

(ii) If A has a matrix representation $\boldsymbol{A} = (a_{ij})_{\substack{i=1,\dots,N\\ j=1,\dots,M}}$ with respect to bases $\mathcal{E} = \{x_1, \dots, x_M\}$ in X and $\mathcal{F} = \{y_1, \dots, y_N\}$ in Y, then the adjoint operator $A^{\#}$ has the representation

$$\boldsymbol{A}^{\#} = (a_{ji})_{\substack{j=1,\dots,M\\ i=1,\dots,N}}$$

(i.e., $\boldsymbol{A}^{\#}$ is the *transpose* of $\boldsymbol{A}$) with respect to the dual bases.

Warning. *We will encounter different adjoint operators in the next section and the adjoint A^* with respect to a scalar product will have a different representation in a complex space!*

Now we turn our attention to a system of linear equations

$$\sum_{j=1}^{M} a_{ij} x_j = b_i, \qquad i = 1, \dots, N. \tag{1.1.6}$$

This system can be written in a more "compact" form, namely as

$$\boldsymbol{A}x = b \tag{1.1.7}$$

where $\boldsymbol{A}$ is a matrix representation of the linear operator A from X into Y. By choosing fixed bases $\mathcal{E} = \{e_1, \dots, e_M\}$ in X and $\mathcal{F} = \{f_1, \dots, f_N\}$ in Y (also $Y = \mathbb{R}^N$ or $\mathbb{C}^N$), A is defined by its matrix representation $\boldsymbol{A} = (a_{ij})_{\substack{i=1,\dots,N\\ j=1,\dots,M}}$ with respect to these bases. In order to formulate results on solvability of (1.1.6) (or, equivalently, of (1.1.7)) the following notation will be useful.

Notation. If $\mathcal{U}$ is a subset of X (not necessarily a subspace of X), then

$$\mathcal{U}^{\perp} = \{f \in X^{\#} : x \in \mathcal{U} \Rightarrow f(x) = 0\}.$$

Similarly,

$$\mathcal{W}_{\perp} = \{x \in X : f \in \mathcal{W} \Rightarrow f(x) = 0\} \qquad \text{for} \quad \mathcal{W} \subset X^{\#}.$$

Proposition 1.1.22.

(i) $(\mathcal{U}^{\perp})_{\perp} = \operatorname{Lin} \mathcal{U}$ *for every* $\mathcal{U} \subset X$.

(ii) *If* $\dim X < \infty$, *then* $(\mathcal{W}_{\perp})^{\perp} = \operatorname{Lin} \mathcal{W}$ *for all* $\mathcal{W} \subset X^{\#}$.

Proof. We include the proof because it contains a construction which should be compared with an analogous one in Section 2.1 (see Proposition 2.1.27 and its proof).

(i) We can assume $\mathcal{U}$ to be a subspace of X since $\mathcal{U}^\perp = (\operatorname{Lin}\mathcal{U})^\perp$. The inclusion $\mathcal{U} \subset (\mathcal{U}^\perp)_\perp$ is obvious. To prove the reverse let us suppose by contradiction that there is an element $x_0 \in (\mathcal{U}^\perp)_\perp \setminus \mathcal{U}$. By the method of proof of Theorem 1.1.3, a subspace Y of X can be found such that

$$X = \operatorname{Lin}\{x_0\} \oplus Y \qquad \text{and} \qquad \mathcal{U} \subset Y.$$

According to Example 1.1.18(ii) there exists $f \in X^\#$ with $\operatorname{Ker} f = Y$. In particular, $f \in \mathcal{U}^\perp$ and $f(x_0) \neq 0$, which contradicts the choice of x_0.

(ii) This part follows from (i) by replacing X by $X^\#$. To repeat the proof we need that $(X^\#)^\#$ could be identified with X. We note that this is possible because $\dim X < \infty$. □

The main idea – separation of a point from a subspace by a linear form (i.e., by a hyperplane) – can be substantially generalized.

Definition 1.1.23. A subset $\mathcal{C}$ of a (real or complex) linear space X is called *convex* if for every $x, y \in \mathcal{C}$, $t \in [0,1]$, the point

$$tx + (1-t)y \qquad \text{belongs to} \quad \mathcal{C}.$$

Proposition 1.1.24. *Let X be a real linear space, $\emptyset \neq \mathcal{C}$ a convex subset of X with a nonempty algebraic interior*

$$\mathcal{C}^0 := \{a \in \mathcal{C} : \forall y \in X \ \exists t_0 > 0 \ \textit{such that} \ a + ty \in \mathcal{C} \ \textit{for all} \ t \in [0, t_0)\}.$$

Let $x_0 \in X \setminus \mathcal{C}$. Then there is $f \in X^\#$ such that

$$f(x) \le f(x_0) \qquad \textit{for all} \quad x \in \mathcal{C}.$$

Proof. It needs a special tool for the treatment of convex sets and a considerably more sophisticated extension procedure,[5] and, therefore, it is omitted. See, e.g., Rockefellar [128, § 11] where the interested reader can find also applications to convex optimization, and also Corollary 2.1.18. □

Theorem 1.1.25. *For $A \in L(X,Y)$ we have*

(i) $\operatorname{Im} A = (\operatorname{Ker} A^\#)_\perp$,

(ii) $\operatorname{Im} A^\# = (\operatorname{Ker} A)^\perp$.

(iii) *If, moreover,* $\dim X = \dim Y < \infty$, *then*

$$\dim \operatorname{Ker} A = \dim \operatorname{Ker} A^\#. \tag{1.1.8}$$

Proof. (i) It is straightforward to prove both the inclusions which lead to the equality $(\operatorname{Im} A)^\perp = \operatorname{Ker} A^\#$. The result follows then from Proposition 1.1.22(i).

[5]See Corollary 2.1.18 for a similar process.

(ii) Let $Y = \operatorname{Im} A \oplus Z$ (Corollary 1.1.8). For $f \in (\operatorname{Ker} A)^{\perp}$ and $y = Ax + z$, $z \in Z$, put $g(y) = f(x)$. This definition does not depend on a concrete choice of x since $f \in (\operatorname{Ker} A)^{\perp}$. This proves that $f = A^{\#}g$ and hence the inclusion $(\operatorname{Ker} A)^{\perp} \subset \operatorname{Im} A^{\#}$ holds. The reverse inclusion is trivial.

(iii) Observe first that $(X|_U)^{\#}$ is isomorphic to $U^{\perp}$ for any subspace U of X, namely,

$$\Phi(F)(x) := F([x]), \qquad F \in (X|_U)^{\#}$$

is the desired isomorphism. If $\dim X < \infty$, then $X|_U$ is isomorphic to $(X|_U)^{\#}$ (both spaces have the same dimension) and, therefore, $X|_U$ is isomorphic to $U^{\perp}$. Now, we apply this observation to $U = \operatorname{Ker} A$. We recall that $\operatorname{Im} A$ is isomorphic to $X|_{\operatorname{Ker} A}$ (Proposition 1.1.14) and therefore to $(\operatorname{Ker} A)^{\perp}$. By (ii), $\operatorname{Im} A$ is isomorphic to $\operatorname{Im} A^{\#}$. The equality (1.1.8) follows from Corollary 1.1.15. □

Remark 1.1.26.

(i) Note that Theorem 1.1.25(i) is an existence result for the equation (1.1.6) (or (1.1.7)) because it can be reformulated as follows:

The equation (1.1.6) *has a solution for* $b = (b_1, \dots, b_N)$ *if and only if*

$$\sum_{i=1}^{N} b_i f_i = 0$$

for all solutions $f = (f_1, \dots, f_N)$ *of the adjoint homogeneous equation*

$$\sum_{i=1}^{N} a_{ji} f_i = 0, \qquad j = 1, \dots, M.$$

In particular, we have also the *alternative* result:

Either the equation (1.1.6) *has a solution for all right-hand sides or*[6] *the adjoint homogeneous equation has a nontrivial solution.*

Theorem 1.1.25(ii) can be reformulated similarly.

(ii) If $\boldsymbol{A}$ is a matrix representation of $A \in L(X, Y)$ (X and Y are finite-dimensional spaces), then $\dim \operatorname{Im} A$ is equal to the number of linearly independent columns of $\boldsymbol{A}$ and is called the *rank* of $\boldsymbol{A}$. If $X = Y$, then $\boldsymbol{A}$ is a square matrix of the type $M \times M$ ($M = \dim X$), and it is called a *regular matrix* provided

$$M = \operatorname{rank} \boldsymbol{A}.$$

Equivalently, $\boldsymbol{A}$ is a regular matrix if and only if its determinant $\det \boldsymbol{A}$ does not vanish. By the proof of Theorem 1.1.25(iii),

$$\dim \operatorname{Im} A = \dim \operatorname{Im} A^{\#}.$$

[6] The conjunction "or" has exclusive character here. This alternative result is sometimes called a *Fredholm alternative* since I. Fredholm proved such a result for linear integral equations. See also Section 2.2.

In particular, this means that the rank of $\boldsymbol{A}$ is equal to the rank of its transpose. The reader is asked to find more matrix formulations of the previous results.

We often do calculations with a matrix representation instead of the operator itself. Since there are plenty of representations of the same operator it would be convenient to work with the simplest possible form. To examine this problem we start with some notions.

Definition 1.1.27. Let X be a complex linear space and $A \in L(X)$. A complex number λ is called an *eigenvalue* of A if there is $x \neq o$ such that

$$Ax = \lambda x.$$

Such an element x is called an *eigenvector* of A (associated with the eigenvalue λ). The set of all eigenvalues of A is called the *spectrum* of A and is denoted by $\sigma(A)$.

Warning. *In infinite dimensions the spectrum of a linear operator can contain also other points than the eigenvalues and is defined in a different way (see page 56)!*

Remark 1.1.28. It is obvious that the following statements are equivalent in a finite-dimensional complex space X:

$$\begin{aligned}\lambda \in \sigma(A) \quad &\Longleftrightarrow \quad \operatorname{Ker}(\lambda I - A) \neq \{o\} \quad \Longleftrightarrow \quad \operatorname{rank}(\lambda \boldsymbol{I} - \boldsymbol{A}) < \dim X \\ &\Longleftrightarrow \quad \det(\lambda \boldsymbol{I} - \boldsymbol{A}) = 0\end{aligned}$$

where $\boldsymbol{A}$ is a representation of A.

Since $P(z) := \det(z\boldsymbol{I} - \boldsymbol{A})$ is a polynomial (the so-called *characteristic polynomial*) of degree $M = \dim X$, the problem of finding $\sigma(A)$ is equivalent to solving an algebraic equation (the so-called *characteristic equation* of A)

$$P(z) = 0. \tag{1.1.9}$$

According to the Fundamental Theorem of Algebra (see Theorem 5.7.13) there exists at least one solution of (1.1.9) in $\mathbb{C}$. The reason for considering complex spaces here is the fact that (1.1.9) need not have a real solution. It is an easy consequence of the Fundamental Theorem of Algebra that the polynomial P can be written in the form

$$P(\lambda) = (\lambda - \lambda_1)^{m_1} \cdots (\lambda - \lambda_k)^{m_k} \tag{1.1.10}$$

where $\sigma(A) = \{\lambda_1, \dots, \lambda_k\}$ ($\lambda_1, \dots, \lambda_k$ are different) and $m_1 + \dots + m_k = \dim X$. The positive integer m_i is called the *multiplicity of the eigenvalue* λ_i.

Definition 1.1.29. Let $A \in L(X)$.

(1) A subspace $Y \subset X$ is said to be *A-invariant* if $A(Y) \subset Y$.

(2) An A-invariant subspace $Y \subset X$ is said to *reduce* A if there is a decomposition $X = Y \oplus Z$ where Z is also A-invariant.

From now on till the end of this section we consider exclusively finite-dimensional spaces.

Example 1.1.30.

(i) Let $X = Y \oplus Z$ where both Y and Z are A-invariant. If $\{e_1, \dots, e_m\}$ is a basis of Y and $\{e_{m+1}, \dots, e_M\}$ is a basis of Z, then the matrix representation $\boldsymbol{A}$ of A with respect to $\{e_1, \dots, e_M\}$ has a block form

$$\boldsymbol{A} = \begin{pmatrix} \boldsymbol{A}_Y & \boldsymbol{O} \\ \boldsymbol{O} & \boldsymbol{A}_Z \end{pmatrix}$$

where $\boldsymbol{A}_Y$ and $\boldsymbol{A}_Z$ are representations of restrictions of A to Y and Z, respectively.

(ii) Assume that there is a basis $\{e_1, \dots, e_M\}$ of X consisting of eigenvectors of $A \in L(X)$ and $Ae_i = \lambda_i e_i$, $i = 1, \dots, M$ ($\lambda_1, \dots, \lambda_M$ are not necessarily distinct). Then the matrix representation of A with respect to this basis is the diagonal matrix

$$\begin{pmatrix} \lambda_1 & 0 & \cdots & 0 \\ 0 & \lambda_2 & \cdots & 0 \\ \vdots & \vdots & \ddots & \vdots \\ 0 & 0 & \cdots & \lambda_M \end{pmatrix}.$$

(iii) The matrix $\boldsymbol{A} = \begin{pmatrix} 1 & 1 \\ 0 & 1 \end{pmatrix}$ is a representation of a linear operator $A \in L(\mathbb{C}^2)$ which has no one-dimensional reducing subspace. Hence A has no diagonal representation. ○

Because of the last example we have to improve our hitherto idea: Choose $\lambda \in \sigma(A)$ and denote

$$N_k := \operatorname{Ker} (\lambda I - A)^k.$$

It is obvious that $N_k \subset N_{k+1}$ and they cannot be all distinct. If $N_k = N_{k+1}$, then $N_i = N_k$ for all $i > k$. Denote by $n(\lambda)$ the least such k and set

$$N(\lambda) := \bigcup_{j=1}^{n(\lambda)} N_j = N_{n(\lambda)}, \qquad R(\lambda) := \operatorname{Im} (\lambda I - A)^{n(\lambda)}.$$

Lemma 1.1.31. *Let $A \in L(X)$ and $\lambda \in \sigma(A)$.*

(i) *Both $N(\lambda)$ and $R(\lambda)$ are A-invariant subspaces and the decomposition*

$$X = N(\lambda) \oplus R(\lambda) \qquad \textit{holds.} \tag{1.1.11}$$

(ii) *Denote by* $A|_N$ *and* $A|_R$ *the restrictions of* A *respectively to* $N(\lambda)$ *and* $R(\lambda)$. *Then*

$$\sigma(A|_N) = \{\lambda\}, \qquad \sigma(A|_R) = \sigma(A) \setminus \{\lambda\}.$$

Moreover, the dimension of $N(\lambda)$ *is equal to the multiplicity of the eigenvalue* λ.

(iii) *If* $\sigma(A) = \{\lambda_1, \dots, \lambda_k\}$, *then*

$$X = N(\lambda_1) \oplus \cdots \oplus N(\lambda_k). \tag{1.1.12}$$

Proof. (i) Since $R(\lambda) \cap N(\lambda) = \{o\}$ (by the definition of $n(\lambda)$) and $\dim X = \dim N(\lambda) + \dim R(\lambda)$ (Corollary 1.1.15), we deduce the decomposition (1.1.11). If

$$y = (\lambda I - A)^{n(\lambda)} x \in R(\lambda),$$

then

$$Ay = -(\lambda I - A)y + \lambda y = -(\lambda I - A)^{n(\lambda)}(\lambda I - A)x + \lambda y \in R(\lambda).$$

The A-invariance of $N(\lambda)$ is also clear.

(ii) Obviously, $\sigma(A|_N) \subset \sigma(A)$. Let $\mu \in \sigma(A) \setminus \{\lambda\}$ and let x be a corresponding eigenvector. By (1.1.11) we have $x = y + z$ where $y \in N(\lambda)$, $z \in R(\lambda)$. Further,

$$o = (\mu I - A)x = (\lambda I - A)y + (\mu - \lambda)y + (\mu I - A)z.$$

By virtue of the uniqueness of the decomposition we have

$$(\lambda I - A)y = (\lambda - \mu)y.$$

This implies that

$$o = (\lambda I - A)^{n(\lambda)} y = (\lambda - \mu)(\lambda I - A)^{n(\lambda)-1} y, \quad \text{i.e,.} \quad y \in \operatorname{Ker}(\lambda I - A)^{n(\lambda)-1}.$$

By repeating this procedure we get

$$y \in \operatorname{Ker}(\lambda I - A)$$

and, therefore, $(\lambda - \mu)y = o$, i.e.,

$$y = o \qquad \text{and} \qquad x = z \in R(\lambda).$$

This shows that $\mu \notin \sigma(A|_N)$ and $\mu \in \sigma(A|_R)$. Since $N(\lambda) \cap R(\lambda) = \{o\}$ the eigenvalue λ does not belong to $\sigma(A|_R)$.

The matrix representation $\boldsymbol{A}$ of A with respect to the basis formed by joining the bases of $N(\lambda)$ and $R(\lambda)$ has the block form

$$\boldsymbol{A} = \begin{pmatrix} \boldsymbol{A}_N & \boldsymbol{O} \\ \boldsymbol{O} & \boldsymbol{A}_R \end{pmatrix}.$$

It follows that

$$\det(z\boldsymbol{I} - \boldsymbol{A}) = \det(z\boldsymbol{I} - \boldsymbol{A}_N)\det(z\boldsymbol{I} - \boldsymbol{A}_R)$$

and hence the characteristic polynomial of $\boldsymbol{A}_N$ is $P_N(z) = (z-\lambda)^{m(\lambda)}$ where $m(\lambda)$ is the multiplicity of the eigenvalue λ of A. Therefore $\dim N(\lambda) = m(\lambda)$.

(iii) This follows by induction with respect to the eigenvalues of A. □

For a polynomial $P(z) = a_n z^n + \cdots + a_1 z + a_0$ and $A \in L(X)$ we put

$$P(A) = a_n A^n + \cdots + a_1 A + a_0 I.$$

Corollary 1.1.32 (Hamilton–Cayley). *Let $A \in L(X)$ and let P be the characteristic polynomial of A. Then $P(A) = O$.*

Proof. Assume that P has the form (1.1.10) and $x = x_1 + \cdots + x_k$ is the decomposition given by (1.1.12). Since $m_k = n(\lambda_k)$,

$$(A - \lambda_k I)^{m_k} x = \sum_{j=1}^{k-1} (A - \lambda_k I)^{m_k} x_j + o.$$

The result follows by induction. □

It remains to compute the representation of the restriction of $\lambda_i I - A$ to $N(\lambda_i)$. Notice that this restriction is nilpotent.[7]

Lemma 1.1.33. *Let $B \in L(X)$ be a nilpotent operator of order n. Then for any $x \in X \setminus \operatorname{Ker} B^{n-1}$ the elements x, Bx, $\dots$, $B^{n-1}x$ are linearly independent and the subspace*

$$Y = \operatorname{Lin}\{x, Bx, \dots, B^{n-1}x\}$$

reduces B. The restriction $B|_Y$ of B to Y has the representation

$$\begin{pmatrix} 0 & 1 & \cdots & 0 \\ 0 & 0 & \cdots & 0 \\ \vdots & \vdots & \ddots & 1 \\ 0 & 0 & \cdots & 0 \end{pmatrix}$$

with respect to the basis $\{B^{n-1}x, \dots, x\}$. There exists a B-invariant direct complement of Y and the restriction of B to such a complement is nilpotent of order $\leq n$.

Proof. It is easy to see the linear independence of the elements $x, \dots, B^{n-1}x$. Indeed, if

$$\sum_{j=0}^{n-1} \alpha_j B^j x = o,$$

[7] An operator $B \in L(X)$ is said to be *nilpotent* if there is such an $n \in \mathbb{N}$ that $B^n = O$. The least such integer n is called the order of nilpotency.

then, by applying B^{n-1}, we get

$$\alpha_0 B^{n-1}x = o, \qquad \text{i.e.,} \qquad \alpha_0 = 0.$$

Repetition shows that $\alpha_j = 0$ for all $j = 0, 1, \dots, n-1$. The form of representation of $B|_Y$ is obvious. The existence of an invariant direct complement of Y can be proved by induction with respect to the order of nilpotency. We omit details and refer to, e.g., Halmos [76, § 57]. □

We are now ready to summarize all information to obtain the following fundamental result.

Theorem 1.1.34 (Jordan Canonical Form). *Let X be a complex linear space of finite dimension and let $A \in L(X)$. Assume that $\sigma(A) = \{\lambda_1, \dots, \lambda_k\}$. Then there exists a basis $\mathcal{F}$ of X in which A has the* canonical block representation

$$\boldsymbol{A}^{\mathcal{F}} = \begin{pmatrix} \boldsymbol{A}_1^{(1)} & & & & \\ & \ddots & & \boldsymbol{O} & \\ & & \boldsymbol{A}_1^{(l_1)} & & \\ & \boldsymbol{O} & & \ddots & \\ & & & & \boldsymbol{A}_k^{(l_k)} \end{pmatrix}$$

where the block matrices (the so-called Jordan cells*) have the form*

$$\boldsymbol{A}_j^{(i)} = \underbrace{\begin{pmatrix} \lambda_j & 1 & 0 & \cdots & 0 \\ 0 & \lambda_j & \ddots & & 0 \\ \vdots & & \ddots & & 1 \\ 0 & & \cdots & & \lambda_j \end{pmatrix}}_{l_j \text{ columns}}, \qquad i = 1, \dots, l_j,\ j = 1, \dots, k. \tag{1.1.13}$$

Remark 1.1.35.

(i) We can also interpret Theorem 1.1.34 as follows. Let $\boldsymbol{A}^{\mathcal{E}}$ be the representation of A with respect to the basis $\mathcal{E}$. By Remark 1.1.12(iii), there is a regular transformation matrix $\boldsymbol{T}$ such that (1.1.2) holds. The canonical matrix $\boldsymbol{A}^{\mathcal{F}}$ may be viewed as a representation of a $B \in L(X)$ with respect to the basis $\mathcal{E}$. Denote by T a linear operator represented in the basis $\mathcal{E}$ by the matrix $\boldsymbol{T}$. Then one has

$$B = T^{-1}AT. \tag{1.1.14}$$

(ii) Assume that $A \in L(X)$ where X is a real linear space. The problem in the application of Theorem 1.1.34 lies in the fact that the spectrum $\sigma(A) \cap \mathbb{R}$ is not sufficient to guarantee the decomposition (1.1.12). This obstacle can be overcome by the complexification $X_{\mathbb{C}}$ of X. Namely, A is extendable to $X_{\mathbb{C}}$ by the formula

$$A_{\mathbb{C}}(x + \mathrm{i}y) = Ax + \mathrm{i}Ay.$$

If $\lambda = \alpha + \mathrm{i}\beta$, $\beta \neq 0$, is an eigenvalue of $A_{\mathbb{C}}$ with an eigenvector $u + \mathrm{i}v$, then u and v are linearly independent in X and the complex conjugate $\overline{\lambda}$ is also an eigenvalue of $A_{\mathbb{C}}$ and $u - \mathrm{i}v$ is the corresponding eigenvector. Moreover, both λ and $\overline{\lambda}$ have the same multiplicity. Rearranging the $A_{\mathbb{C}}$-canonical basis by joining its parts which correspond to λ and $\overline{\lambda}$ we obtain a basis of the real space X in which the representation of A has blocks of the form

$$\begin{pmatrix} \alpha & \beta & 1 & 0 & \cdots & & 0 \\ -\beta & \alpha & 0 & 1 & \cdots & & 0 \\ 0 & 0 & \alpha & \beta & \cdots & & \vdots \\ \vdots & \vdots & -\beta & \alpha & \cdots & 1 & 0 \\ \vdots & \vdots & \vdots & \vdots & \ddots & 0 & 1 \\ & & & & & \alpha & \beta \\ 0 & 0 & 0 & 0 & \cdots & -\beta & \alpha \end{pmatrix}.$$

We drop simple computations which confirm these statements and leave them to the reader.

The simple canonical form is convenient for solving a system of linear differential equations with real constant coefficients. Such a system can be written in the form

$$\dot{x} := \frac{\mathrm{d}x}{\mathrm{d}t} = Ax, \qquad A \in L(X). \tag{1.1.15}$$

If $X = \mathbb{R}^M$ and $\boldsymbol{A} = (a_{ij})$ is the representation of A with respect to the standard basis $e_1, \dots, e_M$, then (1.1.15) is an abstract formulation of the system

$$\dot{x}_i(t) = \sum_{j=1}^{M} a_{ij} x_j(t), \quad i = 1, \dots, M, \quad \text{where } x(t) = \sum_{i=1}^{M} x_i(t) e_i. \tag{1.1.16}$$

In order to find a solution, it is convenient to transform (1.1.16) into a canonical form. If $T \in L(X)$ is invertible, then $x = Ty$ is a solution of (1.1.15) if and only if y solves the equation

$$\dot{y} = By, \qquad \text{where} \quad By := T^{-1}ATy.$$

Theorem 1.1.34 says that T can be chosen in such a way that the representation of B with respect to the standard basis is the Jordan Canonical Form of $\boldsymbol{A}$. Having this form it is easy to solve (1.1.16) (see Exercise 1.1.41).

Qualitative properties of solutions of (1.1.15) are often more interesting than an involved formula for solutions. Therefore it would be convenient to generalize the exponential function solving $\dot{x} = ax$ in $\mathbb{R}$ to $L(X)$. Similarly to the one-dimensional case we put

$$\mathrm{e}^{tA}x := \sum_{n=0}^{\infty} \frac{t^n}{n!} A^n x$$

provided the series is convergent in $L(X)$. We postpone the question of convergence of this series (see Exercise 2.1.35) and give instead an equivalent definition of a function $f(A)$ for $A \in L(X)$ without any use of infinite series.

First we will define $f(B)$ for $B \in L(\mathbb{C}^M)$ which has a representation in the form

$$\boldsymbol{B} = \begin{pmatrix} \lambda & 1 & & \\ & \ddots & \ddots & \\ & & & 1 \\ & & & \lambda \end{pmatrix}. \tag{1.1.17}$$

Assume that f is a polynomial $P\colon z \mapsto a_0 z^n + \cdots + a_n$. Obviously, we define

$$P(B) = a_0 B^n + \cdots + a_n I.$$

It will be convenient to rewrite $P(B)$ in a form which is more adequate for generalization. Since

$$P(z) = \sum_{j=0}^{n} \frac{P^{(j)}(\lambda)}{j!} (z - \lambda)^j,$$

we can write

$$P(z) = \sum_{j=0}^{M-1} \frac{P^{(j)}(\lambda)}{j!} (z - \lambda)^j + (z - \lambda)^M R(z)$$

where R is a polynomial, possibly equal to 0. Since $z \mapsto (z - \lambda)^M$ is the characteristic polynomial of B, we have $(B - \lambda I)^M = O$ (by Corollary 1.1.32). This means that

$$P(B) = \sum_{j=0}^{M-1} \frac{P^{(j)}(\lambda)}{j!} (B - \lambda I)^j. \tag{1.1.18}$$

This shows that we may define

$$f(B) := \sum_{j=0}^{M-1} \frac{f^{(j)}(\lambda)}{j!} (B - \lambda I)^j \tag{1.1.19}$$

for a function f holomorphic on a neighborhood (depending on f) of $\sigma(B) = \{\lambda\}$.[8] We denote by $\mathcal{H}(\sigma(B))$ the collection of such functions.

[8] A weaker assumption on f would be also sufficient but we do not try to obtain an unduly general definition. See also Lemma 1.1.37 below.

It is easy to check that the formula

$$(fg)(B) = f(B)g(B) = g(B)f(B)$$

holds for $f, g \in \mathcal{H}(\sigma(B))$. In particular, for $w \in \mathbb{C} \setminus \{\lambda\}$ and $r_w(z) = (w-z)^{-1}$ we get

$$r_w(B) = (wI - B)^{-1} = \sum_{j=0}^{M-1} \frac{(B-\lambda I)^j}{(w-\lambda)^{j+1}}. \tag{1.1.20}$$

Remark 1.1.36. The following assertion yields another equivalent definition of $f(B)$ which can be used also in a general Banach space for a linear continuous operator $B\colon X \to X$ (see Section 1.2 for the notions of the Banach space and the continuous linear operator). Also Theorem 1.1.38 holds in this more general setting (Dunford Functional Calculus, see Proposition 3.1.19 or Dunford & Schwartz [53]).

Lemma 1.1.37. *Let γ be a positively oriented Jordan curve, $\sigma(B) \subset \operatorname{int}\gamma$, and let f be a holomorphic function on a neighborhood of $\overline{\operatorname{int}\gamma}$. Then*

$$f(B)x = \frac{1}{2\pi \mathrm{i}} \int_\gamma f(w)(wI-B)^{-1}x \,\mathrm{d}w, \qquad x \in X.$$[9]

Proof. By (1.1.20) we have

$$\frac{1}{2\pi \mathrm{i}} \int_\gamma f(w)(wI-B)^{-1}x \,\mathrm{d}w = \sum_{j=0}^{M-1} \left(\frac{1}{2\pi \mathrm{i}} \int_\gamma \frac{f(w)}{(w-\lambda)^{j+1}} \,\mathrm{d}w \right) (B-\lambda I)^j x.$$

The result follows now from the Cauchy Integral Formula.[10] □

Let $A \in L(\mathbb{C}^M)$ have the canonical form (1.1.17), i.e., $A = TBT^{-1}$. Then we define $f(A)$ by (1.1.19) replacing B by A. Notice that

$$f(A) = Tf(B)T^{-1}.$$

We can proceed in the same way for a general $A \in L(X)$ using the decomposition (1.1.12). This leads to the following theorem.

Theorem 1.1.38 (Functional Calculus). *Let X be a complex linear space and let $A \in L(X)$. Then there exists a unique linear operator*

$$\Phi\colon \mathcal{H}(\sigma(A)) \to L(X)$$

with the following properties:

---

[9] Since the integrand is a function $w \in \gamma \to \mathbb{C}^{M\times M}$ (in a matrix representation), the integral is an $M \times M$-tuple of standard curve integrals.

[10] We recall the following result from the theory of functions of a complex variable:

If f and γ are as in Lemma 1.1.37, then

$$f^{(j)}(z) = \frac{j!}{2\pi \mathrm{i}} \int_\gamma \frac{f(w)}{(w-z)^{j+1}} \,\mathrm{d}w \qquad \textit{holds for} \quad z \in \operatorname{int}\gamma \ \textit{and}\ j \in \mathbb{N} \cup \{0\}.$$

(i) $\Phi(fg) = \Phi(f)\Phi(g) = \Phi(g)\Phi(f)$ *for* $f, g \in \mathcal{H}(\sigma(A))$;

(ii) *if* $P(z) = \sum_{j=0}^{n} a_j z^j$, *then* $\Phi(P) = \sum_{j=0}^{n} a_j A^j$;

(iii) *if* $f(z) = \frac{1}{w-z}$ *for* $w \notin \sigma(A)$, *then* $\Phi(f) = (wI - A)^{-1}$.

Remark 1.1.39.

(i) A mapping $\Phi(f)$ can be computed either by Lemma 1.1.37 which is valid also for a general A, or by the formula

$$f(A)x = \sum_{l=1}^{k} \sum_{j=0}^{m(\lambda_l)-1} \frac{f^{(j)}(\lambda_l)}{j!}(A - \lambda_l I)^j \pi_l x$$

where $\sigma(A) = \{\lambda_1, \dots, \lambda_k\}$ and π_l is the projection onto $N(\lambda_l)$ defined by the decomposition (1.1.12). We note that these projections are also functions of A, namely $\pi_l = \chi_l(A)$ where

$$\chi_l(z) = \begin{cases} 1, & z \in B(\lambda_l; \delta), \\ 0, & z \notin \overline{B(\lambda_l; \delta)} \end{cases}$$

and $\delta > 0$ is small enough so that $\sigma(A) \cap B(\lambda_l; \delta) = \{\lambda_l\}$.

(ii) If X is a real linear space of finite dimension and $A \in L(X)$, then we can construct a functional calculus for $X_{\mathbb{C}}$ and $A_{\mathbb{C}}$ (see Remark 1.1.35(ii)).

(iii) We deduced a functional calculus from Theorem 1.1.34. The opposite way is also possible, namely to use functional calculus for finding the canonical form. An important role is played by projections π_l giving the decomposition (1.1.12). The interested reader can find more details, e.g., in Dunford & Schwartz [53, Section VII, 1].

Exercise 1.1.40. Show that

$$\operatorname{sgn} \det \boldsymbol{A} = (-1)^p \qquad \text{where} \quad p = \sum_{\substack{\lambda \in \sigma(A) \\ \lambda < 0}} m(\lambda)$$

for a matrix representation of A with real entries. (The sum over the empty set is defined to be zero.)

Hint. Notice that $\det \boldsymbol{A} = \lambda_1^{m_1} \cdots \lambda_k^{m_k}$.

Exercise 1.1.41. Show that the formula (1.1.19) yields a matrix representation of e^{tB} in the form

$$\begin{pmatrix} e^{t\lambda} & te^{t\lambda} & \cdots & \frac{t^{M-1}}{(M-1)!}e^{t\lambda} \\ 0 & e^{t\lambda} & \cdots & \frac{t^{M-2}}{(M-2)!}e^{t\lambda} \\ \vdots & \vdots & \ddots & \vdots \\ 0 & 0 & \cdots & e^{t\lambda} \end{pmatrix}$$

whenever B has the representation (1.1.17) with respect to the same basis.

Exercise 1.1.42. Use the formula in Remark 1.1.39(i) to estimate $\|e^{tA}x\|_{\mathbb{C}^M}$ for large positive t and large negative t in dependence on $\sigma(A)$.

Hint. Suppose that $\alpha < \operatorname{Re}\lambda < \beta$ for all $\lambda \in \sigma(A)$. Show that there are constants c_1, c_2 such that

$$\|e^{tA}x\| \leq \begin{cases} c_1 e^{\beta t}\|x\| & \text{for} \quad t \geq 0, \\ c_2 e^{\alpha t}\|x\| & \text{for} \quad t \leq 0 \end{cases} \qquad \text{for all} \quad x \in \mathbb{C}^M.$$

In particular, if $\beta < 0$, then all solutions of (1.1.15) tend to zero as $t \to +\infty$ (asymptotic stability).

Exercise 1.1.43. Let $A \in L(\mathbb{C}^M)$ have a regular matrix representation. Show that

(i) all matrix representations of A are regular;

(ii) there exists $B \in L(\mathbb{C}^M)$ such that

$$e^B = A.$$

Is this B unique? How is $\sigma(B)$ related to $\sigma(A)$?

Exercise 1.1.44. Let $A \in L(\mathbb{C}^M)$ and let $\lambda_1, \dots, \lambda_M$ be its eigenvalues repeating according to their algebraic multiplicity. Define the *trace* of A by the formula

$$\operatorname{tr} A := \sum_{j=1}^{M} \lambda_j.$$

Let $f_1, \dots, f_M$ be a basis of $\mathbb{C}^M$ and let $Af_j = \sum\limits_{k=1}^{M} a_{kj} f_k$.

(i) Show that $\operatorname{tr} A = \sum\limits_{j=1}^{M} a_{jj}$ and the expression for the trace does not depend on the choice of basis.

(ii) If $f_1, \dots, f_M$ is an orthonormal basis, then show that

$$\operatorname{tr} A = \sum_{k=1}^{M} (Af_k, f_k).$$

Notice that neither A need not be symmetric nor $f_1, \dots, f_M$ need not be eigenvectors.

1.2 Normed Linear Spaces

In Preface we mentioned that our main attention is focused on the properties of nonlinear mappings defined on various spaces of functions. Besides the linear structure studied in the previous section such spaces also have a topological structure. We will assume that the two structures are tied up (like the distributive law joins multiplication and addition in the notion of the field). A natural requirement of continuity of linear operations leads to the notion of a linear topological space. These spaces are often too general for purposes of nonlinear analysis. For example, basic notions and results of differential calculus in such spaces are not straightforward generalizations of the corresponding notions for functions of several variables and they frequently need profound ideas. Because of that we restrict our interest to cases when a topological structure is given by a metric, especially by a norm. Before starting with this concept we briefly introduce the main topological notions. For more information, the interested reader can consult books like Dugundji [52], Kelley [89].

A set X with a collection $\mathscr{T}$ of its subsets is called a *topological space* if $\mathscr{T}$ possesses the following properties:

(1) $\emptyset, X \in \mathscr{T}$;

(2) an intersection of a finite number of sets of $\mathscr{T}$ belongs to $\mathscr{T}$;

(3) a union of any subcollection of $\mathscr{T}$ belongs to $\mathscr{T}$.

Elements of $\mathscr{T}$ are called *open sets*. A subset $\mathcal{U} \subset X$ is called a *neighborhood* of a point $x \in X$ if there is an open set $\mathcal{G} \subset X$ such that $x \in \mathcal{G} \subset \mathcal{U}$.

An important special case of a topological space is the so-called *metric space*. This is a set X with a real function (*metric*) $\varrho\colon X \times X \to [0, \infty)$ for which

(1) $\varrho(x, y) = 0 \iff x = y$,

(2) $x, y \in X \implies \varrho(x, y) = \varrho(y, x)$ (the so-called *symmetry* of the metric),

(3) $x, y, z \in X \implies \varrho(x, z) \le \varrho(x, y) + \varrho(y, z)$ (the so-called *triangle inequality*).

If ϱ is a metric on X, then

$$B(x; r) := \{y \in X : \varrho(x, y) < r\}$$

is called the *open ball* centered at $x \in X$ with the radius $r > 0$. Open sets in a metric space are defined as subsets $\mathcal{G} \subset X$ which have the following property:

for every $x \in \mathcal{G}$ there is $\delta > 0$ such that $B(x; \delta) \subset \mathcal{G}$.

It is easy to prove that a metric space with this definition of open sets is also a topological space. For the following notions and results see, e.g., Dieudonné [44].

A subset $\mathcal{F}$ of a topological space X is called a *closed set* if $X \setminus \mathcal{F}$ is open. If $\mathcal{A} \subset X$, then the intersection of all closed sets containing $\mathcal{A}$ is called the *closure* of $\mathcal{A}$ and is denoted by $\overline{\mathcal{A}}$, i.e.,

$$\overline{\mathcal{A}} = \bigcap_{\substack{\mathcal{A} \subset \mathcal{F} \\ \mathcal{F} \text{ is closed}}} \mathcal{F}.$$

A dual notion is the *interior* ($\operatorname{int}\mathcal{A}$) of $\mathcal{A}$:

$$\operatorname{int}\mathcal{A} = \bigcup_{\substack{\mathcal{G}\subset\mathcal{A}\\ \mathcal{G}\text{ is open}}} \mathcal{G}.$$

The *boundary* $\partial\mathcal{A}$ of $\mathcal{A}$ is defined by

$$\partial\mathcal{A} := \overline{\mathcal{A}} \cap \overline{X\setminus\mathcal{A}}.$$

A subset $\mathcal{A}$ of X is said to be *dense* if $\overline{\mathcal{A}} = X$.

It is almost obvious that in a metric space X we have the following equivalences:

(i) $x \in \overline{\mathcal{A}} \iff \exists\{x_n\}_{n=1}^{\infty} \subset \mathcal{A} : \lim\limits_{n\to\infty} \varrho(x_n, x) = 0$,[11]

(ii) $x \in \operatorname{int}\mathcal{A} \iff \exists\delta > 0 : B(x;\delta) \subset \mathcal{A}$.

A metric space X is said to be *separable* if there is a countable dense subset of X.

If X, Y are topological spaces and $f\colon X \to Y$, then f is said to be *continuous* on X provided $f_{-1}(\mathcal{G})$ is open in X whenever $\mathcal{G}$ is an open set in Y. If f is injective and surjective, f, f^{-1} are both continuous, then f is called a *homeomorphism* of X onto Y. It is also possible to define continuity at a point $a \in X$ with help of the notion of a neighborhood:

> f is continuous at a if $f_{-1}(\mathcal{U})$ is a neighborhood of a whenever $\mathcal{U}$ is a neighborhood of $f(a)$.

A mapping f is continuous on X if and only if it is continuous at every point of X. The following equivalence holds in metric spaces X and Y:

$$f\colon X \to Y \quad \text{is continuous at } a \in X \quad \iff \quad (x_n \to a \implies f(x_n) \to f(a)).$$

A very important notion is that of compactness: A topological space X is said to be *compact* if for every open *covering* $\{\mathcal{G}_\gamma\}_{\gamma\in\Gamma}$ of X (i.e., $X = \bigcup\limits_{\gamma\in\Gamma} \mathcal{G}_\gamma$) there is a finite subset $\mathcal{K} \subset \Gamma$ such that

$$X = \bigcup_{\gamma\in\mathcal{K}} \mathcal{G}_\gamma$$

---

[11] We also say that the sequence $\{x_n\}_{n=1}^{\infty}$ is convergent to x and write $\lim\limits_{n\to\infty} x_n = x$ or, more simply, $x_n \to x$. The notion of a convergent sequence can be introduced also in topological spaces:

> $\{x_n\}_{n=1}^{\infty}$ is convergent to x if for every neighborhood $\mathcal{U}$ of x there is an index $n_0 \in \mathbb{N}$ such that $x_n \in \mathcal{U}$ for each $n \geq n_0$.

Warning. *In a topological space there need not be enough convergent sequences in order to describe a closure, etc.!* See, e.g., J. von Neumann's example in Dunford & Schwartz [53, Chapter V, 7, Example 38].

(a finite subcovering).

Any subset $\mathcal{A}$ of a topological space X is itself a topological space with the collection of open sets $\{(\mathcal{G} \cap \mathcal{A}) : \mathcal{G} \text{ open in } X\}$. A subset $\mathcal{A}$ of a topological space X is said to be compact in X if $\mathcal{A}$ is a compact topological space in this induced topology. Further, $\mathcal{A} \subset X$ is said to be *relatively compact* provided $\overline{\mathcal{A}}$ is compact.

In metric spaces we have the following characterization.

Proposition 1.2.1. *Let X be a metric space. Then $\mathcal{A} \subset X$ is relatively compact if and only if for any sequence $\{x_n\}_{n=1}^{\infty} \subset \mathcal{A}$ there is a convergent subsequence.*[12]

Beside this proposition, the importance of compactness in analysis is obvious from the next result which will be discussed more deeply in Section 7.2.

Proposition 1.2.2. *Let X be either a compact topological space or a sequentially compact topological space and let f be a continuous real function on X. Then there exist a maximal and a minimal value of f, i.e., there are $x_1, x_2 \in X$ such that*

$$f(x_1) \le f(x) \le f(x_2) \qquad \textit{for all} \quad x \in X.$$

To find a criterion for compactness in a particular space need not be an easy task. To formulate a general result we need one more notion the significance of which goes far beyond our present considerations. A sequence $\{x_n\}_{n=1}^{\infty}$ of elements of a metric space X is called a *Cauchy sequence* if for every $\varepsilon > 0$ there is $n_0 \in \mathbb{N}$ such that

$$\varrho(x_m, x_n) < \varepsilon \qquad \text{for all} \quad m, n \ge n_0.$$

A metric space X is said to be *complete* if any Cauchy sequence in X is convergent (to an element of X). We will encounter complete spaces "almost everywhere" in the subsequent text.

Proposition 1.2.3. *Let X be a complete metric space. Then $\mathcal{A} \subset X$ is relatively compact if and only if for every $\varepsilon > 0$ there is a finite set $\mathcal{K} \subset X$ (the so-called* finite ε-net *for $\mathcal{A}$) such that*

$$\forall a \in \mathcal{A}\ \exists x \in \mathcal{K} : \qquad \varrho(a, x) < \varepsilon.$$

In other words, $\mathcal{A} \subset \bigcup\limits_{x \in \mathcal{K}} B(x; \varepsilon)$.

Proposition 1.2.4. *Let X be a complete metric space and let $f : [\alpha, \beta) \to X$. If f is uniformly continuous on $[\alpha, \beta)$,*[13] *then there exists $\lim\limits_{x \to \beta_-} f(x) \in X$. In particular, if $\beta < \infty$, then f can be continuously extended to $[\alpha, \beta]$.*

---

[12] A topological space X is said to be *sequentially compact* if for any sequence $\{x_n\}_{n=1}^{\infty} \subset X$ there is a subsequence $\{x_{n_k}\}_{k=1}^{\infty}$ which is convergent to a point $x \in X$.

Warning. *These two notions of compactness are different in topological spaces. To be more precise:*

There is a compact topological space which is not sequentially compact and there is a sequentially compact topological space which is not compact!

[13] I.e., $\forall \varepsilon > 0\ \exists \delta > 0\ \forall x, y \in [\alpha, \beta) : |x - y| < \delta \implies \varrho(f(x), f(y)) < \varepsilon$.

Definition 1.2.5. A topological space X is called a *connected space* provided it is not possible to find two disjoint nonempty open sets $\mathcal{G}_1$, $\mathcal{G}_2$ such that

$$X = \mathcal{G}_1 \cup \mathcal{G}_2.$$

For $a \in X$ put

$$\mathcal{C}(a) := \bigcup \{\mathcal{A} \subset X : a \in \mathcal{A} \text{ and } \mathcal{A} \text{ is connected}\}.$$

Then $\mathcal{C}(a)$ is a connected set and it is called the *component* of the point a. If $a, b \in X$, $a \neq b$, then either $\mathcal{C}(a) = \mathcal{C}(b)$ or $\mathcal{C}(a) \cap \mathcal{C}(b) = \emptyset$.

Proposition 1.2.6. *Let X be a connected space, let $f\colon X \to Y$ be continuous. Then $f(X)$ is a connected subset of Y. In particular, if $\gamma\colon [0,1] \to Y$ is continuous, $\mathcal{A} \subset Y$, and $\gamma(0) \in \mathcal{A}$, $\gamma(1) \notin \mathcal{A}$, then there exists $t_0 \in [0,1]$ such that $\gamma(t_0) \in \partial\mathcal{A}$.*

Proposition 1.2.7. *Let X be a complete metric space and let $\mathcal{G}$ be an open subset of X. Then $\mathcal{G}$ is connected if and only if for any two points $a, b \in \mathcal{G}$ there exists a continuous mapping $\gamma\colon [0,1] \to \mathcal{G}$ such that $\gamma(0) = a$, $\gamma(1) = b$.*

For the proof see Kuratowski [97, § 50].

Now we are ready to start with the main subject of this section.

Definition 1.2.8. Let X be a real or complex linear space. A function $\|\cdot\|_X : X \to \mathbb{R}$ is called a *norm* on X if it has the following properties:

(1) $\|x\|_X = 0 \iff x = o$,

(2) $\|\alpha x\|_X = |\alpha|\|x\|_X$ for $\alpha \in \mathbb{R}$ or $\mathbb{C}$ and $x \in X$,

(3) $\|x + y\|_X \leq \|x\|_X + \|y\|_X$ for $x, y \in X$ (the so-called *triangle inequality*).

If a linear space X is endowed with a norm, then X is called a *normed linear space*.

In the sequel we will drop the index of the norm whenever there is no danger of confusion.

It is obvious that $\varrho(x, y) := \|x - y\|$ is a metric on X. Therefore all metric notions and results are transmitted to normed linear spaces. If a normed linear space is complete in this metric, then it is called a *Banach space*. Any metric space can be embedded as a dense set into a complete metric space. For a normed linear space X we get a slightly stronger result:

There exists a Banach space $\tilde{X}$ (the so-called completion *of X) and a linear injection $L\colon X \to \tilde{X}$ such that $\operatorname{Im} L$ is a dense subset of $\tilde{X}$ and*

$$\|x\|_X = \|L(x)\|_{\tilde{X}} \qquad \textit{for all} \qquad x \in X.$$

Example 1.2.9. Let X be an M-dimensional real linear space. Choose a basis $f_1, \dots, f_M$ of X and let $e_1, \dots, e_M$ be the standard basis of $\mathbb{R}^M$. For $x = \sum_{i=1}^{M} x_i f_i \in X$ put $\varphi(x) = \sum_{i=1}^{M} x_i e_i$. Then φ is an isomorphism of X onto $\mathbb{R}^M$. Moreover,

$$\|(x_1, \dots, x_M)\|_1 := \sum_{i=1}^{M} |x_i|, \qquad \|(x_1, \dots, x_M)\|_\infty := \max_{i=1,\dots,M} |x_i|,$$

$$\|(x_1, \dots, x_M)\|_2 := \left[\sum_{i=1}^{M} |x_i|^2\right]^{\frac{1}{2}}$$

are norms on $\mathbb{R}^M$ (for indices $1, \infty$ it is obvious, the triangle inequality for index 2 needs some effort – see also Proposition 1.2.30 below). These norms can be transmitted to X with help of φ, i.e.,

$$\|x\|_\alpha := \|\varphi(x)\|_\alpha, \qquad \alpha = 1, 2, \infty.$$

Similar results are true also for a complex linear space X when $\mathbb{C}^M$ is used instead of $\mathbb{R}^M$. The space X is a Banach space with respect to any of the above norms. ○

The classical Bolzano–Weierstrass result on the compactness of a closed bounded interval in $\mathbb{R}$ has the following generalization:

> *Let X be a finite-dimensional space endowed with an α-norm ($\alpha = 1, 2, \infty$). Then $\mathcal{A} \subset X$ is relatively compact if and only if it is bounded (i.e., there is a constant c such that $\|x\|_\alpha \le c$ for every $x \in \mathcal{A}$).*

We note that this result is true for any norm on X (see Corollary 1.2.11(i) below).

Proposition 1.2.10. *Let X and Y be normed linear spaces and let A be a linear operator from X into Y. Then the following statements are equivalent:*

(i) *A is continuous on X;*

(ii) *A is continuous at $o \in X$;*

(iii) *there is a constant c such that the inequality*

$$\|Ax\|_Y \le c\|x\|_X \qquad \text{is valid for all} \quad x \in X. \tag{1.2.1}$$

Proof. The easy proof is left to the reader as an exercise. □

We denote the collection of all continuous linear operators from X into Y by $\mathcal{L}(X, Y)$ and the least possible constant c in (1.2.1) by $\|A\|_{\mathcal{L}(X,Y)}$. This quantity has all properties of a norm on the linear space $\mathcal{L}(X, Y)$. We will always consider this norm (the so-called *operator norm*) on $\mathcal{L}(X, Y)$. If $X = Y$, we will use the shorter notation $\mathcal{L}(X)$ instead of $\mathcal{L}(X, X)$.

We now return to Example 1.2.9. It is obvious that there are positive constants c_1, c_2 such that

$$c_1\|x\|_1 \le \|x\|_2 \le c_2\|x\|_1 \qquad \text{for all} \quad x \in \mathbb{R}^M \ (\mathbb{C}^M).$$

(Here, e.g., $c_1 = \frac{1}{M}$, $c_2 = \sqrt{M}$.) Such constants exist also for the norms $\|\cdot\|_1$, $\|\cdot\|_\infty$. More generally, two *norms* on a linear space X are called *equivalent* if they satisfy such inequalities. In other words, two norms $\|\cdot\|_\alpha$, $\|\cdot\|_\beta$ on a linear space X are equivalent if the identity map from $(X, \|\cdot\|_\alpha)$ into $(X, \|\cdot\|_\beta)$ is continuous together with its inverse, i.e., it is an *isomorphism*.[14]

Corollary 1.2.11.

(i) *Any two norms on a finite-dimensional linear space X are equivalent. In particular, X is a Banach space.*

(ii) *Let X, Y be normed linear spaces and* $\dim X < \infty$. *Then* $L(X,Y) = \mathcal{L}(X,Y)$, *i.e., any linear operator from X into Y is continuous.*

Proof. (i) Let φ be as in Example 1.2.9 and consider $\mathbb{R}^M$ (or $\mathbb{C}^M$) equipped with the $\|\cdot\|_1$-norm. Then for $x = (x_1, \dots, x_M) \in \mathbb{R}^M$ we have

$$\|\varphi^{-1}(x)\|_X = \left\| \sum_{i=1}^M x_i f_i \right\|_X \le \sum_{i=1}^M |x_i| \, \|f_i\|_X \le c \sum_{i=1}^M |x_i| = c\|x\|_1,$$

i.e., φ^{-1} is continuous. Observe that for proving continuity of φ it is sufficient to show that

$$\inf\{\|\varphi^{-1}(x)\|_X : \|x\|_1 = 1\} > 0.$$

But this is true since the set $\{x \in \mathbb{R}^M : \|x\|_1 = 1\}$ is compact and φ^{-1} is continuous.

Now let $\|\cdot\|$, $\|\cdot\|_\sim$ be two norms on X, $\dim X = M$ and let ι be the identity map from X onto $\tilde{X}$ ($= X$ with the norm $\|\cdot\|_\sim$). The result follows from the commutativity of the diagram in Figure 1.2.1. Since $\mathbb{R}^M$ and $\mathbb{C}^M$ are complete

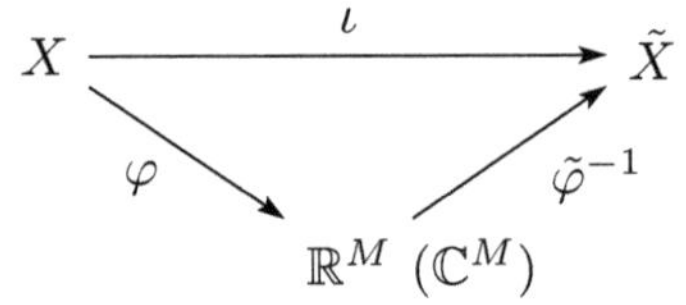

Figure 1.2.1.

spaces with respect to the $\|\cdot\|_1$-norm (the classical Bolzano–Cauchy condition)

[14]Unlike the "algebraic" isomorphism from Definition 1.1.11 here, it is understood in the "topological" sense. In general, $A \in \mathcal{L}(X,Y)$ is an isomorphism if A is injective, surjective and $A^{-1} \in \mathcal{L}(Y,X)$.

and $\{u_n\}_{n=1}^{\infty} \subset X$ is a Cauchy sequence if and only if $\{\varphi(u_n)\}_{n=1}^{\infty}$ is a Cauchy sequence, X is a Banach space.

(ii) It is sufficient to prove continuity with respect to the 1-norm on X. For $x = \sum_{i=1}^{M} x_i f_i \in X$ we have

$$\|Ax\|_Y \leq \sum_{i=1}^{M} |x_i| \, \|Af_i\|_Y \leq c \sum_{i=1}^{M} |x_i| = c\|x\|_1. \qquad \square$$

Example 1.2.12 (spaces of continuous functions). Let T be a compact topological space. Then any continuous real (complex) function f is bounded on T (Proposition 1.2.2), and

$$\|f\|_T = \sup\{|f(x)| : x \in T\}$$

is a norm on the space $C(T)$ of all such functions. Convergence of a sequence in this norm is the *uniform convergence* on T. It follows that $C(T)$ is a Banach space.

If T is not compact, then a continuous function on T need not be bounded. To get a topology on a family of continuous functions on T we can either restrict our attention to the space $BC(T)$ of all bounded, continuous functions on T or assume certain properties of T which are weaker than compactness (the reader can consider $\mathbb{R}^M$ as a model of such T). As a result we wish to obtain a topology on $C(T)$ in which convergence of a sequence is equivalent to the *locally uniform convergence*. This can be done as follows. Let a topological space T be a countable union of open, relatively compact subsets $\mathcal{T}_n$.[15] We leave to the reader to verify that the sum

$$\varrho(f,g) := \sum_{n=1}^{\infty} \frac{1}{2^n} \frac{\|f-g\|_n}{1+\|f-g\|_n} \tag{1.2.2}$$

where

$$\|f-g\|_n := \sup\{|f(x)-g(x)| : x \in \mathcal{T}_n\},$$

defines a metric on $C(T)$ and the convergence of a sequence in this metric is actually the locally uniform convergence, i.e., uniform convergence on any compact subset of T. Since ϱ is bounded it cannot be induced by any norm. Even more is true, namely, there is no norm on $C(T)$ which generates the same system of open sets as the metric ϱ does (provided T itself is not compact). ○

We now state two fundamental results concerning spaces of continuous functions. To formulate the first we need the concept of equicontinuity:

A family $\mathscr{F} \subset C(T)$ is said to be *equicontinuous* if for all $x \in T$ and $\varepsilon > 0$ there is a neighborhood $\mathcal{U}$ of x such that

$$y \in \mathcal{U}, \ f \in \mathscr{F} \qquad \Longrightarrow \qquad |f(y) - f(x)| < \varepsilon.$$

[15] A basic example is $\mathbb{R}^M$ or $\mathbb{C}^M$. Another example is the set $\mathbb{N}$ of natural numbers with the discrete metric: $d(m,n) = 1$ if $m \neq n$ and $d(m,m) = 0$.

Theorem 1.2.13 (Arzelà–Ascoli). *Let T be a topological space which is a union of a sequence of open, relatively compact subsets. Then $\mathcal{F} \subset C(T)$ is relatively compact in the ϱ-metric if and only if the following two conditions are satisfied:*

(i) *$\mathcal{F}$ is equicontinuous;*

(ii) *for each $x \in T$ the set $\{f(x) : f \in \mathcal{F}\}$ is bounded in $\mathbb{R}$ (or $\mathbb{C}$).*[16]

Proof. We omit the proof and refer, e.g., to Dugundji [52, Section XII, 6] or Kelley [89, Chapter 7, Theorem 17] where more general results are proved. □

Since a continuous function can be very strange (e.g., nowhere differentiable) it is often desirable to have an approximation procedure. The first result of this type was the famous Weierstrass Theorem on uniform approximation by polynomials. One of the characteristic features of this approximation consists in the fact that the product of two continuous functions is a continuous function and the same is true for polynomials. In algebraic terms: Both sets are not only linear spaces but also algebras.[17]The following generalization of the Weierstrass Theorem is due to M. H. Stone.

Theorem 1.2.14 (Stone–Weierstrass). *Let T satisfy the assumption of Theorem 1.2.13 and let $C_{\mathbb{R}}(T)$ be an algebra of all real continuous functions on T. Let $\mathcal{A} \subset C_{\mathbb{R}}(T)$ be a subalgebra which contains constant functions and separates points of T (i.e., for any $x, y \in T$, $x \neq y$, there is $f \in \mathcal{A}$ such that $f(x) \neq f(y)$). Then $\mathcal{A}$ is dense in $C_{\mathbb{R}}(T)$ with respect to the ϱ-metric. In particular, $C(\Omega)$ is separable for open $\Omega \subset \mathbb{R}^M$.*

Proof. The proof can be found, e.g., in Dugundji [52, XIII, 3] or Kelley [89, Chapter 7, Exercise T]. In particular, take $\mathcal{A}$ to be the set of all polynomials in $\mathbb{R}^M$. □

We note that Theorem 1.2.14 can be easily extended to the space of complex continuous functions. In this case, $\mathcal{A}$ is assumed to possess the following additional property:

If $f \in \mathcal{A}$, then also $\overline{f} \in \mathcal{A}$.[18]

The reader can ask why certain additional properties are needed for compactness in infinite-dimensional spaces like $C(T)$ in contrast to finite-dimensional spaces. The following theorem explains not only this situation but also the technical difficulties which one meets in the calculus of variations (see Chapter 7).

Proposition 1.2.15 (F. Riesz). *Let X be a normed linear space. Then the closed unit ball $\overline{B(o;1)} := \{x \in X : \|x\| \le 1\}$ is compact (in the norm topology) if and only if X has finite dimension.*

---

[16]If T is compact and (i) holds, then the assumption (ii) is equivalent to the boundedness of $\mathcal{F}$ in $C(T)$.

[17]A linear space X with a binary operation (product) which is associative and distributive with respect to linear operations is called an *algebra*. Further, if X is a normed linear space and for the product the inequality $\|x \cdot y\| \le \|x\|\|y\|$ holds for every $x, y \in X$, then X is called a *normed algebra* and, in the case that X is complete, a *Banach algebra*.

[18]If $z = x + \mathrm{i}y$, $x, y \in \mathbb{R}$, then its complex conjugate $\overline{z}$ is defined by $\overline{z} := x - \mathrm{i}y$.

Proof. Sufficiency is obvious (see Example 1.2.9 and Corollary 1.2.11(i)). It remains to prove necessity. We proceed by contradiction. Assume that $\dim X = \infty$. Choose $0 < \varepsilon < 1$ and suppose that we have $x_1, \dots, x_n \in \overline{B(o;1)}$ such that

$$\|x_i - x_j\| > 1 - \varepsilon \qquad \text{for all} \quad 1 \le i < j \le n.$$

We shall show that we can find another element $x_{n+1} \in \overline{B(o;1)}$ such that $\{x_1, \dots, x_{n+1}\}$ has the same property. Since $X_n = \operatorname{Lin}\{x_1, \dots, x_n\} \ne X$ there is $y \in X \setminus X_n$. Denote

$$d := \inf\{\|y - x\| : x \in X_n\}.$$

Observe that $d > 0$ since X_n is a closed subspace.[19] By the definition of the greatest lower bound, there exists $\tilde{x} \in X_n$ such that

$$d \le \|y - \tilde{x}\| < d(1 + \varepsilon).$$

For $x_{n+1} := \frac{y - \tilde{x}}{\|y - \tilde{x}\|} \in \overline{B(o;1)}$ and $x \in X_n$ we get

$$\|x_{n+1} - x\| = \frac{1}{\|y - \tilde{x}\|}\|y - (\tilde{x} + \|y - \tilde{x}\|x)\| \ge \frac{1}{d(1+\varepsilon)} d > 1 - \varepsilon.$$

Thus an infinite sequence $\{x_n\}_{n=1}^{\infty} \subset \overline{B(o;1)}$ with no convergent subsequence has been constructed, which contradicts compactness of $\overline{B(o;1)}$. □

Example 1.2.16 (spaces of integrable functions). Let Ω be a Lebesgue measurable subset of $\mathbb{R}^M$ and let $\mathrm{d}x$ denote the Lebesgue measure in $\mathbb{R}^M$. For $p \in [1, \infty)$ we denote

$$\mathcal{L}^p(\Omega) := \left\{ f \colon \Omega \to \mathbb{R} \text{ (or } \mathbb{C}) : f \text{ is measurable} \right. \\ \left. \text{and } |f|_{\mathcal{L}^p(\Omega)} := \left[\int_\Omega |f(x)|^p \,\mathrm{d}x \right]^{\frac{1}{p}} < \infty \right\}. \tag{1.2.3}$$

The *Minkowski inequality*

$$|f + g|_{\mathcal{L}^p(\Omega)} \le |f|_{\mathcal{L}^p(\Omega)} + |g|_{\mathcal{L}^p(\Omega)} \tag{1.2.4}$$

implies that $\mathcal{L}^p(\Omega)$ is a linear space. Observe that $|\cdot|_{\mathcal{L}^p(\Omega)}$ is not a norm since $|f|_{\mathcal{L}^p(\Omega)} = 0$ implies only $f = o$ almost everywhere (abbreviation: a.e.) in Ω. Put

$$N(\Omega) = \{f \colon \Omega \to \mathbb{C} : f = o \text{ a.e. in } \Omega\}.$$

Then N is a linear subspace of $\mathcal{L}^p$ and the factor space

$$L^p(\Omega) := \mathcal{L}^p(\Omega)|_N$$

[19] Every finite-dimensional subspace $Y \subset X$ is complete, and therefore closed in X.

is a normed linear space with the norm

$$\|[f]\|_{L^p(\Omega)} = |f|_{\mathcal{L}^p(\Omega)} \qquad \text{for any} \quad f \in [f].$$[20]

For the sake of simplicity we will use the notation f instead of the superfluous $[f]$ for an element of $L^p(\Omega)$ and will call it simply a function. It is also convenient to introduce the space $L^\infty(\Omega)$ of all (classes of) essentially bounded measurable functions. We recall that f is said to be *essentially bounded* on Ω if there is a constant c such that

$$|f(x)| \le c \qquad \text{for a.e.} \quad x \text{ in } \Omega.$$

The least possible c is denoted by $\|f\|_{L^\infty(\Omega)}$. Again $\|\cdot\|_{L^\infty(\Omega)}$ is a norm on $L^\infty(\Omega)$.

We mention another important inequality – the so-called *Hölder inequality*:

If $1 \le p \le \infty$ and p' is the conjugate exponent *($\frac{1}{p} + \frac{1}{p'} = 1$ where $\frac{1}{\infty}$ is here defined to be 0) and $f \in L^p(\Omega)$, $g \in L^{p'}(\Omega)$, then $fg \in L^1(\Omega)$ and*

$$\|fg\|_1 \le \|f\|_p \|g\|_{p'}. \tag{1.2.5}$$

○

Proposition 1.2.17. *$L^p(\Omega)$ is a Banach space for any $1 \le p \le \infty$.*

Proof. We give the proof for $p = 1$ (some small modifications are needed for $1 < p < \infty$, while the proof for $p = \infty$ is similar to the one of completeness of $C(T)$, cf. Example 1.2.12). Let $\{f_n\}_{n=1}^\infty$ be a Cauchy sequence in $L^1(\Omega)$. Then for any $k \in \mathbb{N}$ there is $n_k \in \mathbb{N}$ such that $\|f_n - f_{n_k}\|_1 < \frac{1}{2^k}$ for all $n \ge n_k$. We can assume that the sequence $\{n_k\}_{k=1}^\infty$ is strictly increasing. Put $g_p = \sum\limits_{k=1}^{p} |f_{n_{k+1}} - f_{n_k}|$. Since

$$\int_\Omega g_p(x)\,\mathrm{d}x \le \sum_{k=1}^{p} \int_\Omega |f_{n_{k+1}}(x) - f_{n_k}(x)|\,\mathrm{d}x \le \sum_{k=1}^{p} \frac{1}{2^k},$$

the Monotone Convergence Theorem[21] gives that $g = \lim\limits_{n\to\infty} g_n$ has a finite integral over Ω and therefore g is finite a.e. in Ω. This means that $\sum\limits_{k=1}^{\infty} |f_{n_{k+1}}(x) - f_{n_k}(x)|$ is a.e. convergent, and therefore $f(x) := \lim\limits_{k\to\infty} f_{n_k}(x)$ exists a.e. in Ω. By the Fatou Lemma[22] we have

$$\int_\Omega |f(x) - f_{n_k}(x)|\,\mathrm{d}x \le \liminf_{l\to\infty} \int_\Omega |f_{n_l}(x) - f_{n_k}(x)|\,\mathrm{d}x \le \frac{1}{2^{k-1}}.$$

[20] For the sake of simplicity we will use in the sequel the notation $\|\cdot\|_p$ instead of $\|\cdot\|_{L^p(\Omega)}$.

[21] This theorem reads as follows:

Let $\{g_n\}_{n=1}^\infty$ be an increasing sequence of nonnegative measurable functions on Ω and let $g = \lim\limits_{n\to\infty} g_n$. Then

$$\lim_{n\to\infty} \int_\Omega g_n(x)\,\mathrm{d}x = \int_\Omega g(x)\,\mathrm{d}x.$$

[22] The Fatou Lemma reads:

In particular,

$$f \in L^1(\Omega) \qquad \text{and} \qquad \lim_{k\to\infty} \|f_{n_k} - f\|_1 = 0.$$

The rest of the proof is easy. Indeed, a Cauchy sequence which has a convergent subsequence is itself convergent. □

Remark 1.2.18. The proof shows that the following statement is true:

If $\{f_n\}_{n=1}^{\infty}$ is convergent to f in the L^p-norm, then there is a subsequence $\{f_{n_k}\}_{k=1}^{\infty}$ which converges to f a.e., and there is $g \in L^p(\Omega)$, $g \geq 0$, such that

$$|f_{n_k}(x)| \leq g(x) \qquad \textit{for a.e.} \quad x \in \Omega.$$

Warning. *The whole sequence need not be a.e. convergent!* To see this arrange the characteristic functions of the intervals $\left(\frac{k-1}{2^k}, \frac{k}{2^k}\right)$ into a sequence.

Approximations of integrable functions by more regular functions, like continuous or differentiable ones, are often desirable.

Proposition 1.2.19 (Density Theorem). *Let Ω be an open subset of $\mathbb{R}^M$. Then for any $p \in [1, \infty)$ the subset $C(\Omega) \cap L^p(\Omega)$ is dense in $L^p(\Omega)$. In particular, $L^p(\Omega)$ is separable.*

Proof. It is based on the application of the Luzin Theorem.[23] See also Proposition 1.2.21 below. □

We now show another type of approximations which is more constructive and therefore often more convenient in applications. If f, g are measurable functions on $\mathbb{R}^M$, then we define their *convolution* $f * g$ as

$$(f * g)(x) := \int_{\mathbb{R}^M} f(x-y)g(y)\,\mathrm{d}y \qquad \text{for all} \quad x \in \mathbb{R}^M \tag{1.2.6}$$

for which the integral exists. We note that the properties of the convolution follow from the Fubini Theorem provided measurability of the function $(x, y) \mapsto f(x - y)g(y)$ is established. For details see, e.g., Folland [64], Gripenberg, Londen & Staffans [74, Chapters 2–4], and also Example 2.1.28. The following assertion is a basic result on convolutions.

Let $\{h_n\}_{n=1}^{\infty}$ be a sequence of measurable functions which are uniformly bounded below by an $h \in L^1(\Omega)$. Then

$$\int_\Omega \liminf_{n\to\infty} h_n(x)\,\mathrm{d}x \leq \liminf_{n\to\infty} \int_\Omega h_n(x)\,\mathrm{d}x.$$

The statement holds for $\limsup$ *with the reverse inequality for a sequence bounded above by an integrable function.*

Put here $h_l = |f_{n_l} - f_{n_k}|$.

[23] Roughly speaking, the *Luzin Theorem* says that a bounded measurable function is continuous with respect to sets, measures of which are arbitrarily close to the measure of Ω provided the latter is finite. For a more general formulation and the proof of the Luzin Theorem the reader can consult, e.g., Rudin [132, § 2.23].

Proposition 1.2.20. *Let $f \in L^1(\mathbb{R}^M)$.*

(i) *If $g \in L^p(\mathbb{R}^M)$, $1 \le p \le \infty$, then*

$$f * g \in L^p(\mathbb{R}^M) \qquad \text{and} \qquad \|f * g\|_p \le \|f\|_1 \|g\|_p.$$

(ii) *If $g \in L^\infty(\mathbb{R}^M)$, then $f * g$ is bounded and uniformly continuous on $\mathbb{R}^M$.*

(iii) *If $g \in L^p(\mathbb{R}^M)$ and $\frac{\partial g}{\partial x_i} \in L^p(\mathbb{R}^M)$, then $\frac{\partial}{\partial x_i}(f * g) = f * \frac{\partial g}{\partial x_i}$ a.e. in $\mathbb{R}^M$.*

(iv) *If φ is a nonnegative, measurable function with $\int_{\mathbb{R}^M} \varphi(x)\, \mathrm{d}x = 1$ (the so-called* mollifier*) and $\varphi_n(x) := n^M \varphi(nx)$, then $\varphi_n * g$ converge to g in the L^p-norm for any $g \in L^p(\mathbb{R}^M)$, $1 \le p < \infty$.*

If T is a topological space and $f \colon T \to \mathbb{R}$ $(\mathbb{C})$, then the *support* of f (abbreviation $\operatorname{supp} f$) is the set

$$\overline{\{x \in T : f(x) \ne 0\}}.$$

If $\Omega \subset \mathbb{R}^M$ is an open set, then $\mathcal{D}(\Omega)$ denotes the set of all infinitely differentiable functions on Ω (i.e., their derivatives of arbitrary order are continuous in Ω) which have compact support lying in Ω.

We show that $\mathcal{D}(\Omega)$ contains enough functions. Put

$$\omega(x) = \begin{cases} \mathrm{e}^{-\frac{1}{1-x^2}}, & |x| < 1, \\ 0, & |x| \ge 1. \end{cases}$$

It is a matter of simple calculation to prove that $\omega \in \mathcal{D}(\mathbb{R})$. If $a \in \Omega$, then $B(a;\delta) \subset \Omega$ for a $\delta > 0$ small enough and the function $\varphi(y) := \omega\left(\frac{2}{\delta}\|y - a\|_{\mathbb{R}^M}\right)$ belongs to $\mathcal{D}(\Omega)$. However, much more is true.

Proposition 1.2.21. *Let Ω be an open set in $\mathbb{R}^M$ and let $p \in [1,\infty)$. Then $\mathcal{D}(\Omega)$ is dense in $L^p(\Omega)$.*

Proof. The just defined function φ multiplied by an appropriate constant satisfies the assumptions of Proposition 1.2.20(iv). There is a strictly increasing sequence of compact subsets $\mathcal{C}_m$ of Ω such that

$$\bigcup_{m=1}^{\infty} \mathcal{C}_m = \Omega.$$

Extend $f \in L^p(\Omega)$ by zero outside Ω and put

$$f_m = \chi_m f$$

where χ_m is the characteristic function of the set $\mathcal{C}_m$. Then $f_m \to f$ in the L^p-norm. By Proposition 1.2.20, $\varphi_n * f_m \in \mathcal{D}(\Omega)$ for $n \ge n_m$ and

$$\|\varphi_n * f_m - f\|_p \le \|\varphi_n * (f_m - f)\|_p + \|\varphi_n * f - f\|_p \le \|f_m - f\|_p + \|\varphi_n * f - f\|_p.$$

The result follows from Proposition 1.2.20(iv). □

Remark 1.2.22. If $\operatorname{meas}\Omega < \infty$ and $1 \le \tilde{p} < p \le \infty$, then, by the Hölder inequality,

$$\|f\|_{\tilde{p}} \le (\operatorname{meas}\Omega)^{\frac{1}{\tilde{p}}-\frac{1}{p}} \|f\|_p, \qquad f \in L^p(\Omega). \tag{1.2.7}$$

This means that the identity map of $L^p(\Omega)$ into $L^{\tilde{p}}(\Omega)$ is continuous. We will denote this fact by $L^p(\Omega) \hookrightarrow L^{\tilde{p}}(\Omega)$ and say that $L^p(\Omega)$ is *continuously embedded* into $L^{\tilde{p}}(\Omega)$.

Warning. *Simple examples show that this is not true if* $\operatorname{meas}\Omega = \infty$*!*

The following assertion is an analogue of Theorem 1.2.13.

Proposition 1.2.23 (A. N. Kolmogorov). *Let Ω be an open set in $\mathbb{R}^M$. Then $\mathcal{M} \subset L^p(\Omega)$, $p \in [1,\infty)$, is relatively compact if and only if the following conditions are satisfied:*

(i) *$\mathcal{M}$ is bounded in $L^p(\Omega)$,*

(ii) $\forall \varepsilon > 0\ \exists \delta > 0\ \forall f \in \mathcal{M}$: $\displaystyle\int_\Omega |f(x+y) - f(x)|^p \, dx < \varepsilon$ *for all* $\|y\|_{\mathbb{R}^M} < \delta$,[24]

(iii) $\forall \varepsilon > 0\ \exists \eta > 0\ \forall f \in \mathcal{M}$: $\displaystyle\int_{\{x\in\Omega : \|x\|_{\mathbb{R}^M} \ge \eta\}} |f(x)|^p \, dx < \varepsilon$.

Proof. For the proof based on Proposition 1.2.3 see Yosida [161, Chapter 10, § 1]. □

Remark 1.2.24. All results from 1.2.16–1.2.23 also hold in spaces of sequences

$$l^p := \left\{ x = \{x_n\}_{n=1}^\infty : \|x\|_p = \left(\sum_{n=1}^\infty |x_n|^p \right)^{\frac{1}{p}} < \infty \right\}$$

which can be regarded as $\mathcal{L}^p(\mathbb{N})$ equipped with the counting measure μ ($\mu(\mathcal{A}) = \operatorname{card}\mathcal{A}$).

Example 1.2.25 (spaces of differentiable functions). We can consider either classical derivatives (defined as limits of relative differences) or weak derivatives. We start with the former case.

Let $\alpha = (\alpha_1, \dots, \alpha_M)$ be a *multiindex*, i.e., $\alpha_i \in \mathbb{N} \cup \{0\}$, $i = 1, \dots, M$, and $|\alpha| := \alpha_1 + \dots + \alpha_M$. For a function f on an open set $\Omega \subset \mathbb{R}^M$ we put

$$D^\alpha f(x) := \frac{\partial^{|\alpha|} f(x)}{\partial x_1^{\alpha_1} \cdots \partial x_M^{\alpha_M}}$$

and say that $f \in C^n(\Omega)$ if $D^\alpha f$ are continuous for all multiindices α for which $|\alpha| \le n$. We can use the metric ϱ given by (1.2.2) to define $\varrho_\alpha(f,g) := \varrho(D^\alpha f, D^\alpha g)$ for a multiindex α and set

$$\varrho_n(f,g) := \sum_{|\alpha| \le n} \varrho_\alpha(f,g).$$

[24] If $x + y \notin \Omega$, then we set $f(x+y) := 0$.

Then ϱ_n is a metric on $C^n(\Omega)$ and the convergence in this metric is the locally uniform convergence of all derivatives D^α, $0 \le |\alpha| \le n$ ($D^o f = f$). Another possibility is to consider only such functions $f \in C^n(\Omega)$ for which $D^\alpha f$ is bounded in Ω for all $0 \le |\alpha| \le n$. We denote the collection of such functions by $C^n(\overline{\Omega})$[25] and put

$$\|f\|_{C^n(\overline{\Omega})} := \sum_{|\alpha|\le n} \sup_{x\in\Omega} |D^\alpha f(x)|.$$

This is a norm, $C^n(\overline{\Omega})$ is a Banach space, and the convergence of a sequence $\{f_k\}_{k=1}^\infty \subset C^n(\overline{\Omega})$ to f in this norm means that

$$D^\alpha f_k \rightrightarrows D^\alpha f \qquad \text{uniformly on} \quad \Omega \quad \text{for all} \quad |\alpha| \le n.$$

It is sometimes convenient to have a finer scale of spaces of differentiable functions. We can achieve that by introducing the Hölder continuous functions:

A function $f\colon \Omega \to \mathbb{R}$ (or $\mathbb{C}$) is called *γ-Hölder continuous* ($0 < \gamma \le 1$) if there is a constant c such that the inequality

$$|f(x) - f(y)| \le c\|x - y\|^\gamma \qquad \text{holds for all} \quad x, y \in \Omega.$$[26]

The quantity

$$\|f\|_{C^{0,\gamma}(\overline{\Omega})} := \sup_{x\in\Omega} |f(x)| + \sup_{\substack{x,y\in\Omega\\ x\ne y}} \frac{|f(x) - f(y)|}{\|x - y\|^\gamma}$$

is a norm on the space $C^{0,\gamma}(\overline{\Omega})$ of γ-Hölder continuous, bounded functions on Ω. The space $C^{n,\gamma}(\overline{\Omega})$ is defined similarly.

We note that $C^{n,\gamma}(\overline{\Omega})$ is a Banach space with respect to the above norm (cf. Exercise 5.3.4).

Now we turn our attention to weak derivatives on an open set $\Omega \subset \mathbb{R}^M$. Let

$$f \in L^1_{\mathrm{loc}}(\Omega)$$

(this means that $f \in L^1(\mathcal{K})$ for every compact subset $\mathcal{K} \subset \Omega$), and let α be a multiindex. A function $g \in L^1_{\mathrm{loc}}(\Omega)$ is called an *α-weak derivative* of f if

$$\int_\Omega f(x) D^\alpha \varphi(x)\, \mathrm{d}x = (-1)^{|\alpha|} \int_\Omega g(x)\varphi(x)\, \mathrm{d}x \quad \text{for every } \varphi \in \mathcal{D}(\Omega). \tag{1.2.8}$$

We will denote $g = D^\alpha_{\mathrm{w}} f$ and omit w in case there is no danger of ambiguity.

[25] In connection with this notation observe that for a relatively compact set Ω all derivatives $D^\alpha f$, $|\alpha| \le n - 1$, are uniformly continuous, and therefore continuously extendable to $\overline{\Omega}$.

[26] If $\gamma = 1$, then it is more common to say that f is a *Lipschitz continuous* function. We note that the inequality is satisfied for a $\gamma > 1$ only if f is a constant function (cf. Exercise 5.3.7).

Warning. *Even in the one-dimensional case the ordinary derivative existing almost everywhere need not be the weak derivative!*

For example, the Heaviside function

$$H(x) = \begin{cases} 1, & x \geq 0, \\ 0, & x < 0, \end{cases} \qquad \text{satisfies} \qquad H'(x) = 0 \quad \text{for } x \in \mathbb{R} \setminus \{0\}$$

but the weak derivative does not exist. The distributional derivative of H [27] is the Dirac measure.

We note that an *absolutely continuous function* f on an interval $\mathcal{I} \subset \mathbb{R}$ has a derivative a.e. (the Lebesgue Theorem, see Rudin [132]), and

$$f(x) = f(a) + \int_a^x f'(y)\,dy \qquad \text{for} \quad a, x \in \mathcal{I}.$$

This implies that $D_{\mathrm{w}} f = f'$. The situation in higher dimensions is not so simple since there are several non-equivalent definitions of absolutely continuous functions.

Having a definition of weak derivatives we can define *Sobolev spaces* $W^{k,p}(\Omega)$ for an open set $\Omega \subset \mathbb{R}^M$ as follows:

$$W^{k,p}(\Omega) := \{f \in L^p(\Omega) : \text{derivatives } D^\alpha_{\mathrm{w}} f \text{ exist} \\ \text{and belong to } L^p(\Omega) \text{ for all } |\alpha| \leq k\}$$

with the norm

$$\|f\|_{W^{k,p}(\Omega)} := \sum_{|\alpha| \leq k} \|D^\alpha_{\mathrm{w}} f\|_p. \tag{1.2.9}$$

[28]

Similarly to the definition of L^p spaces, classes of functions are considered here. Since $L^p(\Omega)$ is a Banach space, $W^{k,p}(\Omega)$ is a Banach space, too. ○

As we will see later in this book, Sobolev spaces play an important role in the study of boundary value problems. For this purpose the following assertions are important.

---

[27] A linear form Φ on the linear space $\mathcal{D}(\Omega)$ is called a *distribution* (this notion is due to L. Schwartz) if it has the following continuity property:

If $\varphi_n \in \mathcal{D}(\Omega)$ have their supports in the same compact set $\mathcal{K} \subset \Omega$ and $D^\alpha \varphi_n \rightrightarrows D^\alpha \varphi$ uniformly on Ω for all multiindices α, then

$$\Phi(\varphi_n) \to \Phi(\varphi).$$

Note that any $f \in L^1_{\mathrm{loc}}(\Omega)$ (and even a regular Borel measure on Ω, see, e.g., Rudin [132]) yields a distribution Φ_f by the formula $\Phi_f(\varphi) = \int_\Omega f(x)\varphi(x)\,dx$ for any $\varphi \in \mathcal{D}(\Omega)$. The *distributional derivative* D^α of a distribution Φ is defined as $D^\alpha\Phi(\varphi) := (-1)^{|\alpha|}\Phi(D^\alpha\varphi)$, $\varphi \in \mathcal{D}(\Omega)$. It is easy to prove that $D^\alpha\Phi$ is again a distribution, and an α-weak derivative of $f \in L^1_{\mathrm{loc}}(\Omega)$ is actually equal to the distributional derivative $D^\alpha\Phi_f$. As the Heaviside function shows the converse is not true.

[28] Similarly as for the Lebesgue norm we will use in the sequel the notation $\|\cdot\|_{k,p}$ instead of $\|\cdot\|_{W^{k,p}(\Omega)}$ for the Sobolev norm.

Theorem 1.2.26 (Sobolev Embedding Theorem). *Let $k \in \mathbb{N}$ and let $p \in [1, \infty)$.*

(i) *If $k < \frac{N}{p}$, then $W^{k,p}(\mathbb{R}^N) \hookrightarrow L^{p^*}(\mathbb{R}^N)$ for $\frac{1}{p^*} = \frac{1}{p} - \frac{k}{N}$.*[29]

(ii) *If $k = \frac{N}{p}$, then*

$$W^{k,p}(\mathbb{R}^N) \hookrightarrow L^r(\mathbb{R}^N) \qquad \text{for all} \quad r \in [p, \infty) \qquad \text{and}$$
$$W^{k,p}(\mathbb{R}^N) \subset L^r_{\text{loc}}(\mathbb{R}^N) \qquad \text{for all} \quad r \geq 1.$$

(iii) *If $k > \frac{N}{p}$, then $W^{k,p}(\mathbb{R}^N) \hookrightarrow C^{0,\gamma}(\overline{\mathbb{R}^N})$ for all $0 \leq \gamma < k - \frac{N}{p}$.*[30]

Proof. Proofs of these statements are quite involved and also have a long history. The interested reader can consult, e.g., Adams [2], Kufner, John & Fučík [96], Maz'ja [110], Stein [145, Chapters V, VI]. For a readable account of Sobolev spaces we recommend Evans [59, Chapter 5]. Spaces with fractional derivatives which extend the class of Sobolev spaces can be also defined, e.g., Triebel [153], [154]. □

Remark 1.2.27. The situation for an open set Ω with a nonempty boundary $\partial\Omega$ (in particular, for a bounded Ω) is even more complicated because some techniques from harmonic analysis, like Fourier transform, are not available. One possibility is to extend $f \in W^{k,p}(\Omega)$ to a function $\tilde{f} \in W^{k,p}(\mathbb{R}^N)$. This is possible if the boundary $\partial\Omega$ possesses certain smoothness properties. To explain this more precisely we would need some facts about manifolds (see Section 4.3 and Appendix 4.3A). So we omit details and just state that Theorem 1.2.26 is true provided $\partial\Omega$ is locally Lipschitz (see Section 5.5 for details).

Theorem 1.2.28 (Rellich–Kondrachov). *Let Ω be a bounded open set in $\mathbb{R}^N$ with a locally Lipschitz boundary, $k \in \mathbb{N}$, $p \in [1, \infty)$.*

(i) *Let $k < \frac{N}{p}$ and $q \in [1, p^*)$ where*

$$p^* := \frac{pN}{N - kp}. \tag{1.2.10}$$

Then the embedding $W^{k,p}(\Omega)$ into $L^q(\Omega)$ is compact.[31]

(ii) *If $k = \frac{N}{p}$, then $W^{k,p}(\Omega) \hookrightarrow\hookrightarrow L^q(\Omega)$ for all $q \in [1, \infty)$.*

(iii) *If $0 \leq \gamma < k - \frac{N}{p}$, then $W^{k,p}(\Omega) \hookrightarrow\hookrightarrow C^{0,\gamma}(\overline{\Omega})$.*

Proof. For the proof see references given above. □

---

[29] The exponent $p^* := \frac{pN}{N-kp}$ is sometimes called the *critical Sobolev exponent*.

[30] This means that any function $f \in W^{k,p}(\mathbb{R}^N)$ can be changed on a set of measure zero in such a way that the new function $\tilde{f}$ is γ-Hölder continuous and $\|\tilde{f}\|_{C^{0,\gamma}(\overline{\mathbb{R}^N})} \leq c\|f\|_{W^{k,p}(\mathbb{R}^N)}$. The symbol $\overline{\mathbb{R}^N}$ means that functions from $C^{0,\gamma}(\overline{\mathbb{R}^N})$ are bounded and uniformly γ-Hölder continuous on the whole $\mathbb{R}^N$.

[31] We will use the notation $\hookrightarrow\hookrightarrow$ for compact embeddings. An embedding of X into Y is compact if a ball in X is relatively compact in Y.

Now, we turn our attention to abstract spaces. Proposition 1.2.15 has pointed out the difference between finite-dimensional spaces and (infinite-dimensional) function spaces. Another difference between the finite and infinite dimension lies in the notion of a basis. It can be shown that any algebraic basis in an infinite-dimensional Banach space X has to be uncountable, and therefore the representation of a point by its coordinates can hardly be of any use. This observation leads to the necessity of expressing an element of X by an infinite sum. A sequence $\{e_n\}_{n=1}^{\infty} \subset X$ is called a *Schauder basis* of X if for each $x \in X$ there is a (uniquely determined) sequence $\{\xi_n\}_{n=1}^{\infty}$ of numbers (real or complex according to whether X is real or complex) such that

$$x = \sum_{n=1}^{\infty} \xi_n e_n. \tag{1.2.11}$$

There are several imperfections in this definition. Namely, there are separable[32] Banach spaces which do not possess a Schauder basis. Moreover, the convergence of the sum in (1.2.11) can be understood in several non-equivalent meanings. These problems do not appear in a special class of spaces with an additional structure which is connected with the norm and allows measuring angles.

Definition 1.2.29. Let X be a real (or complex) linear space. A mapping $(\cdot,\cdot)_X :$ $X \times X \to \mathbb{R}$ (or $\mathbb{C}$) is called a *scalar product* on X if the following conditions are satisfied:

(1) for any $y \in X$ the mapping $x \mapsto (x, y)_X$ is linear;

(2) $(x, y)_X = (y, x)_X$ for all $x, y \in X$ in the real case and $(x, y)_X = \overline{(y, x)_X}$ in the complex case;

(3) $(x, x)_X \geq 0$ for every $x \in X$ and $(x, x)_X = 0$ if and only if $x = o$.

Proposition 1.2.30. *Let* $(\cdot,\cdot)$ *be a scalar product on a linear space* X. *Then*

(i) *the so-called* Schwartz inequality

$$|(x, y)|^2 \leq (x, x)(y, y) \qquad \textit{holds for all} \quad x, y \in X; \tag{1.2.12}$$

(ii) *the mapping* $\|\cdot\| : x \mapsto [(x, x)]^{\frac{1}{2}}$ *is a norm on* X.

Proof. Assertion (i). For $x, y \in X$ there exists $c \in \mathbb{C}$, $|c| = 1$, such that for $\hat{y} = cy$ we have $(x, \hat{y}) \in \mathbb{R}$. Hence it suffices to prove (1.2.12) for the real space X. For any $\alpha \in \mathbb{R}$ we have

$$0 \leq (x + \alpha y, x + \alpha y) = (x, x) + 2\alpha(x, y) + |\alpha|^2(y, y),$$

i.e., the discriminant $4|(x, y)|^2 - 4(x, x)(y, y)$ is non-positive. Hence (1.2.12) follows.

[32] If a space X has a Schauder basis, then X is separable. This is not a serious drawback since most function spaces used in analysis are separable.

In assertion (ii), only the triangle inequality has to be checked. For $x, y \in X$ we get[33]

$$\|x+y\|^2 = (x+y, x+y) = (x,x) + 2\operatorname{Re}(x,y) + (y,y) \le \|x\|^2 + 2|(x,y)| + \|y\|^2$$

and the Schwartz inequality completes the proof. □

If X is a linear space with a scalar product we will always consider the norm on X induced by this scalar product. If X is complete with respect to this norm, then X is called a *Hilbert space* and will be usually denoted by H. We note that if X is not complete there exists a Hilbert space $\tilde{H}$ which is a completion of X.

Example 1.2.31.

(i) $\mathbb{R}^M$ with the scalar product

$$(x,y) = \sum_{i=1}^{M} \xi_i \eta_i, \qquad x = \sum_{i=1}^{M} \xi_i e_i, \quad y = \sum_{i=1}^{M} \eta_i e_i,$$

($e_1, \dots, e_M$ the standard basis) is a Hilbert space. Similarly, $\mathbb{C}^M$ is a Hilbert space with respect to the scalar product $(x,y) = \sum\limits_{i=1}^{M} \xi_i \overline{\eta_i}$.

(ii) The norm on $L^2(\Omega)$ given by (1.2.3) is induced by the scalar product

$$(f,g)_{L^2(\Omega)} = \int_\Omega f(x)\overline{g(x)}\,\mathrm{d}x \tag{1.2.13}$$

(in the complex case). Similarly, for $p = 2$ the norm (1.2.9) is equivalent to the norm induced by the scalar product

$$(f,g)_{W^{k,2}(\Omega)} = \sum_{|\alpha| \le k} (D^\alpha f, D^\alpha g)_{L^2(\Omega)}.$$

(iii) The "sup norm" on $BC(\Omega)$ is not induced by any scalar product. This can be seen from the *parallelogram identity*

$$\|x+y\|^2 + \|x-y\|^2 = 2\|x\|^2 + 2\|y\|^2, \qquad x, y \in X \tag{1.2.14}$$

which is valid only in such a space X the norm of which is induced by a scalar product. Indeed, if a norm satisfies (1.2.14), then (in the real case)

$$(x,y) = \frac{1}{4}(\|x+y\|^2 - \|x-y\|^2) \tag{1.2.15}$$

(*polarization identity*) has all properties of a scalar product, and the induced norm coincides with $\|\cdot\|$. It is not difficult to show that the "sup

[33] Notice here a typical procedure with the norm induced by a scalar product, namely using the second power of the norm in calculation.

norm" does not satisfy (1.2.14). Even more is true, namely, the "sup norm" is not equivalent to any norm on $BC(\Omega)$ induced by a scalar product. Since $C[0,1] \subset L^2(0,1)$, the scalar product (1.2.13) is also a scalar product on $C[0,1]$. But the space $C[0,1]$ is not complete in the L^2-norm and, therefore, the L^2-norm on $C[0,1]$ cannot be equivalent to the "sup norm"; only the inequality

$$\|f\|_{L^2(0,1)} \le \|f\|_{C[0,1]}$$

holds. Observe that $L^2(0,1)$ is a completion of $C[0,1]$ with respect to the integral norm given by (1.2.3). ○

The most useful concept in spaces with a scalar product is the following one.

Definition 1.2.32. Let X be a linear space with a scalar product $(\cdot,\cdot)$.

(1) Subsets $\mathcal{A}, \mathcal{B} \subset X$ are said to be *orthogonal* (and denoted by $\mathcal{A} \perp \mathcal{B}$) if $(a,b) = 0$ for every $a \in \mathcal{A}$, $b \in \mathcal{B}$.

(2) A system $\{x_\gamma\}_{\gamma\in\Gamma} \subset X$ is said to be *orthonormal* if

$$(x_\gamma, x_{\tilde{\gamma}}) = \begin{cases} 0, & \gamma \neq \tilde{\gamma}, \\ 1, & \gamma = \tilde{\gamma}. \end{cases}$$

(3) A sequence $\{e_n\}_{n=1}^\infty \subset X$ is called an *orthonormal basis* of X if $\{e_n\}_{n=1}^\infty$ is both an orthonormal system and a Schauder basis of X.

Suppose that $x_1, \dots, x_n$ are linearly independent elements of a space X with a scalar product $(\cdot,\cdot)$. Put $e_1 = \frac{x_1}{\|x_1\|}$ and if orthonormal elements $e_1, \dots, e_k$ $(k < n)$ are constructed in such a way that

$$\operatorname{Lin}\{x_1, \dots, x_k\} = \operatorname{Lin}\{e_1, \dots, e_k\},$$

then define

$$y_{k+1} = x_{k+1} - \sum_{j=1}^{k} (x_{k+1}, e_j) e_j, \qquad e_{k+1} = \frac{y_{k+1}}{\|y_{k+1}\|}.$$

It is obvious that

$$(e_j, e_{k+1}) = 0, \quad j = 1, \dots, k, \qquad \|e_{k+1}\| = 1$$

and

$$\operatorname{Lin}\{x_1, \dots, x_{k+1}\} = \operatorname{Lin}\{e_1, \dots, e_{k+1}\}.$$

This procedure is called the *Schmidt orthogonalization*. For any $x \in Y := \operatorname{Lin}\{x_1, \dots, x_n\}$ we have $x = \sum\limits_{k=1}^{n} \alpha_k e_k$. Taking the scalar product with e_j, we get

$$(x, e_j) = \sum_{k=1}^{n} \alpha_k (e_k, e_j) = \alpha_j,$$

and also

$$\|x\|^2 = \left(\sum_{j=1}^{n}(x,e_j)e_j, \sum_{k=1}^{n}(x,e_k)e_k\right) = \sum_{k=1}^{n}|(x,e_k)|^2.$$

Assume now that $X \neq Y$ and let us look for an approximation of a $y \in X \setminus Y$ by an element $x = \sum_{j=1}^{n} \alpha_j e_j$:

$$\begin{aligned}
\|y-x\|^2 &= \left(y - \sum_{j=1}^{n}\alpha_j e_j, y - \sum_{j=1}^{n}\alpha_j e_j\right) \\
&= \|y\|^2 - \sum_{j=1}^{n}\alpha_j\overline{(y,e_j)} - \sum_{j=1}^{n}\overline{\alpha_j}(y,e_j) + \sum_{j=1}^{n}|\alpha_j|^2 \\
&= \|y\|^2 + \sum_{j=1}^{n}|\alpha_j - (y,e_j)|^2 - \sum_{j=1}^{n}|(y,e_j)|^2 \\
&\geq \|y\|^2 - \sum_{j=1}^{n}|(y,e_j)|^2 = \left\|y - \sum_{j=1}^{n}(y,e_j)e_j\right\|^2 .
\end{aligned} \tag{1.2.16}$$

Two consequences follow from this inequality. First, the best approximation of $y \in X$ by an element of Y is

$$P_n y := \sum_{j=1}^{n}(y,e_j)e_j.$$[34]

Observe also that $(y - P_n y) \perp Y$. Second,

$$\sum_{j=1}^{n}|(y,e_j)|^2 \leq \|y\|^2 \qquad \text{for all} \quad y \in X.$$

Since n is arbitrary (in an infinite-dimensional space) we have obtained the so-called *Bessel inequality*:

If $\{e_n\}_{n=1}^{\infty}$ is an orthonormal system in X, then

[34] We note that this result, namely the linearity of the operator P_n of the best approximation, is typical for spaces with scalar products. In a general normed linear space X and a finite-dimensional subspace Y the best approximation of an arbitrary $x \in X$ by elements of Y exists (by a compactness argument) but a special property of the norm is needed for the uniqueness of the best approximation. Linearity of the best approximation on all subspaces of dimension 2 implies that the norm is induced by a scalar product. More details can be found in the monograph of Singer [141].

$$\sum_{n=1}^{\infty} |(y, e_n)|^2 \le \|y\|^2 \qquad \text{for all} \quad y \in X. \tag{1.2.17}$$

In particular, the sum $\sum_{j=1}^{\infty} |(y, e_j)|^2$ is always convergent.

Proposition 1.2.33. *Let X be a linear space with a scalar product, let X be separable.*[35] *Then there exists an orthonormal basis in X.*

Proof. Let $\{x_1, x_2, x_3, \dots\}$ be a dense set in X. Put

$$Y_n = \operatorname{Lin}\{x_1, \dots, x_n\}, \qquad Y = \bigcup_{n=1}^{\infty} Y_n.$$

Then $\overline{Y} = X$. By omitting linearly dependent elements we can assume that $\dim Y_n = n$. According to the Schmidt orthogonalization there exists an orthonormal sequence $\{e_n\}_{n=1}^{\infty}$ such that $Y_n = \operatorname{Lin}\{e_1, \dots, e_n\}$. Let $x \in X$ and let $\{y_n\}_{n=1}^{\infty}$ be a sequence such that $y_n \in Y_n$ and $\lim_{n\to\infty} y_n = x$ (the density of Y in X). By the inequality (1.2.16),

$$\|x - y_n\| \ge \left\| x - \sum_{j=1}^{n} (x, e_j) e_j \right\|.$$

This means that $x = \sum_{j=1}^{\infty} (x, e_j) e_j$.

To prove uniqueness, suppose that $x = \sum_{j=1}^{\infty} \alpha_j e_j$. Since the scalar product is continuous, we have

$$(x, e_k) = \lim_{n\to\infty} \left(\sum_{j=1}^{n} \alpha_j e_j, e_k \right) = \alpha_k. \qquad \square$$

In order to obtain some useful properties which guarantee that an orthonormal sequence is a basis we need to use completeness. We start with a general approximation result.

Theorem 1.2.34. *Let H be a Hilbert space and let $\mathcal{C}$ be a closed convex subset of H. Then for any $x \in H$ there exists a unique $y \in \mathcal{C}$ such that*

$$\|x - y\| = \inf\{\|x - z\| : z \in \mathcal{C}\}. \tag{1.2.18}$$

This best approximation y is characterized by the following property: $y \in \mathcal{C}$ and

$$\operatorname{Re}(x - y, y - z) \ge 0 \qquad \textit{for all} \quad z \in \mathcal{C} \tag{1.2.19}$$

[35] The assumption on separability is redundant. Without separability an orthonormal basis $\{e_\gamma\}_{\gamma\in\Gamma}$ still exists but Γ is uncountable. Moreover, if $x \in X$, then $(x, e_\gamma) = 0$ for all but countably many γ.

(see Figure 1.2.2 [36]).

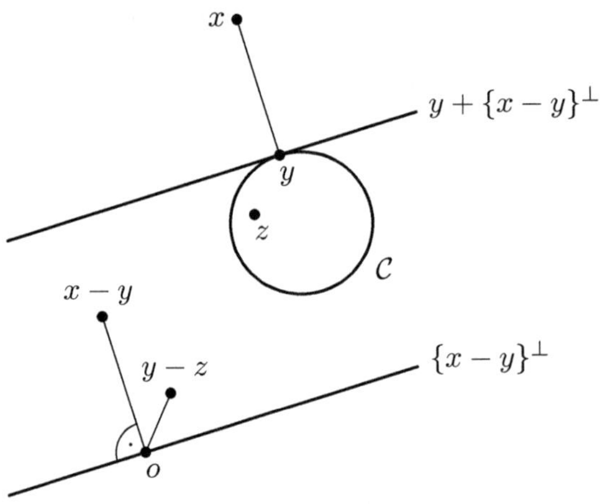

Figure 1.2.2.

Proof.

Step 1 **(Existence)**. Denote the right-hand side in (1.2.18) by d. If $d = 0$, then $x \in \mathcal{C}$ ($\mathcal{C}$ is closed) and $y = x$. Suppose that $d > 0$. Then there are $\{z_n\}_{n=1}^{\infty} \subset \mathcal{C}$ such that

$$d \leq \|x - z_n\| < d + \frac{1}{n}.$$

By (1.2.14) we get

$$\begin{aligned}\|z_n - z_m\|^2 &= \|x - z_m - (x - z_n)\|^2 \\ &= 2(\|x - z_m\|^2 + \|x - z_n\|^2) - 4\left\|x - \frac{z_n + z_m}{2}\right\|^2 \\ &< 2\left(d + \frac{1}{n}\right)^2 + 2\left(d + \frac{1}{m}\right)^2 - 4d^2\end{aligned}$$

(notice that $\frac{z_n+z_m}{2} \in \mathcal{C}$ since $\mathcal{C}$ is convex). This implies that $\{z_n\}_{n=1}^{\infty}$ is a Cauchy sequence, and therefore it is convergent to a $y \in \mathcal{C}$. Obviously, $\|x - y\| = d$.

Step 2 **(Uniqueness)**. Assume that $\|x - y\| = \|x - \tilde{y}\| = d$ for $y, \tilde{y} \in \mathcal{C}$. Using (1.2.14) as above we get $y = \tilde{y}$.

Step 3 **(Characterization)**. Let y be the best approximation of x and let $z \in \mathcal{C}$. Then

$$z_t := tz + (1 - t)y \in \mathcal{C} \qquad \text{for} \quad t \in (0, 1)$$

[36] For $\mathcal{A} \subset H$ we denote $\mathcal{A}^{\perp} := \{x \in H : a \in \mathcal{A} \Longrightarrow (a, x) = 0\}$.

($\mathcal{C}$ is convex) and

$$\begin{aligned}\|x - z_t\|^2 &= \|x - y + t(y - z)\|^2 \\ &= \|x - y\|^2 + t^2\|y - z\|^2 + 2t\,\mathrm{Re}(x - y, y - z) \ge \|x - y\|^2,\end{aligned}$$

i.e.,

$$t\|y - z\|^2 + 2\,\mathrm{Re}(x - y, y - z) \ge 0$$

and taking the limit for $t \to 0_+$, the inequality (1.2.19) follows. If (1.2.19) is satisfied, then

$$\|x - z\|^2 = \|x - y + y - z\|^2 = \|x - y\|^2 + \|y - z\|^2 + 2\,\mathrm{Re}(x - y, y - z) \ge \|x - y\|^2,$$

and therefore y is the best approximation of x. □

Corollary 1.2.35. *Let H be a Hilbert space and M a closed linear subspace of H, $M \neq H$, $M \neq \{o\}$. Then there exists a unique subspace $M^\perp$ (the so-called* orthogonal complement *of M) such that*

$$H = M \oplus M^\perp, \qquad M \perp M^\perp.$$

Moreover, if P denotes the projection to M given by this direct sum[37] *(the so-called* orthogonal projection*), then $P \in \mathcal{L}(H)$, $\|P\|_{\mathcal{L}(H)} = 1$ and*

$$(Px, y) = (x, Py) \qquad \textit{for all} \quad x, y \in H.$$

Proof. A closed linear subspace M is a closed convex set. Denote by $Px \in M$ the best approximation of $x \in H$ in M. Choose $w \in M$ and put

$$z = Px - w \in M.$$

By (1.2.19) we get $\mathrm{Re}(x - Px, w) \ge 0$ and also $\mathrm{Re}(x - Px, \mathrm{i}w) \ge 0$ (by taking $z = Px - \mathrm{i}w$), i.e.,

$$(x - Px, w) \ge 0.$$

Since also $-w \in M$, we finally have

$$(x - Px, w) = 0 \qquad \text{for all} \quad w \in M. \tag{1.2.20}$$

It is easy to see that (1.2.20) is a characterization of Px. By using (1.2.20) for αx, $x_1 + x_2$, we see that P is a linear operator. Since $P^2 = P$, P is a projection onto M. The identity (1.2.20) also shows that

$$\mathrm{Ker}\,P = M^\perp := \{y \in H : x \in M \Longrightarrow (x, y) = 0\}.$$

By the orthogonality of Px and $x - Px$ we have

$$\|x\|^2 = \|Px\|^2 + \|x - Px\|^2 \ge \|Px\|^2, \qquad \text{i.e.,} \qquad \|P\|_{\mathcal{L}(H)} \le 1.$$

Since $Px = x$ for $x \in M$, $\|P\|_{\mathcal{L}(H)} = 1$. By (1.2.20) we get

$$(x, Py) = (x - Px + Px, Py) = (Px, Py) = (Px, Py - y + y) = (Px, y).$$ □

[37] Cf. Example 1.1.13(i).

Corollary 1.2.36. *Let H be a Hilbert space and let $\{e_n\}_{n=1}^{\infty}$ be an orthonormal sequence in H. Then the following statements are equivalent:*

(i) $\{e_n\}_{n=1}^{\infty}$ *is an orthonormal basis;*

(ii) *if* $(x, e_n) = 0$ *for all* n, *then* $x = o$;

(iii) $\operatorname{Lin}\{e_1, e_2, \dots\}$ *is dense in* H;

(iv) *the* Parseval equality

$$\|x\|^2 = \sum_{n=1}^{\infty} |(x, e_n)|^2 \qquad \textit{is valid for all} \quad x \in H. \tag{1.2.21}$$

Proof. The implication (i)⇒(ii) is obvious and follows from the definition of the orthonormal basis.

The implication (ii)⇒(iii): Denote $Y = \operatorname{Lin}\{e_1, e_2, \dots\}$. Assume that Y is not dense, i.e., $\overline{Y} \neq H$. By Corollary 1.2.35 there exists $x \in Y^{\perp} \setminus \{o\}$. In particular, $(x, e_n) = 0$ for all n, a contradiction.

The implication (iii)⇒(iv): The proof of Proposition 1.2.33 shows that the sequence

$$s_n := \sum_{k=1}^{n} (x, e_k) e_k \qquad \text{converges to } x \quad \text{for all } x \in H.$$

Moreover, $s_n \perp (x - s_n)$, and hence

$$\|x\|^2 = \|s_n\|^2 + \|x - s_n\|^2 = \sum_{k=1}^{n} |(x, e_k)|^2 + \|x - s_n\|^2.$$

By taking the limit, the Parseval equality follows.

The implication (iv)⇒(i): Let $x \in H$ be arbitrary. For s_n defined as above and $m > n$ we have

$$\|s_m - s_n\|^2 = \sum_{k=n}^{m} |(x, e_k)|^2.$$

Since the series in (1.2.21) is convergent, the sequence $\{s_n\}_{n=1}^{\infty}$ is Cauchy, and therefore it is convergent to a $y \in H$ since H is complete. Moreover, $(y, e_n) = (x, e_n)$, and by the Parseval equality

$$\|x - y\|^2 = \sum_{n=1}^{\infty} |(x - y, e_n)|^2 = 0.$$ □

Remark 1.2.37. Let H be a Hilbert space and $\{e_n\}_{n=1}^{\infty}$ an orthonormal basis in H. The proof of the last implication shows that for an arbitrary sequence $\{\alpha_n\}_{n=1}^{\infty} \subset \mathbb{R}$ (or $\mathbb{C}$ depending on whether H is a real or complex space) for which $\sum\limits_{n=1}^{\infty} |\alpha_n|^2$ is convergent the series $\sum\limits_{n=1}^{\infty} \alpha_n e_n$ is convergent in H to an $x \in H$ and $(x, e_n) = \alpha_n$. Moreover, the operator

$$U \colon x \in H \mapsto \{(x, e_n)\}_{n=1}^{\infty} \in l^2(\mathbb{N})$$ [38]

is a *unitary operator* (i.e., $(Ux, Uy)_{l^2(\mathbb{N})} = (x, y)$, $x, y \in H$) which is surjective. It implies also that all infinite-dimensional separable Hilbert spaces over the same field of scalars are unitarily equivalent. This statement is known as the *Riesz–Fischer Theorem*. Having this result we can ask why not to restrict our attention only to a single abstract separable Hilbert space. The reason is that in a special function space like $W^{k,2}(\Omega)$ one has more ways of computations since its elements are functions.

Example 1.2.38.

(i) The space $L^2(-\pi, \pi)$ is a Hilbert space. It is separable since continuous 2π-periodic functions are dense in $L^2(-\pi, \pi)$ and any such function can be approximated by trigonometric polynomials of the type $\sum\limits_{k=-n}^{n} a_k \mathrm{e}^{ikt}$ (either the classical Weierstrass Approximation Theorem or Theorem 1.2.14). It is easy to see that

$$e_n \colon t \mapsto \frac{1}{\sqrt{2\pi}} \mathrm{e}^{int}, \qquad t \in (-\pi, \pi), \quad n \in \mathbb{Z},$$

form an orthonormal system in $L^2(-\pi, \pi)$. By Corollary 1.2.36(iii) it is also an orthonormal basis.[39]

(ii) Functions

$$H_n(t)\mathrm{e}^{-\frac{t^2}{2}} \qquad \text{where} \quad H_n(t) = (-1)^n \mathrm{e}^{t^2} \frac{\mathrm{d}^n \mathrm{e}^{-t^2}}{\mathrm{d}t^n}$$

(the so-called *Hermite polynomials*) form (after normalization) an orthonormal basis in $L^2(\mathbb{R})$. For the proof and relevant results in harmonic analysis we recommend the classical book Kaczmarz & Steinhaus [84]. We note that

---

[38] $l^2(\mathbb{N})$ is the space of all (generally complex) sequences $x = \{\xi_n\}_{n=1}^{\infty}$ such that $\sum\limits_{n=1}^{\infty} |\xi_n|^2$ is convergent. The scalar product on $l^2(\mathbb{N})$ is given by $(x, y)_{l^2(\mathbb{N})} = \sum\limits_{n=1}^{\infty} \xi_n \overline{\eta}_n$ for $x = \{\xi_n\}_{n=1}^{\infty}$, $y = \{\eta_n\}_{n=1}^{\infty}$ (see also Remark 1.2.24).

[39] Here this means that

$$f(t) = \sum_{-\infty}^{+\infty} \hat{f}(n) \mathrm{e}^{int} \qquad \text{where} \qquad \hat{f}(n) = (f, e_n)_{L^2(-\pi,\pi)} = \frac{1}{2\pi} \int_{-\pi}^{\pi} f(t) \mathrm{e}^{-int} \, \mathrm{d}t$$

and the series is convergent in the L^2-norm for arbitrary $f \in L^2(-\pi, \pi)$. It is worth noting that the series is actually a.e. convergent to f but this by no means follows from the norm convergence. This result is due to L. Carlesson and it is one of the most difficult and profound results in analysis.

there are many different orthonormal bases in L^2-spaces. We will present one general method of their construction in Theorem 2.2.16. ○

Proposition 1.2.39. *Let $\{e_n\}_{n=1}^{\infty}$ be an orthonormal basis in a Hilbert space H. Then a bounded set $\mathcal{M} \subset H$ is relatively compact if and only if for any $\varepsilon > 0$ there is $k \in \mathbb{N}$ such that*

$$\sum_{n=k}^{\infty} |(x, e_n)|^2 < \varepsilon \qquad \text{for all} \quad x \in \mathcal{M}.$$

Proof. The statement follows from Proposition 1.2.3. □

Theorem 1.2.40 (Riesz Representation Theorem). *Let H be a Hilbert space and let F be a continuous linear form on H. Then there is a unique $f \in H$ such that*

$$F(x) = (x, f) \qquad \text{for all} \quad x \in H.$$

Moreover, $\|F\| = \|f\|$ where $\|F\| = \|F\|_{\mathcal{L}(H,\mathbb{R})}$ or $\|F\| = \|F\|_{\mathcal{L}(H,\mathbb{C})}$ depending on whether H is a real or a complex space.

Proof. Suppose that H is a complex Hilbert space. If $F = o$, then $f = o$. Suppose that $F \neq o$. The idea of constructing f is that f has to be orthogonal to $\operatorname{Ker} F$ which is a closed subspace of H. By Corollary 1.2.35,

$$H = \operatorname{Ker} F \oplus (\operatorname{Ker} F)^{\perp}.$$

Take $x_0 \in (\operatorname{Ker} F)^{\perp}$, $\|x_0\| = 1$, and put $f = \alpha x_0$ where α will be determined later. Let $x = y + \beta x_0$, $y \in \operatorname{Ker} F$, $\beta \in \mathbb{C}$ be arbitrary ($\dim(\operatorname{Ker} F)^{\perp} = 1$, see Example 1.1.18 (ii)). Then

$$(x, f) = \beta\overline{\alpha}, \qquad F(x) = \beta F(x_0).$$

Choose now $\alpha = \overline{F(x_0)}$. If there is another $g \in H$ such that $F(x) = (x, g)$, $x \in H$, then $0 = (x, f - g)$ for all $x \in H$, in particular, for $x = f - g$. Therefore $f = g$. By the Schwartz inequality (1.2.12) we obtain

$$|F(x)| = |(x, f)| \leq \|x\|\|f\|, \qquad \text{i.e.,} \qquad \|F\| \leq \|f\|.$$

Since $F(f) = \|f\|^2$, we have $\|F\| \geq \|f\|$. This shows that $\|F\| = \|f\|$. □

The following variant of the Riesz Representation Theorem is often used in the functional analysis approach to differential equations (see, e.g., Evans [59]).

Proposition 1.2.41 (Lax–Milgram). *Let H be a complex Hilbert space and let $B\colon H \times H \to \mathbb{C}$ be a mapping with the following properties:*

(i) *The mapping $x \mapsto B(x, y)$ is linear for any $y \in H$.*

(ii) *$B(x, \alpha_1 y_1 + \alpha_2 y_2) = \overline{\alpha_1} B(x, y_1) + \overline{\alpha_2} B(x, y_2)$ for every $x, y_1, y_2 \in H$, $\alpha_1, \alpha_2 \in \mathbb{C}$.*

(iii) *There is a constant c such that $|B(x,y)| \le c\|x\|\|y\|$ for every $x, y \in H$.*

Then there is $A \in \mathcal{L}(H)$, $\|A\|_{\mathcal{L}(H)} \le c$, such that

$$B(x,y) = (x, Ay), \qquad x, y \in H.$$

Moreover,

(iv) *if there is a positive constant d such that*

$$B(x,x) \ge d\|x\|^2 \qquad \textit{for each} \quad x \in H,$$

then A is invertible,

$$A^{-1} \in \mathcal{L}(H) \qquad \textit{and} \qquad \|A^{-1}\|_{\mathcal{L}(H)} \le \frac{1}{d}.$$

Proof. The existence of A follows from (i), (iii) and the Riesz Representation Theorem. The property (ii) yields the linearity of A. Since

$$\|Ay\|^2 = (Ay, Ay) = B(Ay, y) \le c\|Ay\|\|y\|,$$

we have $\|Ay\| \le c\|y\|$, i.e., $A \in \mathcal{L}(H)$ and $\|A\|_{\mathcal{L}(H)} \le c$.

The property (iv) means that

$$d\|y\|^2 \le B(y,y) = (y, Ay) \le \|y\|\|Ay\|,$$

i.e.,

$$\|Ay\| \ge d\|y\| \qquad \text{for all} \quad y \in H. \tag{1.2.22}$$

In particular, A is injective. Moreover, $\operatorname{Im} A$ is a closed subspace of H. Indeed, let $Ay_n \to z \in \overline{\operatorname{Im} A}$. By (1.2.22), $\{y_n\}_{n=1}^{\infty}$ is a Cauchy sequence, and hence it is convergent to a $y \in H$. By continuity of A, $Ay = z$, i.e., $z \in \operatorname{Im} A$. In fact,

$$\operatorname{Im} A = H.$$

Indeed, if $w \in (\operatorname{Im} A)^{\perp}$, then

$$d\|w\|^2 \le B(w,w) = (w, Aw) = 0 \qquad \text{and} \qquad w = o.$$

So $\operatorname{Dom}(A^{-1}) = \operatorname{Im} A = H$ and (1.2.22) implies that

$$\|A^{-1}\|_{\mathcal{L}(H)} \le \frac{1}{d}. \qquad \square$$

Exercise 1.2.42. Let X be a normed linear space and let $\mathcal{G}$ be an open subset of X. Then $\mathcal{G}$ is connected if and only if for any two points $a, b \in \mathcal{G}$ there exists a continuous piecewise linear mapping $\gamma\colon [0,1] \to \mathcal{G}$ such that $\gamma(0) = a$, $\gamma(1) = b$. Prove this statement.

Exercise 1.2.43. Let $\{x_n\}_{n=1}^{\infty}$ be a sequence of elements of a Banach space X. If $\sum\limits_{n=1}^{\infty} \|x_n\|$ is convergent (in $\mathbb{R}$) then $\sum\limits_{n=1}^{\infty} x_n$ is convergent in X (in the norm topology). Prove this fact.

Exercise 1.2.44. Let $\{\mathcal{F}_\alpha\}_{\alpha\in\mathcal{A}}$ be a system of closed subsets of a compact space M. Prove the *finite intersection property*:

$$\text{If } \bigcap_{\alpha\in\mathcal{K}} \mathcal{F}_\alpha \neq \emptyset \text{ for any finite } \mathcal{K} \subset \mathcal{A}, \text{ then } \bigcap_{\alpha\in\mathcal{A}} \mathcal{F}_\alpha \neq \emptyset.$$

(This property characterizes compact spaces.)

Hint. Suppose not. Then $\{M \setminus \mathcal{F}_\alpha\}_{\alpha\in\mathcal{A}}$ is an open covering of M.

Exercise 1.2.45. Prove that $\mathcal{F} \subset C[a,b]$ is relatively compact if and only if $\mathcal{F}$ is bounded in $C[a,b]$ and the following equicontinuity condition is satisfied:

$$\forall\varepsilon > 0\ \exists\delta > 0\ \forall f \in \mathcal{F}: \qquad x, y \in [a,b],\ |x-y| < \delta \quad \Longrightarrow \quad |f(x) - f(y)| < \varepsilon.$$

Hint. Use Proposition 1.2.3. Obviously, this statement is also a special case of Theorem 1.2.13.

Exercise 1.2.46. Let $\{e_n\}_{n=1}^{\infty}$ be an orthonormal basis in a Hilbert space H. Define

$$f(x) = \begin{cases} n & \text{if} \quad x = e_n, \\ n(1 - 2\|x - e_n\|) & \text{if} \quad \|x - e_n\| < \dfrac{1}{2}, \\ 0 & \text{otherwise}. \end{cases}$$

Show that f is a well defined continuous functional on H which is not bounded on the closed unit ball.

Exercise 1.2.47. Let $\emptyset \neq \mathcal{M} \subset X$ be a subset of a normed linear space X. For $x \in X$ set

$$\operatorname{dist}(x, \mathcal{M}) = \inf\{\|x - y\| : y \in \mathcal{M}\}.$$

Prove that for any $x_1, x_2 \in X$ we have

$$|\operatorname{dist}(x_1, \mathcal{M}) - \operatorname{dist}(x_2, \mathcal{M})| \leq \|x_1 - x_2\|.$$

Hint. Assume $\operatorname{dist}(x_1, \mathcal{M}) \geq \operatorname{dist}(x_2, \mathcal{M})$. For any $\varepsilon > 0$ there exists $x_\varepsilon \in \mathcal{M}$ such that

$$\|x_2 - x_\varepsilon\| < \operatorname{dist}(x_2, \mathcal{M}) + \varepsilon.$$

Use the triangle inequality for $\|x_1 - x_\varepsilon\|$.

Exercise 1.2.48.[40] Let Ω be a bounded open set in $\mathbb{R}^M$. For $p \in [1, \infty)$ and $k \in \mathbb{N}$ define $W_0^{k,p}(\Omega)$ to be the closure of $\mathcal{D}(\Omega)$ with respect to the $W^{k,p}(\Omega)$-norm (1.2.9).

[40] Supplement to Example 1.2.25.

(i) Prove that

$$W_0^{k,p}(\Omega) \subsetneq W^{k,p}(\Omega)$$

and $W_0^{k,p}(\Omega)$ need not be dense in $W^{k,p}(\Omega)$ (compare it with the statement of Theorem 1.2.28(iii); see also the Trace Theorem (Theorem 5.5.1)).

(ii) Prove the *Poincaré inequality*:

There exists a constant c_p such that for all $u \in W_0^{1,p}(\Omega)$ the inequality

$$\int_\Omega |u(x)|^p \, dx \le c_p \int_\Omega \|\nabla u(x)\|^p \, dx$$ [41] holds.

Hint. It suffices to prove the assertion for $u \in \mathcal{D}(\Omega)$. Consider first $\Omega = (0,1)$ and use the Mean Value Theorem. Then suppose (without loss of generality) $\Omega \subset \tilde{\Omega} := (0,d) \times \mathbb{R}^{M-1}$ and notice that $\mathcal{D}(\Omega) \subset \mathcal{D}(\tilde{\Omega})$.

(iii) Use the Poincaré inequality to prove that

$$|||u|||_{W_0^{1,p}(\Omega)} = \left[\int_\Omega \|\nabla u(x)\|^p \, dx\right]^{\frac{1}{p}}$$

is an equivalent norm on $W_0^{1,p}(\Omega)$ with the norm

$$\|u\|_{W_0^{1,p}(\Omega)} = \left(\int_\Omega |u(x)|^p \, dx\right)^{\frac{1}{p}} + \left(\int_\Omega \|\nabla u(x)\|^p \, dx\right)^{\frac{1}{p}}.$$

Exercise 1.2.49. Let $u \in W^{1,p}(0,1)$, $1 \le p < \infty$. Prove that functions

$$u^+(x) := \max\{u(x), 0\}, \qquad u^-(x) := \max\{-u(x), 0\}$$

also belong to $W^{1,p}(0,1)$. We remark that the corresponding result is false for $W^{k,p}(0,1)$, $k \ge 2$.

[41] Finding the smallest possible value of the constant c_p is a much more difficult problem. See also Exercise 7.4.12 and Example 5.6.4. Here $\nabla u(x) = \left(\frac{\partial u}{\partial x_1}, \dots, \frac{\partial u}{\partial x_M}\right)$ where $\frac{\partial}{\partial x_i}$, $i = 1, \dots, M$, are weak derivatives (see (1.2.8)), is the *gradient* of u.

Chapter 2

Properties of Linear and Nonlinear Operators

2.1 Linear Operators

In this section we point out some fundamental properties of linear operators in Banach spaces. The key assertions presented are the Uniform Boundedness Principle, the Banach–Steinhaus Theorem, the Open Mapping Theorem, the Hahn–Banach Theorem, the Separation Theorem, the Eberlain–Smulyan Theorem and the Banach Theorem. We recall that the collection of all continuous linear operators from a normed linear space X into a normed linear space Y is denoted by $\mathcal{L}(X,Y)$, and $\mathcal{L}(X,Y)$ is a normed linear space with the norm

$$\|A\|_{\mathcal{L}(X,Y)} = \sup\{\|Ax\|_Y : \|x\|_X \le 1\}.$$

Proposition 2.1.1. *Let Y be a Banach space. Then $\mathcal{L}(X,Y)$ is a Banach space, too. In particular, the space X^* of all linear continuous forms on X is complete.*

Proof. Let $\{A_n\}_{n=1}^{\infty}$ be a Cauchy sequence in $\mathcal{L}(X,Y)$. Then for any $\varepsilon > 0$ there is $n_0 \in \mathbb{N}$ such that for all $n, m \ge n_0$ and $x \in X$

$$\|A_n x - A_m x\| \le \|A_n - A_m\|\|x\| \le \varepsilon\|x\|.$$

Since Y is complete, the sequence $\{A_n x\}_{n=1}^{\infty}$ is convergent to a point in Y that can be denoted by Ax. Obviously A is a linear operator from X into Y and

$$\|Ax - A_m x\| = \lim_{n\to\infty} \|A_n x - A_m x\| \le \varepsilon\|x\|, \qquad m \ge n_0, \quad x \in X.$$

This implies (Proposition 1.2.10) that $A \in \mathcal{L}(X,Y)$ and $\|A - A_m\| \to 0$. □

The importance of this result can be seen from the following statement.

Proposition 2.1.2. *Let X be a Banach space and $A \in \mathcal{L}(X)$. If $\|A\| < 1$, then the operator $I - A$ is continuously invertible and*

$$(I - A)^{-1} = \sum_{n=0}^{\infty} A^n$$

where the sum is convergent in the $\mathcal{L}(X)$-norm.

Proof. First we prove the convergence. Let $\varepsilon > 0$ be arbitrary. Put $S_k = \sum\limits_{n=0}^{k} A^n$. Then

$$\|S_l - S_k\| = \left\| \sum_{n=k+1}^{l} A^n \right\| \le \sum_{n=k+1}^{l} \|A^n\| \le \sum_{n=k+1}^{l} \|A\|^n < \varepsilon^{1} \qquad \text{for} \quad l > k$$

provided k is sufficiently large. By Proposition 2.1.1, the limit of S_k exists in the $\mathcal{L}(X)$-norm. Denote $B := \lim\limits_{k\to\infty} S_k = \sum\limits_{n=0}^{\infty} A^n$. We have

$$\begin{aligned}(I - A)B &= \lim_{k\to\infty} (I - A) \sum_{n=0}^{k} A^n = \lim_{k\to\infty} \left(\sum_{n=0}^{k} A^n - \sum_{n=1}^{k+1} A^n \right)\\ &= \lim_{k\to\infty} (I - A^{k+1}) = I\end{aligned}$$

since $\lim\limits_{n\to\infty} A^n = O$. Similarly,

$$B(I - A) = I, \qquad \text{i.e.,} \qquad B = (I - A)^{-1}. \qquad \square$$

If X is a complex Banach space and $A \in \mathcal{L}(X)$, we denote

$$\varrho(A) := \{\lambda \in \mathbb{C} : \lambda I - A \text{ is continuously invertible in } \mathcal{L}(X)\}$$

(the so-called *resolvent set* of A) and

$$\sigma(A) := \mathbb{C} \setminus \varrho(A)$$

(the so-called *spectrum* of A).[2] The operator valued function

$$\lambda \mapsto (\lambda I - A)^{-1}, \qquad \lambda \in \varrho(A),$$

is called the *resolvent* of A.

[1] If $A \in \mathcal{L}(X, Y)$, $B \in \mathcal{L}(Y, Z)$, then $BA \in \mathcal{L}(X, Z)$ and $\|BA\|_{\mathcal{L}(X,Z)} \le \|B\|_{\mathcal{L}(Y,Z)} \|A\|_{\mathcal{L}(X,Y)}$.

[2] The reason for considering only complex spaces consists in the fact that $\sigma(A) \neq \emptyset$ for all $A \in \mathcal{L}(X)$ in this case. This will be proved later in this section (see the discussion following Example 2.1.20).

Corollary 2.1.3. *Let X be a complex Banach space and $A \in \mathcal{L}(X)$. Then $\varrho(A)$ is an open set and*

$$\{\lambda : |\lambda| > \|A\|\} \subset \varrho(A).$$

Proof. If $|\lambda| > \|A\|$, then

$$\lambda I - A = \lambda\left(I - \frac{A}{\lambda}\right)$$

and $\left(I - \frac{A}{\lambda}\right)^{-1} \in \mathcal{L}(X)$ according to Proposition 2.1.2. Hence we have

$$(\lambda I - A)^{-1} = \sum_{n=0}^{\infty} \frac{A^n}{\lambda^{n+1}}.[3]$$

Similarly, if $\lambda_0 \in \varrho(A)$, then

$$\lambda I - A = (\lambda - \lambda_0)I + (\lambda_0 I - A) = (\lambda_0 I - A)[I - (\lambda_0 - \lambda)(\lambda_0 I - A)^{-1}].$$

For a parameter λ such that $\|(\lambda_0 - \lambda)(\lambda_0 I - A)^{-1}\| < 1$, the inverse operator $B = [I - (\lambda_0 - \lambda)(\lambda_0 I - A)^{-1}]^{-1}$ exists and

$$(\lambda I - A)^{-1} = B(\lambda_0 I - A)^{-1}.$$ □

The next theorem together with Theorems 2.1.8 and 2.1.13 is one of the most significant results in linear functional analysis. For the proofs the interested reader can consult textbooks on functional analysis, e.g., Conway [36], Dunford & Schwartz [53], Rudin [131], Yosida [161]. Here we present only quite elementary proof of Theorem 2.1.4 which appeared recently in Sokal [143].

Theorem 2.1.4 (Uniform Boundedness Principle). *Let X be a Banach space and Y a normed linear space. If $\{A_\gamma\}_{\gamma\in\Gamma} \subset \mathcal{L}(X,Y)$ is such that the sets $\{\|A_\gamma x\|_Y : \gamma \in \Gamma\}$ are bounded for all $x \in X$, then $\{\|A_\gamma\|_{\mathcal{L}(X,Y)} : \gamma \in \Gamma\}$ is also bounded.*

Proof. Let $A \in \mathcal{L}(X,Y)$. Then for any $a \in X$ and $r > 0$, we have

$$\sup_{x\in B(a;r)} \|Ax\|_Y \geq r\|A\|_{\mathcal{L}(X,Y)}. \tag{2.1.1}$$

Indeed, using the triangle inequality in the form $\|\alpha - \beta\| \leq \|\alpha\| + \|\beta\|$, we have

$$\max\{\|A(a+x)\|, \|A(a-x)\|\} \geq \frac{1}{2}\left[\|A(a+x)\| + \|A(a-x)\|\right] \geq \|Ax\|.$$

Take supremum over $x \in B(0;r)$ to get (2.1.1). We now proceed to prove the theorem via contradiction. Suppose that $\sup\limits_{\gamma\in\Gamma} \|A_\gamma\|_{\mathcal{L}(X,Y)} = \infty$ and choose $\{A_n\}_{n=1}^{\infty} \subset \{A_\gamma\}_{\gamma\in\Gamma}$ such that $\|A_n\| \geq 4^n$.

[3]This series actually converges for λ such that $|\lambda| > r(A) := \sup\{|\mu| : \mu \in \sigma(A)\}$ but its proof is more involved. The quantity $r(A)$ is called the *spectral radius* of A.

Set $x_0 = 0$, and for $n \geq 1$ use (2.1.1) above to choose inductively $x_n \in X$ such that

$$\|x_n - x_{n-1}\| \leq \frac{1}{3^n} \quad \text{and} \quad \|A_n x_n\| \geq \frac{2}{3} \cdot \frac{1}{3^n} \|A_n\|. \tag{2.1.2}$$

The sequence $\{x_n\}_{n=1}^{\infty}$ is Cauchy and X is a Banach space. Hence, there is $x \in X$ such that

$$\lim_{n \to \infty} \|x_n - x\| = 0.$$

It follows from the first inequality in (2.1.2) that

$$\|x - x_n\| \leq \frac{1}{2} \cdot \frac{1}{3^n}$$

and hence

$$\|A_n x\| \geq \frac{1}{6} \cdot \frac{1}{3^n} \|A_n\| \geq \frac{1}{6} \left(\frac{4}{3}\right)^n \to +\infty,$$

a contradiction. □

The method of construction of sequence $\{x_n\}_{n=1}^{\infty}$ in the proof above is called *the gliding hump method* and the reader can find it in the literature also in different contexts.

Uniform Boundedness Principle is a quintessence of several results on approximation of functions in classical analysis and can be used for "modern" proofs of such results. The following example is typical.

Example 2.1.5. There exists a periodic continuous function the Fourier series of which is divergent at zero.[4] To see this we recall that the nth partial sum of the Fourier series of a function f at 0 is given by

$$s_n(f)(0) = \frac{1}{2\pi} \int_{-\pi}^{\pi} D_n(0 - t) f(t)\, \mathrm{d}t \text{ where } D_n(t) = \frac{\sin\left(n + \frac{1}{2}\right) t}{\sin \frac{t}{2}}, 0 < |t| < \pi$$

(the nth *Dirichlet kernel*). Since $\sigma_n \colon f \mapsto s_n(f)(0)$ are continuous linear forms on the space $C[-\pi, \pi]$, the sequence of their norms $\left\{\|\sigma_n\|_{\mathcal{L}(C[-\pi,\pi],\mathbb{R})}\right\}_{n=1}^{\infty}$ should be bounded provided $\sigma_n(f)$ is convergent for all $f \in C[-\pi, \pi]$ (Theorem 2.1.4). One can calculate that

$$\|\sigma_n\| = \frac{1}{2\pi} \int_{-\pi}^{\pi} |D_n(t)|\, \mathrm{d}t,$$

and a careful estimate shows that $\|\sigma_n\|$ is like $\log n$ for large n. ○

As indicated in the previous example, Theorem 2.1.4 is essentially an approximation result. This is clearer from its next variant.

[4] Even divergent at uncountably many points but always of measure zero. The set of such "bad" functions is dense in $C[-\pi, \pi]$.

Corollary 2.1.6 (Banach–Steinhaus). *Let X and Y be Banach spaces and let $\{A_n\}_{n=1}^{\infty} \subset \mathcal{L}(X,Y)$. Then the limits $\lim_{n\to\infty} A_n x$ exist for every $x \in X$ if and only if the following conditions are satisfied:*

(i) *There is a dense set $\mathcal{M} \subset X$ such that $\lim A_n x$ exists for each $x \in \mathcal{M}$.*

(ii) *The sequence of norms $\{\|A_n\|\}_{n=1}^{\infty}$ is bounded.*

Moreover, under these conditions

$$Ax := \lim_{n\to\infty} A_n x$$

exists for all $x \in X$ and $A \in \mathcal{L}(X,Y)$.[5]

The following proposition is also often useful.

Proposition 2.1.7. *Let X be a Banach space and Y a normed linear space. If $B\colon X \times X \to Y$ is a bilinear operator (i.e., linear in both variables) and*

(i) *for every $y \in X$ the mapping $x \mapsto B(x,y)$ belongs to $\mathcal{L}(X,Y)$;*

(ii) *for every $x \in X$ the mapping $y \mapsto B(x,y)$ belongs to $\mathcal{L}(X,Y)$,*

then there exists a constant c such that

$$\|B(x,y)\|_Y \le c\|x\|_X\|y\|_X, \qquad x, y \in X.$$

In particular, if $x_n \to x$, $y_n \to y$, then $B(x_n, y_n) \to B(x,y)$.

Proof. Denote $B_y\colon x \mapsto B(x,y)$. By (i), $B_y \in \mathcal{L}(X,Y)$ for all $y \in X$, $\|y\| \le 1$. By (ii), $\|B_y(x)\| \le c(x)$. The Uniform Boundedness Principle implies the existence of a constant c such that

$$\sup_{\|x\|\le 1}\ \sup_{\|y\|\le 1} \|B(x,y)\| \le c. \qquad \square$$

Theorem 2.1.8 (Open Mapping Theorem). *Let X, Y be Banach spaces, let $A \in \mathcal{L}(X,Y)$ and let A have a closed range* $\operatorname{Im} A$. *Then for any open set $\mathcal{G} \subset X$ its image $A(\mathcal{G})$ is an open set in* $\operatorname{Im} A$. *In particular, if A is, in addition, injective and surjective, then $A^{-1} \in \mathcal{L}(Y,X)$.*

When applied to linear equations

$$Ax = y,$$

Theorem 2.1.8 says that the continuous dependence of a solution on the right-hand side is a consequence of the existence and uniqueness result. Such continuous dependence is important for any reasonable numerical approximation.

Theorem 2.1.8 can be also used in a “negative” sense:

---

[5] This type of convergence is the so-called *convergence in the strong operator topology*. It is weaker than the norm convergence.

Example 2.1.9. Denote by

$$\hat{f}(n) = \frac{1}{2\pi} \int_{-\pi}^{\pi} f(t) \mathrm{e}^{-int} \, \mathrm{d}t$$

the nth Fourier coefficient of $f \in L^1(-\pi, \pi)$. Since $\hat{f}(n) \to 0$ for $|n| \to \infty$ for all trigonometric polynomials which are dense in $L^1(-\pi, \pi)$, we have

$$\hat{f}(n) \to 0 \qquad \text{for all} \quad f \in L^1(-\pi, \pi)$$

(the so-called Riemann–Lebesgue Lemma). In other words, $A\colon f \mapsto \hat{f}(\cdot)$ is a continuous linear operator from $L^1(-\pi, \pi)$ into

$$c_0(\mathbb{Z}) := \left\{ \{a_n\}_{n \in \mathbb{Z}} : \lim_{|n| \to \infty} |a_n| = 0 \right\}, \qquad \|\{a_n\}\|_{c_0(\mathbb{Z})} = \sup |a_n|.$$

Applications of Fourier series to various problems in analysis (like convolution equations, differential equations, ...) would be much easier if A were a surjective mapping onto $c_0(\mathbb{Z})$. Theorem 2.1.8 shows that this cannot be true for then A^{-1} would be bounded, i.e.,

$$\|f\|_{L^1(-\pi,\pi)} \le c \sup_{n \in \mathbb{Z}} |\hat{f}(n)| \qquad \text{for all} \quad f \in L^1(-\pi, \pi).$$

If $\{D_k\}_{k=1}^{\infty}$ is the sequence of Dirichlet kernels (Example 2.1.5), then

$$\hat{D}_k(n) = \begin{cases} 1, & |n| \le k, \\ 0, & |n| > k, \end{cases} \qquad \text{and} \qquad \|D_k\|_{L^1(-\pi,\pi)} \sim \log k,$$

a contradiction. ○

Theorem 2.1.8 also yields a sufficient condition for a linear operator to be continuous. To formulate it we need the notion of a closed operator:

Let X, Y be normed linear spaces. A linear operator

$$A\colon \operatorname{Dom} A \subset X \to Y$$

is said to be *closed* if

$$\{x_n\}_{n=1}^{\infty} \subset \operatorname{Dom} A, \qquad x_n \to x, \qquad Ax_n \to y$$

implies that

$$x \in \operatorname{Dom} A \qquad \text{and} \qquad Ax = y.$$

Equivalently, A is a closed operator if and only if the graph of A, i.e.,

$$\mathcal{G}(A) := \{(x, Ax) : x \in \operatorname{Dom} A\},$$

is a closed linear subspace of $X \times Y$.

Corollary 2.1.10 (Closed Graph Theorem). *Let X, Y be Banach spaces and let A be a closed operator from* $\operatorname{Dom} A = X$ *into* Y. *Then* A *is continuous.*

Proof. If $\mathcal{G}(A)$ denotes the graph of A, then put

$$T(x, Ax) = x.$$

By Theorem 2.1.8, T^{-1} is continuous, and therefore

$$A = \pi_2 \circ T^{-1}$$

is continuous as well (π_2 is the projection of $X \times Y$ onto the second component Y). □

Example 2.1.11. Many differential operators are either closed or have closed extensions. If they are viewed as operators from X into X, then they are only densely defined. A very simple example:

$$X = C[0,1], \qquad Ax = \dot{x},$$
$$\operatorname{Dom} A = \{x \in X : \dot{x}(t) \text{ exists for all } t \in [0,1] \text{ and } \dot{x} \in X\}.$$

A well-known classical result says that A is a closed operator. But A is not continuous. For $x_n(t) = t^n$ we have $\|x_n\| = 1$, $\|\dot{x}_n\| = n$. ○

Example 2.1.12. Let X be a Banach space and M a linear subspace of X. Let N be an (algebraic) complement of M and let P be the corresponding projection onto M. Then P is continuous if and only if both M and N are closed.

The sufficiency part follows from the Closed Graph Theorem and from an observation that P is closed whenever M and N are closed subspaces. The necessity part is obvious since

$$M = \operatorname{Ker}(I - P), \qquad N = \operatorname{Ker} P.$$

This statement should be compared with the Hilbert space case (Corollary 1.2.35).

An important special case is $\operatorname{codim} M < \infty$. By definition, this means that an algebraic direct complement N has a finite dimension ($\operatorname{codim} M := \dim N$) and therefore N is closed (Corollary 1.2.11(i)). If M is closed as well, then any projection onto M is continuous. We postpone the case of $\dim M < \infty$ to Remark 2.1.19.

We note that if X is a Banach space such that there exists a continuous projection P, $\|P\|_{\mathcal{L}(X)} \le 1$, onto every closed subspace of X, then X has an equivalent norm induced by the scalar product on X (see Kakutani [85]). ○

Now we turn our attention to the *dual space* X^* of all *continuous linear forms* on a normed linear space X. In Section 1.1 we have seen the importance of linear forms. Namely, they allowed us to define an algebraic adjoint operator $A^\#$ and formulate Theorem 1.1.25. The dual space X^* is even more important for a normed linear space X since another topology can be introduced on X with help of X^* which in a certain sense has better properties (Theorem 2.1.25 below).

Surprisingly, the following basic result does not need any topology.

Theorem 2.1.13 (Hahn–Banach). *Let X be a real linear space and let Y be a linear subspace of X. Assume that f is a linear form on Y which is dominated by a sublinear functional p defined on X.*[6] *Then there exists $F \in X^{\#}$ such that*

(i) *$F(y) = f(y)$ for all $y \in Y$ (extension);*

(ii) *$F(x) \leq p(x)$ for all $x \in X$ (dominance).*

Proof. The proof is based on an extension of f to a subspace whose dimension is larger by 1 and such that this extension is dominated by the same p (for details see, e.g., references given on page 58), and the use of Zorn's Lemma as an inductive argument, similarly as in the proof of Theorem 1.1.3. □

Remark 2.1.14. If X is a complex linear space, then we need p to satisfy a stronger condition than (2) in footnote 6, namely

(2') $p(\alpha x) = |\alpha| p(x)$, $\alpha \in \mathbb{C}$, $x \in X$.

In this case p is called a *semi-norm*.[7] The dominance also has to be stronger:

$$|f(x)| \leq p(x).$$

The extension result follows from Theorem 2.1.13 by considering $\operatorname{Re} f$ and $\operatorname{Im} f$ and observing that $\operatorname{Re} f(\mathrm{i}x) = -\operatorname{Im} f(x)$.

Corollary 2.1.15. *Let X be a normed linear space and let Y be a linear subspace of X (not necessarily closed). If $f \in Y^*$, then there exists $F \in X^*$ such that*

(i) *$F(y) = f(y)$ for $y \in Y$;*

(ii) $\|F\|_{X^*} = \|f\|_{Y^*}$.

Proof. Put $p(x) = \|f\|\|x\|$, $x \in X$, and apply Theorem 2.1.13 or Remark 2.1.14, respectively. □

Corollary 2.1.16 (Dual Characterization of the Norm). *Let X be a normed linear space. Then*

$$\|x\|_X = \max\{|f(x)| : f \in X^* \text{ with } \|f\|_{X^*} \leq 1\}. \tag{2.1.3}$$

Proof. Put $g_0(\alpha x) = \alpha\|x\|$, $\alpha \in \mathbb{R}$ (or $\alpha \in \mathbb{C}$). Then g_0 is a continuous linear form on $\operatorname{Lin}\{x\}$ and its norm is 1 (provided $x \neq o$). Let f_0 be its extension from Corollary 2.1.15. Then

$$f_0(x) = \|x\|,\ \|f_0\| = 1, \qquad \text{i.e.,} \qquad \|x\| \leq \sup\{|f(x)| : f \in X^* \text{ with } \|f\| \leq 1\}.$$

The converse inequality follows from the definition of $\|f\|$. □

---

[6] A mapping $p \colon X \to \mathbb{R}$ is called *sublinear* if
(1) $p(x+y) \leq p(x) + p(y)$ for any $x, y \in X$;
(2) $p(\alpha x) = \alpha p(x)$ for any $x \in X$ and $\alpha \geq 0$.

[7] The difference between a norm and a semi-norm is that a semi-norm need not satisfy the condition: $p(x) = 0 \Longrightarrow x = o$.

Remark 2.1.17.

(i) If X is a Hilbert space, then the equality (2.1.3) can be obtained immediately from the Riesz Representation Theorem (Theorem 1.2.40). This theorem can be often used in Hilbert spaces instead of the Hahn–Banach Theorem.

(ii) A slightly weaker form of (2.1.3) is often used:

If $f(x) = 0$ for all $f \in X^*$, then $x = o$.

The equivalent assertion reads as follows:

X^* separates points of X.

Corollary 2.1.18 (Separation Theorem). *Let X be a normed linear space and let $\mathcal{C}$ be a nonempty, closed, convex set. If $x_0 \notin \mathcal{C}$, then there exists $F \in X^*$ such that*

$$\sup\{\operatorname{Re} F(x) : x \in \mathcal{C}\} < \operatorname{Re} F(x_0). \tag{2.1.4}$$

Proof. It is sufficient to give the proof for a real space X and under the additional assumption $o \in \mathcal{C}$. In particular, this assumption means that $x_0 \neq o$. We wish to extend the form f defined on $\operatorname{Lin}\{x_0\}$ by $f(\alpha x_0) = \alpha$, $\alpha \in \mathbb{R}$. To do that we need a suitable dominating functional. Since $d := \operatorname{dist}(x_0, \mathcal{C}) > 0$, there exists a convex neighborhood of $\mathcal{C}$ which does not contain x_0, e.g.,

$$\mathcal{K} = \left\{x + y : x \in \mathcal{C}, \|y\| < \frac{d}{2}\right\}.$$

Put

$$p_{\mathcal{K}}(z) := \inf\left\{\alpha > 0 : \frac{z}{\alpha} \in \mathcal{K}\right\} \qquad \text{for} \quad z \in X.$$[8]

It is a matter of simple calculation to show that $p_{\mathcal{K}}$ is sublinear, $p_{\mathcal{K}}(x_0) > 1$, and $p_{\mathcal{K}}(z) \leq 1$ for $z \in \mathcal{K}$.

Let F be an extension of f given by Theorem 2.1.13. Since $o \in \mathcal{C}$, we have

$$F(\pm y) \leq p_{\mathcal{K}}(\pm y) \leq 1 \qquad \text{for} \quad \|y\| < \frac{d}{2}.$$

This shows that

$$\|F\| \leq \frac{2}{d}, \qquad \text{i.e.,} \qquad F \in X^*.$$

The inequality (2.1.4) follows from domination: namely, we have

$$F(x) + F(y) \leq p_{\mathcal{K}}(x + y) \leq 1 \qquad \text{for} \quad x \in \mathcal{C} \quad \text{and all} \quad \|y\| < \frac{d}{2},$$

i.e.,

$$F(x) \leq 1 - \sup\left\{F(y) : \|y\| < \frac{d}{2}\right\} < 1 = F(x_0). \qquad \square$$

[8] $p_{\mathcal{K}}$ is the so-called *Minkowski functional* of the convex set $\mathcal{K}$.

Remark 2.1.19. If $\mathcal{C}$ from Corollary 2.1.18 is a closed linear subspace of X and $F \in X^*$ satisfies (2.1.4), then $F(x) = 0$ for all $x \in \mathcal{C}$. Notice that $F(x_0) = 1$ for F which has been constructed in the proof.

This observation yields the existence of a continuous projection onto a finite-dimensional subspace Y of X. Namely, suppose that $\{y_1, \dots, y_n\}$ is a basis of Y, and denote by Y_k the span of $y_1, \dots, y_{k-1}, y_{k+1}, \dots, y_n$. Then Y_k is a closed linear subspace of X and $y_k \notin Y_k$. Let $F_k \in X^*$ be such that

$$F_k(y_j) = \begin{cases} 1, & j = k, \\ 0, & j \neq k, \end{cases} \qquad j = 1, \dots, n.$$

Then

$$Px = \sum_{k=1}^{n} F_k(x) y_k$$

is a continuous projection onto Y.

Warning. *It is not true that every projection onto* Y *is continuous even if* codim Y = 1 *but the construction (i.e., the construction of a noncontinuous linear form) is not obvious!*

Example 2.1.20.

(i) By Corollary 1.2.11(ii), $(\mathbb{R}^M)^* = (\mathbb{R}^M)^\#$. This means that $(\mathbb{R}^M)^*$ can be identified with $\mathbb{R}^M$.

(ii) Let $\mathcal{K}$ be a compact subset of $\mathbb{R}^M$. Then for any $F \in [C(\mathcal{K})]^*$ there exists a unique complex Borel measure μ on $\mathcal{K}$ such that

$$F(f) = \int_{\mathcal{K}} f(x)\,\mathrm{d}\mu(x) \qquad \text{for every} \quad f \in C(\mathcal{K}),$$

and $\|F\|_{[C(\mathcal{K})]^*} = |\mu|(\mathcal{K})$ where $|\mu|$ is the total variation of μ. A similar statement holds under a more general assumption on $\mathcal{K}$ – for details and the corresponding notions see Dunford & Schwartz [53, Section IV, 6] or Rudin [132, Chapter 6] and, especially, Bourbaki [17]. In the last book the integration theory is developed on the basis of this representation theorem.

(iii) Let Ω be an open subset of $\mathbb{R}^M$ and let $p \in [1, \infty)$. Then the dual space $[L^p(\Omega)]^*$ can be identified with $L^{p'}(\Omega)$ (p' is the conjugate exponent, i.e., $\frac{1}{p} + \frac{1}{p'} = 1$) in the following sense. For any $F \in [L^p(\Omega)]^*$ there exists a unique $\varphi \in L^{p'}(\Omega)$ such that

$$F(f) = \int_{\Omega} f(x)\varphi(x)\,\mathrm{d}x \qquad \text{for every} \quad f \in L^p(\Omega).$$

Moreover, $\|F\|_{[L^p(\Omega)]^*} = \|\varphi\|_{L^{p'}(\Omega)}$. Details can be found in books cited above.

Warning. *The dual space $[L^\infty(\Omega)]^*$ is much larger than $L^1(\Omega)$!*

(iv) The dual spaces to Sobolev spaces $W^{k,p}(\mathbb{R}^M)$ can be identified with special subspaces of tempered distributions for example via the Fourier transform. We omit details since their description is beyond the scope of this book. ○

The reader can ask why we are so interested in continuous linear forms. One of the reasons is the following. Suppose that φ is a *vector-valued function* (i.e., a mapping from $\mathbb{R}$ or $\mathbb{C}$ into a normed linear space X). For any $f \in X^*$ the composition $f \circ \varphi$ is a real or complex function of a real or complex variable and therefore results of classical analysis can be applied to $f \circ \varphi$. To be more specific, consider the resolvent (see page 56) of $A \in \mathcal{L}(X)$

$$R(\lambda)x := (\lambda I - A)^{-1}x, \qquad \lambda \in \varrho(A),$$

which is an X-valued function for every $x \in X$. Then for any $F \in X^*$, the complex function $\varphi(\lambda) = F[(\lambda I - A)^{-1}x]$ is holomorphic in $\varrho(A)$. For $|\lambda| > \|A\|$ we also have

$$\begin{aligned}|\varphi(\lambda)| &\le \|F\|_{X^*}\|(\lambda I - A)^{-1}\|_{\mathcal{L}(X)}\|x\|_X = \|F\|\|x\|\left\|\sum_{n=0}^{\infty}\frac{A^n}{\lambda^{n+1}}\right\|\\ &\le \|F\|\|x\|\sum_{n=0}^{\infty}\frac{\|A\|^n}{|\lambda|^{n+1}},\end{aligned}$$

and so

$$\lim_{|\lambda|\to\infty}|\varphi(\lambda)| = 0.$$

If $\varrho(A) = \mathbb{C}$, φ would be identically zero (by the Liouville Theorem from the complex functions theory). Since this should be true for all $F \in X^*$, we get $(\lambda I - A)^{-1}x = o$ for all $x \in X$, a contradiction. Therefore, the spectrum $\sigma(A)$ is nonempty for each $A \in \mathcal{L}(X)$. This is a generalization of the existence of an eigenvalue of a linear operator in a finite-dimensional space and therefore also a generalization of the Fundamental Theorem of Algebra (cf. page 15). It is worth mentioning that the Jordan Canonical Form (Theorem 1.1.34) is based on this result.

Warning. *It is not true that any $A \in \mathcal{L}(X)$, $\dim X = \infty$, has an eigenvalue!*
A simple example is

$$X = C[0,1], \qquad Ax(t) := tx(t).$$

Our main reason for considering dual spaces comes from an attempt to find a weaker topology on a normed linear space in which bounded sets would be relatively compact. The importance of this fact will become clear in Chapter 7. We also ask the reader to return to Proposition 1.2.2 for motivation.

Definition 2.1.21. Let $\{x_n\}_{n=1}^{\infty}$ be a sequence of elements in a normed linear space X. We say that $\{x_n\}_{n=1}^{\infty}$ *converges weakly* to $x \in X$ (notation $x_n \rightharpoonup x$ or w-$\lim\limits_{n\to\infty} x_n = x$) if

$$\lim_{n\to\infty} f(x_n) = f(x) \qquad \text{for every} \quad f \in X^*.$$

Proposition 2.1.22.

(i) (*uniqueness*) *If* $x_n \rightharpoonup x$ *and* $x_n \rightharpoonup y$, *then* $x = y$.

(ii) *If* $\lim\limits_{n\to\infty} \|x_n - x\| = 0$, *then* $x_n \rightharpoonup x$.[9]

(iii) *A weakly convergent sequence is bounded. Moreover, if* $x_n \rightharpoonup x$, *then*

$$\|x\| \le \liminf_{n\to\infty} \|x_n\|.$$

(iv) *If* X *is a uniformly convex Banach space,*[10] $x_n \rightharpoonup x$ *and* $\|x_n\| \to \|x\|$, *then* $\{x_n\}_{n=1}^{\infty}$ *converges to* x *in the norm topology.*

Proof. Assertion (i) follows immediately from Remark 2.1.17(ii) since in this case $f(x) = f(y)$ for every $f \in X^*$.

Assertion (ii) is obvious.

Assertion (iii) is basically a consequence of Theorem 2.1.4, but certain preliminaries are needed: Since X^* is a normed linear space, its dual $X^{**} := (X^*)^*$ is defined. Put

$$\kappa(x)\colon f \mapsto f(x), \qquad f \in X^*.$$

Then κ (the so-called *canonical embedding*) is a linear continuous operator from X into X^{**}, and

$$\|\kappa(x)\|_{X^{**}} = \sup_{\|f\|_{X^*}\le 1} |f(x)| = \|x\|_X$$

(Corollary 2.1.16).[11] Since the space X^* is always complete (Proposition 2.1.1), Theorem 2.1.4 can be applied to the sequence $\{\kappa(x_n)\}_{n=1}^{\infty}$. This shows that $\{x_n\}_{n=1}^{\infty}$ is bounded. If $x_n \rightharpoonup x$, we choose $f \in X^*$ such that

$$\|f\| = 1 \qquad \text{and} \qquad f(x) = \|x\|$$

---

[9] **Warning.** *The converse statement is not true in general (see Exercise 2.1.38)!*

[10] A Banach space X is said to be *uniformly convex* if for every $\varepsilon > 0$ there is $\delta > 0$ such that $x, y \in X$, $\|x\| = \|y\| = 1$, $\|x - y\| \ge \varepsilon \Longrightarrow 1 - \left\|\frac{x+y}{2}\right\| \ge \delta$. Every uniformly convex space is reflexive, see Yosida [161, Chapter V, 2]. Hilbert spaces, $L^p(\Omega)$-spaces and $W^{1,p}(\Omega)$-spaces $(1 < p < \infty)$ are uniformly convex (for a Hilbert space this follows from the parallelogram identity (1.2.14), for the other two cases see, e.g., Adams [2, Corollary 2.29 and Theorem 3.5]).

[11] It is not generally true that κ is surjective. A Banach space X is said to be *reflexive* if κ is surjective. Every Hilbert space and spaces $L^p(\Omega), W^{k,p}(\Omega)$, $1 < p < \infty$, are reflexive (the Riesz Representation Theorem and Example 2.1.20(iii)). Spaces $L^1(\Omega)$, $L^\infty(\Omega)$ and $C(\Omega)$ are not reflexive.

(Corollary 2.1.16). Then

$$\|x\| = f(x) = \lim_{n\to\infty} f(x_n) \le \liminf_{n\to\infty} \|x_n\|.$$

Assertion (iv) is obvious for $x = o$. If $x \neq o$, then we may assume that also $x_n \neq o$ and put $y := \frac{x}{\|x\|}$ and $y_n := \frac{x_n}{\|x_n\|}$. Since $x_n \rightharpoonup x$ and $\|x_n\| \to \|x\|$, we have

$$f(y_n) = \frac{1}{\|x_n\|} f(x_n) \to \frac{1}{\|x\|} f(x) = f(y) \quad \text{for any } f \in X^*, \quad \text{i.e.,} \quad y_n \rightharpoonup y.$$

If we prove that $\|y_n - y\| \to 0$, then

$$\|x_n - x\| = \|(y_n\|x_n\| - y\|x\|)\| \le \|x_n\|\|y_n - y\| + \|y\|\big|\|x_n\| - \|x\|\big| \to 0$$

due to the assumption $\|x_n\| \to \|x\|$. To prove $y_n \to y$ we proceed by contradiction using the uniform convexity of X. Suppose that there is $\varepsilon > 0$ such that $\|y_n - y\| \ge \varepsilon$ for infinitely many n. Then, by the uniform convexity of X,

$$\|y_n + y\| \le 2(1 - \delta).$$

Let us choose $f_0 \in X^*$, $\|f_0\| = 1$, $f_0(y) = \|y\| = 1$ (see Corollary 2.1.16). Then

$$2(1-\delta) \ge \limsup_{n\to\infty} \|y_n + y\| \ge \limsup_{n\to\infty} f_0(y_n + y) = 2f_0(y) = 2,$$

a contradiction. □

Remark 2.1.23. The weak convergence is the convergence in the *weak topology*. It is convenient to define this topology by systems of neighborhoods of points. We say that $\mathcal{U} \subset X$ is a *weak neighborhood* of a point $x \in X$ if there are $f_1, \dots, f_n \in X^*$ such that

$$\{y \in X : |f_i(y) - f_i(x)| < 1 \text{ for } i = 1, \dots, n\} \subset \mathcal{U}.$$

A subset $\mathcal{G} \subset X$ is *weakly open* (i.e., open in the weak topology) provided it is a weak neighborhood of each of its points. It is easy to see that a weakly open set is also open in the norm topology. The converse is generally true only in finite-dimensional spaces.

As we have mentioned, our aim is to find compact sets in the weak topology.

Remark 2.1.24. The weak topology in an infinite-dimensional space is not metrizable. Therefore two concepts of compactness, namely the sequential and the covering one (see footnote 12 on page 27) are in principle different. It is surprising that they coincide for weak topologies in Banach spaces. This very deep result is known as the Eberlain–Smulyan Theorem (see Dunford & Schwartz [53, Chapter 5]).

Theorem 2.1.25 (Eberlain–Smulyan). *Let X be a reflexive space. Then any bounded sequence contains a weakly convergent subsequence.*

Proof. We present a simple proof for the case that X is a Hilbert space. A proof for an arbitrary reflexive space can be found, e.g., in Dunford & Schwartz [53], Fabian et al. [60], Yosida [161]. Let $\{x_n\}_{n=1}^{\infty} \subset X$ be a bounded sequence, and put

$$Y = \overline{\operatorname{Lin}\{x_1, x_2, \dots\}}$$

(the closure is taken in the norm topology). Since the sequence of scalar products $\{(x_1, x_n)\}_{n=1}^{\infty}$ is a bounded sequence of numbers (real or complex), there is a subsequence, say $\{x_n^{(1)}\}_{n=1}^{\infty}$, such that $\left\{(x_1, x_n^{(1)})\right\}_{n=1}^{\infty}$ converges. For the same reason there is a subsequence $\{x_n^{(2)}\}_{n=1}^{\infty}$ of $\{x_n^{(1)}\}_{n=1}^{\infty}$ such that $\left\{(x_2, x_n^{(2)})\right\}_{n=1}^{\infty}$ converges, etc. Put $y_k = x_k^{(k)}$ (the diagonal choice). Then $\lim\limits_{k\to\infty} (x_j, y_k)$ exists for all $j \in \mathbb{N}$, and therefore $\lim\limits_{k\to\infty} (x, y_k)$ exists for each $x \in \operatorname{Lin}\{x_1, x_2, \dots\}$.

Since the sequence of linear forms $f_k \colon x \to (x, y_k)$ is bounded in Y^*, the Banach–Steinhaus Theorem (Corollary 2.1.6) implies the existence of $f \in Y^*$ such that

$$\lim_{k\to\infty} f_k(x) = f(x) \qquad \text{for all} \quad x \in Y.$$

Let P be the orthogonal projection onto Y. Put

$$g(x) = f(Px) \qquad \text{for} \quad x \in X.$$

Then $g \in X^*$ and by the Riesz Representation Theorem there is $y \in X$ such that

$$g(x) = (x, y) \qquad \text{for} \quad x \in X.$$

Moreover,

$$\lim_{k\to\infty} (x, y_k) = \lim_{k\to\infty} (Px, y_k) = f(Px) = (x, y) \qquad \text{for all} \quad x \in X.$$

This means that $y_k \rightharpoonup y$. □

Remark 2.1.26. Weak convergence in a dual space X^* is more confusing since two approaches can be used. We say that a sequence $\{f_n\}_{n=1}^{\infty} \subset X^*$

(i) *converges weakly* to $f \in X^*$ (notation $f_n \rightharpoonup f$ or w-$\lim\limits_{n\to\infty} f_n = f$) if

$$\lim_{n\to\infty} F(f_n) = F(f) \qquad \text{for every} \quad F \in X^{**};$$

(ii) *converges weak star* to $f \in X^*$ (notation $f_n \overset{*}{\rightharpoonup} f$ or w*-$\lim\limits_{n\to\infty} f_n = f$) if

$$\lim_{n\to\infty} f_n(x) = f(x) \qquad \text{for every} \quad x \in X.$$

Criteria for weak convergence in L^p-spaces can be found, e.g., in Dunford & Schwartz [53, Chapter IV, 8].

The weak convergence in X^* has obviously the same properties as that in X. Because of the continuous embedding $\kappa\colon X \to X^{**}$ (see the proof of Proposition 2.1.22(iii)) the w-convergence implies the w*-convergence. The converse is true if X is a reflexive space, i.e.,

$$\kappa(X) = X^{**}.$$

Since the w*-topology is generally weaker than the w-topology there can exist more w*-compact sets than the w-compact ones. In fact, the following result (the Alaoglu–Bourbaki Theorem, see Conway [36], Dunford & Schwartz [53], Fabian et al. [60]) holds:

If X is a normed linear space, then any closed ball in X^* is w-compact. If, moreover, X is separable, then the ball is also sequentially w*-compact.*

For example, this theorem can be applied to balls in $L^p(\Omega)$, $1 < p \leq \infty$.

In the rest of this section we will examine *adjoint operators*. Suppose that X and Y are normed linear spaces and $A \in \mathcal{L}(X,Y)$. If $g \in Y^*$, then

$$A^*g := g(A) \in X^*.$$

The operator $A^*\colon Y^* \to X^*$ is obviously linear, and it is also continuous since

$$|A^*g(x)| = |g(Ax)| \leq \|g\|_{Y^*}\|Ax\|_Y \leq \|g\|_{Y^*}\|A\|_{\mathcal{L}(X,Y)}\|x\|_X.$$

If H_1, H_2 are Hilbert spaces and $A \in \mathcal{L}(H_1,H_2)$ we have another approach to the definition of an adjoint operator, namely the one based on the Riesz Representation Theorem: For $y \in H_2$ the mapping $f\colon x \mapsto (Ax,y)_{H_2}$ is a continuous linear form on H_1, and hence there is $z \in H_1$ for which $f(x) = (x,z)_{H_1}$. This z is uniquely determined by y, and we denote for a moment $z = A^+y$, i.e.,

$$(Ax,y)_{H_2} = (x,A^+y)_{H_1}.$$

There is a very slight difference between A^* and A^+, e.g., $(\alpha A)^* = \alpha A^*$ and $(\alpha A)^+ = \overline{\alpha}A^+$ (see also Example 2.1.28 below). So we will use the same notation, namely A^*, for both concepts. Symmetric matrices have certain special properties (e.g., their canonical forms are diagonal). The same can be expected for their generalization in the Hilbert space setting which is defined as follows:

An operator $A \in \mathcal{L}(H)$ is said to be *self-adjoint* if $A = A^*$, i.e.,

$$(Ax,y) = (x,Ay) \qquad \text{for all} \quad x,y \in H.$$

In order to generalize Theorem 1.1.25 to continuous linear operators on infinite-dimensional normed linear spaces we will use the same notation but with a slightly different meaning:

If $\mathcal{M} \subset X$, then

$$\mathcal{M}^{\perp} := \{f \in X^* : x \in \mathcal{M} \Rightarrow f(x) = 0\}.$$

If $\mathcal{N} \subset X^*$, then

$$\mathcal{N}_{\perp} := \{x \in X : f \in \mathcal{N} \Rightarrow f(x) = 0\}.$$

We invite the reader to compare these symbols with that for orthogonal complements in Hilbert spaces.

Proposition 2.1.27. *Let X, Y be normed linear spaces and let $A \in \mathcal{L}(X,Y)$. Then*

(i) *if $x_n \rightharpoonup x$, then $Ax_n \rightharpoonup Ax$;*

(ii) *if A is, moreover, continuously invertible, then A^* is also continuously invertible and $(A^*)^{-1} = (A^{-1})^*$;*

(iii) $\operatorname{Ker} A = (\operatorname{Im} A^*)_{\perp}$;

(iv) $\overline{\operatorname{Im} A} = (\operatorname{Ker} A^*)_{\perp}$.

Proof. (i) It is easy with the use of A^*.

(ii) It is sufficient to show that $(A^{-1})^* A^* = I_{Y^*}$ and $A^*(A^{-1})^* = I_{X^*}$. This follows from the more general result

$$(AB)^* = B^* A^*$$

which is easily verified.

(iii) The inclusion $\subset$ is obvious from the definition, for the converse inclusion $\supset$ it is sufficient to use the fact that Y^* separates the points of Y.

(iv) It is easy to see that $(\operatorname{Im} A)^{\perp} = \operatorname{Ker} A^*$. To get (iv) it suffices to prove that

$$(\mathcal{M}^{\perp})_{\perp} = \overline{\operatorname{Lin} \mathcal{M}} \qquad \text{for} \quad \mathcal{M} \subset X.$$

If x_0 belonged to $(\mathcal{M}^{\perp})_{\perp} \setminus \overline{\operatorname{Lin} \mathcal{M}}$, x_0 would be separated from $\overline{\operatorname{Lin} \mathcal{M}}$ by a linear form $f \in X^*$ (Corollary 2.1.18). Since $\overline{\operatorname{Lin} \mathcal{M}}$ is a subspace of X, this separating f would be in $(\overline{\operatorname{Lin} \mathcal{M}})^{\perp} = \mathcal{M}^{\perp}$. Therefore $f(x_0) = 0$, and a contradiction is obtained. The converse inclusion $\overline{\operatorname{Lin} \mathcal{M}} \subset (\mathcal{M}^{\perp})_{\perp}$ is obvious. □

Notice that the statement (iv) is not a sufficient condition for solvability of the equation

$$Ax = y$$

since only the closure of $\operatorname{Im} A$ is characterized. There are many operators the range of which is not closed. A simple example is $Ax(t) = \int_0^t x(s)\, ds$ considered either in $C[0,1]$ or in $L^2(0,1)$. It is not an easy task to decide whether an operator has a closed range or not. The following statement is useful in applications.

If X, Y are Banach spaces and $A \in \mathcal{L}(X,Y)$ is injective, then $\operatorname{Im} A$ *is closed if and only if there is a positive constant c such that*

$$\|Ax\| \geq c\|x\| \qquad \textit{for all} \quad x \in X.$$

Sufficiency is easy, the necessity part follows from the Open Mapping Theorem.

There is an important subclass of operators with a closed range, namely the so-called *Fredholm operators.* An operator $A \in \mathcal{L}(X)$ is said to be Fredholm if

$$\dim \operatorname{Ker} A < \infty, \qquad \operatorname{Im} A \quad \text{is closed,} \qquad \text{and} \qquad \operatorname{codim} \operatorname{Im} A < \infty$$

(i.e., the dimension of any direct complement of $\operatorname{Im} A$ is finite). We note that

$$\operatorname{codim} \operatorname{Im} A = \dim \operatorname{Ker} A^*$$

(this is basically Proposition 2.1.27(iv)). We define

$$\operatorname{ind} A := \dim \operatorname{Ker} A - \dim \operatorname{Ker} A^*$$

and call it the *index of the Fredholm operator*. A special class of Fredholm operators will be examined in the next section.

We have not yet introduced any sufficiently broad family of continuous linear operators. The next example fills this gap.

Example 2.1.28 (Integral operators). Let Ω and $\tilde{\Omega}$ be open subsets of $\mathbb{R}^M$ and $\mathbb{R}^{\tilde{M}}$, respectively. Assume that $k\colon \tilde{\Omega} \times \Omega \to \mathbb{C}$ is a measurable function for which there are constants c_1, c_2 such that

$$\int_\Omega |k(t,s)| \, \mathrm{d}s \leq c_1 \quad \text{for a.a. } t \in \tilde{\Omega}, \qquad \int_{\tilde{\Omega}} |k(t,s)| \, \mathrm{d}t \leq c_2 \quad \text{for a.a. } s \in \Omega.$$

Then the operator A defined by

$$Ax(t) = \int_\Omega k(t,s)x(s) \, \mathrm{d}s \tag{2.1.5}$$

is a linear bounded operator from $L^p(\Omega)$ into $L^p(\tilde{\Omega})$ for $1 \leq p \leq \infty$.[12]

To prove this assertion we have to show that $Ax(t)$ exists for a.a. $t \in \tilde{\Omega}$, is measurable on $\tilde{\Omega}$ and belongs to $L^p(\tilde{\Omega})$. For $1 \leq p < \infty$,[13] by the Hölder inequality, we get for $\frac{1}{p'} = 1 - \frac{1}{p}$:

$$|Ax(t)| \leq \int_\Omega |k(t,s)|^{\frac{1}{p'}} |k(t,s)|^{\frac{1}{p}} |x(s)| \, \mathrm{d}s \leq c_1^{\frac{1}{p'}} \left[\int_\Omega |k(t,s)||x(s)|^p \, \mathrm{d}s \right]^{\frac{1}{p}}.$$

[12] Conditions on the kernel k which guarantee that $A \in \mathcal{L}(L^p(\Omega), L^r(\tilde{\Omega}))$ can be found in Dunford & Schwartz [53, Chapter VI, 11A].

[13] The case $p = \infty$ is left to the reader.

Set

$$\varphi(t) := \left[\int_{\Omega} |k(t,s)||x(s)|^p \,\mathrm{d}s\right]^{\frac{1}{p}}.$$

Since the measurable function $(t,s) \mapsto |k(t,s)||x(s)|^p$ can be approximated by step functions (consider first $\tilde{\Omega} \times \Omega$ bounded), the function $t \mapsto \varphi(t)$ is measurable on $\tilde{\Omega}$. The Fubini Theorem yields

$$\begin{aligned}\int_{\tilde{\Omega}} |\varphi(t)|^p \,\mathrm{d}t &= \int_{\tilde{\Omega}} \left[\int_{\Omega} |k(t,s)||x(s)|^p \,\mathrm{d}s\right] \mathrm{d}t \\ &= \int_{\Omega} \left[\int_{\tilde{\Omega}} |k(t,s)| \,\mathrm{d}t\right] |x(s)|^p \,\mathrm{d}s \le c_2 \|x\|^p_{L^p(\Omega)}.\end{aligned}$$

In particular, φ is finite a.e. Since $t \mapsto Ax(t)$ is measurable on $\tilde{\Omega}$ (by the same argument as above), we also have

$$\|Ax\|_{L^p(\tilde{\Omega})} = \left[\int_{\tilde{\Omega}} |Ax(t)|^p \,\mathrm{d}t\right]^{\frac{1}{p}} \le c_1^{\frac{1}{p'}} c_2^{\frac{1}{p}} \|x\|_{L^p(\Omega)}.$$

The Fubini Theorem also yields (we identify $g \in [L^p(\tilde{\Omega})]^*$, $1 \le p < \infty$, with a function from $L^{p'}(\tilde{\Omega})$ – see Example 2.1.20(iii))

$$\begin{aligned}\int_{\tilde{\Omega}} Ax(t)g(t) \,\mathrm{d}t &= \int_{\tilde{\Omega}} \left[\int_{\Omega} k(t,s)x(s) \,\mathrm{d}s\right] g(t) \,\mathrm{d}t \\ &= \int_{\Omega} \left[\int_{\tilde{\Omega}} k(t,s)g(t) \,\mathrm{d}t\right] x(s) \,\mathrm{d}s = \int_{\Omega} (A^*g)(s)x(s) \,\mathrm{d}s,\end{aligned}$$

i.e.,

$$A^*g \colon s \mapsto \int_{\tilde{\Omega}} k(t,s)g(t) \,\mathrm{d}t, \qquad g \in L^{p'}(\tilde{\Omega}).$$

We note that the adjoint operator to A for $p = 2$ in the sense of the Riesz Representation Theorem is of the form

$$A^*g(s) = \int_{\tilde{\Omega}} \overline{k(t,s)} g(t) \,\mathrm{d}t.$$

In particular, A is self-adjoint if $\Omega = \tilde{\Omega}$ and $k(t,s) = \overline{k(s,t)}$. We will continue the study of integral operators in the next section (Example 2.2.5). ○

In Example 2.1.11 we have mentioned that differential operators on a function space are not continuous and are only densely defined. Therefore we wish to extend the notion of the adjoint operator to this case. Assume that A is a linear operator defined on a dense subspace $\operatorname{Dom} A$ of X with values in Y. Put

$$\begin{aligned}D^* = \{g \in Y^* :\ &\text{a linear form } x \in \operatorname{Dom} A \mapsto g(Ax) \\ &\text{has a continuous extension } f \text{ to the whole of } X\}.\end{aligned}$$

Obviously, D^* is a linear subspace of Y^* containing o and the extension f is uniquely determined by g. We denote

$$A^*g := f, \qquad \mathrm{Dom}\,(A^*) = D^*$$

and call A^* the *adjoint operator* to A.

Example 2.1.29. The simplest differential operator is defined by

$$Ax(t) = \dot{x}(t).$$

This relation can be considered in various function spaces and also with different domains. If we are interested in its adjoint we should have a good representation of the dual space. This leads to an observation that spaces of integrable functions would be more convenient than spaces of continuous functions. Therefore let $X = L^p(0,1)$, $1 \le p < \infty$ and $\mathrm{Dom}\,A = C^1[0,1]$. Consider $A\colon \mathrm{Dom}\,A \subset X \to X$. We wish to compute A^*. Assume $g \in \mathrm{Dom}\,(A^*) \subset L^{p'}(0,1)$ and $A^*g = f$, i.e.,

$$g(Ax) = \int_0^1 \dot{x}(t)g(t)\,\mathrm{d}t = \int_0^1 x(t)f(t)\,\mathrm{d}t = A^*g(x) \qquad \text{for all} \quad x \in \mathrm{Dom}\,A.$$

In particular, for

$$x \in \mathcal{V} = \{x \in \mathrm{Dom}\,A : x(1) = 0\} \qquad \text{and} \qquad F(t) = \int_0^t f(s)\,\mathrm{d}s,$$

the integration by parts[14] yields

$$\int_0^1 x(t)f(t)\,\mathrm{d}t = x(t)F(t)|_0^1 - \int_0^1 \dot{x}(t)F(t)\,\mathrm{d}t = -\int_0^1 \dot{x}(t)F(t)\,\mathrm{d}t.$$

Since the restriction $A|_{\mathcal{V}}$ of A to $\mathcal{V}$ has a dense range in $L^p(0,1)$ ($\mathrm{Im}\,A|_{\mathcal{V}} = C[0,1]$), we have $F + g = o$ in $L^{p'}(0,1)$. This means that g can be changed on a set of measure zero to have g absolutely continuous and

$$\dot{g} = -f \in L^{p'}(0,1), \qquad \text{i.e.,} \qquad g \in W^{1,p'}(0,1).$$

Moreover, $g(0) = -F(0) = 0$. Taking $F(t) = -\int_t^1 f(s)\,\mathrm{d}s$ we see that also $g(1) = 0$. This proves that

$$\mathrm{Dom}\,(A^*) \subset \{g \in W^{1,p'}(0,1) : g(0) = g(1) = 0\} = W_0^{1,p'}(0,1)\,^{15}$$

[14] If you are not familiar with integration by parts for the Lebesgue integral (notice that $f \in L^{p'}(0,1) \subset L^1(0,1)$), you can approximate f by a continuous function to get a standard situation for integration by parts.

and

$$A^*g = -\dot{g}.$$

Integration by parts yields also the converse inclusion, i.e.,

$$\mathrm{Dom}(A^*) = W_0^{1,p'}(0,1).$$

Notice that $\operatorname{Im} A$ is dense in $L^p(0,1)$ but not closed while A^* is injective and

$$\operatorname{Im} A^* = \left\{ f \in L^{p'}(0,1) : \int_0^1 f(t)\,\mathrm{d}t = 0 \right\}$$

is closed but not dense in $L^{p'}(0,1)$.

Notice also that $\varrho(A) = \varrho(A^*) = \emptyset$ and any $\lambda \in \mathbb{C}$ is an eigenvalue of A. To the contrary, A^* has no eigenvalues. ○

A more general result (due to S. Banach) is stated in the following proposition (see, e.g., Yosida [161]).

Proposition 2.1.30. *Let X, Y be Banach spaces and let A be a closed densely defined linear operator from X into Y. Then $\operatorname{Im} A$ is closed if and only if $\operatorname{Im} A^*$ is closed. Moreover,*

$$\overline{\operatorname{Im} A} = (\operatorname{Ker} A^*)_\perp \qquad \textit{and} \qquad \operatorname{Ker} A = (\operatorname{Im} A^*)_\perp.$$

Nevertheless, notice that A is not closed in our example. Proposition 2.1.30 can be applied to A^* (A^* is always closed); $\mathrm{Dom}\,(A^{**}) = W^{1,p}(0,1)$, $A^{**}x = \dot{x}$.[16] This simple example shows how the domain of a (linear) noncontinuous operator affects its properties.

Example 2.1.31. Put

$$Ax = -\ddot{x} \qquad \text{with} \qquad \mathrm{Dom}\, A = \{x \in C^2(a,b) : x(a) = x(b) = 0\}.$$

If the equation

$$Ax = \lambda x$$

has a nonzero solution w ($\in \mathrm{Dom}\, A$), then λ is called an eigenvalue and w a corresponding eigenfunction of A. Simple calculation shows that $\frac{k^2\pi^2}{(b-a)^2}$ are all eigenvalues of A,[17] and $\sin \frac{k\pi}{b-a}(t-a)$ are the corresponding eigenfunctions. Consider now the boundary value problem

$$\begin{cases} -\ddot{x}(t) = \lambda x(t) + f(t), & t \in (a,b), \\ x(a) = x(b) = 0. \end{cases} \tag{2.1.6}$$

[15] The last equality should be proved. A deeper insight into these Sobolev spaces will be given in Chapter 7, cf. also Exercise 1.2.48.

[16] Notice that A^{**} is an extension of A and, moreover, the graph of A^{**} is the closure of the graph of A (it is also said that A^{**} is the closure of A).

[17] The sign minus in the definition of A is conventional; it is introduced to obtain positive eigenvalues.

Let φ_1, φ_2 be a fundamental system for the differential equation $-\ddot{x} - \lambda x = 0$. The Variation of Constants Formula shows that

$$x(t) = c_1\varphi_1(t) + c_2\varphi_2(t) + \int_a^t \frac{\varphi_1(s)\varphi_2(t) - \varphi_1(t)\varphi_2(s)}{W(s)} f(s)\,\mathrm{d}s \tag{2.1.7}$$

is a solution to $-\ddot{x} - \lambda x = f$. Here W is the Wronski determinant of φ_1, φ_2 (notice that for this equation we always can choose φ_1, φ_2 such that $W \equiv 1$). We wish to find constants c_1, c_2 such that x given by (2.1.7) satisfies the boundary conditions $x(a) = x(b) = 0$. The number λ is not an eigenvalue if and only if

$$\det \begin{pmatrix} \varphi_1(a) & \varphi_2(a) \\ \varphi_1(b) & \varphi_2(b) \end{pmatrix} \neq 0.$$

In this case the formula (2.1.7) shows that for any $f \in C[a,b]$ the problem (2.1.6) has a unique solution in $\operatorname{Dom} A$ [18] which is called a *classical solution*. This means that $\lambda \in \varrho(A)$. Suppose now that λ is an eigenvalue. Then we can take φ_1 as a corresponding eigenfunction and get

$$x(a) = c_2\varphi_2(a), \qquad \text{i.e.,} \qquad c_2 = 0$$

($\varphi_2(a) \neq 0$ since φ_1, φ_2 are linearly independent), and

$$x(b) = \varphi_2(b) \int_a^b \varphi_1(s) f(s)\,\mathrm{d}s = 0, \qquad \text{i.e.,} \qquad \int_a^b \varphi_1(s) f(s)\,\mathrm{d}s = 0 \tag{2.1.8}$$

since $\varphi_2(b) \neq 0$ (by the same argument as above). Notice that (2.1.8) is also a necessary condition for solvability of (2.1.6).

We will return to this example in the next section (see Example 2.2.17). ○

Example 2.1.32. Linear differential operators of the second order with nonconstant coefficients are more complicated. To simplify our exposition we consider a differential expression

$$Lx := p_0\ddot{x} + p_1\dot{x} + p_2 x$$

where $\ddot{p}_0$, $\dot{p}_1$, p_2 are continuous functions on a closed bounded interval $[a,b]$ and $p_0 < 0$ on this interval (the so-called regular case). Let $X = L^p(a,b)$, $1 \le p < \infty$ and

$$\mathcal{D} = \{x \in W^{2,p}(a,b) : x(a) = x(b) = 0\}.$$

Put

$$Ax = Lx, \qquad x \in \mathcal{D} = \operatorname{Dom} A$$

and consider

$$A\colon \operatorname{Dom} A \subset X \to X.$$

[18] If $f \in L^p(a,b)$, then it is possible to show that the function $x = x(t)$ given by (2.1.7) belongs to $W^{2,2}(a,b)$, $x(a) = x(b) = 0$, and the equation in (2.1.6) is satisfied a.e. in (a,b). Such a solution is called a *strong solution*.

A solution of $Ax = f$ is therefore a strong solution of

$$\begin{cases} Lx(t) = f(t), & t \in (a,b), \\ x(a) = x(b) = 0. \end{cases}$$

It can be proved that A is injective provided $p_2 > 0$ in $[a,b]$. (Assume by contradiction that $\operatorname{Ker} A \neq \{o\}$ and show that there is $x_0 \in \operatorname{Ker} A$ which has a negative minimum at an interior point $c \in (a,b)$. Deduce that $Lx_0(c) < 0$.) The Variation of Constants Formula shows that the operator A is also surjective and A^{-1} is an integral operator

$$A^{-1}f(t) = \int_a^b G(t,s)f(s)\,\mathrm{d}s \tag{2.1.9}$$

where G is the so-called *Green function* of L. The Green function is nonnegative on $[a,b] \times [a,b]$ and satisfies the estimates from Example 2.1.28. Therefore $A^{-1} \in \mathcal{L}(X)$. In order to calculate the adjoint A^* it is convenient to consider the so-called formal adjoint expression to L, i.e.,

$$My = (p_0 y)^{\cdot\cdot} - (p_1 y)^{\cdot} + p_2 y$$

which is obtained by integrating by parts in the integral $\displaystyle\int_a^b Lx(t)y(t)\,\mathrm{d}t$ and omitting the boundary terms. Put

$$By = My \qquad \text{for} \quad y \in \mathcal{D} = \operatorname{Dom} B.$$

The same integration as above shows that $B \subset A^*$. The proof of the equality $A^* = B$ needs a more careful calculation.

The interested reader can consult the books Coddington & Levinson [35, Chapter 9], Edmunds & Evans [55] or Dunford & Schwartz [54], in particular Chapter XIII, for details and also for more complicated singular cases which are important in applications, e.g., in Quantum Mechanics (the Schrödinger equation). ○

Exercise 2.1.33. Let X, Y be Banach spaces. If $A \in \mathcal{L}(X,Y)$ has a continuous inverse $A^{-1} \in \mathcal{L}(Y,X)$ and $B \in \mathcal{L}(X,Y)$ is such that

$$\|B - A\| < \frac{1}{\|A^{-1}\|},$$

then B is also continuously invertible and

$$\|B^{-1}\| \le \frac{\|A^{-1}\|}{1 - \|A^{-1}\|\|B-A\|}, \quad \|B^{-1} - A^{-1}\| \le \frac{\|A^{-1}\|^2}{1 - \|A^{-1}\|\|B-A\|}\|B - A\|.$$

Hint. Examine the proof of Corollary 2.1.3 and write $A^{-1}B = A^{-1}(B-A) + I$.

Exercise 2.1.34. Let X be a Banach space. Prove that the spectrum depends continuously on an operator in the following sense: For any $A \in \mathcal{L}(X)$ and an open set $\mathcal{G}$ containing $\sigma(A)$ there is $\delta > 0$ such that for $B \in \mathcal{L}(X)$, $\|B - A\| < \delta$, the spectrum $\sigma(B) \subset \mathcal{G}$.

Hint. First show that $\|(\lambda I - A)^{-1}\|$ is bounded on $\mathbb{C}\backslash\mathcal{G}$ and then use Exercise 2.1.33.

Exercise 2.1.35. Show that

$$e^{tA} = \sum_{n=0}^{\infty} \frac{t^n A^n}{n!}$$

is well defined for all $t \in \mathbb{R}$, $A \in \mathcal{L}(X)$, provided X is a Banach space, and, moreover, the vector function

$$\varphi\colon t \mapsto e^{tA} x_0$$

solves the differential equation

$$\dot{x}(t) = Ax(t)$$

and satisfies the initial condition $\varphi(0) = x_0$. (See also the end of Section 1.1.)

Exercise 2.1.36. Let K be a continuous real function on $[a, b] \times [a, b]$ and let $h \in C[a, b]$ be fixed. Let

$$M = \max_{(t,\tau)\in[a,b]\times[a,b]} |K(t, \tau)|$$

and let $\lambda \in \mathbb{R}$ be such that

$$|\lambda| < \frac{1}{M(b-a)}.$$

Prove that the integral equation

$$x(t) = \lambda \int_a^b K(t, \tau)x(\tau)\,d\tau + h(t)$$

has a unique solution $x \in C[a, b]$.

Exercise 2.1.37. Let $\{x_n\}_{n=1}^{\infty}$, $\{y_n\}_{n=1}^{\infty}$ be sequences in a Hilbert space H such that $x_n \rightharpoonup x$, $y_n \to y$. Then

$$(x_n, y_n) \to (x, y).$$

Hint. Use Proposition 2.1.22(iii).

Exercise 2.1.38. Let $\{e_n\}_{n=1}^{\infty}$ be an orthonormal sequence in a Hilbert space. Show that

$$e_n \rightharpoonup o.$$

Hint. Use the Bessel inequality (1.2.17).

Exercise 2.1.39. Prove assertion (iv) of Proposition 2.1.22 for a Hilbert space X.

Hint. Use the relation between the scalar product and the norm in X.

Exercise 2.1.40. Show that a convex set (in particular a subspace) of a normed linear space is weakly closed if and only if it is closed in the norm topology.

Hint. Suppose by contradiction that $\mathcal{C}$ is a norm-closed convex set which is not weakly closed. Then there is $x_0 \in \overline{\mathcal{C}}^{\,w} \setminus \mathcal{C}$. Use the Separation Theorem (Corollary 2.1.18) to obtain a contradiction.

Exercise 2.1.41. Prove that actually

$$\|A^*\|_{\mathcal{L}(Y^*,X^*)} = \|A\|_{\mathcal{L}(X,Y)}.$$

Hint. The inequality $\|A^*\| \le \|A\|$ follows from the calculation after Remark 2.1.26. For the converse inequality use the dual characterization of the norm $\|Ax\|$.

2.2 Compact Linear Operators

In this section we present a class of continuous linear operators the properties of which are closely related to the properties of finite-dimensional linear operators. The key assertions presented concern the Riesz–Schauder Theory and the Hilbert–Schmidt Theorem.

Definition 2.2.1. Let X and Y be normed linear spaces. A linear operator $A \in L(X,Y)$ is called a *compact operator* if the image of a ball in X is relatively compact in Y. The set of all compact operators from X into Y is denoted by $\mathscr{C}(X,Y)$.

Remark 2.2.2.

(i) Every compact linear operator is continuous.

(ii) The compactness condition is mostly used in the following equivalent form:

For any bounded sequence $\{x_n\}_{n=1}^{\infty} \subset X$ there is a subsequence $\{x_{n_k}\}_{k=1}^{\infty}$ such that Ax_{n_k} converge in the norm topology of Y.

(iii) Replacing the norm topology in Y by the weak topology a *weakly compact operator* can be defined. If either X or Y is reflexive, then any $A \in \mathcal{L}(X,Y)$ is weakly compact. This follows from the Eberlain–Smulyan Theorem (Remark 2.1.24) and the observation that $A \in \mathcal{L}(X,Y)$ maps a weakly convergent sequence into a weakly convergent one (cf. Proposition 2.1.27(i)).

Example 2.2.3.

(i) If $A \in \mathcal{L}(X,Y)$ and $\dim \operatorname{Im} A < \infty$ (the so-called *operator of finite rank*), then $A \in \mathscr{C}(X,Y)$.

(ii) Let $\{e_n\}_{n=1}^{\infty}$ be an orthonormal basis in a Hilbert space H. Put

$$Ae_n = \lambda_n e_n$$

and extend A by linearity to the dense set $\mathcal{D} := \operatorname{Lin}\{e_1, \dots\}$ in H. The operator A is bounded on $\mathcal{D}$ (and therefore it can be uniquely extended to a continuous operator on H) if and only if $\{\lambda_n\}_{n=1}^{\infty}$ is a bounded sequence. In addition,

$$\|A\| = \sup_n |\lambda_n|.$$

This follows immediately from the identity

$$\|Ax\|^2 = \sum |\lambda_n|^2 |(x, e_n)|^2 \qquad \text{for every} \quad x \in H.$$

Moreover, A is a compact operator on H if and only if

$$\lim_{n\to\infty} \lambda_n = 0.$$

This is an easy consequence of Proposition 1.2.39. ○

Proposition 2.2.4. *Let X, Y and Z be normed linear spaces. Then*

(i) *if $A \in \mathscr{C}(X,Y)$, $B \in \mathcal{L}(Y,Z)$, then $BA \in \mathscr{C}(X,Z)$;*

(ii) *if $A \in \mathscr{C}(Y,Z)$, $B \in \mathcal{L}(X,Y)$, then $AB \in \mathscr{C}(X,Z)$;*

(iii) *if $A \in \mathscr{C}(X,Y)$ and a sequence $\{x_n\}_{n=1}^{\infty} \subset X$ converges weakly to $x \in X$, then*

$$\lim_{n\to\infty} \|Ax_n - Ax\| = 0.$$

(iv) *Assume that Y is a Banach space and a sequence $\{A_n\}_{n=1}^{\infty} \subset \mathscr{C}(X,Y)$ converges to $A \in \mathcal{L}(X,Y)$ in the norm operator topology. Then $A \in \mathscr{C}(X,Y)$.*

Proof. The assertions (i) and (ii) are obvious.

To prove (iii) assume by contradiction that there is a subsequence $\{x_{n_k}\}_{k=1}^{\infty}$ such that

$$\|Ax_{n_k} - Ax\| \geq c > 0.$$

The sequence $\{x_n\}_{n=1}^{\infty}$ is bounded (Proposition 2.1.22(iii)), and hence there exists a subsequence $\left\{x_{n_{k_l}}\right\}_{l=1}^{\infty}$ and $y \in Y$ such that

$$\|Ax_{n_{k_l}} - y\| \to 0.$$

Since

$$f(Ax_n) = A^*f(x_n) \to A^*f(x) = f(Ax) \qquad \text{for every} \quad f \in X^*,$$

we have $y = Ax$, and hence a contradiction.

(iv) Let $B(o;1)$ be the unit ball. By Proposition 1.2.3 it suffices to show that for any $\varepsilon > 0$ there is a finite ε-net of $A(B(o;1))$. We choose n such that $\|A_n - A\| < \frac{\varepsilon}{2}$, and a finite $\frac{\varepsilon}{2}$-net for $A_n(B(o;1))$. By the triangle inequality, this is the desired ε-net for $A(B(o;1))$. □

Example 2.2.5.

(i) Let k be a continuous function on the Cartesian product $[a,b] \times [a,b]$. Then the operator

$$Ax\colon t \in [a,b] \mapsto \int_a^b k(t,s)x(s)\,\mathrm{d}s$$

is compact as an operator from $C[a,b]$ into itself.[19] We give two proofs of this assertion.

The first is based on the use of the Arzelà–Ascoli Theorem (Theorem 1.2.13). Its assumptions are satisfied for $\mathcal{F} = A(B(o;1))$ where $B(o;1)$ is the unit ball in $C[a,b]$. The equicontinuity of $\mathcal{F}$ follows from the uniform continuity of k on $[a,b] \times [a,b]$.

The second proof uses Proposition 2.2.4(iv). Put

$$\mathcal{A} = \{(t,s) \mapsto x(t)y(s) : x, y \in C[a,b]\}.$$

It is easy to see that $\mathcal{A}$ is a subalgebra of $C([a,b] \times [a,b])$ which satisfies the assumptions of the real or complex Stone–Weierstrass Theorem (Theorem 1.2.14). Hence there are sequences $\{q_n\}_{n=1}^{\infty}$, $\{r_n\}_{n=1}^{\infty}$ in $C[a,b]$ such that

$$q_n(t)r_n(s) \rightrightarrows k(t,s) \qquad \text{uniformly in} \quad [a,b] \times [a,b].$$

In particular, this means that the operators

$$A_n x\colon t \mapsto q_n(t) \int_a^b r_n(s)x(s)\,\mathrm{d}s$$

converge in the operator norm to A. Since $\operatorname{Im} A_n \subset \operatorname{Lin}\{q_n\}$, all A_n are compact and, therefore, A is compact.

(ii) Let Ω be a measurable subset of $\mathbb{R}^M$ and let $k \in L^2(\Omega \times \Omega)$. Then the operator

$$Ax(t) = \int_\Omega k(t,s)x(s)\,\mathrm{d}s$$

is compact as an operator from $L^2(\Omega)$ into itself.

We present again two proofs of this statement. The first will be a typical Hilbert space proof, the second will use the reflexivity of $L^2(\Omega)$ and we will show how it could be used to get compactness of an integral operator on $L^p(\Omega)$.

The first proof is based on the following observation:

---

[19] This is true under more general assumptions, e.g., if the interval $[a,b]$ is replaced by a compact topological space K, μ is a Borel measure on K and A is defined by $Ax(t) = \displaystyle\int_K k(t,s)x(s)\,\mathrm{d}\mu(s)$.

Let $\{e_k\}_{k=1}^{\infty}$, $\{f_k\}_{k=1}^{\infty}$ *be two orthonormal bases in a separable Hilbert space* H. *Let* $B \in \mathcal{L}(H)$. *By the Parseval equality we have*

$$\|B\|_2^2 := \sum_{k,n=1}^{\infty} |(Be_k, f_n)|^2 = \sum_{k=1}^{\infty} \|Be_k\|^2 = \sum_{n=1}^{\infty} \|B^* f_n\|^2 \le \infty.$$

This shows that the quantity $\|B\|_2^2$ depends only on B and not on the particular choice of bases. Moreover, if $\|B\|_2^2 < \infty$, then B is called a *Hilbert–Schmidt operator* and $B \in \mathscr{C}(H)$.

To see this take $n_\varepsilon \in \mathbb{N}$ such that $\sum\limits_{n=n_\varepsilon+1}^{\infty} \|B^* f_n\|^2 < \varepsilon$ and define

$$B_\varepsilon x = \sum_{n=1}^{n_\varepsilon} (Bx, f_n) f_n.$$

Then $\dim \operatorname{Im} B_\varepsilon < \infty$ and

$$\|B_\varepsilon x - Bx\|^2 = \sum_{n=n_\varepsilon+1}^{\infty} |(Bx, f_n)|^2 \le \|x\|^2 \sum_{n=n_\varepsilon+1}^{\infty} \|B^* f_n\|^2 \le \varepsilon \|x\|^2.$$

The compactness of B follows from Proposition 2.2.4(iv).

In order to apply this statement to the integral operator A choose an orthonormal basis $\{e_n\}_{n=1}^{\infty}$ in $L^2(\Omega)$ and notice that

$$\varphi_{m,n}(t,s) := e_m(t)\overline{e_n(s)}$$

is an orthonormal set in $L^2(\Omega \times \Omega)$ (use Corollary 1.2.36). Since

$$(Ae_n, e_m)_{L^2(\Omega)} = (k, \varphi_{m,n})_{L^2(\Omega\times\Omega)},$$

we have

$$\|A\|_2 = \|k\|_{L^2(\Omega\times\Omega)}.$$

Now we give the second proof. Let $\{x_n\}_{n=1}^{\infty}$ be a bounded set in $L^2(\Omega)$. Since $L^2(\Omega)$ as a Hilbert space is reflexive, there is a subsequence – denote it again by $\{x_n\}_{n=1}^{\infty}$ – which is weakly convergent to an x in $L^2(\Omega)$. In particular,

$$\int_\Omega k(t,s)x_n(s)\,\mathrm{d}s \to \int_\Omega k(t,s)x(s)\,\mathrm{d}s \qquad \text{for a.a.} \quad t \in \Omega$$

(the Fubini Theorem shows that $k(t,\cdot) \in L^2(\Omega)$ for a.a. $t \in \Omega$). Since

$$\begin{aligned}|Ax_n(t) - Ax(t)| &\le \int_\Omega |k(t,s)|\,|x_n(s) - x(s)|\,\mathrm{d}s \\ &\le \|x_n - x\|_{L^2(\Omega)} \left[\int_\Omega |k(t,s)|^2\,\mathrm{d}s\right]^{\frac{1}{2}} \le c\left[\int_\Omega |k(t,s)|^2\,\mathrm{d}s\right]^{\frac{1}{2}},\end{aligned}$$

the Lebesgue Dominated Convergence Theorem yields

$$\|Ax_n - Ax\|_{L^2(\Omega)} \to 0. \qquad \bigcirc$$

Proposition 2.2.6. *Let H be a Hilbert space and $A \in \mathcal{L}(H)$. Then A is a compact operator if and only if there is a sequence $\{A_n\}_{n=1}^{\infty} \subset \mathcal{L}(H)$ of operators of finite rank which converges to A in the operator norm topology.*

Proof. Because of Proposition 2.2.4 only the necessity part is left to be proved. Let $B(o;1)$ be the unit ball in H. Since $\overline{A(B(o;1))}$ is compact, it is a separable metric space, and therefore

$$Y = \overline{\operatorname{Lin} A(B(o;1))}$$

is a separable Hilbert space. Let $\{e_n\}_{n=1}^{\infty}$ be an orthonormal basis in Y. Put

$$A_n x = \sum_{k=1}^{n} (Ax, e_k)e_k.$$

Then A_n has finite rank and

$$\|A_n x - Ax\|^2 = \sum_{k=n+1}^{\infty} |(Ax, e_k)|^2 < \varepsilon \qquad \text{for every} \quad x \in B(o;1)$$

provided n is sufficiently large (Proposition 1.2.39). □

Remark 2.2.7. The proof of the preceding proposition indicates that the result holds also in a Banach space X with a Schauder basis $\{e_n\}_{n=1}^{\infty}$ (see page 41). The famous conjecture of S. Banach was that any separable Banach space has a Schauder basis. The first counterexample was constructed by P. Enflo. He found a compact operator in a separable Banach space which cannot be approximated by operators of finite rank. We notice that separable Banach spaces of functions like $C(\overline{\Omega})$, $L^p(\Omega)$, $W^{k,p}(\Omega)$ $(1 \le p < \infty)$ have a Schauder basis.

One of our goals in this section is to generalize the Fredholm alternative (see footnote 6 on page 14). As we have seen in Section 1.1 the notion of the adjoint operator is very important.

Proposition 2.2.8 (Schauder). *Let X, Y be Banach spaces and assume that $A \in \mathcal{L}(X,Y)$. Then A is compact if and only if A^* is compact.*

Proof.

Step 1 **(the "only if" part)**. Suppose that $A \in \mathscr{C}(X,Y)$ and $\{g_n\}_{n=1}^{\infty} \subset Y^*$, $\|g_n\|_{Y^*} \le 1$. It is easy to verify the assumptions of the Arzelà–Ascoli Theorem (Theorem 1.2.13) for the sequence of functions

$$g_n \colon \mathcal{K} := \overline{A(B(o;1))} \to \mathbb{R} \quad (\text{or } \mathbb{C})$$

($B(o;1)$ is the unit ball in X). By this theorem there is a subsequence $\{g_{n_k}\}_{k=1}^{\infty}$ which is uniformly convergent on $\mathcal{K}$. Since

$$|A^*g_{n_k}(x) - A^*g_{n_l}(x)| \le \sup_{y\in\mathcal{K}} |g_{n_k}(y) - g_{n_l}(y)| \qquad \text{for each} \quad x \in B(o;1)$$

and X^* is complete, the sequence $\{A^*g_{n_k}\}_{k=1}^{\infty}$ is convergent in X^*.

Step 2 **(the "if" part)**. Assume now that $A^* \in \mathscr{C}(Y^*, X^*)$. We embed X into X^{**} and Y into Y^{**} with help of the canonical isometrical embeddings κ_X and κ_Y (see the proof of Proposition 2.1.22(iii)). Since A^* is compact, A^{**} is compact by the first part of the proof. It suffices to show that

$$\kappa_Y(Ax) = A^{**}\kappa_X(x) \qquad \text{for} \quad x \in X$$

and we leave that to the reader. □

If $A \in \mathscr{C}(X,Y)$, then the equation

$$Ax = y \tag{2.2.1}$$

is scarcely ever well-posed[20] as follows from the first part of the next theorem. This is the reason why we are interested rather in equations of the type

$$x - Ax = y. \tag{2.2.2}$$

Theorem 2.2.9 (Riesz–Schauder Theory). *Let X be a Banach space and $A \in \mathscr{C}(X)$. Then*

(i) *if* $\operatorname{Im} A$ *is closed, then* $\dim \operatorname{Im} A < \infty$;

(ii) $\dim \operatorname{Ker}(I - A) < \infty$;

(iii) $\operatorname{Im}(I - A)$ *is closed;*

(iv) *(the Fredholm alternative)*

$$\operatorname{Im}(I-A) = X \qquad \textit{if and only if} \qquad \operatorname{Ker}(I-A) = \{o\};$$

(v) $\dim \operatorname{Ker}(I - A) = \dim \operatorname{Ker}(I^* - A^*)$.

Proof. (i) If $Y = \operatorname{Im} A$ is closed, then $A\colon X \to Y$ is an open mapping (Theorem 2.1.8). This means that a certain ball $B(o;\delta)$ in Y is contained in the relatively compact set $A(B(o;1))$, i.e., $B(o;\delta)$ itself is relatively compact. By Proposition 1.2.15, $\dim Y < \infty$.

[20] An equation (2.2.1) is said to be *well-posed* if A is injective and A^{-1} is continuous. If A is an integral operator, then (2.2.1) is called an *integral equation of the first kind*. The equation (2.2.2) is called an *integral equation of the second kind*. The research of these equations carried out by I. Fredholm is supposed to be one of the starting points in the development of functional analysis.

(ii) For the rest of the proof we put

$$T := I - A \qquad \text{and} \qquad Y := \operatorname{Ker} T.$$

Then the restriction of A to the Banach space Y maps Y onto Y. By (i), $\dim Y < \infty$.

(iii) Because of (ii) there exists a continuous projection P of X onto Y (Remark 2.1.19). Denote

$$Z := \operatorname{Ker} P, \qquad \text{i.e.,} \qquad X = Y \oplus Z$$

and both Y and Z are Banach spaces. Since T is injective on Z, $\operatorname{Im} T$ is closed provided there is a positive constant c such that

$$\|Tz\|_Y \geq c\|z\|_Z \qquad \text{for each} \quad z \in Z,$$

see page 71. Suppose by contradiction that such c does not exist, i.e., there are $z_n \in Z$ such that

$$\|z_n\|_Z = 1 \qquad \text{and} \qquad \|Tz_n\|_Y < \frac{1}{n}\|z_n\|_Z.$$

Then one can find a subsequence $\{z_{n_k}\}_{k=1}^{\infty}$ for which Az_{n_k} converges to a y. Since $Tz_{n_k} \to o$, we have $\lim\limits_{n\to\infty} z_{n_k} = y \in Z$. This means that

$$Ty = o, \qquad \text{i.e.,} \qquad y \in Y \cap Z, \qquad \text{and thus} \qquad y = o.$$

This is a contradiction since $z_{n_k} \to y$ implies that $\|y\|_Y = 1$.

(iv) We will prove the necessity part by way of contradiction. Put

$$Y_k := \operatorname{Ker} T^k.$$

Then

$$Y_1 \subsetneq Y_2 \subsetneq \cdots \subsetneq Y_k \subsetneq \cdots$$

since for $x_1 \in \operatorname{Ker}(I - A)$, $x_1 \neq o$, there is x_2 such that $x_1 = Tx_2$, i.e., $x_2 \in Y_2 \setminus Y_1$, etc. It follows from the construction in the proof of Proposition 1.2.15 that there are $y_k \in Y_k$, $\|y_k\|_{Y_k} = 1$, such that $\operatorname{dist}(y_{k+1}, Y_k) \geq \frac{1}{2}$. For $k > l$ we have

$$\|Ay_k - Ay_l\|_{Y_k} = \|y_k - (y_l - Ty_l + Ty_k)\|_{Y_k} \geq \operatorname{dist}(y_k, Y_{k-1}) \geq \frac{1}{2}.$$

This means that there is no convergent subsequence of $\{Ay_k\}_{k=1}^{\infty}$, a contradiction.

The sufficiency part is now easy: It follows from Proposition 2.2.8 and the previous part (iii) that $\operatorname{Im} T^*$ is closed. Assume that $\operatorname{Ker} T = \{o\}$. By Proposition 2.1.27(iii),

$$\operatorname{Im} T^* = (\operatorname{Ker} T)^{\perp} = X^*.$$

According to the first part of this proof, $\operatorname{Ker} T^* = \{o\}$ and, again by (iii) and Proposition 2.1.27(iv),

$$\operatorname{Im} T = (\operatorname{Ker} T^*)_\perp = X.$$

(v) As in the proof of (iii), $X = Y \oplus Z$ and the corresponding projection P of X onto Y is continuous. It can be shown that a direct complement W of $\operatorname{Im} T$ in X is isomorphic to $\operatorname{Ker} T^*$.[21] This means that

$$\dim W = \dim \operatorname{Ker} T^* < \infty.$$

Denote

$$\dim \operatorname{Ker} T = n \qquad \text{and} \qquad \dim \operatorname{Ker} T^* = n^*.$$

We shall prove that $n = n^*$. Assume that

$$n > n^*.$$

In particular, this means that there is a surjective linear operator $\Phi \in \mathcal{L}(Y, W)$. Such Φ cannot be injective (see Corollary 1.1.15), i.e., there is $x_0 \in Y$, $x_0 \neq o$, for which $\Phi(x_0) = o$. Put now

$$B := A + \Phi P.$$

Since $P \in \mathscr{C}(X)$, we have $B \in \mathscr{C}(X)$ and

$$Bx_0 = Ax_0 + o = x_0, \qquad \text{i.e.,} \qquad \operatorname{Ker}(I - B) \neq \{o\}.$$

By the Fredholm alternative (iv), $\operatorname{Im}(I - B) \neq X$. But

$$(I - B)(Z) = \operatorname{Im} T \qquad \text{and} \qquad (I - B)(Y) = \Phi(Y) = W,$$

i.e.,

$$\operatorname{Im}(I - B) = \operatorname{Im} T + W = X,$$

a contradiction. This proves the inequality $n \leq n^*$.

By interchanging T and T^* we similarly obtain $n^* \leq n$.[22] □

Remark 2.2.10. The proof of the following statement is similar to that of Lemma 1.1.31(i).

If $A \in \mathscr{C}(X)$ and $1 \in \sigma(A)$, then there is $k \in \mathbb{N}$ such that

$$X = \operatorname{Ker}(I - A)^k \oplus \operatorname{Im}(I - A)^k.$$

Moreover, both the spaces on the right-hand side are A-invariant, and $\dim \operatorname{Ker}(I - A)^k < \infty$.[23]

---

[21] This is clear for X being a Hilbert space, since $\operatorname{Im} T$ is closed and the orthogonal complement $(\operatorname{Im} T)^\perp$ is equal to $\operatorname{Ker} T^*$ (Proposition 2.1.27(iv)). In a general Banach space we can use the factor space $X|_{\operatorname{Im} T}$ which is algebraically isomorphic to a direct complement W of $\operatorname{Im} T$ and for $g \in \left(X|_{\operatorname{Im} T}\right)^*$ put $f(x) = g([x])$. It remains to show that the correspondence $g \to f$ is an (isometric) isomorphism onto $(\operatorname{Im} T)^\perp = \operatorname{Ker} T^*$.

[22] We recommend to the reader to do that carefully to see that no reflexivity of X is needed.

[23] This dimension is called the *multiplicity of the eigenvalue* 1.

Remark 2.2.11. Theorem 2.2.9 can be generalized to operators $A \in \mathcal{L}(X)$ for which there is $k \in \mathbb{N}$ such that $A^k \in \mathscr{C}(X)$.

Another way of generalization is connected with perturbations of Fredholm operators. Notice that the statement (v) of Theorem 2.2.9 says that $I - A$ is a Fredholm operator of index zero provided $A \in \mathscr{C}(X)$. The following theorem states the stability of index.

Theorem 2.2.12. *Let X, Y be Banach spaces and let $A \in \mathcal{L}(X,Y)$ be a Fredholm operator. Then*

(i) *if $B \in \mathscr{C}(X,Y)$, then $A + B$ is Fredholm and*

$$\operatorname{ind} A = \operatorname{ind}(A + B); \tag{2.2.3}$$

(ii) *the set of Fredholm operators in $\mathcal{L}(X,Y)$ is an open subset of $\mathcal{L}(X,Y)$; furthermore,* ind *is a continuous function on this open set.*

Proof. The proofs and further results can be found, e.g., in Kato [87, § IV.5.]. □

Corollary 2.2.13. *Let X be a complex Banach space and let $A \in \mathscr{C}(X)$. Then*

(i) *$\sigma(A) \setminus \{0\}$ is a countable set of eigenvalues of finite multiplicity;*

(ii) *if $\dim X = \infty$, then $0 \in \sigma(A)$, and if λ is an accumulation point of $\sigma(A)$, then $\lambda = 0$.*

Proof. (i) If $\lambda \neq 0$, then $\lambda I - A = \lambda\left(I - \frac{A}{\lambda}\right)$ and Theorem 2.2.9 can be applied. In particular, if such λ belongs to $\sigma(A)$, then λ is an eigenvalue of finite multiplicity. It remains to show that for any $r > 0$ the set $\varrho = \{\lambda \in \sigma(A) : |\lambda| > r\}$ is finite.

Assume by way of contradiction that there is a sequence of mutually different points $\{\lambda_n\}_{n=1}^{\infty} \subset \varrho$ and let x_n be the corresponding nonzero eigenvectors. Put

$$W_n = \operatorname{Lin}\{x_1, \dots, x_n\}.$$

It is easy to see by induction that $x_1, \dots, x_n$ are linearly independent. So we can find $y_{n+1} \in W_{n+1}$ such that

$$\|y_{n+1}\| = 1 \qquad \text{and} \qquad \operatorname{dist}(y_{n+1}, W_n) \geq \frac{1}{2}.$$

Now for $k > l$ we have

$$\|Ay_k - Ay_l\| = \|\lambda_k y_k - [(\lambda_k I - A)y_k + (\lambda_l I - A)y_l - \lambda_l y_l]\| \geq |\lambda_k| \operatorname{dist}(y_k, W_{k-1}) \geq \frac{r}{2}$$

and this contradicts the compactness of A.

(ii) The statement on accumulation points follows immediately from the proof of (i). To see that 0 is a point of $\sigma(A)$ provided $\dim X = \infty$ it is sufficient to realize that $\sigma(A)$ cannot be a finite set of nonzero numbers $\lambda_1, \dots, \lambda_n$. Indeed, with help of Remark 2.2.10 we get

$$X = \operatorname{Ker}(\lambda_1 I - A)^{k_1} \oplus \cdots \oplus \operatorname{Ker}(\lambda_n I - A)^{k_n} \oplus V \tag{2.2.4}$$

where V is a nontrivial closed A-invariant subspace of X. Therefore the spectrum $\sigma(A|_V)$ of the restriction $A|_V$ of A to V is a subset of $\sigma(A)$. Since $\sigma(A|_V) \neq \emptyset$ (see the discussion following Example 2.1.20), we have

$$\{\lambda_1, \ldots, \lambda_n\} \neq \sigma(A). \qquad \square$$

Example 2.2.14. Consider

$$Ax(t) := \int_0^t x(s)\,\mathrm{d}s \qquad \text{on the space} \quad L^2(0,1).$$

This is a special class of operators which have been examined in Example 2.2.5(ii):

$$k(t,s) = \begin{cases} 1 & \text{for} \quad 0 \le s \le t \le 1, \\ 0 & \text{for} \quad 0 \le t < s \le 1. \end{cases}$$

Therefore $A \in \mathscr{C}(L^2(0,1))$.

If $\lambda \neq 0$ were an eigenvalue of A with an eigenfunction x, then

$$x(t) = \frac{1}{\lambda}\int_0^t x(s)\,\mathrm{d}s,$$

i.e., x is absolutely continuous and

$$\dot{x} = \frac{1}{\lambda}x, \qquad x(0) = 0.$$

This implies that $x = o$ in $[0,1]$. Since $\sigma(A)$ cannot be empty, $\sigma(A) = \{0\}$, and 0 is no eigenvalue of A.

We notice that the same statement (with a more complicated proof) is valid for any *Volterra integral operator*

$$Ax(t) = \int_0^t k(t-s)x(s)\,\mathrm{d}s, \qquad x \in L^2(0,1),$$

provided, e.g., $k \in L^2(0,1)$. See also Example 2.3.7. ○

Corollary 2.2.13 can be significantly strengthened in the case that X is a Hilbert space and A is a compact, self-adjoint operator. To see this we need some technicalities.

Proposition 2.2.15. *Let H be a Hilbert space and A a self-adjoint continuous operator on H. Then*

(i) $\|A\| = \sup\limits_{\|x\|=1} |(Ax,x)|$;

(ii) $m := \inf\limits_{\|x\|=1} (Ax,x)$ *and* $M := \sup\limits_{\|x\|=1} (Ax,x)$ *belong to the spectrum of* A;

(iii) $\|A\| = \sup\{|\lambda| : \lambda \in \sigma(A)\}$;

(iv) $\sigma(A) \subset \mathbb{R}$;

(v) *if* $Ax = \lambda x$, $Ay = \mu y$, $\lambda \neq \mu$, *then* $(x, y) = 0$.

Proof. (i) Denote the right-hand side by α. Obviously $\alpha \leq \|A\|$. To prove the converse inequality take

$$o \neq x \in H, \qquad y = Ax.$$

Then for any $t > 0$, using (1.2.14), we have

$$\begin{aligned}
\|Ax\|^2 &= \left(A(tx), \frac{1}{t}Ax\right) = \left(A(tx), \frac{1}{t}y\right) \\
&= \frac{1}{4}\left[\left(A\left(tx + \frac{1}{t}y\right), tx + \frac{1}{t}y\right) - \left(A\left(tx - \frac{1}{t}y\right), tx - \frac{1}{t}y\right)\right] \\
&\leq \frac{\alpha}{4}\left[\left\|tx + \frac{1}{t}y\right\|^2 + \left\|tx - \frac{1}{t}y\right\|^2\right] = \frac{\alpha}{2}\left[t^2\|x\|^2 + \frac{1}{t^2}\|y\|^2\right].
\end{aligned}$$

Now we choose t such that

$$t^2\|x\|^2 + \frac{1}{t^2}\|y\|^2 = 2\|x\|\|y\|, \qquad \text{i.e.,} \qquad \left(t\|x\| - \frac{1}{t}\|y\|\right)^2 = 0.$$

Hence

$$t = \left(\frac{\|y\|}{\|x\|}\right)^{\frac{1}{2}} \qquad \text{and} \qquad \|Ax\|^2 \leq \alpha\|x\|\|y\| \qquad \text{follows.}$$

(ii) By taking $A + \|A\|I$ instead of A, we can assume that $0 \leq m \leq M = \|A\|$ (the last equality follows from (i)). Let $\{x_n\}_{n=1}^{\infty}$ be a sequence such that

$$\|x_n\| = 1 \qquad \text{and} \qquad \lim_{n\to\infty} (Ax_n, x_n) = M.$$

Then

$$\begin{aligned}
\limsup_{n\to\infty} \|Ax_n - Mx_n\|^2 &= \limsup_{n\to\infty} [(Ax_n, Ax_n) - 2M(Ax_n, x_n) + M^2] \\
&\leq \limsup_{n\to\infty} [2M^2 - 2M(Ax_n, x_n)] = 0.
\end{aligned}$$

If $M \in \varrho(A)$, then there is a constant $c > 0$ such that

$$\|Ax - Mx\| \geq c\|x\|.$$

The previous calculation shows that this cannot be true.

The assertion on m is obtained by replacing A by $-A$.

(iii) This is a consequence of (i) and (ii) and Corollary 2.1.3.

(iv) Let $\lambda = \alpha + \mathrm{i}\beta$, $\beta \neq 0$. A simple calculation yields that

$$\|\lambda x - Ax\|^2 \geq |\beta|^2\|x\|^2 \qquad \text{for every} \qquad x \in H.$$

This inequality shows that both

$$\lambda I - A \qquad \text{and} \qquad \overline{\lambda} I - A^* = \overline{\lambda} I - A$$

are injective and $\mathrm{Im}\,(\lambda I - A)$ is closed. By Proposition 2.1.27(iv) and Corollary 1.2.35,

$$\mathrm{Im}\,(\lambda I - A) = [\mathrm{Ker}\,(\lambda I - A)^*]^\perp = [\mathrm{Ker}\,(\overline{\lambda} I - A)]^\perp = H.$$

Therefore $\lambda \in \varrho(A)$.

(v) We have

$$\lambda(x, y) = (Ax, y) = (x, Ay) = (x, \mu y) = \mu(x, y)$$

(by (iv), $\mu \in \mathbb{R}$). Since $\lambda \neq \mu$, we conclude that $(x, y) = 0$. □

Theorem 2.2.16 (Hilbert–Schmidt). *Let H be a separable Hilbert space and A a self-adjoint compact operator. Then there exists an orthonormal basis $\{e_n\}_{n=1}^\infty$ where e_n are the eigenvectors of A.*

If

$$Ae_n = \lambda_n e_n \qquad \text{and} \qquad x = \sum_{n=1}^\infty (x, e_n)e_n,$$

then

$$Ax = \sum_{n=1}^\infty \lambda_n (x, e_n)e_n.$$

Proof. Let $\{\lambda_n\}_{n=1}^\infty$ be the sequence of all nonzero and pairwise distinct eigenvalues of A. Choose an orthonormal basis $e_1^{(k)}, \dots, e_{n_k}^{(k)}$ of

$$N_k := \mathrm{Ker}\,(\lambda_k I - A).$$

Remember that $N_k \perp N_{k+1}$ (Proposition 2.2.15(v)). Let us align the collection

$$\bigcup_k \{e_1^{(k)}, \dots, e_{n_k}^{(k)}\}$$

into a sequence $\{e_1, e_2, \dots\}$. This sequence is an orthonormal basis of

$$H_1 := \overline{\mathrm{Lin}\{e_1, e_2, \dots\}}.$$

If $H_1 = H$, the proof is complete. Assume therefore that $H \neq H_1$. The orthogonal complement $H_1^\perp$ is A-invariant. This means that the restriction

$$B := A|_{H_1^\perp}$$

is a self-adjoint operator on the Hilbert space $H_1^\perp$. Since $\sigma(B) \subset \sigma(A)$, $\sigma(B)$ cannot contain any nonzero number (Corollary 2.2.13(i)). As $\sigma(B) \neq \emptyset$, we have $\sigma(B) = \{0\}$ and, by Proposition 2.2.15(iii),

$$B = O \qquad \text{on} \quad H_1^\perp.$$

Hence 0 is an eigenvalue of B as well as of A. By adding an orthonormal basis of $H_1^\perp$ to $\{e_1, e_2, \dots\}$ we obtain an orthonormal basis of H. □

Example 2.2.17.[24] We have found that the inverse operator to

$$Ax = -(p\dot{x})^{\cdot} + qx,^{25} \qquad x \in \operatorname{Dom} A = \{x \in W^{2,2}(a,b) : x(a) = x(b) = 0\},$$

exists provided $\dot{p}, q \in C[a,b]$ and $p, q > 0$ on $[a,b]$. Moreover, A^{-1} is an integral operator

$$A^{-1}f(t) = \int_a^b G(t,s)f(s)\,\mathrm{d}s$$

where G is the Green function of the differential expression. From the construction of G it follows that $G \in C([a,b] \times [a,b])$, in particular, $G \in L^2(a,b)$, and G is a real symmetric function ($G(t,s) = G(s,t)$), see, e.g., Walter [156].

By Example 2.2.5(ii), A^{-1} is a compact, self-adjoint[26] operator in the real space $L^2(a,b)$ and Theorem 2.2.16 can be applied to obtain an orthonormal basis of $L^2(a,b)$ formed by the eigenfunctions $\{e_n\}_{n=1}^\infty$ of A^{-1}, i.e., by the eigenfunctions of A. Since

$$(Ax,x)_{L^2(a,b)} = \int_a^b [p(t)\dot{x}^2(t) + q(t)|x(t)|^2]\,\mathrm{d}t > 0 \quad \text{for all } x \in \operatorname{Dom} A, x \neq o,$$

all eigenvalues are positive. If λ is an eigenvalue of A (equivalently $\frac{1}{\lambda}$ is an eigenvalue of A^{-1}), then

$$\dim \operatorname{Ker}(\lambda I - A) = 1$$

since the equation

$$(p\dot{x})^{\cdot} + (q - \lambda)x = 0$$

cannot have two linearly independent solutions satisfying the initial condition $x(a) = 0$. Let the eigenvalues λ_n of A be arranged into a sequence so that

$$0 < \lambda_1 < \lambda_2 < \cdots$$

From the properties of compact operators (Corollary 2.2.13) it follows that $\lambda_n \to \infty$. It is sometimes important to know how quickly λ_n tend to infinity. A simple

[24] A continuation of Example 2.1.32.
[25] This operator is called a *Sturm–Liouville operator*.
[26] We restrict our attention to a special differential operator A in contrast to the general operator from Example 2.1.32 in order to get a self-adjoint inverse A^{-1}.

estimate can be obtained with help of the quantity $\|A^{-1}\|_2$ (Example 2.2.5(ii)), namely

$$\sum_{n=1}^{\infty} \frac{1}{\lambda_n^2} = \|A^{-1}\|_2^2 < \infty$$

(see Excercise 2.2.22). However, this result is far from being optimal. We remark here that a variational approach to an eigenvalue problem for compact, self-adjoint operators will be briefly described in Section 7.8.

Consider now the equation

$$Ax = \lambda x + f \tag{2.2.5}$$

or, equivalently (cf. Exercise 2.2.21),

$$\sum_{n=1}^{\infty} (\lambda_n - \lambda)(x, e_n)e_n = \sum_{n=1}^{\infty} (f, e_n)e_n, \quad \text{i.e.,} \quad (\lambda_n - \lambda)(x, e_n) = (f, e_n) \quad \text{for } n \in \mathbb{N}.$$

If λ is no eigenvalue of A, then $\inf\limits_n |\lambda_n - \lambda| > 0$ (since $\lambda_n \to \infty$) and

$$x = \sum_{n=1}^{\infty} \frac{(f, e_n)}{\lambda_n - \lambda} e_n$$

is a unique solution of (2.2.5). (Notice that this series is convergent.) If $\lambda = \lambda_n$, then the condition

$$(f, e_n) = 0$$

is a necessary and sufficient condition for solvability of (2.2.5) (see also Example 2.1.31).

If we examined singular differential operators, e.g., on the interval $[0, \infty)$, we would meet with many difficulties arising for example from the fact that A^{-1} is no more compact and, therefore, its spectrum is more complicated. The interested reader can consult the book Dunford & Schwartz [54]. ○

Remark 2.2.18. The Hilbert–Schmidt Theorem allows us to introduce a functional calculus for compact, self-adjoint operators similarly as it has been done for matrices in Theorem 1.1.38:

Let A be a compact, self-adjoint operator on a Hilbert space H. Then there exists a unique mapping

$$\Phi \colon C(\sigma(A)) \to \mathcal{L}(H)$$ [27]

with the following properties:

(i) Φ is an algebra homomorphism (i.e., Φ preserves operations on $C(\sigma(A))$ and $\mathcal{L}(H)$);

[27] If $\sigma(A) = \{0\} \cup \{\lambda_n\}_{n=1}^{\infty}$, then $f \in C(\sigma(A))$ if and only if $\lim\limits_{n\to\infty} f(\lambda_n) = f(0)$.

(ii) Φ is a continuous mapping from $C(\sigma(A))$ into $\mathcal{L}(H)$ with the operator topology;

(iii) if $P(x) = \sum_{k=0}^{m} a_k x^k$, then $\Phi(P) = \sum_{k=0}^{m} a_k A^k$;

(iv) if $w \notin \sigma(A)$ and $f(x) = \frac{1}{w-x}$, then $\Phi(f) = (wI - A)^{-1}$;

(v) $\sigma(\Phi(f)) = f(\sigma(A))$ for every $f \in C(\sigma(A))$.

If $Ax = \sum_{n=1}^{\infty} \lambda_n (x, e_n) e_n$, then it is easy to verify properties (i)–(v) for

$$\Phi(f)x := \sum_{n=1}^{\infty} f(\lambda_n)(x, e_n)e_n.$$

In particular if $A \in \mathscr{C}(H)$ then $|A|(x) := \sum_{n=1}^{\infty} |\lambda_n|(x, e_n)e_n$, where $\{\lambda_n^2\}_{n=1}^{\infty}$ is a sequence of eigenvalues of A^*A.

We omit the proof of uniqueness.

It is worth mentioning that we can introduce a functional calculus for a linear operator A which has a compact, self-adjoint resolvent $(\lambda_0 I - A)^{-1}$. We leave this easy construction to the interested reader. Example 2.2.17 shows a class of such operators.

Exercise 2.2.19. Consider a special case of the Sturm–Liouville operator

$$Ax = -\ddot{x}$$

in the space $L^2(0, \pi)$ with the boundary conditions

(i) $x(0) = x(\pi) = 0$ (Dirichlet boundary conditions),

(ii) $\dot{x}(0) = \dot{x}(\pi) = 0$ (Neumann boundary conditions),

(iii) $\alpha_0 x(0) + \beta_0 \dot{x}(0) = 0$, $\alpha_1 x(\pi) + \beta_1 \dot{x}(\pi) = 0$ (mixed or Newton–Robin boundary conditions),

(iv) $x(0) = x(\pi)$, $\dot{x}(0) = \dot{x}(\pi)$ (periodic conditions).

Find Green functions, eigenvalues and eigenfunctions. What follows from the Hilbert–Schmidt Theorem? Compare this result with that of Example 1.2.38(i).

Exercise 2.2.20. Define e^{tA} for the operator A from Exercise 2.2.19 (see Remark 2.2.18). Take $x \in \operatorname{Dom} A$ and show that the function

$$u(t, \xi) := \left(\mathrm{e}^{tA} x\right)(\xi), \qquad t \geq 0,$$

is a solution to the heat equation

$$\frac{\partial u}{\partial t} = \frac{\partial^2 u}{\partial \xi^2}$$

satisfying the initial condition $u(0, \cdot) = x(\cdot)$ and the boundary conditions given by $u(t, \cdot) \in \operatorname{Dom} A$. Do not forget to define the notion of a solution.

Exercise 2.2.21. Let A be as in Example 2.2.17. Prove that

$$\operatorname{Dom} A = \left\{ x = \sum_{n=1}^{\infty} (x, e_n) e_n : \sum_{n=1}^{\infty} |\lambda_n|^2 |(x, e_n)|^2 < \infty \right\}$$

and

$$Ax = \sum_{n=1}^{\infty} \lambda_n (x, e_n) e_n.$$

Exercise 2.2.22. Denote $\mathcal{L}_2(H)$ the set of all Hilbert-Schmidt operators on a Hilbert space H. Prove that

(i) $\mathcal{L}_2(H)$ is an *ideal* in $\mathcal{L}(H)$, i.e., for $A \in \mathcal{L}(H)$, $B \in \mathcal{L}_2(H)$ we have AB and BA belong to $\mathcal{L}_2(H)$;

(ii) $A \in \mathcal{L}_2(H) \Leftrightarrow |A| := (A^*A)^{\frac{1}{2}} \in \mathcal{L}_2(H)$. (For the definition of $|A|$ see Remark 2.2.18). If $\lambda_1, \dots$ are eigenvalues of $|A|$ then $\|A\|_2^2 = \sum_n \lambda_n^2$. For more information see, e.g., Schatten [137].

Exercise 2.2.23. Suppose that $A \in \mathcal{L}(X, Y)$ maps a weakly convergent sequence into a strongly convergent one. Prove that A is compact provided X is reflexive.

Exercise 2.2.24. Prove the assertion from Remark 2.2.10 and the decomposition (2.2.4).

2.3 Contraction Principle

The previous four sections have been devoted to some basic facts in the linear theory. It is now time to start with nonlinear problems, especially with the solution of the nonlinear equation

$$f(x) = a \qquad \text{for} \qquad f\colon X \to X. \tag{2.3.1}$$

The basic assertions in this section are fixed point theorems for contractible and non-expansive mappings. If X is a linear space, (2.3.1) is equivalent to the equation

$$F(x) := a - f(x) + x = x.$$

The solution of this equation is called a *fixed point* of F. In the case that

$$f(x) = x - Ax \qquad (F(x) = Ax + a)$$

where $A \in \mathcal{L}(X)$, we succeeded in solving this equation in Section 2.1 (cf. Proposition 2.1.2) by applying the iteration process

$$x_0 = a, \qquad x_n = a + Ax_{n-1} \qquad \text{provided} \quad \|A\| < 1.$$

This idea can be easily generalized to the following result which is often attributed to S. Banach. Notice that in the forthcoming assertion we do not need the linear structure of the space at all.

Theorem 2.3.1 (Contraction Principle). *Let M be a complete metric space and let $F\colon M \to M$ be a* contraction, *i.e., there is $q \in [0, 1)$ such that*

$$\varrho(F(x), F(y)) \le q\varrho(x, y) \qquad \textit{for every} \quad x, y \in M.$$

Then there exists a unique fixed point $\tilde{x}$ of F in M. Moreover, if

$$x_0 \in M, \qquad x_n = F(x_{n-1}),$$

then the sequence $\{x_n\}_{n=1}^{\infty}$ converges to $\tilde{x}$ and the estimates

$$\varrho(x_n, \tilde{x}) \le \frac{q^n}{1-q}\varrho(x_1, x_0) \qquad \textit{(a priori estimate)}, \tag{2.3.2}$$

$$\varrho(x_n, \tilde{x}) \le \frac{q}{1-q}\varrho(x_n, x_{n-1}) \qquad \textit{(a posteriori estimate)} \tag{2.3.3}$$

hold.

Proof. We prove that $\{x_n\}_{n=1}^{\infty}$ is a Cauchy sequence. Indeed, for $m > n$ we have

$$\begin{aligned}\varrho(x_m, x_n) &\le \varrho(x_m, x_{m-1}) + \cdots + \varrho(x_{n+1}, x_n)\\ &= \varrho(F(x_{m-1}), F(x_{m-2})) + \cdots + \varrho(F(x_n), F(x_{n-1}))\\ &\le q[\varrho(x_{m-1}, x_{m-2}) + \cdots + \varrho(x_n, x_{n-1})]\\ &\le (q^{m-1} + \cdots + q^n)\varrho(x_1, x_0) \le \frac{q^n}{1-q}\varrho(x_1, x_0).\end{aligned}$$

Since $q < 1$, the right-hand side is arbitrarily small for sufficiently large n. The Cauchy sequence $\{x_n\}_{n=1}^{\infty}$ has a limit $\tilde{x}$ in the complete space M, and for this limit the estimate (2.3.2) holds. Being a contraction, F is a continuous mapping, and therefore

$$\tilde{x} = \lim_{n\to\infty} x_n = \lim_{n\to\infty} F(x_{n-1}) = F\left(\lim_{n\to\infty} x_{n-1}\right) = F(\tilde{x}).$$

Uniqueness of a fixed point is even easier: If $\tilde{x} = F(\tilde{x})$, $\tilde{y} = F(\tilde{y})$, then

$$\varrho(\tilde{x}, \tilde{y}) = \varrho(F(\tilde{x}), F(\tilde{y})) \le q\varrho(\tilde{x}, \tilde{y}), \qquad \text{i.e.,} \quad \varrho(\tilde{x}, \tilde{y}) = 0 \qquad (q < 1).$$

The a posteriori estimate also follows from the above estimate of $\varrho(x_m, x_n)$. □

The fixed point of F the existence of which has been just established often depends on a parameter. The following result is useful in investigating this dependence.

Corollary 2.3.2. *Let M be a complete metric space and A a topological space. Assume that $F\colon A \times M \to M$ possesses the following properties:*

(i) *There is $q \in [0, 1)$ such that*

$$\varrho(F(a, x), F(a, y)) \le q\varrho(x, y) \qquad \textit{for all} \quad a \in A \quad \textit{and} \quad x, y \in M.$$

(ii) *For every $x \in M$ the mapping $a \mapsto F(a,x)$ is continuous on A.*

Then for each $a \in A$ there is a unique $\varphi(a) := \tilde{x}$ such that

$$F(a, \tilde{x}) = \tilde{x}.$$

Moreover, φ is continuous on A.

Proof. The existence of φ follows directly from Theorem 2.3.1. The estimates

$$\begin{aligned}\varrho(\varphi(a), \varphi(b)) &= \varrho(F(a, \varphi(a)), F(b, \varphi(b)))\\ &\le \varrho(F(a, \varphi(a)), F(b, \varphi(a))) + \varrho(F(b, \varphi(a)), F(b, \varphi(b)))\\ &\le \varrho(F(a, \varphi(a)), F(b, \varphi(a))) + q\varrho(\varphi(a), \varphi(b))\end{aligned}$$

yield

$$\varrho(\varphi(a), \varphi(b)) \le \frac{1}{1-q}\varrho(F(a, \varphi(a)), F(b, \varphi(a))),$$

and the continuity of φ follows. □

Remark 2.3.3. Notice that φ is Lipschitz continuous provided $a \mapsto F(a,x)$ is Lipschitz continuous uniformly with respect to x (and, of course, A is a metric space).

There is an enormous amount of applications of the Contraction Principle. The proof of the existence theorem for the initial value problem for ordinary differential equations belongs to standard applications. However, the historical development went in the opposite direction. The following theorem had been proved (by iteration) about thirty years before the Contraction Principle was formulated in its full generality. Another application will be given in Section 4.1.

Theorem 2.3.4 (Picard). *Let $\mathcal{G}$ be an open set in $\mathbb{R} \times \mathbb{R}^N$ and let $f\colon (t, x_1, \dots, x_N) \in \mathcal{G} \to \mathbb{R}^N$ be continuous and locally Lipschitz continuous with respect to the x-variables, i.e., for every $(s, y) \in \mathcal{G}$ there exist $\delta > 0$, $\hat{\delta} > 0$, $L > 0$ such that*

$$\|f(t, x_1) - f(t, x_2)\| \le L\|x_1 - x_2\| \quad \textit{whenever } |t - s| < \delta,\ \|x_i - y\| < \hat{\delta},\ i = 1, 2.$$

Then for any $(t_0, \xi_0) \in \mathcal{G}$ there exists $\delta > 0$ such that the equation

$$\dot{x} = f(t, x) \tag{2.3.4}$$

has a unique solution on the interval $(t_0 - \delta, t_0 + \delta)$ satisfying the initial condition

$$x(t_0) = \xi_0. \tag{2.3.5}$$

Proof. First we rewrite the initial value problem (2.3.4), (2.3.5) into an equivalent fixed point problem for an integral operator F defined by

$$F(x)\colon t \mapsto \xi_0 + \int_{t_0}^{t} f(s, x(s))\,\mathrm{d}s, \qquad t \in (t_0 - \delta, t_0 + \delta).^{28} \tag{2.3.6}$$

This equivalence is easy to establish (by integration and by differentiation with respect to t). Therefore we wish to solve the equation

$$F(x) = x$$

in a complete metric space M. We choose M to be a closed subset of the Banach space $C[t_0 - \delta, t_0 + \delta]$ for a certain small $\delta > 0$.

We need two properties of F and M, namely that F maps M into M and F is a contraction on M. Choose first δ_1, $\hat{\delta}_1$ such that

$$\mathcal{R}_1 := [t_0 - \delta_1, t_0 + \delta_1] \times \{x \in \mathbb{R}^N : \|x - \xi_0\| \le \hat{\delta}_1\} \subset \mathcal{G}.$$

This set $\mathcal{R}_1$ is compact, and therefore f is bounded and uniformly Lipschitz continuous on it, i.e., there are constants K, L such that

$$\|f(s,x)\| \le K, \qquad \|f(s,x) - f(s,y)\| \le L\|x - y\| \qquad \text{for} \quad (s,x),(s,y) \in \mathcal{R}_1.$$

Put

$$M = \{x \in C[t_0 - \delta, t_0 + \delta] : \|x(t) - \xi_0\| \le \hat{\delta}_1 \, \forall t \in [t_0 - \delta, t_0 + \delta]\} \text{ for a } \delta \le \delta_1.$$

Then

$$\sup_{t \in \mathcal{I}_\delta} \|F(x(t)) - \xi_0\| \le \delta K, \quad \sup_{t \in \mathcal{I}_\delta} \|F(x(t)) - F(y(t))\| \le \delta L \sup_{t \in \mathcal{I}_\delta} \|x(t) - y(t)\|$$

where

$$\mathcal{I}_\delta := [t_0 - \delta, t_0 + \delta].$$

If we choose δ so small that $\delta K \le \hat{\delta}_1$ and $\delta L \le \frac{1}{2}$, then F maps M into itself (the first condition) and is a contraction with $q = \frac{1}{2}$ (the second condition). By the Contraction Principle, F has a unique fixed point y in M and this is a solution of (2.3.4), (2.3.5) on the interval $(t_0 - \delta, t_0 + \delta)$. If $\tilde{x}$ is a solution of (2.3.4), (2.3.5) on the interval $(t_0 - \delta, t_0 + \delta)$, then $\tilde{x} \in M$ (prove it!), i.e., $y = \tilde{x}$, and the uniqueness follows. □

Remark 2.3.5. The mapping F defined by (2.3.6) depends actually not only on x but also on t_0, ξ_0. By taking smaller δ we can prove that F is also Lipschitz continuous with respect to the initial conditions and Corollary 2.3.2 yields that the solution $x(t;\cdot,\cdot)$ of (2.3.4), (2.3.5) is also Lipschitz continuous with respect to the initial conditions. Moreover, if $f \in C^1(\mathcal{G})$ then the solution $x(t;.,.)$ is also continuously differentiable with respect to the initial conditions. We can use for $F(x;t_0,\xi_0)$ the Implicit Function Theorem to obtain that $x(t;t_0,\xi_0)$ is differentiable with respect to all variables (Example 4.2.5).

---

[28] If $t < t_0$, then we define $\int_{t_0}^{t} f(s,x(s))\,\mathrm{d}s = -\int_{t}^{t_0} f(s,x(s))\,\mathrm{d}s$, and $\int_{t_0}^{t_0} f(s,x(s))\,\mathrm{d}s = 0$.

Remark 2.3.6. A simple example $\dot{x} = x^2$ shows that solutions need not be global. On the other hand, if $\mathcal{G} = (\alpha, \beta) \times \mathbb{R}^N$ and the right-hand side f is *sublinear*, i.e., there are nonnegative functions $a \in L^1_{\text{loc}}(\alpha, \beta)$, $b \in C[\alpha, \beta]$ such that

$$\|f(t,x)\| \le a(t) + b(t)\|x\|, \quad (t,x) \in (\alpha, \beta) \times \mathbb{R}^N, \tag{2.3.7}$$

then for any initial condition the equation (2.3.4) has a unique solution on (α, β). To prove this fact we have to show that the local solution x given by Theorem 2.3.4 cannot "explode" inside (α, β) and, therefore, it can be continued to the entire interval (α, β). This method is based on the estimate of x which follows from the Gronwall inequality (see Proposition 2.3.9 below):

$$\|x(t)\| \le \|\xi_0\| + \int_{t_0}^{t} \|f(s, x(s))\| ds \le \|\xi_0\| + \int_{t_0}^{t} a(s) ds$$

$$+ \int_{t_0}^{t} b(s)\|x(s)\| ds, \quad t \ge t_0.$$

By this inequality,

$$\|x(t)\| \le A(t) + \int_{t_0}^{t} A(s)b(s)e^{\int_s^t b(\sigma)d\sigma},$$

where $A(t) = \|\xi_0\| + \int_{t_0}^{t} a(s)ds$. If x were defined on $[t_0, \tau)$, $\tau < \beta$, then the last inequality shows that x is bounded and therefore $t \mapsto f(t, x(t))$ is bounded on $[t_0, \tau)$ as well. Hence, x is uniformly continuous on this interval and, in particular, the limit $\lim_{t \to \tau-} x(t) = \eta$ exists and is finite. So, we can "glue" together the solution x with a solution satisfying the initial condition $x(\tau) = \eta$.

We emphasize the special case of a system of linear differential equations

$$\dot{x} = \boldsymbol{A}(t)x + g(t)$$

with a continuous matrix $\boldsymbol{A}$ and a continuous vector function g on some interval (α, β) that satisfies the estimate (2.3.7). However, in this case we can avoid the use of Contraction Principle at all and prove the convergence of iterations directly. The following example demonstrates this approach.

Example 2.3.7. Let k be a bounded measurable function on the set

$$\mathcal{M} = \{(s,t) \in \mathbb{R}^2 : 0 \le s \le t \le 1\}.$$

Then for any $f \in L^1(0,1)$ and $\lambda \neq 0$ there is a unique solution to the integral equation

$$x(t) - \lambda \int_0^t k(t,s)x(s)\,\mathrm{d}s = f(t). \tag{2.3.8}$$

To prove this assertion, denote

$$Ax(t) = \int_0^t k(t,s)x(s)\,\mathrm{d}s.$$

Then $A \in \mathcal{L}(L^1(0,1))$ (Example 2.1.28). Put

$$x_0 = f, \qquad x_n = f + \lambda A x_{n-1}.$$

Due to the completeness of $L^1(0,1)$ the sequence $\{x_n\}_{n=1}^{\infty}$ is convergent in $L^1(0,1)$ provided the sum $\sum\limits_{n=1}^{\infty} \|x_n - x_{n-1}\|_{L^1(0,1)}$ is convergent. We have

$$x_n - x_{n-1} = \lambda^n A^n f \qquad \text{and} \qquad A^n f(t) = \int_0^t k_n(t,s)f(s)\,\mathrm{d}s$$

where

$$k_1 = k \qquad \text{and} \qquad k_n(t,s) = \int_s^t k_{n-1}(t,\sigma)k(\sigma,s)\,\mathrm{d}\sigma$$

(check this relation). It is easy to prove by induction that

$$|k_n(t,s)| \le \|k\|^n_{L^\infty(\mathcal{M})} \frac{(t-s)^{n-1}}{(n-1)!}, \qquad (t,s) \in \mathcal{M},$$

and hence

$$\begin{aligned}\|x_n - x_{n-1}\|_{L^1(0,1)} &\le |\lambda|^n \int_0^1 \left| \int_0^t k_n(t,s)f(s)\,\mathrm{d}s \right| \mathrm{d}t \\ &\le |\lambda|^n \int_0^1 |f(s)| \left[\int_s^1 |k_n(t,s)|\,\mathrm{d}t \right] \mathrm{d}s \le \frac{|\lambda|^n \|k\|^n}{n!} \|f\|_{L^1(0,1)}.\end{aligned}$$

Since the series $\sum\limits_{n=1}^{\infty} \frac{a^n}{n!}$ is convergent for any $a \in \mathbb{R}$ the limit $\lim\limits_{n\to\infty} x_n = \tilde{x} \in L^1(0,1)$ exists and $\tilde{x}$ is a solution to (2.3.8). In fact $\tilde{x}$ is a unique solution (see Exercise 2.3.21). Moreover, $\tilde{x}$ depends continuously on f, which means that $\sigma(A) = \{0\}$.[29] This result holds also for $k \in C(\mathcal{M})$ in the space $C[0,1]$. The proof is the same. ○

[29] Actually, we have proved that $\mathbb{C} \setminus \{0\} \subset \varrho(A)$, i.e., $\sigma(A) \subset \{0\}$. Since $A \in \mathcal{L}(L^1(0,1))$, $\sigma(A) \neq \emptyset$, we have $\sigma(A) = \{0\}$.

Example 2.3.8. Assume $E \in C^1(\mathbb{R}^N)$, $\lim\limits_{\|x\|\to\infty} E(x) = \infty$, and denote the vector of its partial derivatives at x as $\nabla E(x)$ (gradient of E). Let $f \in C\left([0,T];\mathbb{R}^N\right)$. Consider the following *gradient system*

$$\dot{x}(t) + \nabla E\left(x(t)\right) = f(t), \quad x(0) = x_0. \tag{2.3.9}$$

This system need not satisfy the assumptions of Theorem 2.3.4 but the weaker existence theorem (Proposition 5.2.7 below) can be used to yield a maximal solution to (2.3.9) on certain interval $[0,\tau)$. We show that this solution is actually global, i.e., $\tau = T$. To this end assume $\tau < T$ and multiply (2.3.9) by $\dot{x}$ and integrate over $[0,t)$:

$$\int_0^t \|\dot{x}(s)\|^2 ds + \int_0^t \left(\nabla E\left(x(s)\right), \dot{x}(s)\right) ds = \int_0^t \left(f(s), \dot{x}(s)\right) ds$$

$$\leq \frac{1}{2}\int_0^t \|f(s)\|^2 ds + \frac{1}{2}\int_0^t \|\dot{x}(s)\|^2 ds.$$

Since

$$\left(\nabla E\left(x(s)\right), \dot{x}(s)\right) = \frac{d}{ds} E\left(x(s)\right),$$

we have

$$E\left(x(t)\right) \leq \frac{1}{2}\int_0^t \|\dot{x}(s)\|^2 ds + E\left(x(t)\right) \leq E(x_0) + \frac{1}{2}\int_0^\tau \|f(s)\|^2 ds.$$

This inequality shows that x is bounded on $[0,\tau)$ due to the assumption

$$\lim_{\|x\|\to\infty} E(x) = \infty.$$

Similarly as in Remark 2.3.6 we deduce that there exists a finite $\lim\limits_{t\to\tau-} x(t)$ and x can be continued as a solution beyond τ. But this contradicts the maximality of x. ○

Proposition 2.3.9 (Gronwall Lemma). *Let $a \in C^1[\alpha,\beta)$, $b \in C[\alpha,\beta)$ and let $b \geq 0$ on the interval $[\alpha,\beta)$. If z is a continuous function satisfying the inequality*

$$z(t) \leq a(t) + \int_\alpha^t b(s)z(s)ds, \qquad t \in [\alpha,\beta) \tag{2.3.10}$$

then

$$z(t) \leq a(t) + \int_\alpha^t a(s)b(s)e^{\int_s^t b(\xi)d\xi} ds \tag{2.3.11}$$

for $t \in [\alpha,\beta)$.

Proof. Denote the right-hand side in (2.3.10) as $y(t)$. Then

$$\dot{y}(t) = \dot{a}(t) + b(t)z(t) \le \dot{a}(t) + b(t)y(t).$$

Multiplying by $e^{-B(t)}$, where $B(t) = \int\limits_{\alpha}^{t} b(s)ds$, we obtain

$$\left[y(t)e^{-B(t)}\right]^{\cdot} \le \dot{a}(t)e^{-B(t)} = \left[\int\limits_{\alpha}^{t} \dot{a}(s)e^{-B(s)}ds\right]^{\cdot}.$$

Therefore $y(\alpha) \ge y(t)e^{-B(t)} - \int\limits_{\alpha}^{t} \dot{a}(s)e^{-B(s)}ds$. Integrating by parts we get (2.3.11).

□

Using this proposition for $z(t) := \|x(t)\|$, where x is a local solution of the linear equation $\dot{x} = \boldsymbol{A}(t)x + g(t)$, we find a function φ which depends only on $\boldsymbol{A}$ and g and it is defined on the interval (a,b) such that the inequality $z(t) \le \varphi(t)$ holds on the domain of a solution x. This means that the continuation process can be prolonged (by the uniform continuity of x) to the whole interval (a,b). Similarly, this continuation process works for the equation (2.3.4) provided f is globally Lipschitz continuous with respect to the x-variables and $\mathcal{G} = \mathbb{R} \times \mathbb{R}^N$.

Example 2.3.10. Find sufficient conditions for the existence of a classical solution (cf. Example 2.1.31) of the boundary value problem

$$\begin{cases} \ddot{x}(t) = \varepsilon f(t, x(t)), & t \in (0,1), \\ x(0) = x(1) = 0. \end{cases} \tag{2.3.12}$$

Theorem 2.3.4 suggests the assumption that f is continuous with respect to t and Lipschitz continuous with respect to the x-variable on a certain rectangle $[0,1] \times [-r,r]$. Denote a Lipschitz constant on this interval by $L(r)$. We wish to rewrite the problem (2.3.12) as a fixed point problem. To reach this goal suppose that we have a solution y and let $g(t) := \varepsilon f(t, y(t))$. Then y solves also the equation

$$\ddot{y} = g$$

and satisfies $y(0) = y(1) = 0$.

It is easy to see that this problem has exactly one solution which is given by

$$y(t) = \int_0^1 G(t,s)g(s)\,\mathrm{d}s := \int_0^t (t-1)sg(s)\,\mathrm{d}s + \int_t^1 t(s-1)g(s)\,\mathrm{d}s$$

(G is the Green function – see Example 2.1.32). Therefore, we are looking for a continuous function x which solves the integral equation

$$x(t) = \varepsilon \int_0^1 G(t,s)f(s,x(s))\,\mathrm{d}s. \tag{2.3.13}$$

Denote

$$F(\varepsilon, x) := \varepsilon \int_0^1 G(t,s) f(s, x(s))\,\mathrm{d}s.$$

We can solve (2.3.13) by applying the Contraction Principle in

$$M := \{x \in C[0,1] : \|x\| \le r\} \qquad \text{for an appropriate choice of } r.$$

For $x \in M$ we have

$$|f(s, x(s))| \le |f(s,0)| + |f(s,x(s)) - f(s,0)| \le K + L(r)r$$

where $K > 0$ is a constant such that $|f(s,0)| \le K$, $s \in [0,1]$, and

$$\|F(\varepsilon, x)\| \le \frac{|\varepsilon|}{8}(K + L(r)r).$$

This estimate shows that F maps M into itself if

$$q := \frac{|\varepsilon|}{8} L(r) < 1 \qquad \text{and} \qquad r \ge \frac{|\varepsilon| K}{8} \frac{1}{1-q}.$$

Then F is also a contraction on M with the constant q. We can conclude that for a given r there is $\varepsilon_0 > 0$ such that for $|\varepsilon| \le \varepsilon_0$ both the above conditions[30] are satisfied and (2.3.13) has a solution.

Now we have to show that a continuous solution x of (2.3.13) is actually a classical solution of the boundary value problem (2.3.12). Since we know the explicit form of the Green function G, it is obvious that $x(0) = x(1) = 0$ and it is also easy to differentiate twice the right-hand side of (2.3.13) (taking into account that x is continuous).

We remark that we have not used all properties of the integral operator with the kernel G. In particular, such an operator is compact (Example 2.2.5(i)) and this property has not been used. This property will be significant in Chapter 5.

○

The a posteriori estimate (2.3.3) shows that the convergence of iterations may be rather slow. It can be sometimes desirable to have faster convergence at the expense of more restrictive assumptions. The classical Newton Method for solving an equation

$$f(x) = 0, \qquad f\colon \mathbb{R} \to \mathbb{R},$$

is illustrated in Figure 2.3.1.

In order to generalize this method we need the notion of a derivative of $f\colon X \to X$. This will be the main subject of the next chapter.

---

[30]Notice that for a fixed ε these conditions are antagonistic, namely the first requires small r and the other large r. This situation is typical in applications of the Contraction Principle.

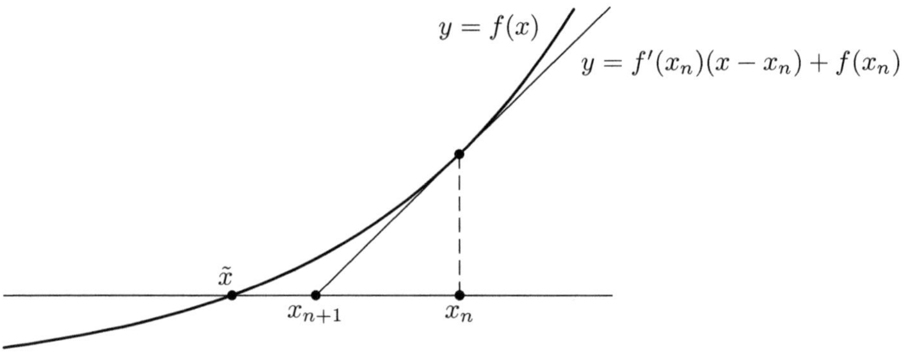

Figure 2.3.1.

There are many generalizations of the Contraction Principle. One of them concerns the assumption $q < 1$. A mapping $F\colon M \to M$ is called *non-expansive* if

$$\varrho(F(x), F(y)) \le \varrho(x, y) \qquad \text{for all} \quad x, y \in M.$$

A simple example $F(x) = x + 1$, $x \in \mathbb{R}$, shows that F may have no fixed point. This can be caused by the fact that F does not map any bounded set into itself. However, there are non-expansive mappings which map the unit ball into itself and do not possess any fixed point either. See the following example or Exercise 2.3.20.

Example 2.3.11 (Beals). Let M be the space of all sequences with zero limit with the sup norm (this space is usually denoted by c_0) and let

$$F(x) = (1, x_1, x_2, \dots) \qquad \text{for} \quad x = (x_1, x_2, \dots) \in M.$$

Then F is a non-expansive map of the unit ball into itself without any fixed point. ○

This example indicates that some special properties of the space are needed. We formulate the following assertion in a Hilbert space and use the Hilbert structure essentially in its proof. The statement is true also in uniformly convex spaces but the proof is more involved (see, e.g., Goebel [72]). Let us note an interesting fact that the validity of Proposition 2.3.12 in a reflexive Banach space is an open problem.

Proposition 2.3.12 (Browder). *Let $\mathcal{M}$ be a bounded closed and convex set in a Hilbert space H. Let F be a non-expansive mapping from $\mathcal{M}$ into itself. Then there is a fixed point of F in $\mathcal{M}$. Moreover, if*

$$x_0 \in \mathcal{M}, \qquad x_n = F(x_{n-1}) \qquad \textit{and} \qquad y_n = \frac{1}{n}\sum_{k=0}^{n-1} x_k,$$

then the sequence $\{y_n\}_{n=1}^{\infty}$ is weakly convergent to a fixed point.

Proof. The existence result is not difficult to prove.[31] So we will prove a more interesting result which yields also a numerical method for finding a fixed point. The proof consists of four steps, the last one is crucial and has a variational character.

Step 1. Since $\mathcal{M}$ is bounded, closed and convex, $y_n \in \mathcal{M}$ and there is a subsequence $\{y_{n_k}\}_{k=1}^{\infty}$ weakly convergent to an $\tilde{x} \in \mathcal{M}$ (Theorem 2.1.25 and Exercise 2.1.40). Fix such a weakly convergent subsequence $\{y_{n_k}\}_{k=1}^{\infty}$ and its weak limit $\tilde{x} \in \mathcal{M}$.

Step 2. We have $\lim\limits_{n\to\infty} \|F(y_n) - y_n\| = 0$. Indeed,

$$\|F^k(x_0) - F(y_n) + F(y_n) - y_n\|^2 = \|F^k(x_0) - F(y_n)\|^2 + \|F(y_n) - y_n\|^2 \\ + 2\operatorname{Re}(F^k(x_0) - F(y_n), F(y_n) - y_n)$$

where

$$F^k(x_0) = F(F^{k-1}(x_0)).$$

Summing up this equality from $k = 0$ to $k = n-1$ and dividing by n we get

$$\begin{aligned}\frac{1}{n}\sum_{k=0}^{n-1}\|F^k(x_0) - y_n\|^2 &= \frac{1}{n}\sum_{k=0}^{n-1}\big(\|F^k(x_0) - F(y_n)\|^2 \\ &\quad + \|F(y_n) - y_n\|^2 + 2\operatorname{Re}(y_n - F(y_n), F(y_n) - y_n)\big) \\ &= \frac{1}{n}\sum_{k=0}^{n-1}\|F^k(x_0) - F(y_n)\|^2 - \|F(y_n) - y_n\|^2.\end{aligned}$$

Since F is non-expansive, we conclude from this equality that

$$\begin{aligned}\|F(y_n) - y_n\|^2 &\le \frac{1}{n}\sum_{k=1}^{n-1}\|F^{k-1}(x_0) - y_n\|^2 + \frac{1}{n}\|x_0 - F(y_n)\|^2 \\ &\quad - \frac{1}{n}\sum_{k=0}^{n-1}\|F^k(x_0) - y_n\|^2 \\ &= \frac{1}{n}\|x_0 - F(y_n)\|^2 - \frac{1}{n}\|F^{n-1}(x_0) - y_n\|^2 - \frac{1}{n}\|x_0 - y_n\|^2 \to 0\end{aligned}$$

(all sequences belong to $\mathcal{M}$, and hence they are bounded).

Step 3. The element $\tilde{x}$ is a fixed point of F. To see this, observe that the inequality

$$(z - F(z) - (y_{n_k} - F(y_{n_k})), z - y_{n_k}) \\ = (z - y_{n_k}, z - y_{n_k}) - (F(z) - F(y_{n_k}), z - y_{n_k}) \ge \|z - y_{n_k}\|^2 - \|z - y_{n_k}\|^2 = 0$$

[31] It is possible to assume that $o \in \mathcal{M}$. For any $t \in (0,1)$ the mapping $F_t(x) := tF(x)$ is a contraction. Letting $t \to 1$ we obtain a sequence $\{x_n\}_{n=1}^{\infty} \subset \mathcal{M}$ for which $x_n - F(x_n) \to o$. Therefore it is sufficient to show that $(I - F)(\mathcal{M})$ is closed. This needs a trick which is typical for monotone operators (Section 6.1). Notice that $I - F$ is monotone provided F is non-expansive.

holds for any $z \in \mathcal{M}$. By Exercise 2.1.37 and Step 2, the limit of the left-hand side is $(z - F(z), z - \tilde{x})$, i.e., the inequality

$$(z - F(z), z - \tilde{x}) \geq 0 \tag{2.3.14}$$

is also true. Now take $t \in (0, 1)$ and put

$$z = (1 - t)\tilde{x} + tF(\tilde{x}) \qquad (z \in \mathcal{M}).$$

For $t \to 0$, the inequality (2.3.14) divided by t yields

$$\|\tilde{x} - F(\tilde{x})\|^2 \leq 0.$$

Step 4. If x is a fixed point of F, then

$$\|x_n - x\|^2 = \|F(x_{n-1}) - F(x)\|^2 \leq \|x_{n-1} - x\|^2$$

and, therefore, the limit $\varphi(x) := \lim\limits_{n\to\infty} \|x_n - x\|^2$ exists. By Step 3, $\tilde{x}$ is also a fixed point, and we get

$$\varphi(\tilde{x}) \leq \|\tilde{x} - x_k\|^2 = \|\tilde{x} - v\|^2 + \|v - x_k\|^2 + 2\,\mathrm{Re}(\tilde{x} - v, v - x_k) \qquad \text{for any} \quad v \in H.$$

Summing up from $k = 0$ to $k = n - 1$ and dividing by n we arrive at

$$\varphi(\tilde{x}) \leq \|\tilde{x} - v\|^2 + \frac{1}{n}\sum_{k=0}^{n-1} \|v - x_k\|^2 + 2\,\mathrm{Re}(\tilde{x} - v, v - y_n). \tag{2.3.15}$$

Let v be a weak limit of a subsequence $\{y_{n_l}\}_{l=1}^{\infty} \subset \{y_n\}_{n=1}^{\infty}$, possibly different from $\{y_{n_k}\}_{k=1}^{\infty}$. Then v is a fixed point of F by virtue of the previous steps. Set $n = n_k$ and take the limit for $k \to \infty$ in (2.3.15). We finally obtain[32]

$$\varphi(\tilde{x}) \leq \varphi(v) - \|\tilde{x} - v\|^2,$$

and $v = \tilde{x}$ follows. In particular, the limit of any weakly convergent subsequence of $\{y_n\}_{n=1}^{\infty}$ coincides with $\tilde{x}$, and therefore the whole sequence $\{y_n\}_{n=1}^{\infty}$ weakly converges to $\tilde{x}$. □

Remark 2.3.13. We have mentioned in footnote 31 on page 103 that $I - F$ is a monotone operator whenever F is non-expansive. The converse statement is not true even in $\mathbb{R}$. Consider, e.g., $F(x) = -2x$. Proposition 2.3.12 should be compared with Theorem 6.1.4.

Exercise 2.3.14. Let $\{a_n\}_{n=1}^{\infty}$ be a sequence. Prove that

$$\lim_{n\to\infty} \frac{1}{n}\sum_{k=1}^{n} a_k = \lim_{n\to\infty} a_n$$

provided the right-hand side exists.

---

[32]Observe that $\lim\limits_{k\to\infty} \frac{1}{n_k} \sum\limits_{j=0}^{n_k-1} \|v - x_j\|^2 = \lim\limits_{n\to\infty} \|v - x_n\|^2 = \varphi(v)$, cf. Excercise 2.3.14.

In the following three exercises we briefly show other modifications of the Contraction Principle.

Exercise 2.3.15. If M is a complete metric space, $F\colon M \to M$ and there is a function $V\colon M \to \mathbb{R}^+$ such that

$$V(F(x)) + \varrho(x, F(x)) \le V(x), \qquad x \in M, \tag{2.3.16}$$

then for arbitrary

$$x_0 \in M, \qquad x_n = F(x_{n-1}),$$

the sequence $\{x_n\}_{n=1}^\infty$ is convergent in M to an $\tilde{x}$. Moreover, if the graph of F is closed in $M \times M$, then

$$F(\tilde{x}) = \tilde{x}.$$

Hint. Show that $\{V(x_n)\}_{n=1}^\infty$ is a decreasing sequence; this implies that $\{x_n\}_{n=1}^\infty$ is a Cauchy sequence.

Remark 2.3.16. The condition (2.3.16) is suitable for a vector-valued mapping F and plays an important role in the game theory. For details see, e.g., Aubin & Ekeland [11, Chapter VI].

Exercise 2.3.17. Let M be a complete metric space and let $F\colon M \to M$. If there is $n \in \mathbb{N}$ such that F^n is a contraction, then F has a unique fixed point in M.

Hint. Let $\tilde{x}$ be a fixed point of

$$G := F^n, \qquad \tilde{x} = \lim_{k\to\infty} G^k(x_0).$$

Estimate $\varrho(F(G^k(x_0)), G^k(x_0))$. It is possible to show that

$$\tilde{x} = \lim_{k\to\infty} F^k(x_0).$$

Remark 2.3.18. The power $n \in \mathbb{N}$ need not be the same for all $x, y \in M$, i.e., if there is $q \in [0,1)$ such that for every $x, y \in M$ there exist $n(x), n(y) \in \mathbb{N}$ and

$$\varrho(F^{n(x)}(x), F^{n(y)}(y)) \le q\varrho(x, y),$$

then F also has a unique fixed point (Sehgal [139]). The proof is similar to the previous one.

Exercise 2.3.19 (Edelstein). Let M be a compact metric space and let $F\colon M \to M$ satisfy the condition

$$\varrho(F(x), F(y)) < \varrho(x, y) \qquad \text{for all} \quad x, y \in M,\ x \ne y.$$

Then F has a unique fixed point in M.

Hint. Only existence has to be proved: By compactness there is a convergent subsequence $F^{n_k}(x_0) \to \tilde{x}$. Now show that the sequence

$$\alpha_n := \varrho(F^n(x_0), F^{n+1}(x_0))$$

is decreasing and

$$\lim_{n\to\infty} \alpha_n = \varrho(\tilde{x}, F(\tilde{x})) = \varrho(F(\tilde{x}), F^2(\tilde{x})), \qquad \text{i.e.,} \qquad F(\tilde{x}) = \tilde{x}.$$

Exercise 2.3.20. Let

$$\mathcal{K} = \{x \in C[0,1] : 0 \le x(t) \le 1, x(0) = 0, x(1) = 1\}, \qquad F\colon x(t) \mapsto tx(t).$$

Then $F(\mathcal{K}) \subset \mathcal{K}$, F is non-expansive and there is no fixed point of F in $\mathcal{K}$! Prove these facts and explain their relation to Proposition 2.3.12.

Exercise 2.3.21. Let $x \in L^1(0,1)$ be a solution of

$$x(t) = \lambda \int_0^t k(t,s)x(s)\,\mathrm{d}s$$

where λ and k are as in Example 2.3.7. Prove that $x = 0$ a.e. in $(0,1)$.

Hint. First show that $x \in L^\infty(0,1)$. From the equation we have

$$\|x\|_{L^\infty(0,t)} \le |\lambda| t \|k\|_{L^\infty(\mathcal{M})} \|x\|_{L^\infty(0,t)}, \qquad t \in (0,1).$$

Now deduce $x = 0$ a.e. in $(0,1)$.

Exercise 2.3.22. Prove Corollary 2.1.3 using Theorem 2.3.1.

Exercise 2.3.23. Let $f \in C[0,1]$. Prove that there exists $\varepsilon_0 > 0$ such that for any $\varepsilon \in [0, \varepsilon_0]$ the boundary value problem

$$\begin{cases} \ddot{x}(t) - x(t) + \varepsilon \arctan x(t) = f(t), & t \in (0,1), \\ x(0) = x(1) = 0, \end{cases}$$

has a unique solution $x \in C^2[0,1]$.

Exercise 2.3.24. Let K be a continuous real function on $[a,b] \times [a,b] \times \mathbb{R}$ and assume there exists a constant $N > 0$ such that for any $t, \tau \in [a,b]$, $z_1, z_2 \in \mathbb{R}$, we have

$$|K(t,\tau,z_1) - K(t,\tau,z_2)| \le N|z_1 - z_2|.$$

Let $h \in C[a,b]$ be fixed and let $\lambda \in \mathbb{R}$ be such that

$$|\lambda| < \frac{1}{N(b-a)}.$$

Prove that the integral equation

$$x(t) = \lambda \int_a^b K(t,\tau,x(\tau))\,\mathrm{d}\tau + h(t)$$

has a unique solution $x \in C[a,b]$.

Exercise 2.3.25. Let $\boldsymbol{A}\colon (a,b) \to \mathbb{R}^{M\times M}$ be a continuous matrix-valued function and let $\alpha \in (a,b)$, $\xi \in \mathbb{R}^M$.

(i) Modify the procedure from Example 2.3.7 to prove that the initial value problem

$$\begin{cases} \dot{x}(t) = \boldsymbol{A}(t)x(t), \\ x(\alpha) = \xi, \end{cases}$$

has a unique solution which is defined on (a,b).

(ii) Prove that the equation

$$\dot{x}(t) = \boldsymbol{A}(t)x(t) \tag{2.3.17}$$

has M linearly independent solutions $\varphi^1,\dots,\varphi^M$ on the interval (a,b) and any solution of (2.3.17) is a linear combination of $\varphi^1,\dots,\varphi^M$. The matrix $\boldsymbol{\Phi} = (\varphi_i^j)_{i,j=1,\dots,M}$ is called a fundamental matrix of (2.3.17).

(iii) Let $\boldsymbol{A}$ be continuous on $\mathbb{R}$ and T-periodic ($T>0$). Denote $\boldsymbol{C} = \boldsymbol{\Phi}(T)$ where $\boldsymbol{\Phi}$ is a fundamental matrix, $\boldsymbol{\Phi}(0) = \boldsymbol{I}$. Suppose that $\boldsymbol{B}$ is a solution of the equation

$$\mathrm{e}^{T\boldsymbol{B}} = \boldsymbol{C}$$

(see Exercise 1.1.43). Prove that

$$\boldsymbol{Q}(t) := \boldsymbol{\Phi}(t)\mathrm{e}^{-t\boldsymbol{B}}$$

is regular for all $t \in \mathbb{R}$ and T-periodic. Moreover, x is a solution to (2.3.17) if and only if

$$y(t) := \boldsymbol{Q}^{-1}(t)x(t)$$

is a solution of the equation

$$\dot{y} = \boldsymbol{B}y$$

which has constant coefficients. Find a condition in terms of $\sigma(\boldsymbol{C})$ for the existence of a nontrivial kT-periodic solution to (2.3.17) ($k \in \mathbb{N}$).

(iv) Let $f\colon \mathbb{R} \to \mathbb{R}^M$ be a continuous and T-periodic mapping. Is there any relation between the existence of a nontrivial T-periodic solution to (2.3.17) and the existence of a T-periodic solution to the equation

$$\dot{x}(t) = \boldsymbol{A}(t)x(t) + f(t)?$$

Hint. Use the Variation of Constant Formula and (iii).

Exercise 2.3.26. Which of equations $\dot{x} = \pm x^3$ agrees with Example 2.3.8.

Exercise 2.3.27. If E from Example 2.3.8 is convex on $\mathbb{R}^N$ then the problem (2.3.9) has a unique solution on $[0,T)$. Prove this statement.

Hint. Write the equation which satisfies the difference of possible two solutions, $z(t) = x_1(t) - x_2(t)$. Multiply this equation by z. Use the convexity of E to show that $\|z(\cdot)\|$ is decreasing.

Exercise 2.3.26.

Chapter 3

Abstract Integral and Differential Calculus

3.1 Integration of Vector Functions

This section is devoted to the integration of mappings which take values in a Banach space X. We will consider two types of domains of such mappings: either compact intervals or measurable spaces. For scalar functions the former case leads to the Riemann integral and the latter to the Lebesgue integral with respect to a measure.

Definition 3.1.1. Let $f\colon [a,b] \to X$. Let there exist $x \in X$ with the following property: For every $\varepsilon > 0$ there is $\delta > 0$ such that for all divisions

$$\mathcal{D} := \{a = t_0 < \cdots < t_n = b\}$$

for which $|\mathcal{D}| := \max\limits_{i=1,\dots,n} (t_i - t_{i-1}) < \delta$ and for all choices $\tau_i \in [t_{i-1}, t_i]$, $i = 1, \dots, n$, the inequality

$$\left\| \sum_{i=1}^{n} f(\tau_i)(t_i - t_{i-1}) - x \right\|_X < \varepsilon \tag{3.1.1}$$

is satisfied. Then x is called the *Riemann integral* of f over $[a,b]$ and it is denoted by

$$\int_a^b f(t)\,\mathrm{d}t.$$

A basic existence theorem is a straightforward generalization of the classical (Riemann's) result.

Theorem 3.1.2 (Graves). *Let X be a Banach space and let $f\colon [a,b] \to X$ be continuous. Then the Riemann integral $\int_a^b f(t)\,\mathrm{d}t$ exists.*

Proof. Since f is continuous on the compact interval $[a,b]$, f is uniformly continuous on it. Take an equidistant division $\mathcal{D}_n = \{a = t_0^n < \cdots < t_n^n = b\}$ of the interval $[a,b]$, i.e.,

$$t_i^n = a + \frac{i}{n}(b-a), \quad i = 1, \dots, n, \qquad \text{and} \qquad |\mathcal{D}_n| = \frac{b-a}{n}.$$

Then

$$s_n = \sum_{i=1}^{n} f(t_i^n)(t_i^n - t_{i-1}^n)$$

forms a Cauchy sequence (by the uniform continuity of f). Let $x = \lim_{n\to\infty} s_n$. It is easy to see, again by the uniform continuity of f, that condition (3.1.1) is satisfied whenever $|\mathcal{D}|$ is sufficiently small. □

Since the Riemann integral is linear and the estimate

$$\left\| \int_a^b f(t)\,\mathrm{d}t \right\|_X \le \int_a^b \|f(t)\|_X\,\mathrm{d}t \le \|f\|_{C([a,b],X)}(b-a)\,^{1} \tag{3.1.2}$$

holds for each $f \in C([a,b],X)$, the integral is a linear continuous operator. Its commutativity with linear operators is important.

Proposition 3.1.3. *Let X, Y be Banach spaces and let $f\colon [a,b] \to X$ be Riemann integrable.*

(i) *If $A \in \mathcal{L}(X,Y)$, then Af is also integrable and*

$$A\int_a^b f(t)\,\mathrm{d}t = \int_a^b Af(t)\,\mathrm{d}t \qquad \textit{holds.} \tag{3.1.3}$$

(ii) *If $A\colon X \to Y$ is a linear closed operator and Af is Riemann integrable, then*

$$\int_a^b f(t)\,\mathrm{d}t \in \operatorname{Dom} A$$

and (3.1.3) is true.

Proof. The verification of (i) is straightforward, for (ii) we choose a sequence $\{s_n\}_{n=1}^{\infty}$ of Riemann sums such that

$$\lim_{n\to\infty} s_n = \int_a^b f(t)\,\mathrm{d}t \qquad \text{and} \qquad \lim_{n\to\infty} As_n = \int_a^b Af(t)\,\mathrm{d}t.$$

The statement follows from the definition of a closed operator. □

[1] Here $\|f\|_{C([a,b],X)} = \max_{t\in[a,b]} \|f(t)\|_X$.

With help of this generalization of the Riemann integral we can also prove a basic result on the existence and uniqueness of a solution of a differential equation in a Banach space. Assume that

$$f\colon \mathcal{I} \times \mathcal{G} \to X$$

where $\mathcal{I}$ is an open interval in $\mathbb{R}$, $\mathcal{G}$ is an open subset of a Banach space X. By a *solution of a differential equation*

$$\dot{x} = f(t, x) \tag{3.1.4}$$

we mean a mapping $x\colon \mathcal{J} \to X$ where $\mathcal{J}$ is an open interval, $\mathcal{J} \subset \mathcal{I}$, such that $x(\mathcal{J}) \subset \mathcal{G}$ and for every $t \in \mathcal{J}$ the limit

$$\dot{x}(t) := \lim_{\tau \to 0} \frac{x(t+\tau) - x(t)}{\tau}$$

exists and

$$\dot{x}(t) = f(t, x(t)).$$

Theorem 3.1.4. *Let $\mathcal{I}$ be an open interval in $\mathbb{R}$ and let $\mathcal{G}$ be an open subset of a Banach space X. Assume that $f\colon \mathcal{I} \times \mathcal{G} \to X$ is continuous and locally satisfies the Lipschitz condition with respect to the second variable, i.e., for every $s \in \mathcal{I}$, $y \in \mathcal{G}$ there are $\delta > 0$, $\hat{\delta} > 0$, $L > 0$ such that*

$$\|f(t, x_1) - f(t, x_2)\| \le L\|x_1 - x_2\|$$

whenever $|t - s| < \delta$ and $\|x_i - y\| < \hat{\delta}$, $i = 1, 2$.

Then for each $t_0 \in \mathcal{I}$, $x_0 \in \mathcal{G}$ there exists $h > 0$ such that the equation (3.1.4) *has a unique solution on the interval $\mathcal{J} = (t_0 - h, t_0 + h)$ which satisfies the initial condition*

$$x(t_0) = x_0. \tag{3.1.5}$$

The proof of this theorem is based on the use of the Contraction Principle for the equivalent integral equation (see also the proof of Theorem 2.3.4)

$$x(t) = x_0 + \int_{t_0}^{t} f(s, x(s))\,\mathrm{d}s,^{2} \qquad t \in \mathcal{J}, \tag{3.1.6}$$

where the integral is the Riemann integral. The equivalence of (3.1.4), (3.1.5) and (3.1.6) is established in the following lemma.

Lemma 3.1.5. *Suppose that f is continuous on $\mathcal{I} \times \mathcal{G}$ and $(t_0, x_0) \in \mathcal{I} \times \mathcal{G}$. Then a continuous function $x\colon \mathcal{J} \to \mathcal{G}$ is a solution of* (3.1.4) *on an interval $\mathcal{J} \subset \mathcal{I}$ and satisfies the condition* (3.1.5) *if and only if x solves on $\mathcal{J}$ the integral equation* (3.1.6).

---

[2]Recall that $\int_{t_0}^{t} g(s)\,\mathrm{d}s = -\int_{t}^{t_0} g(s)\,\mathrm{d}s$ for $t < t_0$ (see footnote 28 on page 96).

Proof.

Step 1. Assume first that x is a solution of (3.1.4). Then x as well as the mapping $t \in \mathcal{J} \mapsto f(t, x(t))$ are continuous on $\mathcal{J}$. Choose $\tau \in \mathcal{J}$ and integrate both sides of (3.1.4) over the interval $[t_0, \tau]$ (or $[\tau, t_0]$). Notice that both sides are Riemann integrable (Theorem 3.1.2). Moreover,

$$\varphi\left(\int_{t_0}^{\tau} \dot{x}(t)\,\mathrm{d}t\right) = \int_{t_0}^{\tau} \varphi(\dot{x}(t))\,\mathrm{d}t = \int_{t_0}^{\tau} \frac{\mathrm{d}}{\mathrm{d}t}\varphi(x(t))\,\mathrm{d}t = \varphi(x(\tau)) - \varphi(x_0)$$

for all $\varphi \in X^*$ (the last equality follows from the so-called Basic Theorem of Calculus). By the Hahn–Banach Theorem, in particular Remark 2.1.17(ii), we have

$$\int_{t_0}^{\tau} \dot{x}(t)\,\mathrm{d}t = x(\tau) - x_0,$$

i.e., x satisfies (3.1.6).

Step 2. Suppose now that $x\colon \mathcal{J} \to X$ is a continuous solution of (3.1.6). Then x satisfies (3.1.5) and it remains to check that

$$\frac{\mathrm{d}}{\mathrm{d}t}\int_{t_0}^{t} f(s, x(s))\,\mathrm{d}s = f(t, x(t)).$$

This can be done by copying the proof for the scalar case. □

Proof of Theorem 3.1.4. Choose $\delta > 0$, $\hat{\delta} > 0$ small enough and $K > 0$ such that

$$\|f(s, x_1)\| \le K, \qquad \|f(s, x_1) - f(s, x_2)\| \le L\|x_1 - x_2\|$$

for $s \in [t_0 - \delta, t_0 + \delta]$, $\|x_i - x_0\| \le \hat{\delta}$, $i = 1, 2$. Let $0 < h \le \min\left\{\frac{\hat{\delta}}{K}, \frac{1}{2L}, \delta\right\}$ and

$$M_h = \{x \in C([t_0 - h, t_0 + h], X) : \|x(s) - x_0\| \le \hat{\delta} \text{ for } s \in [t_0 - h, t_0 + h]\}.$$

Then M_h is a complete metric space (with respect to the induced metric) and the operator

$$F(x)\colon t \mapsto x_0 + \int_{t_0}^{t} f(s, x(s))\,\mathrm{d}s, \quad t \in [t_0 - h, t_0 + h],$$

is well defined on M_h, $F(M_h) \subset M_h$ (by (3.1.2)), and

$$\begin{aligned}\|F(x_1) - F(x_2)\| &= \sup_{t \in [t_0 - h, t_0 + h]} \left\|\int_{t_0}^{t} [f(s, x_1(s)) - f(s, x_2(s))]\,\mathrm{d}s\right\| \\ &\le Lh\|x_1 - x_2\| \le \frac{1}{2}\|x_1 - x_2\| \qquad \text{for} \quad x_1, x_2 \in M_h.\end{aligned}$$

By the Contraction Principle (Theorem 2.3.1), there is a unique $x \in M_h$ such that

$$F(x) = x.$$

Using Lemma 3.1.5 we conclude that x is a solution of (3.1.4), (3.1.5) on the interval $\mathcal{J} = (t_0 - h, t_0 + h)$.

Let y be another solution of the same problem on the interval $\mathcal{J}$. Then $y \in M_k$ for a $k \leq h$. Because of the uniqueness in the Contraction Principle, $y(t) = x(t)$ for $t \in [t_0 - k, t_0 + k]$. Taking $(t_0 \pm k, x(t_0 \pm k))$ as new initial conditions we can extend the equality $y(t) = x(t)$ to a larger closed interval $[t_0 - \tilde{k}, t_0 + \tilde{k}]$, i.e., $y \in M_{\tilde{k}}$. This argument shows that

$$y(t) = x(t) \qquad \text{for} \quad t \in \mathcal{J}. \qquad \square$$

Corollary 3.1.6. *Let f satisfy the assumptions of Theorem 3.1.4 where $\mathcal{I} = (\alpha, \infty)$, $\mathcal{G} = X$. If, moreover, f is bounded on $\mathcal{I} \times X$, then for each $t_0 \in \mathcal{I}$ and $x_0 \in X$,* (3.1.4) *has a unique solution satisfying the initial condition* (3.1.5) *which is defined on the whole interval $\mathcal{I}$.*

Proof. The problem (3.1.4), (3.1.5) has a solution x_γ on an interval $(\beta, \gamma) \subset \mathcal{I}$. Such an interval exists due to Theorem 3.1.4 and the solution x_γ is unique on this interval by a similar argument as in the proof of uniqueness. Let

$$\tilde{\gamma} = \sup\{\gamma > \beta : \text{ there is a solution } x_\gamma \text{ on } (\beta, \gamma)\}.$$

If $\gamma_1 < \gamma_2$ and x_{γ_i} is a corresponding solution on (β, γ_i), $i = 1, 2$, then

$$x_{\gamma_1}(t) = x_{\gamma_2}(t) \qquad \text{for} \quad t \in (\beta, \gamma_1)$$

(by uniqueness). This allows us to define the solution $x = x(t)$ on the entire interval $(\beta, \tilde{\gamma})$. Since

$$\|x(t) - x(s)\| \leq \int_s^t \|f(\sigma, x(\sigma))\| \,\mathrm{d}\sigma \leq |t - s| \sup_{(\tau, y) \in \mathcal{I} \times X} \|f(\tau, y)\|$$

for any $\beta < s < t < \tilde{\gamma}$, the solution x is uniformly continuous on $(\beta, \tilde{\gamma})$ and, therefore, continuously extendable at $\tilde{\gamma}$ provided $\tilde{\gamma} < \infty$ (see Proposition 1.2.4). The local Theorem 3.1.4 allows us to continue x as a solution beyond the value of $\tilde{\gamma}$, a contradiction. Hence $\tilde{\gamma} = \infty$. Similarly, we prove $\inf \beta = \alpha$. $\square$

We note that the same result holds provided that f satisfies the Lipschitz condition on $(\alpha, \infty) \times X$. This can be shown similarly as for a system of ordinary differential equations with help of the Gronwall inequality (see Proposition 2.3.9).

Remark 3.1.7. Under the assumptions of Corollary 3.1.6 the solution x depends continuously on the initial data. In order to formulate this result we denote by $x = x(\cdot; t_0, x_0)$ the solution of (3.1.4), (3.1.5) on the interval $\mathcal{I}$. The continuous dependence now reads as follows:

For any compact interval $\mathcal{J} \subset \mathcal{I}$, $t_0 \in \mathcal{J}$, and any $\varepsilon > 0$ there is $\delta > 0$ such that

$$\|x(t; t_0, x_0) - x(t; t_1, x_1)\| < \varepsilon \qquad \textit{for all} \quad t \in \mathcal{J}$$

provided $t_1 \in \mathcal{J}$, $|t_0 - t_1| < \delta$ and $\|x_0 - x_1\| < \delta$.

Cf. Remark 2.3.5 and the use of the Implicit Function Theorem as in Example 4.2.5.

Remark 3.1.8. Another existence theorem for the scalar differential equation (3.1.4) (i.e., $X = \mathbb{R}^N$) is based on the continuity of f only (cf. Proposition 5.2.7).

Warning. *A generalization to an infinite-dimensional space does not hold, e.g.,*

$$f(x) = \left\{ |x_n|^{\frac{1}{2}} + \frac{1}{n+1} \right\}_{n=1}^{\infty} \qquad \textit{for} \quad x = (x_1, x_2, \dots) \in c_0$$

where c_0 is the space of sequences which converge to zero. As a norm on c_0 the "sup norm" is taken. It is not difficult to see that the equation $\dot{x} = f(x)$ has no solution satisfying the initial condition $x(0) = o$! See [44, *Exercise* 10.5.5].

We now turn to the integration of vector functions defined on a measurable space (M, Σ, μ) where Σ is a σ-algebra of subsets of M and μ is a (positive) measure defined on Σ. A generalization of the abstract Lebesgue integral[3] can be done in two different ways: Either by integrating $\varphi \circ f$ over M for all $\varphi \in X^*$ or by approximating f by step functions for which the integral is naturally defined, and then passing to the limit. The former approach leads to a *weak integral* and the latter one to the so-called *Bochner integral*. Since an existence theorem for the weak integral (i.e., the existence of $x \in X$ such that $\varphi(x) = \displaystyle\int_M \varphi(f)\, \mathrm{d}\mu$ for all $\varphi \in X^*$) is complicated, we only briefly describe the less general Bochner integral.

Definition 3.1.9. Let (M, Σ, μ) be a measurable space and let X be a Banach space.

(1) A function $s\colon M \to X$ is called a *step function* if there are pairwise disjoint sets $\mathcal{M}_1, \dots, \mathcal{M}_n$ in Σ with $\mu(\mathcal{M}_k) < \infty$, $k = 1, \dots, n$, such that s is constant (say, equal to x_k) on each $\mathcal{M}_k$ and

$$s(t) = o \qquad \text{for} \quad t \in M \setminus \bigcup_{k=1}^{n} \mathcal{M}_k.$$

The integral of s is defined by

$$\int_M s\, \mathrm{d}\mu = \sum_{k=1}^{n} x_k \mu(\mathcal{M}_k).$$

(2) A function $f\colon M \to X$ is said to be *strongly measurable* if there is a sequence $\{s_m\}_{m=1}^{\infty}$ of step functions such that

$$\lim_{m\to\infty} s_m(t) = f(t) \qquad \text{exists for } \mu\text{-a.a. } t \in M.$$

[3]The reader who is not acquainted with the measure theory and abstract Lebesgue integral can assume that M is an open subset of $\mathbb{R}^N$, μ is a Lebesgue measure and Σ is a collection of all Lebesgue measurable subsets of M.

(3) A strongly measurable function $f\colon M \to X$ is said to be *Bochner integrable* if there is a sequence $\{s_m\}_{m=1}^{\infty}$ of step functions which converges to f μ-a.e. and

$$\lim_{m\to\infty} \int_M \|f - s_m\|_X \,\mathrm{d}\mu = 0. \tag{3.1.7}$$

In this case we put

$$\int_M f \,\mathrm{d}\mu = \lim_{m\to\infty} \int_M s_m \,\mathrm{d}\mu. \tag{3.1.8}$$

Remark 3.1.10. In order to show that this definition is correct we need to prove that the norm of a strongly measurable function is a measurable function (this is obvious) and, therefore, the condition (3.1.7) makes sense. From (3.1.7) it also immediately follows that the limit in (3.1.8) does not depend on any special choice of $\{s_n\}_{n=1}^{\infty}$.

The following statement offers a very useful criterion for Bochner integrability.

Proposition 3.1.11 (Bochner). *Let X be a Banach space and let (M, Σ, μ) be a measurable space. A strongly measurable vector function $f\colon M \to X$ is Bochner integrable if and only if the norm $\|f\|_X$ is Lebesgue integrable. Moreover,*

$$\left\| \int_M f \,\mathrm{d}\mu \right\|_X \le \int_M \|f\|_X \,\mathrm{d}\mu. \tag{3.1.9}$$

Proof.
Step 1. Let f be Bochner integrable and let $\{s_n\}_{n=1}^{\infty}$ be a sequence of step functions from the definition. Then $\left\{ \int_M \|s_n\| \,\mathrm{d}\mu \right\}_{n=1}^{\infty}$ is a Cauchy sequence (by (3.1.7)), in particular, its limit, say $\alpha \in \mathbb{R}$, exists. Then

$$\left\| \int_M f \,\mathrm{d}\mu \right\| = \lim_{n\to\infty} \left\| \int_M s_n \,\mathrm{d}\mu \right\| \le \lim_{n\to\infty} \int_M \|s_n\| \,\mathrm{d}\mu = \alpha.$$

It is easy to see that α does not depend on any special choice of $\{s_n\}_{n=1}^{\infty}$ from the definition and, moreover,

$$\int_M \|f\| \,\mathrm{d}\mu \le \int_M \|f - s_n\| \,\mathrm{d}\mu + \int_M \|s_n\| \,\mathrm{d}\mu, \qquad \text{i.e.,} \qquad \int_M \|f\| \,\mathrm{d}\mu \le \alpha.$$

Step 2. Suppose now that $\|f\|$ is Lebesgue integrable and that $\{s_n\}_{n=1}^{\infty}$ is a sequence of step functions from the definition of strong measurability. Put

$$\sigma_n(t) = \begin{cases} s_n(t) & \text{if } \|s_n(t)\| \le \left(1 + \dfrac{1}{n}\right) \|f(t)\|, \\ o & \text{otherwise.} \end{cases}$$

Then $\sigma_n \to f$ μ-a.e. and, by the Lebesgue Dominated Convergence Theorem,

$$\int_M \|\sigma_n - f\| \,\mathrm{d}\mu \to 0 \qquad \text{since} \qquad \|\sigma_n(t) - f(t)\| \le \left(2 + \frac{1}{n}\right) \|f(t)\|.$$

It follows from the independence of α of the special choice of approximating step functions that $\alpha \le \int_M \|f\| \,\mathrm{d}\mu$. This proves the inequality (3.1.9). □

Proposition 3.1.12. *Let X, Y be Banach spaces, let (M, Σ, μ) be a measurable space and let $f\colon M \to X$ be Bochner integrable.*

(i) *If $A \in \mathcal{L}(X, Y)$, then Af is also Bochner integrable and*

$$\int_M Af \,\mathrm{d}\mu = A \int_M f \,\mathrm{d}\mu. \tag{3.1.10}$$

(ii) *If A is a closed linear operator from X into Y and Af is Bochner integrable, then*

$$\int_M f \,\mathrm{d}\mu \in \operatorname{Dom} A$$

and (3.1.10) *holds.*

Proof. The proof of statement (i) is straightforward. To prove (ii) let

$$Z = \{(x, Ax) : x \in \operatorname{Dom} A\}$$

be equipped with the graph norm

$$\|(x, Ax)\|_Z := \|x\|_X + \|Ax\|_Y.$$

Since A is closed and X, Y are Banach spaces, Z is a Banach space as well. The crucial point of the proof is to show that

$$g(t) := (f(t), Af(t)), \qquad t \in M,$$

is strongly measurable. Achieving this[4] the rest of the proof is easy: By Proposition 3.1.11, g is Bochner integrable and

$$\int_M g \,\mathrm{d}\mu = \left[\int_M f \,\mathrm{d}\mu, \int_M Af \,\mathrm{d}\mu\right] \in Z$$

[4]We sketch the proof of this result: Let $\varphi \in Z^*$. According to the Hahn–Banach Theorem there is an extension $\Phi = (\Phi_1, \Phi_2)$ of φ to $(X \times Y)^*$. Since f and Af are strongly measurable, we conclude that $t \mapsto \varphi(g(t)) = \Phi_1(f(t)) + \Phi_2(Af(t))$ is measurable. It can be also shown that there is $\mathcal{N} \subset M$, $\mu(M \setminus \mathcal{N}) = 0$ such that $g(\mathcal{N})$ is separable. The result now follows from the *Pettis Theorem* (see, e.g., Dunford & Schwartz [53, Chapter III, 6], Yosida [161]):

A function $g\colon M \to Z$ (Banach space) is strongly measurable if and only if the following two conditions are satisfied:

(i) *For every $\varphi \in Z^*$ the function $t \mapsto \varphi(g(t))$ is a measurable function.*

(ii) *There is $\mathcal{N} \subset M$ such that $\mu(M \setminus \mathcal{N}) = 0$ and $g(\mathcal{N})$ is a separable subspace of Z.*

(since g maps into Z, its integral has to belong to Z, too). In particular, $\int_M f \,\mathrm{d}\mu \in \operatorname{Dom} A$ and (3.1.10) holds. □

Remark 3.1.13. If $f\colon M \to X$ is a Bochner integrable function and $\varphi \in X^*$, then, by the previous proposition,

$$\varphi(f)\colon M \to \mathbb{R} \quad (\text{or } \mathbb{C})$$

is integrable (in this case the Bochner and the Lebesgue integrals coincide). This shows that the Bochner integral is a restriction of any notion of a weak integral.

The notion of the Bochner integral and Proposition 3.1.11 allows us to define the Sobolev space also for vector-valued functions. As in the scalar case we start with the notion of vector-valued distribution. Let $\mathcal{I}$ be an open interval, $\mathcal{D}(\mathcal{I})$ the space of all C^∞-real functions with compact support in $\mathcal{I}$ with the standard notion of convergence.[5] The space of all linear continuous operators from $\mathcal{D}(\mathcal{I})$ into X is denoted by $\mathcal{D}^*(\mathcal{I}, X)$ and its elements are vector-valued distributions (with values in X). For example, if $f \in L^1_{\mathrm{loc}}(\mathcal{I}, X)$ then the corresponding distribution $T_f \in \mathcal{D}^*(\mathcal{I}, X)$ is defined by

$$T_f(\varphi) := \int_{\mathcal{I}} \varphi(t) f(t) dt, \quad \varphi \in \mathcal{D}(\mathcal{I}).$$

Any vector-valued distribution $T \in \mathcal{D}^*(\mathcal{I}, X)$ has the derivative DT defined by

$$DT(\varphi) = -T(\varphi'), \quad \varphi \in \mathcal{D}(\mathcal{I}).$$

Using the convergence in $\mathcal{D}(\mathcal{I})$ one can see that $DT \in \mathcal{D}^*(\mathcal{I}, X)$.

Definition 3.1.14. Let $1 \le p \le \infty$, $\mathcal{I}$ an open interval in $\mathbb{R}$ and X be a Banach space. Then the *Sobolev space* $W^{1,p}(\mathcal{I}, X)$ is the space of all functions $f \in L^p(\mathcal{I}, X)$ the distributional derivative Df of which belongs to $L^p(\mathcal{I}, X)$. The norm on $W^{1,p}(\mathcal{I}, X)$ is given by

$$\|f\|_{W^{1,p}(\mathcal{I},X)} = \|f\|_{L^p(\mathcal{I},X)} + \|Df\|_{L^p(\mathcal{I},X)}.$$

Proposition 3.1.15.

(i) *$W^{1,p}(\mathcal{I}, X)$ is a Banach space.*

(ii) *If $f \in W^{1,p}(\mathcal{I}, X)$ then there exists a sequence $\{f_n\}_{n=1}^\infty \subset D(\mathcal{I}, X)$ such that $f_n \to f$ in the norm of $W^{1,p}(\mathcal{I}, X)$ and also in the norm of X for almost all $t \in \mathcal{I}$.*

Proof. Both statements are similar as in the scalar case $X = \mathbb{R}$. Indeed, a sequence $\{f_n\}$ can be constructed by convolution as in Proposition 1.2.20(iv) using mollifiers $\varphi_n \in \mathcal{D}(\mathbb{R})$, extending f as a zero function outside $\mathcal{I}$ and truncating this extension (cf. the proof of Proposition 1.2.21). □

[5] See, e.g., footnote 27 on page 39.

In order to characterize elements of $W^{1,p}(\mathcal{I},X)$, we prove some simple properties of vector-valued distributions.

Lemma 3.1.16. *Let $T \in \mathcal{D}^*(\mathcal{I},X)$. If $DT = 0$ then there exists $x_0 \in X$ such that*

$$T(\varphi) = x_0 \int_{\mathcal{I}} \varphi(t)dt \quad \text{for any} \quad \varphi \in \mathcal{D}(\mathcal{I}).$$

Proof. Choose some $\varphi_0 \in \mathcal{D}(\mathcal{I})$ with $\int_{\mathcal{I}} \varphi_0(t)dt = 1$ and put $x_0 = T(\varphi_0)$. For $\varphi \in \mathcal{D}(\mathcal{I})$ take $t_0 \in \mathcal{I}$ to the left from the supports of both φ, φ_0 and consider

$$\psi(t) = \int_{t_0}^{t} \left[\varphi(s) - \varphi_0(s) \int_{\mathcal{I}} \varphi(\sigma)d\sigma\right] ds, \quad t \in \mathcal{I}.$$

By the choice of t_0, ψ has also a compact support in $\mathcal{I}$, i.e., $\psi \in \mathcal{D}(\mathcal{I})$. Since

$$0 = DT(\psi) = -T(\psi') = -T(\varphi) + x_0 \int_{\mathcal{I}} \varphi(\sigma)d\sigma,$$

the result follows. □

Lemma 3.1.17. *Let $g \in L^1(\mathcal{I},X)$ and let $t_0 \in \mathcal{I}$. Then*

$$h(t) = \int_{t_0}^{t} g(s)ds, \quad t \in \mathcal{I}$$

is bounded and uniformly continuous function on $\mathcal{I}$, $h' = g$ a.e. in $\mathcal{I}$ and $Dh = g$.

Proof. For $t, s \in \mathcal{I}$, $s < t$, we have

$$\|h(t) - h(s)\|_X = \|\int_s^t g(\sigma)d\sigma\|_X \le \int_s^t \|g(\sigma)\|_X d\sigma$$

and the uniform continuity of h follows from the standard properties of scalar-valued Lebesgue integral.

Moreover, we get

$$\|h\|_{C(\mathcal{I},\mathbb{R})} = \sup_{t\in\mathcal{I}} \|h(t)\|_X \le \|g\|_{L^1(\mathcal{I},X)}.$$

For $s \neq 0$, t, $t+s \in \mathcal{I}$, we also have

$$\left\|\frac{h(t+s) - h(t)}{s} - g(t)\right\|_X = \left\|\frac{1}{s}\int_t^{t+s} (g(\sigma) - g(t))\, d\sigma\right\|_X \le \left|\frac{1}{s}\int_t^{t+s} \|g(\sigma) - g(t)\|_X d\sigma\right|. \tag{3.1.11}$$

Since the function $\sigma \mapsto \|g(\sigma) - g(t)\|_X$ is locally integrable in $\mathcal{I}$ and almost every point in $\mathcal{I}$ is its Lebesgue point (i.e., we have that $\frac{h(t+s)-h(t)}{s} \to g(t)$ for $s \to 0$ and a.a. $t \in \mathcal{I}$), we obtain from (3.1.11) that $h' = g$ a.e. in $\mathcal{I}$.

By the definition of distributional derivative,

$$DT_h(\varphi) = -T_h(\varphi') = -\int_{\mathcal{I}} \varphi'(t)h(t)dt, \quad \varphi \in \mathcal{D}(\mathcal{I}),$$

and, integrating by parts (see Exercise 3.1.24 (ii)).

$$DT_h(\varphi) = \int_{\mathcal{I}} \varphi(t)h'(t)dt = \int_{\mathcal{I}} \varphi(t)g(t)dt = T_g(\varphi),$$

i.e., $Dh = g$. □

We have the following characterization of vector-valued functions from the Sobolev space.

Proposition 3.1.18. *For $f \in L^p(\mathcal{I}, X)$ the following properties are equivalent:*

(i) $f \in W^{1,p}(\mathcal{I}, X)$.

(ii) *For any $t_0 \in \mathcal{I}$ there is $x_0 \in X$ and $g \in L^p(\mathcal{I}, X)$ such that*

$$f(t) = x_0 + \int_{t_0}^{t} g(s)ds \quad \text{for} \quad \text{a.a. } t \in \mathcal{I}. \tag{3.1.12}$$

(iii) *After changing f on a set of zero measure, f is bounded and uniformly continuous on $\mathcal{I}$, and $f' \in L^p(\mathcal{I}, X)$. Moreover,*

$$f(t) = f(t_0) + \int_{t_0}^{t} f'(s)ds \quad \text{for all} \quad t_0, t \in \mathcal{I}. \tag{3.1.13}$$

Proof. (i)⇒(ii) Choose $t_0 \in \mathcal{I}$ and set

$$k(t) := f(t) - \int_{t_0}^{t} Df(s)ds.$$

By Lemma 3.1.17, $Dk = 0$ and there is $x_0 \in X$ such that

$$\int_{\mathcal{I}} k(t)\varphi(t)dt = x_0 \int_{\mathcal{I}} \varphi(t)dt$$

for any $\varphi \in \mathcal{D}(\mathcal{I})$, see Lemma 3.1.16. Since φ is arbitrary, we get

$$f(t) - x_0 - \int_{t_0}^{t} Df(s)ds = 0$$

for a.a. $t \in \mathcal{I}$.

(ii)$\Rightarrow$(iii) Since $h(t) = \int_{t_0}^{t} g(s)ds$ is bounded and uniformly continuous on $\mathcal{I}$ (Lemma 3.1.17), we can change f such that (3.1.12) holds everywhere in $\mathcal{I}$. Lemma 3.1.17 shows that $Df = f' = g$.

(iii)$\Rightarrow$(iv) This implication is obvious since from (3.1.13) we get that $Df = f'$. □

We now return to the functional calculus given for matrices (see Theorem 1.1.38). Let $B \in \mathcal{L}(X)$ and let $\mathcal{H}(\sigma(B))$ be a collection of holomorphic functions on a neighborhood of $\sigma(B)$ (this neighborhood can depend on a function). If $f \in \mathcal{H}(\sigma(B))$, then there exists a positively oriented Jordan curve γ such that $\sigma(B) \subset \operatorname{int}\gamma$ and f is holomorphic on a neighborhood of $\overline{\operatorname{int}\gamma}$. Hence the following integral

$$f(B)x := \frac{1}{2\pi \mathrm{i}} \int_\gamma f(w)(wI - B)^{-1}x \, \mathrm{d}w, \qquad x \in X,$$

exists. Its properties are collected in the following assertions.

Proposition 3.1.19 (Dunford Functional Calculus). *Let X be a complex Banach space and let $B \in \mathcal{L}(X)$. There exists a unique linear mapping $\Phi : \mathcal{H}(\sigma(B)) \to \mathcal{L}(X)$ with the following properties:*

(i) $\Phi(fg) = \Phi(f)\Phi(g) = \Phi(g)\Phi(f)$ *for* $f, g \in \mathcal{H}(\sigma(B))$*;*

(ii) *if* $P(w) = \sum_{j=0}^{n} a_j w^j$*, then* $P(B) = \sum_{j=0}^{n} a_j B^j$*;*

(iii) *if* $f(w) = \frac{1}{\lambda - w}$ *for* $w \neq \lambda$ *and* $\lambda \notin \sigma(B)$*, then* $f(B) = (\lambda I - B)^{-1}$*;*

(iv) *if* $\{f_n\}_{n=1}^{\infty} \subset \mathcal{H}(\sigma(B))$*,* $f_n \rightrightarrows f$ *on a neighborhood of* $\overline{\operatorname{int}\gamma}$*, then we have* $f_n(B) \to f(B)$ *in the norm topology;*

(v) *if* $f \in \mathcal{H}(\sigma(B))$*, then* $\sigma(f(B)) = f(\sigma(B))$*.*

Proof. The proof can be found, e.g., in Dunford & Schwartz [53, Section VII.3]. □

Suppose that $\lambda_0 \in \sigma(B)$ is an isolated point of the spectrum of B, i.e., there exist disjoint neighborhoods $\mathcal{U}_0$ of λ_0 and $\mathcal{U}$ of $\sigma(B) \setminus \{\lambda_0\}$. The function

$$f(w) = \begin{cases} 1 & \text{for} \quad w \in \mathcal{U}_0, \\ 0 & \text{for} \quad w \in \mathcal{U}, \end{cases}$$

belongs to the collection $\mathcal{H}(\sigma(B))$ and the operator $P_0 := f(B)$ is a projection of X onto the subspace $X_0 := P_0(X)$ since $P_0^2 = P_0$. The operator $P_1 := I - P_0$ is a projection onto the complementary subspace X_1, i.e., $X = X_0 \oplus X_1$. Denote by B_0 and B_1 the restrictions of B onto X_0 (i.e., $B_0 \in \mathcal{L}(X_0)$) and X_1, respectively. Proposition 3.1.19(v) implies that

$$\sigma(B_0) = \{\lambda_0\}, \qquad \sigma(B_1) = \sigma(B) \setminus \{\lambda_0\}.$$

Put

$$\gamma_0 := \{\lambda_0 + r\mathrm{e}^{\mathrm{i}\varphi} : \varphi \in [0, 2\pi]\} \qquad \text{for a (small) positive } r.$$

Using Proposition 3.1.19 we get (see Exercise 3.1.27)

$$\begin{aligned}(\mu I_0 - B_0)^{-1} x_0 &= \frac{1}{2\pi\mathrm{i}} \int_{\gamma_0} \frac{(\lambda I_0 - B_0)^{-1} x_0}{\mu - \lambda} \,\mathrm{d}\lambda \\ &= \frac{1}{2\pi\mathrm{i}} \int_{\gamma_0} \frac{1}{\mu - \lambda_0} \sum_{n=0}^{\infty} \left(\frac{\lambda - \lambda_0}{\mu - \lambda_0} \right)^n (\lambda I_0 - B_0)^{-1} x_0 \,\mathrm{d}\lambda \\ &= \frac{x_0}{\mu - \lambda_0} + \sum_{n=1}^{\infty} \frac{(-1)^n}{(\mu - \lambda_0)^{n+1}} (\lambda_0 I_0 - B_0)^n x_0 \end{aligned} \tag{3.1.14}$$

for $x_0 \in X_0$, $|\mu - \lambda_0| > r$. The Taylor series for the function $\mu \mapsto (\mu I_1 - B_1)^{-1} x_1$ has the form (see Exercise 3.1.27)

$$\begin{aligned}(\mu I_1 - B_1)^{-1} x_1 &= \sum_{n=0}^{\infty} \frac{(\mu - \lambda_0)^n}{n!} \frac{\mathrm{d}^n}{\mathrm{d}\lambda^n} (\lambda I_1 - B_1)^{-1} x_1 \Big|_{\lambda = \lambda_0} \\ &= \sum_{n=0}^{\infty} (-1)^n (\mu - \lambda_0)^n (\lambda_0 I_1 - B_1)^{-(n+1)} x_1, \end{aligned} \tag{3.1.15}$$

$x_1 \in X_1$, $|\mu - \lambda_0| < r_0 := \|(\lambda_0 I_1 - B_1)^{-1}\|^{-1}$.

Proposition 3.1.20. *If λ_0 is an isolated point of the spectrum $\sigma(B)$, $B \in \mathcal{L}(X)$, then there exist operators $A_n \in \mathcal{L}(X)$, $n \in \mathbb{Z}$, and $r > 0$ such that*

$$(\mu I - B)^{-1} x = \sum_{n=-\infty}^{+\infty} (\mu - \lambda_0)^n A_n x \tag{3.1.16}$$

for all $x \in X$ and $0 < |\mu - \lambda_0| < r$. Moreover, if $k \in \mathbb{N}$ is such that

$$A_{-n} = O \qquad \text{for every} \quad n > k \qquad \text{and} \qquad z := A_{-k} x \neq o,$$

then

$$Bz = \lambda_0 z.$$

On the other hand, if λ_0 is a nonzero eigenvalue of a compact operator B, then λ_0 is a pole of the resolvent of B, i.e., there is $k \in \mathbb{N}$ such that

$$A_{-n} = O \qquad \text{for all} \quad n > k.$$

Proof. Let λ_0 be an isolated point of $\sigma(B)$ and $B \in \mathcal{L}(X)$. If P_0, P_1 are the above projections onto X_0, X_1, then

$$(\mu I - B)^{-1}x = (\mu I - B_0)^{-1}x_0 + (\mu I - B_1)^{-1}x_1, \qquad P_0 x = x_0, \quad P_1 x = x_1,$$

and (3.1.16) follows from (3.1.14) and (3.1.15). Since

$$A_{-(k+1)}x = (B - \lambda_0 I)A_{-k}x,$$

the second statement holds as well.

Suppose now that B is compact and $\lambda_0 \neq 0$ is an eigenvalue of B. By Corollary 2.2.13, λ_0 is an isolated point of $\sigma(B)$. Since the restriction B_0 of B onto the subspace X_0 has the spectrum $\sigma(B_0)$ consisting of λ_0 only, B_0^{-1} exists and is continuous. Therefore, the unit ball $B(x_0;1) = B_0(B_0^{-1}(B(x_0;1)))$ is a compact set. Proposition 1.2.15 says that $M = \dim X_0$ is finite. It follows from Lemma 1.1.31 that

$$X_0 = \operatorname{Ker}(\lambda_0 I_0 - B_0)^k \qquad \text{for a certain} \quad k \in \mathbb{N}$$

and (see (1.1.20))

$$(\mu I_0 - B_0)^{-1}x_0 = \sum_{n=0}^{k-1} \frac{(-1)^n}{(\mu - \lambda_0)^{n+1}}(\lambda_0 I_0 - B_0)^n x_0 = \sum_{n=1}^{k} \frac{A_{-n}x}{(\mu - \lambda_0)^n},$$

where $P_0 x = x_0$. The proof is complete. □

Exercise 3.1.21. Let X be a Banach space. Assume that $f\colon [a,b] \to X$ has the Riemann integral over the interval $[a,b]$. Show that then the Bochner integral $\int_a^b f(t)\,dt$ also exists and the two integrals are equal. In particular, Proposition 3.1.3 is a special case of Proposition 3.1.12. However, the proof of Proposition 3.1.3(ii) is much simpler.

Exercise 3.1.22. Prove that $W^{1,2}(0,T;X)$ is continuously embedded into

$$C([0,T];X) \text{ for } T < \infty.$$

Hint. The set inclusion $W^{1,2} \subset C$ follows from Proposition 3.1.18. To prove that the map $\kappa : W^{1,2} \to C$ is continuous, it is enough to prove that κ is closed (use the Closed Graph Theorem).

Exercise 3.1.23.

(i) Prove the following form of the Fubini Theorem:
Let $M = \{(t,s) \in \mathbb{R}^2 : 0 \le s \le t \le a\}$ and let $f \in L^1(M;X)$ where X is a Banach space.
Then

$$\int_0^a \left[\int_0^t f(t,s)ds\right] dt = \int_0^a \left[\int_s^a f(t,s)dt\right] ds = \int_M f(t,s)dtds.$$

Hint. Apply $\varphi \in X^*$ to $\int\limits_M f$ and use the classical Fubini theorem.

(ii) Let $T(t)$, $t \in [0,a]$ be a family of continuous linear operators on a Banach space X such that the mapping $t \mapsto T(t)x$ is continuous on $[0,a]$ for every $x \in X$. Let $g \in L^1(0,a;X)$. Prove that

$$(t,s) \mapsto T(t)g(s) \in L^1(M;X)$$

where M is as in (i) and, moreover,

$$\int_0^a T(t)\left[\int_0^t g(s)ds\right]dt = \int_0^a \left[\int_s^a T(t)g(s)dt\right]ds.$$

Hint. To prove that $T(t)g(s) \in L^1(M;X)$ show that this function is measurable and then, by the Uniform Boundedness Principle (Theorem 2.1.4), $\sup\limits_{t\in[0,a]} \|T(t)\| < \infty$. Follow now carefully the hint to (i). Notice that it is not possible to write $\int\limits_s^a T(t)g(s)dt = \left[\int\limits_s^a T(t)dt\right] g(s)$, since, in general, $T \notin L^1(0,a;\mathcal{L}(X))$ (the map $t \mapsto T(t)$ need not be measurable as a function into $\mathcal{L}(X)$).

Exercise 3.1.24. Let $f \in W^{1,2}(a,b)$ and $g \in W^{1,2}(a,b;X)$.

(i) Prove that $fg \in W^{1,2}(a,b;X)$ and

$$(fg)^{\cdot}(t) = \dot{f}(t)g(t) + f(t)\dot{g}(t) \quad \text{a.e. in } (a,b).$$

(ii) Prove that for a bounded interval (a,b) the following integration by parts formula holds:

$$\int_a^b \dot{f}(t)g(t)dt = f(b)g(b) - f(a)g(a) - \int_b^a f(t)\dot{g}(t)dt.$$

Exercise 3.1.25. Let X be a Banach space and let $f\colon (a,b)\times(a,b) \to X$ satisfy the following conditions:

(i) $f(t,\cdot) \in L^1(a,b;X)$ for all $t \in (a,b)$.

(ii) For a.a. $s \in (a,b)$ there exists the partial derivative $\frac{\partial f}{\partial t}(\cdot,s)$ and it is measurable in (a,b).

(iii) There is $g \in L^1(a,b)$ such that $\|\frac{\partial f}{\partial t}(t,s)\|_X \le g(s)$ for a.a. $t \in (a,b)$.

Denote $g(t) = \int\limits_a^t f(t,s)ds$ and prove that

$$\dot{g}(t) = \int_a^t \frac{\partial f}{\partial t}(t,s)ds + f(t,t) \quad \text{for a.a.}\, t \in (a,b).$$

Hint. Use Lemma 3.1.17.

Exercise 3.1.26. Compare the functional calculus from Proposition 3.1.19 with that of Remark 2.2.18. More precisely, show that for a compact, self-adjoint operator B the functional calculus given in Remark 2.2.18 is an extension of that of Proposition 3.1.19.

Exercise 3.1.27. Give details to confirm the formulae of resolvent (3.1.14) and (3.1.15).

Hint. For (3.1.14) replace the sum and the integral and use Proposition 3.1.19(ii). For (3.1.15) use the resolvent identity

$$(\lambda I - B)^{-1} - (\mu I - B)^{-1} = (\mu - \lambda)(\lambda I - B)^{-1}(\mu I - B)^{-1}$$

and induction.

Exercise 3.1.28. Let $A\colon \operatorname{Dom} A \subset H \to H$ be a densely defined linear operator on a Hilbert space H. Assume that A has a compact resolvent that is also self-adjoint.

(i) Extend the functional calculus (Remark 2.2.18 and Proposition 3.1.19) to such A. In particular, show that the formula for $\Phi(f)x$ still holds provided that

$$\sum_{n=1}^{\infty} |f(\lambda_n)|^2 |(x, e_n)|^2 < \infty$$

(here $\sigma(A) = \{\lambda_1, \dots\}$). Notice that $\Phi(f)$ need not be bounded if $\dim H = \infty$ since $\sigma(A)$ is unbounded in this case. Also, $\Phi(f)$, $\Phi(g)$ do not commute in general.

Hint. If $\dim H = \infty$ and the spectrum $\sigma\left((\lambda_0 I - A)^{-1}\right) = \{\mu_n\}_{n=1}^{\infty} \subset \mathbb{R}$ $(\lambda_0 \in \mathbb{R})$ then $\lim \mu_n = 0$. Note that A and $(\lambda_0 I - A)^{-1}$ have the same eigenvectors $\{e_n\}_{n=1}^{\infty}$ which form an orthonormal basis of H. Moreover, $\sigma(A) = \{\lambda_n := \lambda_0 - \frac{1}{\mu_n}\}_{n=1}^{\infty}$. If $\Phi(f) \in \mathcal{L}(H)$ and $\Phi(g)$ is only a closed operator then it can happen that $\operatorname{Im}(\Phi(f)) \cap \operatorname{Dom}(\Phi(g)) = \{o\}$. This shows that $\Phi(f), \Phi(g)$ need not commute.

(ii) From now till the end of this exercise we suppose that the spectrum $\sigma(A)$ is bounded above. Show that for the functions $f_t(\lambda) = e^{t\lambda}$, $t \geq 0$, the operators $\Phi(f_t)x := e^{tA}x$ are defined on H and have the following properties:

(1) $e^{0A} = I$ and $e^{tA} \in \mathcal{L}(H)$ for all $t \geq 0$.

(2) For all $x \in H$ the vector function $t \mapsto e^{tA}x$ is continuous on the interval $[0, +\infty)$.

(3) $e^{(t+s)A} = e^{tA}e^{sA}$ for all $t, s \geq 0$.

(4) e^{tA} maps $\operatorname{Dom} A$ into $\operatorname{Dom} A$ for all $t \geq 0$.

A system of operators $\{e^{tA}\}_{t \geq 0}$ which satisfies properties (1) – (3) is called a C_0-*semigroup*. For more information on C_0-*semigroup* see Section 8.1.

Hint. By the definition we have

$$e^{tA}x = \sum_{n=1}^{\infty} e^{\lambda_n t}(x, e_n)e_n$$

provided this series is convergent in H (λ_n and e_n are as in (i)). Use Exercise 1.2.43 to prove the convergence. For the continuity (see (2)) estimate $\|e^{tA}x - e^{sA}x\|^2$. Notice that you can get a good estimate of $\|e^{tA} - e^{sA}\|$ only if $A \in \mathcal{L}(H)$.

(iii) Show that the function

$$u(t) := \mathrm{e}^{tA}x_0 \qquad \text{with} \quad x_0 \in \operatorname{Dom} A$$

is a classical solution to the initial value problem

$$\begin{cases} \dot{x}(t) = Ax(t), \\ x(0) = x_0. \end{cases} \tag{3.1.17}$$

(iv) Prove that

$$\int_0^t \mathrm{e}^{sA}x \,\mathrm{d}s \in \operatorname{Dom} A \qquad \text{and} \qquad A\int_0^t \mathrm{e}^{sA}x \,\mathrm{d}s = \mathrm{e}^{tA}x - x \quad \text{for } x \in H.$$

In other words, $\mathrm{e}^{tA}x$ is a continuous solution of the integral form of (3.1.17).

(v) Prove that

$$\int_0^\infty \mathrm{e}^{-\lambda t}\mathrm{e}^{tA}x \,\mathrm{d}t = (\lambda - A)^{-1}x \qquad \text{for all} \quad x \in H$$

and sufficiently large $\operatorname{Re}\lambda$ (actually for $\operatorname{Re}\lambda > \sup\{(Ax, x) : x \in \operatorname{Dom} A, \|x\| = 1\}$).

Hint. Estimate $\|\mathrm{e}^{tA}x_0\|$ to prove the existence of the (Bochner) integral for sufficiently large $\operatorname{Re}\lambda$. Let $B(\lambda)$ denote the integral. Use Proposition 3.1.12 (ii) to show that $B(\lambda)\ (\lambda I - A)x = (\lambda I - A)B(\lambda)x$ for all $x \in \operatorname{Dom} A$.

(vi) Let $g\colon [0,\infty) \to H$ be a continuous mapping and let $u\colon [0,\infty) \to H$ be a classical solution to the initial value problem

$$\begin{cases} \dot{x}(t) = Ax(t) + g(t), \\ x(0) = x_0. \end{cases}$$

Show that

$$u(t) = \mathrm{e}^{tA}x_0 + \int_0^t \mathrm{e}^{(t-s)A}g(s)\,\mathrm{d}s \qquad \text{for} \quad t \geq 0.$$

This formula is called the *variation of constants formula*.

Hint. Let u be a solution on an interval $[0,t]$ and let $v(s) = e^{(t-s)A}u(s)$, $s \in [0,t]$. Prove that $\dot{v}(s) = e^{(t-s)A}g(s)$ and integrate from $s = 0$ to $s = t$.

(vii) Find conditions on a continuous mapping $h\colon H \to H$ such that the existence of a continuous solution to the integral equation

$$x(t) = \mathrm{e}^{tA}x_0 + \int_0^t \mathrm{e}^{(t-s)A}h(x(s))\,\mathrm{d}s$$

follows from the Contraction Principle. Such a solution is called a *mild solution* of the problem

$$\begin{cases} \dot{x}(t) = Ax(t) + h(x(t)), \\ x(0) = x_0. \end{cases} \tag{3.1.18}$$

If $-A$ is as in Exercise 2.2.19(i), (ii), (iv), or, more generally, $A = \Delta$ with suitable boundary conditions, then (3.1.18) is a semilinear partial differential equation of parabolic type.

Hint. Cf. the proof of Theorem 2.3.4.

Exercise 3.1.29. Let B be a compact, self-adjoint operator on a Hilbert space H and let $\lambda_0 \in \sigma(B) \setminus \{0\}$. Compute A_n in the expression (3.1.16).

3.2 Differential Calculus in Normed Linear Spaces

We suppose that the reader is acquainted with partial derivatives and the differential of functions of two real variables. Our goal in this section is to extend these notions to mappings between normed linear spaces. Many infinite-dimensional spaces vary from $\mathbb{R}^N$ by the lack of any natural basis. In particular this means that there is no way of generalizing partial derivatives. We define a directional derivative instead.

Definition 3.2.1. Let X, Y be normed linear spaces (both over the same scalar field) and let $f\colon X \to Y$. If for $a, h \in X$ the limit (in the norm of Y)

$$\lim_{\substack{t\to 0\\ t\in\mathbb{R}}} \frac{f(a+th) - f(a)}{t}$$

exists, then its value is called the *derivative of f at the point a and in the direction h* (or *directional derivative* or *Gâteaux variation*) and is denoted by $\delta f(a;h)$.

If $\delta f(a;h)$ exists for all $h \in X$ and the mapping $Df(a)\colon h \mapsto \delta f(a;h)$ is linear and continuous, then $Df(a)$ is called the *Gâteaux derivative* of f at the point a.[6]

[6]The terminology concerning Gâteaux differentiability is not fixed. Some authors do not assume linearity of $Df(a)$.

Remark 3.2.2. Simple examples of functions of two variables show that the directional derivative need not be linear in h and not even the existence of $Df(a)$ guarantees the continuity of f at the point a.

Example 3.2.3. Consider the standard bases $e_1^M, \dots, e_M^M$ and $e_1^N, \dots, e_N^N$ of $\mathbb{R}^M$ and $\mathbb{R}^N$, respectively. Then we can write $f\colon \mathbb{R}^M \to \mathbb{R}^N$ in the form

$$f(x) = \sum_{i=1}^{N} f^i(x) e_i^N \qquad \text{(or briefly } f = (f^1, \dots, f^N)\text{)}.$$

It is easy to see that $\delta f(a; h)$ exists if and only if $\delta f^i(a; h)$ exists for all $i = 1, \dots, N$. In particular, for $h = e_j^M$, the directional derivative $\delta f^i(a; h)$ is nothing else than $\frac{\partial f^i}{\partial x_j}(a)$. This means that the Gâteaux derivative $Df(a)$ has the matrix representation with respect to the standard bases in the form

$$\begin{pmatrix} \dfrac{\partial f^1}{\partial x_1}(a) & \cdots & \dfrac{\partial f^1}{\partial x_M}(a) \\ \vdots & \ddots & \vdots \\ \dfrac{\partial f^N}{\partial x_1}(a) & \cdots & \dfrac{\partial f^N}{\partial x_M}(a) \end{pmatrix}.$$

This matrix is called the *Jacobi matrix* of f at the point a. If $M = N$, then its determinant is denoted by

$$\frac{\partial(f^1, \dots, f^M)}{\partial(x_1, \dots, x_M)} = J_f$$

and is called the *Jacobian* of f at a. ○

Example 3.2.4. Suppose that H is a Hilbert space and $f\colon H \to \mathbb{R}$ (or $\mathbb{C}$) has a Gâteaux derivative $Df(a)$ at $a \in H$. Then, by the Riesz Representation Theorem, there exists a unique point $\nabla f(a) \in H$ such that

$$Df(a)h = (h, \nabla f(a))_H.$$

The element $\nabla f(a)$ is called the *gradient* of f at a. Notice that the gradient ∇f is a mapping from H into itself. ○

Remark 3.2.5. One of the most important applications of the notion of derivative is in extremal problems of classical analysis. The well-known theorem (due to Fermat) asserts that the derivative is zero at an extremal point provided this derivative exists. The same result obviously holds for $f\colon X \to \mathbb{R}$ also in an infinite-dimensional space X.[7]

[7] A simple reason for this observation comes from the fact that the directional derivative $\delta f(a; h)$ describes the behavior of the functional f along the straight line $\{a + th : t \in \mathbb{R}\}$, i.e., the behavior of the real function $t \mapsto f(a + th)$ near zero.

The previous remark indicates the use of the notion of derivative for solving the equation

$$F(x) = o \qquad \text{for} \quad F\colon H \to H.$$

Namely, suppose that there is a functional $f\colon H \to \mathbb{R}$ such that

$$\nabla f = F.$$

Then it is sufficient to show that f has a local maximum or minimum. However, it is a very nontrivial problem to find such f (the so-called *potential* of F) or to find conditions to ensure its existence. See Chapter 7 for more details. A discussion of the finite-dimensional case ($H = \mathbb{R}^M$) is given in Appendix 4.3B (Remark 4.3.66 and Theorem 4.3.68).

We postpone examples since various properties of the derivative will be needed to introduce them.

Theorem 3.2.6 (Mean Value Theorem). *Let X be a normed linear space and Y a Banach space. Let $f\colon X \to Y$ have the directional derivative at all points of the segment joining points $a, b \in X$ in the direction of this segment, i.e., $\delta f(a + t(b - a); b - a)$ exists for all $t \in [0, 1]$. If the mapping $t \mapsto \delta f(a + t(b - a); b - a)$ is continuous on $[0, 1]$, then*

$$f(b) - f(a) = \int_0^1 \delta f(a + t(b - a); b - a)\,\mathrm{d}t. \tag{3.2.1}$$

Proof. Take a $\varphi \in Y^*$ and denote

$$g(t) = \varphi(f(a + t(b - a))), \qquad t \in [0, 1].$$

By the definition of the directional derivative, we have

$$g'(t) = \varphi[\delta f(a + t(b - a); b - a)]$$

and g' is continuous on $[0, 1]$. It follows from the Basic Theorem of Calculus that

$$\int_0^1 \varphi[\delta f(a + t(b - a); b - a)]\,\mathrm{d}t = g(1) - g(0) = \varphi(f(b)) - \varphi(f(a)).$$

The Riemann integral $\displaystyle\int_0^1 \delta f(a + t(b - a); b - a)\,\mathrm{d}t$ exists (see Theorem 3.1.2) and, by Proposition 3.1.12(i), we get

$$\int_0^1 \varphi[\delta f(a + t(b - a); b - a)]\,\mathrm{d}t = \varphi\left[\int_0^1 \delta f(a + t(b - a); b - a)\,\mathrm{d}t\right].$$

Since $\varphi \in Y^*$ has been chosen arbitrary, the Hahn–Banach Theorem (in particular, Remark 2.1.17(ii)) implies the equality (3.2.1). $\square$

The following result offers another possible formulation.

Theorem 3.2.7 (Mean Value Theorem). *Let X, Y be normed linear spaces and let $f\colon X \to Y$. If for given $a, b \in X$ the directional derivative $\delta f(a + t(b-a); b-a)$ exists for all $t \in [0,1]$, then*

$$\|f(b) - f(a)\|_Y \le \sup_{t\in[0,1]} \|\delta f(a + t(b-a); b-a)\|_Y \tag{3.2.2}$$

and

$$\|f(b) - f(a) - \delta f(a; b-a)\|_Y \le \sup_{t\in[0,1]} \|\delta f(a + t(b-a); b-a) - \delta f(a; b-a)\|_Y. \tag{3.2.3}$$

Moreover, if $Df(a + t(b-a))$ exists for all $t \in [0,1]$, then

$$\|f(b) - f(a)\|_Y \le \sup_{t\in[0,1]} \|Df(a + t(b-a))\|_{\mathcal{L}(X,Y)} \|b-a\|_X. \tag{3.2.4}$$

Proof. An idea similar to the previous proof is used. By the dual characterization of the norm (Corollary 2.1.16) there is $\varphi \in X^*$, $\|\varphi\| = 1$, such that

$$\|f(b) - f(a)\| = \varphi(f(b) - f(a)).$$

Define now

$$g(t) = \varphi(f(a + t(b-a))), \qquad t \in [0,1].$$

Then

$$g'(t) = \varphi(\delta f(a + t(b-a); b-a))$$

and, therefore, the function g satisfies all assumptions of the classical Mean Value Theorem. Consequently, if X is a real space, we get

$$\begin{aligned}\|f(b) - f(a)\| &= g(1) - g(0) = g'(\vartheta) = \varphi(\delta f(a + \vartheta(b-a); b-a)) \\ &\le \|\delta f(a + \vartheta(b-a); b-a)\| \qquad \text{for a} \quad \vartheta \in (0,1).\end{aligned}$$

If X is a complex space, we consider $\operatorname{Re} g$ and obtain

$$\|f(b) - f(a)\| \le \sup_{\vartheta\in(0,1)} |g'(\vartheta)|$$

(see the next remark) and the assertion also follows. The proof of (3.2.3) is similar and (3.2.4) is an easy consequence of (3.2.2). $\square$

Remark 3.2.8. The Mean Value Theorem for functions from $\mathbb{R} \to \mathbb{R}$ is often stated in the following form:

There is $\vartheta \in (0,1)$ such that

$$f(b) - f(a) = f'(a + \vartheta(b-a))(b-a)$$

provided f is continuous on the interval $[a,b]$ and $f'(x)$ exists for every $x \in (a,b)$.

Warning. *This equality does not hold even for* $f\colon \mathbb{R} \to \mathbb{C}$ ($\sim \mathbb{R}^2$) *(e.g.,* $f(x) = \mathrm{e}^{\mathrm{i}x}$*,* $a = 0$*,* $b = 2\pi$*)!*

Example 3.2.9. Differentiability of the norm is connected with the properties of the corresponding space (see, e.g., Fabian et al. [60, Chapter 5]). As a simple example we will show the relation between the uniqueness of the supporting hyperplane at a given point $a \in X$, $\|a\| = 1$, and the Gâteaux differentiability of the norm at the point a. For the sake of simplicity we assume that X is a real space. It follows from Corollary 2.1.16 that there is $\varphi \in X^*$, $\|\varphi\| = 1$, such that $\varphi(a) = \|a\| = 1$ and

$$\varphi(x) \le 1 \qquad \text{for all} \quad x \in X,\ \|x\| \le 1.$$[8]

Put $f(x) = \|x\|$. Fix $h \in X$ and let $g(t) = \|a + th\|$, $t \in \mathbb{R}$. The function g is a convex real function, and therefore there exist the right and the left derivatives at zero and $g'_-(0) \le g'_+(0)$. Further, we have

$$\frac{g(t) - g(0)}{t} \ge \frac{\varphi(a + th) - \varphi(a)}{t} = \varphi(h) \qquad \text{for} \quad t > 0.$$

In particular, $g'_+(0) \ge \varphi(h)$ and similarly $g'_-(0) \le \varphi(h)$. This means that φ is uniquely determined provided the directional derivative of the norm exists at a for all $h \in X$. In particular,

$$\delta f(a; h) = \varphi(h),$$

i.e., the norm is Gâteaux differentiable at a. The converse is also true. Indeed, suppose by contradiction that $\delta f(a; h)$ does not exist for an h, i.e., $g'_+(0) > g'_-(0)$. Choose $\alpha \in [g'_-(0), g'_+(0)]$ and define $\psi(\gamma a + th) = \gamma + t\alpha$ for scalars γ, t. Then

$$\alpha \le g'_+(0) \le \frac{g(t) - g(0)}{t} = \frac{\|a + th\| - \|a\|}{t} \qquad \text{for} \quad t > 0,$$

and therefore

$$\psi(a + th) = 1 + t\alpha \le \|a + th\|.$$

The same inequality holds for $t \le 0$. As an easy consequence we get

$$|\psi(\gamma a + th)| \le \|\gamma a + th\|, \qquad \psi(a) = 1.$$

This means that

$$\psi \in Y^*, \qquad \|\psi\| = 1 \qquad \text{where} \quad Y = \operatorname{Lin}\{a, h\}.$$

The Hahn–Banach Theorem yields an extension φ of ψ which determines a supporting hyperplane. Since for a different α we get a different φ there is no uniqueness of supporting hyperplanes at a and the duality mapping[9] is not single-valued at a. ○

[8]The hyperplane $\mathcal{M} = \{x \in X : \varphi(x) = 1\}$ is then called a *supporting hyperplane* to the unit ball of X at the point a. Such a $\varphi \in X^*$ need not be uniquely determined.

[9] The map $\kappa\colon X \to \exp X^* : \kappa(x) := \{f \in X^* : \|f\|_{X^*} = \|x\|_X,\ f(x) = \|x\|_X^2\}$ is called the *duality mapping*. It is a multi-valued mapping and belongs to the fundamental concepts in the Banach space theory.

Similarly to partial derivatives, the Gâteaux derivative is also unsuitable for the Chain Rule for differentiability. We recommend to the reader to construct examples of $f\colon \mathbb{R}^2 \to \mathbb{R}$, $g\colon \mathbb{R} \to \mathbb{R}^2$ such that $f(g)$ has no derivative at o in spite of the fact that

$$Df(o) = 0, \qquad g(0) = o, \qquad g'(0) = (0,0).$$

For this purpose a stronger notion of differentiability is needed. The following definition is a straightforward generalization of the differential of a function of two variables.

Definition 3.2.10. Let X, Y be normed linear spaces (both over the same scalar field). A mapping $f\colon X \to Y$ is said to be *Fréchet differentiable* at a point $a \in X$ if there exists $A \in \mathcal{L}(X,Y)$ such that

$$\lim_{h \to o} \frac{\|f(a+h) - f(a) - Ah\|_Y}{\|h\|_X} = 0. \tag{3.2.5}$$

In this case A is called the *Fréchet derivative* of f at the point a and is denoted by $f'(a)$.

Remark 3.2.11.

(i) If $f'(a)$ exists, then also $Df(a)$ exists. Moreover,

$$f'(a)h = Df(a)h \qquad \text{for all} \quad h \in X.$$

(ii) Suppose that a linear operator $A\colon X \to Y$ has the property (3.2.5). It is easy to see that A is continuous if and only if f is continuous at a.

(iii) A basic analytical approach to the investigation of nonlinear problems involves their approximation by simpler objects. Among them linear approximations are more appropriate from the local point of view. The classical notion of the derivative as the best local linear approximation is the most transparent confirmation of this phenomenon (e.g., the Fermat Theorem for local extremal points). The notion of Fréchet derivative is a genuine generalization to infinite-dimensional spaces.

Theorem 3.2.12 (Chain Rule). *Let X, Y, Z be normed linear spaces and let there exist $\delta g(a;h)$ for $g\colon X \to Y$. If $g(a) = b$ and for $f\colon Y \to Z$ the Fréchet derivative $f'(b)$ exists, then*

$$\delta(f \circ g)(a;h) = f'(b)[\delta g(a;h)].^{10} \tag{3.2.6}$$

[10]For more transparent notation we will often use the symbol $f \circ g$ instead of $f(g)$ for the composition of f and g.

Proof. Choose $\varepsilon > 0$ and $h \in X$. By (3.2.5) there is $\eta > 0$ such that

$$\|f(b+k) - f(b) - f'(b)k\|_Z \le \varepsilon \|k\|_Y \qquad \text{for} \quad \|k\|_Y < \eta.$$

Put

$$\omega(t) := g(a+th) - g(a) - t\delta g(a;h).$$

By Definition 3.2.1, there is $\hat{\delta} > 0$ such that $\|\omega(t)\|_Y \le \varepsilon |t|$ for $|t| < \hat{\delta}$. For

$$k := g(a+th) - g(a) = g(a+th) - b$$

we have

$$\|k\|_Y \le |t| \|\delta g(a;h)\|_Y + \|\omega(t)\|_Y \le |t| [\|\delta g(a;h)\|_Y + \varepsilon].$$

We may choose $\hat{\delta}$ so small that the right-hand side in this inequality is less than η whenever $|t| < \hat{\delta}$. Using all the information and $\delta g(a;h) = \frac{k - \omega(t)}{t}$ we obtain

$$\begin{aligned} &\left\| \frac{f(g(a+th)) - f(g(a))}{t} - f'(b)[\delta g(a;h)] \right\|_Z \\ &\qquad = \left\| \frac{f(b+k) - f(b) - f'(b)k}{t} + f'(b)\frac{\omega(t)}{t} \right\|_Z \\ &\qquad \le \frac{\varepsilon \|k\|_Y}{|t|} + \|f'(b)\|_{\mathcal{L}(Y,Z)} \frac{\|\omega(t)\|_Y}{|t|} \le \varepsilon [\varepsilon + \|\delta g(a;h)\|_Y + \|f'(b)\|_{\mathcal{L}(Y,Z)}] \end{aligned}$$

for $0 < |t| < \hat{\delta}$. The formula (3.2.6) follows. □

Corollary 3.2.13. *Let the hypotheses of Theorem* 3.2.12 *be satisfied. If, moreover,* $Dg(a)$ *exists, then also* $D(f \circ g)(a)$ *does exist and the analogue of* (3.2.6) *is true. A similar assertion is true for* $(f \circ g)'(a)$ *provided* $g'(a)$ *exists.*

Proof. The assertion on $D(f \circ g)(a)$ follows from (3.2.6). The proof for $(f \circ g)'(a)$ is similar to that given above. □

Corollary 3.2.14. *Let* $A \in \mathcal{L}(Y,Z)$ *and let* $\delta f(a;h)$ *exist for* $f \colon X \to Y$. *Then*

$$\delta(Af)(a;h) = A\delta f(a;h)$$

and similarly for $D(Af)(a)$ *and* $(Af)'(a)$.

Proof. It is sufficient to show that

$$A'(y) = A \qquad \text{for all} \quad y \in Y,$$

but this follows immediately from the definition. □

The verification of the degree of linear approximation needed in (3.2.5) is not always an easy task. The following condition can be of use in such situations.

Proposition 3.2.15. *Let $Df(x)$ exist for all x in a neighborhood of a point $a \in X$. If $x \mapsto Df(x)$ is continuous at a (as a mapping $X \to \mathcal{L}(X,Y)$), then $f'(a)$ exists and it is equal to $Df(a)$.*

Proof. According to the estimate (3.2.3) we have for small h

$$\|f(a+h) - f(a) - Df(a)h\|_Y \le \sup_{t\in[0,1]} \|Df(a+th) - Df(a)\|_{\mathcal{L}(X,Y)}\|h\|_X$$

and the continuity of Df yields (3.2.5). □

Definition 3.2.16. Let $\mathcal{G}$ be an open set in X and let $f: X \to Y$. If the Gâteaux derivative $Df: \mathcal{G} \to \mathcal{L}(X,Y)$ is continuous on $\mathcal{G}$ (or equivalently, f' is continuous on $\mathcal{G}$), then we write $f \in C^1(\mathcal{G})$.

One of the convenient conditions for the existence of the differential (i.e., Fréchet derivative) of $f: \mathbb{R}^2 \to \mathbb{R}$ is the continuity of partial derivatives. These can be interpreted also as derivatives with respect to one-dimensional subspaces. A generalization leads to the following definition.

Definition 3.2.17. Let $f: X \to Y$ where $X = X_1 \times X_2$ and X_1, X_2, Y are normed linear spaces.[11] Let $a_2 \in X_2$ and let $f_1: x_1 \mapsto f(x_1, a_2)$. If f_1 has the Gâteaux (or Fréchet) derivative at $a_1 \in X_1$, then $Df_1(a_1)$ (or $f_1'(a_1)$) is called the *partial Gâteaux* (or *Fréchet*) *derivative* of f at (a_1, a_2) with respect to the first variable and is denoted by $D_1f(a_1,a_2)$ (or $f_1'(a_1,a_2)$).

Similarly the partial derivative with respect to the second variable (D_2f or f_2') is defined.

If $Df(a_1,a_2)$ exists, then also $D_1f(a_1,a_2)$, $D_2f(a_1,a_2)$ exist and

$$Df(a_1,a_2)(h_1,h_2) = D_1f(a_1,a_2)h_1 + D_2f(a_1,a_2)h_2. \tag{3.2.7}$$

For the converse assertion we need more assumptions:

Proposition 3.2.18. *Assume that D_2f exists on a neighborhood of a point (a_1,a_2) and the mapping*

$$D_2f: X_1 \times X_2 \to \mathcal{L}(X_2, Y)$$

is continuous at (a_1,a_2). Assume, moreover, that D_1f exists at the point (a_1,a_2). Then $Df(a_1,a_2)$ exists and (3.2.7) *holds.*

[11]Then X is a normed linear space, too. A norm on X is, for example, defined by $\|x\|_X = \|x_1\|_{X_1} + \|x_2\|_{X_2}$, $x = (x_1,x_2) \in X_1 \times X_2$.

Proof. Choose sufficiently small h_1, h_2. Then, by (3.2.3),

$$\begin{aligned}
&\|f(a_1+th_1,a_2+th_2)-f(a_1,a_2)-tD_1f(a_1,a_2)h_1-tD_2f(a_1,a_2)h_2\| \\
&\quad\le \|f(a_1+th_1,a_2+th_2)-f(a_1+th_1,a_2)-tD_2f(a_1+th_1,a_2)h_2\| \\
&\quad\quad+\|D_2f(a_1+th_1,a_2)-D_2f(a_1,a_2)\||t|\|h_2\| \\
&\quad\quad\quad+\|f(a_1+th_1,a_2)-f(a_1,a_2)-tD_1f(a_1,a_2)h_1\| \\
&\quad\le \sup_{0\le\tau\le1}\|D_2f(a_1+th_1,a_2+t\tau h_2)-D_2f(a_1+th_1,a_2)\||t|\|h_2\| \\
&\quad\quad+\|D_2f(a_1+th_1,a_2)-D_2f(a_1,a_2)\||t|\|h_2\| \\
&\quad\quad\quad+\left\|\frac{f(a_1+th_1,a_2)-f(a_1,a_2)}{t}-D_1f(a_1,a_2)h_1\right\||t|=o(|t|)
\end{aligned}$$

as $t\to0$, and the result follows. □

Remark 3.2.19. If, in addition to the assumptions of Proposition 3.2.18, $f_1'(a_1,a_2)$ exists, then $f'(a_1,a_2)$ exists, too. The proof then follows the same lines as that above.

Corollary 3.2.20. *Let $\mathcal{G}$ be an open subset of $X=X_1\times X_2$ and $f\colon X\to Y$. Then $f\in C^1(\mathcal{G})$ if and only if both f_1, f_2 belong to $C^1(\mathcal{G})$.*

Example 3.2.21. One of the most important nonlinear mappings is the so-called *Nemytski operator* which is sometimes also called the *substitution* (or *superposition*) *operator*. As the latter term indicates it arises by the substitution of a function $\varphi\colon\mathcal{G}\subset\mathbb{R}^M\to\mathbb{R}$ into the function $f\colon\mathcal{G}\times\mathbb{R}\to\mathbb{R}$. This leads to a new operator

$$F\colon\varphi\mapsto f(\cdot,\varphi(\cdot))$$

which acts on a space X of functions φ. We wish to find conditions on f for F to be a mapping from X into X and to have some derivatives. We start with the case $X=C[0,1]$.

It is clear that the continuity of f on $[0,1]\times\mathbb{R}$ is sufficient to guarantee that $F\colon X\to X$. Since f is uniformly continuous on compact sets of the form

$$\{(x,y)\in[0,1]\times\mathbb{R}:|y-\varphi(x)|\le1\}\qquad\text{for}\quad\varphi\in C[0,1],$$

F is also continuous on X.

Suppose now that the partial derivative $\frac{\partial f}{\partial y}$ is continuous on $[0,1]\times\mathbb{R}$. For $\varphi,h\in X$ we have, by the classical Mean Value Theorem,

$$\frac{f(x,\varphi(x)+th(x))-f(x,\varphi(x))}{t}=\frac{\partial f}{\partial y}(x,\varphi(x)+\vartheta(t,x)th(x))h(x)$$

for a $\vartheta(t,x) \in (0,1)$ and

$$\sup_{x\in[0,1]} \left| \frac{f(x,\varphi(x)+th(x)) - f(x,\varphi(x))}{t} - \frac{\partial f}{\partial y}(x,\varphi(x))h(x) \right|$$
$$\leq \sup_{x\in[0,1]} \sup_{0\leq\vartheta\leq 1} \left| \frac{\partial f}{\partial y}(x,\varphi(x)+\vartheta th(x)) - \frac{\partial f}{\partial y}(x,\varphi(x)) \right| |h(x)| \leq \varepsilon \|h\|_{C[0,1]}$$

for all sufficiently small $|t|$ (again by uniform continuity of $\frac{\partial f}{\partial y}$ on compact sets). This means that the Gâteaux derivative $DF(\varphi)$ exists and

$$DF(\varphi)h : x \mapsto \frac{\partial f}{\partial y}(x,\varphi(x))h(x).$$

Moreover, DF is continuous as a mapping $X \to \mathcal{L}(X)$ (again by the uniform continuity of $\frac{\partial f}{\partial y}$) and, by Proposition 3.2.15, $F'(\varphi)$ exists for any $\varphi \in X$.

Warning. *It is not always true that the existence of* $\frac{\partial f}{\partial y}$ *implies the existence of* DF*!*

For example, let

$$X = \{\varphi \in C[0,\infty) : \varphi(x)\mathrm{e}^{-x} \text{ is bounded on } [0,\infty)\}$$

with the norm

$$\|\varphi\| := \sup_{x\in[0,\infty)} |\varphi(x)\mathrm{e}^{-x}|$$

and let $f(y) = \sin y$. Since f is Lipschitz continuous with constant 1 we obtain

$$\|F(\varphi_1) - F(\varphi_2)\| \leq \|\varphi_1 - \varphi_2\|.$$

In particular, F is a continuous mapping from X into itself. But

$$\delta F(0;h) \neq \sin'(0)h, \qquad h \in X,$$

as could be erroneously supposed by analogy. Namely, for $h(x) = \mathrm{e}^x \in X$,

$$\sup_{x\in[0,\infty)} \left| \mathrm{e}^{-x} \left[\frac{\sin(t\mathrm{e}^x) - 0}{t} - \mathrm{e}^x \right] \right| = \sup_{y\in[t,\infty)} \left| \frac{\sin y}{y} - 1 \right| \geq 1 \quad \text{for any } t > 0.\text{[12]}$$

Similar calculations yield that $\delta F(0;h)$, $h \neq o$, does not exist at all. ○

The study of the Nemytski operator in spaces of integrable functions is much more complicated. First it has to be proved that $F(\varphi)$ is a measurable function on Ω for $\varphi \in L^p(\Omega)$. The following notion is crucial for this purpose.

[12]The lack of differentiability of the Nemytski operators in weighted spaces causes big problems in the use of the Implicit Function Theorem, cf. the discussion in Example 4.2.7.

Definition 3.2.22. Let Ω be an open set in $\mathbb{R}^N$. A function $f\colon \Omega \times \mathbb{R} \to \mathbb{R}$ is said to have the *Carathéodory property* (notation: $f \in \mathrm{CAR}(\Omega \times \mathbb{R})$) if

(M) for all $y \in \mathbb{R}$ the function $x \mapsto f(x,y)$ is (Lebesgue) measurable on Ω;

(C) for a.a. $x \in \Omega$ the function $y \mapsto f(x,y)$ is continuous on $\mathbb{R}$.

Proposition 3.2.23. *Let Ω be an open set in $\mathbb{R}^N$. Then*

(i) *if $f\colon \Omega \times \mathbb{R} \to \mathbb{R}$ is continuous on $\Omega \times \mathbb{R}$, then $f \in \mathrm{CAR}(\Omega \times \mathbb{R})$;*

(ii) *if $f \in \mathrm{CAR}(\Omega \times \mathbb{R})$ and $\varphi\colon \Omega \to \mathbb{R}$ is (Lebesgue) measurable on Ω, then*

$$F(\varphi)\colon x \mapsto f(x, \varphi(x)), \qquad x \in \Omega,$$

is a measurable function on Ω.

Proof. (i) Since a continuous function $f(\cdot, y)$ is Lebesgue measurable, the assertion is obvious.

(ii) Let φ be a measurable function on Ω. Then there is a sequence $\{s_n\}_{n=1}^{\infty}$ of step functions which converge to φ a.e. in Ω. If

$$s(x) = \sum_{i=1}^{k} \alpha_i \chi_{\Omega_i}(x)$$

is a step function on Ω, i.e., there are pairwise disjoint $\Omega_1, \dots, \Omega_k$ which are measurable,

$$\Omega = \bigcup_{i=1}^{k} \Omega_i \qquad \text{and} \qquad \chi_{\Omega_i}(x) = \begin{cases} 1, & x \in \Omega_i, \\ 0, & x \notin \Omega_i, \end{cases}$$

then

$$f(x, s(x)) = \sum_{i=1}^{k} f(x, \alpha_i) \chi_{\Omega_i}(x)$$

is a measurable function (property (M) in Definition 3.2.22). By property (C),

$$\lim_{n \to \infty} f(x, s_n(x)) = f(x, \varphi(x)) \quad \text{for a.a. } x \in \Omega, \quad \text{i.e.,} \quad F(\varphi)(x) = f(x, \varphi(x))$$

is measurable. □

Having measurability of $F(\varphi)$ we can ask when $F(\varphi) \in L^q(\Omega)$. It is plausible that a certain growth condition for f is needed.

Theorem 3.2.24. *Let $f \in \mathrm{CAR}(\Omega \times \mathbb{R})$ and $p, q \in [1, \infty)$. Let there exist $g \in L^q(\Omega)$ and $c \in \mathbb{R}$ such that*

$$|f(x,y)| \le g(x) + c|y|^{\frac{p}{q}} \qquad \text{for a.a.} \quad x \in \Omega \quad \text{and all} \quad y \in \mathbb{R}. \tag{3.2.8}$$

Then

(i) *$F(\varphi) \in L^q(\Omega)$ for all $\varphi \in L^p(\Omega)$;*[13]

(ii) *F is a continuous mapping from $L^p(\Omega)$ into $L^q(\Omega)$;*

(iii) *F maps bounded sets in $L^p(\Omega)$ into bounded sets in $L^q(\Omega)$.*

Proof. The proof of (i) is based on Proposition 3.2.23 and the use of the Minkowski inequality (Example 1.2.16) and it is straightforward.

The proof of (ii) is quite involved and its crucial step consists in the fact that F maps sequences converging in measure into sequences with the same property. We omit details (see, e.g., Krasnoselski [92, § I.2] or Appell & Zabreiko [9]).

The property (iii) follows from the growth condition (3.2.8). □

Remark 3.2.25. The Carathéodory property can be generalized to functions $f\colon \Omega \times \mathbb{R}^M \to \mathbb{R}$. Proposition 3.2.23 and Theorem 3.2.24 hold similarly for

$$F(\varphi_1, \dots, \varphi_M)(x) := f(x, \varphi_1(x), \dots, \varphi_M(x)).$$

Remark 3.2.26. Let $\Omega \subset \mathbb{R}^N$ be an open subset of $\mathbb{R}^N$ and $f\colon \Omega \times \mathbb{R}^{N+1} \to \mathbb{R}$ satisfy the Carathéodory property. Assume, moreover, there exist $g \in L^q(\Omega)$ and $c \in \mathbb{R}$ such that

$$|f(x,y)| \le g(x) + c\sum_{i=0}^{N} |y_i|^{\frac{p}{q}}$$

for a.a. $x \in \Omega$ and all $y = (y_0, y_1, \dots, y_N) \in \mathbb{R}^{N+1}$. Then F defined by

$$F(u)(x) := f(x, u(x), \nabla u(x))$$

is a continuous mapping from $W^{1,p}(\Omega)$ into $L^q(\Omega)$ and maps bounded sets into bounded sets.

The growth condition with respect to y_0 can be relaxed according to the Embedding Theorem for $W^{1,p}(\Omega)$ (cf. Fučík & Kufner [66]).

Now we turn our attention to the directional derivative of the Nemytski operator in the space $L^2(\Omega)$. The exponents $p = q = 2$ are considered for simplicity only.

In accordance with the computation in Example 3.2.21 we could expect

$$\delta F(\varphi; h)(x) = \frac{\partial f}{\partial y}(x, \varphi(x))h(x) \tag{3.2.9}$$

provided the right-hand side belongs to $L^2(\Omega)$. This is true if $\frac{\partial f}{\partial y}(\cdot, \varphi(\cdot)) \in L^\infty(\Omega)$, e.g., whenever $\frac{\partial f}{\partial y}$ is a bounded continuous function on $\Omega \times \mathbb{R}$. But this is not the whole story since we have to show that

$$\int_\Omega \left| \frac{f(x, \varphi(x) + th(x)) - f(x, \varphi(x))}{t} - \frac{\partial f}{\partial y}(x, \varphi(x))h(x) \right|^2 \mathrm{d}x \to 0 \quad \text{for } t \to 0.$$

[13] Actually, it can be proved that this property implies that (3.2.8) is satisfied for $g \in L^q(\Omega)$ and $c \in \mathbb{R}$, cf. Appell & Zabreiko [9].

For a.a. $x \in \Omega$ the function under the integral sign can be estimated by the Mean Value Theorem (the formula (3.2.1)):

$$\left| \frac{f(x,\varphi(x)+th(x)) - f(x,\varphi(x))}{t} - \frac{\partial f}{\partial y}(x,\varphi(x))h(x) \right|$$
$$\leq \frac{1}{|t|} \sup_{0\leq\vartheta\leq 1} \left| \frac{\partial f}{\partial y}(x,\varphi(x)+t\vartheta h(x)) - \frac{\partial f}{\partial y}(x,\varphi(x)) \right| |t|\,|h(x)|.^{14}$$

The right-hand side converges to zero for $t \to 0$ for a.a. $x \in \Omega$ (by the continuity of $\frac{\partial f}{\partial y}$). In order to justify the use of the Lebesgue Dominated Convergence Theorem we need a square integrable majorant. In particular, boundedness of $\frac{\partial f}{\partial y}$ on $\Omega \times \mathbb{R}$ is sufficient.[15]

In the case when F depends also on the gradient of φ the situation is only technically slightly more complicated. Nemytski operators appear often under the integral (see, e.g., Chapter 7). Since the integral is a continuous linear form, in particular, it is Fréchet differentiable, we can use the Chain Rule to get

$$D\Phi(\varphi)h = \int_\Omega \left[\frac{\partial f}{\partial y_0}(x,\varphi(x),\nabla\varphi(x))h(x) + \sum_{i=1}^{N} \frac{\partial f}{\partial y_i}(x,\varphi(x),\nabla\varphi(x))\frac{\partial h}{\partial x_i}(x) \right] \mathrm{d}x \tag{3.2.10}$$

for

$$\Phi(\varphi) := \int_\Omega f(x,\varphi(x),\nabla\varphi(x))\,\mathrm{d}x,$$

under appropriate assumptions on f.

Now we are turning our attention to higher derivatives. We restrict our attention to the second derivatives and believe that the reader will be able to define the third and higher order derivatives as well. Higher order derivatives of functions are defined by induction. We will do the same for abstract mappings.

Let $f\colon X \to Y$, and $a, h, k \in X$. Put $g(t,s) = f(a+th+sk)$. Then

$$\frac{\partial g}{\partial t}(0,s) = \delta f(a+sk;h),$$

which is a mapping from $\mathbb{R}$ (of variable s) into Y and can be differentiated again:

$$\frac{\partial^2 g}{\partial t \partial s}(0,0) := \frac{\partial}{\partial s}\left(\frac{\partial g}{\partial t}(0,s)\right)\bigg|_{s=0}.$$

[14]It is worth noticing how the classical Mean Value Theorem is used here: to avoid problems with measurability of $x \mapsto \frac{\partial f}{\partial y}(x,\varphi(x)+\vartheta(x)th(x))$ "the inequality form" of the theorem is employed.

[15]The reader should notice problems in finding conditions which ensure that (3.2.9) is also the Fréchet derivative. The situation is even much worse than one would expect. The function f has to be linear for $F\colon L^2(\Omega) \to L^2(\Omega)$ to be Fréchet differentiable (see, e.g., Ambrosetti & Prodi [6, Chapter 1, Proposition 2.8]). See also Exercise 3.2.41 and Remark 3.2.42.

If these derivatives exist, then

$$\delta^2 f(a; h, k) := \frac{\partial}{\partial s}\left(\frac{\partial g}{\partial t}(0, s)\right)\Big|_{s=0}$$

is called the *second directional derivative* (in the directions h, k). Notice that generally $\delta^2 f(a; h, k) \neq \delta^2 f(a; k, h)$. (Find an example for $f\colon \mathbb{R}^2 \to \mathbb{R}$!) It is easy to see that for $f\colon \mathbb{R}^M \to \mathbb{R}$ we have

$$\delta^2 f(a; e_i, e_j) = \frac{\partial^2 f}{\partial x_i \partial x_j}(a)$$

if e_i, e_j are the unit coordinate vectors in $\mathbb{R}^M$.

It may occur that the operator

$$(h, k) \in X \times X \mapsto \delta^2 f(a; h, k)$$

is linear in both variables (i.e., it is the so-called *bilinear operator*) and is continuous on $X \times X$.[16] In that case $\delta^2 f(a; \cdot, \cdot)$ is called the *second Gâteaux derivative* and is denoted by $D^2 f(a)$.

Proposition 3.2.27 (Taylor Formula). *Let X be a normed linear space and Y a Banach space. Assume that $a, h \in X$ and that $\delta^2 f(x; h, h)$ exists for all $x \in \mathcal{M} := \{a + th : t \in [0, 1]\}$ and is continuous as a mapping from $\mathcal{M}$ into Y. Then*

$$f(a + h) = f(a) + \delta f(a; h) + \int_0^1 (1 - t)\delta^2 f(a + th; h, h)\,\mathrm{d}t.^{17} \qquad (3.2.11)$$

Proof. Put $g(t) = (1 - t)\delta f(a + th; h)$. Then we have

$$g'(t) = -\delta f(a + th; h) + (1 - t)\delta^2 f(a + th; h, h), \qquad t \in [0, 1].$$

Since both terms on the right-hand side are continuous we get,

$$g(1) - g(0) = \int_0^1 g'(t)\,\mathrm{d}t = -\int_0^1 \delta f(a + th; h)\,\mathrm{d}t + \int_0^1 (1 - t)\delta^2 f(a + th; h, h)\,\mathrm{d}t.$$

Using Theorem 3.2.6 we obtain (3.2.11). □

---

[16] Equivalently, it is continuous at the point (o, o) if there is a constant c such that $\|\delta^2 f(a; h, k)\|_Y \leq c\|h\|_X\|k\|_X$ for all $h, k \in X$ (see a similar assertion in Proposition 1.2.10 for a linear operator). Denoting the space of all continuous bilinear operators from $X \times X$ into Y by $\mathcal{B}_2(X, Y)$ we see that the least possible constant c in the above inequality is a norm on $\mathcal{B}_2(X, Y)$. See also the important Proposition 2.1.7.

[17] If for $n \in \mathbb{N}$ the nth directional derivative $\delta^n f(a; h, \dots, h)$ exists for all $h \in X$, then the mapping $h \mapsto f(a) + \sum\limits_{k=1}^{n} \frac{1}{k!}\delta^k f(a; \underbrace{h, \dots, h}_{k\text{-times}})$ is called the *Taylor polynomial* of the degree n of f at the point a.

If we wanted to define the second Fréchet derivative also by induction we should differentiate $f' \colon X \to \mathcal{L}(X, Y)$ at $a \in X$ to obtain $f''(a) \in \mathcal{L}(X, \mathcal{L}(X, Y))$. But this space seems to be rather strange and the space $\mathcal{L}(X, \mathcal{L}(X, \mathcal{L}(X, Y)))$ (for $f'''(a)$) really awkward. Because of that we identify $\mathcal{L}(X, \mathcal{L}(X, Y))$ with the space of continuous bilinear operators $\mathcal{B}_2(X, Y)$ (see footnote 16) and define the *second Fréchet derivative* $f''(a)$ to be an element of $\mathcal{B}_2(X, Y)$ with the approximation property

$$\lim_{h \to o} \frac{\|f'(a+h) - f'(a) - f''(a)(h, \cdot)\|_{\mathcal{L}(X,Y)}}{\|h\|_X} = 0. \tag{3.2.12}$$

The careful reader can ask why we have written $f''(a)(h, \cdot)$ and not $f''(a)(\cdot, h)$ in (3.2.12). The reason is that the mapping $(h, k) \mapsto f''(a)(h, k)$ is actually symmetric.

Proposition 3.2.28. *If $f''(a)$ exists, then*

$$f''(a)(h, k) = f''(a)(k, h) \qquad \textit{for all} \quad h, k \in X.$$

Proof. Similarly to the proof of the classical result on mixed partial derivatives we express the difference

$$f(a + h + k) - f(a + h) - f(a + k) + f(a)$$

which is equal to $g_i(1) - g_i(0)$ for

$$\begin{aligned} g_1(t) &:= f(a + th + k) - f(a + th), && t \in [0, 1], \\ g_2(s) &:= f(a + h + sk) - f(a + sk), && s \in [0, 1]. \end{aligned}$$

Since $f''(a)$ exists, both the mappings f and f' are defined on a neighborhood $\mathcal{U}$ of a. Elements h, k are chosen so small that all variables belong to $\mathcal{U}$.

We can express

$$g_i(1) - g_i(0) = A_i + g_i'(0) \qquad \text{where} \quad A_i := g_i(1) - g_i(0) - g_i'(0),$$

and

$$\begin{aligned} g_1'(0) &= e_1(h, k) + f''(a)(k, h), & e_1(h, k) &:= f'(a + k)h - f'(a)h - f''(a)(k, h), \\ g_2'(0) &= e_2(k, h) + f''(a)(h, k), & e_2(k, h) &:= f'(a + h)k - f'(a)k - f''(a)(h, k). \end{aligned}$$

Since $g_1(1) - g_1(0) = g_2(1) - g_2(0)$, we have

$$f''(a)(h, k) - f''(a)(k, h) = A_1 - A_2 + e_1(h, k) - e_2(k, h). \tag{3.2.13}$$

Now we estimate all terms on the right-hand side of this equality.

By Theorem 3.2.7,

$$\|A_i\| \le \sup_{t \in [0,1]} \|g_i'(t) - g_i'(0)\|.$$

Since $f''(a)$ is bilinear, we have

$$\begin{aligned} g_1'(t) - g_1'(0) &= f'(a+th+k)h - f'(a+th)h - f'(a+k)h + f'(a)h \\ &= [f'(a+th+k)h - f'(a)h - f''(a)(th+k,h)] \\ &\quad - [f'(a+th)h - f'(a)h - f''(a)(th,h)] \\ &\quad - [f'(a+k)h - f'(a)h - f''(a)(k,h)]. \end{aligned} \tag{3.2.14}$$

Choose now $\varepsilon > 0$ and $\delta > 0$ corresponding to the definition of $f''(a)$ such that

$$\|f'(a+u)v - f'(a)v - f''(a)(u,v)\| \le \varepsilon\|u\|\|v\| \quad \text{for} \quad \|u\| < \delta \text{ and any } v \in X.$$

Then every term on the right-hand side of (3.2.14) is bounded by $\varepsilon(\|h\| + \|k\|)^2$ provided $\|h\|, \|k\| < \delta$. The same estimate holds for $\|e_1(h,k)\|$ and similarly also for $\|g_2'(t) - g_2'(0)\|$, $\|e_2(k,h)\|$.

By (3.2.13) we obtain

$$\|f''(a)(h,k) - f''(a)(k,h)\| \le \|A_1\| + \|A_2\| + \|e_1(h,k)\| + \|e_2(k,h)\| \le 8\varepsilon[\|h\| + \|k\|]^2 \tag{3.2.15}$$

provided $\|h\|, \|k\| < \delta$. Choose $h_0, k_0 \in X$ and put

$$h = \alpha h_0, \qquad k = \alpha k_0.$$

For a sufficiently small α the estimate (3.2.15) holds. Because of the bilinearity of $f''(a)$ we get

$$\|f''(a)(h_0,k_0) - f''(a)(k_0,h_0)\| \le 8\varepsilon[\|h_0\|^2 + \|k_0\|^2] \qquad \text{for any} \quad \varepsilon > 0.$$

This completes the proof. □

Remark 3.2.29.

(i) It is not difficult to see that the existence of $f''(a)$ implies the existence of $D^2 f(a)$ and the equality

$$f''(a)(h,k) = D^2 f(a)(h,k).$$

It is also possible to prove that the continuity of $D^2 f$ on an open set $\mathcal{G} \subset X$ (as a mapping from $\mathcal{G}$ into $\mathcal{B}_2(X,Y)$) is equivalent to the continuity of f'' on $\mathcal{G}$. In this case we write $f \in C^2(\mathcal{G})$.

(ii) If $X = \mathbb{R}^M$, $Y = \mathbb{R}$ and $D^2 f(a)$ exists for $f\colon \mathbb{R}^M \to \mathbb{R}$, then it is sufficient to know the values

$$D^2 f(a)(e_i, e_j), \qquad i,j = 1,\dots,M,$$

to determine $D^2 f(a)$. This means that $D^2 f(a)$ (and also $f''(a)$) can be represented by the matrix (the so-called *Hess matrix*)

$$\left(\frac{\partial^2 f}{\partial x_i \partial x_j}(a)\right).$$

Exercise 3.2.30. Let $A \in \mathcal{L}(X, Y)$ and $B \in \mathcal{B}_2(X, Y)$. Compute A'', B' and B''!

Exercise 3.2.31. Let $f\colon X \to Y$ be injective on an open set $\mathcal{G} \subset X$. Denote $(f|_{\mathcal{G}})^{-1} = g$. Suppose that $f'(a)$ and $g'(b)$ exist for an $a \in \mathcal{G}$, $f(a) = b$. Is it true that

$$g'(b) = [f'(a)]^{-1}\,?$$

(For conditions which guarantee the existence of $g'(b)$ see Section 4.1.)

Exercise 3.2.32. Put $\Phi(A) = A^{-1}$ for an invertible $A \in \mathcal{L}(X, Y)$ (here X, Y are Banach spaces). Show that

$$\Phi'(A)(B) = -A^{-1}BA^{-1}, \qquad B \in \mathcal{L}(X, Y) \quad \text{for all } A \in \operatorname{Dom}\Phi.$$

Hint. Use the same method as in Exercise 2.1.33 and Corollary 2.1.3 to show that $\operatorname{Dom}\Phi$ is open in $\mathcal{L}(X, Y)$ and $(A + B)^{-1} - A^{-1} = A^{-1}\left[(I + BA^{-1})^{-1} - I\right] = A^{-1}\sum\limits_{n=1}^{\infty}(-1)^n(BA^{-1})^n$ for $\|B\|$ small.

Exercise 3.2.33. Let X be either $C[0, 1]$ or $L^p(0, 1)$, $1 \le p < \infty$. Compute $\delta f(x, h)$, $Df(x)$ and $f'(x)$ for $f(x) = \|x\|$, $x \in X$.

Exercise 3.2.34.

(i) Compute the duality mapping (for the definition see the footnote 9 on page 130) for the space $L^p(\Omega)$, $1 \le p < \infty$.

(ii) Show that the duality mapping for the space $C[0, 1]$ need not be single-valued.

Hint. Note that the dual $(C\,[0, 1])^*$ can be identified with regular Borel measures on $[0, 1]$ (see, e.g., Rudin [132]) in such a way that $\varphi(x) = \int\limits_0^1 x(t)d\mu(t)$ for $\varphi \in (C[0, 1])^*$. It is easy to see that if $|x|$ takes its maximum at $t_0 \in [0, 1]$ then $|x(t_0)|\delta_{t_0} \in \kappa(x)$, where δ_{t_0} is the Dirac measure concentrated at t_0.

Exercise 3.2.35. Let $p > 1$, $\Omega \subset \mathbb{R}^N$ and let

$$f(u) = \frac{1}{p}\int_\Omega \|\nabla u(x)\|^p \,\mathrm{d}x, \qquad g(u) = \frac{1}{p}\int_\Omega |u(x)|^p \,\mathrm{d}x$$

be functionals defined on $W^{1,p}(\Omega)$ (here $\|\nabla u(x)\| := \left[\sum\limits_{i=1}^{N}\left(\frac{\partial u(x)}{\partial x_i}\right)^2\right]^{\frac{1}{2}}$). Prove that f and g are Fréchet differentiable at each $u \in W^{1,p}(\Omega)$, and

$$f'(u)v = \int_\Omega \|\nabla u(x)\|^{p-2}(\nabla u(x), \nabla v(x))\,\mathrm{d}x, \quad g'(u)v = \int_\Omega |u(x)|^{p-2}u(x)v(x)\,\mathrm{d}x.$$

Hint. Let

$$\varphi(t) = |t|^{p-2}t, \qquad t \neq 0, \qquad \varphi(0) = 0.$$

Then φ is continuous and $\frac{\mathrm{d}}{\mathrm{d}t}\left(\frac{1}{p}|t|^p\right) = \varphi(t)$, $t \in \mathbb{R}$. Similarly, for $y \in \mathbb{R}^N$, $\|y\| = \left(\sum\limits_{i=1}^{N} y_i^2\right)^{\frac{1}{2}}$, set

$$\psi(y) = \|y\|^{p-2}y, \qquad y \neq o, \qquad \psi(o) = o.$$

Then $\nabla\left(\frac{1}{p}\|y\|^p\right) = \psi(y)$ for all $y \in \mathbb{R}^N$.

Exercise 3.2.36. Find conditions on k and f for the so-called *Hammerstein operator*

$$H\varphi(t) = \int_a^b k(t,s)f(s,\varphi(s))\,\mathrm{d}s$$

to map $L^2(a,b)$ into itself, and then differentiate H!

Exercise 3.2.37. Differentiate the following operators:

(i) $F(\varphi) = \displaystyle\int_0^1 \left[\int_0^t |\varphi(s)|^2\,\mathrm{d}s\right]^3 \mathrm{d}t$, $\varphi \in C[0,1]$ or $\varphi \in L^2(0,1)$;

(ii) $F(\varphi)(t) = \left[\displaystyle\int_0^t \varphi(s)\,\mathrm{d}s\right]^2$ as

- $F\colon L^1(0,1) \to L^1(0,1)$,
- $F\colon C[0,1] \to C[0,1]$,
- $F\colon C[0,1] \to C^1[0,1]$.

Exercise 3.2.38. Let $f\colon [0,1] \times \mathbb{R} \to \mathbb{R}$ and

$$F(\varphi) = \int_0^1 f(t,\varphi(t))\,\mathrm{d}t.$$

Under which conditions on f there exist $D^2F(\varphi)$ and $F''(\varphi)$ if we consider

$$F\colon C[0,1] \to \mathbb{R} \qquad \text{and} \qquad F\colon L^2(0,1) \to \mathbb{R}?$$

Remark 3.2.39. The following assertion is due to I. V. Skrypnik:

If $\frac{\partial^2 f}{\partial y^2}$ is continuous and bounded on $(0,1) \times \mathbb{R}$, then $F \in C^2(L^2(0,1))$ (F is defined as in Exercise 3.2.38) if and only if

$$f(t,y) = a(t) + b(t)y + c(t)y^2 \qquad \textit{where} \quad a,b,c \in L^\infty(0,1).$$

Exercise 3.2.40. Let $f\colon [0,1] \times \mathbb{R} \times \mathbb{R} \to \mathbb{R}$ and

$$F(\varphi) = \int_0^1 f(t,\varphi(t),\varphi'(t))\,\mathrm{d}t, \qquad \varphi \in C^1[0,1].$$

Under which conditions does $D^2F(\varphi)$ exist?

Exercise 3.2.41. Suppose that Ω is a bounded open subset of $\mathbb{R}^N$, function $f\colon \Omega \times \mathbb{R} \to \mathbb{R}$ and its partial derivatives $\frac{\partial f}{\partial y}$ are continuous on $\Omega \times \mathbb{R}$ (or both satisfy the Carathéodory property). Let $p > 2$ and let there exist constants a, b such that

$$\left|\frac{\partial f}{\partial y}(x,y)\right| \le a + b|y|^{p-2}, \qquad x \in \Omega,\ y \in \mathbb{R}.$$

If $f(\cdot,0) \in L^{p'}(\Omega)$ where $p' = \frac{p}{p-1}$ (the conjugate exponent) and $F(\varphi)(x) = f(x,\varphi(x))$, show the following facts:

(i) F maps $L^p(\Omega)$ into $L^{p'}(\Omega)$.
Hint. Integrate $\frac{\partial f}{\partial y}$ and use the above estimate and Theorem 3.2.24.

(ii) $\delta F(\varphi)h\colon x \mapsto \frac{\partial f}{\partial y}(x,\varphi(x))h(x)$ for all $h \in L^p(\Omega)$.
Hint. Proceed similarly to the main text. Use the Hölder inequality to show that $F_y(\varphi)(x) := \frac{\partial f}{\partial y}(x,\varphi(x))$ maps $L^p(\Omega)$ into $L^q(\Omega)$, $q = \frac{p}{p-2}$.

(iii) The Fréchet derivative $F'(\varphi)$ exists for all $\varphi \in L^p(\Omega)$.

Remark 3.2.42. If differentiability properties of F are also needed for $p \le 2$, one should replace $L^p(\Omega)$ by more sophisticated spaces like Besov or Triebel–Lizorkin ones. See, e.g., Runst & Sickel [134, Chapter 5].

3.2A Newton Method

The Contraction Principle offers a very effective method for solving nonlinear equations, either to prove the existence of a solution or to find them numerically. Since the speed of convergence is not always satisfactory, various modifications have appeared. One of these modifications is even much older than the Contraction Principle itself and goes back to I. Newton. An idea of this method can be seen from Figure 3.2.1 where the iterations for solving the equation

$$f(x) = o$$

are shown.

Suppose that we have found an approximate solution a. We wish to construct a correction $\tilde{y}$ such that

$$f(a + \tilde{y}) = o.$$

By the Taylor expansion,

$$f(a) = f(a+\tilde{y}) - f'(a+\tilde{y})\tilde{y} + r(\tilde{y}) = -f'(a+\tilde{y})\tilde{y} + r(\tilde{y}),$$

i.e.,

$$\tilde{y} = -[f'(a+\tilde{y})]^{-1}[f(a) - r(\tilde{y})] = -[f'(a+\tilde{y})]^{-1}[f(a+\tilde{y}) - f'(a+\tilde{y})\tilde{y}] =: F(\tilde{y}) \quad (3.2.16)$$

provided $[f'(a+\tilde{y})]^{-1}$ exists.

The idea is to solve the equation

$$\tilde{y} = F(\tilde{y}) \quad (3.2.17)$$

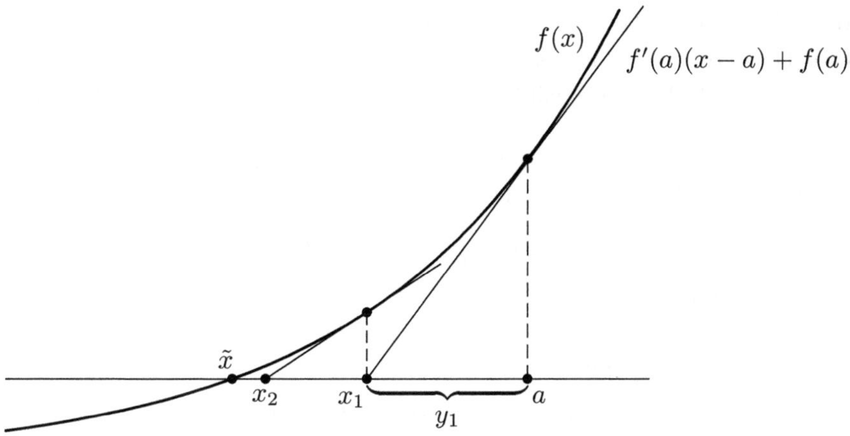

Figure 3.2.1.

in a certain closed ball $B(o;\varrho)$ around o by iterations

$$y_{n+1} = F(y_n), \qquad y_0 = o.$$

Denoting $x_n = a + y_n$ we can rewrite these iterations in the form

$$x_{n+1} - a = -[f'(x_n)]^{-1} f(x_n) + x_n - a, \tag{3.2.18}$$

which are exactly the iterations from Figure 3.2.1. If the sequence of iterations $\{y_n\}_{n=1}^{\infty}$ converges to $\tilde{y}$, then

$$f(a + \tilde{y}) = o$$

as follows from (3.2.16). Our goal is to show:

(A1) There is $\delta > 0$ such that F maps $B(o;\delta)$ into itself and it is a contraction on this ball.

(A2) The convergence of $\{x_n\}_{n=1}^{\infty}$ is faster than the convergence of iterations given by the Contraction Principle (cf. Theorem 2.3.1), actually there is a constant c such that

$$\|x_{n+1} - x_n\| \le c\|x_n - x_{n-1}\|^2.[18] \tag{3.2.19}$$

We apparently need some assumptions to reach this goal. We assume that X is a Banach space, $f\colon X \to X$, and, moreover:

(H1) There is a ball $B(a;\hat{\delta})$ such that $f \in C^1(B(a;\hat{\delta}))$ and f' satisfies the Lipschitz condition on this ball:

there exists L such that

$$\|f'(x) - f'(y)\|_{\mathcal{L}(X)} \le L\|x - y\|_X \qquad \text{for} \quad x, y \in B(a;\hat{\delta}).$$

[18] Compare this quadratic estimate (which yields an exponential one for $\|\tilde{x} - x_n\|$) with an estimate from the Contraction Principle $\|\tilde{x} - x_n\| \le q^n \|x_1 - x_0\|$ for a $0 < q < 1$.

(H2) The value $f(a)$ is sufficiently small.[19]

(H3) The derivative $f'(a)$ has a continuous inverse $[f'(a)]^{-1} \in \mathcal{L}(X)$.

The proof of (A1), (A2) will be done in several steps. For the sake of simplicity we denote

$$A := f'(a), \qquad A(x) := f'(x), \qquad \alpha := \|[f'(a)]^{-1}\|.$$

Step 1. If $\delta < \frac{1}{\alpha L}$, $\delta \le \hat{\delta}$, then $A^{-1}(x)$ exists and

$$\|A^{-1}(x)\| \le \frac{\alpha}{1 - \alpha L\delta} \qquad \text{for} \quad x \in B(a;\delta).$$

Indeed, we can write

$$A(x) = A[I + A^{-1}(A(x) - A)].$$

Since $\|A(x) - A\| \le L\|x - a\|$, we get

$$\|A^{-1}(A(x) - A)\| \le \alpha L\|x - a\|$$

and $A^{-1}(x)$ exists for $x \in B(a;\delta)$ (by Proposition 2.1.2), and

$$A^{-1}(x) = \sum_{n=0}^{\infty} (-1)^n [A^{-1}(A(x) - A)]^n A^{-1}.$$

The estimate for $\|A^{-1}(x)\|$ follows.

Step 2. If $w, x \in B(a;\delta)$, then

$$\|A^{-1}(w) - A^{-1}(x)\| \le \left(\frac{\alpha}{1 - \alpha L\delta}\right)^2 L\|w - x\|.$$

This estimate follows from the identity

$$A^{-1}(w) - A^{-1}(x) = A^{-1}(w)[A(x) - A(w)]A^{-1}(x)$$

and Step 1.

Step 3. We have

$$\|r(y)\| \le \frac{L}{2}\delta^2 \qquad \text{and} \qquad \|r(y) - r(z)\| \le 3L\delta\|y - z\| \qquad \text{for} \quad y, z \in B(o;\delta)$$

where

$$r(y) := f(a) - f(a + y) + A(a + y)y$$

(see (3.2.16)). Indeed, by Theorem 3.2.6, we get

$$r(y) = \int_0^1 [A(a + y) - A(a + (1 - t)y)]y \,\mathrm{d}t$$

and

$$\begin{aligned} r(y) - r(z) &= f(a + z) - f(a + y) + A(a + y)y - A(a + z)z \\ &= \int_0^1 [A(a + t(z - y)) - A(a + y)](z - y)\,\mathrm{d}t + [A(a + y) - A(a + z)]z. \end{aligned}$$

The estimates now follow from (H1) and Step 2.

[19] This assumption means that we actually need a good approximation of a solution of the equation $f(x) = o$ (see Step 4 for the estimate of $\|f(a)\|$).

Step 4. The assertion (A1) holds. Indeed, we have

$$F(y) - F(z) = A^{-1}(a+y)[r(y) - r(z)] + [A^{-1}(a+z) - A^{-1}(a+y)][f(a) - r(z)].$$

From (H1) and Steps 1–3 we get

$$\|F(y) - F(z)\| \le c(\delta + \|f(a)\|L)\|y - z\|$$

with a c which is a bounded function of $\delta \in [0, \delta_0]$ (δ_0 small enough). This means that we can choose δ and the estimate of $\|f(a)\|$ in (H2) such that

$$\|F(y) - F(z)\| \le q\|y - z\|, \qquad y, z \in B(o;\delta) \qquad \text{for a} \quad q \in (0,1).$$

Moreover,

$$\|F(y)\| \le \|F(y) - F(o)\| + \|F(o)\| \le q\delta + \alpha\|f(a)\| \le \delta,$$

provided $\|f(a)\|$ is sufficiently small.

Step 5. We can now prove the assertion (A2). By (3.2.18) and Theorem 3.2.6,

$$\begin{aligned} f(x_n) &= f(x_n) - f(x_{n-1}) - f'(x_{n-1})(x_n - x_{n-1}) \\ &= \int_0^1 [f'(x_{n-1} + t(x_n - x_{n-1})) - f'(x_{n-1})](x_n - x_{n-1})\, dt. \end{aligned}$$

Hence

$$\|f(x_n)\| \le \frac{L}{2}\|x_n - x_{n-1}\|^2,$$

and also

$$\|x_{n+1} - x_n\| \le \|A^{-1}(x_n)\|\|f(x_n)\| \le c\|x_n - x_{n-1}\|^2.$$

Remark 3.2.43. The drawback of the iteration procedure (3.2.18) consists in the requirement to compute the inverse to the derivative at each step. This is the price for fast convergence. One can assume that by replacing $[f'(x)]^{-1}$ by the fixed inverse $[f'(a)]^{-1}$ we should avoid this disadvantage. This idea is also due to I. Newton. Conditions for convergence of these iterations were found by Kantorovich (see Kantorovich [86]). Serious problems appear when the derivative $f'(x)$ is injective but not continuously invertible.

In applications, e.g., to nonlinear partial differential equations, we have many possibilities of the choice of Banach spaces X_α, Y_α such that $f\colon X_\alpha \to Y_\alpha$ (see, e.g., Example 1.2.25 and Example 2.1.29). It can happen that

$$[f'(x)]^{-1} \in \mathcal{L}(Y_\alpha, X_\beta) \qquad \text{where} \quad X_\alpha \subset X_\beta.$$

This means that the equation

$$f'(x_n)(x - x_n) = -f(x_n)$$

(see (3.2.18)) which has to be solved to obtain the $(n+1)^{\text{st}}$-iteration x_{n+1}, has a solution in a larger space X_β provided $x_n \in X_\alpha$. Therefore the iterations belong to larger and larger spaces and, after a finite number of steps, there is no solution at all. This can be also expressed by an observation that x_{n+1} is less smooth than x_n, or that "derivatives are lost" during iterations. One way to overcome these difficulties consists in the approximation of $[f'(x)]^{-1}$ by a "better" operator $L(x)$ in the sense that

$$\|f'(x)L(x) - I\|_{\mathcal{L}(Y_\alpha)}$$

is smaller and smaller when x approaches a solution of

$$f(x) = o.$$

Precise conditions under which new iterations

$$w_{n+1} = w_n - L(w_n)f(w_n)$$

converge to a solution can be found, e.g., in Moser [115]. A similar idea appeared earlier in Nash [116]. See also Remark 4.1.6 for a slightly different explanation.

Exercise 3.2.44. Let $f \in C^1(\mathbb{R})$ be a convex real function.

(i) Using only the results of elementary calculus prove the convergence of Newton approximations under appropriate assumptions. Give the reccurence formula for

$$f(x) = x^2 - a, \qquad a > 0.$$

(ii) The same as in (i) for the Kantorovich approximations (Remark 3.2.43).

Chapter 4

Local Properties of Differentiable Mappings

4.1 Inverse Function Theorem

In this section we are looking for conditions which allow us to invert a map $f\colon X \to Y$, especially $f\colon \mathbb{R}^M \to \mathbb{R}^N$. The simple case of a linear operator f indicates that a reasonable assumption is that $M = N$.

Let us start with the simplest case $M = N = 1$. The well-known theorem says that if f is continuous and strictly monotone on an open interval $\mathcal{I}$, then f is injective and $f(\mathcal{I})$ is an open interval $\mathcal{J}$. Moreover, the inverse function f^{-1} is continuous on $\mathcal{J}$.

It is not clear how to generalize the monotonicity assumption to $\mathbb{R}^M$ (cf. Section 6.1), and without it the theorem is not true even in $\mathbb{R}$. Since the monotonicity of a differentiable function $f\colon \mathbb{R} \to \mathbb{R}$ is a consequence of the sign of the derivative of f, we take into consideration also f'. The example $f(x) = x^2$ where f is not injective in any neighborhood of the origin shows that we have to assume $f'(x) \neq 0$. In fact, if f is continuous on an open interval $\mathcal{I}$, $f'(x)$ exists (possibly infinite) at all points of $\mathcal{I}$, and f' does not vanish at any point of $\mathcal{I}$, then f is injective (actually strictly monotone since f' is either strictly positive or strictly negative in $\mathcal{I}$), and f^{-1} is continuous and differentiable on the open interval $f(\mathcal{I})$.

Therefore, we are looking for a generalization of the assumption $f'(x) \neq 0$ for maps $f\colon \mathbb{R}^M \to \mathbb{R}^M$. Since we are interested in a (unique) solution of the equation

$$f(x) = y,$$

the case of a linear function $f\colon \mathbb{R}^M \to \mathbb{R}^M$ (then $f'(x) = f$) suggests to assume that $f'(x)$ is either an injective or, equivalently because of the finite dimension, a surjective linear map. In both cases, $f'(x)$ is an isomorphism of $\mathbb{R}^M$ onto $\mathbb{R}^M$ (for the case of Banach spaces see Theorem 2.1.8).

However, there is still one more problem: Let

$$f(r,\vartheta) = (r\cos\vartheta, r\sin\vartheta), \ (r,\vartheta) \in (0,\infty)\times\mathbb{R}, \quad \text{or} \quad g(z) = \mathrm{e}^z, \ z \in \mathbb{C}.$$

Both the functions are infinitely many times differentiable on their domains,

$$\det f'(r,\varphi) = r \neq 0, \qquad g'(z) \neq 0,$$

and $f(r,\cdot)$ is 2π-periodic and g is $2\pi\mathrm{i}$-periodic, i.e., f and g are not injective.

Therefore, we cannot expect more than only local invertibility. The philosophy of that is simple. Since the notion of derivative is a local one, we can deduce only local information from it.

After these preliminary considerations we can state the main theorem. Since there is no simplification in the case of finite dimension, we formulate it for general Banach spaces.

Theorem 4.1.1 (Local Inverse Function Theorem). *Let X, Y be Banach spaces, $\mathcal{G}$ an open set in X, $f\colon X\to Y$ continuously differentiable on $\mathcal{G}$. Let the derivative $f'(a)$ be an isomorphism of X onto Y for $a\in\mathcal{G}$. Then there exist neighborhoods $\mathcal{U}$ of a, $\mathcal{V}$ of $f(a)$ such that f is injective on $\mathcal{U}$, $f(\mathcal{U}) = \mathcal{V}$. If g denotes the inverse to the restriction $f|_{\mathcal{U}}$, then $g\in C^1(\mathcal{V})$.*

Proof. We will solve the equation

$$f(x) = y$$

for a fixed y near the point $b = f(a)$ by the iteration process. To do that we have to rewrite the equation $f(x) = y$ as an equation in X. We denote by A the inverse map $[f'(a)]^{-1}\in\mathcal{L}(Y,X)$. Then

$$f(x) = y \qquad \Longleftrightarrow \qquad F_y(x) := x - A[f(x) - y] = x. \tag{4.1.1}$$

The simplest condition for the convergence of iterations is given by the Contraction Principle (see Theorem 2.3.1). We have

$$\begin{aligned}&\|F_y(x_1) - F_y(x_2)\| = \|x_1 - x_2 - A[f(x_1) - f(x_2)]\|\\ &\quad\le \|A\|\|f(x_2) - f(x_1) - f'(a)(x_2 - x_1)\| \le \|A\| \sup_{\xi\in\overline{B(a;r)}} \|f'(\xi) - f'(a)\|\|x_1 - x_2\|,\end{aligned}$$

$x_1, x_2 \in \overline{B(a;r)}$. Here we have used the Mean Value Theorem (see formula (3.2.3)). In other words, we can choose $r > 0$ so small that

$$\|F_y(x_1) - F_y(x_2)\| \le \frac{1}{2}\|x_1 - x_2\| \tag{4.1.2}$$

for $x_1, x_2 \in \overline{B(a;r)} \subset \mathcal{G}$, $y \in Y$. Further,

$$\|F_y(x) - a\| = \|F_y(x) - F_y(a) + F_y(a) - a\| \le \frac{1}{2}\|x - a\| + \|A\|\|b - y\|.$$

If $\delta > 0$ is such that $\|A\|\delta \le \frac{r}{2}$, then $F_y(x) \in B(a;r)$ provided $x \in \overline{B(a;r)}$, $y \in B(b;\delta)$. By the Contraction Principle, the equation (4.1.1) has a unique solution in $\overline{B(a;r)}$,

$$x := g(y) \in B(a;r) \qquad \text{for any} \quad y \in B(b;\delta).$$

Moreover, if $g(y_i) = x_i$, $i = 1, 2$, then

$$\begin{aligned}\|g(y_1) - g(y_2)\| &= \|F_{y_1}(x_1) - F_{y_2}(x_2)\| \\ &\le \|F_{y_1}(x_1) - F_{y_1}(x_2)\| + \|F_{y_1}(x_2) - F_{y_2}(x_2)\| \le \frac{1}{2}\|x_1 - x_2\| + \|A\|\|y_1 - y_2\|,\end{aligned}$$

i.e.,

$$\|g(y_1) - g(y_2)\| \le 2\|A\|\|y_1 - y_2\|. \tag{4.1.3}$$

In particular, g is a Lipschitz continuous map on $B(b;\delta)$. To prove the differentiability of g, fix a $y \in B(b;\delta)$ and choose $\hat{\delta} > 0$ so small that $B(y;\hat{\delta}) \subset B(b;\delta)$.

A candidate for $g'(y)$ is the inverse $C(x) := [f'(x)]^{-1}$ for $x = g(y)$. By (4.1.3), $x \in B(a;r)$ and hence (since $r > 0$ was chosen so small) we have

$$\|f'(x) - f'(a)\| \le \frac{1}{2\|[f'(a)]^{-1}\|}.$$

This means that $C(x)$ exists and $C(x) \in \mathcal{L}(Y, X)$ (cf. Exercise 2.1.33). So we wish to estimate the expression

$$\alpha(k) := g(y + k) - g(y) - C(x)k \qquad \text{for} \quad k \in Y, \ \|k\| < \hat{\delta}.$$

Put

$$h = g(y + k) - g(y), \qquad \text{i.e.,} \qquad k = f(x + h) - f(x).$$

We have

$$\alpha(k) = h - C(x)k = -C(x)[f(x + h) - f(x) - f'(x)h].$$

By the definition of the Fréchet derivative, for any $\varepsilon > 0$ there is $\eta > 0$ such that

$$\|f(x + h) - f(x) - f'(x)h\| \le \varepsilon\|h\| \qquad \text{provided} \quad \|h\| < \eta.$$

But (see (4.1.3)) $\|h\| \le 2\|A\|\|k\|$. This means that

$$\|\alpha(k)\| = o(\|k\|), \qquad \text{i.e.,} \qquad g'(y) = C(x) = [f'(x)]^{-1}.$$

This also implies the continuity of $g'(y)$ since the inverse $[f'(x)]^{-1}$ depends continuously on x (see Exercise 2.1.33).

To complete the proof it remains to put

$$\mathcal{V} = B(b;\delta) \qquad \text{and} \qquad \mathcal{U} = f_{-1}(\mathcal{V}) \cap B(a;r). \qquad \square$$

Corollary 4.1.2. *Let X, Y be Banach spaces, $\mathcal{G}$ an open subset of X, $f \in C^1(\mathcal{G}, Y)$. If $f'(x)$ is an isomorphism of X onto Y for all $x \in \mathcal{G}$, then $f(\mathcal{G})$ is an open subset of Y.*

Proof. Use the definition of an open set and Theorem 4.1.1. □

Example 4.1.3. If $f \in C^k(\mathcal{G})$, $k \in \mathbb{N}$, then $g \in C^k(\mathcal{V})$. This follows easily from the formula

$$g'(y) = [f'(x)]^{-1}, \qquad x = g(y),$$

the Chain Rule and Exercise 3.2.32. ○

Definition 4.1.4. Let X, Y be Banach spaces. Then $f\colon X \to Y$ is called a *diffeomorphism* of $\mathcal{G} \subset X$ (or a diffeomorphism of $\mathcal{G}$ onto $\mathcal{H} = f(\mathcal{G})$) if the following conditions are satisfied:

(1) $\mathcal{G}$ is an open set in X, $f \in C^1(\mathcal{G})$,

(2) $f(\mathcal{G}) = \mathcal{H}$ is an open set in Y,

(3) f is injective on $\mathcal{G}$ and the inverse $g = (f|_{\mathcal{G}})^{-1}$ belongs to $C^1(\mathcal{H})$.

If, moreover, $f \in C^k(\mathcal{G})$ for some $k \in \mathbb{N}$, and (therefore) $g \in C^k(\mathcal{H})$, then f is called a *C^k-diffeomorphism.*

A diffeomorphism in $\mathbb{R}^M$ can be viewed as a nonlinear generalization of a linear invertible operator $A\colon \mathbb{R}^M \to \mathbb{R}^M$. Such A yields a linear transformation of coordinates

$$y = Ax.$$

If φ is a diffeomorphism of $\mathcal{G}$ onto $\mathcal{H}$ and $a \in \mathcal{G}$, we can suppose without loss of generality that

$$\varphi(a) = o$$

(if this is not true consider a new diffeomorphism on $\mathcal{G}$: $\tilde{\varphi}(x) = \varphi(x) - \varphi(a)$). Then the Cartesian coordinates $y_1, \dots, y_M$ of $y = \varphi(x)$ can be taken as (generalized or nonlinear or non-Cartesian) coordinates of a point x in the neighborhood $\mathcal{G}$ of a. Such coordinates play an important role in problems where we have to work on non-flat domains (e.g., on nonlinear manifolds – see Appendix 4.3A).

Notice that we can also interpret Theorem 4.1.1 in the finite-dimensional case as follows:

> *The Cartesian coordinates $(y_1, \dots, y_M)$ of the point $y = f(x)$ are nonlinear coordinates of the point x. In these nonlinear coordinates the diffeomorphism f of $\mathcal{U}$ is equal to the identity map.*

Example 4.1.5. Standard examples of nonlinear coordinates:

(i) Polar coordinates in $\mathbb{R}^2$:

$$x = r\cos\varphi, \qquad y = r\sin\varphi$$

($\psi(r,\varphi) = (x,y)$ is a diffeomorphism of $(0,\infty) \times (\alpha, \alpha + 2\pi)$ onto $\mathbb{R}^2$ without a half line);

(ii) Spherical coordinates in $\mathbb{R}^3$:

$$x = r\cos\varphi_1\cos\varphi_2, \qquad y = r\sin\varphi_1\cos\varphi_2, \qquad z = r\sin\varphi_2;$$

(iii) Spherical coordinates in $\mathbb{R}^M$:

$$\begin{aligned} x_1 &= r\cos\varphi_1\cos\varphi_2\cdots\cos\varphi_{M-1},\\ x_2 &= r\sin\varphi_1\cos\varphi_2\cdots\cos\varphi_{M-1},\\ x_3 &= r\sin\varphi_2\cos\varphi_3\cdots\cos\varphi_{M-1},\\ &\ \vdots\\ x_{M-1} &= r\sin\varphi_{M-2}\cos\varphi_{M-1},\\ x_M &= r\sin\varphi_{M-1}. \end{aligned}$$

Before using the Local Inverse Function Theorem we have to check that functions

$$\psi_i(r,\varphi_1,\dots,\varphi_{M-1}) = x_i, \qquad i = 1,\dots,M,$$

have continuous partial derivatives (obvious) and their Jacobi matrix is regular. Equivalently, the determinant J_ψ of the Jacobi matrix is nonzero at a point $(\tilde{r},\tilde{\varphi}_1,\dots,\tilde{\varphi}_{M-1})$, $\psi(\tilde{r},\tilde{\varphi}_1,\dots,\tilde{\varphi}_{M-1}) = a$. Here

$$J_\psi = r^{M-1}\prod_{k=1}^{M-2}\cos^k\varphi_{k+1}, \qquad M \geq 2.^{1}$$ ○

Example 4.1.6. The following question concerning the assumptions of Theorem 4.1.1 naturally arises: "What happens if $f'(a)$ is not an isomorphism?"

In the case of finite dimension, $f'(a)$ cannot be an isomorphism for $f\colon \mathbb{R}^M \to \mathbb{R}^N$ whenever $M \neq N$. If $M > N$, i.e., the number of equations is smaller than the number of variables, then we can expect (we recommend to consider the case of a linear f) to compute some of the variables. The simplest case is solved in the next Section 4.2 (the Implicit Function Theorem). If $M < N$, then $f(\mathcal{G})$ will probably be a "thin" subset of $\mathbb{R}^N$. This case leads to the notion of a (differentiable) manifold (see the first part of Section 4.3 (Definition 4.3.4) and Appendix 4.3A).

If both X and Y have infinite dimension, it can occur that $f'(a)$ is injective but $\operatorname{Im} f'(a)$ is a dense subset of Y, different from Y. In this case, $A = [f'(a)]^{-1}$ exists but it is not continuous into X. We can also explain this situation as follows. If there is a constant $c > 0$ such that

$$\|f'(a)h\|_Y \geq c\|h\|_X \qquad \text{for all} \quad h \in X, \tag{4.1.4}$$

[1] We use the notation $\prod_{j=1}^{p} a_j := a_1\cdots a_p$ ($\prod_{\emptyset} := 1$).

then $f'(a)$ is injective, $Y_1 := \operatorname{Im} f'(a)$ is a closed subspace of Y. Moreover, if we know that Y_1 is dense in Y, then $Y_1 = Y$ and Theorem 4.1.1 can be applied. But sometimes we are able to prove only a weaker estimate, namely that there is a constant $c > 0$ such that

$$\|f'(a)h\|_Y \geq c\|h\|_{\tilde{X}} \qquad \text{for all} \quad h \in X$$

where $\|\cdot\|_{\tilde{X}}$ is a weaker norm than $\|\cdot\|_X$. By this we mean that only the estimate

$$\|h\|_{\tilde{X}} \leq d\|h\|_X \qquad \text{holds for all} \quad h \in X$$

(e.g., $X = C^1[0,1]$, $\|h\|_X = \sup\limits_{t\in[0,1]} |h(t)| + \sup\limits_{t\in[0,1]} |h'(t)|$, $\|h\|_{\tilde{X}} = \sup\limits_{t\in[0,1]} |h(t)|$). Then $A := [f'(a)]^{-1}$ maps Y continuously into the completion of X with respect to the norm $\|\cdot\|_{\tilde{X}}$ (remember that we need complete spaces for the Contraction Principle).

In the above example "one derivative is lost" in an iteration. An idea how to overcome this problem is to use an approximation of A and a more rapid iteration process (e.g., the Newton iteration – see Appendix 3.2A) to compensate errors in the approximations of A (results of this type are the so-called "Hard" Local Inverse/Implicit Function Theorems) see, e.g., Deimling [42], Hamilton [77], Moser [115], Nash [116] or Nirenberg [118]. ○

We now turn to a global version of the Inverse Function Theorem.

Theorem 4.1.7 (Global Inverse Function Theorem). *Let X, Y be Banach spaces and let $f\colon X \to Y$ be continuously differentiable on X. Suppose that $f'(x)$ is continuously invertible for all $x \in X$ and there is a constant $c > 0$ such that*

$$\|[f'(x)]^{-1}\|_{\mathcal{L}(Y,X)} \leq c \qquad \textit{for all} \quad x \in X.$$

Then f is a diffeomorphism of X onto Y.

Proof. It is sufficient to prove that f is injective and surjective. The statement on the diffeomorphism follows then from Theorem 4.1.1. Fix an $a \in X$ and denote $b = f(a)$.

Step 1. The map f is surjective, i.e., $f(X) = Y$. To see this choose $y \in Y$ and put

$$\varphi(t) = (1-t)b + ty, \qquad t \in [0,1].$$

We wish to show that there is a curve $\psi\colon [0,1] \to X$ such that

$$f(\psi(t)) = \varphi(t), \qquad \text{in particular,} \qquad y = \varphi(1) = f(\psi(1)).$$

Since f is locally invertible at $a \in X$ (Theorem 4.1.1), there is a neighborhood $\mathcal{U}$ of a and $\delta > 0$ such that $\psi(t) = (f|_{\mathcal{U}})^{-1}\varphi(t)$ is well defined for $t \in [0,\delta)$ and $\psi \in C^1([0,\delta), X)$. Let

$$\mathcal{A} := \{\tau \in [0,1] : \exists \omega \in C^1([0,\tau], X), f(\omega(t)) = \varphi(t),\ t \in [0,\tau]\}, \tag{4.1.5}$$

and $\alpha = \sup \mathcal{A}$. Notice that ω is uniquely determined by (4.1.5) (this follows from the local invertibility of f), and therefore there is $\omega \in C^1([0,\alpha), X)$ such that$f(\omega(t)) = \varphi(t)$, $t \in [0,\alpha)$. Since we have

$$\|\omega(t_1) - \omega(t_2)\| \leq \sup_{t\in[t_1,t_2]} \|\omega'(t)\| |t_1 - t_2| \leq c\|y - b\| |t_1 - t_2|$$

for all $t_1, t_2 \in [0,\alpha)$, the mapping ω is uniformly continuous on the interval $[0,\alpha)$, hence

$$\lim_{t\to\alpha_-} \omega(t) := \omega(\alpha)$$

exists (X is a complete space) and the equality (4.1.5) holds for all $t \in [0,\alpha]$.

Now we are ready to prove that $\alpha = 1$. Indeed, if $\alpha < 1$, then we can apply Theorem 4.1.1 at the point $\omega(\alpha)$ to obtain a contradiction with the definition of α.

Step 2. The map f is injective. Suppose by contradiction that there are different $x_1, x_2 \in X$ for which

$$f(x_1) = f(x_2).$$

Put

$$y := f(x_2), \quad \psi_i(t) = (1-t)a + tx_i, \quad \varphi_i(t) = f(\psi_i(t)), \quad t \in [0,1],\ i = 1,2.$$

By a slight modification of the above procedure it is possible to prove the existence of a mapping $G\colon [0,1] \times [0,1] \to X$ such that

$$f(G(t,s)) = (1-s)\varphi_1(t) + s\varphi_2(t), \qquad (t,s) \in [0,1] \times [0,1].$$

Then

$$f(G(1,s)) = (1-s)f(x_1) + sf(x_2) = y \qquad \text{for every} \quad s \in [0,1].$$

This contradicts the local invertibility of f at x_1 ($\neq x_2$). □

Exercise 4.1.8. A complex function $f\colon \mathbb{C} \to \mathbb{C}$ is called *holomorphic* in an open set $\mathcal{G} \subset \mathbb{C}$ if $f'(z)$ exists for every $z \in \mathcal{G}$. If $f'(a) \neq 0$ for an $a \in \mathcal{G}$, then f is locally invertible (Theorem 4.1.1). Prove that $(f|_{\mathcal{U}})^{-1}$ is holomorphic and apply this result to $f(z) = \mathrm{e}^z$ to obtain a power series expression of a continuous branch of the "multi-valued function" log. (For the "complex function" proof see, e.g., Rudin [132, Theorem 10.30].)

Exercise 4.1.9. Let

$$f(x) = x + 2x^2 \sin\frac{1}{x}, \quad x \neq 0, \qquad f(0) = 0.$$

Show that f is not injective on any neighborhood of zero. Which assumption of Theorem 4.1.1 is not satisfied?

Hint. If $\mathcal{U}$ is a neighborhood of 0, show that $f'(x) = 0$ has a solution in $\mathcal{U}$ and $f''(x) \neq 0$ at any such solution. Hence f is not injective on $\mathcal{U}$. Note also that f' is not continous at 0.

Exercise 4.1.10. Find the form of the Laplace operator

$$\Delta u := \frac{\partial^2 u}{\partial x^2} + \frac{\partial^2 u}{\partial y^2}$$

in the polar coordinates in the set

$$\mathcal{G} = \{(x,y) \in \mathbb{R}^2 : x^2 + y^2 > 0\} \qquad \text{for} \quad u \in C^2(\mathcal{G}).$$

Hint. If $v(r,\varphi) = u(r\cos\varphi, r\sin\varphi)$, then

$$(\Delta u) \circ \Phi = \frac{\partial^2 v}{\partial r^2} + \frac{1}{r^2}\frac{\partial^2 v}{\partial \varphi^2} + \frac{1}{r}\frac{\partial v}{\partial r} \qquad \text{where} \quad \Phi(r,\varphi) = (r\cos\varphi, r\sin\varphi)$$

is the transformation. Note that we have

$$\left(\frac{\partial u}{\partial x}, \frac{\partial u}{\partial y}\right) = \left(\frac{\partial v}{\partial r}, \frac{\partial v}{\partial \varphi}\right) \circ (\Phi')^{-1}.$$

It is more comfortable to use this formula once again to compute $\frac{\partial^2 u}{\partial x^2}, \frac{\partial^2 u}{\partial y^2}$.

Exercise 4.1.11. Show that the estimate

$$\|[f'(x)]^{-1}\|_{\mathcal{L}(Y,X)} \leq c + d\|x\|_X$$

is sufficient in Theorem 4.1.7.

Hint. Use the Gronwall Lemma (Proposition 2.3.9) to estimate $\|\omega'(t)\|$.

4.2 Implicit Function Theorem

Let us start with a simple example of $f\colon \mathbb{R}^2 \to \mathbb{R}$, e.g.,

$$f(x,y) := x^2 + y^2 - 1.$$

Denote

$$\mathcal{M} = \{(x,y) \in \mathbb{R}^2 : f(x,y) = 0\},$$

i.e., $\mathcal{M}$ is the unit circle in $\mathbb{R}^2$. We would like to solve the equation

$$f(x,y) = 0$$

for the unknown variable y or to express $\mathcal{M}$ as the graph of a certain function

$$y = \varphi(x).$$

We immediately see that for any $x \in (-1, 1)$ there is a couple of y's ($y_{1,2} = \pm\sqrt{1-x^2}$) such that $(x, y) \in \mathcal{M}$. In particular, $\mathcal{M}$ is not a graph of any function $y = \varphi(x)$. We can only obtain that $\mathcal{M}$ is locally a graph, i.e., for $(a, b) \in \mathcal{M}$, $a \in (-1, 1)$, there is a neighborhood $\mathcal{U}$ of (a, b) such that $\mathcal{M} \cap \mathcal{U}$ is the graph of a function $y = \varphi(x)$. On the other hand, for $x = \pm 1$ there is a unique y ($y = 0$) for which $(x, y) \in \mathcal{M}$. But there is no neighborhood $\mathcal{U}$ of $(1, 0)$ such that $\mathcal{M} \cap \mathcal{U}$ is the graph of a function $y = \varphi(x)$. What is the difference between these two cases?

In the former case the tangent line to $\mathcal{M} \cap \mathcal{U}$ exists at the point (a, b) with the slope $\varphi'(a)$. Since

$$f(x, \varphi(x)) = 0 \qquad \text{for} \quad x \in (a - \delta, a + \delta),$$

we have (formally by the Chain Rule)

$$\frac{\partial f}{\partial x}(a, b) + \frac{\partial f}{\partial y}(a, b)\varphi'(a) = 0, \tag{4.2.1}$$

i.e., $\varphi'(a) = -\frac{a}{b}$, since $\frac{\partial f}{\partial y}(a, b) = 2b \neq 0$.

In the latter case, where $(a, b) = (1, 0)$, we have $\frac{\partial f}{\partial y}(1, 0) = 0$, and $\varphi'(1)$ cannot be determined from (4.2.1). The tangent line to $\mathcal{M}$ at the point $(0, 1)$ is parallel to the y-axis, which indicates some problems with determining a solution, i.e., the (implicit) function φ. The reader is invited to sketch a figure.

This discussion shows the importance of the assumption

$$\frac{\partial f}{\partial y}(a, b) \neq 0.$$

How this assumption can be generalized to $f : \mathbb{R}^{M+N} \to \mathbb{R}^N$? A brief inspection of the linear case leads to the observation that we can compute the unknowns $y_{M+1}, \dots, y_{M+N}$ from the equations

$$f_i(y_1, \dots, y_{M+N}) = \sum_{j=1}^{M+N} a_{ij} y_j = 0, \qquad i = 1, \dots, N,$$

uniquely as functions of $y_1, \dots, y_M$ if and only if

$$\det (a_{ij})_{\substack{i=1,\dots,N \\ j=M+1,\dots,M+N}} \neq 0.$$

Nevertheless, $a_{ij} = \frac{\partial f_i}{\partial y_j}$, and the condition on the regularity of the matrix $(a_{ij})_{\substack{i=1,\dots,N \\ j=M+1,\dots,M+N}}$ means that the partial (Fréchet) derivative of f (see Definition 3.2.17) with respect to the last N variables is an isomorphism of $\mathbb{R}^N$.

Theorem 4.2.1 (Implicit Function Theorem). *Let X, Y, Z be Banach spaces, $f : X \times Y \to Z$. Let $(a, b) \in X \times Y$ be such a point that*

$$f(a, b) = o.$$

Let $\mathcal{G}$ be an open set in $X \times Y$ containing the point (a,b). Let $f \in C^1(\mathcal{G})$ and let the partial Fréchet derivative $f_2'(a,b)$ be an isomorphism of Y onto Z.

Then there are neighborhoods $\mathcal{U}$ of a and $\mathcal{V}$ of b such that for any $x \in \mathcal{U}$ there exists a unique $y \in \mathcal{V}$ for which

$$f(x,y) = o.$$

Denote this y by $\varphi(x)$. Then $\varphi \in C^1(\mathcal{U})$. Moreover, if $f \in C^k(\mathcal{G})$, $k \in \mathbb{N}$, then $\varphi \in C^k(\mathcal{U})$.

Proof. We denote $A := [f_2'(a,b)]^{-1}$ and define

$$F(x,y) = (x, Af(x,y)), \qquad (x,y) \in \mathcal{G}.$$

Then $F\colon X \times Y \to X \times Y$, $F \in C^1(\mathcal{G})$ and

$$F'(a,b)(h,k) = (h, Af'(a,b)(h,k)).$$

One can verify that $F'(a,b)$ is an isomorphism of $X \times Y$ onto itself. Hence we can apply Theorem 4.1.1 to get neighborhoods $\mathcal{U} \times \mathcal{V}$ of (a,b) and $\tilde{\mathcal{U}} \times \tilde{\mathcal{V}}$ of (a,o) such that for any $\xi \in \tilde{\mathcal{U}}$ and $\eta = o \in \tilde{\mathcal{V}}$ there exists a unique $(x,y) \in \mathcal{U} \times \mathcal{V}$ such that

$$F(x,y) = (x, Af(x,y)) = (\xi, o), \qquad \text{i.e.,} \qquad x = \xi, \quad \mathcal{U} = \tilde{\mathcal{U}},$$

and, denoting $y = \varphi(x)$,

$$f(x, \varphi(x)) = o.$$

This means that

$$F^{-1}(x,o) = (x, \varphi(x)).$$

Since the inverse F^{-1} is differentiable, by Theorem 4.1.1 we conclude that $\varphi \in C^1(\mathcal{U})$. □

Remark 4.2.2. We can also deduce a formula for $\varphi'(x)$: Indeed, since

$$f(x, \varphi(x)) = o \qquad \text{for every} \quad x \in \mathcal{U}$$

and both the functions f and φ are differentiable, we get from the Chain Rule

$$f_1'(x, \varphi(x)) + f_2'(x, \varphi(x)) \circ \varphi'(x) = o,$$

and therefore

$$\varphi'(x) = -[f_2'(x, \varphi(x))]^{-1} \circ f_1'(x, \varphi(x)) \qquad \text{for} \quad x \in \mathcal{U}_1 \tag{4.2.2}$$

where $\mathcal{U}_1 \subset \mathcal{U}$ may be smaller if necessary in order to guarantee the existence of the inverse

$$[f_2'(x, \varphi(x))]^{-1}$$

(see Exercise 2.1.33).

Remark 4.2.3. The statement of Theorem 4.2.1 is by no means the best one. If we have used the Contraction Principle directly we would obtain the existence of a solution $y = \varphi(x)$ under weaker assumptions. Namely, $f(x, y) = o$ is equivalent to

$$y = y - Af(x, y)$$

and since x is a parameter here we do not need to assume the differentiability with respect to x if we content ourselves just with the existence of φ (and give up its differentiability). We recommend to the reader to use directly the Contraction Principle to obtain the following statement:

Let X be a normed linear space, Y, Z be Banach spaces and let $f\colon X \times Y \to Z$ be continuous at the point (a, b) where

$$f(a, b) = o.$$

Assume that the partial Fréchet derivative $f_2'(a, b)$ is an isomorphism of Y onto Z and $f_2'\colon X \times Y \to \mathcal{L}(Y, Z)$ is continuous at (a, b).

Then there are neighborhoods $\mathcal{U}$ of a and $\mathcal{V}$ of b such that for any $x \in \mathcal{U}$ there is a unique $y = \varphi(x) \in \mathcal{V}$ for which

$$f(x, \varphi(x)) = o.$$

Moreover, φ is continuous at a.

It is also possible to avoid partly the requirement of invertibility of $f_2'(a, b)$ (see Remark 4.1.6 and references given there).

There are many examples in Calculus where the Implicit Function Theorem is used. We give one in Exercise 4.2.9, see also exercises in Dieudonné [44]. Our attention is turned mainly towards more theoretical applications.

Example 4.2.4. Let

$$P(z) = z^n + a_{n-1}z^{n-1} + \cdots + a_0$$

be a polynomial with real or complex coefficients $a_0, \dots, a_{n-1}$. The famous Basic Theorem of Algebra (Theorem 5.7.13) says that if $n \geq 1$, then the equation $P(z) = 0$ has at least one solution $\tilde{z} \in \mathbb{C}$ and actually n solutions if all of them are counted with their multiplicity. This means that P can be factorized as follows:

$$P(z) = (z - z_1)^{k_1} \cdots (z - z_l)^{k_l}, \qquad k_1 + \cdots + k_l = n,$$

where $z_1, \dots, z_l$ are different. A natural question arises: How do these solutions $z_1, \dots, z_l$ depend on the coefficients $a_0, \dots, a_{n-1}$ of P? Let

$$F(z, y_0, \dots, y_{n-1}) = z^n + y_{n-1}z^{n-1} + \cdots + y_0 \colon \mathbb{C} \times \mathbb{C}^n \to \mathbb{C}.$$

Then

$$F(z_1, a_0, \dots, a_{n-1}) = P(z_1) = 0 \qquad \text{and} \qquad F \in C^\infty(\mathbb{C} \times \mathbb{C}^n).$$

If z_1 is a simple root, i.e., $k_1 = 1$, then

$$\frac{\partial F}{\partial z}(z_1, a_0, \dots, a_{n-1}) \neq 0,$$

and the Implicit Function Theorem says that z_1 depends continuously on $a_0, \dots, a_{n-1}$ (also in the real case). But what happens if $k_1 > 1$? Notice that the cases of real and complex roots are different. In the former case, the real root can disappear ($x^2 + \varepsilon = 0$ for $\varepsilon > 0$), and in the latter case, the uniqueness can be lost. Since the solution z_1 ramifies or bifurcates at $a_0, \dots, a_{n-1}$, this phenomenon is called a *bifurcation*. We postpone a basic information on this very important nonlinear phenomenon till the end of the next section. See also Proposition 5.7.16. ○

Example 4.2.5 (dependence of solutions on initial conditions). Suppose that $f \colon \mathbb{R} \times \mathbb{R}^N \to \mathbb{R}^N$ is continuous in an open set $\mathcal{G} \subset \mathbb{R} \times \mathbb{R}^N$ and has continuous partial derivatives with respect to the last N variables in $\mathcal{G}$. Denote by $\varphi(\cdot; \tau, \xi)$ a (unique) solution of the initial value problem

$$\begin{cases} \dot{x} = f(t, x), \\ x(\tau) = \xi \end{cases}$$

(see Theorem 2.3.4). We are now interested in the properties of φ with respect to the variables $(\tau, \xi) \in \mathcal{G}$, cf. Remark 2.3.5. Let us define

$$\Phi(\tau, \xi, \varphi)(t) := \xi + \int_\tau^t f(s, \varphi(s))\, \mathrm{d}s - \varphi(t). \tag{4.2.3}$$

For a fixed $(t_0, x_0) \in \mathcal{G}$ the solution $\varphi(\cdot; t_0, x_0)$ of $\Phi(t_0, x_0, \varphi) = o$ is defined on an open interval $\mathcal{J}$. Choose a compact interval $\mathcal{I} \subset \mathcal{J}$ such that $t_0 \in \operatorname{int} \mathcal{I}$. Then the mapping Φ given by (4.2.3) is defined on a certain open subset $\mathcal{H} \subset \mathbb{R} \times \mathbb{R}^N \times C(\mathcal{I}, \mathbb{R}^N)$ and takes its values in $C(\mathcal{I}, \mathbb{R}^N)$. Further,

$$\Phi(t_0, x_0, \varphi(\cdot; t_0, x_0)) = o$$

and

$$[\Phi_1'(\tau, \xi, \varphi)](t) = -f(\tau, \varphi(\tau)), \qquad [\Phi_2'(\tau, \xi, \varphi)\eta](t) = \eta,$$
$$[\Phi_3'(\tau, \xi, \varphi)\psi](t) = \int_\tau^t f_2'(s, \varphi(s))\psi(s)\, \mathrm{d}s - \psi(t),$$
$$t \in \mathcal{I}, \quad \eta \in \mathbb{R}^N, \quad \psi \in C(\mathcal{I}, \mathbb{R}^N) \quad \text{whenever } (\tau, \xi, \varphi) \in \mathcal{H}.$$

Since these partial Fréchet derivatives are continuous, $\Phi \in C^1(\mathcal{H})$ (see Proposition 3.2.18). The crucial assumption of the Implicit Function Theorem is the continuous invertibility of $\Phi_3'(t_0, x_0, \varphi(\cdot; t_0, x_0))$ in the space $C(\mathcal{I}, \mathbb{R}^N)$. Put

$$B\psi(t) = \int_{t_0}^t f_2'(s, \varphi(s; t_0, x_0))\psi(s)\, \mathrm{d}s, \qquad \psi \in C(\mathcal{I}, \mathbb{R}^N).$$

We have proved in Example 2.3.7 that $\sigma(B) = \{0\}$. In particular,

$$B - I = \Phi_3'(t_0, x_0, \varphi(\cdot; t_0, x_0))$$

is continuously invertible. By Theorem 4.2.1, there exist neighborhoods $\mathcal{U}$ of (t_0, x_0) and $\mathcal{V}$ of $\varphi(\cdot; t_0, x_0)$ such that for any $(\tau, \xi) \in \mathcal{U}$ there is a unique $\varphi \in \mathcal{V}$ such that

$$\Phi(\tau, \xi, \varphi) = o.$$

Moreover, this φ is continuously differentiable with respect to τ and ξ, and for the continuous mappings

$$\Theta(\cdot) := \frac{\partial \varphi}{\partial \tau}(\cdot; t_0, x_0) \qquad \text{and} \qquad \Xi(\cdot) := \frac{\partial \varphi}{\partial \xi}(\cdot; t_0, x_0)$$

we have, by Remark 4.2.2,

$$-f(t_0, x_0) + \int_{t_0}^{t} f_2'(s, \varphi(s; t_0, x_0))\Theta(s)\, \mathrm{d}s - \Theta(t) = o,$$

$$\eta + \int_{t_0}^{t} f_2'(s, \varphi(s; t_0, x_0))\Xi(s)\eta\, \mathrm{d}s - \Xi(t)\eta = o, \qquad \eta \in \mathbb{R}^N.$$

This means that Θ and Ξ solve the so-called *equation in variations*

$$\dot{y}(t) = f_2'(t, \varphi(t; t_0, x_0))y(t) \tag{4.2.4}$$

(this is a system of N linear equations for Θ and a system of $N \times N$ equations for Ξ) and fulfil the initial conditions

$$\Theta(t_0) = -f(t_0, x_0), \qquad \Xi(t_0) = \boldsymbol{I}. \tag{4.2.5}$$

In particular, $\Xi(\cdot)$ is a fundamental matrix of (4.2.4). ○

As an application of differentiability with respect to initial conditions we briefly sketch the approach to orbital stability of periodic solutions.

Example 4.2.6. Assume that we know a non-constant T-periodic solution φ_0 of an autonomous system

$$\dot{x} = f(x),$$

and that we are interested in the behavior of other solutions which start at time $t = 0$ near $\varphi_0(0) = x_0$. We assume that $f \in C^1(\mathcal{G})$, $\mathcal{G}$ is an open set in $\mathbb{R}^N$, and denote by $\varphi(\cdot, \xi)$ the solution satisfying $\varphi(0, \xi) = \xi$. Let

$$\mathcal{M} = \{x \in \mathbb{R}^N : (x - x_0, f(x_0))_{\mathbb{R}^N} = 0\}.$$

In order to show that a solution $\varphi(\cdot, \xi)$ exists on such an interval $[0, t(\xi)]$ that it meets $\mathcal{M} \cap \mathcal{U}$ ($\mathcal{U}$ is a neighborhood of x_0) for the first positive time $t(\xi)$ near T (T is the period of φ_0), see Figure 4.2.1, we can solve the equation

$$\Phi(t, \xi) := (\varphi(t, \xi) - x_0, f(x_0)) = 0$$

in the vicinity of the point (T, x_0). We have

$$\Phi_1'(T, x_0) = (f(x_0), f(x_0)) > 0$$

($f(x_0) \neq 0$ since φ_0 is non-constant) and

$$\Phi_2'(t, x_0)\eta = \left(\frac{\mathrm{d}\varphi}{\mathrm{d}\xi}(t, x_0)\eta, f(x_0)\right) = (\Xi(t, x_0)\eta, f(x_0))$$

(see the previous example) where $\Xi(t, x_0)$ is a fundamental matrix of the linear T-periodic equation $\dot{y}(t) = f'(\varphi_0(t))y(t)$ (cf. (4.2.4)). So, we may use the Implicit Function Theorem to get a function $t(\xi)$ such that

$$\Phi(t(\xi), \xi) = 0, \qquad t(x_0) = T, \qquad \xi \in \mathcal{U}(x_0).$$

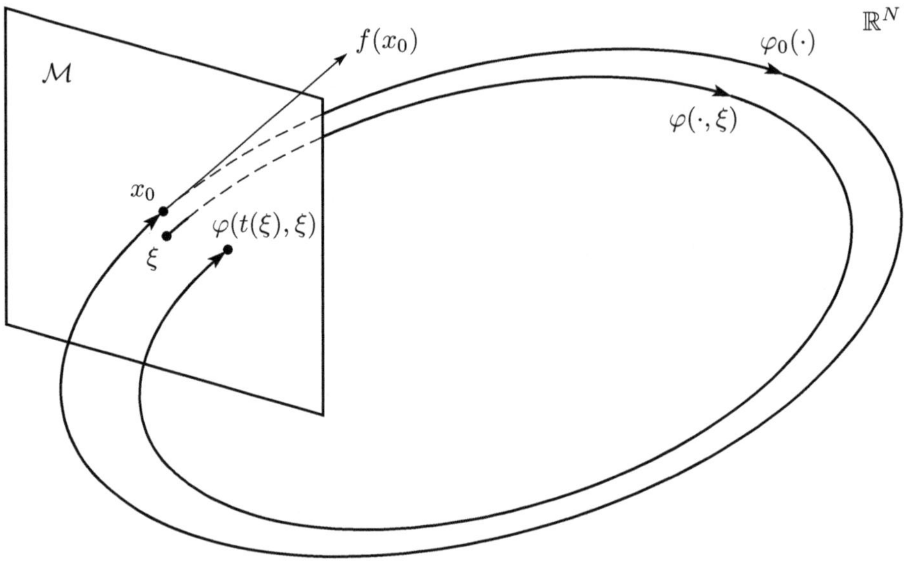

Figure 4.2.1.

By (4.2.2) we also have

$$\frac{\mathrm{d}t}{\mathrm{d}\xi}(T, x_0)\eta = -\frac{1}{\|f(x_0)\|^2}(\Xi(t, x_0)\eta, f(x_0)), \qquad \eta \in \mathbb{R}^N.$$

This allows us to investigate the behavior of the so-called *Poincaré mapping*

$$P(\xi) := \varphi(t(\xi), \xi), \qquad \xi \in \mathcal{U} \cap \mathcal{M}.$$

The asymptotic orbital stability of φ_0 can be defined by the requirement

$$\lim_{n\to\infty} P^n(\xi) = x_0.$$

For more details the interested reader can consult, e.g., Amann [4, Section 23].○

We are often interested in asymptotic behavior of solutions of a system of ordinary differential equations (linear or nonlinear), e.g., boundedness of solutions or its convergence to some special solutions (constant, periodic, etc.). In the following example we briefly sketch a method which can be used.

Example 4.2.7. Consider the equation

$$\dot{x} = \boldsymbol{A}x + f \tag{4.2.6}$$

where $\boldsymbol{A}$ is a constant $N \times N$ matrix and $f\colon \mathbb{R} \to \mathbb{R}^N$ is bounded and continuous on $\mathbb{R}$ ($f \in BC(\mathbb{R}, \mathbb{R}^N)$). We are interested in bounded solutions of (4.2.6) on $\mathbb{R}$. Let us assume $\sigma(\boldsymbol{A}) \cap \mathrm{i}\mathbb{R} = \emptyset$. With help of Functional Calculus (Theorem 1.1.38, in particular, Remark 1.1.39(i), (ii)) we can construct two projections P^+, P^- onto complementary subspaces X^+, X^- of $\mathbb{R}^N$ which commute with $A \in \mathcal{L}(\mathbb{R}^N)$ ($\boldsymbol{A}$ is the matrix representation of A in the standard basis) and such that

$$\sigma(A^+) = \sigma(A) \cap \{z \in \mathbb{C} : \operatorname{Re} z > 0\}, \qquad \sigma(A^-) = \sigma(A) \cap \{z \in \mathbb{C} : \operatorname{Re} z < 0\}$$

(A^+, A^- are the restrictions of A to X^+, X^-, respectively). With help of the Variation of Constants Formula it can be proved that for any $f \in BC(\mathbb{R}, \mathbb{R}^N)$ there is a unique solution x of (4.2.6) in the space $BC(\mathbb{R}, \mathbb{R}^N)$, and this solution is given by the formula

$$x(t) = \int_{-\infty}^{t} \mathrm{e}^{(t-s)A^-} P^- f(s)\,\mathrm{d}s - \int_{t}^{+\infty} \mathrm{e}^{(t-s)A^+} P^+ f(s)\,\mathrm{d}s.^2 \tag{4.2.7}$$

If we are interested in bounded solutions only on $\mathbb{R}^+ := [0, \infty)$, a similar computation shows that all such solutions for $f \in BC(\mathbb{R}^+, \mathbb{R}^N)$ are given by

$$x(t) = \mathrm{e}^{tA^-} x^- + \int_{0}^{t} \mathrm{e}^{(t-s)A^-} P^- f(s)\,\mathrm{d}s - \int_{t}^{\infty} \mathrm{e}^{(t-s)A^+} P^+ f(s)\,\mathrm{d}s \tag{4.2.8}$$

where x^- is an arbitrary point in X^-.

[2]The interested reader can check this formula and also (4.2.8) as an exercise on the use of the Variation of Constants Formula.
Hint. Use the estimates $\|\mathrm{e}^{tA}x\| \le c\mathrm{e}^{-\alpha t}\|x\|$ for $x \in X^-$, $t > 0$, and $\|\mathrm{e}^{tA}x\| \le c\mathrm{e}^{\alpha t}\|x\|$ for $x \in X^+$, $t < 0$, where the positive constants α, c are independent of t and x, α is such that $\sigma(A) \cap \{\lambda \in \mathbb{C} : |\operatorname{Re}\lambda| \le \alpha\} = \emptyset$ and c depends on α only. These estimates follow from Functional Calculus (see Exercise 1.1.42) and they ensure that integrals in (4.2.7) do exist. Apply P^+ to both sides of the Variation of Constants Formula and send $t \to \infty$ to obtain $P^+x(t) = -\int_t^\infty \mathrm{e}^{(t-s)A^+} P^+ f(s)\,\mathrm{d}s$ provided x is a bounded solution. Similarly $P^-x(t)$.

Both formulae (4.2.7) and (4.2.8) may be used for finding bounded solutions to a semilinear equation

$$\dot{x} = \boldsymbol{A}x + f(x) \qquad \text{where} \quad f(o) = o, \ f'(o) = o, \ f \in C^1(\mathcal{U}) \tag{4.2.9}$$

($\mathcal{U}$ is a neighborhood of $o \in \mathbb{R}^N$). To do that we solve the corresponding nonlinear equations (4.2.7), (4.2.8) where $f(\cdot)$ is replaced by $g(x(\cdot))$ where g is bounded and $g(y) = f(y)$ in a neighborhood of 0. For details see Hale [75, Sections III.6 and IV.3]. A solution in (4.2.8) depends on the parameter x^-, so we have the equation

$$\Phi(\xi, \varphi)(t)$$
$$:= \varphi(t) - \mathrm{e}^{tA^-}\xi - \int_0^t \mathrm{e}^{(t-s)A^-} P^- g(\varphi(s))\,\mathrm{d}s + \int_t^\infty \mathrm{e}^{(t-s)A^+} P^+ g(\varphi(s))\,\mathrm{d}s = o$$

with $\Phi\colon X^- \times BC(\mathbb{R}^+, \mathbb{R}^N) \to BC(\mathbb{R}^+, \mathbb{R}^N)$ (check it – you have to use the estimates given in footnote 2 on page 163). This formulation is suitable for the use of the Implicit Function Theorem. We leave details to the interested reader. The graph of the mapping

$$\kappa\colon \xi \in X^- \mapsto P^+\varphi(0, \xi)$$

is the so-called *local stable manifold* $\mathscr{W}^{\mathrm{s}}_{\mathrm{loc}}(x_0)$ of the equation (4.2.9) ($\varphi(\cdot, \xi)$ is a solution of $\Phi(\xi, \varphi) = o$).[3] It follows from the formula (4.2.2) that

$$\kappa'(o) = o,$$

i.e., $\mathscr{W}^{\mathrm{s}}_{\mathrm{loc}}(x_0)$ is tangent to the stable manifold X^- of the linear equation $\dot{x} = \boldsymbol{A}x$, see Figure 4.2.2. ○

Remark 4.2.8. It is sometimes convenient to define a solution of nonlinear, in particular, partial differential equations, more generally, not assuming that a solution has all classical derivatives which appear in the equation (see Chapters 5, 6, 7 and 8). Actually, we have seen one such possibility in the reformulation of a differential equation as an integral equation

$$x = F(x)$$

[3]The so-called *stable manifold* $\mathscr{W}^{\mathrm{s}}(x_0)$ of the stationary point x_0 of the equation $\dot{x} = g(x)$ ($g(x_0) = o$) is defined as follows: Let $\psi(\cdot, \xi)$ be a solution of this differential equation satisfying the initial condition $\psi(0, \xi) = \xi$. Then the stable manifold is $\mathscr{W}^{\mathrm{s}}(x_0) = \left\{\xi : \lim\limits_{t\to\infty} \psi(t, \xi) = x_0\right\}$ and a *local stable manifold* is defined by $\mathscr{W}^{\mathrm{s}}_{\mathrm{loc}}(x_0) := \{\xi \in \mathscr{W}^{\mathrm{s}}(x_0) : \psi(t, \xi) \in \mathcal{U} \text{ for } t \geq 0\}$ where $\mathcal{U}$ is a neighborhood of x_0.
Notice the crucial assumption $\sigma(\boldsymbol{A}) \cap \mathrm{i}\mathbb{R} = \emptyset$ (i.e., o is the so-called *hyperbolic stationary point* of the equation (4.2.9)) in the above argument. Figure 4.2.2 shows also the distinction between stable and local stable manifolds. It is worth mentioning that a similar approach cannot be used in the case $\sigma(A) \cap \mathrm{i}\mathbb{R} \neq \emptyset$. Since there can exist eigenvalues on the imaginary axis of the multiplicity greater than 1, we cannot expect a manifold consisting of bounded solutions. To get the so-called *central manifold* we are forced to solve a nonlinear version of the equations (4.2.7) in a weighted space instead of $BC(\mathbb{R}, \mathbb{R}^N)$. However, this problem is more difficult due to the lack of differentiability of the Nemytski operator (see footnote 12 on page 135). For details see, e.g., Chow, Li & Wang [33, Chapter 1] and references given there.

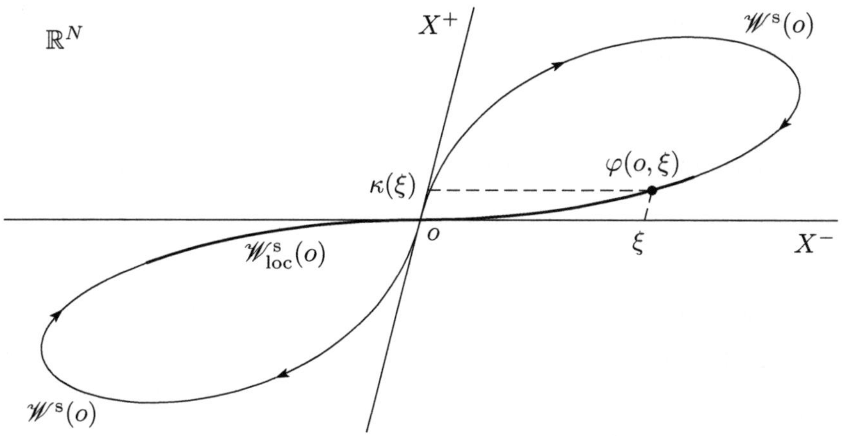

Figure 4.2.2.

where F is given by the formula (2.3.6). Having a more general notion of solution a natural question arises: Under what conditions is this solution smoother, in particular, is it a "classical" solution? Such results are known as regularity assertions. The Implicit Function Theorem can be occasionally used to prove such statements. See Theorem 7.1.14.

Exercise 4.2.9. Let $f\colon \mathbb{R}^M \to \mathbb{R}^N$ and let Φ be a diffeomorphism defined on a neighborhood $\mathcal{U}$ of the graph of f onto $\mathcal{V} \subset \mathbb{R}^{M+N}$. Write

$$\Phi^{-1}(\xi,\eta) = (\psi^1(\xi,\eta), \psi^2(\xi,\eta)) \qquad \text{for} \quad (\xi,\eta) \in \mathcal{V}.$$

This means the graph of f is isomorphic to

$$\Gamma = \{(\xi,\eta) \in \mathbb{R}^{M+N} : \psi^2(\xi,\eta) - f(\psi^1(\xi,\eta)) = o\}.$$

The Implicit Function Theorem yields conditions for Γ to be the graph of a function $\eta = g(\xi)$.

(i) Formulate these conditions!

(ii) Express the derivative of f in terms of the derivative of g.
Hint. $f'(\Phi^{-1}) = [(\psi^2)_2' g' + (\psi^2)_1'] \circ [(\psi^1)_1' + (\psi^1)_2' g']^{-1}$.
Control question: Have you checked that the second term on the right-hand side is an isomorphism of $\mathbb{R}^M$ onto $\mathbb{R}^M$?

(iii) Without using the general result from (ii) transform the equation

$$\frac{\mathrm{d}y}{\mathrm{d}x} = f(x,y)$$

into polar coordinates!

Exercise 4.2.10. Let M be a metric space and $f\colon M \times \mathbb{R} \to \mathbb{R}$ a continuous map. Let $c > 0$ be such that for all $x \in M$, $y_1, y_2 \in \mathbb{R}$, we have

$$(f(x,y_1) - f(x,y_2))(y_1 - y_2) \ge c|y_1 - y_2|^2.$$

Prove that for any $x \in M$ there exists a unique $y(x) \in \mathbb{R}$ such that

$$f(x,y(x)) = 0$$

and, moreover, $y\colon x \mapsto y(x)$ is a continuous map from M into $\mathbb{R}$.

Hint. Use the properties of real functions of one real variable.

Exercise 4.2.11. Let M be a normed linear space, let f be as in Exercise 4.2.10 and, moreover, $f \in C^k(M \times \mathbb{R})$ with some $k \in \mathbb{N}$. Then the implicit function $y = y(x)$ from Exercise 4.2.10 is of the class $C^k(M)$. Prove it!

Hint. Use Theorem 4.2.1.

Exercise 4.2.12. Give details which are omitted in Example 4.2.7.

Exercise 4.2.13. Let A be a densely defined linear operator in a Hilbert space. Assume that A has a compact self-adjoint resolvent. Extend the construction of the local stable manifold (Example 4.2.7) to the equation (4.2.6). See Exercise 3.1.28 for the properties of this equation.

Exercise 4.2.14. Assume that

$$f(x,y) = \sum_{j,k=0}^{\infty} a_{jk}(x-x_0)^j(y-y_0)^k, \qquad |x-x_0| < \alpha, \quad |y-y_0| < \beta.$$

Moreover, let $a_{00} = 0$, $a_{01} \ne 0$. Apply the Implicit Function Theorem and show that the implicit function $y(x)$ is the sum of a power series in a neighborhood of x_0.

Note that for complex variables the result follows directly from the properties of holomorphic functions and Theorem 4.2.1. In the real case one has to prove that the formal power series for $y(x)$ has a positive radius of convergence.

4.3 Local Structure of Differentiable Maps, Bifurcations

We now revert to the topic of Example 4.1.6, i.e., to the case when the assumptions of the Local Inverse Function Theorem (Theorem 4.1.1) are violated. In particular, it has been mentioned there that the assumptions of the Local Inverse Function Theorem are never satisfied for $f\colon \mathbb{R}^M \to \mathbb{R}^N$ provided $M \ne N$. In the first part we will study local behavior of such mappings. In the second part we stress the main idea of the Lyapunov–Schmidt Reduction and the approach to bifurcation phenomena (Crandall–Rabinowitz Bifurcation Theorem).

Definition 4.3.1. Let $f\colon X \to Y$ be a differentiable map in a neighborhood of a point $a \in X$. If $f'(a)$ is neither injective nor surjective, then a is called a *singular point* of f.

The following proposition deals with the first non-singular case for the mapping $f\colon \mathbb{R}^M \to \mathbb{R}^N$, $M < N$. For the second one see Proposition 4.3.8.

Proposition 4.3.2. *Let $f\colon \mathbb{R}^M \to \mathbb{R}^N$ be a differentiable map on an open set $\mathcal{G} \subset \mathbb{R}^M$. Let $a \in \mathcal{G}$ and let $f'(a)$ be injective. Let Q be a (linear) projection of $\mathbb{R}^N$ onto $Y_1 := \operatorname{Im} f'(a)$. Then there exist neighborhoods $\mathcal{U}$ of a, $\mathcal{V}$ of $Qf(a)$ in Y_1, a diffeomorphism φ of $\mathcal{U}$ onto $\mathcal{V}$ and a differentiable map $g\colon \mathcal{V} \to \mathbb{R}^N$ such that*

$$f = g \circ \varphi$$

(*see Figure* 4.3.1).

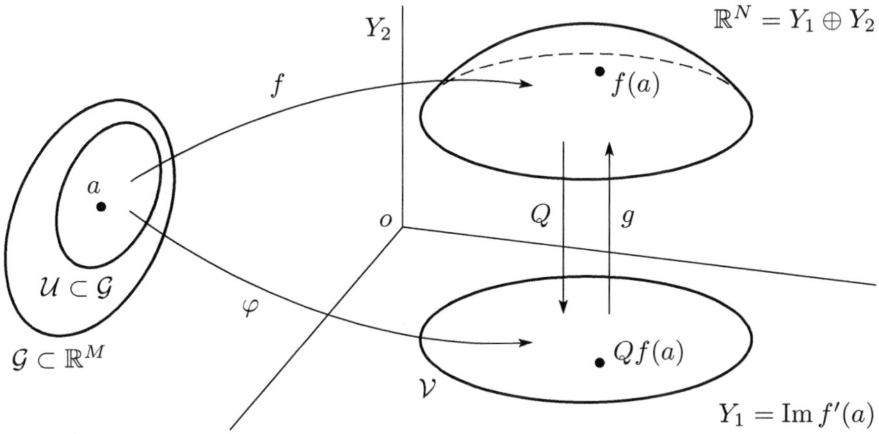

Figure 4.3.1.

Proof. The proof is almost obvious from Figure 4.3.1. Put $\varphi = Q \circ f$. Then

$$\varphi'(a) = Qf'(a)$$

is an isomorphism of $\mathbb{R}^M$ onto Y_1. Since $\dim Y_1 = M$ is finite, Y_1 is a Banach space (as a closed subspace of the Banach space $\mathbb{R}^N$) and, by Theorem 4.1.1, φ is a diffeomorphism of a neighborhood $\mathcal{U}$ of a onto a neighborhood (in Y_1) $\mathcal{V}$ of $Qf(a)$. It suffices to put

$$g = f \circ \varphi^{-1}. \qquad \square$$

Remark 4.3.3.

(i) We have used the finite dimension of $Y := \mathbb{R}^N$ to ensure both the existence of a continuous linear projection Q and the closedness of the range $\operatorname{Im} f'(a)$. If $f\colon X \to Y$, X, Y are Banach spaces, then neither of these two conditions has to be satisfied. It follows from the proof that Proposition 4.3.2 holds under these two additional assumptions. We notice that these assumptions are superfluous provided X has a finite dimension (see Remark 2.1.19).

(ii) It is also easy to prove that

$$\Psi(y) := g(Qy) - (I - Q)y - Qf(a)$$

is a diffeomorphism of a neighborhood $\mathcal{W}$ of $b = f(a)$ onto a neighborhood $\tilde{\mathcal{W}}$ of o in $\mathbb{R}^N$. Indeed,

$$\Psi(b) = o, \qquad \Psi'(b)k = f'(a)h - (I - Q)k$$

where $h \in \mathbb{R}^M$ is such that

$$Qk = \varphi'(a)h.$$

Moreover, $y \in f(\mathcal{G}) \cap \mathcal{W}$ if and only if there is an $x \in \mathcal{G}$ such that

$$y = f(x) \qquad \text{and} \qquad (I - Q)\Psi(f(x)) = o.$$

This means that there exists a local (nonlinear) transformation of coordinates in $\mathcal{W}$ (given by Ψ) such that $f(\mathcal{G}) \cap \mathcal{W}$ is expressed by

$$z_{M+1} = \cdots = z_N = 0$$

in these new coordinates.

(iii) An interpretation similar to (ii) follows:

$$(I - Q)f = (I - Q)g(\varphi) = (I - Q)g(Qf), \qquad \Phi(Qf) := (I - Q)g(Qf).$$

This means that after a linear transformation of coordinates the last $N - M$ components of f (i.e., $(I - Q)f$) depend (via Φ) on the first M components of f in a neighborhood of a. Compare this local nonlinear result to the linear one for the equation

$$Ax = y, \qquad A \in \mathcal{L}(\mathbb{R}^M, \mathbb{R}^N).$$

(iv) A map f which satisfies the assumptions of Proposition 4.3.2 at each point $a \in \mathcal{G}$ is often called an *immersion* of $\mathcal{G}$ into $\mathbb{R}^N$. An injective immersion which is also a homeomorphism of $\mathcal{G}$ onto $f(\mathcal{G})$ (in the induced topology from $\mathbb{R}^N$) is called an *embedding*. Some examples of immersions which are not embeddings are shown in Figures 4.3.2 and 4.3.3. We note that we have already used the term embedding for an injective continuous linear operator.

Figure 4.3.2. Immersion

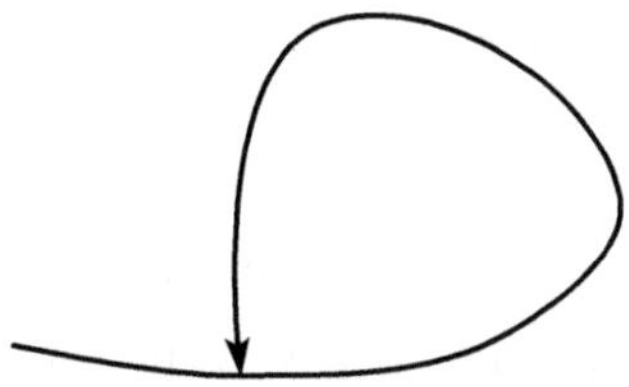

Figure 4.3.3. Injective immersion

Further examination of Proposition 4.3.2 leads to the following definition of a differentiable manifold. This notion is basic for differential geometry and global nonlinear analysis. In this textbook we will mostly use it for purposes of terminology only. Some basic facts on manifolds are given in Appendix 4.3A and will be used for developing the notion of degree (Appendix 5.7A) and study of certain differential equations on manifolds (Section 8.4) .

Definition 4.3.4. A *differentiable manifold* of dimension M and of the class C^k is a subset $\mathscr{M}$ of $\mathbb{R}^N$ $(N \geq M)$ with the following property:

For each $x \in \mathscr{M}$ there is a neighborhood $\mathcal{W}$ of x (in $\mathbb{R}^N$) and a C^k-diffeomorphism ψ of $\mathcal{W}$ into $\mathbb{R}^N$ such that

$$\psi(\mathscr{M} \cap \mathcal{W}) = \{y = (y_1, \dots, y_N) \in \mathbb{R}^N : y_{M+1} = \cdots = y_N = 0\} \cap \psi(\mathcal{W}).$$

A relative neighborhood $\mathcal{W} \cap \mathscr{M}$ together with ψ is called a *(local) chart* at the point $x \in \mathscr{M}$. The first M coordinates $(y_1, \dots, y_M)$ are called the *local coordinates* of x on $\mathscr{M}$. The collection of all charts of $\mathscr{M}$ is called an *atlas* of $\mathscr{M}$.

Example 4.3.5.

(i) An open subset $\mathcal{G} \subset \mathbb{R}^M$ is an M-dimensional differentiable manifold of the class C^k for any $k \in \mathbb{N}$ (i.e., of the class C^∞).

(ii) The graph of a function $f\colon \mathbb{R}^M \to \mathbb{R}$, $f \in C^k(\mathcal{G})$, $\mathcal{G}$ an open subset of $\mathbb{R}^M$, is an M-dimensional differentiable manifold of the class C^k in $\mathbb{R}^N$, $N \geq M+1$.

(iii) Let

$$\mathcal{S}^2 = \{(x, y, z) \in \mathbb{R}^3 : x^2 + y^2 + z^2 = 1\}$$

be the 2-dimensional sphere. Then $\mathcal{S}^2$ is a 2-dimensional differentiable manifold of the class C^∞ in $\mathbb{R}^N$, $N \geq 3$. Indeed, a chart for the upper open half-sphere can be constructed as follows: let

$$\psi(x, y, z) = (x, y, z - \sqrt{1 - x^2 - y^2}),$$
$$\mathcal{W} = \{(x, y, z) \in \mathbb{R}^3 : x^2 + y^2 < 1, \ z > 0\}.$$

Then ψ is a diffeomorphism of $\mathcal{W}$ into $\mathbb{R}^3$ and

$$\psi(\mathcal{W} \cap \mathcal{S}^2) = \{(u, v, w) \in \mathbb{R}^3 : u^2 + v^2 < 1, w = 0\}.$$

We will see a more comfortable proof in Example 4.3.10. ○

Definition 4.3.6. Let X, Y be Banach spaces, $f\colon X \to Y$ a differentiable map on a neighborhood of a point $a \in X$. If $f'(a)$ is a surjective map onto Y, then the point a is called a *regular point*. If a is not a regular point, then it is called a *critical point*. A value $b \in Y$ is called a *critical value* of f provided the set

$$f_{-1}(b) := \{x \in X : f(x) = b\}$$

contains a critical point. In the other case, b is a *regular value*.

Remark 4.3.7. There is a difference between the notion of a singular point (Definition 4.3.1) and a critical point. For example, if $f\colon \mathbb{R}^M \to \mathbb{R}^N$, $M < N$, then all points in $\mathbb{R}^M$ are critical (but some of them can be non-singular). The importance of the notion of a critical point will be more apparent in connection with the Sard Theorem (Theorem 5.7.3) and its applications.

Proposition 4.3.8. *Let $\mathcal{G}$ be an open subset of $\mathbb{R}^M$, $f\colon \mathcal{G} \to \mathbb{R}^N$, $f \in C^k(\mathcal{G})$. Let $a \in \mathcal{G}$ be a regular point of f. Then there are neighborhoods $\mathcal{U}$ of $o \in \mathbb{R}^M$, $\mathcal{V}$ of a, and a diffeomorphism $\varphi \in C^k$ of $\mathcal{U}$ onto $\mathcal{V}$ such that*

$$\{x \in \mathcal{V} : f(x) = f(a)\} = \varphi(\mathcal{U} \cap \operatorname{Ker} f'(a))$$

(see Figure 4.3.4).

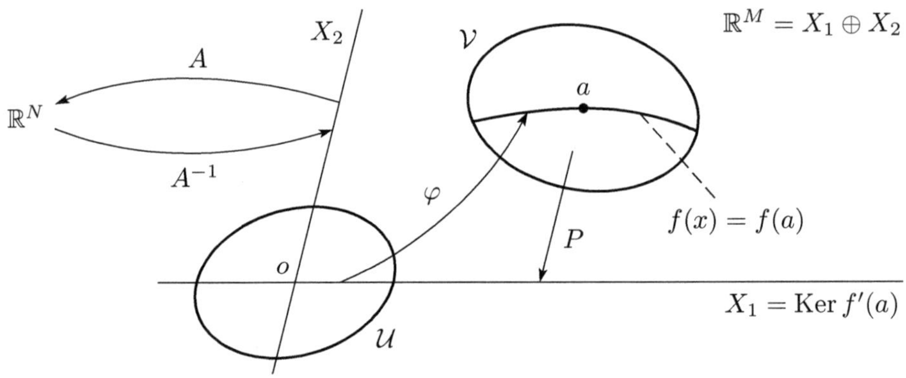

Figure 4.3.4.

Proof. By Remark 4.3.7, $M \geq N$. If $M = N$, then Theorem 4.1.1 can be applied. Therefore, we assume that $M > N$. Denote by P a (linear continuous) projection of $X := \mathbb{R}^M$ onto $X_1 := \operatorname{Ker} f'(a)$ and by X_2 the complementary subspace given by $X_2 = \operatorname{Im}(I - P)$. If A is the restriction of $f'(a)$ to X_2, then A is an isomorphism of X_2 onto $\mathbb{R}^N$ (A is both injective and surjective). Denote by A^{-1} the inverse isomorphism of $\mathbb{R}^N$ onto X_2 (A^{-1} is also called a *right inverse* of $f'(a)$). We can rewrite f in the following way:

$$f(x) = f(a) + f'(a)[A^{-1}(f(x) - f(a)) + P(x - a)].$$

Let us denote

$$\psi(x) = A^{-1}(f(x) - f(a)) + P(x - a).$$

A simple calculation shows that

$$\psi'(a)h = A^{-1}f'(a)h + Ph = (I - P)h + Ph = h \qquad \text{for any} \quad h \in X.$$

Since $\psi(a) = o$, ψ is a diffeomorphism of a neighborhood $\mathcal{V} \subset \mathcal{G}$ of a onto a neighborhood $\mathcal{U}$ of o (Theorem 4.1.1). Further, $x \in \{y \in \mathcal{V} : f(y) = f(a)\}$ if and only if $x \in \mathcal{V}$ and $\psi(x) = P(x - a)$, i.e.,

$$\psi(x) \in \mathcal{U} \cap \operatorname{Ker} f'(a).$$

The desired diffeomorphism φ is the inverse of ψ. □

Remark 4.3.9.

(i) Proposition 4.3.8 together with its proof also holds for $f\colon X \to Y$, X, Y Banach spaces provided there exists a linear continuous projection P of X onto $\operatorname{Ker} f'(a)$. The continuity of A^{-1} follows in this case from the Open Mapping Theorem (Theorem 2.1.8). The existence of such a projection P can be shown in two important cases, namely, when Y has finite dimension (and therefore $\operatorname{Ker} f'(a)$ has finite codimension – Example 2.1.12) or $\operatorname{Ker} f'(a)$ has finite dimension (Remark 2.1.19).

(ii) Notice that φ can be viewed as a local (nonlinear) transformation of coordinates in which f is a linear map, namely

$$f(\varphi(y)) = f(a) + f'(a)y, \qquad y \in \mathcal{U}.$$

This formula also shows that all points in $\mathcal{V}$ are regular. Moreover, if z is sufficiently close to $b = f(a)$, then

$$y = A^{-1}(z - b) \in \mathcal{U} \qquad \text{and} \qquad f(\varphi(y)) = z.$$

This shows that $f(\mathcal{G})$ is an open set in $\mathbb{R}^N$ provided all points of $\mathcal{G}$ are regular.

(iii) In the terms of differentiable manifolds (Definition 4.3.4) the statement of Proposition 4.3.8 can be formulated as follows:

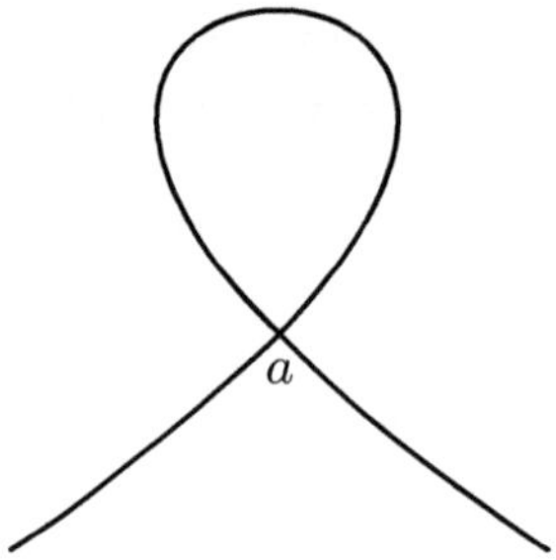

Figure 4.3.5.

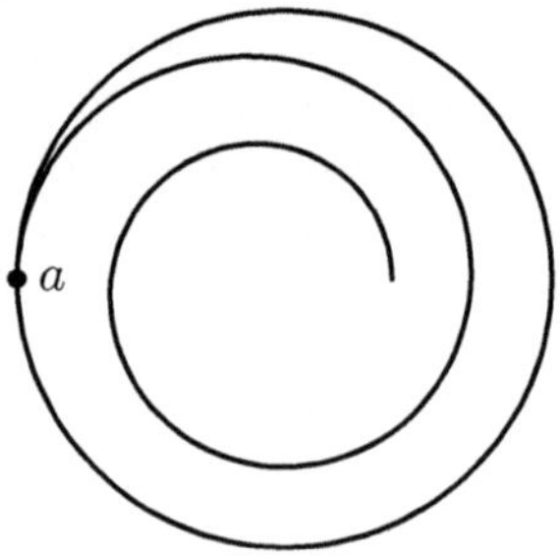

Figure 4.3.6.

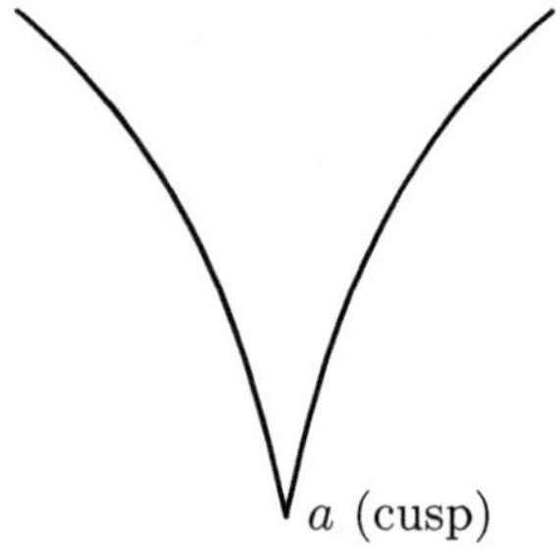

Figure 4.3.7.

If $f\colon \mathbb{R}^M \to \mathbb{R}^N$ *is a differentiable map in an open set* $\mathcal{G} \subset \mathbb{R}^M$, $b \in \mathbb{R}^N$, *then the set*

$$\{x \in \mathcal{G} : f(x) = b\}$$

is a differentiable manifold (either empty or of dimension $M - N$*) provided* b *is a regular value of* f.

(iv) Proposition 4.3.8 imposes certain restrictions on the set

$$\{x \in \mathbb{R}^M : f(x) = f(a)\}.$$

In Figures 4.3.5–4.3.7 there are some cases in which a is not a regular point (i.e., it is a critical point). The value $f(a)$ is critical in all cases.

Example 4.3.10. The sphere $\mathcal{S}^2$ is a C^∞-differentiable manifold. To see this it is sufficient to use Remark 4.3.9(iii) for

$$f(x, y, z) = x^2 + y^2 + z^2 - 1, \qquad b = 0. \qquad \bigcirc$$

The assertions of the last two propositions are part of the following more general result.

Theorem 4.3.11 (Rank Theorem). *Let* $f\colon \mathbb{R}^M \to \mathbb{R}^N$ *be a differentiable map on an open subset* $\mathcal{G} \subset \mathbb{R}^M$ *and let the dimension of* $\operatorname{Im} f'(x)$ *be constant for* $x \in \mathcal{G}$ *(and equal to* $L \in \mathbb{N}$*). Then for any* $a \in \mathcal{G}$ *there exist neighborhoods* $\mathcal{U}$ *of* a, $\mathcal{W}$ *of* $b = f(a)$, *cubes* $\mathcal{C}$ *in* $\mathbb{R}^M$, $\mathcal{D}$ *in* $\mathbb{R}^N$ *and diffeomorphisms* $\Phi\colon \mathcal{C} \to \mathcal{U}$, $\Psi\colon \mathcal{W} \to \mathcal{D}$ *such that the map* F *defined by* $F = \Psi \circ f \circ \Phi$ *has the form*

$$F(z_1, \dots, z_M) = (z_1, \dots, z_L, 0, \dots, 0) \qquad \textit{for all} \quad z = (z_1, \dots, z_M) \in \mathcal{C}$$

(see Figure 4.3.8).

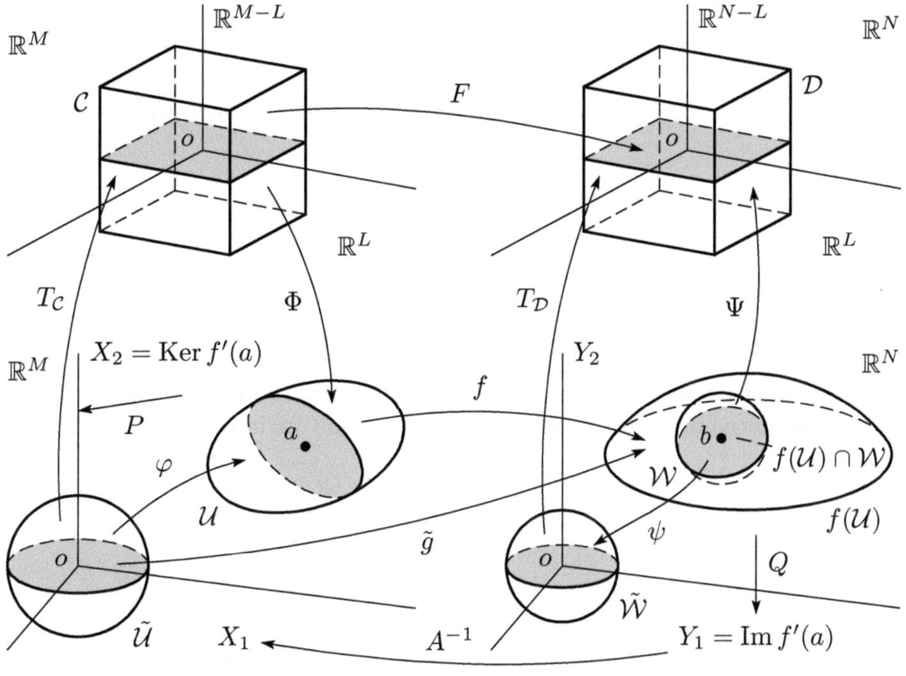

Figure 4.3.8.

Proof. Denote $X_2 = \operatorname{Ker} f'(a)$, P a (linear) projection in $\mathbb{R}^M$ onto X_2, $X_1 = \operatorname{Ker} P$ and, similarly, $Y_1 = \operatorname{Im} f'(a)$, Q a (linear) projection in $\mathbb{R}^N$ onto Y_1, $Y_2 = \operatorname{Ker} Q$. Then the restriction A of $f'(a)$ to X_1 is an isomorphism of X_1 onto Y_1. Let A^{-1} be the inverse isomorphism, $A^{-1}\colon Y_1 \to X_1$. By the proof of Proposition 4.3.8,

$$\alpha(x) = A^{-1}Q(f(x) - f(a)) + P(x - a)$$

is a diffeomorphism of the neighborhood $\mathcal{U}$ of $a \in \mathbb{R}^M$ onto the neighborhood $\tilde{\mathcal{U}}$ of $o \in \mathbb{R}^M$. Denote by φ the inverse to α. For $h_1 \in X_1$ and $x \in \mathcal{U}$ we have

$$\alpha'(x)h_1 = A^{-1}Qf'(x)h_1.$$

This implies that $f'(x)$ is injective on X_1 ($\alpha'(x)$ is an isomorphism). Since

$$\dim X_1 = \dim \operatorname{Im} f'(x) = L,$$

the restriction of $f'(x)$ to X_1 is an isomorphism of X_1 onto $\operatorname{Im} f'(x)$. We can express this fact in the commutative diagram (Figure 4.3.9).
Using the decomposition $\mathbb{R}^M = X_1 \oplus X_2$, we write

$$u = u_1 + u_2 \qquad \text{for} \quad u \in \tilde{\mathcal{U}},\ u_i \in X_i,\ i = 1, 2,$$

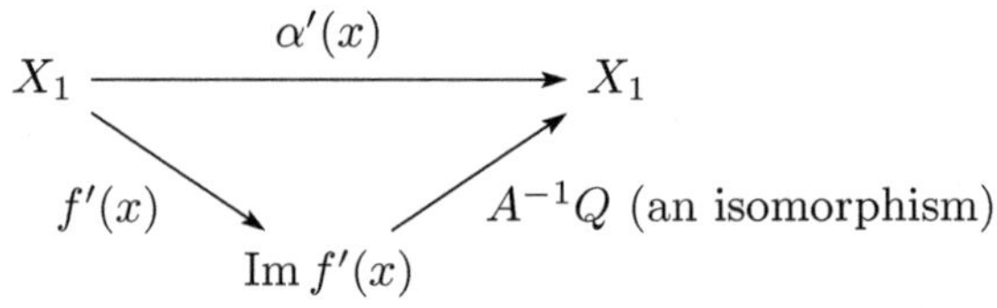

Figure 4.3.9.

and define

$$g(u_1, u_2) = f(\varphi(u_1 + u_2)).$$

Now, we show that g actually depends on the first variable only. To see this we compute the derivate of g with respect to the second variable:

$$g_2'(u_1, u_2)h_2 = f'(\varphi(u))\varphi'(u)h_2.$$

For $k := \varphi'(u)h_2$ and $\varphi(u) = x$ we have

$$h_2 = \alpha'(x)k = A^{-1}Qf'(x)k + Pk.$$

This means that $A^{-1}Qf'(x)k = o$. Since $A^{-1}Q$ is an isomorphism of $\operatorname{Im} f'(x)$ onto X_1 (see Figure 4.3.9), we have $f'(x)k = o$, i.e.,

$$g_2'(u_1, u_2)h_2 = o \qquad \text{for any} \quad h_2 \in X_2.$$

The Mean Value Theorem (Theorem 3.2.7) implies that

$$g(u_1, u_2) = g(u_1, o) \qquad \text{for} \quad (u_1, u_2), (u_1, o) \in \tilde{\mathcal{U}}.$$ [4]

This result is shown in Figure 4.3.8 by shaded areas. Put $\tilde{g}(u_1) := g(u_1, o)$.

We employ Proposition 4.3.2, in particular Remark 4.3.3(ii) to complete the proof. Replacing there $\tilde{g}$ for f, we obtain a diffeomorphism ψ of a neighborhood $\mathcal{W}$ of $b = f(a)$ onto a neighborhood $\tilde{\mathcal{W}}$ of $o \in \mathbb{R}^N$ such that

$$(I - Q)\psi(f(\mathcal{U}) \cap \mathcal{W}) = o$$

(see the right lower corner of Figure 4.3.8). We get cubes $\mathcal{C}$ and $\mathcal{D}$ by diffeomorphisms $T_{\mathcal{C}}$, $T_{\mathcal{D}}$ in $\mathbb{R}^M$, $\mathbb{R}^N$, respectively, which transform non-Cartesian coordinates α in $X_1 \oplus X_2$ or ψ in $Y_1 \oplus Y_2$ into Cartesian coordinates in $\mathbb{R}^M = \mathbb{R}^L \times \mathbb{R}^{M-L}$ ($T_{\mathcal{C}}(X_1) = \mathbb{R}^L$), or in $\mathbb{R}^N = \mathbb{R}^L \times \mathbb{R}^{N-L}$ ($T_{\mathcal{D}}(Y_1) = \mathbb{R}^L$), respectively (see the upper part of Figure 4.3.8 and page 152). □

[4]In fact, the use of Theorem 3.2.7 requires the segment joining (u_1, o) to (u_1, u_2) to lie in $\tilde{\mathcal{U}}$. Taking a smaller $\tilde{\mathcal{U}}$ if necessary we can assume that $\tilde{\mathcal{U}}$ is convex.
Notice that we have got a similar result at the end of the proof of Proposition 4.3.8 where we have considered only one fiber, namely $\{x : f(x) = f(a)\}$.

Remark 4.3.12. The assertion of the Rank Theorem can be formulated in a slightly less informative way as follows:

Under the hypotheses of Theorem 4.3.11, $f(\mathcal{G})$ *is a differentiable manifold of dimension* L.

Definition 4.3.13. Functions $f_1, \dots, f_N \colon \mathbb{R}^M \to \mathbb{R}^N$ are said to be *independent* in an open set $\mathcal{G} \subset \mathbb{R}^M$ if any point $x \in \mathcal{G}$ is regular for $f = (f_1, \dots, f_N)$. In the other case, the functions are called *dependent*.

The following assertion explains the notions of the dependent and independent functions.

Suppose the assumptions of the Rank Theorem are satisfied for

$$f = (f_1, \dots, f_L, f_{L+1}, \dots, f_N) \colon \mathbb{R}^M \to \mathbb{R}^N$$

where functions $f_1, \dots, f_L$ *are independent in a neighborhood of a point* $a \in \mathbb{R}^M$. *Then there is a smooth function* $G \colon \mathbb{R}^L \to \mathbb{R}^{N-L}$ *such that*

$$(f_{L+1}(x), \dots, f_N(x)) = G(f_1(x), \dots, f_L(x))$$

for x *in a certain neighborhood of* a.

To prove this assertion notice first that $\operatorname{Im} f'(a)$ is an L-dimensional subspace of $\mathbb{R}^N$ and can be identified with $\mathbb{R}^L \times \{0\}$. This means that

$$Qf(x) = H_1(x) := (f_1(x), \dots, f_L(x))$$

and, in the notation of the proof of Theorem 4.3.11,

$$f(x) = \tilde{g}(u_1) \qquad \text{where} \qquad u_1 = A^{-1}(H_1(x) - H_1(a)).$$

In particular, $f_{L+1}, \dots, f_N$ are smooth functions of $f_1, \dots, f_L$.

The notion of independent functions plays an important role also in the theory of ordinary differential equations. Indeed, let

$$\dot{x} = v(x)$$

be a system of M differential equations. A smooth non-constant function $f \colon \mathbb{R}^M \to \mathbb{R}$ is called the *first integral* of this system in an open set $\mathcal{G} \subset \mathbb{R}^M$ if for any $a \in \mathcal{G}$ there is an interval $\mathcal{I}_a$ such that for a solution $\varphi(\cdot, a)$ of the system such that

$$\varphi(0, a) = a,$$

we have that

$$\varphi(t, a) \in \mathcal{G} \qquad \text{and} \qquad \frac{\mathrm{d}}{\mathrm{d}t} f(\varphi(t, a)) = 0 \qquad \text{hold for} \quad t \in \mathcal{I}_a.$$

It has been proved in the theory of ordinary differential equations that a system $\dot{x} = v(x)$ ($v\colon \mathcal{G} \subset \mathbb{R}^M \to \mathbb{R}^M$ is smooth) has $M-1$ independent first integrals $f_1, \dots, f_{M-1}$ in a neighborhood $\mathcal{U}$ of any non-stationary point $a \in \mathcal{G}$. A smooth function $g\colon \mathcal{U} \to \mathbb{R}$ is the first integral if and only if $g, f_1, \dots, f_{M-1}$ are dependent on $\mathcal{U}$.

We remark that the knowledge of the first integrals reduces the original system. For example, let $f_1, \dots, f_{M-1}$ be independent first integrals in a neighborhood $\mathcal{U}$ of a non-stationary point a and assume, e.g., that $v_M \neq 0$. Then the transformation of coordinates

$$y_i = f_i(x), \quad i = 1, \dots, M-1, \qquad y_M = x_M$$

leads to a new system

$$\dot{y}_i = 0, \quad i = 1, \dots, M-1, \qquad \dot{y}_M = w(y) \qquad \text{for a certain function } w,$$

and after rescaling in time $\left(t(s) = \int\limits_0^s \frac{d\sigma}{w(\varphi(\sigma,a))} \right)$ to

$$\frac{dy_i}{ds} = 0, \quad i = 1, \dots, M-1, \qquad \frac{dy_M}{ds} = 1.$$

For another interpretation and a generalization of the notion of the first integral see Exercise 4.3.26 and the end of Appendix 4.3A.

Remark 4.3.14. A result similar to the Rank Theorem holds also for a differentiable map $f\colon X \to Y$ where X, Y are Banach spaces. The delicate question is the existence of continuous linear projections P of X (onto $\operatorname{Ker} f'(a)$) and Q of Y (onto $\operatorname{Im} f'(a)$). Such projections exist provided $f'(a)$ is a Fredholm operator, i.e., $\operatorname{Ker} f'(a)$ has finite dimension and $\operatorname{Im} f'(a)$ is a closed subspace of finite codimension in Y (see page 71). Notice that the equation $f(x) = y$ can be solved by the following procedure which is often called the *Lyapunov–Schmidt Reduction*:

The equation

$$f(x) = y$$

is equivalent to the couple of equations

$$y_1 := Qy = Qf(x_1 + x_2), \qquad y_2 := (I-Q)y = (I-Q)f(x_1+x_2)$$

where

$$x = x_1 + x_2, \qquad x_2 = Px.$$

Suppose that the first equation may be solved[5] for x_1 assuming x_2 to be fixed (looking at x_2 as to a parameter). We obtain

$$x_1 = g(y_1, x_2).$$

[5]E.g., by the Implicit Function Theorem (Theorem 4.2.1) in the vicinity of a known solution $b = f(a)$ since $f'(a)$ is an isomorphism of X_1 onto Y_1 or, more generally, by an iteration process.

The second equation is now an equation (it is called the *bifurcation equation* or the *alternative problem*) of the form

$$(I-Q)f(x_2+g(y_1,x_2)) = y_2 \qquad \text{for an unknown } x_2.$$

If $f'(a)$ is a Fredholm map, then this equation is an equation in finite-dimensional spaces:

$$x_2 \in \operatorname{Ker} f'(a),\ y_2 \in Y_2,\ \dim \operatorname{Ker} f'(a) < \infty,\ \dim Y_2 = \operatorname{codim} \operatorname{Im} f'(a) < \infty.$$

Notice that the Implicit Function Theorem ensures a unique local solution to the first equation for y sufficiently close to $b = f(a)$. In this situation we also obtain

$$g_2'(b_1, a_2) = o,$$

i.e., the point a_2 is a critical point for

$$F(x_2) := (I-Q)f(x_2 + g(b_1, x_2)) - b_2.$$

The simplest case for the local study of F is that

$$\operatorname{codim} \operatorname{Im} f'(a) = 1, \qquad \text{i.e.,} \qquad F\colon X_2 = \operatorname{Ker} f'(a) \to \mathbb{R}$$

(see Example 4.3.20). Notice that $\dim X_2$ is finite for $f'(a)$ being a Fredholm map.

Example 4.3.15. As an application we will investigate the existence of a solution of the following boundary value problem for a system of ordinary differential equations

$$\begin{cases} \dot{x}(t) = f(t, x(t)), & t \in (0,1), \\ x(0) = x(1). \end{cases} \tag{4.3.1}$$

We suppose (see Theorem 2.3.4) that f together with its partial derivatives with respect to the variables $x = (x_1, \dots, x_N)$ are continuous on $[0,1] \times \mathbb{R}^N$. We know that any solution starting at $t = 0$ satisfies the integral equation

$$x(t) - x(0) = \int_0^t f(s, x(s))\,\mathrm{d}s$$

for all t from the interval of its existence. This means that x satisfies the boundary value problem (4.3.1) if and only if

$$G(x_0) := \int_0^1 f(s, x(s, x_0))\,\mathrm{d}s = o.$$

Here $x(\cdot, x_0)$ denotes a (unique) solution of $\dot{x}(t) = f(t, x(t))$ such that $x(0, x_0) = x_0$. The problem of solving the equation

$$G(x_0) = o \qquad \text{for} \quad G\colon \mathbb{R}^N \to \mathbb{R}^N$$

is a nontrivial topological task which we will deal with in Chapter 5.

Notice that we cannot use the Implicit Function Theorem directly since there is no parameter in (4.3.1). Therefore we modify the problem by adding a multiplicative parameter ε to (4.3.1), i.e., we investigate the problem

$$\begin{cases} \dot{x}(t) = \varepsilon f(t, x(t)), & t \in (0,1), \\ x(0) = x(1). \end{cases} \tag{4.3.2}$$

Notice that for $\varepsilon = 0$ any N-dimensional constant a solves (4.3.2). To be able to use the abstract approach described above we rewrite (4.3.2) in an operator form. To do this we define Banach spaces

$$X = \{x \in C([0,1], \mathbb{R}^N) : x(0) = x(1)\}, \qquad Y = \{y \in C([0,1], \mathbb{R}^N) : y(0) = o\}$$

and operators $L, N \colon X \to Y$:

$$Lx \colon t \mapsto x(t) - x(0), \qquad N(x) \colon t \mapsto \int_0^t f(s, x(s))\, \mathrm{d}s, \qquad t \in [0,1].$$

Then the system (4.3.2) is equivalent to the operator equation

$$G(x, \varepsilon) := Lx - \varepsilon N(x) = o. \tag{4.3.3}$$

The operator L is linear and continuous, therefore differentiable:

$$L'(x)h = Lh \qquad \text{for} \qquad h \in X.$$

The operator N is also continuously differentiable and

$$N'(x)h \colon t \mapsto \int_0^t f_2'(s, x(s))h(s)\, \mathrm{d}s, \qquad t \in [0,1],\ h \in X.$$

Check this expression yourself, see also Example 3.2.21. This means that

$$G_1'(a, 0)h = Lh$$

is not injective and $X_2 := \operatorname{Ker} L$ consists of N-dimensional constant functions. Moreover,

$$Y_1 := \operatorname{Im} L = \{y \in Y : y(1) = y(0) = o\}.$$

There are continuous linear projections P, Q onto closed subspaces X_2 and Y_1, respectively, given by

$$Px \colon t \mapsto x(0), \qquad Qy \colon t \mapsto y(t) - ty(1).$$

Having the decompositions

$$X = X_1 \oplus X_2, \qquad Y = Y_1 \oplus Y_2,$$

we can use the Lyapunov–Schmidt Reduction, i.e.,

$$x = x_1 + a, \qquad x_1 \in X_1, \quad a \in X_2,$$

solves (4.3.3) if and only if it solves the couple of equations

$$G^1(x_1, a, \varepsilon) := Lx_1 - \varepsilon QN(x_1 + a) = o, \tag{4.3.4}$$

$$G^2(x_1, a, \varepsilon) := (I - Q)N(x_1 + a) = o. \tag{4.3.5}$$

Since $G^1(o, a, 0) = o$ and

$$(G^1)'_1(o, a, 0)h = Lh$$

is an isomorphism of X_1 onto Y_1 (it is both injective and surjective), the inverse is continuous by the Open Mapping Theorem (Theorem 2.1.8). The Implicit Function Theorem yields a solution

$$x_1 := \varphi(b, \varepsilon)$$

of (4.3.4) in a neighborhood of $(a, 0)$ for a given $a \in X_2$. We also have

$$\varphi(a, 0) = o \qquad \text{and} \qquad \varphi'_1(a, 0) = o$$

(check it again). This means that it is sufficient to solve

$$H(b, \varepsilon) := (I - Q)N(\varphi(b, \varepsilon) + b) = o$$

with respect to b. Since $\dim X_2 = \dim Y_2 = N < \infty$ and $H\colon X_2 \times \mathbb{R} \to Y_2$ we can try to use the Implicit Function Theorem once more. To this end we need an $\tilde{a} \in X_2$ for which

$$(I - Q)N(\tilde{a}) := t\int_0^1 f(s, \tilde{a})\,\mathrm{d}s = o, \qquad \text{i.e.,} \qquad \int_0^1 f(s, \tilde{a})\,\mathrm{d}s = o,$$

and the equation

$$(I - Q)N'(\tilde{a})d := t\left(\int_0^1 f'_2(s, \tilde{a})\,\mathrm{d}s\right)d = tc$$

has a unique solution for every $c \in \mathbb{R}^N$. The last requirement means that the $N \times N$-matrix $\int_0^1 f'_2(s, \tilde{a})\,\mathrm{d}s$ has to be regular. ○

To summarize the considerations of the previous example, we get the following conclusion.

Proposition 4.3.16. *Let $f = (f^1, \dots, f^N)\colon [0,1] \times \mathbb{R}^N \to \mathbb{R}^N$ be continuous and have continuous partial derivatives $\frac{\partial f^i}{\partial x_j}$ $(i, j = 1, \dots, N)$. Let the function f satisfy the conditions*

$$\int_0^1 f(s, \tilde{a})\,\mathrm{d}s = o, \qquad \det\left(\int_0^1 \frac{\partial f^i}{\partial x_j}(s, \tilde{a})\,\mathrm{d}s\right) \neq 0$$

for a certain constant $\tilde{a} \in \mathbb{R}^N$. *Then there exist* $\delta > 0$ *and a differentiable map* $\varepsilon \mapsto x(\cdot, \varepsilon)$, $|\varepsilon| < \delta$, *such that* $x(\cdot, 0) = \tilde{a}$ *and the functions* $x(\cdot, \varepsilon)$ *satisfy the boundary value problem* (4.3.2).

Remark 4.3.17. Let us give some remarks on this result. If the function f in (4.3.1) is 1-periodic in the variable t, then x is a solution of (4.3.1) if and only if

$$\tilde{x}(t) = x(t-n), \qquad n = [t],\ t \in \mathbb{R},$$

is a 1-periodic solution of $\dot{x} = f(t, x)$. Only technical difficulties appear when one generalizes the just described approach to a more general equation

$$\dot{x}(t) = \boldsymbol{A}(t)x + \varepsilon f(t, x)$$

with more general boundary conditions

$$\boldsymbol{B}x(0) - \boldsymbol{C}x(1) = o$$

($\boldsymbol{B}$, $\boldsymbol{C}$ are $N \times N$ matrices). Notice also that having a result for a system of differential equations we can investigate boundary value problems for second order equations. For example, we put

$$x = \begin{pmatrix} y \\ \dot{y} \end{pmatrix}, \quad f(t,x) = \begin{pmatrix} 0 \\ g(t, y, \dot{y}) \end{pmatrix}, \quad \boldsymbol{B} = \begin{pmatrix} b_1 & b_2 \\ 0 & 0 \end{pmatrix}, \quad \boldsymbol{C} = \begin{pmatrix} 0 & 0 \\ c_1 & c_2 \end{pmatrix},$$

to rewrite

$$\begin{cases} \ddot{y}(t) = a(t)y(t) + \varepsilon g(t, y(t), \dot{y}(t)), & t \in (0,1), \\ b_1 y(0) + b_2 \dot{y}(0) = 0, \quad c_1 y(1) + c_2 \dot{y}(1) = 0 \end{cases}$$

into the form

$$\begin{cases} \dot{x}(t) = \begin{pmatrix} 0 & 1 \\ a(t) & 0 \end{pmatrix} x(t) + \varepsilon f(t, x(t)), & t \in (0,1), \\ \boldsymbol{B}x(0) + \boldsymbol{C}x(1) = o. \end{cases}$$

Many other examples of the use of the Implicit Function Theorem can be found in Vejvoda et al. [155]. We will return to the problem (4.3.1) in Example 5.8.7.

We now turn to the study of the behavior of a differentiable function in the vicinity of a critical point. We recommend to the reader to consider the cases

$$f(x) = x^n, \quad n > 1, \qquad \text{and} \qquad f(x) = \sum_{i,j=1}^{n} a_{ij} x_i x_j, \quad a_{ij} = a_{ji},$$

first.

Definition 4.3.18. Let $\mathcal{G}$ be an open set in a Banach space X, $f\colon X \to \mathbb{R}$, $f \in C^2(\mathcal{G})$. A critical point $a \in \mathcal{G}$ of f is said to be *non-degenerate* if for any $h \in X$, $h \neq o$, the linear form $f''(a)(h,\cdot)$ does not vanish.

The following basic result holds also in a Hilbert space but its finite-dimensional version is more transparent.

Theorem 4.3.19 (Morse). *Let $\mathcal{G}$ be an open set in $\mathbb{R}^M$, $f\colon \mathbb{R}^M \to \mathbb{R}$, $f \in C^2(\mathcal{G})$. Let $a \in \mathcal{G}$ be a non-degenerate critical point of f. Then there exists a diffeomorphism φ of a neighborhood $\mathcal{U}$ of a onto a neighborhood $\mathcal{V}$ of $o \in \mathbb{R}^M$ such that for $x \in \mathcal{U}$, $y = \varphi(x)$, the function f can be expressed in the form*

$$f(x) = f(a) + \frac{1}{2}\sum_{i=1}^{M} \lambda_i y_i^2$$

where $\lambda_1, \dots, \lambda_M$ are the eigenvalues of the symmetric matrix $f''(a)$.

Proof. We identify a bilinear operator with its matrix representation in the standard basis in $\mathbb{R}^M$ (Remark 3.2.29(ii)) and denote the collection of all $M \times M$ matrices by $\mathscr{M}$. Then we can write

$$\boldsymbol{B}(x)(x-a, x-a) := (\boldsymbol{B}(x)(x-a), x-a)_{\mathbb{R}^M}.$$

We choose a norm $\|\cdot\|_{M\times M}$ on $\mathscr{M}$ and keep it fixed throughout the proof. A subset of $\mathscr{M}$ consisting of symmetric matrices is denoted by $\mathscr{S}$. We also denote by F and $F_{\mathscr{S}}$ the sets of all bounded continuous maps of $\mathcal{G}$ into $\mathscr{M}$ and $\mathscr{S}$, respectively. The space F equipped with the norm

$$\|\boldsymbol{A}\|_F := \sup_{x\in\mathcal{G}} \|\boldsymbol{A}(x)\|_{M\times M}$$

is a Banach space and $F_{\mathscr{S}}$ is its closed subspace. Without loss of generality we can assume that $\mathcal{G}$ is a convex neighborhood of the point a so small that f'' is bounded on $\mathcal{G}$.

After these preliminaries we start with the proof. Since $f'(a) = o$, the Taylor Formula (Proposition 3.2.27) gives

$$\begin{aligned} f(x) &= f(a) + \int_0^1 (1-t) f''(a + t(x-a))(x-a, x-a)\,\mathrm{d}t \\ &= f(a) + \boldsymbol{B}(x)(x-a, x-a) \quad \text{with} \quad \boldsymbol{B}(x) := \int_0^1 (1-t) f''(a + t(x-a))\,\mathrm{d}t \end{aligned}$$

(the Riemann integral of a function with values in $\mathbb{R}^{M\times M}$). Note that we have $\boldsymbol{B}(\cdot) \in F_{\mathscr{S}}$. Our aim is to show that we can choose $\boldsymbol{C}(\cdot) \in F$ such that

$$\boldsymbol{B}(x) = \boldsymbol{C}^*(x)\boldsymbol{J}\boldsymbol{C}(x)$$

where $\boldsymbol{J}$ is the canonical form of $\boldsymbol{B}(a) = \frac{1}{2}f''(a)$, i.e.,

$$\boldsymbol{J} = \frac{1}{2}\begin{pmatrix} \lambda_1 & & 0 \\ & \ddots & \\ 0 & & \lambda_M \end{pmatrix}.[6]$$

Here $\boldsymbol{C}^*$ stands for the adjoint matrix to $\boldsymbol{C}$, i.e., $\boldsymbol{C}^* = (c_{ji})$ provided $\boldsymbol{C} = (c_{ij})$. The transformation of coordinates

$$y = \boldsymbol{C}(x)(x-a)$$

then yields

$$f(x) = f(a) + (\boldsymbol{J}(y), y)_{\mathbb{R}^M} = f(a) + \frac{1}{2}\sum_{i=1}^{M} \lambda_i y_i^2.$$

To achieve this goal we will use the Implicit Function Theorem (Theorem 4.2.1). We put

$$\Phi(\boldsymbol{B}, \boldsymbol{C}) = \boldsymbol{C}^*(x)\boldsymbol{J}\boldsymbol{C}(x) - \boldsymbol{B}(x) \colon F_{\mathscr{S}} \times F \to F_{\mathscr{S}}.$$

In particular,

$$\Phi(\boldsymbol{B}(a), \boldsymbol{T}) = \boldsymbol{T}^*\boldsymbol{J}\boldsymbol{T} - \boldsymbol{B}(a) = o,$$

provided $\boldsymbol{T}$ is a unitary matrix which transforms $\boldsymbol{B}(a)$ into its canonical form $\boldsymbol{J}$. Put $\boldsymbol{A} := \boldsymbol{J}\boldsymbol{T}$. The partial differential of Φ with respect to the second variable has the form

$$\Phi_2'(\boldsymbol{B}, \boldsymbol{C})\boldsymbol{M} \colon x \mapsto \boldsymbol{M}^*(x)\boldsymbol{J}\boldsymbol{C}(x) + \boldsymbol{C}^*(x)\boldsymbol{J}\boldsymbol{M}(x), \qquad x \in \mathcal{G}.$$

Then

$$\operatorname{Ker} \Phi_2'(\boldsymbol{B}(a), \boldsymbol{T}) = \{\boldsymbol{M} \in F : \boldsymbol{M}^*(\cdot)\boldsymbol{A} + \boldsymbol{A}^*\boldsymbol{M}(\cdot) = o\}$$

and

$$Q \colon \boldsymbol{M} \mapsto \frac{1}{2}(\boldsymbol{M} - (\boldsymbol{A}^*)^{-1}\boldsymbol{M}^*\boldsymbol{A})$$

is a continuous linear projection of F onto $\operatorname{Ker} \Phi_2'(\boldsymbol{B}(a), \boldsymbol{T})$. By the assumption on the point a, $\boldsymbol{J}$ is injective. Further, $\boldsymbol{T}$ is a unitary matrix, i.e., $\boldsymbol{T}^* = \boldsymbol{T}^{-1}$. This means that $(\boldsymbol{A}^*)^{-1} = \boldsymbol{J}^{-1}\boldsymbol{T}$ exists. It can be seen that $I - Q$ is a projection onto

$$F_1 := (\boldsymbol{A}^*)^{-1}(F_{\mathscr{S}}).$$

The partial differential $\Phi_2'(\boldsymbol{B}(a), \boldsymbol{T})$ is an isomorphism of F_1 onto $F_{\mathscr{S}}$. Namely,

$$\boldsymbol{M} := \frac{1}{2}\boldsymbol{J}^{-1}\boldsymbol{T}\boldsymbol{S} \in F_1 \qquad \text{and} \qquad \Phi_2'(\boldsymbol{B}(a), \boldsymbol{T})\boldsymbol{M} = \boldsymbol{S} \in F_{\mathscr{S}}.$$

[6] A symmetric matrix has a diagonal canonical form – see Proposition 7.8.8.

We can now apply the Implicit Function Theorem to $\Phi\colon F_{\mathscr{S}} \times F_1 \to F_{\mathscr{S}}$ ($\boldsymbol{T} \in F_1$) and obtain positive numbers ε and δ such that for any $\boldsymbol{B} \in F_{\mathscr{S}}$, $\|\boldsymbol{B}(\cdot) - \boldsymbol{B}(a)\|_F < \varepsilon$ there is a unique $\boldsymbol{C} \in F_1$, $\|\boldsymbol{C}(\cdot) - \boldsymbol{T}\|_F < \delta$ for which

$$\Phi(\boldsymbol{B}, \boldsymbol{C}) = \boldsymbol{C}^*(x)\boldsymbol{J}\boldsymbol{C}(x) - \boldsymbol{B}(x) = o \qquad \text{for all} \quad x \in \mathcal{G}.$$

To finish the proof we have to show that there is a neighborhood $\mathcal{U}$ of a such that

$$\|\boldsymbol{B}(x) - \boldsymbol{B}(a)\|_F < \varepsilon \qquad \text{for all} \quad x \in \mathcal{U}.$$

By the definition of $\boldsymbol{B}$,

$$\begin{aligned}\|\boldsymbol{B}(\cdot) - \boldsymbol{B}(a)\|_F &= \sup_{x \in \mathcal{G}} \left\| \int_0^1 (1-t)[f''(a + t(x-a)) - f''(a)]\, \mathrm{d}t \right\|_{M \times M} \\ &\leq \frac{1}{2} \sup_{x \in \mathcal{G}} \|f''(x) - f''(a)\|_{M \times M}.\end{aligned}$$

This means that we can find the desired neighborhood $\mathcal{U}$. □

Example 4.3.20. Let X, Y be Banach spaces and let $f\colon X \to Y$. Consider the equation

$$f(x) = o \tag{4.3.6}$$

in the vicinity of a known solution $x = a$. Let f be a C^2-mapping in a neighborhood of a. Suppose that $f'(a)$ is a Fredholm operator (Remark 4.3.14) and, moreover, that the above equation can be reduced to the bifurcation equation

$$(I - Q)f(g(x_2) + x_2) = o.$$

Here Q is a projection of Y onto $\operatorname{Im} f'(a)$, $X = X_1 \oplus \operatorname{Ker} f'(a)$ and $g(x_2)$ is a (unique) solution of

$$Qf(x_1 + x_2) = o \qquad \text{for} \quad x_2 \in \operatorname{Ker} f'(a),$$

and x_2 is in a neighborhood of $a_2 \in \operatorname{Ker} f'(a)$ ($a = a_1 + a_2$). We also assume that this g is given by the Implicit Function Theorem. In particular, this means that

$$g'(a_2) = o.$$

Suppose now that

$$\operatorname{codim} \operatorname{Im} f'(a) = 1,$$

i.e., $I - Q$ is a projection onto a 1-dimensional subspace Y_2 of Y. Let

$$Y_2 = \operatorname{Lin}\{y_2\}.$$

By Corollary 2.1.18 and Remark 2.1.19, there is $\varphi \in Y^*$, $\varphi \in [\operatorname{Im} f'(a)]^{\perp}$, and we may assume that

$$\varphi(y_2) = 1.$$

In other words,

$$(I - Q)y = \varphi(y)y_2,$$

and the bifurcation equation has the form

$$F(x_2) := \varphi(f(g(x_2) + x_2)) = 0.$$

We have

$$F'(a_2)h = \varphi[f'(a)(g'(a_2)h + h)] = 0, \qquad h \in \operatorname{Ker} f'(a),$$

i.e., a_2 is a critical point of F. Further,

$$\begin{aligned} F''(a_2)(h,k) &= \varphi[f''(a)(g'(a_2)h + h, g'(a_2)k + k)] + \varphi[f'(a)(g''(a_2)(h,k))] \\ &= \varphi[f''(a)(h,k)] \end{aligned}$$

since

$$\varphi \circ f'(a) = 0 \qquad \text{and} \qquad g'(a_2) = o.$$

If, for example,

$$\dim \operatorname{Ker} f'(a) = 2$$

(this can occur for $f\colon \mathbb{R}^{N+1} \to \mathbb{R}^N$) and the matrix of $F''(a_2)$ is regular, i.e., a_2 is a non-degenerate critical point of F, then after a suitable transformation of coordinates we get

$$F(x_2) = \frac{1}{2}(\lambda_1 \xi^2 + \lambda_2 \eta^2)$$

(the Morse Theorem) and the following conclusion:

If $\operatorname{sgn} \lambda_1 = \operatorname{sgn} \lambda_2$, *then the equation* (4.3.6) *has an isolated solution*

$$x = a;$$

if $\operatorname{sgn} \lambda_1 = -\operatorname{sgn} \lambda_2$, *then there are two curves of solutions given by*

$$\xi = \pm \sqrt{-\frac{\lambda_2}{\lambda_1}}\,\eta.$$

○

The previous example can be generalized. The following problem is a standard one in the bifurcation theory: A differentiable map $f\colon \mathbb{R} \times X \to Y$ is given where X, Y are Banach spaces.[7] A smooth curve $x = \alpha(\lambda)$, $\lambda \in (-\delta, \delta)$, of solutions of the equation

$$f(\lambda, x) = o \tag{4.3.7}$$

is known. After the transformation $\xi = x - \alpha(\lambda)$, we can suppose that

$$f(\lambda, o) = o \tag{4.3.8}$$

for λ in a neighborhood of (e.g.) $0 \in \mathbb{R}$.

[7] The set of parameters $\mathbb{R}$ can be replaced by a normed linear space in general.

Definition 4.3.21. Let (4.3.8) be satisfied for the equation (4.3.7). The point $(0, o) \in \mathbb{R} \times X$ is called a *bifurcation point* provided in any neighborhood of $(0, o)$ there is a solution (λ_0, x_0) of (4.3.7) such that $x_0 \neq o$.

Notice that whenever f is differentiable in a neighborhood $\mathcal{U}$ of $(0, o)$ and $f_2'(0, o)$ is an isomorphism, then $(0, o)$ is not a bifurcation point (the Implicit Function Theorem). In order to find a sufficient condition for bifurcation suppose that $f \in C^2(\mathcal{U})$ and $A = f_2'(o, o)$ is not an isomorphism. More precisely, let $\operatorname{Ker} A$ be nontrivial, i.e., let 0 be an eigenvalue of A. The simplest case occurs when 0 is a *simple eigenvalue*, i.e.,

$$\operatorname{Ker} A = \operatorname{Lin}\{x_0\}, \qquad x_0 \neq o.$$

The following result is a classical one (see Crandall & Rabinowitz [37]).

Theorem 4.3.22 (Local Bifurcation Theorem). *Let X, Y be Banach spaces, $f \colon \mathbb{R} \times X \to Y$ a twice continuously differentiable map on a neighborhood of $(0, o)$. Let f satisfy the assumptions*

(i) *$f(\lambda, o) = o$ for all $\lambda \in (-\delta, \delta)$ for some $\delta > 0$,*

(ii) $\dim \operatorname{Ker} f_2'(0, o) = \operatorname{codim} \operatorname{Im} f_2'(0, o) = 1$,

(iii) *if $f_2'(0, o)x_0 = o$, $x_0 \neq o$, then $f_{1,2}''(0, o)(1, x_0) \notin \operatorname{Im} f_2'(0, o)$.*

Denote by X_1 the topological complement[8] *of $\operatorname{Ker} f_2'(0, o)$ in X.*

Then there is a C^1-curve $(\varphi, \psi) : (-\eta, \eta) \to \mathbb{R} \times X_1$ (for some $\eta > 0$) such that

$$\varphi(0) = 0, \qquad \psi(0) = o, \qquad f(\varphi(t), t(x_0 + \psi(t))) = o.$$

Moreover, there is a neighborhood $\mathcal{U}$ of $(0, o)$ in $\mathbb{R} \times X$ such that

$$f(\lambda, x) = o \qquad \text{for} \quad (\lambda, x) \in \mathcal{U}$$

if and only if either $x = o$ or

$$\lambda = \varphi(t), \quad x = t(x_0 + \psi(t)) \quad \text{for a certain } t$$

– see Figure 4.3.10. *Such a picture is called a* bifurcation diagram.

Proof. We are giving two proofs. The first one for a finite-dimensional case when $X = Y = \mathbb{R}^M$ is based on the Morse Theorem. The second one which is due to M. Crandall and P. Rabinowitz is based on the Implicit Function Theorem and will be only sketched.

The first proof We choose a direct complement X_1 to $\operatorname{Ker} f_2'(0, o) = \operatorname{Lin}\{x_0\}$ in the space X and a continuous projection Q in the space Y onto $\operatorname{Im} f_2'(0, o)$. The

[8] Let $X = X_1 \oplus X_2$ and the corresponding projection P of X onto X_1 be continuous. Then X_1 is called a *topological complement* of X_2 and vice versa. See also Remark 2.1.19, Remark 4.3.9 (i) and Remark 4.3.14.

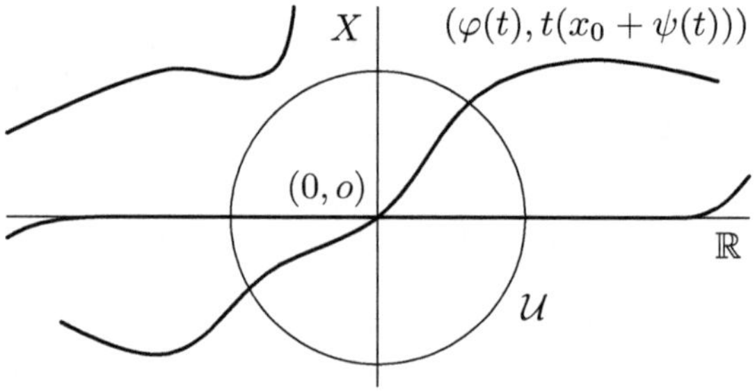

Figure 4.3.10.

complementary projection $I - Q$ is a projection onto one-dimensional space, i.e., there are $\omega \in Y^*$, $\omega \neq o$, and $y \in Y$, $Qy_1 = 0$, $y_1 \neq o$, such that

$$(I - Q)y = \omega(y)y_1.$$

In particular, $y \in \operatorname{Im} f_2'(0, o)$ if and only if $\omega(y) = 0$.

By the Implicit Function Theorem, the equation

$$G(\lambda, t, x_1) := Qf(\lambda, tx_0 + x_1) = 0$$

has a unique C^2-solution $x_1 = g(\lambda, t) \in X_1$ in a neighborhood of $(0, o) \in \mathbb{R} \times X$. Since $g(\lambda, 0) = o$, we have $g_1'(\lambda, 0) = o$ and, by the Implicit Function Theorem, $g_2'(0, 0) = o$. We can define C^1-function h as follows

$$h(\lambda, t) = \begin{cases} \dfrac{g(\lambda, t)}{t}, & \text{for} \quad t \neq 0, \\ \frac{\partial g}{\partial t}(\lambda, 0), & \text{for} \quad t = 0. \end{cases}$$

i.e., $g(\lambda, t) = th(\lambda, t)$.

Using now the Lyapunov-Schmidt reduction (Remark 4.3.14) the equation $f(\lambda, x) = o$ is equivalent to the equation

$$F(\lambda, t) := \omega\left[f\left(\lambda, tx_0 + g(\lambda, t)\right)\right] = 0$$

in a neighborhood of $(0, o) \in \mathbb{R} \times X$. To solve this new equation we want to use the Morse Theorem. To do this we have to show that $(0, 0) \in \mathbb{R}^2$ is a non-degenerate critical point of the C^2-function F. Since $F(\lambda, 0) = 0$ we get

$$\frac{\partial F}{\partial \lambda}(\lambda, 0) = 0 \qquad \text{and} \qquad \frac{\partial^2 F}{\partial \lambda^2}(0, 0) = 0.$$

Further,

$$\frac{\partial F}{\partial t}(\lambda, 0) = \omega\left[f_2'(\lambda, o)\left(x_0 + g_2'(\lambda, 0)\right)\right], \qquad \frac{\partial F}{\partial t}(0, 0) = 0,$$

and

$$\beta := \frac{\partial F}{\partial\lambda\partial t}(0,0) = \frac{\partial^2 F}{\partial t\partial\lambda}(0,0) = \omega[f''_{2,1}(0,o)(1,x_0) + f'_2(0,o)g''_{2,1}(0,0)]$$
$$= \omega[f''_{2,1}(0,o)(1,x_0)],$$

since $\omega(z) = 0$ for every $z \in \operatorname{Im} f'_2(0,o)$. This proves that $(0,0)$ is a critical point of F which is non-degenerate (see assumption (iii)). The matrix representation of $F''(0,0)$ has nonzero eigenvalues $\pm\beta$.

The rest of the proof follows from the Morse Theorem. This yields a transformation of coordinates (i.e., diffeomorphism) $(\lambda,t) = (\kappa(\mu,s),\omega(\mu,s))$ in a neighborhood of $(0,0) \in \mathbb{R}^2$ such that $F(\lambda,t) = \frac{\beta}{2}(\mu^2 - s^2)$. This means that there are locally two (smooth) branches of solutions of $F(\lambda,t) = 0$, namely $\lambda = \kappa(\pm s,s), t = \omega(\pm s,s)$. One of these branches is $t \equiv 0$ – assume, e.g., that $\omega(-s,s) = 0$. In particular, $\omega'_1(0,0) = \omega'_2(0,0)$. The second, non-trivial one, is given by $\lambda = \kappa(s,s) =: \widetilde{\kappa}(s)$, $t = \omega(s,s) =: \widetilde{\omega}(s)$. Since (κ,ω) is a diffeomorphism, $\frac{d\widetilde{\omega}}{ds}(0) = 2\omega'_1(0,0) \neq 0$. This shows that the function $\widetilde{\omega}$ is locally invertible, i.e., $s = \gamma(t)$. So we have proved that locally $f(\lambda,x) = 0$ if and only if either $x = 0$ or $\lambda = \widetilde{\kappa}(\gamma(t))$, $x = tx_0 + g(t,\widetilde{\kappa}(\gamma(t))) = t[x_0 + h(t,\widetilde{\kappa}(\gamma(t)))]$. □

The second proof proceeds by using the Implicit Function Theorem for the function $\Phi\colon \mathbb{R}\times\mathbb{R}\times X_1 \to Y$ defined by

$$\Phi(\lambda,t,x_1) = \begin{cases} \frac{1}{t}f(\lambda,t(x_0+x_1)) & \text{for} \quad t\neq 0,\\ f'_2(\lambda,o)(x_0+x_1) & \text{for} \quad t = 0.\end{cases}$$

Notice that $\Phi(0,0,o) = o$, and

$$(\lambda,h) \mapsto \Phi'_1(0,0,o)\lambda + \Phi'_3(0,0,o)h$$

is an isomorphism of $\mathbb{R}\times X_1$ onto Y (assumptions (ii) and (iii)). For details see Crandall & Rabinowitz [37]. □

Example 4.3.23. The following two functions offer very simple illustrative examples:

$$f(\lambda,x) = \lambda x - x^2, \qquad g(\lambda,x) = \lambda x - x^3.$$

Their bifurcation diagrams are shown in Figures 4.3.11 and 4.3.12.

We use these functions to point out the typical examples of the changing of stability of a differential equation when the so-called non-hyperbolic stationary point is crossed.[9] In these figures branches of stationary solutions of equations

$$\dot{x} = f(\lambda,x), \qquad \dot{x} = g(\lambda,x)$$

are shown with an indication of their stability (s for stable, u for unstable). ○

[9] A stationary point $a \in \mathbb{R}^M$ is called *hyperbolic* for the equation $\dot{x} = f(x)$ provided $f(a) = o$ and $\sigma(f'(a)) \cap \mathrm{i}\mathbb{R} = \emptyset$. See also footnote 3 on page 164.

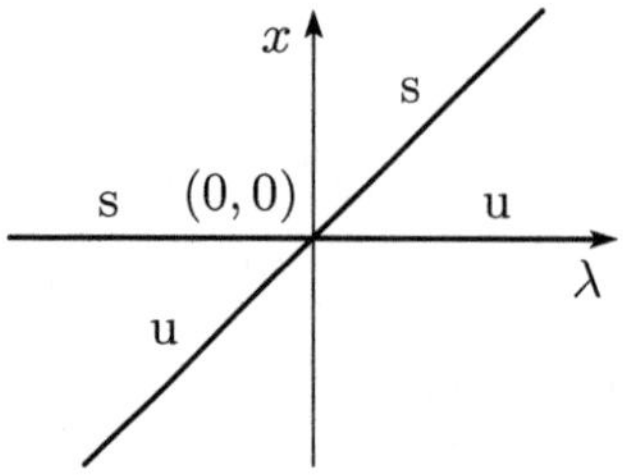

Figure 4.3.11. Transcritical bifurcation

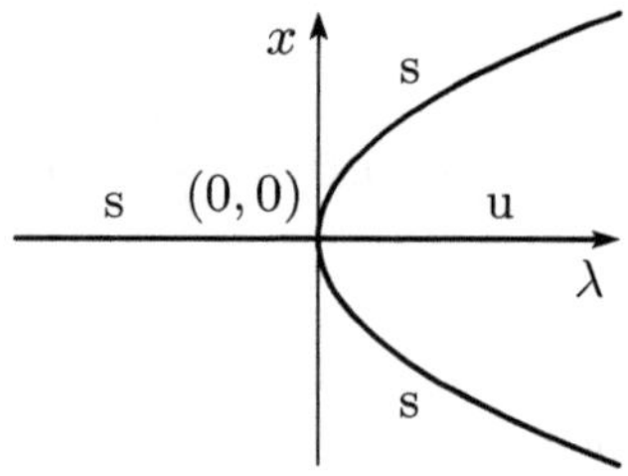

Figure 4.3.12. Pitchfork bifurcation

Example 4.3.24. We wish to find a nontrivial 2π-periodic solution of the nonlinear pendulum equation

$$\ddot{x}(t) + \lambda \sin x(t) = 0. \tag{4.3.9}$$

We put

$$f(\lambda, x) : t \mapsto \ddot{x}(t) + \lambda \sin x(t),$$

and

$$X = \{x \in C^2(\mathbb{R}) : x \text{ is } 2\pi\text{-periodic}\},$$
$$\|x\|_X = \max_{t\in[0,2\pi]} |x(t)| + \max_{t\in[0,2\pi]} |\dot{x}(t)| + \max_{t\in[0,2\pi]} |\ddot{x}(t)|,$$
$$Y = \{y \in C(\mathbb{R}) : y \text{ is } 2\pi\text{-periodic}\}, \quad \|y\|_Y = \max_{t\in[0,2\pi]} |y(t)|.$$

It is easy to show that

$$f_2'(\lambda, o)h = \ddot{h}(t) + \lambda h$$

and, therefore, $\operatorname{Ker} {f_2}'(\lambda, o)$ is nontrivial if and only if $\lambda = n^2$, $n \in \mathbb{N} \cup \{0\}$, and

$$\operatorname{Ker} f_2'(0, o) = \{\text{constant functions}\},$$

$$\operatorname{Ker} f_2'(n^2, o) = \operatorname{Lin}\{\sin nt, \cos nt\} \qquad \text{for} \quad n \in \mathbb{N}.$$

In the former case, i.e., $n = 0$, we can apply Theorem 4.3.22. Since

$$\operatorname{Im} f_2'(0, o) = \left\{ y \in Y : \int_0^{2\pi} y(s)\,\mathrm{d}s = 0 \right\}$$

and

$$f''_{1,2}(0, o)(1, c) = c \qquad \text{for} \quad c \in \operatorname{Ker} f_2'(0, o),$$

the assumptions of Theorem 4.3.22 are satisfied.

What can we do in the latter case when $n \in \mathbb{N}$ and the dimension of $\operatorname{Ker} f_2'(n^2, o)$ is equal to 2? In spite of the fact that Theorem 4.3.22 cannot be used we still may proceed with the Lyapunov–Schmidt Reduction: Denote

$$A := f_2'(n^2, o), \qquad Y = \operatorname{Im} A \oplus Z,$$

$$(I - Q)y\colon t \mapsto \frac{\sin nt}{\pi} \int_0^{2\pi} y(s) \sin ns \, \mathrm{d}s + \frac{\cos nt}{\pi} \int_0^{2\pi} y(s) \cos ns \, \mathrm{d}s, \quad y \in Y.$$

Then $I - Q$ is the projection onto Z such that $\operatorname{Ker}(I - Q) = \operatorname{Im} A$. Similarly, let

$$X = \operatorname{Ker} A \oplus V \qquad \text{where} \quad V = \{v \in X : (I - Q)v = o\}.$$

The operator f can be expressed by

$$f(\mu + n^2, u + v) = Av + \mu(u + v) + h(\mu, u, v), \qquad u \in \operatorname{Ker} A, v \in V,$$

where

$$h(\mu, u, v) = \mu[\sin(u + v) - (u + v)] + n^2[\sin(u + v) - (u + v)].$$

Because of this special form of h we will try to find a solution of (4.3.9) in the form $x = \mu(u + v)$. The equality

$$f(\mu + n^2, \mu(u + v)) = 0$$

holds if and only if

$$Av + \mu(u + v) + \mu^2 g(\mu, u, v) = 0 \tag{4.3.10}$$

where

$$g(\mu, u, v) = \begin{cases} \dfrac{h(\mu, \mu u, \mu v)}{\mu^3}, & \mu \neq 0, \\ -\dfrac{n^2}{6}(u + v)^3, & \mu = 0. \end{cases}$$

For solving (4.3.10) we use the Lyapunov–Schmidt Reduction. According to it, the equation (4.3.10) is equivalent to the following couple of equations:

$$\begin{aligned} 0 &= Av + \mu Q(u + v) + \mu^2 Q g(\mu, u, v) && (= Av + \mu v + \mu^2 Q g(\mu, u, v)), \\ 0 &= (I - Q)(u + v) + \mu(I - Q) g(\mu, u, v) && (= u + \mu(I - Q) g(\mu, u, v)). \end{aligned}$$

By the Implicit Function Theorem, the first equation has a unique solution $v = \varphi(\mu, u)$ in a neighborhood of the point $(0, u^*, o)$ for any u^*. We insert φ into the bifurcation equation obtaining

$$\Phi(\mu, u) = u + \mu(I - Q) g(\mu, u, \varphi(\mu, u)) = 0. \tag{4.3.11}$$

Since $\Phi(0, u^*) = u^*$, we take $u^* = o$ and solve (4.3.11) in a neighborhood of $(0, o)$. This can be done with help of the Implicit Function Theorem since $\Phi_2'(0, o)$ is an isomorphism of $\operatorname{Ker} A$ onto Z. Denoting this solution by $u = \omega(\mu)$ we can come to the following conclusion:

Any point (n^2, o) is a bifurcation point of the equation (4.3.9) *and a nontrivial branch of* 2π*-periodic solutions of* (4.3.9) *has the form*

$$x = \mu(\omega(\mu) + \varphi(\mu, \omega(\mu))), \qquad \mu \in (-\delta, \delta).$$

The reader is invited to generalize this procedure to obtain sufficient conditions for a bifurcation for the equation

$$f(\lambda, x) = o$$

assuming that

$$f(\lambda, o) = o \qquad \text{for} \quad |\lambda - \lambda^*| < \delta, \qquad f \in C^2(\mathcal{U}),$$
$$\dim \operatorname{Ker} f_2'(\lambda^*, o) = \operatorname{codim} \operatorname{Im} f_2'(\lambda^*, o) = 2$$

where $\mathcal{U}$ is a neighborhood of (λ^*, o).

We notice that no uniqueness of the bifurcation branch was proved even in our concrete example. Compare this with the assertion given in Theorem 4.3.22. This is due to our special choice of the form of the bifurcation branch, namely $x = \mu(u + v)$. ○

Example 4.3.25 (Application of Theorem 4.3.22). We will study the bifurcation points of the periodic problem

$$\begin{cases} \ddot{x}(t) + \lambda x(t) + g(\lambda, t, x(t), \dot{x}(t)) = 0, & t \in (0, 2\pi), \\ x(0) = x(2\pi), \quad \dot{x}(0) = \dot{x}(2\pi). \end{cases} \tag{4.3.12}$$

In this example we will concentrate on the point $\lambda = 0$ which is an eigenvalue of the associated eigenvalue problem

$$\begin{cases} \ddot{x}(t) + \lambda x(t) = 0, & t \in (0, 2\pi), \\ x(0) = x(2\pi), \quad \dot{x}(0) = \dot{x}(2\pi), \end{cases} \tag{4.3.13}$$

of multiplicity 1. We consider the same function spaces X, Y as in the previous example (Example 4.3.24).

Let us define $f \colon X \times \mathbb{R} \to Y$ by

$$f(\lambda, x) : t \mapsto \ddot{x}(t) + \lambda x(t) + g(\lambda, t, x(t), \dot{x}(t))$$

where the function

$$g = g(\lambda, t, x, y)$$

satisfies the following hypotheses:

(i) g is 2π-periodic in t and continuous with respect to all four variables (as a function from $\mathbb{R}^4$ into $\mathbb{R}$);

(ii) the derivatives of g with respect to x, y, λ up to the order p ($p \geq 2$) are continuous functions from $\mathbb{R}^4$ into $\mathbb{R}$;

(iii) $g(\lambda, t, 0, 0) = 0$ for all $t, \lambda \in \mathbb{R}$;

(iv) $g_3'(\lambda, t, 0, 0) = g_4'(\lambda, t, 0, 0) = 0$ for all $t, \lambda \in \mathbb{R}$.

It follows from (iii) that $f(\lambda, o) = o$ for all $\lambda \in \mathbb{R}$. Moreover, thanks to (iv) we have

$$f_2'(\lambda, o)h = \ddot{h} + \lambda h,$$

and so we conclude

$$\dim \operatorname{Ker} f_2'(0, o) = 1.$$

It follows from (ii) that $f \in C^p(X \times [0, 2\pi], Y)$. By Proposition 2.1.27(iv),

$$Y_1 = \operatorname{Im} f_2'(0, o) = \left\{ y \in Y : \int_0^{2\pi} y(t)\, dt = 0 \right\}$$

is a closed subspace of Y of codimension 1. Set

$$x_0 = 1, \qquad \operatorname{Ker} f_2'(0, o) = \operatorname{Lin}\{1\}, \qquad X_1 = \left\{ x \in X : \int_0^{2\pi} x(t)\, dt = 0 \right\}.$$

Since

$$f_{1,2}''(0, o)1 = 1 \qquad \text{and} \qquad 1 \notin \operatorname{Im} f_2'(0, o),$$

the condition (iii) in Theorem 4.3.22 is verified, too.

It follows from the Crandall–Rabinowitz Bifurcation Theorem (see Theorem 4.3.22) that $(0, o)$ is a bifurcation point of (4.3.12). In particular, the point $(0, o) \in \mathbb{R} \times X$ belongs to the branch of trivial solutions (λ, o), but also to the branch

$$\Gamma = \{(\hat{\lambda}(s), s + x(s)) : s \in (-\varepsilon, \varepsilon)\}, \quad \hat{\lambda}(0) = 0, \qquad x(0) = o,\ \dot{x}_s(0) = o.$$

Hence for any $s \in (-\varepsilon, \varepsilon)$, $s \neq 0$, the nontrivial solution $s + x(s)$ is the sum of a constant function (with respect to t) and the perturbed function $x(s)$ (which depends on t) such that $x(s)$ belongs to X_1. ○

Exercise 4.3.26.

(i) Let $\mathcal{G}$ be an open subset of $\mathbb{R}^M$ and let $v \in C^1(\mathcal{G}, \mathbb{R}^M)$. Assume that $a \in \mathcal{G}$ is not a stationary point of v, i.e., $v(a) \neq o$. Prove that there exists a diffeomorphism F of a neighborhood $\mathcal{U}$ of a onto a neighborhood $\mathcal{V}$ of o such that F maps solutions to the equation

$$\dot{x}(t) = v(x(t)) \tag{4.3.14}$$

which lie in $\mathcal{U}$ to solutions in $\mathcal{V}$ of the system of equations

$$\begin{aligned} \dot{y}_1(t) &= 1, \\ \dot{y}_i(t) &= 0, \qquad i = 2, \dots, M. \end{aligned}$$

Hint. Choose a subspace Y of $\mathbb{R}^M$ for which $\mathbb{R}^M = Y \oplus \operatorname{Lin}\{v(a)\}$. Define

$$G(z) = \varphi(t; a + y) \qquad \text{for} \quad z = y + tv(a),\ y \in Y,$$

where $\varphi(t;\xi)$ is a solution to (4.3.14) such that

$$\varphi(0;\xi) = \xi.$$

Prove that G is a local diffeomorphism and $F = G^{-1}$ has the desired property.

(ii) Deduce from (i) that the equation (4.3.14) has $M - 1$ independent first integrals in a neighborhood of a nonstationary point.

(iii) Is there any relation between the first integral of (4.3.14) and the linear partial differential equation

$$v_1(x)\frac{\partial u}{\partial x_1} + \cdots + v_M(x)\frac{\partial u}{\partial x_M} = 0\,?$$

Exercise 4.3.27. Apply Theorem 4.3.22 to the (Dirichlet) boundary value problem

$$\begin{cases} \ddot{x}(t) + \lambda x(t) + g(\lambda, t, x(t), \dot{x}(t)) = 0, & t \in (0,\pi),\\ x(0) = x(\pi) = 0, \end{cases}$$

and show that every (k^2, o), $k \in \mathbb{N}$, is a bifurcation point!

Exercise 4.3.28. Replace the Dirichlet boundary condition in Exercise 4.3.27 by the Neumann boundary condition

$$\dot{x}(0) = \dot{x}(\pi) = 0$$

and prove that every (k^2, o), $k \in \mathbb{N} \cup \{0\}$, is a bifurcation point!

Exercise 4.3.29. Why can the approach used in Example 4.3.25 not be applied to prove that the points (k^2, o), $k \in \mathbb{N}$, are bifurcation points of (4.3.12) even if k^2 is an eigenvalue of the associated eigenvalue problem (4.3.13)? Can you modify the method from Example 4.3.24?

Exercise 4.3.30. Apply Theorem 4.3.22 to the boundary value problem

$$\begin{cases} \dfrac{\mathrm{d}^4 x}{\mathrm{d}t^4}(t) - \lambda x(t) + g(\lambda, t, x(t), \dot{x}(t), \ddot{x}(t)\dddot{x}(t)) = 0, & t \in (0,\pi),\\ x(0) = \ddot{x}(0) = x(\pi) = \ddot{x}(\pi), \end{cases}$$

and show that, under appropriate assumptions on g, every (k^2, o), $k \in \mathbb{N}$, is a bifurcation point.

4.3A Differentiable Manifolds, Tangent Spaces and Vector Fields

We have defined a differentiable manifold in the basic text (Definition 4.3.4) and have also shown some examples (Examples 4.3.5 and 4.3.10). In the following Appendices 4.3A–4.3C we will provide more information about this object in order to develop a geometric approach to one of the most powerful tools of nonlinear analysis, namely to the Brouwer degree (See Sections 5.7 and 5.7A).

There is no doubt as to the importance of the notion of the derivative (or differential) for local study of functions of one or more variables. Therefore a notion of the differential will be a rudiment of analysis on differentiable manifolds, too. We have learned that it is convenient to define the differential of

$$f\colon \mathbb{R}^M \to \mathbb{R} \qquad \text{at a point} \quad a \in \mathbb{R}^M$$

as a linear form $f'(a)$ on $\mathbb{R}^M$ which approximates f locally at a in a given precise way. To extend such an approach to functions on differentiable manifolds we have to say what is a linear form on a (nonlinear) manifold. This is done by the important notion of the *tangent space*

$$T_a\mathscr{M}$$

of a differentiable manifold $\mathscr{M}$ at a point $a \in \mathscr{M}$. Roughly speaking, $T_a\mathscr{M}$ is the collection of all tangent vectors to $\mathscr{M}$ at the point a. We can imagine a tangent vector v at the point $a \in \mathscr{M}$ with help of the following physical interpretation.

Consider a force F which acts on a material point P. Suppose that there are certain rigid constraints which make P move along a smooth curve γ which lies on a manifold $\mathscr{M} \subset \mathbb{R}^N$ (this manifold is determined by these constraints). As a concrete example, imagine that we move on the globe (because of gravitation) which can be approximated by the smooth two-dimensional sphere $\mathcal{S}^2$. Let the point P be at a certain instant, say $t = 0$, at the position $a = \gamma(0) \in \mathscr{M}$, and let the force F and also all constraints stop to operate suddenly at this time. What will happen then? According to the First Newton Law the point P will continue to move with a constant speed

$$v = |\dot\gamma(0)| \qquad \left(\dot\gamma(0) := \frac{\mathrm{d}\gamma}{\mathrm{d}t}(0)\right)$$

along the line with the directional vector $\dot\gamma(0)$. This vector is the tangent vector to the curve γ at the point a. The collection of all these tangent vectors (which are given by all possible motions through a fixed point $a \in \mathscr{M}$) forms the tangent space $T_a\mathscr{M}$. More precisely, we give the following definition.

Definition 4.3.31. Let $\mathscr{M}$ be an M-dimensional differentiable manifold in $\mathbb{R}^N$ and let $a \in \mathscr{M}$. The *tangent space* $T_a\mathscr{M}$ is the collection of tangent vectors $\dot\gamma(0)$ to all smooth curves

$$\begin{aligned}\gamma \in \Gamma_a := \{\gamma\colon \mathbb{R} \to \mathbb{R}^N &: \text{there is an open interval } \mathcal{I}_\gamma \ni 0\\ &\text{such that } \gamma \in C^1(\mathcal{I}_\gamma),\ \gamma(\mathcal{I}_\gamma) \subset \mathscr{M},\ \gamma(0) = a\}.\end{aligned}$$

The method for computation of tangent vectors to a "parametrized" M-dimensional manifold $\mathscr{M} \subset \mathbb{R}^N$ is based on the use of local coordinates: Let $a \in \mathscr{M}$, let ψ be a diffeomorphism of a neighborhood $\mathcal{W} \subset \mathbb{R}^N$ of the point a into $\mathbb{R}^N$ (see Definition 4.3.4). If

$$\mathcal{V} = \mathcal{W} \cap \mathscr{M},$$

then $(\mathcal{V}, \psi)$ is a local chart of $\mathscr{M}$ at the point $a \in \mathscr{M}$. Denote the inverse of the restriction of $P\psi$ to $\mathcal{V}$ by φ where $Py = (y_1, \dots, y_M) \in \mathbb{R}^M$ for $y = (y_1, \dots, y_N) \in \mathbb{R}^N$. Then φ maps the neighborhood $\mathcal{U} = P\psi(\mathcal{V})$ of the point $b = P\psi(a) \in \mathbb{R}^M$ into $\mathscr{M}$ and we will consider φ also as an embedding (see Remark 4.3.3(iv)) of $\mathcal{U}$ into $\mathbb{R}^N$. We call φ the *local parametrization* of $\mathcal{V} \subset \mathscr{M}$. The main reason for introducing φ is that φ can be differentiated, but $\psi|_\mathcal{V}$ cannot, and the whole ψ does not describe $\mathscr{M}$. See Figure 4.3.13.

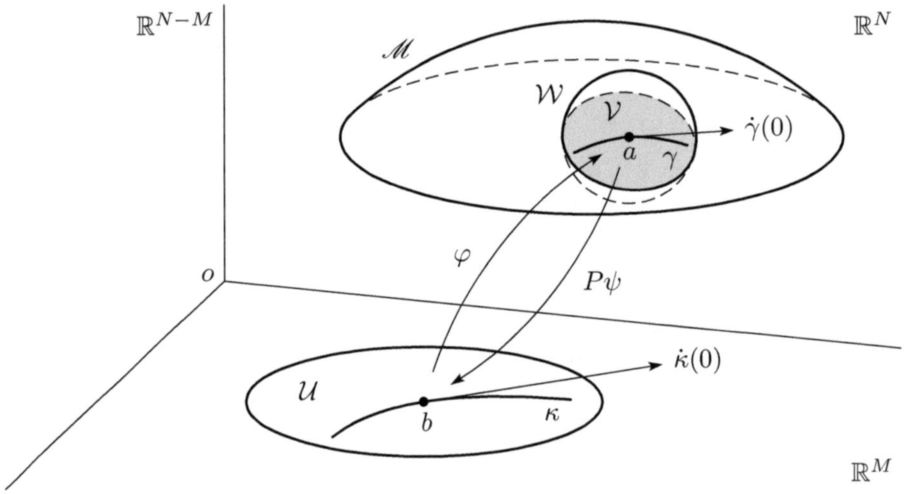

Figure 4.3.13. Manifold

Consider now a smooth curve $\gamma \in \Gamma_a$ (see Figure 4.3.13). We can choose $\mathcal{I}_\gamma$ so small that $\gamma(\mathcal{I}_\gamma) \subset \mathcal{V}$. Then

$$\kappa(t) = (P\psi)(\gamma(t))$$

is a smooth curve in $\mathcal{U} \subset \mathbb{R}^M$. We have $\varphi \circ \kappa = \gamma$ and, consequently,

$$\dot{\gamma}^j(0) = \sum_{i=1}^{M} \frac{\partial \varphi^j}{\partial y_i}(b)\dot{\kappa}^i(0), \text{[10]} \qquad j = 1, \dots, N,$$

or, more briefly,

$$\dot{\gamma}(0) = \varphi'(b)(\dot{\kappa}(0)). \tag{4.3.15}$$

[10]The Einstein summation convention is often used in differential geometry. According to it, the sum is taken with respect to all indices which appear simultaneously in upper and lower positions. For example, if $e_1, \dots, e_M$ is a basis in $\mathbb{R}^M$, then the coordinates of a point $x \in \mathbb{R}^M$ with respect to this basis should be denoted by $x^1, \dots, x^M$, since $x = \sum_{i=1}^{M} x^i e_i =: x^i e_i$ by this convention. Similarly, $\varphi\colon \mathbb{R}^M \to \mathbb{R}^N$ has components $\varphi^1, \dots, \varphi^N$ and values $\varphi(x^1, \dots, x^M)$ $(= \varphi(x))$ and partial derivatives $\frac{\partial \varphi^j}{\partial x_i}$. Moreover, $\varphi'(a)(h^i e_i) = \frac{\partial \varphi^j}{\partial x_i}(a) h^i e_j$ $\left(= \sum_{j=1}^{N} \left[\sum_{i=1}^{M} \frac{\partial \varphi^j}{\partial x_i}(a) h^i\right] e_j\right)$. Since this notation is not too common in analysis we do not use it.

Since also

$$\dot{\kappa}(0) = P\psi'(a)(\dot{\gamma}(0)),$$

there is a correspondence between the tangent vector $v = (\dot{\gamma}^1(0), \dots, \dot{\gamma}^N(0)) \in T_a\mathscr{M}$ and the tangent vector $w = (\dot{\kappa}^1(0), \dots, \dot{\kappa}^M(0))$ to a curve

$$\kappa = (P\psi) \circ \gamma \qquad \text{at} \quad b = \kappa(0) = \psi(a).$$

Obviously, for any $w = (w^1, \dots, w^M) \in \mathbb{R}^M$ there is a smooth curve κ (e.g., $\kappa(t) = b + t \sum_{i=1}^{M} w^i e_i$; $e_1, \dots, e_M$ is the standard coordinate basis in $\mathbb{R}^M$) such that $w = \dot{\kappa}(0)$. This means that

$$T_a\mathscr{M} = \operatorname{Im} \varphi'(b)$$

and the linear operations in $\mathbb{R}^M$ induce those in $T_a\mathscr{M}$. Therefore, $T_a\mathscr{M}$ is a linear space of dimension M ($M = \dim \mathscr{M}$) and $\varphi'(b)e_i$, $i = 1, \dots, M$, form a basis of $T_a\mathscr{M}$. Since

$$(y^1, \dots, y^M) := (\psi^1(x), \dots, \psi^M(x))$$

can be viewed as local (nonlinear) coordinates of a point $x \in \mathcal{V}$, the vector $\varphi'(b)e_i$ is also denoted by $\frac{\partial}{\partial y_i}$. This means that

$$\dot{\gamma}(0) = \varphi'(b)(\dot{\kappa}(0)) = \varphi'(b)\left(\sum_{i=1}^{M} \dot{\kappa}^i(0)e_i\right) = \sum_{i=1}^{M} \dot{\kappa}^i(0)\frac{\partial}{\partial y_i}. \tag{4.3.16}$$

Example 4.3.32. Let us compute the tangent space to the 2-dimensional sphere

$$\mathcal{S}^2 = \{(x, y, z) \in \mathbb{R}^3 : x^2 + y^2 + z^2 = 1\} \qquad \text{at the point} \quad a = \left(\frac{1}{2}, \frac{1}{2}, \frac{\sqrt{2}}{2}\right).$$

As local coordinates we choose the spherical coordinates

$$x = \cos\alpha \cos\beta, \quad y = \sin\alpha \cos\beta, \quad z = \sin\beta,$$

i.e., $(x, y, z) = \varphi(\alpha, \beta)$, $b = \left(\frac{\pi}{4}, \frac{\pi}{4}\right)$ and $\varphi\left(\frac{\pi}{4}, \frac{\pi}{4}\right) = a$. Then

$$\frac{\partial}{\partial\alpha} := \frac{\partial\varphi}{\partial\alpha}\left(\frac{\pi}{4}, \frac{\pi}{4}\right) = \left(-\frac{1}{2}, \frac{1}{2}, 0\right), \qquad \frac{\partial}{\partial\beta} := \frac{\partial\varphi}{\partial\beta}\left(\frac{\pi}{4}, \frac{\pi}{4}\right) = \left(-\frac{1}{2}, -\frac{1}{2}, \frac{\sqrt{2}}{2}\right)$$

is a basis of $T_a\mathcal{S}^2$. Choosing a perpendicular vector v to both $\frac{\partial}{\partial\alpha}$ and $\frac{\partial}{\partial\beta}$, e.g., $v = (1, 1, \sqrt{2})$, we get the following expression for $T_a\mathcal{S}^2$:

$$T_a\mathcal{S}^2 = \{(x, y, z) \in \mathbb{R}^3 : x + y + \sqrt{2}z = 0\}.$$

(If you have drawn a picture, you get a slightly better insight.) ○

It was shown in Remark 4.3.9(iii) that a manifold can also be given implicitly, i.e., as the set of solutions of the equation

$$f(x) = o.$$

Proposition 4.3.33. *Let $f\colon \mathbb{R}^M \to \mathbb{R}^N$ have continuous partial derivatives in an open set $\mathcal{G} \subset \mathbb{R}^M$ and let o be a regular value of f (Definition 4.3.6). Then*

$$\mathscr{M} = \{x \in \mathcal{G} : f(x) = o\}$$

is an $(M-N)$-dimensional differentiable manifold provided $\mathscr{M}$ is not empty, and for $a \in \mathscr{M}$ the tangent space $T_a\mathscr{M}$ is equal to $\operatorname{Ker} f'(a)$.

Proof. The first part is exactly Proposition 4.3.8 and Remark 4.3.9(iii). If a map $\gamma\colon \mathcal{I}_\gamma \to \mathscr{M}$ is a smooth curve, $\gamma(0) = a$, then

$$f(\gamma(t)) = o \quad \text{for } t \in \mathcal{I}_\gamma \qquad \text{and} \qquad f'(a)\dot\gamma(0) = o, \quad \text{i.e.,} \quad \dot\gamma(0) \in \operatorname{Ker} f'(a).$$

Since $T_a\mathscr{M} \subset \operatorname{Ker} f'(a)$ and both the spaces have the same (finite) dimension (the assumption on regularity of o), we have

$$T_a\mathscr{M} = \operatorname{Ker} f'(a). \qquad \square$$

Since the same geometric object $\mathscr{M}$ can be viewed as a manifold with different local parametrizations or as solutions of different equations, we would like to know how the notion of the tangent space (and other notions to be introduced later on) depends on the way it is introduced. As the implicit definition of manifold leads to local parametrizations (see the proof of Proposition 4.3.8) we can consider only the definition given by parametrizations. First of all we should say when two atlases of $\mathscr{M}$ define the same structure on $\mathscr{M}$.

Definition 4.3.34. Two C^k-atlases $(\mathcal{V}_\alpha, \psi_\alpha)_{\alpha\in\mathcal{A}}$, $(\tilde{\mathcal{V}}_\beta, \tilde\psi_\beta)_{\beta\in\mathcal{B}}$ of $\mathscr{M}$ are said to be *equivalent* if for every $a \in \mathscr{M}$ and any $\alpha \in \mathcal{A}$, $\beta \in \mathcal{B}$ for which $a \in \mathcal{V}_\alpha \cap \tilde{\mathcal{V}}_\beta$ the mapping

$$\Phi = (P\tilde\psi_\beta) \circ \varphi_\alpha$$

is a C^k-diffeomorphism of $(\varphi_\alpha)_{-1}(\mathcal{V}_\alpha \cap \tilde{\mathcal{V}}_\beta)$ onto $(\tilde\varphi_\beta)_{-1}(\mathcal{V}_\alpha \cap \tilde{\mathcal{V}}_\beta)$ (see Figure 4.3.14).

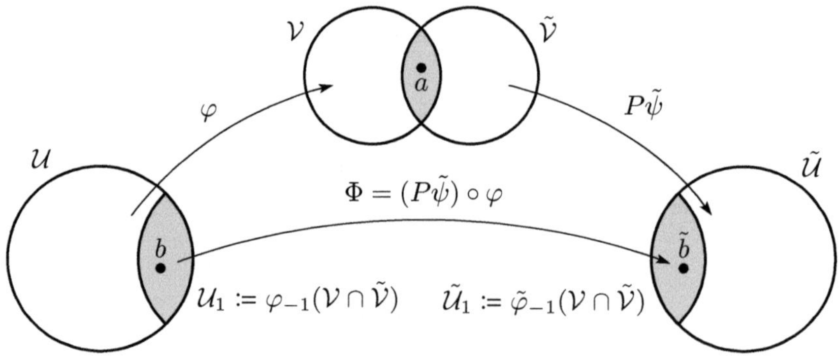

Figure 4.3.14.

Example 4.3.35. Let $(\mathcal{V}, \psi)$, $(\tilde{\mathcal{V}}, \tilde{\psi})$ be two local charts at a point $a \in \mathcal{M}$ which belong to equivalent atlases of $\mathcal{M}$. Let $v \in T_a\mathcal{M}$, $v = \dot{\gamma}(0)$ for a smooth curve γ. Then

$$\gamma = \varphi \circ \kappa = \tilde{\varphi} \circ \tilde{\kappa} \qquad \text{where} \quad \tilde{\kappa} = \Phi \circ \kappa$$

and Φ is defined as in Definition 4.3.34. It follows that

$$\varphi'(b)(\dot{\kappa}(0)) = \tilde{\varphi}'(\tilde{b})(\Phi'(b)\dot{\kappa}(0)).$$

In particular, for $e_i = \dot{\kappa}(0)$, denoting $\frac{\partial}{\partial y_i} := \varphi'(b)e_i$ (as above) and $\frac{\partial}{\partial z_j} := \tilde{\varphi}'(\tilde{b})e_j$, we get the transformation rule for the tangent vectors

$$\frac{\partial}{\partial y_i} = \tilde{\varphi}'(b)\left(\sum_{j=1}^{M} \frac{\partial \Phi^j}{\partial y_i}(b)e_j\right) = \sum_{j=1}^{M} \frac{\partial \Phi^j}{\partial y_i}(b)\frac{\partial}{\partial z_j}, \qquad i = 1, \dots, M. \tag{4.3.17}$$

○

We will now examine a more general situation. Let $\mathcal{M}$, $\tilde{\mathcal{M}}$ be differentiable manifolds in $\mathbb{R}^N$ and $\mathbb{R}^{\tilde{N}}$, respectively. Suppose that $g\colon \mathcal{M} \to \tilde{\mathcal{M}}$ and let $(\mathcal{V}, \psi)$, $(\tilde{\mathcal{V}}, \tilde{\psi})$ be local charts in $a \in \mathcal{M}$ and $\tilde{a} = g(a) \in \tilde{\mathcal{M}}$. Put

$$G = (\tilde{P}\tilde{\psi}) \circ g \circ \varphi$$

(see Figure 4.3.15). Then G is called a *local realization* of g.

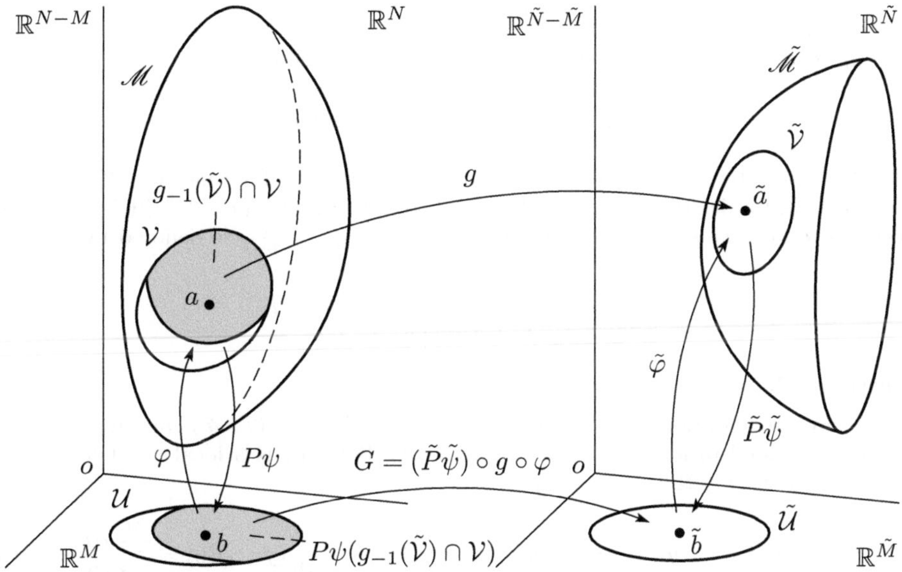

Figure 4.3.15.

We say that a mapping $g\colon \mathcal{M} \to \tilde{\mathcal{M}}$ *is of the class* C^k (or k-times continuously differentiable) if $G \in C^k(\mathcal{U}, \tilde{\mathcal{U}})$ for all charts $(\mathcal{V}, \psi)$ and $(\tilde{\mathcal{V}}, \tilde{\psi})$. In this case g maps a smooth curve $\gamma \in \Gamma_a$ onto a smooth curve $g \circ \gamma \in \Gamma_{\tilde{a}}$. Namely,

$$g \circ \gamma = \tilde{\varphi} \circ G \circ \kappa \qquad \text{where} \quad \gamma = \varphi \circ \kappa,$$

and

$$\frac{\mathrm{d}}{\mathrm{d}t}(g \circ \gamma)(0) = \tilde{\varphi}'(b)(G'(b)\dot{\kappa}(0)). \tag{4.3.18}$$

We say that g "pushes forward" the tangent vector $\dot{\gamma}(0) \in T_a\mathscr{M}$ to the tangent vector

$$\frac{\mathrm{d}}{\mathrm{d}t}(g \circ \gamma)(0) \in T_{g(a)}\tilde{\mathscr{M}}$$

which is denoted by $g_*(\dot{\gamma}(0))$ and which is called a *push-forward*. In particular, g pushes forward the tangent vector $\frac{\partial}{\partial y_i}$ to $g_*\left(\frac{\partial}{\partial y_i}\right)$ where

$$g_*\left(\frac{\partial}{\partial y_i}\right) = \sum_{j=1}^{\tilde{M}} \frac{\partial G^j}{\partial y_i}(b)\frac{\partial}{\partial z_j}; \qquad \frac{\partial}{\partial z_j} = \tilde{\varphi}'(\tilde{b})e_j. \tag{4.3.19}$$

We wish to point out that g_* is the generalization of $g'(a)$ for $g\colon \mathbb{R}^M \to \mathbb{R}^{\tilde{M}}$.

The transformation rule (4.3.17) is a special case of (4.3.18) where $g = I$.

An important special case of a smooth mapping is a differentiable function on a manifold. For such a function we define the notion of the differential.

Definition 4.3.36. A function $f\colon \mathscr{M} \to \mathbb{R}$ is said to be *differentiable* at $a \in \mathscr{M}$ if

$$f \circ \gamma \in C^1(\mathcal{I}_\gamma) \qquad \text{on some} \quad \gamma \in \Gamma_a.$$

The *differential* $\mathrm{d}f(a)$ of f at the point a is defined by the relation

$$\mathrm{d}f(a)(\dot{\gamma}(0)) := \frac{\mathrm{d}}{\mathrm{d}t}(f \circ \gamma)(0) \qquad \text{on some} \quad \gamma \in \Gamma_a.$$

We denote by $\mathcal{F}(\mathscr{M})$ the collection of all differentiable functions on $\mathscr{M}$.

The (algebraic) dual space to $T_a\mathscr{M}$ will be denoted by $(T_a\mathscr{M})^*$ (it is sometimes called the *cotangent space*) and the dual basis to $\frac{\partial}{\partial y_1}, \dots, \frac{\partial}{\partial y_M}$ is denoted by $\mathrm{d}y^1, \dots, \mathrm{d}y^M$, i.e.,

$$\mathrm{d}y^j\left(\frac{\partial}{\partial y_i}\right) = \delta_i^j = \begin{cases} 1 & \text{for} \quad i = j, \\ 0 & \text{for} \quad i \neq j. \end{cases}$$

Remark 4.3.37.

(i) From the definition of $\mathrm{d}f(a)$ it is obvious that $\mathrm{d}f(a) \in (T_a\mathscr{M})^*$ and its values can be expressed in local coordinates as follows: If $(\psi, \mathcal{V})$ is a local chart at $a \in \mathscr{M}$, $F = f \circ \varphi\colon \mathcal{U} \subset \mathbb{R}^M \to \mathbb{R}$, then

$$f \circ \gamma = F \circ \kappa \qquad \text{for} \quad \gamma = \varphi \circ \kappa \in \Gamma_a, \qquad \text{i.e.,} \qquad \frac{\mathrm{d}}{\mathrm{d}t}(f \circ \gamma)(0) = \sum_{i=1}^{M} \frac{\partial F}{\partial y_i}(b)\dot{\kappa}^i(0).$$

In other words,

$$\mathrm{d}f(a) = \sum_{i=1}^{M} \frac{\partial F}{\partial y_i}(b)\,\mathrm{d}y^i. \tag{4.3.20}$$

In particular, for $f(x) = \psi^i(x)$ we have

$$\mathrm{d}\psi^i(a) = \mathrm{d}y^i, \qquad i = 1, \dots, M.$$

Observe that the formula (4.3.20) allows us to define continuity of the mapping

$$x \in \mathscr{M} \mapsto \mathrm{d}f(x) \in (T_x\mathscr{M})^*$$

by the requirement that all F (corresponding to all charts of $\mathscr{M}$) have continuous partial derivatives.[11]

(ii) Suppose now that there are two local charts $(\mathcal{V},\psi)$, $(\tilde{\mathcal{V}},\tilde{\psi})$ at a point $a \in \mathscr{M}$. Let $f\colon \mathscr{M} \to \mathbb{R}$ be differentiable at $a \in \mathscr{M}$. Put

$$F := f \circ \varphi, \qquad \tilde{F} := f \circ \tilde{\varphi}, \qquad \text{i.e.,} \qquad F = \tilde{F} \circ \Phi$$

where Φ is defined in Definition 4.3.34. By virtue of (4.3.20) we have

$$\begin{aligned} \mathrm{d}f(a) &= \sum_{i=1}^{M} \frac{\partial F}{\partial y_i}(b)\,\mathrm{d}y^i = \sum_{i=1}^{M} \left(\sum_{j=1}^{M} \frac{\partial \tilde{F}}{\partial z_j}(\tilde{b}) \frac{\partial \Phi^j}{\partial y_i}(b) \right) \mathrm{d}y^i \\ &= \sum_{j=1}^{M} \frac{\partial \tilde{F}}{\partial z_j}(\tilde{b}) \left[\sum_{i=1}^{M} \frac{\partial \Phi^j}{\partial y_i}(b)\,\mathrm{d}y^i \right] = \sum_{j=1}^{M} \frac{\partial \tilde{F}}{\partial z_j}(\tilde{b})\,\mathrm{d}z^j. \end{aligned} \tag{4.3.21}$$

The equality

$$\mathrm{d}z^j = \sum_{i=1}^{M} \frac{\partial \Phi^j}{\partial y_i}(b)\,\mathrm{d}y^i \tag{4.3.22}$$

follows from this calculation applied to the jth coordinate function

$$z^j = \tilde{\psi}^j(x), \qquad x \in \mathcal{V} \cap \tilde{\mathcal{V}}$$

and Remark 4.3.37(i).[12]

According to the last remark, the existence of the differential $\mathrm{d}f(a)$ does not depend on the choice of local chart. The fact that the differentiability of functions, and similarly other notions, do not depend on a particular choice of local coordinates is crucial for differential geometry and global analysis. Namely, these parts of mathematics study objects in their own geometric nature. The invariance of geometric objects with respect to various groups of transformations (in our case this group is the group of diffeomorphisms) plays a very important role in various applications, mainly in physics (e.g., in general relativity). It is also worth mentioning that the emphasis on invariance is a certain kind of the philosophy dual to Descartes. Analytic geometry and classical differential geometry transform geometric properties into the language of analysis. Local coordinates which introduce analytic tools into geometry are used mainly for computations.

---

[11]It is possible to define the structure of a differentiable manifold on the collection $T\mathscr{M} := \{(x,v) : x \in \mathscr{M}, v \in T_x\mathscr{M}\}$ of all tangent spaces together with their "base" points. This set $T\mathscr{M}$ is called a *tangent bundle* of $\mathscr{M}$. The structure of a differentiable manifold on $T\mathscr{M}$ is given by the local charts which are constructed as follows: If $(\mathcal{V},\psi)$ is a local chart at $a \in \mathscr{M}$, then $\mathcal{V}_T = \{\{x\} \times T_x\mathscr{M} : x \in \mathcal{V}\} \subset \mathbb{R}^{N+M}$ and $\psi_T(x,w) = (\psi(x), P\psi'(x)w) \subset \mathbb{R}^{N+M}$. In a similar way the *cotangent bundle* is the collection $T^*\mathscr{M} := \{(x,\kappa) : x \in \mathscr{M}, \kappa \in (T_x\mathscr{M})^*\}$. The continuity of $\mathrm{d}f$ which has just been defined is then the continuity of $\mathrm{d}f\colon \mathscr{M} \to T^*\mathscr{M}$.

[12]Notice the difference between the transformation rules (4.3.17) and (4.3.22). The reader who is acquainted with tensor analysis can realize that tangent vectors are transformed as contravariant tensors and differentials as covariant ones.

Remark 4.3.38. The transformation rule (4.3.21) can be generalized. Consider a function $f\colon \tilde{\mathscr{M}} \to \mathbb{R}$ and suppose that a mapping $g\colon \mathscr{M} \to \tilde{\mathscr{M}}$ is given. We can look at g as a "generalized" transformation (we do not assume that g is injective). We are interested in the relation between $\mathrm{d}f(\tilde{a})$ and the differential of the "transformed" function $f\circ g\colon \mathscr{M} \to \mathbb{R}$. The desired chain rule can be obtained again with help of local charts $(\mathcal{V},\psi)$ at $a \in \mathscr{M}$ and $(\tilde{\mathcal{V}},\tilde{\psi})$ at $\tilde{a} = g(a) \in \tilde{\mathscr{M}}$ (see Figure 4.3.15). This means that we investigate $F\circ G$ instead of $f\circ g$. Here

$$G = (\tilde{P}\tilde{\psi})\circ g\circ\varphi \qquad \text{and} \qquad F = f\circ\tilde{\varphi}.$$

According to (4.3.19)–(4.3.21) we have

$$\begin{aligned}\mathrm{d}(f\circ g)(a)\left(\frac{\partial}{\partial y_i}\right) &= \frac{\partial(F\circ G)}{\partial y_i}(b) = \sum_{j=1}^{\tilde{M}}\frac{\partial F}{\partial z_j}(\tilde{b})\frac{\partial G^j}{\partial y_i}(b)\\ &= \mathrm{d}f(\tilde{a})\left(g_*\left(\frac{\partial}{\partial y_i}\right)\right) =: g^*(\mathrm{d}f(\tilde{a}))\left(\frac{\partial}{\partial y_i}\right).\end{aligned} \tag{4.3.23}$$

The linear form $g^*(\mathrm{d}f(\tilde{a})) \in (T_a\mathscr{M})^*$ is called a *pull-back* of the linear form $\mathrm{d}f(\tilde{a}) \in (T_{\tilde{a}}\tilde{\mathscr{M}})^*$. This operation will play an important role in the definition of the degree (see Proposition 5.7.18).

We will return to pull-back in Exercise 4.3.75.

Remark 4.3.39. The notion of the differentiable manifold can be generalized in such a way that it is not a priori assumed that $\mathscr{M}$ is a subset of $\mathbb{R}^N$. In fact, we have needed in Definition 4.3.4 that $\mathscr{M}$ has a topological structure (inherited from $\mathbb{R}^N$) such that the neighborhood $\mathcal{V} = \mathcal{W}\cap\mathscr{M}$ is homeomorphic via $P\psi|_{\mathcal{V}}$ ($(P\psi|_{\mathcal{V}})^{-1} = \varphi$) with $\mathcal{U}\subset\mathbb{R}^M$. A differential structure on $\mathscr{M}$ is introduced with help of differentiability properties of the mappings $\Phi_{1,2} := \varphi_2^{-1}\circ\varphi_1$ for different neighborhoods $\mathcal{V}_1$, $\mathcal{V}_2$ of $\mathscr{M}$ (cf. Definition 4.3.34). This is sufficient to correctness of the definition of the smooth function $f\colon\mathscr{M}\to\mathbb{R}$. (It is smooth provided $f\circ\varphi\colon\mathcal{U}\to\mathbb{R}$ is smooth for all φ. Namely,

$$f(x) = (f\circ\varphi_1)(y) = (f\circ\varphi_2)[\Phi_{1,2}(y)] \qquad \text{for} \quad x\in\mathcal{V}_1\cap\mathcal{V}_2.)$$

These considerations allow us to say that the topological space $\mathscr{M}$[13] is called the M-*dimensional differentiable manifold* if $\mathscr{M}$ is locally homeomorphic to open sets of $\mathbb{R}^M$ in such a way that all composite mappings $\Phi_{1,2}$ belong to the class C^k.[14]

Remark 4.3.40. We can define an *infinite-dimensional differentiable* (e.g, of the class C^k) *manifold* by replacing $\mathbb{R}^M$ by a Banach space X. As an important example consider a mapping $f\in C^k(X,\mathbb{R})$, $k\geq 1$, and define

$$\mathscr{M} = \{x\in X : f(x) = 1\}.$$

[13]If $\mathscr{M} = \bigcup_\alpha \mathcal{V}_\alpha$ is a set such that there are injective and surjective mappings $\varphi_\alpha\colon\mathcal{U}_\alpha\to\mathcal{V}_\alpha$ where $\mathcal{U}_\alpha$ are open subsets of $\mathbb{R}^M$, then the sets $\mathcal{V}_\alpha$ form a subbase of a topology in $\mathscr{M}$.

[14]The deep theorem due to the H. Whitney roughly says that a connected M-dimensional differentiable manifold can be embedded (Remark 4.3.3(iv)) into $\mathbb{R}^{2M+1}$ (see, e.g., Whitney [158, Chapter IV], Aubin [13, Theorem 1.22] or Sternberg [146, Theorem 2.4.4]). This means that our previous approach was not too restrictive.

If $\mathcal{M} \neq \emptyset$ and all its points are regular (i.e., $f'(x) \neq 0$ for all $x \in \mathcal{M}$), then $\mathcal{M}$ is an infinite-dimensional manifold of the class C^k (this follows from Proposition 4.3.8 and Remark 4.3.9(i)). Moreover, the tangent space $T_a\mathcal{M}$ is equal to $\operatorname{Ker} f'(a)$ for $a \in \mathcal{M}$. Indeed, the inclusion $T_a\mathcal{M} \subset \operatorname{Ker} f'(a)$ can be proved as in Proposition 4.3.33. To get the reverse inclusion let $h \in \operatorname{Ker} f'(a)$ and let φ be the diffeomorphism from Proposition 4.3.8. For $\gamma(t) := \varphi(th)$ there exists $\delta > 0$ such that $\gamma(t) \in \mathcal{M}$ for $|t| < \delta$ and $\dot\gamma(0) = \varphi'(o)h = h$ (cf. the proof of Proposition 4.3.8), i.e., $h \in T_a\mathcal{M}$. In order to define the differential $\mathrm{d}f(a)$ for $f\colon \mathcal{M} \to \mathbb{R}$ we need a generalization of the notion of the tangent space $T_a\mathcal{M}$ in this more general setting. For $a \in \mathcal{V} \subset \mathcal{M}$ we define

$$\Gamma_a := \{\gamma\colon \mathbb{R} \to \mathcal{M} : \exists \text{ open interval } \mathcal{I}_\gamma \ni 0 : \gamma(0) = a \in \mathcal{V}, \varphi^{-1} \circ \gamma \in C^1(\mathcal{I}_\gamma),$$
$$\text{where } \varphi\colon \mathcal{U} \to \mathcal{V} \text{ is a local parametrization of } \mathcal{V}\}.$$

Similarly as above, $T_a\mathcal{M}$ is the collection of all vectors $\frac{\mathrm{d}}{\mathrm{d}t}(\varphi^{-1} \circ \gamma)\big|_{t=0}$ with a linear structure of $\mathbb{R}^M$. (Actually, $T_a\mathcal{M}$ now coincides with $\mathbb{R}^M$. We remark that previously $T_a\mathcal{M}$ was an M-dimensional subset of $\mathbb{R}^N$.) Further,

$$\mathrm{d}f(a)\colon T_a\mathcal{M} \to \mathbb{R}$$

is defined as follows

$$\mathrm{d}f(a)v := \frac{\mathrm{d}}{\mathrm{d}t}(f \circ \gamma)\bigg|_{t=0} = \frac{\mathrm{d}}{\mathrm{d}t}(F \circ \kappa)\bigg|_{t=0} \quad \text{where } v = \dot\kappa(o), \kappa = \varphi^{-1} \circ \gamma, F = f \circ \varphi^{-1},$$

see Figure 4.3.16.

The following notion is also useful in nonlinear analysis.

Definition 4.3.41. A *vector field* on a differentiable manifold $\mathcal{M}$ is such a mapping $v\colon \mathcal{M} \to T\mathcal{M}$ that

$$v(x) \in T_x\mathcal{M} \qquad \text{for all} \quad x \in \mathcal{M}.$$

The set of all smooth vector fields on $\mathcal{M}$ is denoted by $\mathcal{V}(\mathcal{M})$.

A vector field v determines the differential equation

$$\dot x = v(x). \tag{4.3.24}$$

A *solution* of (4.3.24) is a curve $\gamma\colon \mathcal{I}_\gamma \to \mathcal{M}$ ($\mathcal{I}_\gamma$ is an open interval in $\mathbb{R}$) such that

$$\dot\gamma(t) = v(\gamma(t)) \qquad \text{for all} \quad t \in \mathcal{I}_\gamma.$$

If we choose a chart $(\mathcal{V}, \psi)$ at a point $a \in \mathcal{M}$, we can try to find a solution γ of (4.3.24) which passes through the point a, i.e., we are looking for a curve $\gamma\colon \mathcal{I}_\gamma \ni 0 \to \mathcal{M}$ which is a solution of (4.3.24) and $\gamma(0) = a$. If

$$v(x) = \sum_{i=1}^{M} v^i(x)\frac{\partial}{\partial y_i} \qquad \text{for} \quad x \in \mathcal{V},$$

then the equation (4.3.24) has the form of a system

$$\dot y^i = v^i(\varphi \circ y), \qquad i = 1, \dots, M. \tag{4.3.25}$$

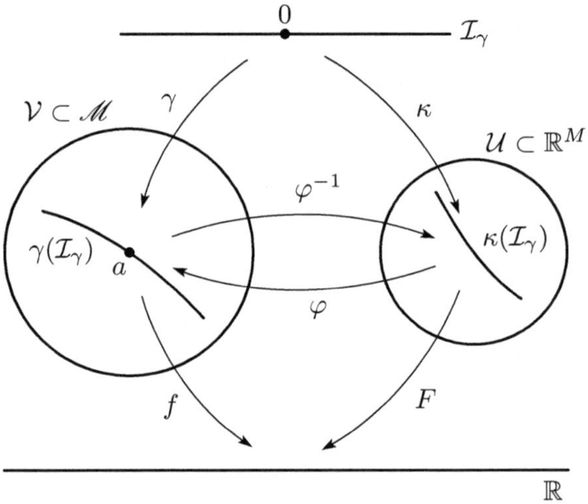

Figure 4.3.16.

The local existence (and uniqueness) theorem for the system (4.3.25) can be used assuming that the vector field v is continuous (and all partial derivatives $\frac{\partial(v^i \circ \varphi)}{\partial y_j}$ are continuous). A standard continuation process then yields a solution γ which is defined on a maximal "time" interval $\mathcal{I}_a$ ($\gamma(0) = a$). It is well known that even very simple differential equations in $\mathbb{R}$ need not have any solution defined on the whole of $\mathbb{R}$ (e.g., $\dot{x} = x^2 + 1$). The situation is better in the case of a compact differentiable manifold $\mathcal{M}$ (i.e., when $\mathcal{M}$ is a compact subset of $\mathbb{R}^N$). If v is continuous on $\mathcal{M}$, $a \in \mathcal{M}$, then there exists a solution γ of (4.3.24), $\gamma(0) = a$, which is defined on the whole of $\mathbb{R}$. Because of the compactness of $\mathcal{M}$ there is a finite number of charts $(\mathcal{V}_i, \psi_i)$, $i = 1, \dots, k$ such that $\mathcal{M} = \bigcup_{i=1}^{k} \mathcal{V}_i$. By the continuity of v there is also a constant K for which

$$|v(x)| \le K \qquad \text{for all} \quad x \in \mathcal{M}.$$

Any solution γ is therefore uniformly continuous on $\mathcal{I}_\gamma$. If $\mathcal{I}_\gamma \neq \mathbb{R}$, then the limits (in $\mathcal{M}$) of γ at the terminal point(s) of $\mathcal{I}_\gamma$ exist, and γ could be continued. The reader is invited to fill in all details of this proof using local coordinates.

The global existence of solutions means that the map $\sigma \colon \mathbb{R} \times \mathcal{M} \to \mathcal{M}$:

$$\sigma(t, a) = \gamma_a(t)$$

where γ_a is a solution of (4.3.24) on $\mathbb{R}$ which satisfies

$$\gamma_a(0) = a,$$

is a smooth (provided v is smooth) dynamical system on $\mathcal{M}$. In particular, for fixed t the mapping $a \mapsto \sigma(t, a)$ is a diffeomorphism on $\mathcal{M}$ (the derivative of a solution of a differential equation with respect to the initial conditions - Example 4.2.5). On the other hand, with any smooth dynamical system σ on $\mathscr{M}$ we can associate a smooth vector field v on $\mathscr{M}$. The reader who is interested in dynamical systems on manifolds can consult, e.g., Chillingworth [31], Ruelle [133] for brief information, Palis & De Melo [119] or Katok & Hasselblatt [88] (this is actually much more than an introduction). If $f \in \mathcal{F}(\mathscr{M})$ and $v \in \mathcal{V}(\mathscr{M})$ then the function $x \mapsto \mathrm{d}f(x)v(x)$ (the derivative of f at the point x in the direction $v(x)$) is called the *Lie derivative* of f with respect to v and it is denoted by $\mathcal{L}_v f(x)$. It is not difficult to see that $\mathcal{L}_v f(x) = \lim\limits_{t\to 0} \frac{f(\sigma(t,x))-f(x)}{t}$, where σ is the dynamical system determined by v. In order to define a derivative of a vector field w with respect to the given vector field v we have to overcome the problem that $w(\sigma(t,x))$ and $w(x)$ belong to different tangent spaces and therefore their difference is not well defined. But we can push-forward the vector $w(\sigma(t,x))$ by $\sigma(-t,\cdot)$ to $T_x\mathscr{M}$, i.e., we consider $\sigma_*(-t,\cdot)w(\sigma(t,x))$ and define

$$\mathcal{L}_v w(x) := \lim_{t\to 0} \frac{\sigma_*(-t,\cdot)w(\sigma(t,x)) - w(x)}{t}$$

provided that this limit exists. This *Lie derivative of a vector field* is again a vector field and it has local coordinates at x given by

$$\mathcal{L}_v w(x) = \sum_{i=1}^{M} z^i \frac{\partial}{\partial y_i}, \quad z^i = \sum_{j=1}^{M} \left(w^j \frac{\partial v^i}{\partial y_j} - v^j \frac{\partial w^i}{\partial y_j} \right),$$

where w^i and v^i are local coordinates of w and v with respect to the basis $\frac{\partial}{\partial y_1}, \dots, \frac{\partial}{\partial y_M}$. We will see another way to define a derivative of w with respect to v in Section 8.4 (see Definition 8.4.1).

We have mentioned the role of the first integrals for (autonomous) systems of ordinary differential equations on page 175. The notion of the first integral is connected with partial differential equations, since $f\colon \mathscr{M} \to \mathbb{R}$ is the first integral of (4.3.24) if f satisfies (in local coordinates) the linear partial differential equation of the first order

$$\frac{\mathrm{d}}{\mathrm{d}t} f(\gamma(t)) = \frac{\mathrm{d}}{\mathrm{d}t} F(\kappa(t)) = \sum_{i=1}^{n} \frac{\partial F}{\partial y_i}(\kappa(t)) v^i(\gamma(t)) = \mathrm{d}f(x)(v(x)) = o$$

where

$$x = \gamma(t) \qquad \text{and} \qquad \gamma(t) = \varphi(\kappa(t)).$$

The solutions of (4.3.24) are called the *characteristics* of this partial differential equation for an unknown function F.

We obtain a system of partial differential equations by considering a family of vector fields. Let $v_1, \dots, v_k$ be vector fields on a manifold $\mathscr{M}$ and denote by

$$V(x) = \operatorname{Lin}\{v_1(x), \dots, v_k(x)\}$$

the subspace of $T_x\mathscr{M}$. The first integral of the system $v_1, \dots, v_k$ (or the collection of subspaces $\{V(x)\}_{x\in\mathscr{M}}$) is a function $f\colon \mathscr{M} \to \mathbb{R}$ for which

$$\mathrm{d}f(x)(v_i(x)) = 0, \qquad i = 1, \dots, k, \quad x \in \mathscr{M}, \tag{4.3.26}$$

or equivalently, $\mathrm{d}f$ annihilates all $V(x)$, $x \in \mathscr{M}$, i.e., $\mathcal{L}_w f = o$ for all vector fields w such that $w(x) \in V(x)$. From this formulation it is clear that we can suppose that the vector fields $v_1, \dots, v_k$ are linearly independent at each $x \in \mathscr{M}$. Contrary to the case of one equation, the system (4.3.26) need not have a solution.

The following problem is similar to the preceding one: Let $\mathcal{G}$ be an open subset of $\mathbb{R}^M$ and let $g = (g^1, \dots, g^M)$ be a smooth mapping from $\mathcal{G} \times \mathbb{R}$ into $\mathbb{R}^M$. Since $\mathbb{R}^M$ can be also interpreted as the dual space to $\mathbb{R}^M$, the mapping g determines a system of partial differential equations

$$u'(x) = g(x, u(x)), \qquad x \in \mathcal{G} \subset \mathbb{R}^M. \tag{4.3.27}$$

Expressing the Frechét derivative (i.e., the differential) $u'(x)$ in terms of partial derivatives we get the system

$$\frac{\partial u}{\partial x_i}(x) = g^i(x, u(x)), \qquad i = 1, \dots, M, \quad x \in \mathcal{G}.$$

If this system has a solution u, then $u \in C^2(\mathcal{G})$ (since g is supposed to be smooth), which implies a necessary condition for the existence of a solution given by mixed derivatives $\left(\frac{\partial^2 u}{\partial x_i \partial x_j} = \frac{\partial^2 u}{\partial x_j \partial x_i}, i, j = 1, \dots, M\right)$:

$$\frac{\partial g^i}{\partial x_j} + \frac{\partial g^i}{\partial u} g^j = \frac{\partial g^j}{\partial x_i} + \frac{\partial g^j}{\partial u} g^i, \qquad i, j = 1, \dots, M. \tag{4.3.28}$$

It is a question how to formulate this *integrability condition* for the system (4.3.26).

The system (4.3.26) is said to be *completely integrable* in $\mathscr{M}$ if for any $x \in \mathscr{M}$ there is a submanifold $\mathscr{N}(x)$ of (the *integral manifold*) $\mathscr{M}$ containing x such that

$$i_*(T_y \mathscr{N}(x)) = V(y) \qquad \text{for all} \quad y \in \mathscr{N}(x)$$

(i is the natural embedding of $\mathscr{N}(x)$ into $\mathscr{M}$). Notice that for one vector field $v \neq o$ (i.e., $\dim V(x) = 1$) the integral manifold is the integral curve of the system $\dot{x} = v(x)$ and, in general, it contains the integral curves of all equations

$$\dot{x} = v^i(x), \qquad i = 1, \dots, M.$$

The gluing of all integral curves need not be a manifold. A possible problem is shown in Remark 4.3.43 below.

The basic result on complete integrability is the next theorem (Frobenius Theorem) which we state without proof. This theorem is an important tool in differential geometry, and the reader can find its proof in textbooks on this subject, e.g., Aubin [13], Sternberg [146, § 3.5]. To formulate the theorem in a compact form we need[15] the notion of the *Lie brackets* (they are sometimes, mainly in applications to Hamiltonian mechanics, called the *Poisson brackets*). If v, w are two smooth vector fields with local representations

$$v = \sum_{i=1}^{M} v^i \frac{\partial}{\partial y_i}, \qquad w = \sum_{j=1}^{M} w^j \frac{\partial}{\partial y_j},$$

then $[v, w]$ is the vector field with the local representation

$$[v, w] = \sum_{j=1}^{M} \left(\sum_{k=1}^{M} \left(v^k \frac{\partial w^j}{\partial y_k} - w^k \frac{\partial v^j}{\partial y_k} \right) \right) \frac{\partial}{\partial y_j}. \tag{4.3.29}$$

For another interpretation of this operation see Remark 4.3.43 below or Exercise 4.3.48.

[15] We will give another formulation of this theorem at the end of Appendix 4.3B.

Theorem 4.3.42 (Frobenius). *Let $v_1, \dots, v_k$ be smooth vector fields on a manifold $\mathscr{M}$. Then this system is completely integrable if and only if*

$$[v_i(x), v_j(x)] \in V(x) \qquad \textit{for all} \quad x \in \mathscr{M}, \quad i, j = 1, \dots, k.$$

Remark 4.3.43. Suppose that two smooth vector fields v, w are given on a (compact) manifold $\mathscr{M}$ and let σ_v, σ_w be the corresponding dynamical systems on $\mathscr{M}$. There is no reason to expect that these systems commute, i.e.,

$$\sigma_v(t, \sigma_w(s, x)) = \sigma_w(s, \sigma_v(t, x)),$$

and it is not difficult to construct a counterexample which confirms that (see Figure 4.3.17).

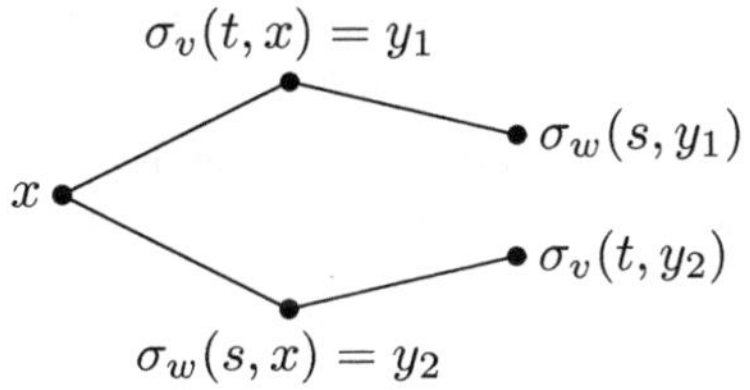

Figure 4.3.17.

It can be shown that a necessary and sufficient condition for commutativity is that

$$[v, w] = o$$

(see (4.3.29)).

Exercise 4.3.44. Let $g_*(x) \colon \mathcal{T}_x\mathscr{M} \to \mathcal{T}_{g(x)}\widetilde{\mathscr{M}}$ be the push-forward defined on p. 198. If $g_*(x)$ is an isomorphism then there is an open neighborhood $\mathcal{U}$ of x such that $g(\mathcal{U})$ is open in $\widetilde{\mathscr{M}}$ and $g|\mathcal{U}$ is a diffeomorphism. Prove this generalisation of Theorem 4.1.1.

Exercise 4.3.45. We say that a function $f \colon \mathscr{M} \to \mathbb{R}$ is an integral of the equation (4.3.24) if $f(\gamma(\cdot))$ is constant for any solution γ of (4.3.24). If this is true only locally, f is called a *local integral.* Suppose that $\dim \mathscr{M} = 2$ and $(\mathcal{V}, \psi)$ is a local chart of $\mathscr{M}$, f is the integral of (4.3.24) and $\mathrm{d}f(x) \neq 0$ for all $x \in \mathcal{V}$. Prove the following assertions:

(i) There is no stationary point of (4.3.24) in $\mathcal{V}$ ($a \in \mathcal{V}$ is a stationary point of (4.3.24) if $\gamma(t) = a$, $t \in \mathbb{R}$, is a solution of (4.3.24)).

(ii) Take functions $g \colon \mathcal{U} \to \mathbb{R}$, $\varphi(\mathcal{U}) = \mathcal{V}$, such that

$$z = \Phi(y) = (F(y), g(y)) \qquad \text{where} \quad F = f \circ \varphi,$$

is a diffeomorphism of $\mathcal{U}$ onto $\tilde{\mathcal{U}}$ (why does such a function exist?). Then there exists $h \colon \tilde{\mathcal{U}} \to \mathbb{R}$ such that the vector field v has the form

$$v(x) = 0 \cdot \frac{\partial}{\partial z_1} + h(z) \frac{\partial}{\partial z_2}$$

in these new local coordinates $z = (z_1, z_2)$ (use the transformation rule (4.3.17) and the fact that f is an integral of (4.3.24)). Notice that $h(z) \neq 0$ for all $z \in \tilde{\mathcal{U}}$.

(iii) Put

$$H(z_1, z_2) = \int_{z_2^0}^{z_2} \frac{\mathrm{d}\eta}{h(z_1, \eta)}$$

(what is the relation of H to a solution of (4.3.24) in the z-coordinates?). Consider another transformation of coordinates

$$\xi = (z_1, H(z_1, z_2)).$$

Then the vector field v has the form

$$v(x) = 0 \cdot \frac{\partial}{\partial \xi_1} + 1 \cdot \frac{\partial}{\partial \xi_2}$$

in the local coordinates $\xi = (\xi_1, \xi_2)$. Cf. Exercise 4.3.26.

Can you formulate the result in terms of ordinary differential equations? Can you generalize this result to higher-dimensional manifolds?

Exercise 4.3.46. Let $v_1, \dots, v_k$ be smooth vector fields on a differentiable manifold $\mathscr{M}$ which are linearly independent on a neighborhood $\mathcal{U}$ of $a \in \mathscr{M}$. Assume that

$$[v_i, v_j] = o, \qquad i, j = 1, \dots, k, \qquad \text{on} \quad \mathcal{U}.$$

Prove that there exist local coordinates $(y_1, \dots, y_M)$ such that

$$v_i = \frac{\partial}{\partial y_i}, \qquad i = 1, \dots, k,$$

in a neighborhood of a.

Hint. Cf. Exercises 4.3.45 and 4.3.26.

Exercise 4.3.47. Let $\mathscr{M}$ be a differentiable manifold of the class C^∞ and let X be the set of all C^∞ functions on $\mathscr{M}$. Let $D \colon X \to X$ satisfy

(D1) D is linear,

(D2) $D(fg) = gDf + fDg$ for all $f, g \in X$ (pointwise multiplication).

Show that there is a vector field v on $\mathscr{M}$ such that

$$\mathcal{L}_v f := \mathrm{d}f(x)(v(x)) = Df(x), \qquad x \in \mathscr{M}, \quad f \in X.$$

Here $\mathcal{L}_v f$ is the directional derivative (in the direction of the vector field v) which is also called the Lie derivative (cf. page 203).[16]

Hint. Put

$$v = \sum_{i=1}^{M} a_i \frac{\partial}{\partial y_i} \qquad \text{where} \quad a_i = Dy_i.$$

Show that

$$Df - \mathcal{L}_v f = o$$

holds for polynomials of degree ≥ 1 on $\mathcal{U}$. Then use the Taylor polynomial. It remains to extend this result from local charts to the whole manifold – use a partition of unity. See Definition 4.3.80 and Theorem 4.3.82.

[16]Sophus Lie was one of the promoters of geometric methods in analysis. A topological group (i.e., the group with a topological structure such that group operations are continuous) with the structure of a differentiable manifold (e.g., S^N, groups of regular or orthogonal matrices) is called a *Lie group*.

The converse statement, i.e., the fact that $\mathcal{L}_v$ satisfies (D1), (D2) for smooth v is easy to prove. (Do that.) Is there any difference between the differential and the Lie derivative?

Exercise 4.3.48. Let v, w be two smooth vector fields on a differentiable manifold. Define the vector field $[v, w]$, the so-called *commutator* (or Lie bracket) of v, w, by the formula

$$\mathcal{L}_{[v,w]}f = \mathcal{L}_v(\mathcal{L}_w f) - \mathcal{L}_w(\mathcal{L}_v f) \qquad \text{for every} \quad f \in X$$

(see (4.3.29) and Exercise 4.3.47). Show that this definition is correct, i.e., $[v, w]$ is a vector field, and show that the Jacobi identity

$$[u, [v, w]] + [v, [w, u]] + [w, [u, v]] = o$$

holds for any smooth vector fields u, v, w.[17]

Exercise 4.3.49. Show that $\mathcal{L}_v w = [v, w]$.

Hint. Compute coordinates of $[v, w]$.

Exercise 4.3.50. Prove that the mapping $\mathcal{L} : \mathcal{V}(\mathscr{M}) \times \mathcal{V}(\mathscr{M}) \to \mathcal{V}(\mathscr{M})$ has the following properties:

(i) $\mathcal{L}$ is bilinear.

(ii) If $f \in \mathcal{F}(\mathscr{M})$ then $\mathcal{L}_v(fw) = \mathcal{L}_v(f)w + f\mathcal{L}_v(w)$. (Leibniz rule for the derivative of a product.)

4.3B Differential Forms

Before starting with the notion of a differential form we need to summarize some basic facts from multilinear algebra. Let X be a (real) linear space. A *bilinear form* is a map $A\colon X \times X \to \mathbb{R}$ which is linear in both variables. A typical example of a bilinear form is the scalar product in a real Hilbert space.[18] A p-linear form $A\colon \underbrace{X \times \cdots \times X}_{p} \to \mathbb{R}$ is defined in a similar way.

Definition 4.3.51. A p-linear form A is said to be *skew-symmetric* if

$$A(x_{\pi(1)}, \dots, x_{\pi(p)}) = \operatorname{sgn} \pi A(x_1, \dots, x_p)$$

holds for any permutation π of the set $\{1, \dots, p\}$ and all $x_1, \dots, x_p \in X$. Here $\operatorname{sgn} \pi = 1$ if the number of sign changes in π (a sign change occurs whenever $i < j$ and $\pi(i) > \pi(j)$) is even and $\operatorname{sgn} \pi = -1$ if this number is odd. The collection of all skew-symmetric p-linear forms is denoted by $\Lambda_p(X)$.

---

[17] Let $\mathcal{A}$ be a set with two binary operations $+$, $\cdot$ such that

(A1) $\mathcal{A}$ with operations $+$, $\cdot$ is a ring,

(A2) $a \cdot a = o$ for all $a \in \mathcal{A}$,

(A3) $(a \cdot b) \cdot c + (b \cdot c) \cdot a + (c \cdot a) \cdot b = o$ for all $a, b, c \in \mathcal{A}$.

Then $\mathcal{A}$ is said to be a *Lie ring.* If $\mathcal{A}$ is, moreover, a linear space, then $\mathcal{A}$ is called a *Lie algebra.* (If $\mathcal{A}$ is an associative ring and $[a, b] = a \cdot b - b \cdot a$, then $(\mathcal{A}, +, [\cdot, \cdot])$ is a Lie ring.) For more information see, e.g., Adams [1], Bourbaki [18], Bröcker & Dieck [21], Helgason [78].

[18] This is not true for a complex Hilbert space since $(x, \alpha y) = \overline{\alpha}(x, y)$ for $\alpha \in \mathbb{C}$.

Remark 4.3.52.

(i) Let $e_1, \dots, e_M$ be a basis of X and $f^1, \dots, f^M$ its dual basis, i.e., a basis of the space X^* of all linear forms on X for which

$$f^i(e_j) = \delta_j^i = \begin{cases} 1, & i = j, \\ 0, & i \neq j. \end{cases}$$

Then any element $x \in X$ can be expressed in the from

$$x = \sum_{i=1}^{M} f^i(x) e_i.$$

For $A \in \Lambda_p(X)$, $p \leq M = \dim X$, we have

$$\begin{aligned} A(x_1, \dots, x_p) &= \sum_{i_1, \dots, i_p = 1}^{M} f^{i_1}(x_1) \cdot \dots \cdot f^{i_p}(x_p) A(e_{i_1}, \dots, e_{i_p}) \\ &= \sum_{1 \leq i_1 < \dots < i_p \leq M} \left[\sum_{\pi \in \{1, \dots, p\}} \operatorname{sgn} \pi f^{i_{\pi(1)}}(x_1) \cdot \dots \cdot f^{i_{\pi(p)}}(x_p) \right] \\ &\qquad \times A(e_{i_1}, \dots, e_{i_p}) \\ &= \sum_{1 \leq i_1 < \dots < i_p \leq M} \det \left(f^{i_j}(x_k) \right)_{j,k=1,\dots,p} A(e_{i_1}, \dots, e_{i_p}). \end{aligned}$$

In particular, if $p = M$, then

$$A(x_1, \dots, x_M) = \det \left(f^i(x_k) \right)_{i,k=1,\dots,M} A(e_1, \dots, e_M), \tag{4.3.30}$$

i.e., $\dim \Lambda_M(X) = 1$. Notice also that $\Lambda_p(X) = \{o\}$ for $p > M$.

(ii) Elements $x_1, \dots, x_p$ of X are linearly dependent if and only if

$$A(x_1, \dots, x_p) = 0 \qquad \text{for all} \quad A \in \Lambda_p(X).$$

This follows easily from the formula given above.

The product operation can be defined in the family of skew-symmetric forms.

Definition 4.3.53. Let $A \in \Lambda_p(X)$, $B \in \Lambda_q(X)$ be skew-symmetric forms. Then their *exterior product* $A \wedge B$ is the $(p+q)$-skew-symmetric form defined by the formula

$$A \wedge B(x_1, \dots, x_{p+q}) = \sum_{\substack{\pi \in \{1, \dots, p+q\} \\ \pi(1) < \dots < \pi(p) \\ \pi(p+1) < \dots < \pi(p+q)}} \operatorname{sgn} \pi A(x_{\pi(1)}, \dots, x_{\pi(p)}) B(x_{\pi(p+1)}, \dots, x_{\pi(p+q)}).$$

Remark 4.3.54.

(i) The exterior product of three and more skew-symmetric forms is defined by induction and the associative law holds, i.e.,

$$A \wedge B \wedge C := (A \wedge B) \wedge C = A \wedge (B \wedge C).$$

(ii) The exterior product is not commutative. Namely,

$$B \wedge A = (-1)^{pq} A \wedge B \qquad \text{for} \quad A \in \Lambda_p(X), \quad B \in \Lambda_q(X).$$

Example 4.3.55.

(i) If A, B are one-forms (i.e., linear forms), then

$$A \wedge B(x_1, x_2) = A(x_1)B(x_2) - A(x_2)B(x_1).$$

More generally, by induction,

$$A_1 \wedge \cdots \wedge A_n(x_1, \ldots, x_n) = \det\left(A_i(x_j)\right)_{i,j=1,\ldots,n} \qquad \text{for one-forms} \quad A_1, \ldots, A_n.$$

(ii) If $e_1, \ldots, e_M$ and $f^1, \ldots, f^M$ are mutually dual bases of X and X^*, respectively, then

$$A = A(e_1, \ldots, e_M) f^1 \wedge \cdots \wedge f^M \qquad \text{for any} \quad A \in \Lambda_M(X).$$

In other words, $f^1 \wedge \cdots \wedge f^M$ generates $\Lambda_M(X)$ ($\dim X = M$). More generally, the products

$$(f^{i_1} \wedge \cdots \wedge f^{i_n})_{1 \le i_1 < \cdots < i_n \le M}$$

form a basis of $\Lambda_n(X)$, i.e., for any $A \in \Lambda_n(X)$ there are scalars $a_{\ldots}$ such that

$$A = \sum_{1 \le i_1 < \cdots < i_n \le M = \dim X} a_{i_1,\ldots,i_n} f^{i_1} \wedge \cdots \wedge f^{i_n} \tag{4.3.31}$$

where $a_{i_1,\ldots,i_n} = A(e_{i_1}, \ldots, e_{i_n})$ (see Remark 4.3.52(i)). ○

The main goal of this appendix is to investigate skew-symmetric forms on manifolds which are continuous (smooth) with respect to the topological (differential) structure of the manifold. The basic definition is the following:

Definition 4.3.56. Let $\mathscr{M}$ be a differentiable manifold of dimension M. A mapping

$$\omega : x \in \mathscr{M} \mapsto \omega(x) \in \Lambda_p(T_x\mathscr{M})$$

is said to be a *p-differential form* on $\mathscr{M}$ if ω is continuous (or smooth) in the following sense:

Let $(\mathcal{V}, \psi)$ be a local chart of $\mathscr{M}$ and let

$$\omega(x) = \sum_{1 \le i_1 \le \cdots \le i_p \le M} a_{i_1,\ldots,i_p}(x)\, \mathrm{d}y^{i_1} \wedge \cdots \wedge \mathrm{d}y^{i_p} \tag{4.3.32}$$

be the representation of ω in this chart (see (4.3.31)). Then all functions $a_{i_1,\ldots,i_p}$ are continuous (or smooth) in $\mathcal{V}$.

Remark 4.3.57.

(i) A smooth function $f\colon \mathscr{M} \to \mathbb{R}$ is sometimes called a differential form of order 0. Its differential $\mathrm{d}f$ is the one-form with the local representation

$$\mathrm{d}f(x) = \sum_{i=1}^{M} \frac{\partial (f \circ \varphi)(y)}{\partial y_i}\,\mathrm{d}y^i, \qquad \varphi(y) = x$$

(see (4.3.20)).

(ii) Let ω be a p-differential form in $\mathbb{R}^N$ with the representation

$$\omega(x) = \sum_{1 \le i_1 < \cdots < i_p \le N} a_{i_1,\dots,i_p}(x)\,\mathrm{d}f^{i_1} \wedge \cdots \wedge \mathrm{d}f^{i_p}$$

($f^1,\dots,f^N$ is the dual basis to the standard one $e_1,\dots,e_N$ in $\mathbb{R}^N$). In accordance with the notation of coordinates in $\mathbb{R}^N$ we can also write

$$\omega(x) = \sum_{1 \le i_1 < \cdots < i_p \le N} a_{i_1,\dots,i_p}(x)\,\mathrm{d}x^{i_1} \wedge \cdots \wedge \mathrm{d}x^{i_p}.$$

If $\mathscr{M}$ is a differentiable manifold in $\mathbb{R}^N$ of dimension $M \ge p$, then ω can be restricted to $\mathscr{M}$. Since $T_x\mathscr{M} \subset \mathbb{R}^N$ (see (4.3.16)) we have

$$\omega(x)\left(\frac{\partial}{\partial y_{j_1}},\dots,\frac{\partial}{\partial y_{j_p}}\right) = \sum_{1 \le i_1 < \cdots < i_p \le N} a_{i_1,\dots,i_p}(x)\det\left(f^{i_k}\left(\frac{\partial}{\partial y_{j_l}}\right)\right)_{k,l=1,\dots,p}.$$

(iii) If $(\mathcal{V},\psi)$ and $(\tilde{\mathcal{V}},\tilde{\psi})$ are local charts at the same point, then we have two representations for a p-differential form ω in $\mathcal{V} \cap \tilde{\mathcal{V}}$:

$$\omega(x) = \sum_{i \le i_1 < \cdots < i_p \le M} f_{i_1,\dots,i_p}\,\mathrm{d}y^{i_1} \wedge \cdots \wedge \mathrm{d}y^{i_p},$$

$$\omega(x) = \sum_{1 \le j_1 < \cdots < j_p \le M} g_{j_1,\dots,j_p}\,\mathrm{d}z^{j_1} \wedge \cdots \wedge \mathrm{d}z^{j_p}.$$

(Here $\mathrm{d}y^1,\dots,\mathrm{d}y^M$ is the basis of $(T_x\mathscr{M})^*$ with respect to the local chart $(\mathcal{V},\psi)$ and similarly $\mathrm{d}z^1,\dots,\mathrm{d}z^M$ for $(\tilde{\mathcal{V}},\tilde{\psi})$.) The Transformation Rule (4.3.22) yields a relation between the coefficients $f_{\dots}$ and $g_{\dots}$. This relation is simple for M-forms ($M = \dim\mathscr{M}$), namely

$$\omega(x) = g(x)\,\mathrm{d}z^1 \wedge \cdots \wedge \mathrm{d}z^M = g(x)\left[\det \frac{\partial \Phi_j}{\partial y_i}(\psi(x))\right]\mathrm{d}y^1 \wedge \cdots \wedge \mathrm{d}y^M$$

where $\left(\frac{\partial \Phi_j}{\partial y_i}(y)\right)_{i,j=1,\dots,M}$ is the Jacobi matrix of the transformation $z = \Phi(y)$ of local coordinates (see Figure 4.3.14). The determinant of the Jacobi matrix will be called the Jacobian and denoted by J_Φ (Example 4.1.5). This Transformation Rule can be generalized to mappings between manifolds in a way similar as in (4.3.19) and (4.3.23). If $g\colon \mathscr{M} \to \tilde{\mathscr{M}}$ is a smooth map and ω is a p-differential form on $\tilde{\mathscr{M}}$, then the formula

$$g^*\omega(x)(v_1,\dots,v_p) = \omega(g(x))(g_*v_1,\dots,g_*v_p), \tag{4.3.33}$$

$x \in \mathscr{M}$, $v_1, \dots, v_p \in T_x\mathscr{M}$, where $g_* v_i$ is the push-forward of the tangent vector v_i (see (4.3.19)), defines the *pull-back* of ω. To obtain a local representation of the type (4.3.33) we choose local coordinates at x, put $v_k = \frac{\partial}{\partial y_{j_k}}$ and use the Transformation Rule (4.3.19). However, the final formula is rather cumbersome and we will not need it with the exception of the case when $\dim \mathscr{M} = \dim \tilde{\mathscr{M}} = M$ and ω is an M-form,

$$\omega(z) = f(z)\, \mathrm{d}z^1 \wedge \cdots \wedge \mathrm{d}z^M.$$

Then

$$g^*\omega(x) = f(g(x)) J_G(\psi(x))\, \mathrm{d}y^1 \wedge \cdots \wedge \mathrm{d}y^M$$

where G is the local realization of g (see Figure 4.3.15). An important special case is $\varphi^*\omega$ where φ is a coordinate mapping $\mathcal{U} \subset \mathbb{R}^M \to \mathcal{V} \subset \mathscr{M}$. The next example shows how to compute the pull-back of $(M-1)$-forms for small M. These formulae are often used in vector calculus — see also special cases of integration in Appendix 4.3C.

Example 4.3.58.

(i) Let

$$\omega(x, y) = f(x, y)\, \mathrm{d}x + g(x, y)\, \mathrm{d}y$$

be a 1-form in $\mathbb{R}^2$ and $\gamma = (\gamma_1, \gamma_2)\colon (a, b) \to \mathbb{R}^2$ a smooth curve. Then

$$\begin{aligned} \gamma^*\omega(t)\left(\frac{\partial}{\partial t}\right) &= \omega(\gamma(t))\left(\gamma_* \frac{\partial}{\partial t}\right) \\ &= f(\gamma_1(t), \gamma_2(t))\dot{\gamma}_1(t)\, \mathrm{d}t + g(\gamma_1(t), \gamma_2(t))\dot{\gamma}_2(t)\, \mathrm{d}t. \end{aligned}$$

(ii) Let

$$\omega(x, y, z) = f(x, y, z)\, \mathrm{d}y \wedge \mathrm{d}z + g(x, y, z)\, \mathrm{d}z \wedge \mathrm{d}x + h(x, y, z)\, \mathrm{d}x \wedge \mathrm{d}y$$

be a 2-form in $\mathbb{R}^3$ and $\varphi\colon (u, v) \in \mathcal{U} \subset \mathbb{R}^2 \to \mathbb{R}^3$ a smooth parametrization of a surface $\mathcal{S}$ in $\mathbb{R}^3$. Then

$$[\varphi^*\omega(u, v)](e_1, e_2) = \omega(\varphi(u, v))(\underbrace{\varphi'(u, v)e_1}_{\frac{\partial}{\partial u}}, \underbrace{\varphi'(u, v)e_2}_{\frac{\partial}{\partial v}})$$

and, if

$$\frac{\partial}{\partial u} = u_1 \frac{\partial}{\partial x} + u_2 \frac{\partial}{\partial y} + u_3 \frac{\partial}{\partial z}, \qquad \frac{\partial}{\partial v} = v_1 \frac{\partial}{\partial x} + v_2 \frac{\partial}{\partial y} + v_3 \frac{\partial}{\partial z}$$

(here $\frac{\partial}{\partial x}$ is actually the first vector of the standard basis in $\mathbb{R}^3$ — see Remark 4.3.57(ii)), then

$$\mathrm{d}y \wedge \mathrm{d}z\left(\frac{\partial}{\partial u}, \frac{\partial}{\partial v}\right) = u_2 v_3 - u_3 v_2, \qquad \text{etc.},$$

and eventually

$$\varphi^*\omega\left(\frac{\partial}{\partial u}, \frac{\partial}{\partial v}\right) = \left((f, g, h), \frac{\partial}{\partial u} \times \frac{\partial}{\partial v}\right)_{\mathbb{R}^3}$$

where the brackets $(\cdot,\cdot)_{\mathbb{R}^3}$ denote the scalar product in $\mathbb{R}^3$ and $\frac{\partial}{\partial u}\times\frac{\partial}{\partial v}$ is the so-called *cross* (or vector) *product* of vectors $\frac{\partial}{\partial u}=(u_1,u_2,u_3)$, $\frac{\partial}{\partial v}=(v_1,v_2,v_3)$ in $\mathbb{R}^3$, i.e.,

$$\frac{\partial}{\partial u}\times\frac{\partial}{\partial v}=(u_2v_3-u_3v_2,u_3v_1-u_1v_3,u_1v_2-u_2v_1).$$ ○

Remark 4.3.59. The reader can ask why it is necessary (or reasonable) to introduce differential forms even though vectors and vector fields have been defined. Actually, there is only a technical difference for one-forms, since $T_a\mathscr{M}$ is isomorphic to its dual $(T_a\mathscr{M})^*$. For example, $\mathrm{d}f(a)\in(T_a\mathscr{M})^*$ and therefore it can be represented by a scalar product in $T_a\mathscr{M}$. Since $T_a\mathscr{M}$ is a linear subspace of $\mathbb{R}^N$ (for $\mathscr{M}\subset\mathbb{R}^N$) we may define the scalar product in $T_a\mathscr{M}$ as

$$(v,w)_{T_a\mathscr{M}}:=(v,w)_{\mathbb{R}^N}\qquad\text{for}\quad v,w\in T_a\mathscr{M}.$$[19]

In particular, this means that there is a vector $\nabla f(a)$ — the so-called *gradient* of f — such that

$$\mathrm{d}f(a)(v)=(v,\nabla f(a))_{T_a\mathscr{M}}.$$

If

$$\mathrm{d}f(a)=\sum f_i(a)\,\mathrm{d}y_i,$$

then

$$f_i(a)=\left(\frac{\partial}{\partial y_i},\nabla f(a)\right)_{T_a\mathscr{M}}.$$[20]

The reason for distinguishing between differential forms and vector fields lies in the richer structure of the collection of all differential forms — there are operations like the exterior product and the exterior differential (Definition 4.3.60). Moreover, the differential forms

$$\omega^1=f\,\mathrm{d}x+g\,\mathrm{d}y+h\,\mathrm{d}z\qquad\text{and}\qquad\omega^2=f\,\mathrm{d}y\wedge\mathrm{d}z+g\,\mathrm{d}z\wedge\mathrm{d}x+h\,\mathrm{d}x\wedge\mathrm{d}y$$

can be attached to the vector field

$$F=f\frac{\partial}{\partial x}+g\frac{\partial}{\partial y}+h\frac{\partial}{\partial z}.$$

We will see in Appendix 4.3C that the integral of ω^1 along a curve γ can be interpreted as work done by the force field F along γ and the integral of ω^2 along a surface $\mathcal{S}$ has the meaning of the rate at which a fluid flow represented by the velocity field F crosses $\mathcal{S}$. Another reason consists in a simplification of various notions and results of classical vector analysis and differential geometry. Examples like orientation, elementary volume and the Stokes Theorem will be shown in Appendix 4.3C.

Definition 4.3.60. Let $\mathscr{M}$ be a differentiable manifold of dimension M and let ω be a smooth p-differential form on $\mathscr{M}$ which has the local representation (4.3.32). Then the *differential* of ω is the $(p+1)$-differential form $\mathrm{d}\omega$ with the local representation

$$\mathrm{d}\omega(x)=\sum_{1\le i_1<\dots<i_p\le M}\mathrm{d}a_{i_1,\dots,i_p}(x)\wedge\mathrm{d}y^{i_1}\wedge\dots\wedge\mathrm{d}y^{i_p}.\tag{4.3.34}$$

[19] In this connection see the notion of a Riemann metric on p. 227.

[20] We warn that vectors $\frac{\partial}{\partial y_1},\dots,\frac{\partial}{\partial y_M}$ need not be orthogonal in $T_a\mathscr{M}$!

Example 4.3.61.

(i) If $f\colon \mathscr{M} \to \mathbb{R}$ is a differentiable function, i.e., a 0-form, then the differential $\mathrm{d}f$ given by Remark 4.3.57(i) is the same as that in (4.3.34).

(ii) Let

$$\omega(x) = f_1(x)\,\mathrm{d}x^1 + f_2(x)\,\mathrm{d}x^2 + f_3(x)\,\mathrm{d}x^3$$

be a 1-form on an open set $\mathcal{G}$ in $\mathbb{R}^3$ and (x_1, x_2, x_3) the Cartesian coordinates of a point x. If f_1, f_2, f_3 are smooth functions on $\mathcal{G}$, then

$$\begin{aligned}
\mathrm{d}\omega(x) &= \mathrm{d}f_1(x) \wedge \mathrm{d}x^1 + \mathrm{d}f_2(x) \wedge \mathrm{d}x^2 + \mathrm{d}f_3(x) \wedge \mathrm{d}x^3 \\
&= \frac{\partial f_1}{\partial x_1} \underbrace{\mathrm{d}x^1 \wedge \mathrm{d}x^1}_{=0} + \frac{\partial f_1}{\partial x_2}\,\mathrm{d}x^2 \wedge \mathrm{d}x^1 + \frac{\partial f_1}{\partial x_3}\,\mathrm{d}x^3 \wedge \mathrm{d}x^1 + \cdots \\
&= \left(\frac{\partial f_2}{\partial x_1} - \frac{\partial f_1}{\partial x_2}\right) \mathrm{d}x^1 \wedge \mathrm{d}x^2 + \left(\frac{\partial f_3}{\partial x_2} - \frac{\partial f_2}{\partial x_3}\right) \mathrm{d}x^2 \wedge \mathrm{d}x^3 \\
&\quad + \left(\frac{\partial f_1}{\partial x_3} - \frac{\partial f_3}{\partial x_1}\right) \mathrm{d}x^3 \wedge \mathrm{d}x^1.
\end{aligned}$$

If we interpret the components (f_1, f_2, f_3) of the form ω as those of a vector field v (Remark 4.3.59), then the components of $\mathrm{d}\omega$, more precisely

$$\left(\frac{\partial f_3}{\partial x_2} - \frac{\partial f_2}{\partial x_3}, \frac{\partial f_1}{\partial x_3} - \frac{\partial f_3}{\partial x_1}, \frac{\partial f_2}{\partial x_1} - \frac{\partial f_1}{\partial x_2}\right),$$

form the so-called *curl* of v (notation $\operatorname{curl} v$ or $\nabla \times v$; for the cross product see Example 4.3.58(ii)). ○

Remark 4.3.62. By computing the differentials $\mathrm{d}f_{i_1\dots i_p}$ as in the previous example and rearranging the sum in (4.3.34) to get rid of the zero terms $\mathrm{d}y^{i_1} \wedge \cdots \wedge \mathrm{d}y^{i_{p+1}}$ where two indices coincide, we obtain, e.g., for an $(M-1)$-form,

$$\omega(x) = \sum_{i=1}^{M} f_i(x)\,\mathrm{d}y^1 \wedge \cdots \wedge \widehat{\mathrm{d}y^i} \wedge \cdots \wedge \mathrm{d}y^M$$

(here $\widehat{\mathrm{d}y^i}$ means that $\mathrm{d}y^i$ is missing),

$$\mathrm{d}\omega(x) = \sum_{i=1}^{M} (-1)^i \frac{\partial (f_i \circ \varphi)}{\partial y_i}(\psi(x))\,\mathrm{d}y^1 \wedge \cdots \wedge \mathrm{d}y^M.$$

Here φ, ψ are given by a local chart at $x \in \mathscr{M}$.

Remark 4.3.63. If $v \in \mathcal{V}(\mathscr{M})$ we can define the *Lie derivative of a form* $\omega \in \Lambda_p(\mathscr{M})$ similarly as we did for a vector field, namely,

$$\mathcal{L}_v\omega(x) := \lim_{t\to 0} \frac{\sigma^*(t,\cdot)\omega(\sigma(t,x)) - \omega(x)}{t},$$

where σ is a solution of the equation $\dot{x} = v(x)$ (see p. 202). Notice that $\mathcal{L}_v\omega \in \Lambda_p(\mathscr{M})$.

Example 4.3.61(i) leads to the following question: Are all one-differential forms differentials of smooth functions? In other words, has any (continuous) one-form ω a "primitive function" f, i.e., is there f such that $\mathrm{d}f = \omega$? A short speculation on one-forms in $\mathbb{R}^2$ suggests obstacles caused by mixed partial derivatives. We investigate this problem in a more general way.

Proposition 4.3.64.

(i) *Let ω be a differential form of the class C^2. Then*

$$\mathrm{d}^2\omega := \mathrm{d}(\mathrm{d}\omega) = 0.$$

(ii) *Let ω and κ be a p-differential form and q-differential form, respectively, then*

$$\mathrm{d}(\omega \wedge \kappa) = (\mathrm{d}\omega) \wedge \kappa + (-1)^q \omega \wedge (\mathrm{d}\kappa).$$

Proof. An easy proof is left to the reader. Notice however that the exchangeability of mixed partial derivatives of C^2-functions is the crucial point in the statement (i). □

Definition 4.3.65. A differential form ω is said to be

(1) *closed* if $\mathrm{d}\omega = 0$,

(2) *exact* if there is a differential form κ such that $\omega = \mathrm{d}\kappa$.

Remark 4.3.66. The concept of exact differential forms is a generalization of the classical notion of the potential of a mapping $f\colon \mathbb{R}^M \to \mathbb{R}^M$: A function $F\colon \mathcal{G} \to \mathbb{R}$ is called a *potential* of f in an open set $\mathcal{G} \subset \mathbb{R}^M$ if

$$F'(x)h = (f(x), h)_{\mathbb{R}^M}, \qquad x \in \mathcal{G}, \quad h \in \mathbb{R}^M.$$

In particular, if F is a potential of a C^1-function, then

$$\frac{\partial f_i}{\partial x_j}(x) = \frac{\partial f_j}{\partial x_i}(x), \qquad i, j = 1, \dots, M, \quad x \in \mathcal{G}.$$

The following example shows that this necessary condition is not sufficient.

Example 4.3.67. Let $\mathcal{G} = \mathbb{R}^2 \setminus \{(0,0)\}$ and let

$$\omega(x,y) = -\frac{y}{x^2+y^2}\,\mathrm{d}x + \frac{x}{x^2+y^2}\,\mathrm{d}y$$

be a 1-form in $\mathcal{G}$. This form is closed in $\mathcal{G}$. Suppose now that ω is exact, i.e., there is a function $f\colon \mathcal{G} \to \mathbb{R}$ such that $\mathrm{d}f = \omega$, in particular,

$$\frac{\partial f}{\partial x} = -\frac{y}{x^2+y^2}, \qquad \frac{\partial f}{\partial y} = \frac{x}{x^2+y^2} \qquad \text{in} \quad \mathcal{G}.$$

Integrating, we obtain

$$f(x,y) = \begin{cases} -\arctan\dfrac{x}{y} + C(y) & \text{for} \quad (x,y) \in \mathcal{G},\ y \neq 0, \\ \arctan\dfrac{y}{x} + D(x) & \text{for} \quad (x,y) \in \mathcal{G},\ x \neq 0. \end{cases}$$

Since $\arctan z + \arctan \frac{1}{z} = \frac{\pi}{2}$ for $z \neq 0$, we have $C(y) - D(x) = \frac{\pi}{2}$, i.e., C and D are constant functions in all quadrants. Taking limits for $x \to 0_\pm$, $y \to 0_\pm$ we arrive at a contradiction, i.e., ω is not exact.

The reader can ask how we have found this example. The problem is more transparent if $\mathbb{R}^2$ is identified with the complex plane. If

$$F(z) = \frac{1}{z}, \qquad z = x + \mathrm{i}y,$$

then

$$\operatorname{Re} F(x + \mathrm{i}y) = \frac{x}{x^2 + y^2}, \qquad \operatorname{Im} F(x + \mathrm{i}y) = -\frac{y}{x^2 + y^2}.$$

It is well known that there is no (holomorphic) function Φ such that

$$\Phi'(z) = \frac{1}{z} \qquad \text{for all} \quad z \in \mathbb{C} \setminus \{0\}$$

(consider $\displaystyle\int_{S^1} \frac{\mathrm{d}z}{z}$). In the theory of functions of a complex variable a primitive function can be constructed by a curve integral. We will use the same approach in constructing a "primitive form" to a differential form. This is the main idea of the proof of the following basic result. ○

Theorem 4.3.68 (H. Poincaré). *Any closed differential form on a differentiable manifold is locally exact.*

Proof. Let ω be a closed p-form on an M-dimensional manifold $\mathscr{M}$ $(1 \leq p \leq M)$. We choose a local chart $(\mathcal{V}, \psi)$ such that $P\psi(\mathcal{V}) = \mathcal{U}$ is an open ball in $\mathbb{R}^M$ with center at the origin. The pull-back (see Remark 4.3.57(iii)) $\Omega := \varphi^*\omega$ $(\varphi = (P\psi)^{-1})$ is a p-form in $\mathcal{U}$. We define a $(p-1)$-form σ on $\mathcal{U}$ by the formula

$$\sigma(y)(v_1, \dots, v_{p-1}) = \int_0^1 t^{p-1}\Omega(ty)(y, v_1, \dots, v_{p-1})\,\mathrm{d}t$$

for $y \in \mathcal{U}$, $v_1, \dots, v_{p-1} \in T_y\mathcal{U}$.[21] We have to show that

(i) the integral exists (this fact follows from the continuity of $t \mapsto \Omega(ty)$);

(ii) σ is a $(p-1)$-differential form on $\mathcal{U}$ (the skew-symmetry of σ follows from the same property of Ω);

(iii) $\mathrm{d}\sigma(y) = \Omega(y)$ for $y \in \mathcal{U}$.

Verification of the last statement is technically complicated. The case $p = 1$ is more transparent, and therefore we will give the computation only for this case. For the induction step the reader can consult, e.g., Sternberg [146, Theorem III.4.1], Cartan [25, Theorem II.3.2.12.1] or Taylor [149, Theorem 1.13.2].

Suppose that Ω has in $\mathcal{U}$ the form

$$\Omega(y) := (\varphi^*\omega)(y) = \sum_{i=1}^{M} g_i(y)\,\mathrm{d}f^i, \qquad y \in \mathcal{U},$$

[21] Here we identify the point $y \in \mathcal{U}$ with the vector $y \in T_y\mathcal{U} = \mathbb{R}^M$.

where $f^1, \dots, f^M$ is the dual basis to the standard one $e_1, \dots, e_M$ in $\mathbb{R}^M$. We wish to show that the function

$$\sigma(y) := \sum_{i=1}^{M} \int_0^1 g_i(ty) y_i \, \mathrm{d}t, \qquad y = (y_1, \dots, y_M) \in \mathcal{U}, \tag{4.3.35}$$

has the differential

$$\mathrm{d}\sigma(y) = \Omega(y), \qquad \text{i.e.,} \qquad \frac{\partial \sigma}{\partial y_j}(y) = g_j(y), \quad j = 1, \dots, M.$$

By differentiating the integral (4.3.35) with respect to the parameter y_j we obtain

$$\frac{\partial \sigma}{\partial y_j}(y) = \int_0^1 g_j(ty) \, \mathrm{d}t + \sum_{i=1}^{M} \int_0^1 \frac{\partial g_i}{\partial y_j}(ty) t y_i \, \mathrm{d}t = \int_0^1 g_j(ty) \, \mathrm{d}t + \sum_{i=1}^{M} \int_0^1 \frac{\partial g_j}{\partial y_i}(ty) t y_i \, \mathrm{d}t.$$

For the last equality we have used the assumption $\mathrm{d}\omega = 0$ and Exercise 4.3.75(iv):

$$\mathrm{d}\Omega = \mathrm{d}(\varphi^* \omega) = \varphi^* (\mathrm{d}\omega) = 0$$

and, consequently,

$$\frac{\partial g_i}{\partial y_j} = \frac{\partial g_j}{\partial y_i}, \qquad i, j = 1, \dots, M.$$

Using integration by parts we get

$$\int_0^1 g_j(ty) \, \mathrm{d}t = \big[t g_j(ty) \big]_{t=0}^{t=1} - \int_0^1 t \frac{\mathrm{d}}{\mathrm{d}t} g_j(ty) \, \mathrm{d}t = g_j(y) - \sum_{i=1}^{M} \int_0^1 t \frac{\partial g_j}{\partial y_i}(ty) y_i \, \mathrm{d}t.$$

If we put $f(x) = \sigma(y)$ for $x = \varphi(y)$, then

$$\mathrm{d}f = \omega. \qquad \square$$

Remark 4.3.69. The proof of the case $p = 1$ shows that there exists a potential σ of a smooth mapping $g = (g^1, \dots, g^M) \colon \mathcal{U} \to \mathbb{R}^M$ in a ball $\mathcal{U}$ provided the symmetry conditions

$$\frac{\partial g^i}{\partial y_j} = \frac{\partial g^j}{\partial y_i}, \qquad i, j = 1, \dots, M, \qquad \text{hold.}$$

Example 4.3.67 suggests that certain topological properties of $\mathcal{U}$ are necessary if $\mathcal{U}$ is not a ball. In the proof of the previous theorem the potential σ was defined by the curve integral

$$\sigma(y) = \int_0^1 (g(\gamma(t)), \dot{\gamma}(t))_{\mathbb{R}^M} \, \mathrm{d}t =: \int_{\gamma_{o,y}} \Omega \text{ [22]}$$

along the curve $\gamma_{o,y} = \{ty : t \in [0, 1]\}$. The crucial point in the direct computation of the Fréchet derivative of σ is an estimate of the difference $\sigma(y + h) - \sigma(y)$. If the curve integral depends only on the initial and terminal points and not on the path which joins these points, then

$$\sigma(y + h) - \sigma(y) = \int_{\gamma_{y,y+h}} \Omega = \int_0^1 (g(y + th), h)_{\mathbb{R}^M} \, \mathrm{d}t = (g(y), h)_{\mathbb{R}^M} + o(\|h\|_{\mathbb{R}^M})$$

[22] The definition of the curve integral and of the integral of a differential form is given in the next Appendix 4.3C.

provided g is continuously differentiable. Example 4.3.67 can be easily adapted to show that the independence of the curve integral on the path in $\mathcal{U}$ implies that $\mathcal{U}$ is not punctured. There is another way to express this observation. It consists in considering the obstacles preventing a closed form from being exact.

Assume that $\mathscr{M}$ is a connected manifold and denote the group (with respect to pointwise addition) of closed p-differential forms on $\mathscr{M}$ by $Z^p(\mathscr{M})$ and the subgroup of exact p-forms by $B^p(\mathscr{M})$. The quotient

$$H^p(\mathscr{M}) := Z^p(\mathscr{M})|_{B^p(\mathscr{M})}$$

is called the p-(de Rham) *cohomology group* of $\mathscr{M}$. If $H^1(\mathscr{M})$ is trivial, i.e., any closed one-form in $\mathscr{M}$ is exact, then $\mathscr{M}$ is said to be *simply connected*. More details on the role of cohomological groups in the study of differentiable manifolds can be found, e.g., in Whitney [158, Chapter IV]. The calculation of cohomology groups is by no means trivial.

Example 4.3.70.

(i) Let $f\colon \mathscr{M} \to \mathscr{N}$ be a smooth map. If $\omega \in Z^p(\mathscr{N})$, then

$$\mathrm{d}(f^*\omega) = f^*(\mathrm{d}\omega) = 0, \qquad \text{i.e.,} \qquad f^*\omega \in Z^p(\mathscr{M}).$$

(Exercise 4.3.75(iv)). Similarly, if $\omega \in B^p(\mathscr{N})$, then $f^*\omega \in B^p(\mathscr{M})$. This means that f induces a linear map

$$f^* : H^p(\mathscr{N}) \to H^p(\mathscr{M}).$$

In particular, if f is a diffeomorphism of $\mathscr{M}$ onto $\mathscr{N}$, then $H^p(\mathscr{M})$ is isomorphic to $H^p(\mathscr{N})$.

(ii) Using the previous example we can show that $H^1(\mathcal{S}^1)$ is isomorphic to $\mathbb{R}$. Instead of $H^1(\mathcal{S}^1)$ it is sufficient to compute $H^1(\mathbb{R}|_{\mathbb{Z}})$: Denote by i the natural projection of $\mathbb{R}$ onto $\mathbb{R}|_{\mathbb{Z}}$ and consider a closed one-form ω on $\mathbb{R}|_{\mathbb{Z}}$. Then

$$f(x)\,\mathrm{d}x := i^*\omega(x)$$

where f is a 1-periodic function. Define

$$\vartheta(\omega) = \int_0^1 f(x)\,\mathrm{d}x.$$

It is easy to see that $\vartheta(\omega) = 0$ if and only if ω is exact, and also that ϑ maps $B^1(\mathbb{R}|_{\mathbb{Z}})$ onto $\mathbb{R}$. This shows that ϑ induces the isomorphism of $H^1(\mathbb{R}|_{\mathbb{Z}})$ onto $\mathbb{R}$. ○

Now we explain the notion of a simply connected domain in another way which will be important in the sequel (e.g., in the degree theory — Appendix 5.7A).

Definition 4.3.71. Let X, Y be metric (topological) spaces. Continuous maps $f, g\colon X \to Y$ are called *homotopic* if there exists a continuous map $\Phi\colon X \times [0,1] \to Y$ such that

$$\Phi(\cdot, 0) = f(\cdot), \qquad \Phi(\cdot, 1) = g(\cdot).$$

Such Φ is said to be a *homotopy* between f and g.

Remark 4.3.72. The relation between two continuous maps to be homotopic is clearly an equivalence relation. The set of all continuous maps $C(X,Y)$ is therefore divided into disjoint classes of mutually homotopic maps. We denote the class containing f by $[f]$.

Here we are using the homotopy concept mainly for curves. The reader can imagine that curves $\gamma_0, \gamma_1 \colon [0,1] \to \mathscr{M}$ are homotopic if γ_0 may be continuously deformed (in $\mathscr{M}$!) into γ_1. A curve γ is called *null-homotopic* if γ is homotopic to a constant curve (point) $\tilde{\gamma} \colon t \in [0,1] \mapsto a \in \mathscr{M}$. In particular, this is important for closed curves ($\gamma(0) = \gamma(1)$). To see this, choose a fixed point $a \in \mathscr{M}$ and define

$$H_1(\mathscr{M}) := \{[\gamma] : \gamma \colon [0,1] \to \mathscr{M} \text{ is continuous, } \gamma(0) = \gamma(1) = a\}.$$

$H_1(\mathscr{M})$ forms a group — the so-called *fundamental group* of $\mathscr{M}$ — under "multiplication" $\gamma = \gamma_2 \cdot \gamma_1$ defined by

$$\gamma(t) = \begin{cases} \gamma_1(2t), & t \in \left[0, \frac{1}{2}\right], \\ \gamma_2(2t-1), & t \in \left[\frac{1}{2}, 1\right] \end{cases}$$

(notice that the definition $[\gamma_2] \cdot [\gamma_1] := [\gamma_2 \cdot \gamma_1]$ is correct). If $\mathscr{M}$ is path-connected, i.e., for any $a, b \in \mathscr{M}$ there exists a continuous curve γ in $\mathscr{M}$ such that $\gamma(0) = a$, $\gamma(1) = b$, then the fundamental group of $\mathscr{M}$ does not depend on the choice of the point a. Whenever the fundamental group is trivial, i.e., any closed curve can be continuously deformed into a point, then there are no holes in $\mathscr{M}$ and the integral of a one-form along a closed curve is zero, i.e., this integral does not depend on the path (Remark 4.3.69).

Cohomology and homotopy groups belong to the main tools in algebraic topology. The reader who is interested in these techniques can consult the corresponding textbooks, e.g., Adams [1], Dold [45], Greenberg [73], Kosniowski [91] (here, in Chapter 26, you can find applications of fundamental groups to the classification of two-dimensional compact connected surfaces; for example, such a surface is simply connected if and only if it is homeomorphic to $\mathcal{S}^2$), Spanier [144].

At the end of this appendix we link differential one-forms and systems of differential equations and continue the discussion from Appendix 4.3A. To simplify it we assume that

$$\alpha = \sum_{i=1}^{M} \gamma_i \, \mathrm{d}x^i$$

is a non-vanishing smooth one-form in an open set $\mathscr{G} \subset \mathbb{R}^M$. The form α is uniquely determined by its kernel up to a multiplication factor. Let $v^1, \dots, v^{M-1}$ be a basis of this kernel, i.e., $v^1, \dots, v^{M-1}$ are vector fields on $\mathscr{G}$ which are linearly independent at each point $x \in \mathscr{G}$ and annihilate α. The equation

$$\alpha = 0 \tag{4.3.36}$$

is called the *exterior differential equation* in $\mathscr{G}$ and its solution is a mapping

$$T \colon x \in \mathscr{G} \mapsto \text{ a subspace } T(x) \subset T_x\mathscr{G}$$

such that

$$\alpha(v) = 0 \qquad \text{for all} \quad v \in T(x).$$

A submanifold $\mathscr{S}$ of $\mathscr{G}$ is said to be an *integral manifold* for the equation (4.3.36) if

$$\dim \mathscr{S} = M - 1 \qquad \text{and} \qquad \varphi^* \alpha(y) = 0 \quad \text{for all} \quad y \in \mathscr{S}$$

where φ is a local parametrization of $\mathscr{S}$ in a neighborhood of y. This integral manifold is the same object as that at the end of Appendix 4.3A, i.e.,

$$i_*(T_y\mathscr{S}) = \operatorname{Lin}\{v^1(y), \dots, v^{M-1}(y)\}$$

where $i\colon \mathscr{S} \to \mathscr{G}$ is an embedding.

The notion of the exterior differential equation can be generalized to a system

$$\alpha^j = 0, \qquad j = 1, \dots, k, \tag{4.3.37}$$

where $\alpha^1, \dots, \alpha^k$ are differential forms on a manifold $\mathscr{M}$ not necessarily of the same order. In the special case when $\alpha^1, \dots, \alpha^k$ are linearly independent one-forms ((4.3.37), it is the so-called *Pfaff system*), the intersection of their kernels has a basis formed by $M - k$ vector fields. The Frobenius Theorem for the existence of an integral manifold for (4.3.37) has the following form.

Theorem 4.3.73 (Frobenius, the differential forms version). *Let $\alpha^1, \dots, \alpha^k$ be smooth differential one-forms on a differentiable manifold $\mathscr{M}$. The necessary and sufficient condition for the existence of an integral manifold in a vicinity of any point of $\mathscr{M}$ is that*

$$\mathrm{d}\alpha^i \wedge \alpha^1 \wedge \dots \wedge \alpha^k = 0, \qquad i = 1, \dots, k.$$

Proof. The equivalence of Theorem 4.3.42 and Theorem 4.3.73 is not difficult to prove, and the case $k = 1$, $\dim \mathscr{M} = 3$ is rather instructive. In this case a connection to the Poincaré Theorem 4.3.68 and its proof should also be recognized. □

Exercise 4.3.74. Denote by $\mathcal{O}(N)$ the set of all regular linear mappings $A\colon \mathbb{R}^N \to \mathbb{R}^N$ for which $A^{-1} = A^*$ and by $\mathcal{SO}(N)$ the set

$$\{A \in \mathcal{O}(N) : \det A = 1\}.$$

(i) As a subset of $\mathbb{R}^{N\times N}$ the set $\mathcal{O}(N)$ is a differentiable manifold. Find its dimension.
Hint. Consider $A \to A^*A$. The dimension is $\frac{N(N-1)}{2}$.

(ii) Show that $\mathcal{SO}(N)$ is the component of $\mathcal{O}(N)$ containing the identity.

(iii) Show that $A \in \mathcal{O}(N)$ induces a mapping of $\mathcal{S}^{N-1}$ into itself.

(iv) Let ω be a one-form on $\mathcal{S}^2$ which is invariant under $\mathcal{SO}(3)$, i.e.,

$$A^*\omega = \omega \qquad \text{for all} \quad A \in \mathcal{SO}(3).$$

Prove that $\omega = 0$.

(v) Does a result analogous to (iv) hold for a two-form ω on $\mathbb{R}^3$?

Exercise 4.3.75. Prove the following properties of the pull-back operation:

(i) $g^*(\omega \wedge \kappa) = (g^*\omega) \wedge (g^*\kappa)$,

(ii) $(h \circ g)^*\omega = g^*(h^*\omega)$,

(iii) $g^*(\mathrm{d}f) = \mathrm{d}(f \circ g)$ for $f \colon \mathscr{M} \to \mathbb{R}$,[23]

(iv) $g^*(\mathrm{d}\omega) = \mathrm{d}(g^*\omega)$ where d denotes the differential.

Exercise 4.3.76. Let f be a differentiable function on $\mathscr{M}$ (i.e., $f \in \mathcal{F}(\mathscr{M})$) and $\omega \in \Lambda_p(\mathscr{M})$. Then

$$\mathrm{d}(f\omega) = \mathrm{d}f \wedge \omega + f\,\mathrm{d}\omega.$$

(Leibniz rule for the differential.) Prove this formula.

Exercise 4.3.77. Let ω be a 1-form and let $v \in \mathcal{V}(\mathscr{M})$. In order to determine $\mathcal{L}_v\omega$ it is sufficient to compute $(\mathcal{L}_v\omega)(w)$ for any $w \in \mathcal{V}(\mathscr{M})$.

(i) Show that $(\mathcal{L}_v\omega)(w) = \mathcal{L}_v(\omega(w)) - \omega(\mathcal{L}_v w)$.

(ii) If $\omega = \sum \omega_i\,\mathrm{d}y^i$, $v = \sum v^j \frac{\partial}{\partial y_j}$, deduce that

$$\mathcal{L}_v(\omega)\left(\frac{\partial}{\partial y_i}\right) = \sum_j v^j \frac{\partial \omega_i}{\partial y_j} + \sum_j \omega_j \frac{\partial v^j}{\partial y_i}.$$

Exercise 4.3.78. Let $\mathscr{M}$ be the open unit ball in $\mathbb{R}^2$ without its origin. Show that $H_1(\mathscr{M})$ is isomorphic to $\mathbb{Z}$. Is it also true in $\mathbb{R}^3$?

Exercise 4.3.79. Show that $H_1(\mathcal{S}^1)$ is isomorphic to $\mathbb{Z}$.

Hint. Use an approach similar to that in Example 4.3.70(ii), and instead of the mapping ϑ show that there is a lifting $\tilde{\gamma}$ of a continuous closed curve $\gamma\colon [0,1] \to \mathbb{R}|_{\mathbb{Z}}$, i.e., $\tilde{\gamma}\colon [0,1] \to \mathbb{R}$ continuous such that

$$i(\tilde{\gamma}) = \gamma, \qquad \tilde{\gamma}(0) = 0.$$

Now consider $\tilde{\gamma}(1)$ (actually this is the degree of γ — see Appendix 5.7A). For details see, e.g., Kosniowski [91, Chapter 16].

4.3C Integration on Manifolds

We have met the curve integral in the previous Appendix 4.3B. There are two objects which can be integrated along curves: functions and differential one-forms.

The situation with functions is simple. If $\mathscr{M}$ is an M-dimensional differentiable manifold in $\mathbb{R}^N$, $f\colon \mathscr{M} \to \mathbb{R}$ is a continuous function and $\gamma\colon [a,b] \to \mathscr{M}$ is a smooth curve, then we define

$$\int_\gamma f\,\mathrm{d}\gamma := \int_a^b f(\gamma(t))\|\dot{\gamma}(t)\|\,\mathrm{d}t. \tag{4.3.38}$$

The Eucledian norm $\|\dot{\gamma}(\cdot)\|$ of tangent vectors expresses here a quantity which could be viewed as the "infinitesimal" length of γ. Recall in this context the formula for the length of γ:

$$l(\gamma) := \int_a^b \|\dot{\gamma}(t)\|\,\mathrm{d}t.$$

We will return to the length and area of a nonlinear object later in this appendix.

The integral on the right-hand side of (4.3.38) is the Riemann integral and consequently it has reasonable properties. It could be generalized to some noncontinuous

[23] If we interpret f as a 0-form, then the notation g^*f instead of $f \circ g$ is more agreeable.

functions (via the Lebesgue integral) and/or to certain non-smooth curves (pairwise smooth or with bounded variation via the Riemann–Stieltjes integral). Since we are not interested in these generalizations we always assume that all objects are so smooth as we need (manifolds at least of the class C^1, functions, vector fields, differential forms at least continuous).

The situation with integration of one-forms is different. Namely, differential forms are defined only on manifolds (recall that an open subset of $\mathbb{R}^M$ is also a manifold) and curves need not be manifolds (see Figure 4.3.2). There are two possibilities how to avoid these obstacles: either to assume that γ lies on a manifold where the one-form is defined or to restrict integration to curves which are themselves manifolds. We now examine the first possibility and postpone the other one to Definition 4.3.92.

For the definition of the integral of a one-form given on page 216 we have assumed that the whole curve γ lies in one chart to get the same representation of the form at all points of γ. If more charts are needed to cover the curve we have to be careful not to integrate over some parts of the curve several times. To eliminate this risk the following tool is very useful. In order to build it up we need a topological interlude.

Definition 4.3.80. Let $(\mathcal{V}_n, \psi_n)_{n\in\mathbb{N}}$ be an atlas of a differentiable manifold $\mathscr{M}$. Let $\{\alpha_n\}_{n\in\mathbb{N}}$ be a collection of smooth (often C^∞) nonnegative functions on $\mathscr{M}$ which have the following properties:

(1) for all $n \in \mathbb{N}$ the *support* of α_n defined by

$$\operatorname{supp}\alpha_n := \overline{\{x \in \mathscr{M} : \alpha_n(x) \neq 0\}}$$

is a compact subset of $\mathcal{V}_n$;

(2) $\sum\limits_{n=1}^{\infty} \alpha_n(x) = 1$ for all $x \in \mathscr{M}$.

Then $\{\alpha_n\}_{n\in\mathbb{N}}$ is said to be a *partition of unity* subordinate to $\{\mathcal{V}_n\}_{n\in\mathbb{N}}$.[24]

Since $\mathscr{M} \subset \mathbb{R}^N$ is separable in the induced topology a countable atlas always exists. It is also possible to construct a sequence $\{\mathcal{G}_n\}_{n=1}^{\infty}$ of open subsets of $\mathscr{M}$ such that

$$\mathcal{G}_n \subset \operatorname{int}\overline{\mathcal{G}_{n+1}},^{25} \qquad \overline{\mathcal{G}_n} \quad \text{is compact,} \qquad \text{and} \qquad \mathscr{M} = \bigcup_{n=1}^{\infty} \mathcal{G}_n.$$

For example, $\mathcal{G}_n$ can be chosen as the intersection of $\mathscr{M}$ with the open ball centered at o with radius n. For the construction of a partition of unity the following topological device is convenient. We will need various types of balls so $B(a; r)$ will denote the open ball in $\mathbb{R}^M$ ($M = \dim \mathscr{M}$).

Lemma 4.3.81. *Let $\{\mathcal{W}_\alpha\}_{\alpha\in\mathcal{A}}$ be an open covering of an M-dimensional manifold $\mathscr{M}$ in $\mathbb{R}^N$. Then there is a countable open covering $\{\mathcal{V}_m\}_{m\in\mathbb{N}}$ of $\mathscr{M}$ with the following properties:*

(i) *$\{\mathcal{V}_m\}_{m\in\mathbb{N}}$ is subordinate to $\{\mathcal{W}_\alpha\}_{\alpha\in\mathcal{A}}$, i.e., for each $m \in \mathbb{N}$ there is an index $\alpha_m \in \mathcal{A}$ such that $\mathcal{V}_m \subset \mathcal{W}_{\alpha_m}$;*

[24] A partition of unity is defined in topology in a more general way; see the corresponding textbooks, e.g., Dugundji [52, Chapter VIII].

[25] We wish to point out that topological notions (like interior) are taken here with respect to the topology of $\mathscr{M}$, i.e., $\mathcal{G} \subset \mathscr{M}$ is open provided there is an open set $\mathcal{H} \subset \mathbb{R}^N$ such that $\mathcal{G} = \mathscr{M} \cap \mathcal{H}$.

(ii) *there are smooth mappings* $\varphi_m \colon B(o;2) \to \mathcal{V}_m$ *such that* $\bigcup\limits_{m=1}^{\infty} \varphi_m(B(o;1)) = \mathscr{M}$*;*

(iii) *the collection* $\{\mathcal{V}_m\}_{m\in\mathbb{N}}$ *forms a* locally finite *system, i.e., any point* $x \in \mathscr{M}$ *has a neighborhood which intersects only a finite number of* $\{\mathcal{V}_m\}_{m\in\mathbb{N}}$.

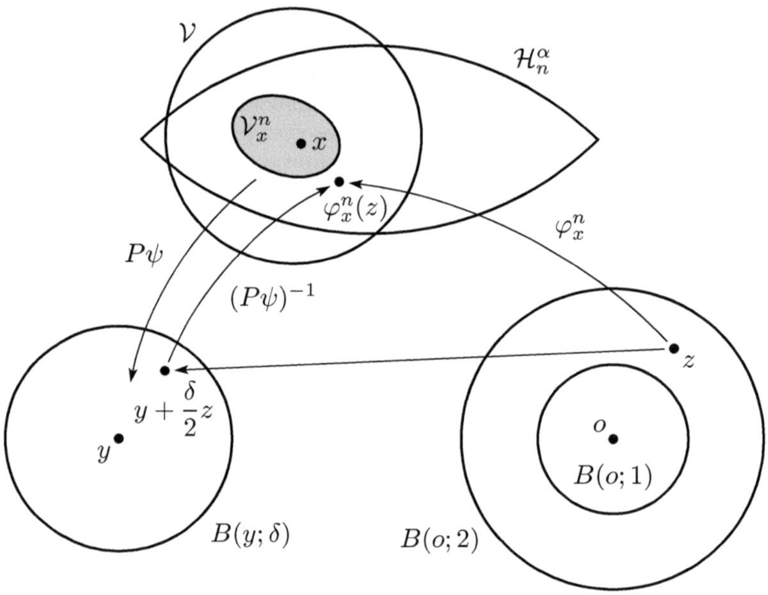

Figure 4.3.18.

Proof. Choose a sequence $\{\mathcal{G}_n\}_{n=1}^{\infty}$ of open subsets of $\mathscr{M}$ which has the property stated prior to this lemma. Put in addition $\mathcal{G}_0 = \mathcal{G}_{-1} = \emptyset$. The main idea behind the forthcoming construction is that the compact sets

$$\mathcal{K}_n := \overline{\mathcal{G}_n} \setminus \mathcal{G}_{n-1}, \qquad n \in \mathbb{N},$$

cover $\mathscr{M}$ and the larger open sets

$$\mathcal{H}_n := \mathcal{G}_{n+1} \setminus \overline{\mathcal{G}_{n-2}}, \qquad n \in \mathbb{N},$$

form a locally finite system. Fix $n \in \mathbb{N}$. Let $(\mathcal{V}, \psi)$ be a local chart at

$$x \in \mathcal{H}_n^\alpha := \mathcal{H}_n \cap \mathcal{W}_\alpha.$$

Put

$$y = P\psi(x).$$

There is a ball $B(y;\delta) \subset P\psi(\mathcal{V} \cap \mathcal{H}_n^\alpha)$. We now shift the center y to the origin and expand the ball appropriately, namely we put

$$\varphi_x^n(z) = (P\psi)^{-1}\left(y + \frac{\delta}{2}z\right) \qquad \text{for} \quad z \in B(o;2).$$

With help of these smooth maps φ_x^n we return back to the manifold by setting

$$\mathcal{V}_x^n = \varphi_x^n(B(o;1))$$

(see Figure 4.3.18). Notice that $\mathcal{V}_x^n$ is open in $\mathscr{M}$. Open sets $\{\mathcal{V}_x^n\}_{x\in\mathcal{H}_n^\alpha,\alpha\in\mathcal{A}}$ cover the compact set $\mathcal{K}_n$. We choose a finite subcovering

$$\mathcal{V}_{x_1}^n, \dots, \mathcal{V}_{x_{k_n}}^n.$$

The collection $\{\mathcal{V}_{x_j}^n\}_{j=1,\dots,k_n,\, n\in\mathbb{N}}$, covers $\mathscr{M}$, and

$$\{\varphi_{x_j}^n(B(o;2))\}_{j=1,\dots,k_n,\, n\in\mathbb{N}},$$

is the desired locally finite countable system $\{\mathcal{V}_m\}_{m\in\mathbb{N}}$. □

Theorem 4.3.82. *For any atlas $(\mathcal{W}_\alpha, \psi_\alpha)_{\alpha\in\mathcal{A}}$ of a manifold $\mathscr{M}$ there exists a subordinate partition of unity.*

Proof. According to the previous Lemma 4.3.81 we choose a locally finite subordinate covering $\{\mathcal{V}_k\}_{k\in\mathbb{N}}$ of $\mathscr{M}$ and the corresponding functions $\{\varphi_k\}_{k=1}^\infty$. It is easy to show that the function

$$\eta(y) = \begin{cases} \mathrm{e}^{-\frac{1}{1-\|y\|^2}}, & \|y\| < 1, \\ 0, & \|y\| \ge 1, \end{cases} \qquad y \in \mathbb{R}^M,$$

is a C^∞-function in $\mathbb{R}^M$. Put

$$\beta_k(x) = \begin{cases} \eta(y), & x \in \mathcal{V}_k,\ x = \varphi_k(y), \\ 0, & x \in \mathscr{M} \setminus \mathcal{V}_k. \end{cases}$$

Then β_k is smooth (of the same order as $\mathscr{M}$) and

$$\beta_k(x) > 0 \qquad \text{for} \quad x \in \varphi_k(B(o;1)).$$

Since $\{\mathcal{V}_k\}_{k\in\mathbb{N}}$ is a locally finite system, the series $\sum\limits_{k=1}^\infty \beta_k(x)$ has only a finite number of nonzero terms. Moreover, $\sum\limits_{k=1}^\infty \beta_k(x) > 0$ for all $x \in \mathscr{M}$ due to $\bigcup\limits_{k=1}^\infty \varphi_k(B(o;1)) = \mathscr{M}$. It is now sufficient to put

$$\alpha_k(x) = \frac{\beta_k(x)}{\sum\limits_{n=1}^\infty \beta_n(x)}, \qquad k \in \mathbb{N},$$

to obtain the desired partition of unity. □

We can now return to the definition of the integral of a one-form ω. If $\{\alpha_n\}_{n\in\mathbb{N}}$ is a partition of unity which is subordinate to a covering $\{\mathcal{V}_n\}_{n\in\mathbb{N}}$ of $\mathscr{M}$ where $(\mathcal{V}_n, \psi_n)_{n\in\mathbb{N}}$ is an atlas of $\mathscr{M}$, then

$$\omega(x) = \sum_{n=1}^\infty \alpha_n(x)\omega(x), \qquad x \in \mathscr{M}.$$

Notice that $\alpha_n\omega$ is a one-form and $\operatorname{supp}\alpha_n\omega \subset \mathcal{V}_n$. This decomposition of ω allows us to define the integral locally. If γ is a smooth curve in $\mathscr{M}$ and $\gamma(t) \in \mathcal{V}_n$, then $\dot\gamma(t) \in T_{\gamma(t)}\mathscr{M}$ and it can be written in the form

$$\dot\gamma(t) = \sum_{i=1}^{M} \dot\gamma^i(t)\frac{\partial}{\partial y_i}.$$

Definition 4.3.83. Let $\mathscr{M}$ be an M-dimensional differentiable manifold and denote $(\mathcal{V}_n, \psi_n)_{n\in\mathbb{N}}$ its atlas. Let $\{\alpha_n\}_{n\in\mathbb{N}}$ be a partition of unity subordinate to $\{\mathcal{V}_n\}_{n\in\mathbb{N}}$. Let $\gamma\colon \mathcal{I} \to \mathscr{M}$ be a smooth curve and ω a one-form on $\mathscr{M}$. If

$$\omega(x) = \sum_{i=1}^{M} f_i(x)\,\mathrm{d}y^i, \qquad x \in \mathcal{V}_n,$$

then we define

$$\int_\gamma \omega = \sum_{n=1}^{\infty}\int_\gamma \alpha_n\omega := \sum_{n=1}^{\infty}\int_{\mathcal{I}} \alpha_n(\gamma(t))\sum_{i=1}^{M} f_i(\gamma(t))\dot\gamma^i(t)\,\mathrm{d}t \tag{4.3.39}$$

provided the integrals on the right-hand side exist and the sum is absolutely convergent.

Remark 4.3.84.

(i) If γ is a smooth curve defined on a compact interval $\mathcal{I} = [a, b]$ and $\{\mathcal{V}_n\}_{n\in\mathbb{N}}$ is a locally finite covering of $\mathscr{M}$, then γ lies in a finite number of $\{\mathcal{V}_n\}_{n\in\mathbb{N}}$ only. If, moreover, the form ω is continuous, then the integrals in (4.3.39) exist and the sum is absolutely convergent, since it contains only finitely many nonzero terms. We require absolute convergence of the series because we do not want the value of the integral to depend on the arrangement of charts $(\mathcal{V}_n, \psi_n)_{n\in\mathbb{N}}$ into a sequence.

(ii) It can be proved (do it as an exercise!) that the formula (4.3.39) does not depend on the choice of partition of unity. It should be also proved that the right-hand side in (4.3.39) is the same for all equivalent atlases on $\mathscr{M}$ (see Definition 4.3.34). This follows from the transformation rules for tangent vectors (4.3.17) and for differential forms (4.3.22).

Remark 4.3.85. We can interpret the local coordinates $(f_1, \dots, f_M)$ of a one-form ω as the local coordinates of a vector field

$$F(x) = \sum_{i=1}^{M} f_i(x)\frac{\partial}{\partial x_i}, \qquad x \in \mathcal{V}_n$$

(and vice-versa – Remark 4.3.59). If we define

$$\int_\gamma F = \int_\gamma \omega,$$

then the integral $\displaystyle\int_\gamma F$ expresses the work done by the vector field F along the curve γ. The special cases $\mathscr{M} = \mathbb{R}^2, \mathbb{R}^3$ are known from introductory courses in mechanics (see, e.g., Kittel, Knight & Ruderman [90, Chapter 5]).

Remark 4.3.86. Figures 4.3.2 and 4.3.3 show that a smooth curve need not be a differentiable manifold in $\mathbb{R}^N$. In order to avoid such cases it is sufficient to assume that the curve γ has a parametrization which is an embedding (Remark 4.3.3(iv)). If, moreover, γ lies on a manifold $\mathscr{M}$, then it is the so-called *submanifold* of $\mathscr{M}$ in the following sense: A subset $\mathscr{P}$ of a differentiable manifold $\mathscr{M}$ is said to be a P-dimensional *submanifold* of $\mathscr{M}$ if there is an atlas $(\mathcal{V}_n, \psi_n)_{n\in\mathbb{N}}$ of $\mathscr{M}$ such that

$$\psi_n(x) = (y_1, \dots, y_P, 0, \dots, 0) \in \mathbb{R}^N \qquad \text{for all} \qquad x \in \mathcal{V}_n \cap \mathscr{P}.$$

The proof of Proposition 4.3.2 shows that the image of an embedding is a submanifold.

In order to integrate functions over a surface in $\mathbb{R}^3$, or more generally over a manifold, we need to generalize the notion of area of a parallelogram to a non-flat domain. Let us recall here the definition of the multiple Riemann integral. The notion of (normalized) area or volume is based on the fact that the unit cube

$$\mathcal{C} := \left\{ \sum_{i=1}^{M} x_i e_i : 0 \le x_i \le 1 \right\}$$

($e_1, \dots, e_M$ is the standard basis in $\mathbb{R}^M$) has the M-dimensional volume (i.e., the Lebesgue measure) equal to 1.

Let $\mathcal{A}$ be a parallelepiped in $\mathbb{R}^M$ spanned by vectors $v_1, \dots, v_M$, i.e.,

$$\mathcal{A} := \left\{ \sum_{j=1}^{M} \alpha_j v_j : 0 \le \alpha_j \le 1 \right\}.$$

Then the volume $V(\mathcal{A})$ of $\mathcal{A}$ is defined by

$$V(\mathcal{A}) := \int_{\mathcal{A}} 1 \,\mathrm{d}x.$$

This integral can be calculated with help of the linear operator $T\colon \mathbb{R}^M \to \mathbb{R}^M$ which sends the vectors $e_1, \dots, e_M$ of the standard basis to $v_1, \dots, v_M$ $\left(Te_j = v_j = \sum\limits_{i=1}^{M} t_{ij} e_i\right)$. It is well-known that

$$\int_{\mathcal{A}} 1 \,\mathrm{d}x = \int_{\mathcal{C}} |\det \boldsymbol{T}| \,\mathrm{d}y = |\det \boldsymbol{T}| := |\det (t_{ij})_{i,j=1,\dots,M}|.\text{[26]} \tag{4.3.40}$$

There is a problem with the generalization to a manifold since a manifold is bent. Nonetheless, a manifold can be supposed to be locally flat provided it is smooth. This basic principle of analysis allows us to define an infinitesimal area or volume via these notions for flat tangent spaces. Another problem now arises since there is no natural unit cube in $T_a\mathscr{M}$. To overcome this obstacle we want to express the M-volume of the parallelepiped, spanned by the coordinate vectors $\frac{\partial}{\partial y_1}, \dots, \frac{\partial}{\partial y_M}$, without using the standard basis. This can be done for the parallelepiped $\mathcal{A}$ given above with help of the scalar

[26] More generally: If T is a (nonlinear) diffeomorphism which maps $\mathcal{C}$ onto $\mathcal{A} := T(\mathcal{C})$, then the formula $V(\mathcal{A}) := \int_{\mathcal{A}} 1 \,\mathrm{d}x = \int_{\mathcal{C}} |\det \boldsymbol{T}'(y)| \,\mathrm{d}y$ holds for the Lebesgue measure $V(\mathcal{A})$ of $\mathcal{A}$. The proof of this nonlinear version is based on (4.3.40).

product (in which the standard basis is an orthonormal basis). If $\boldsymbol{G}(v_1,\dots,v_M)$ is the so-called *Gramm matrix* of vectors $v_1,\dots,v_M$, i.e.,

$$\boldsymbol{G}(v_1,\dots,v_M)=\begin{pmatrix}(v_1,v_1)_{\mathbb{R}^N} & \cdots & (v_1,v_M)_{\mathbb{R}^N}\\ \vdots & \ddots & \vdots\\ (v_M,v_1)_{\mathbb{R}^N} & \cdots & (v_M,v_M)_{\mathbb{R}^N}\end{pmatrix},$$

then the formula

$$V(\mathcal{A})=[\det \boldsymbol{G}(v_1,\dots,v_M)]^{\frac{1}{2}} \tag{4.3.41}$$

holds (see Exercise 4.3.109).

Since $T_a\mathcal{M}\subset\mathbb{R}^N$, the scalar product in $\mathbb{R}^N$ can be restricted to $T_a\mathcal{M}$ to get a natural scalar product in $T_a\mathcal{M}$. This justifies the next definition. We wish to point out that the Gramm matrix $\boldsymbol{G}\left(\frac{\partial}{\partial y_1},\dots,\frac{\partial}{\partial y_M}\right)$ consists of scalar products of vectors $\frac{\partial}{\partial y_i},\frac{\partial}{\partial y_j}$ in $\mathbb{R}^N$ (the differential structure of $\mathcal{M}$ is inherited from $\mathbb{R}^N$). If $\mathcal{M}$ is two-dimensional differentiable manifold in $\mathbb{R}^3$, i.e., a *surface*, then the quadratic form which corresponds to the Gramm matrix G is called a *first fundamental form* of $\mathcal{M}$.

Definition 4.3.87. Let $\mathcal{M}$ be a differentiable manifold with an atlas $(\mathcal{V}_n,\psi_n)_{n\in\mathbb{N}}$ and let

$$\mathcal{U}_n=P_n\psi_n(\mathcal{V}_n),\qquad \varphi_n=(\psi_n|_{\mathcal{V}_n})^{-1}.$$

Let $\{\alpha_n\}_{n\in\mathbb{N}}$ be a partition of unity in $\mathcal{M}$ subordinate to $\{\mathcal{V}_n\}_{n\in\mathbb{N}}$. If f is a continuous function on $\mathcal{M}$, then we define

$$\int_{\mathcal{M}} f\,\mathrm{d}V:=\sum_{n=1}^{\infty}\int_{\mathcal{U}_n}(\alpha_n f)(\varphi_n(y))\left[\det \boldsymbol{G}\left(\frac{\partial}{\partial y_1},\dots,\frac{\partial}{\partial y_M}\right)\right]^{\frac{1}{2}}\mathrm{d}y_1\cdots\mathrm{d}y_M \tag{4.3.42}$$

provided the right-hand side exists and the sum is absolutely convergent.

Remark 4.3.88. It is possible to show that this definition does not depend on the partition of unity and on the choice of an atlas. The right-hand side in (4.3.42) exists whenever f has compact support or, in particular, if $\mathcal{M}$ is a compact manifold.

The scalar product on $T_x\mathcal{M}$ inherited from $\mathbb{R}^N$ (it will be called the natural Riemann metric on $\mathcal{M}$) leads to uniformly distributed mass or currents. However, it is sometimes unrealistic in physical applications. To cover further applications in a more realistic manner we can consider different scalar products at different points of a manifold. Since any positive definite symmetric bilinear form in $\mathbb{R}^M\times\mathbb{R}^M$ determines a scalar product in $\mathbb{R}^M$, we can introduce a metric structure on a manifold $\mathcal{M}$ by a (smooth) mapping $g\colon x\in\mathcal{M}\mapsto S_2^+(T_x\mathcal{M})$ (positive definite symmetric bilinear forms on $T_x\mathcal{M}$). Such g is called a *Riemann metric* on $\mathcal{M}$ and the couple $(\mathcal{M},g)$ is called a *Riemann manifold* . If $\frac{\partial}{\partial y_1},\dots,\frac{\partial}{\partial y_M}$ is a basis of $T_x\mathcal{M}$ and $\mathrm{d}y^1,\dots,\mathrm{d}y^M$ is the dual basis of $(T_x\mathcal{M})^*$ then g can be written in the form

$$g=\sum_{i,j=1}^{M} g_{ij}\,\mathrm{d}y^i\,\mathrm{d}y^j,$$

where $\mathrm{d}y^i\,\mathrm{d}y^j$ is tensor product [27]

$$\mathrm{d}y^i \otimes \mathrm{d}y^j \in (T_x\mathscr{M})^* \otimes (T_x\mathscr{M})^*.$$

We have defined the derivative $\mathrm{d}f(x)$ of $f : \mathscr{M} \to \mathbb{R}$ as an element of $(T_x\mathscr{M})^*$. By the Riesz Representation Theorem (Theorem 1.2.40) there exists a unique vector $\nabla_g f(x) \in T_x\mathscr{M}$ (the so-called *Rieman gradient* of f at the point x) such that

$$\mathrm{d}f(x)v = (v, \nabla_g f(x))_g$$

for all $v \in T_x\mathscr{M}$. If g is the natural Riemann metric on $\mathscr{M}$ then $\nabla_g f$ is the same as the gradient defined in Remark 4.3.59.

Example 4.3.89. Compute the surface area $V(\mathcal{S}^2)$ of the unit sphere $\mathcal{S}^2$ in $\mathbb{R}^3$.

It is obvious that the two-dimensional surface area of the "Greenwich meridian $\mathcal{G}$" is zero. The rest $\mathcal{S}^2 \setminus \mathcal{G}$ is covered by one chart with

$$\varphi(\alpha, \vartheta) = (\cos\alpha\cos\vartheta, \sin\alpha\cos\vartheta, \sin\vartheta), \qquad \alpha \in (0, 2\pi), \quad \vartheta \in \left(-\frac{\pi}{2}, \frac{\pi}{2}\right).$$

Since $n = 1$, $\mathcal{U}_1 = (0, 2\pi) \times \left(-\frac{\pi}{2}, \frac{\pi}{2}\right)$, $\alpha_1 = 1$ and $\det \boldsymbol{G}\left(\frac{\partial}{\partial\alpha}, \frac{\partial}{\partial\vartheta}\right) = \cos^2\vartheta$,

$$\frac{\partial}{\partial\alpha} = (-\sin\alpha\cos\vartheta, \cos\alpha\cos\vartheta, 0), \qquad \frac{\partial}{\partial\vartheta} = (-\cos\alpha\sin\vartheta, -\sin\alpha\sin\vartheta, \cos\vartheta),$$

we have

$$V(\mathcal{S}^2) = \int_{\mathcal{S}^2} 1\,\mathrm{d}S = \iint_{(0,2\pi)\times\left(-\frac{\pi}{2},\frac{\pi}{2}\right)} \cos\vartheta\,\mathrm{d}\alpha\,\mathrm{d}\vartheta = 4\pi.$$

(It is more common to denote the integration symbol in the two-dimensional case by $\mathrm{d}S$ instead of $\mathrm{d}V$.) ○

It follows from Definition 4.3.83 (see also Exercise 4.3.106) that the integral of a one-form along a curve γ depends on the orientation of γ. Namely, if

$$\tilde{\gamma}(t) = \gamma(1 - t), \qquad t \in (0, 1),$$

then

$$\dot{\tilde{\gamma}}(t) = -\dot{\gamma}(1 - t), \qquad \text{and hence} \qquad \int_{\tilde{\gamma}} \omega = -\int_{\gamma} \omega.$$

This dependence on an orientation is crucial for the generalization of the curve integral to an integral of a differential form over a manifold. What can the orientation of a manifold be? Let us start with simple examples like $\mathbb{R}$, $\mathbb{R}^2$, $\mathbb{R}^3$. It is the common understanding that the "standard" (equivalently "positive") orientation on $\mathbb{R}$ is from the left to the right, in $\mathbb{R}^2$ anticlockwise and by the right-thumb rule in $\mathbb{R}^3$. These slightly vague formulations can be made precise by taking fixed bases in $\mathbb{R}$, $\mathbb{R}^2$, $\mathbb{R}^3$, e.g., the standard bases. Then

[27] If $\eta, \omega \in T_x\mathscr{M}^*$ and $v, w \in T_x\mathscr{M}$ then

$$\eta \otimes \omega(v, w) = \eta(v)\omega(w).$$

In particular, for $v = \Sigma v^i \frac{\partial}{\partial y_i}$, $w = \sum w^j \frac{\partial}{\partial y_j}$ we have $\mathrm{d}y^i \otimes \mathrm{d}y^j(v, w) = v^i w^j$ and $g(v, w) = \sum\limits_{i,j=1}^{M} g_{ij} v^i w^j$.

all bases are divided into two disjoint classes according to the sign of the determinant of the transformation matrix $\boldsymbol{T}$ which sends the fixed (e.g., standard) basis into a new one. We say that $\tilde{e}_1, \dots, \tilde{e}_M$ is a positive basis if $\det \boldsymbol{T} > 0$. We want to remind the reader that

$$\det \boldsymbol{T} = f^1 \wedge \dots \wedge f^M(\tilde{e}_1, \dots, \tilde{e}_M) \qquad \text{if} \qquad \boldsymbol{T} e_i = \tilde{e}_i, \quad i = 1, \dots, M,$$

and $f^1, \dots, f^M$ is the dual basis to $e_1, \dots, e_M$ (Example 4.3.55(i)). This indicates that the choice of a fixed nowhere vanishing continuous M-form $\overline{\omega}$ (i.e., $\overline{\omega}(x) \neq 0$ for all $x \in \mathscr{M}$) on the M-dimensional manifold $\mathscr{M}$ makes it possible to introduce an orientation on $\mathscr{M}$. If $(\mathcal{V}, \psi)$ is a local chart at a point $x \in \mathscr{M}$, then the basis $\frac{\partial}{\partial y_1}, \dots, \frac{\partial}{\partial y_M}$ of $T_x\mathscr{M}$ is said to be a *positive basis* of $T_x\mathscr{M}$ provided

$$\overline{\omega}(x)\left(\frac{\partial}{\partial y_1}, \dots, \frac{\partial}{\partial y_M}\right) > 0.$$

It can be proved that a continuous non-vanishing form exists on $\mathscr{M}$ if and only if there is an atlas $(\mathcal{V}_n, \psi_n)$ of $\mathscr{M}$ such that for any two charts $(\mathcal{V}, \psi)$, $(\tilde{\mathcal{V}}, \tilde{\psi})$ of this atlas the transformation matrix $((P\tilde{\psi}) \circ \varphi)'(y)$ (see (4.3.17)) has a positive determinant for all $y \in \psi(\mathcal{V} \cap \tilde{\mathcal{V}})$ (provided $\mathcal{V} \cap \tilde{\mathcal{V}} \neq \emptyset$).

Definition 4.3.90. A differentiable manifold $\mathscr{M}$ of dimension M is said to be *orientable* if there exists a continuous nowhere vanishing M-form on $\mathscr{M}$. If such a form $\overline{\omega}$ is fixed, then $(\mathscr{M}, \overline{\omega})$ is called an *oriented manifold*.

Example 4.3.91.

(i) Suppose that $\mathscr{M}$ is a two-dimensional orientable connected manifold in $\mathbb{R}^3$ (i.e., a surface) and $\overline{\omega}$ is a nowhere vanishing two-form on $\mathscr{M}$. The question is how these orientations of $T_x\mathscr{M}$ cohere with the natural orientation of $\mathbb{R}^3$. To find an answer we choose a point $x \in \mathscr{M}$ and local coordinates at x such that $\frac{\partial}{\partial y_1}, \frac{\partial}{\partial y_2}$ form a positive basis in $T_x\mathscr{M}$. It is obvious that there is a vector $n \in \mathbb{R}^3$ which is perpendicular to $T_x\mathscr{M} \subset \mathbb{R}^3$ and such that

$$\left(n, \frac{\partial}{\partial y_1}, \frac{\partial}{\partial y_2}\right)$$

is a positive basis of $\mathbb{R}^3$. The vector $\frac{n}{\|n\|}$ is called a (unit) *outer normal vector* to $\mathscr{M}$ at the point x. It is easy to prove that

$$n = c\left(\frac{\partial}{\partial y_1} \times \frac{\partial}{\partial y_2}\right), \qquad c > 0.$$

For the definition of the cross product see Example 4.3.58(ii).

(ii) The Möbius strip $\mathscr{S} \subset \mathbb{R}^3$ is an example of a non-orientable manifold. An argument to prove that can be based on the above consideration. Choose a point $a \in \mathscr{S}$, a basis $\frac{\partial}{\partial y_1}, \frac{\partial}{\partial y_2}$ in $T_a\mathscr{S}$ and find the outer normal vector n_a. Now move the point a together with the basis $\left(n, \frac{\partial}{\partial y_1}, \frac{\partial}{\partial y_2}\right)$ along the whole strip to come back to the initial position. The vector n_a will end at $-n_a$ (see Figure 4.3.19). ○

Definition 4.3.92. Let $(\mathscr{M}, \overline{\omega})$ be an oriented M-dimensional differentiable manifold. Let $(\mathcal{V}_n, \psi_n)_{n \in \mathbb{N}}$ be an atlas of $\mathscr{M}$ for which the coordinate vectors $\frac{\partial}{\partial y_1}, \dots, \frac{\partial}{\partial y_M}$ form a

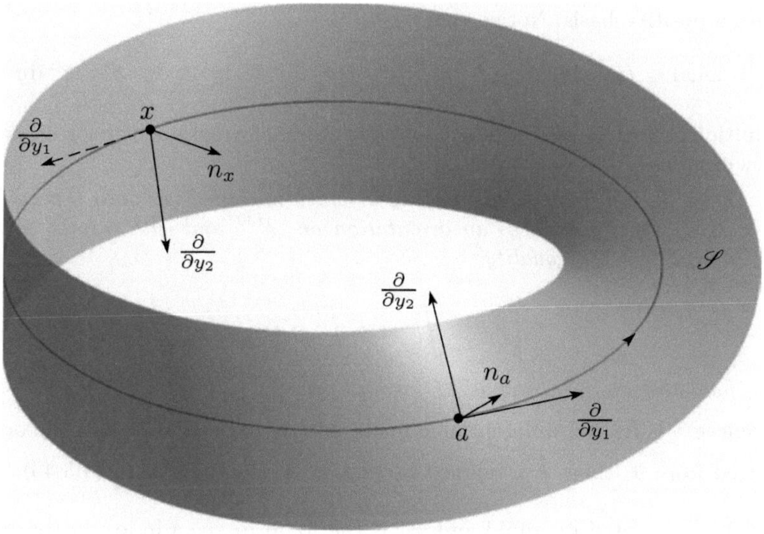

Figure 4.3.19.

positive basis of $T_x\mathscr{M}$ for all $x \in \mathcal{V}_n$ and all $n \in \mathbb{N}$. Let $\{\alpha_n\}_{n\in\mathbb{N}}$ be a partition of unity subordinate to this atlas. If ω is a continuous M-form with the local representation

$$\omega(x) = f_n(x)\,\mathrm{d}y^1 \wedge \cdots \wedge \mathrm{d}y^M, \qquad x \in \mathcal{V}_n,$$

then we define the *integral of* ω *over* $\mathscr{M}$ as

$$\begin{aligned}\int_{\mathscr{M}} \omega = \sum_{n=1}^{\infty} \int_{\mathscr{M}} \alpha_n \omega &:= \sum_{n=1}^{\infty} \int_{\mathcal{U}_n} \varphi_n^*(\alpha_n\omega)(y) \\ &= \sum_{n=1}^{\infty} \int_{\mathcal{U}_n} (\alpha_n f_n)(\varphi_n(y))\,\mathrm{d}y_1 \cdots \mathrm{d}y_M \text{ [28]}\end{aligned} \tag{4.3.43}$$

provided the right-hand side exists and the sum is absolutely convergent.

Remark 4.3.93.

(i) If a form ω has compact support, in particular, if $\mathscr{M}$ is a compact set in $\mathbb{R}^N$, then the integral $\displaystyle\int_{\mathscr{M}} \omega$ exists.

(ii) Definition 4.3.92 does not depend on the concrete choice of an atlas and on a partition of unity. If the coordinate vectors $\frac{\partial}{\partial y_1}, \ldots, \frac{\partial}{\partial y_M}$ determine a negative basis in $T_x\mathscr{M}$, then we change the order, e.g., to

$$\frac{\partial}{\partial y_2}, \frac{\partial}{\partial y_1}, \frac{\partial}{\partial y_3}, \ldots, \frac{\partial}{\partial y_M},$$

[28] For $\eta(y) = g(y)\,\mathrm{d}f^1 \wedge \cdots \wedge \mathrm{d}f^M$ a continuous M-form on a measurable set $\mathcal{U} \subset \mathbb{R}^M$ (cf. Remark 4.3.57(ii)) we define $\displaystyle\int_{\mathcal{U}} g(y)\,\mathrm{d}f^1 \wedge \cdots \wedge \mathrm{d}f^M = \int_{\mathcal{U}} g(y)\,\mathrm{d}y_1 \cdots \mathrm{d}y_M$.

to get a positive basis. Notice that

$$\omega(x) = f(x)\,\mathrm{d}y^1 \wedge \cdots \wedge \mathrm{d}y^M = -f(x)\,\mathrm{d}y^2 \wedge \mathrm{d}y^1 \wedge \mathrm{d}y^3 \wedge \cdots \wedge \mathrm{d}y^M.$$

(iii) Definition 4.3.92 is also independent on a transformation of coordinates in the following sense:

Let g be a diffeomorphism of an oriented manifold $\mathscr{M}$ onto a manifold $\tilde{\mathscr{M}}$.[29] *Then g induces an orientation on $\tilde{\mathscr{M}}$*[30] *and with respect to this orientation the equality*

$$\int_{\tilde{\mathscr{M}}} \omega = \int_{\mathscr{M}} g^* \omega \qquad (4.3.44)$$

holds for any continuous M-form ω on $\tilde{\mathscr{M}}$.

(iv) If a curve γ is itself a differentiable manifold (Remark 4.3.86) with the orientation induced from $\mathbb{R}$, then $\int_\gamma \omega$ defined by (4.3.39) is the same as in (4.3.43).

(v) If $\mathscr{M}$ is an oriented manifold and ω_V is the M-form given in local coordinates by

$$\omega_V = \left[\det \boldsymbol{G}\left(\frac{\partial}{\partial y_1}, \dots, \frac{\partial}{\partial y_M}\right)\right]^{\frac{1}{2}} \mathrm{d}y^1 \wedge \cdots \wedge \mathrm{d}y^M$$

where $\frac{\partial}{\partial y_1}, \dots, \frac{\partial}{\partial y_M}$ is a positive basis, then the volume $V(\mathscr{M}) := \int_{\mathscr{M}} \mathrm{d}V$ (see (4.3.42)) is given by

$$V(\mathscr{M}) = \int_{\mathscr{M}} \omega_V.$$

Example 4.3.94. Let $\mathcal{M}$ be part of the hemisphere given by

$$\left\{(x, y, z) \in \mathbb{R}^3 : x^2 + y^2 + z^2 = 1,\ z > -\frac{1}{2}\right\}$$

and

$$\omega(x, y, z) = y^2\,\mathrm{d}x \wedge \mathrm{d}y + yz\,\mathrm{d}x \wedge \mathrm{d}z + x^2\,\mathrm{d}y \wedge \mathrm{d}z.$$

We choose the spherical coordinates φ:

$$x = \cos\alpha\cos\vartheta, \quad y = \sin\alpha\cos\vartheta, \quad z = \sin\vartheta, \quad (\alpha, \vartheta) \in \mathcal{U} := (-\pi, \pi) \times \left(-\frac{\pi}{6}, \frac{\pi}{2}\right)$$

and the orientation such that $\left(\frac{\partial}{\partial\alpha}, \frac{\partial}{\partial\vartheta}\right)$ is a positive basis. We wish to compute the integral $\int_{\mathscr{M}} \omega$.

---

[29] We have not defined this notion yet, but it is almost evident how to generalize the well-known case $\mathbb{R}^M \to \mathbb{R}^M$. One has to overcome certain difficulties which are caused by the local definition of a manifold and the global notion of diffeomorphism.

[30] Let $(\mathcal{V}_n, \psi_n)_{n\in\mathbb{N}}$ be an atlas of $\mathscr{M}$ such that the coordinate vectors $\frac{\partial}{\partial y_1}, \dots, \frac{\partial}{\partial y_M}$ form a positive basis of $T_x\mathscr{M}$ for $x \in \mathcal{V}_n$. Then $g_*\frac{\partial}{\partial y_1}, \dots, g_*\frac{\partial}{\partial y_M}$ determine a positive basis of $T_{g(x)}\tilde{\mathscr{M}}$.

The curve $\gamma = \{(-\cos\vartheta, 0, \sin\vartheta) : \vartheta \in \left(-\frac{\pi}{6}, \frac{\pi}{2}\right)\}$ has two-dimensional surface measure equal to zero, and therefore

$$\int_{\mathcal{M}} \omega = \int_{\mathcal{U}} \varphi^* \omega \, \mathrm{d}\alpha \, \mathrm{d}\vartheta.$$

We have (see Example 4.3.58(ii) or (4.3.33) and (4.3.19))

$$\begin{aligned} \varphi^*(\mathrm{d}x \wedge \mathrm{d}y) &= \sin\vartheta \cos\vartheta \, \mathrm{d}\alpha \wedge \mathrm{d}\vartheta, \\ \varphi^*(\mathrm{d}y \wedge \mathrm{d}z) &= \cos\alpha \cos^2\vartheta \, \mathrm{d}\alpha \wedge \mathrm{d}\vartheta, \qquad \text{i.e.,} \qquad \varphi^*\omega = \cos^3\alpha \cos^4\vartheta \, \mathrm{d}\alpha \wedge \mathrm{d}\vartheta. \\ \varphi^*(\mathrm{d}z \wedge \mathrm{d}x) &= \sin\alpha \cos^2\vartheta \, \mathrm{d}\alpha \wedge \mathrm{d}\vartheta, \end{aligned}$$

An easy computation gives

$$\int_{\mathcal{M}} \omega = \int_{(-\pi,\pi)\times\left(-\frac{\pi}{6}, \frac{\pi}{2}\right)} \cos^3\alpha \cos^4\vartheta \, \mathrm{d}\alpha \, \mathrm{d}\vartheta = 0. \qquad \circ$$

It is evident that the computation of $\int_{\mathscr{M}} \omega$ need not be an easy task when several charts have to be used to cover the support of ω. The main reason is that a partition of unity must be constructed and this is technically difficult. Because of that, we would like to have such useful tools like the Fubini Theorem and the Fundamental Theorem of Calculus. The later theorem, i.e.,

$$\int_a^b f'(x) \, \mathrm{d}x = f(b) - f(a),$$

can be interpreted in the manifold language as follows. The closed interval $[a, b]$ is a manifold $\mathscr{M}$ positively oriented from the left to the right. The one-form $f'(x)\,\mathrm{d}x$ is the differential of the zero-form (i.e., the function f), and the (oriented) boundary of $\mathscr{M}$ consists of the points b, a (in this order). The Fundamental Theorem of Calculus reduces the integral of the form $\mathrm{d}f(x) = f'(x)\,\mathrm{d}x$ over $\mathscr{M}$ to an "integral" of f over $\partial\mathscr{M}$. This observation is essential for the generalization to manifolds with boundaries. To do that we have to define the boundary of $\mathscr{M}$ first, and then to show how this boundary inherits the orientation of $\mathscr{M}$.

Definition 4.3.95. Let $\mathscr{N}$ be an M-dimensional differentiable manifold in $\mathbb{R}^N$. A closed subset $\mathscr{M}$ of $\mathscr{N}$ is said to be an M-dimensional *differentiable manifold with boundary*[31] if

$$\overline{\operatorname{int} \mathscr{M}} = \mathscr{M}$$

(the interior and closure are taken in the topology of $\mathscr{N}$) and for any point $x \in \mathscr{M}$ there is a chart $(\mathcal{V}, \psi)$ of an atlas of $\mathscr{N}$ such that either

(i) $\mathcal{V} \subset \mathscr{M}$

or

(ii) $P\psi(x) = (0, y_2, \ldots, y_M)$ and $P\psi(\mathcal{V} \cap \mathscr{M}) = \{(y_1, \ldots, y_M) \in \mathbb{R}^M : y_1 \le 0\} \cap P\psi(\mathcal{V})$.

[31] This boundary can be an empty set (see, e.g., Remark 4.3.96(i)). If this boundary is nonempty, then the manifold $\mathscr{M}$ is not a differentiable manifold in the sense of Definition 4.3.4. See also Remark 4.3.96(ii).

A point x is called an *interior point* of $\mathscr{M}$ in the case (i). If x is not an interior point, then x is called a *boundary point* of $\mathscr{M}$. The collection of all boundary points is called the *boundary* of $\mathscr{M}$ and denoted by $\partial\mathscr{M}$ (see Figure 4.3.20).

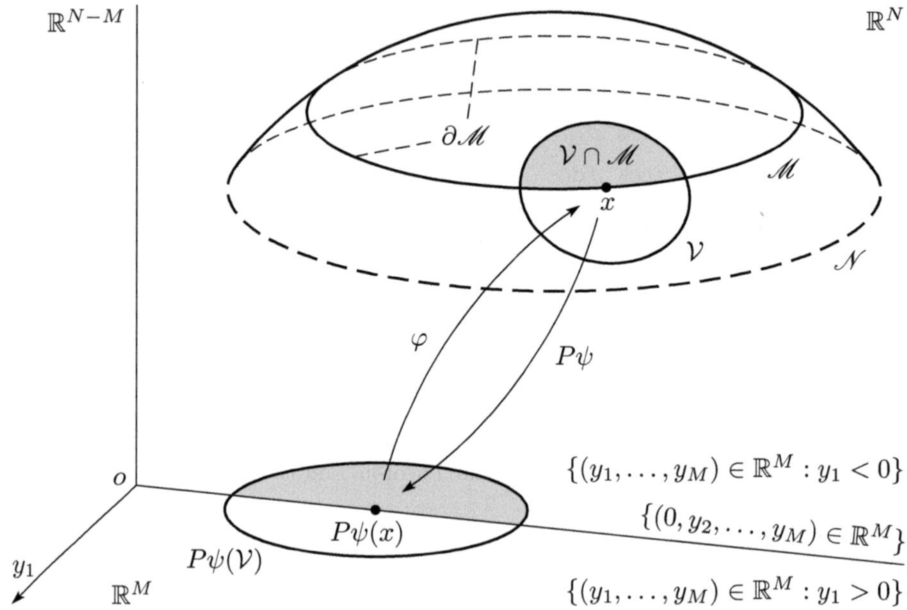

Figure 4.3.20.

Remark 4.3.96.

(i) A manifold can have empty boundary. The sphere $\mathcal{S}^2$ in $\mathbb{R}^3$ is an example of this fact. Such a manifold is also called a manifold without boundary.

(ii) If $\mathscr{M}$ is a manifold with nonempty boundary $\partial\mathscr{M}$, then the boundary $\partial\mathscr{M}$ is itself a differentiable manifold of dimension $M-1$ in $\mathbb{R}^N$. An atlas is given by the restriction of the original atlas. We notice that $\partial\mathscr{M}$ has empty boundary, i.e., $\partial(\partial\mathscr{M})=\emptyset$.

(iii) Let $\mathscr{M}$ be a manifold with boundary. The tangent space $T_a\mathscr{M}$ for an interior point $a\in\mathscr{M}$ is defined as in Definition 4.3.31 and $T_a\mathscr{M}=T_a\mathscr{N}$. If $a\in\partial\mathscr{M}$, then we take all smooth curves in $\overline{\mathbb{R}^M_-}$ ($\mathbb{R}^M_-=\{(y_1,\dots,y_M)\in\mathbb{R}^M : y_1<0\}$) through the point $b=\psi(a)$ and transfer them into $\mathbb{R}^N$ by applying

$$\varphi=(P\psi)^{-1}.$$

The tangent vectors at the point a of these transfered curves form the tangent space $T_a(\partial\mathscr{M})$. If $\omega\colon(-1,1)\to\mathbb{R}^M$ is a smooth curve such that

$$\omega(0)=b,\qquad \omega(t)\in\begin{cases}\mathbb{R}^M_- & \text{for}\quad t<0,\\ \mathbb{R}^M_+ & \text{for}\quad t>0,\end{cases}$$

then

$$\varphi'(b)[\dot{\omega}(0)]$$

is the so-called *outer vector* to $\partial\mathscr{M}$. The *outer normal* n to $\partial\mathscr{M}$ at the point $a \in \partial\mathscr{M}$ is the unit outer vector which is perpendicular in $\mathbb{R}^N$ to $T_a(\partial\mathscr{M})$ (see Figure 4.3.21).

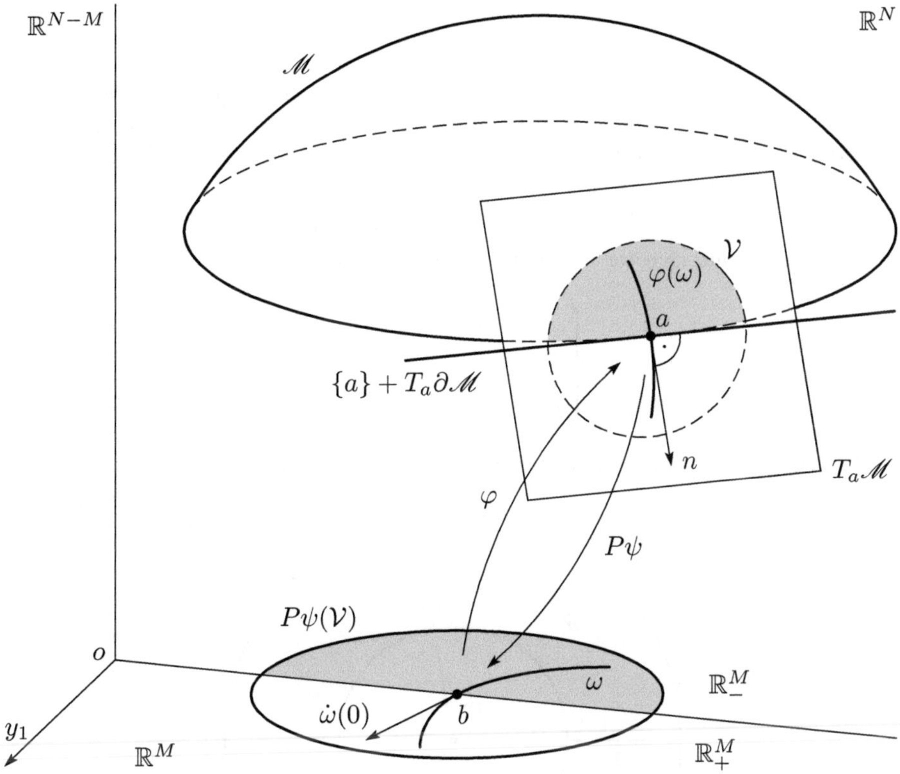

Figure 4.3.21.

(iv) If $\mathscr{M} \subset \mathscr{N}$ is a differentiable manifold with boundary and $(\mathscr{N}, \overline{\omega})$ is an oriented manifold, then $\overline{\omega}$ induces an orientation on $\partial\mathscr{M}$ as follows. Choose $a \in \partial\mathscr{M}$ and

$$\overline{v}_1 := \frac{\mathrm{d}}{\mathrm{d}t}\varphi(t, b_2, \ldots, b_M)\Big|_{t=0}$$

(a special case of an outer vector), and define

$$\overline{\omega}_\partial(a)(v_2, \ldots, v_M) = \overline{\omega}(a)(\overline{v}_1, v_2, \ldots, v_M).$$

The form $\overline{\omega}_\partial$ defines the *induced orientation* on $\partial\mathscr{M}$. In other words: $\overline{v}_2, \ldots, \overline{v}_M$ is a positive basis of $T_a(\partial\mathscr{M})$ provided $\overline{v}_1, \ldots, \overline{v}_M$ is a positive basis of $T_a\mathscr{N}$.

Example 4.3.97.

(i) Let $\mathscr{M}$ be the closed ring

$$\{(x,y) \in \mathbb{R}^2 : 1 \le x^2 + y^2 \le 4\}.$$

Then $\mathscr{M} \subset \mathbb{R}^2$ is a two-dimensional manifold with boundary

$$\partial\mathscr{M} = \mathcal{S}_1^1 \cup \mathcal{S}_2^1$$

($\mathcal{S}_r^1$ denotes the circle with radius r and center at the origin). As the local coordinates we take the polar coordinates (r, ϑ). The (standard) orientation on $\mathscr{M}$ is given either by the standard Euclidean basis $e_1 = (1,0)$, $e_2 = (0,1)$ in $T_a\mathscr{M} = \mathbb{R}^2$ or by the form

$$\overline{\omega}(a) = f^1 \wedge f^2 =: \mathrm{d}x \wedge \mathrm{d}y = r(a)\, \mathrm{d}r \wedge \mathrm{d}\vartheta.$$

The special outer vector $\overline{v}_1$ mentioned in Remark 4.3.96(iv) is

$$\overline{v}_1 = \begin{cases} -\dfrac{\partial}{\partial r} & \text{at the point} \quad a_1 = (1, \vartheta_1), \\ \dfrac{\partial}{\partial r} & \text{at the point} \quad a_2 = (2, \vartheta_2). \end{cases}$$

The completion to a positive basis in $T_a\mathbb{R}^2$ is shown in Figures 4.3.22 and 4.3.23.

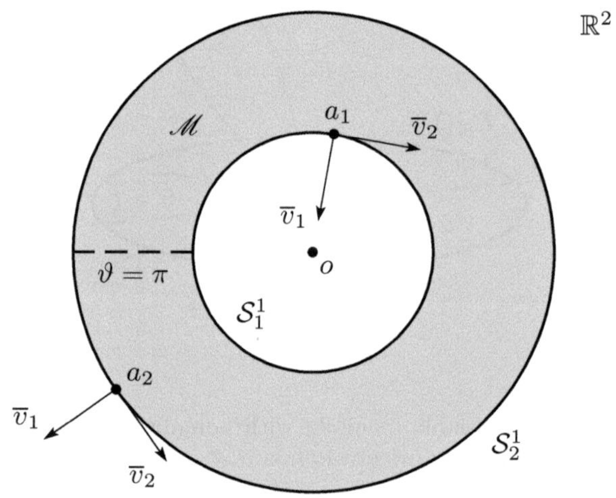

Figure 4.3.22.

The form $\overline{\omega}_\partial$ on $\mathcal{S}_1^1$ is given (in the polar coordinates r, ϑ) by

$$\overline{\omega}_\partial(a)\left(\frac{\partial}{\partial\vartheta}\right) = r(a)\, \mathrm{d}r \wedge \mathrm{d}\vartheta \left(-\frac{\partial}{\partial r}, \frac{\partial}{\partial\vartheta}\right) = -r(a).$$

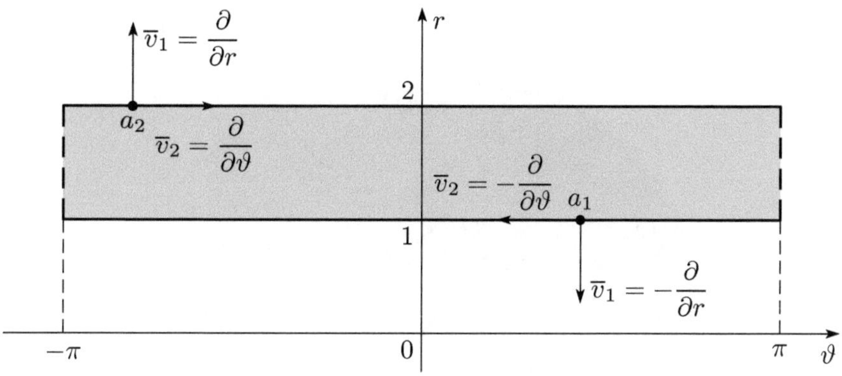

Figure 4.3.23.

(ii) Let $\mathcal{B}$ be the closed unit ball in $\mathbb{R}^3$. Then $\mathcal{B}$ is a three-dimensional manifold (included in $\mathbb{R}^3$) with boundary $\partial\mathcal{B} = \mathcal{S}^2$ (the two-dimensional sphere). The standard orientation in $\mathbb{R}^3 = T_a\mathcal{B}$ gives the orientation in $\operatorname{int}\mathcal{B}$. The induced orientation on $\partial\mathcal{B}$ is obtained by Remark 4.3.96(iv). Namely, we take a normal vector $n = \frac{\partial}{\partial r}$ at a point $a \in \partial\mathcal{B}$ and independent vectors $\overline{v}_2, \overline{v}_3 \in T_a(\partial\mathcal{B})$ in the order given by the right-thumb rule for $(n, \overline{v}_2, \overline{v}_3)$ (see Figure 4.3.24).[32] Notice that the orientation on $\mathcal{B}$ is given, e.g., by

$$\overline{\omega} = \mathrm{d}x \wedge \mathrm{d}y \wedge \mathrm{d}z = (r^2 \cos\vartheta)\,\mathrm{d}r \wedge \mathrm{d}\alpha \wedge \mathrm{d}\vartheta$$

in the spherical coordinates. Similarly,

$$\overline{\omega}_\partial(v_2, v_3) = \overline{\omega}\left(\frac{\partial}{\partial r}, v_2, v_3\right). \qquad \bigcirc$$

The next theorem is a basic result on differential forms and it is the promised generalization of the fundamental theorem of calculus.

Theorem 4.3.98 (Stokes, abstract version). *Let $\mathscr{M}$ be an M-dimensional oriented manifold with boundary $\partial\mathscr{M}$. Let ω be a smooth $(M-1)$-differential form on $\mathscr{M}$ with compact support. Then*

$$\int_{\mathscr{M}} \mathrm{d}\omega = \int_{\partial\mathscr{M}} i^*\omega\ ^{33}$$

provided $\partial\mathscr{M}$ has the induced orientation and $i\colon \partial\mathscr{M} \to \mathscr{M}$ is the canonical injection.

Proof. Let $(\mathcal{V}_n, \psi_n)_{n\in\mathbb{N}}$ be an atlas of $\mathscr{M}$ such that the coordinate vectors given by $\frac{\partial}{\partial y_1}, \dots, \frac{\partial}{\partial y_M}$ form positive bases in the corresponding tangent spaces. Let $\{\alpha_n\}_{n\in\mathbb{N}}$ be

[32] More precisely, for $n := \overline{v}_2 \times \overline{v}_3$ (the cross product is defined in Example 4.3.58(ii)) the vectors n, $\overline{v}_2$, $\overline{v}_3$ form a positive basis in $\mathbb{R}^3$ and n is perpendicular to $\overline{v}_2$, $\overline{v}_3$.

[33] Here $\int_{\mathscr{M}} \mathrm{d}\omega$ is defined as in Definition 4.3.92 where $\mathcal{U}_n$ need not be an open set – see footnote 28 on page 229.

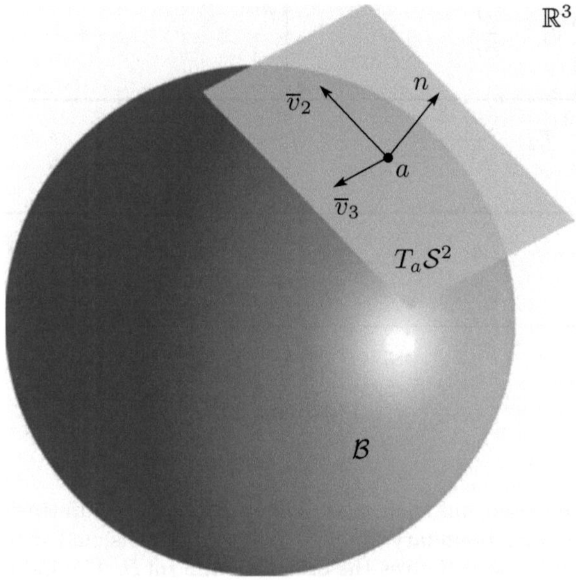

Figure 4.3.24.

a partition of unity subordinate to $\{\mathcal{V}_n\}_{n\in\mathbb{N}}$. Since ω has compact support, we have

$$\int_{\mathscr{M}} \mathrm{d}\omega = \sum_{j=1}^{k} \int_{\mathscr{M}} \alpha_j \,\mathrm{d}\omega \qquad \text{where} \quad \operatorname{supp}(\alpha_j\,\mathrm{d}\omega) \subset \mathcal{V}_j .$$

Therefore, it is sufficient to prove the Stokes Theorem only for the case when $\operatorname{supp}\omega$ is contained in one coordinate neighborhood, say $\mathcal{V}$. Suppose that ω has the representation

$$\omega(x) = \sum_{i=1}^{M} (-1)^{i-1} f_i(y)\,\mathrm{d}y^1 \wedge \cdots \wedge \widehat{\mathrm{d}y^i} \wedge \cdots \wedge \mathrm{d}y^M, \qquad x = \varphi(y) \in \mathcal{V},$$

where the hat denotes that the term $\mathrm{d}y^i$ is missing. Then

$$\mathrm{d}\omega(x) = \left(\sum_{i=1}^{M} \frac{\partial f_i}{\partial y_i}(y) \right) \mathrm{d}y^1 \wedge \cdots \wedge \mathrm{d}y^M.$$

There are two cases for $\mathcal{V}$:

Case 1 ($\mathcal{V} \cap \partial\mathscr{M} = \emptyset$). Then $\int_{\partial\mathscr{M}} i^*\omega = 0$ and

$$\int_{\mathscr{M}} \mathrm{d}\omega = \sum_{i=1}^{M} \int_{\mathcal{U}} \frac{\partial f_i}{\partial y_i}(y)\,\mathrm{d}y_1 \ldots \mathrm{d}y_M = 0$$

by the Fubini Theorem, since $f_i = 0$ outside of a compact subset of $\mathcal{U}$.

Case 2 ($\mathcal{V} \cap \partial\mathcal{M} \neq \emptyset$). According to the definition of the boundary we can assume that $\mathcal{V}$ and $\psi(\mathcal{V}) = \mathcal{U}$ have the form as in Figure 4.3.20. Then

$$i^*\omega(x) = f_1(y)\,\mathrm{d}y^2 \wedge \cdots \wedge \mathrm{d}y^M \qquad \text{for} \quad x = \varphi(y) \in \mathcal{V} \cap \partial\mathcal{M}$$

and

$$\int_{\partial\mathcal{M}} i^*\omega = \int_{\partial\mathcal{M}} f_1(y)\,\mathrm{d}y^2 \wedge \cdots \wedge \mathrm{d}y^M = \int_{\mathcal{U}\cap\mathbb{R}^{M-1}} f_1(0, y_2, \dots, y_M)\,\mathrm{d}y_2 \dots \mathrm{d}y_M.$$

On the other hand,

$$\int_{\mathcal{M}} \mathrm{d}\omega = \sum_{i=1}^{M} \int_{\mathcal{U}} \frac{\partial f_i}{\partial y_i}\,\mathrm{d}y_1 \dots \mathrm{d}y_M = \int \cdots \int_{\mathcal{U}\cap\mathbb{R}^{M-1}} \left(\int_{-\infty}^{0} \frac{\partial f_1}{\partial y_1}(y)\,\mathrm{d}y_1 \right) \mathrm{d}y_2 \dots \mathrm{d}y_M$$
$$= \int_{\mathcal{U}\cap\mathbb{R}^{M-1}} f_1(0, y_2, \dots, y_M)\,\mathrm{d}y_2 \dots y_M$$

since the integrals for $i = 2, \dots, M$ vanish because of the compact support of the restriction of f to $\mathcal{U} \cap \mathbb{R}^{M-1}$.

This completes the proof. □

Several special cases of the Stokes Theorem are worth mentioning. We say that a curve $\gamma\colon [a, b] \to \mathbb{R}^N$ is *simple* if

$$\gamma(t_1) = \gamma(t_2), \quad t_1 < t_2, \qquad \text{implies} \qquad a = t_1, \quad b = t_2.$$

Corollary 4.3.99 (Green). *Let Ω be a bounded open subset of $\mathbb{R}^2$ the boundary of which is the image of a simple closed smooth curve γ which is oriented so that Ω is on the left-hand side when we move along the curve. Let $F = (f, g)\colon \mathbb{R}^2 \to \mathbb{R}^2$ be a C^1-mapping in a neighborhood of the closure $\overline{\Omega}$. Then*

$$\int_{\Omega} \left(\frac{\partial g}{\partial x} - \frac{\partial f}{\partial y} \right) \mathrm{d}x\,\mathrm{d}y = \int_{\gamma} (f\,\mathrm{d}x + g\,\mathrm{d}y). \tag{4.3.45}$$

(See Definition 4.3.83 *for the integral on the right-hand side.)*

Proof. Notice that $\overline{\Omega}$ is an oriented manifold (the positive orientation of $\mathbb{R}^2$) with boundary $\partial\Omega$ (smoothness of γ) and the above mentioned orientation of γ agrees with the induced orientation on $\partial\Omega$. Put

$$\omega = f\,\mathrm{d}x + g\,\mathrm{d}y$$

and use the Stokes Theorem. □

Remark 4.3.100. For $F(x, y) = \frac{1}{2}(-y, x)$ the formula (4.3.45) gives

$$V(\Omega) = \frac{1}{2} \int_{\gamma} -y\,\mathrm{d}x + x\,\mathrm{d}y,$$

which can be used for computation of the area of planar sets.

Corollary 4.3.101 (Gauss–Ostrogradski). *Let Ω be a bounded open subset of $\mathbb{R}^3$ such that $\overline{\Omega}$ is a differentiable manifold with boundary $\partial\Omega$ (cf. page 265 and Exercise 5.3.10). Assume that $F = (f^1, f^2, f^3)\colon \mathbb{R}^3 \to \mathbb{R}^3$ is a smooth mapping in a neighborhood of $\overline{\Omega}$. Then*

$$\int_\Omega \sum_{i=1}^{3} \frac{\partial f^i}{\partial x_i}\, \mathrm{d}x_1\, \mathrm{d}x_2\, \mathrm{d}x_3 = \int_{\partial\Omega} (F, n)\, \mathrm{d}S \tag{4.3.46}$$

where n is the unit outer normal vector to $\partial\Omega$ and (F, n) is the scalar product in $\mathbb{R}^3$. The integral on the right-hand side is defined in Definition 4.3.87.

Proof. Put

$$\omega = f^1\, \mathrm{d}y \wedge \mathrm{d}z + f^2\, \mathrm{d}z \wedge \mathrm{d}x + f^3\, \mathrm{d}x \wedge \mathrm{d}y$$

and choose local coordinates (u, v) on $\partial\Omega$ such that $\left(n, \frac{\partial}{\partial u}, \frac{\partial}{\partial v}\right)$ forms a positive orthonormal basis in $\mathbb{R}^3$. The pull-back $i^*\omega$ has been computed in Example 4.3.58(ii):

$$i^*\omega = \left(F, \frac{\partial}{\partial u} \times \frac{\partial}{\partial v}\right) \mathrm{d}u \wedge \mathrm{d}v.$$

The cross product $\frac{\partial}{\partial u} \times \frac{\partial}{\partial v}$ is the unit vector which is perpendicular in $\mathbb{R}^3$ to $\frac{\partial}{\partial u}, \frac{\partial}{\partial v}$. So, $\frac{\partial}{\partial u} \times \frac{\partial}{\partial v} = n$. □

Corollary 4.3.102 (Stokes, classical version). *Let $\mathscr{M}$ be a bounded two-dimensional oriented manifold in $\mathbb{R}^3$ (i.e., a surface) with boundary which is described by a simple smooth curve γ. Let F be a C^1-vector field on $\mathscr{M}$. Then*

$$\int_{\mathscr{M}} (\operatorname{curl} F, n)\, \mathrm{d}S = \int_\gamma F$$

where γ has the orientation induced from $\mathscr{M}$, the vector $\operatorname{curl} F$ *is defined in Example* 4.3.61(ii) *and n is the unit outer normal vector to $\mathscr{M}$ in $\mathbb{R}^3$ (if $\frac{\partial}{\partial y_1}, \frac{\partial}{\partial y_2}$ is a positive basis at $a \in \mathscr{M}$, then n is perpendicular in $\mathbb{R}^3$ to $T_a\mathscr{M}$ and $\left(n, \frac{\partial}{\partial y_1}, \frac{\partial}{\partial y_2}\right)$ is a positive basis in $\mathbb{R}^3$).*

Proof (a hint). Rewrite the abstract Stokes theorem for this special case using the corresponding definitions of integrals and Example 4.3.61(ii). □

Remark 4.3.103. Considering F in Corollary 4.3.101 as a velocity field of a fluid flow (Remark 4.3.59) we can interpret the right-hand side in (4.3.46) as the amount of the fluid which flows out of a region Ω per a unit time. In particular, if the *divergence* of F ($\operatorname{div} F := \sum_{i=1}^{3} \frac{\partial f^i}{\partial x_i}$) vanishes everywhere in Ω, then this amount is zero for any subregion of Ω. In other words, the fluid is incompressible in this case.

Remark 4.3.104. Using an "infinitesimal" ball Ω centered at a in Corollary 4.3.101 it is possible to interpret the value of the function $\operatorname{div} F = \sum_{i=1}^{3} \frac{\partial f^i}{\partial x_i}$ at the point a physically. From the mathematical point of view it is more interesting that this is a starting point for the generalization of basic differential operators to non-flat domains. We now briefly describe this procedure. Let F be a vector field on a manifold $\mathscr{M}$ and let ω be an M-form. Define an $(M-1)$-form $\omega \rfloor F$ by

$$(\omega \rfloor F)(v_1, \dots, v_{M-1}) := \omega(F, v_1, \dots, v_{M-1}).$$

If $(\mathscr{M},\overline{\omega})$ is an oriented manifold, then $\mathrm{d}(\omega\rfloor F)$ has to be a multiple of $\overline{\omega}$:

$$\mathrm{d}(\omega\rfloor F) =: (\operatorname{div} F)\overline{\omega}.$$

We strongly recommend the reader to compute $\operatorname{div} F$, e.g., on the sphere $\mathcal{S}^2$. One of the most important partial differential operators is the *Laplacian* Δ: if $\mathcal{G}$ is an open set in $\mathbb{R}^N$, $f\colon \mathcal{G} \to \mathbb{R}$ is smooth, then

$$\Delta f := \sum_{i=1}^{M} \frac{\partial^2 f}{\partial x_i^2}$$

and it is easy to see that

$$\Delta f = \operatorname{div}(\nabla f) \qquad \text{where} \quad \nabla f := \left(\frac{\partial f}{\partial x_1}, \dots, \frac{\partial f}{\partial x_N}\right)$$

is the gradient of f. Since the notion of the gradient has been defined for functions on manifolds (Remark 4.3.59), we are able to generalize the Laplacian to functions defined on a manifold $\mathscr{M}$:

$$\Delta_{\mathscr{M}} f = \operatorname{div}(\nabla f).$$

This operator $\Delta_{\mathscr{M}}$ is often called the *Laplace–Beltrami operator*. See also Section 8.4. For more information on the significance of this operator the reader can consult, e.g., Chavel [28], Davies & Safarov [39], Robinson [127] or Rosenberg [129].

Remark 4.3.105. In weak formulations of boundary problems for elliptic differential equations (see Section 5.5) the following *Green formula* is frequently used:

> *Let a manifold $\mathscr{M}$ satisfy the assumptions of the abstract Stokes Theorem and let $f, g \in C^2(\mathscr{M})$. Then*

$$\int_{\mathscr{M}} (\Delta_{\mathscr{M}} f) g\overline{\omega} + \int_{\mathscr{M}} (\nabla f, \nabla g)_{T_x\mathscr{M}}\overline{\omega} = \int_{\partial\mathscr{M}} g(\nabla f, n)_{T_x(\partial\mathscr{M})}\overline{\omega}_\partial. \tag{4.3.47}$$

The proof is based on a generalization of (4.3.46):

$$\int_{\mathscr{M}} (\operatorname{div} F)\overline{\omega} = \int_{\partial\mathscr{M}} (F, n)\overline{\omega}_\partial$$

which follows from the abstract Stokes Theorem. Another ingredient is the formula for the divergence of the product of a function and a vector field

$$\operatorname{div}(gF) = g \operatorname{div} F + (\nabla g, F).$$

This formula follows from the definition of the divergence by computation.

Exercise 4.3.106. Prove that the definition of $\displaystyle\int_\gamma \omega$ does not depend on the choice of an atlas and a partition of unity. What can be said about the dependence on a parametrization of γ?

Exercise 4.3.107. Let ω be an exact one-form and f its "primitive function", i.e., $\mathrm{d}f = \omega$. Compute $\displaystyle\int_\gamma \omega$! What is the result if γ is a closed curve?

Exercise 4.3.108. Let ω be a closed one-form in $\mathscr{M}$. Show that ω is exact if and only if $\int_\gamma \omega = 0$ for any smooth closed curve in $\mathscr{M}$.

Hint. Consult the proof of Theorem 4.3.68.

Exercise 4.3.109. Prove (4.3.41)!

Hint. Use induction. To prove the induction step it is convenient to compute the distance δ of v_M from the span of $v_1, \dots, v_{M-1}$. Show that

$$\delta^2 = \frac{\det \boldsymbol{G}(v_1, \dots, v_M)}{\det \boldsymbol{G}(v_1, \dots, v_{M-1})}$$

provided $v_1, \dots, v_{M-1}$ are linearly independent.

Exercise 4.3.110. Check that the definition (4.3.38) is a special case of (4.3.42)! Express the length of a curve $\gamma \subset \mathbb{R}^3$ in spherical coordinates!

Exercise 4.3.111. Let a surface $\mathcal{S}$ be determined by the graph of a smooth function $f\colon \mathcal{U} \subset \mathbb{R}^2 \to \mathbb{R}^3$. Find a formula for the area of $\mathcal{S}$!

Hint. $\iint_{\mathcal{U}} \sqrt{1 + \left(\frac{\partial f}{\partial x}\right)^2 + \left(\frac{\partial f}{\partial y}\right)^2}\, \mathrm{d}x\, \mathrm{d}y$

Exercise 4.3.112. Let $\mathscr{M}$ be the graph of a smooth function $f\colon \mathbb{R}^M \to \mathbb{R}$ defined on an open set $\mathcal{U} \subset \mathbb{R}^M$. Show that $\mathscr{M}$ is an orientable manifold!

Exercise 4.3.113. Let $f\colon \mathbb{R}^N \to \mathbb{R}$ be a smooth mapping for which o is a regular value. Show that

$$\mathscr{M} = \{x \in \mathbb{R}^N : f(x) = 0\}$$

is an orientable manifold of dimension $N - 1$ (provided $\mathscr{M} \neq \emptyset$)!

Exercise 4.3.114. Deduce a version of the Cauchy Theorem, i.e.,

$$\int_\gamma f\, \mathrm{d}z = 0,$$

from the Green Theorem (Corollary 4.3.99)!

Hint. Interpret $f(z)\, \mathrm{d}z$ as the couple of differential 1-forms

$$(g\, \mathrm{d}x - h\, \mathrm{d}y, h\, \mathrm{d}x + g\, \mathrm{d}y).$$

Exercise 4.3.115. Let Ω be a bounded open subset of $\mathbb{R}^N$ such that $\overline{\Omega}$ is a differentiable manifold with boundary $\partial\Omega$ (cf. Exercise 5.3.10).

(i) Generalize the Gauss-Ostrogradski Theorem (Corollary 4.3.101) for a smooth $F\colon \mathbb{R}^N \to \mathbb{R}^N$, i.e., prove that

$$\int_\Omega \operatorname{div} F\, \mathrm{d}x = \int_{\partial\Omega} (F, n)\, \mathrm{d}S$$

holds. Here n is the unit outer normal vector to $\partial\Omega$.

(ii) Deduce from (i) the particular case for $F\colon \mathbb{R}^N \to \mathbb{R}$, i.e.,

$$\int_{\Omega} \nabla F \,\mathrm{d}x = \int_{\partial\Omega} F n \,\mathrm{d}S.$$

Results (i) and (ii) are often known as the *Divergence Theorem.*

Exercise 4.3.116. Find the formula[34] for Δf

(i) in polar coordinates;

Hint. You should get

$$\Delta f = \frac{\partial^2 f}{\partial r^2} + \frac{1}{r}\frac{\partial f}{\partial r} + \frac{1}{r^2}\frac{\partial^2 f}{\partial \vartheta^2};$$

(ii) in spherical coordinates in $\mathbb{R}^3$;

Hint. You should get

$$\Delta f = \frac{\partial^2 f}{\partial r^2} + \frac{2}{r}\frac{\partial f}{\partial r} + \frac{1}{r^2 \cos\vartheta}\left(\frac{\partial^2 f}{\partial \alpha^2} + \cos\vartheta \frac{\partial^2 f}{\partial \vartheta^2} + \sin\vartheta \frac{\partial f}{\partial \vartheta}\right);$$

(iii) on $\mathcal{S}^2$;

(iv) on the Riemann manifold (see page 226).

Exercise 4.3.117. Let $\mathscr{M}$ be a connected differentiable manifold.

(i) Prove that there exists a Riemann metric g on $\mathscr{M}$.

Hint. $\mathscr{M}$ is embedded into $\mathbb{R}^N$.

(ii) Prove that any two points $x, y \in \mathscr{M}$ can be connected by a C^1-curve, i.e.,

$$\text{there is} \quad \gamma\colon [a,b] \to \mathscr{M}, \quad \gamma \in C^1, \quad \gamma(a) = x,\ \gamma(b) = y.$$

Hint. Use the assumption that $\mathscr{M}$ is connected.

(iii) Define the length of a C^1-curve $\gamma\colon [a,b] \to \mathscr{M}$ by the formula

$$l_g(\gamma) := \int_a^b \sqrt{g_{\gamma(t)}(\dot\gamma(t), \dot\gamma(t))}\,\mathrm{d}t.$$

Put

$$\varrho_g(x,y) = \inf\{l_g(\gamma) : \gamma\colon [a,b] \to \mathscr{M}, \gamma(a) = x, \gamma(b) = y\}$$

and show that ϱ_g is a metric on $\mathscr{M}$.

(iv) How is the topology on $\mathscr{M}$ given by ϱ_g related to the topology induced from $\mathbb{R}^N$?

[34] Such formulae are convenient if one is looking for a solution with some symmetries, i.e., invariant with respect to some group actions.

Chapter 5

Topological Methods

5.1 Brouwer Fixed Point Theorem

One of the most frequent problems in analysis, especially in its applications, consists in solving the equation

$$F(x) = y$$

where F is a mapping from a Banach space X into a Banach space Y.[1] Such an equation can be reduced to the equation $F(x) = o$, or, provided $X \subset Y$, to the equation

$$F(x) = x. \tag{5.1.1}$$

In this section we present two basic results on the solvability of (5.1.1) in a special case, namely, for a continuous mapping F and a finite-dimensional X, and a compact mapping F in a general Banach space of infinite dimension – the Brouwer and the Schauder Fixed Point Theorems.

We start with the finite-dimensional case. A brief inspection of $F\colon \mathbb{R} \to \mathbb{R}$ indicates that reasonable assumptions on F are continuity on a closed interval $\mathcal{I}$ and $F(\mathcal{I}) \subset \mathcal{I}$. Moreover, the interval $\mathcal{I}$ should also be bounded. The Intermediate Value Theorem from Calculus applied to $g(x) = F(x) - x$ says that there is a solution of (5.1.1) in $\mathcal{I}$ provided these assumptions are satisfied. Notice that these assumptions are too weak to say anything about the number of solutions. Having no appropriate ordering in $\mathbb{R}^2$ standard proofs of the above result fail in $\mathbb{R}^2$ and, therefore, a generalization is far from being simple.

Instead of an interval we consider the closed unit ball $\mathcal{B} := \overline{B(o;1)}$ in $\mathbb{R}^N$, $N \geq 2$, and a continuous mapping $F\colon \mathcal{B} \to \mathcal{B}$. Suppose that the equation (5.1.1) has no solution in $\mathcal{B}$. Define a map G as indicated in Figure 5.1.1, i.e.,

$$G(x) = \tau(x)F(x) + (1 - \tau(x))x$$

[1] Spaces X, Y are assumed to have linear and topological structure since we discuss problems of analysis and not only of algebra or topology. Banach space structures are supposed mainly for simplification here, but sometimes they can be crucial.

where $\tau(x) \le 0$ is a solution of the quadratic equation

$$\|\tau F(x) + (1-\tau)x\|^2 = 1.$$

The mapping G is well defined (remember our assumption that $F(x) \neq x$ for

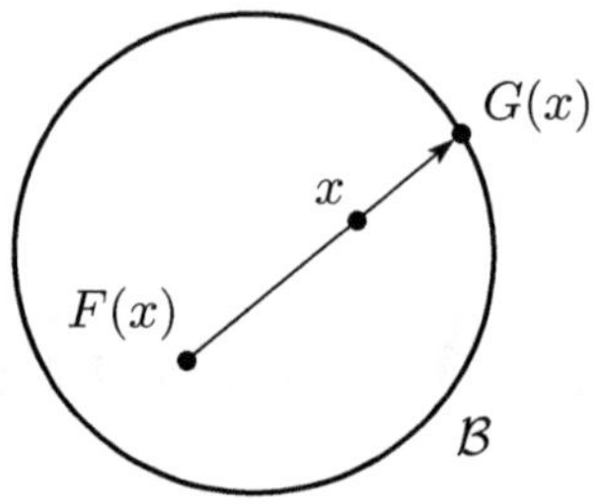

Figure 5.1.1.

$x \in \mathcal{B}$), it maps $\mathcal{B}$ continuously onto the unit sphere $\mathcal{S}^{N-1}$ in $\mathbb{R}^N$ and

$$G(x) = x \qquad \text{for} \quad \|x\| = 1.$$

However, this seems to be impossible as our experience says that if the ball is continuously deformed (by G) onto the sphere, this ball has to be punctured. Nevertheless, a rigorous proof of this fact is far from being obvious. Such a proof could be based on introducing certain topological notions which are preserved under continuous deformation. If we show that some of these notions are different for the ball and its boundary, we obtain a contradiction with the existence of G. Algebraic topology is devoted to the study of topological invariants of an algebraic nature (homotopic groups, homological groups, etc.). However, such methods are beyond the scope of this book.

Instead we will give an analytic proof of the existence of a fixed point of a continuous mapping $F\colon \mathcal{B} \to \mathcal{B}$. This proof which is due to Milnor [112] is based on the idea of approximating a "bad" nonlinear mapping F by a simpler one. A smooth approximation is possible by the Weierstrass Approximation Theorem (see Theorem 1.2.14 and the discussion there).

Suppose, by contradiction, that F has no fixed point in $\mathcal{B}$. For any $\varepsilon > 0$ there are polynomials $P_1, P_2, \dots, P_N$ of N variables, such that for $P = (P_1, \dots, P_N)$ we have

$$\sup_{\|x\| \le 1} \|F(x) - P(x)\| < \varepsilon$$

and also that $\tilde{P} := \frac{P}{1+\varepsilon}\colon \mathcal{B} \to \mathcal{B}$. Moreover, if $\varepsilon > 0$ is small enough, $\tilde{P}$ has no fixed points in $\mathcal{B}$, either. This follows from the estimate

$$\begin{aligned}\|F(x) - x\| &\le \|F(x) - P(x)\| + (1+\varepsilon)\left\|\tilde{P}(x) - x + \frac{\varepsilon}{1+\varepsilon}x\right\| \\ &< 2\varepsilon + (1+\varepsilon)\|\tilde{P}(x) - x\|\end{aligned}$$

and the fact that $\inf_{x \in \mathcal{B}} \|F(x) - x\| > 0$, since $\mathcal{B}$ is compact. Now we construct the mapping $G\colon \mathcal{B} \to \mathcal{S}^{N-1}$ corresponding to $\tilde{P}$ as has been shown in Figure 5.1.1 for F, i.e., $G(x) = \tau(x)\widetilde{P}(x) + (1 - \tau(x))\, x$. We put

$$H(t,x) := (1-t)x + tG(x), \qquad x \in \mathcal{B}.$$

The most important properties of H are given in the next lemma.

Lemma 5.1.1.

(i) *$H(t,\cdot)$ maps $\mathcal{B}$ into itself for every $t \in [0,1]$.*

(ii) *$H(t,\cdot)$ maps $\operatorname{int} \mathcal{B}$ into itself for every $t \in [0,1)$.*

(iii) *The partial Frechét derivative $H_2'(t,x)$ exists on $[0,1] \times \operatorname{int} \mathcal{B}$ and is bounded on this set.*

(iv) *For small $t \geq 0$ the mapping $H(t,\cdot)$ is a diffeomorphism of* $\operatorname{int} \mathcal{B}$ onto *itself.*

Proof. The first two statements are obvious, the third follows from differentiation of $\tau(x)$ (see Exercise 5.1.7). Let us prove the fourth statement. For a small positive t the derivative

$$H_2'(t,x) = (1-t)I + tG'(x)$$

is an isomorphism (Proposition 2.1.2) and, by the Local Inverse Function Theorem, $H(t,\cdot)$ is a local diffeomorphism. Since, by the Mean Value Theorem, applied to G,

$$\|H(t,x) - H(t,y)\| \geq (1-t)\|x-y\| - t \sup_{\|z\|<1} \|G'(z)\|\, \|x-y\| \geq (1-ct)\|x-y\|$$

for a constant c and all $x, y \in \operatorname{int} \mathcal{B}$, the mapping $H(t,\cdot)$ is injective for small t. It remains to prove that

$$H(t,\cdot)(\operatorname{int} \mathcal{B}) = \operatorname{int} \mathcal{B}.$$

Notice that $\operatorname{int} \mathcal{B}$ is a connected set and thus it is sufficient to show that

$$\mathcal{M} := H(t,\cdot)(\operatorname{int} \mathcal{B})$$

is open and relatively closed in $\operatorname{int} \mathcal{B}$. The former property is a consequence of the local continuous invertibility of $H(t,\cdot)$. To see the latter property we point out that

$$\mathcal{M} = H(t,\cdot)(\mathcal{B}) \cap \operatorname{int} \mathcal{B}$$

($\|H(t,x)\| = 1$ for every $\|x\| = 1$). Since $\mathcal{B}$ is compact, $H(t,\cdot)(\mathcal{B})$ is also compact and therefore closed, i.e., $\mathcal{M}$ is relatively closed in $\operatorname{int} \mathcal{B}$. □

Having Lemma 5.1.1 we continue in the proof of our main statement on the existence of a fixed point of F. To reach a contradiction with the assumption of non-existence, we will use the substitution theorem for the Lebesgue integral:

If $\operatorname{meas}\mathcal{A}$ *denotes the* N*-dimensional Lebesgue measure of* $\mathcal{A} \subset \mathbb{R}^N$*, we have*

$$\operatorname{meas}(\operatorname{int}\mathcal{B}) = \int_{\operatorname{int}\mathcal{B}} \mathrm{d}x = \int_{\operatorname{int}\mathcal{B}} \det H_2'(t,y)\,\mathrm{d}y \tag{5.1.2}$$

for small positive t*.*

Notice that $H_2'(0,y) = I$ and thus $\det H_2'(t,y)$ is positive for small $t > 0$. The second equality follows from the substitution $x = H(t,y)$ (Lemma 5.1.1(iv)). The last integral in (5.1.2) is defined for all $t \in [0,1]$, and it is a polynomial $Q(t)$ of the variable $t \in [0,1]$. Since $Q(t)$ is a constant for small t, we also obtain that

$$Q(1) = \operatorname{meas}(\operatorname{int}\mathcal{B}).$$

The substitution $G(y) = H(1,y) =: z$ yields

$$Q(1) = \int_{\operatorname{int}\mathcal{B}} \det G'(y)\,\mathrm{d}y = \int_{G(\operatorname{int}\mathcal{B})} \mathrm{d}z = \operatorname{meas} G(\operatorname{int}\mathcal{B}).^2$$

But $G(\mathcal{B}) = \mathcal{S}^{N-1}$ and $\operatorname{meas}\mathcal{S}^{N-1} = 0$. Hence

$$0 = Q(1) = Q(0) = \operatorname{meas}(\operatorname{int}\mathcal{B}),$$

a contradiction.

In order to get a fixed point theorem in reasonable generality we prove the following simple topological result.

Lemma 5.1.2. *Let* $\mathcal{K}$ *be a convex, closed and bounded subset of* $\mathbb{R}^N$ *which contains at least two different points. Then* $\mathcal{K}$ *is homeomorphic to the unit ball* $\mathcal{B}^M := \overline{B(o;1)}$ *in* $\mathbb{R}^M$ *for some* $M \le N$*.*

Proof. Assume without loss of generality that $o \in \mathcal{K}$. Choose linearly independent elements $x_1, \dots, x_M$ of $\mathcal{K}$ such that

$$X := \operatorname{Lin}\{x_1, \dots, x_M\}$$

contains $\mathcal{K}$. The fact that for any $x \in X$ there is $\alpha > 0$ such that $\frac{1}{\alpha}x \in \mathcal{K}$ can be proved by induction with respect to the dimension of X. Define the Minkowski functional of $\mathcal{K}$ (see p. 63), i.e.,

$$p(x) = \inf\left\{\alpha > 0 : \frac{1}{\alpha}x \in \mathcal{K}\right\}, \quad \varphi(x) = p(x)\frac{x}{\|x\|_{\mathbb{R}^N}}, \quad x \in X \setminus \{o\}, \varphi(o) = o.$$

It is not difficult to prove that φ is a homeomorphism of $\mathcal{K}$ onto $\mathcal{B}^N \cap X$. Since X with the induced $\mathbb{R}^N$-norm is isomorphic to $\mathbb{R}^M$ (Corollary 1.2.11(i)), $\mathcal{K}$ is also homeomorphic to $\mathcal{B}^M$. □

[2] Notice that for this substitution we do not need G to be a diffeomorphism.

The first main result of this section is the following *Brouwer Fixed Point Theorem.*

Theorem 5.1.3 (Brouwer Fixed Point Theorem). *Let $\mathcal{K}$ be a nonempty, convex, closed and bounded subset of $\mathbb{R}^N$. Assume that $F\colon \mathcal{K} \to \mathcal{K}$ is continuous. Then F has a fixed point in $\mathcal{K}$.*

Proof. If $\mathcal{K}$ has exactly one point then the statement is obvious. In other cases choose a homeomorphism Φ of $\mathcal{B}^M = \overline{B(o;1)} \subset \mathbb{R}^M$ onto $\mathcal{K}$ (Lemma 5.1.2). According to the above discussion, the mapping $\Phi^{-1} \circ F \circ \Phi\colon \mathcal{B}^M \to \mathcal{B}^M$ has a fixed point $\tilde{x} \in \mathcal{B}^M$. Then

$$F(\Phi(\tilde{x})) = \Phi(\tilde{x}) \in \mathcal{K}.$$

Another proof is based on the Brouwer degree (see Example 5.7.9). □

The following example shows an interesting application of the Brouwer Fixed Point Theorem in linear algebra.

Example 5.1.4. Let $\boldsymbol{A} = (a_{ij})_{i,j=1,\dots,N}$ be a matrix such that

$$a_{ij} \geq 0 \qquad \text{for all} \quad i,j = 1,\dots,N.$$

Then there exists a nonnegative eigenvalue λ of $\boldsymbol{A}$ with an eigenvector $x = (x_1,\dots,x_N)$ having all its components

$$x_i \geq 0, \qquad i = 1,\dots,N.$$

Indeed, consider the l^1-norm on $\mathbb{R}^N$, i.e.,

$$\|x\|_1 = \sum_{i=1}^{N} |x_i|, \quad \text{and let} \quad \mathcal{D} := \{x \in \mathbb{R}^N : \|x\|_1 = 1,\ x_i \geq 0,\ i = 1,\dots,N\}.$$

Then $\mathcal{D}$ is a nonempty, closed, convex and bounded subset of $\mathbb{R}^N$. Let $A\colon \mathbb{R}^N \to \mathbb{R}^N$ be a linear operator with the representation in the standard basis given by the matrix $\boldsymbol{A}$. If A vanishes at an $x \in \mathcal{D}$, then such x is an eigenvector for the eigenvalue $\lambda = 0$. If this is not the case, put

$$f(x) = \frac{Ax}{\|Ax\|_1} \qquad \text{for} \quad x \in \mathcal{D}.$$

Since f maps $\mathcal{D}$ continuously into $\mathcal{D}$, it has a fixed point x in $\mathcal{D}$. Then

$$Ax = \lambda x \qquad \text{where} \quad \lambda = \|Ax\|_1. \qquad ○$$

Let us mention now the standard application of the Brouwer Fixed Point Theorem to the existence of periodic solutions of ordinary differential equations.

The basic idea goes back to H. Poincaré: Denote by $x(\cdot;\xi)$ a solution of the initial value problem

$$\dot{x}(t) = f(t, x(t)), \qquad x(0) = \xi. \tag{5.1.3}$$

Assume that f satisfies conditions which ensure the existence and uniqueness of (5.1.3) (see, e.g., Theorem 2.3.4) and, moreover, that $f(\cdot, x)$ is T-periodic. Then $x(\cdot;\xi)$ is a T-periodic solution of (5.1.3) if and only if $x(\cdot;\xi)$ is defined on the interval $[0,T]$ and

$$P(\xi) := x(T;\xi) = \xi$$

(P is called the *Poincaré mapping*). Since $x(\cdot;\xi)$ depends continuously on the initial condition ξ (under reasonable assumptions on f, see Remark 2.3.5 and Example 4.2.5), the Poincaré mapping is continuous and its fixed points can be found with help of the Brouwer Fixed Point Theorem as the following example suggests.

Example 5.1.5. Assume in addition that there exists $r > 0$ such that

$$(x, f(t,x))_{\mathbb{R}^N} \le 0 \qquad \text{for all} \quad t \in [0,T] \quad \text{and for all} \quad \|x\|_{\mathbb{R}^N} \le r.$$

Then there exists a T-periodic solution of (5.1.3).

To be able to apply the Brouwer Fixed Point Theorem to the Poincaré mapping P it is sufficient to show that P maps the closed ball $\overline{B(o;r)}$ into itself. Since

$$\frac{\mathrm{d}}{\mathrm{d}t}\frac{1}{2}\|x(t)\|^2 = (x(t), f(t,x(t))) \le 0 \qquad \text{whenever} \quad \|x(t)\| \le r,$$

the function $t \mapsto \|x(t)\|$ is decreasing provided $\|x(0)\| \le r$. In particular, P is well defined (i.e., a solution $x(\cdot, x_0)$ exists on the interval $[0,T]$ provided $\|x(0)\| < r$) and P maps $\overline{B(o;r)}$ into itself. ○

Example 5.1.6. Assume that the right-hand side in (5.1.3) is asymptotically linear in $\mathbb{R}^N$, i.e., there exist a T-periodic continuous matrix $\boldsymbol{A}(t)$ and the function $g\colon \mathbb{R} \times \mathbb{R}^N \to \mathbb{R}^N$ continuous, T-periodic in t, and locally Lipschitz with respect to the x-variables such that for

$$f(t,x) = \boldsymbol{A}(t)x + g(t,x) \tag{5.1.4}$$

the following condition is satisfied:

(H) $\forall \varepsilon > 0\ \exists b > 0: \quad \|g(t,x)\| \le b + \varepsilon\|x\| \qquad \text{for all} \quad t \in \mathbb{R},\ x \in \mathbb{R}^N$.[3]

We are again interested in periodic solutions to (5.1.3) with f given by (5.1.4).

First we have to show that the Poincaré mapping is well defined for all $\xi \in \mathbb{R}^N$. Denote by

$$\boldsymbol{\Phi}(t,s) := \boldsymbol{\Phi}(t)\boldsymbol{\Phi}^{-1}(s)$$

[3] Roughly speaking: g has a uniformly (with respect to t) "vanishing derivative at infinity".

where $\mathbf{\Phi}(t)$ is the fundamental matrix of the linear equation

$$\dot{x} = \boldsymbol{A}(t)x \tag{5.1.5}$$

such that $\mathbf{\Phi}(s,s) = \boldsymbol{I}$.[4] Then the solution of (5.1.3) satisfies the integral equation

$$x(t;\xi) = \mathbf{\Phi}(t,0)\xi + \int_0^t \mathbf{\Phi}(t,s)g(s,x(s;\xi))\,\mathrm{d}s \tag{5.1.6}$$

(the Variation of Constants Formula) whenever it exists on the interval $[0,t]$. The fundamental matrix $\mathbf{\Phi}$ is continuous on $[0,T]\times[0,T]$ and therefore it is bounded:

$$\|\mathbf{\Phi}(t,s)\|_{\mathbb{R}^{N\times N}} \le K, \qquad (t,s)\in[0,T]\times[0,T].$$

By (H), we get the estimate

$$\|x(t;\xi)\| \le K\|\xi\| + KbT + K\varepsilon\int_0^t \|x(s;\xi)\|\,\mathrm{d}s$$

and, with help of the Gronwall inequality (see Proposition 2.3.9),

$$\|x(t;\xi)\| \le K(bT + \|\xi\|)\mathrm{e}^{K\varepsilon T} = L_1 + L_2\|\xi\| \tag{5.1.7}$$

whenever $x(\cdot;\xi)$ is defined on the interval $[0,t]\subset[0,T]$. If the maximal interval of the existence of the solution $x(\cdot;\xi)$ is $[0,\tau)$ with $\tau\le T$, then the boundedness of $x(\cdot;\xi)$, (5.1.7) and the condition (H) imply that $x(\cdot;\xi)$ is uniformly continuous on $[0,\tau)$, and therefore it can be extended to a larger interval (see Proposition 1.2.4 and cf. a similar idea in Corollary 3.1.6). This implies that $x(\cdot;\xi)$ is defined on $[0,T]$ (actually on $\mathbb{R}$) and the mapping P is defined for all $\xi\in\mathbb{R}^N$.

To apply the Brouwer Fixed Point Theorem (the problem is to show that P maps a ball into itself) we assume that 1 is not the *Floquet multiplier* of the linear equation (5.1.5), i.e., $1\notin\sigma(\mathbf{\Phi}(T,0))$ or, equivalently, the equation (5.1.5) possesses only the trivial T-periodic solution. Then the equation $P(\xi)=\xi$ is equivalent to the equation

$$F(\xi) := [\boldsymbol{I} - \mathbf{\Phi}(T,0)]^{-1}[x(T;\xi) - \mathbf{\Phi}(T,0)\xi] = \xi.$$

From (5.1.6) and (5.1.7) we obtain

$$\|F(\xi)\| \le \|[\boldsymbol{I} - \mathbf{\Phi}(T,0)]^{-1}\|[Kb + K\varepsilon(L_1 + L_2\|\xi\|)]T = c_1(\varepsilon) + c_2\varepsilon\|\xi\|$$

where c_2 does not depend on ε. Choose ε small enough to satisfy $c_2\varepsilon<1$. Keeping such ε fixed there is $r>0$ such that $c_1(\varepsilon)+c_2\varepsilon r\le r$. It follows that F maps the ball $\overline{B(o;r)}\subset\mathbb{R}^N$ of the radius r into itself and the Brouwer Fixed Point Theorem yields a T-periodic solution of (5.1.4). ○

[4] This means that $x\colon t\mapsto\mathbf{\Phi}(t,s)\xi$ is a (unique) solution of the equation (5.1.5) which satisfies $x(s)=\xi$.

Exercise 5.1.7. Let τ be as in the proof of Lemma 5.1.1. Prove that

$$x \mapsto \tau(x), \qquad x \in \operatorname{int} \mathcal{B},$$

has a bounded Fréchet derivative.

Hint. Use the Implicit Function Theorem for

$$\Phi(\tau, x) := \|\tau \tilde{P}(x) + (1-\tau)x\|^2 - 1.$$

5.1A Contractible Sets

This appendix has solely an auxiliary character and will be used in the proof of the Krasnoselski Potential Bifurcation Theorem in Appendix 7.8B. The proofs of the assertions from this appendix rely on the Brouwer Fixed Point Theorem (Theorem 5.1.3).

Definition 5.1.8. Let $\mathcal{A}$ and $\mathcal{B}$ be subsets of a topological space Y. Then by definition $\mathcal{A}$ is *contractible* into $\mathcal{B}$ in the space Y, briefly

$$\mathcal{A} \prec \mathcal{B} \quad \text{in} \quad Y,$$

if there exists a *homotopy* $h \in C([0,1] \times \mathcal{A}, Y)$ such that for any $u \in \mathcal{A}$,

$$h(0,u) = u, \qquad h(1,u) \in \mathcal{B}.$$

The next assertion shows that "$\prec$" is a transitive relation.

Lemma 5.1.9. *Let $\mathcal{A}$, $\mathcal{B}$ and $\mathcal{C}$ be subsets of Y. If $\mathcal{A} \prec \mathcal{B}$ and $\mathcal{B} \prec \mathcal{C}$ in Y, then also $\mathcal{A} \prec \mathcal{C}$ in Y.*

Proof. Let us assume that $\mathcal{A} \prec \mathcal{B}$ and $\mathcal{B} \prec \mathcal{C}$ by means of homotopies h and g. Define a homotopy $f \in C([0,1] \times \mathcal{A}, Y)$ by

$$f(t,u) = \begin{cases} h(2t,u), & 0 \le t \le \frac{1}{2}, \quad u \in \mathcal{A}, \\ g(2t-1, h(t,u)), & \frac{1}{2} < t \le 1, \quad u \in \mathcal{A}. \end{cases}$$

Then $f \in C([0,1] \times \mathcal{A}, Y)$ and Definition 5.1.8 yields $\mathcal{A} \prec \mathcal{C}$. □

Let H_1 and H_2 be two closed subspaces of a Hilbert space H such that

$$H = H_1 \oplus H_2.$$

Let $P_i \colon H \to H_i$, $i = 1, 2$, be projections (cf. Example 1.1.13(i)), and assume that

$$\dim H_1 < \infty.$$

Set

$$R = \{x \in H : P_1 x \neq o\}.$$

The set R equipped with the metric induced by the norm in H is a metric space.

Lemma 5.1.10. *The set $\mathcal{S}_{1,r} := \partial B(o;r) \cap H_1$ is not contractible to a point in R.*[5]

[5] I.e., there is no $x \in R$ such that $\mathcal{S}_{1,r} \prec \{x\}$ in R.

Proof. It is enough to prove this assertion for the sphere with radius $r = 1$. Let us denote it by $\mathcal{S}_1$. We proceed in two steps. We prove first that if $\mathcal{S}_1$ were contractible to a point in R, then it would have to be contractible to a point in $\mathcal{S}_1$. In the second step we show that this fact contradicts the Brouwer Fixed Point Theorem (Theorem 5.1.3).

Step 1. If $\mathcal{S}_1$ is contractible to a point in R, then there exists a continuous mapping $f\colon [0,1] \times \mathcal{S}_1 \to R$ and $x_0 \in R$ such that

$$f(0,x) = x, \qquad f(1,x) = x_0 \qquad \text{for all} \quad x \in \mathcal{S}_1.$$

For $t \in [0,1]$, $x \in \mathcal{S}_1$ set

$$g(t,x) = \frac{P_1 f(t,x)}{\|P_1 f(t,x)\|}.$$

Then g deforms the set $\mathcal{S}_1$ continuously to the point $\frac{P_1 x_0}{\|P_1 x_0\|}$ in $\mathcal{S}_1$.

Step 2. Let the unit sphere $\mathcal{S}_1 \subset H_1$ be contractible to a point in $\mathcal{S}_1$, i.e., there exists a continuous map $g\colon [0,1] \times \mathcal{S}_1 \to \mathcal{S}_1$ and a point $x_0 \in \mathcal{S}_1$ such that

$$g(0,x) = x, \qquad g(1,x) = x_0 \qquad \text{for all} \quad x \in \mathcal{S}_1.$$

Now, we define $h\colon \overline{B(o;1)} \cap H_1 \to S_1$ by

$$h\colon x \mapsto \begin{cases} -g\left(1 - \|x\|, \dfrac{x}{\|x\|}\right) & \text{for} \quad x \neq o, \\ -x_0 & \text{for} \quad x = o. \end{cases}$$

Then h is continuous. Since $\dim H_1 < \infty$, the Brouwer Fixed Point Theorem (Theorem 5.1.3) implies that there exists $y \in \overline{B(o;1)} \cap H_1$ such that

$$h(y) = y.$$

Since h assumes only values from $\mathcal{S}_1$, we have $y \in \mathcal{S}_1$, $\|y\| = 1$. On the other hand,

$$h(y) = -g(0,y) = -y,$$

which is a contradiction. □

Lemma 5.1.11. *Let $\mathcal{F}$ be a subset of R. If there exists $x_0 \in H_1$, $\|x_0\| = 1$, such that*

$$P_1(\mathcal{F}) \cap \{y \in H_1 : y = ax_0,\ a \in \mathbb{R}\} = \emptyset,$$

then $\mathcal{F}$ is contractible to a point in R.

Proof. Define $f\colon [0,1] \times \mathcal{F} \to H$ as

$$f(t,x) = \begin{cases} x + 2tx_0[1 - (x,x_0)] & \text{for} \quad t \in \left[0, \frac{1}{2}\right],\ x \in \mathcal{F}, \\ x_0 + 2(1-t)[x - (x,x_0)x_0] & \text{for} \quad t \in \left(\frac{1}{2}, 1\right],\ x \in \mathcal{F}. \end{cases}$$

The mapping f is continuous and deforms $\mathcal{F}$ to the point $x_0 \in R$. It is sufficient to verify that for any $t \in [0,1]$, $x \in \mathcal{F}$ we have

$$P_1 f(t,x) \neq o.$$

Indeed, for any $t \in \left[0, \frac{1}{2}\right]$ we have

$$P_1 f(t,x) = 2t[1 - (x, x_0)]x_0 + P_1 x,$$

for $t \in \left(\frac{1}{2}, 1\right]$ we have

$$P_1 f(t,x) = [1 - 2(1-t)(x, x_0)]x_0 + 2(1-t)P_1 x.$$

For $t \in [0,1)$ we have then $P_1 f(t,x) \neq o$ due to the assumption $P_1(\mathcal{F}) \cap \operatorname{Lin}\{x_0\} = \emptyset$. For $t = 1$ we have

$$P_1 f(t,x) = x_0 \neq o.$$

□

5.2 Schauder Fixed Point Theorem

The Brouwer Fixed Point Theorem is a very strong device for solving finite-dimensional nonlinear equations. Unfortunately, it does not hold in infinite dimensions as the following example shows.

Example 5.2.1 (Kakutani). Let H be a separable Hilbert space with an orthonormal basis $\{e_n\}_{n=1}^{\infty}$. Denote by $A \in \mathcal{L}(H)$ the right shift given by

$$Ae_n = e_{n+1}, \quad \text{i.e.,} \quad A\left(\sum_{n=1}^{\infty} x_n e_n\right) = \sum_{n=1}^{\infty} x_n e_{n+1},$$

and

$$F(x) = (1 - \|x\|^2)^{\frac{1}{2}} e_1 + Ax.$$

Then F is continuous and

$$\|F(x)\|^2 = 1 - \|x\|^2 + \|Ax\|^2 = 1 \qquad \text{for} \quad \|x\| \leq 1.$$

If $x = \sum\limits_{n=1}^{\infty} x_n e_n$ is a fixed point of F, then

$$x_n = x_{n+1} \qquad \text{and} \qquad x_1 = (1 - \|x\|^2)^{\frac{1}{2}}.$$

The series $\sum\limits_{n=1}^{\infty} x_n^2$, with $x_n = x_{n+1}$, is convergent only if $x_n = 0$ for all n, i.e., $\|x\| = 0$. Then

$$x_1 = 1,$$

a contradiction. ○

Notice that in the previous example the apparently simple linear operator A is perturbed by a nonlinear operator with a one-dimensional range. Continuous operators with the range in a finite-dimensional subspace form an important special subclass of the so-called *(nonlinear) compact operators*.

Definition 5.2.2. Let X, Y be normed linear spaces and let $\mathcal{M} \subset X$. A mapping $F\colon \mathcal{M} \to Y$ is called a *compact operator* on $\mathcal{M}$ into Y if F is continuous on $\mathcal{M}$ ($\mathcal{M}$ being a metric space with the metric induced by the norm of X) and $F(\mathcal{M} \cap \mathcal{K})$ is a relatively compact set in Y for any bounded set $\mathcal{K} \subset X$.

The set of all compact operators from $\mathcal{M}$ into Y is denoted by $\mathscr{C}(\mathcal{M}, Y)$. If the range of $F \in \mathscr{C}(\mathcal{M}, Y)$ is a subset of a finite-dimensional subspace of Y, then we say that F is a *finite-dimensional operator* and write $F \in \mathscr{C}_{\mathrm{f}}(\mathcal{M}, Y)$.

We recall that linear compact operators have been investigated in Section 2.2.

Warning. *In contrast to the linear case the continuity of a nonlinear operator F is not a consequence of the fact that F maps bounded sets onto relatively compact ones!* A simple example can be constructed for $F\colon \mathbb{R} \to \mathbb{R}$.

Our interest in compact operators arises from the observation that they are close to finite-dimensional ones. The precise formulation follows.

Theorem 5.2.3. *Let X be a normed linear space, Y a Banach space and let $\mathcal{M}$ be a bounded subset of X.*

(i) *If $F \in \mathscr{C}(\mathcal{M}, Y)$, then there is a sequence $\{F_n\}_{n=1}^{\infty} \subset \mathscr{C}_{\mathrm{f}}(\mathcal{M}, Y)$ which converges to F uniformly on $\mathcal{M}$.*

(ii) *If $\{F_n\}_{n=1}^{\infty} \subset \mathscr{C}(\mathcal{M}, Y)$ and $\lim\limits_{n\to\infty} F_n(x) = F(x)$ uniformly for $x \in \mathcal{M}$, then $F \in \mathscr{C}(\mathcal{M}, Y)$.*

Proof. (i) Since $\overline{F(\mathcal{M})}$ is compact there is a finite $\frac{1}{n}$-net $y_1, \dots, y_m \in F(\mathcal{M})$ of $F(\mathcal{M})$ (Proposition 1.2.3). Functions $\varphi_k(x) = \max\left\{0, \frac{1}{n} - \|F(x) - y_k\|\right\}$ are continuous on $\mathcal{M}$ and $\sum\limits_{i=1}^{m} \varphi_i(x) > 0$ for every $x \in \mathcal{M}$. Therefore the functions

$$\mu_k(x) := \frac{\varphi_k(x)}{\sum\limits_{i=1}^{m} \varphi_i(x)}, \qquad k = 1, \dots, m,$$

form a continuous partition of unity on $\mathcal{M}$. Put $F_n(x) = \sum\limits_{k=1}^{m} \mu_k(x) y_k$, $x \in \mathcal{M}$. Then $F_n \in \mathscr{C}_{\mathrm{f}}(\mathcal{M}, Y)$ and

$$\|F(x) - F_n(x)\| \le \sum_{k=1}^{m} \mu_k(x) \|F(x) - y_k\| < \frac{1}{n} \qquad \text{for every} \quad x \in \mathcal{M}.$$

(ii) If we literally translate the classical proof for vector functions we see that F is continuous on $\mathcal{M}$. Let $n \in \mathbb{N}$ be such that

$$\sup_{x \in \mathcal{M}} \|F(x) - F_n(x)\| < \varepsilon$$

and $y_1, \dots, y_k$ is an ε-net for $F_n(\mathcal{M})$. Then it is also a 2ε-net for $F(\mathcal{M})$. Since Y is a Banach space, Proposition 1.2.3 shows that $\overline{F(\mathcal{M})}$ is compact. □

Remark 5.2.4. The assertion (i) of Theorem 5.2.3 obviously holds for linear compact operators, but generally we cannot guarantee linearity of the approximating sequence $\{F_n\}_{n=1}^{\infty}$ (see Remark 2.2.7).

The following theorem is a generalization of the Brouwer Fixed Point Theorem into the infinite-dimensional setting.

Theorem 5.2.5 (Schauder Fixed Point Theorem). *Let $\mathcal{K}$ be a nonempty, closed, convex and bounded subset of a normed linear space X. Assume that $F \in \mathscr{C}(\mathcal{K}, X)$ and $F(\mathcal{K}) \subset \mathcal{K}$. Then there is a fixed point of F in $\mathcal{K}$.*

Proof. Let $\{F_n\}_{n=1}^{\infty}$ be the sequence constructed in the proof of Theorem 5.2.3(i). Denote

$$X_n := \operatorname{Lin}\{y_1, \dots, y_m\}.$$

Since $y_1, \dots, y_m \in F(\mathcal{K})$ and $\mathcal{K}$ is convex, we have

$$F_n(\mathcal{K}) \subset \mathcal{K} \cap X_n.$$

The restriction of F_n to $\mathcal{K} \cap X_n$ satisfies the assumptions of the Brouwer Fixed Point Theorem and hence there is $x_n \in \mathcal{K} \cap X_n$ such that

$$F_n(x_n) = x_n.$$

By the compactness of F there is a subsequence $\{F(x_{n_k})\}_{k=1}^{\infty}$ which converges to an $x \in \overline{F(\mathcal{K})} \subset \overline{\mathcal{K}} = \mathcal{K}$. The estimate

$$\|F(x_{n_k}) - x_{n_k}\| = \|F(x_{n_k}) - F_{n_k}(x_{n_k})\| < \frac{1}{n_k}$$

implies that also $\lim\limits_{k\to\infty} x_{n_k} = x$. Since F is continuous, we conclude that

$$\lim_{k\to\infty} F(x_{n_k}) = F(x) \qquad \text{and} \qquad F(x) = x. \qquad \square$$

Remark 5.2.6. The above proof of Theorem 5.2.5 is based on the approximation of F by $F_n \in \mathscr{C}_{\mathrm{f}}(\mathcal{K}, X)$. The construction in the proof of Theorem 5.2.3(i) is surely not unique. We recommend to the reader to think about a possible simplification when F acts on a separable Hilbert space.

Another possibility occurs when $\mathcal{K}$ is a compact convex set. We obtain a typical situation as soon as X is a reflexive Banach space and $\mathcal{K}$ is a closed, convex and bounded subset of X. Then $\mathcal{K}$ is compact in the weak topology (Remark 2.1.24 and Theorem 2.1.25)[6] and the continuity of $F\colon \mathcal{K} \to \mathcal{K}$ in the weak topology (it sends weakly convergent sequences into weakly convergent ones) is sufficient to justify application of Theorem 5.2.5.

A slightly more general statement was proved by A. N. Tikhonov (for a proof see, e.g., Dugundji [52, Appendix 1] and Deimling [42, § 10.3]).

[6] We also have to use that a convex set which is closed in the norm topology is also weakly closed (cf. Exercise 2.1.40).

We now show how the Schauder Fixed Point Theorem can be applied to differential equations. To avoid technical details we restrict ourselves to ordinary differential equations. For partial differential equations see Section 5.6. Their solutions are generally smooth which suggests a relation to compact operators.

Proposition 5.2.7. *Let $\mathcal{G}$ be an open subset of $\mathbb{R}^{N+1}$ and let $f\colon \mathcal{G} \to \mathbb{R}^N$ be continuous on $\mathcal{G}$. Then for any $(t_0, x_0) \in \mathcal{G}$ there exists $\delta > 0$ such that the equation*

$$\dot{x} = f(t, x)$$

has a solution on the interval $(t_0-\delta, t_0+\delta)$ which satisfies the condition $x(t_0) = x_0$.

Proof. It has been shown in Lemma 3.1.5 that the initial value problem is equivalent to the integral equation

$$F(x)(t) := x_0 + \int_{t_0}^{t} f(s, x(s))\, \mathrm{d}s = x(t) \tag{5.2.1}$$

in the space $C[t_0 - \delta, t_0 + \delta]$. Choose $\delta > 0$, $r > 0$ such that

$$\mathcal{M} = [t_0 - \delta, t_0 + \delta] \times \overline{B(x_0; r)} \subset \mathcal{G}$$

($\overline{B(x_0; r)}$ is the closed ball in $\mathbb{R}^N$ of radius r centered at x_0). Then $\mathcal{M}$ is a compact set in $\mathbb{R}^{N+1}$, and therefore f is bounded on $\mathcal{M}$, say

$$\|f(t, x)\|_{\mathbb{R}^N} \le c \qquad \text{for} \quad (t, x) \in \mathcal{M}.$$

Then

$$\|F(x) - x_0\|_{C[t_0-\delta, t_0+\delta]} \le c\delta \le r$$

for

$$x \in \mathcal{K} := \{y \in C[t_0 - \delta, t_0 + \delta] : \|y - x_0\|_{C[t_0-\delta, t_0+\delta]} \le r\}$$

provided δ is sufficiently small. This proves that $F(\mathcal{K}) \subset \mathcal{K}$. Since f is also uniformly continuous on $\mathcal{M}$, the operator F is continuous on $\mathcal{K}$ (the convergence on $\mathcal{K}$ is the uniform convergence). Further, for $t, s \in [t_0 - \delta, t_0 + \delta]$, $t < s$, $x \in \mathcal{K}$, we have

$$\|F(x)(t) - F(x)(s)\|_{\mathbb{R}^N} \le \int_t^s \|f(\sigma, x(\sigma))\|_{\mathbb{R}^N}\, \mathrm{d}\sigma \le c|s - t|.$$

This means that $F(\mathcal{K})$ is equicontinuous. By Theorem 1.2.13, $F(\mathcal{K})$ is relatively compact on $C[t_0-\delta, t_0+\delta]$. It follows from Theorem 5.2.5 that the equation (5.2.1) has a solution. □

Our second example concerns a boundary value problem for an ordinary differential equation.

Example 5.2.8. Let f be a continuous function on $[0,1] \times \mathbb{R}$. We wish to solve the equation

$$\ddot{x}(t) = f(t, x(t)) \tag{5.2.2}$$

with the Dirichlet boundary conditions

$$x(0) = x(1) = 0. \tag{5.2.3}$$

We have dealt with this problem already in Example 2.3.10. It has been proved there that y is a solution of this problem if and only if it is continuous and satisfies the integral equation (f is assumed to be continuous)

$$F(y)(t) := \int_0^1 G(t,s) f(s, y(s))\, \mathrm{d}s = y(t) \tag{5.2.4}$$

where the Green function G is given by

$$G(t,s) = \begin{cases} s(t-1), & 0 \le s \le t \le 1, \\ t(s-1), & 0 \le t \le s \le 1. \end{cases}$$

The operator F maps $C[0,1]$ into itself (actually into $C^2[0,1]$) and is compact. This can be proved by two types of argument:

(i) For any $R > 0$ there is $c(R)$ such that

$$|f(s,y)| \le c(R) \qquad \text{for} \quad s \in [0,1],\ |y| \le R.$$

Since

$$\frac{\mathrm{d}^2}{\mathrm{d}t^2} F(y)(t) = f(t, y(t)),$$

F maps the ball $\overline{B(o;R)}$ in $C[0,1]$ into the set of functions which have uniformly bounded second derivatives. Thus $F(\overline{B(o;R)})$ is relatively compact in $C[0,1]$ (see Theorem 1.2.13).

(ii) The operator F is a composition of a linear integral operator and a Nemytski operator (see Example 3.2.21). The Nemytski operator

$$\Phi \colon y \mapsto f(\cdot, y(\cdot))$$

is continuous from $C[0,1]$ into itself, and the integral operator

$$K \colon x \mapsto \int_0^1 G(\cdot, s) x(s)\, \mathrm{d}s$$

is compact from $C[0,1]$ into itself (Example 2.2.5). Therefore $F = K \circ \Phi$ is also compact.

It remains to prove that F maps a ball $\overline{B(o;R)} \subset C[0,1]$ into itself. For this purpose some growth assumptions on f are needed. If

$$|f(s,y)| \leq a + b|y| \qquad \text{for} \quad s \in [0,1],\ y \in \mathbb{R},$$

then

$$\|F(y)\|_{C[0,1]} \leq [a + b\|y\|_{C[0,1]}] \sup_{t\in[0,1]} \int_0^1 |G(t,s)|\,\mathrm{d}s \leq \frac{a + b\|y\|_{C[0,1]}}{8}.$$

Whenever $b < 8$, $R \geq \frac{a}{8-b}$, then F maps $\overline{B(o;R)}$ into itself and (5.2.4) has a solution. ○

Exercise 5.2.9. If f has a sublinear growth in y, i.e., there is $\alpha \in [0,1)$ such that

$$|f(s,y)| \leq a + b|y|^{\alpha},$$

then no restriction on b is needed. Prove this fact!

Exercise 5.2.10. Regard the operator F given by (5.2.4) as an operator on a space of integrable functions. Repeat the argument from Example 5.2.8.

Exercise 5.2.11. Let f in (5.2.2) depend also on $\dot{x}(t)$, i.e., $f = f(t, x(t), \dot{x}(t))$. Formulate assumptions on $f(x,y,z)$ to get the existence of a solution of the boundary value problem (5.2.2), (5.2.3). See also Example 5.8.5.

Exercise 5.2.12. Let K be a bounded continuous real function on $[a,b] \times [a,b] \times \mathbb{R}$ and let $h \in C[a,b]$. Prove that the integral equation

$$x(t) = \int_a^b K(t,\tau,x(\tau))\,\mathrm{d}\tau + h(t)$$

has at least one solution $x \in C[a,b]$.

5.2A Fixed Point Theorems for Noncompact Operators

There are many generalizations of the Schauder Fixed Point Theorem. We mention here one which shows that the assumption of compactness of the operator can be relaxed. However, having in mind Example 5.2.1 this must be done carefully and more than continuity of the operator must be required. To this purpose we need a tool which will measure "how much non-compact" the operator actually is.

Definition 5.2.13. Let $\mathcal{M}$ be a bounded set in a metric space (X,ϱ). The *Kuratowski measure of noncompactness* $\chi(\mathcal{M})$ is defined to be the infimum of the set of all numbers $d > 0$ with the property that

(KM) $\mathcal{M}$ can be covered by finitely many sets, each of whose diameters[7] is less than or equal to d.

[7] The diameter of $\mathcal{M}$ is defined as $\operatorname{diam} \mathcal{M} := \sup \varrho(x,y)$ where the supremum is taken over all $x, y \in \mathcal{M}$.

If X is complete, then it follows from Proposition 1.2.3 that $\mathcal{M}$ is relatively compact if and only if (KM) holds for every $d > 0$. Therefore $\chi(\mathcal{M}) = 0$ is equivalent to relative compactness of $\mathcal{M}$. If the value of $\chi(\mathcal{M})$ increases, $\mathcal{M}$ deviates more strongly (in the sense of condition (KM)) from relatively compact sets.

Proposition 5.2.14 (Properties of the Kuratowski measure of noncompactness). *Let X be a (real or complex) Banach space. Then for all bounded subsets $\mathcal{M}$, $\mathcal{M}_1, \dots, \mathcal{M}_n$, $\mathcal{N}$ of X the following assertions hold:*

(i) $\chi(\emptyset) = 0$;

(ii) $\chi(\mathcal{M}) = 0 \iff \mathcal{M}$ *is relatively compact;*

(iii) $0 \le \chi(\mathcal{M}) \le \operatorname{diam} \mathcal{M}$;

(iv) $\mathcal{M} \subset \mathcal{N} \implies \chi(\mathcal{M}) \le \chi(\mathcal{N})$;

(v) $\chi(\mathcal{M} + \mathcal{N}) \le \chi(\mathcal{M}) + \chi(\mathcal{N})$;[8]

(vi) $\chi(\beta\mathcal{M}) = |\beta|\chi(\mathcal{M})$ *for all* $\beta \in \mathbb{R}$ *(or* $\mathbb{C}$*);*

(vii) $\chi(\mathcal{M}) = \chi(\overline{\mathcal{M}})$;

(viii) $\chi\left(\bigcup_{i=1}^{n} \mathcal{M}_i\right) = \max\{\chi(\mathcal{M}_1), \dots, \chi(\mathcal{M}_n)\}$;

(ix) $\chi(\mathcal{M}) = \chi(\operatorname{Co} \mathcal{M})$, *where* $\operatorname{Co} \mathcal{M}$ *(the* convex hull *of* $\mathcal{M}$*) is the least convex set containing* $\mathcal{M}$.

Proof. The properties (i)–(vii) follow directly from Definition 5.2.13, and so the proof is left to the reader. Let us prove (viii). Set

$$\mathcal{M} = \bigcup_{i=1}^{n} \mathcal{M}_i \qquad \text{and} \qquad a = \max\{\chi(\mathcal{M}_1), \dots, \chi(\mathcal{M}_n)\}.$$

Then it follows from $\mathcal{M}_i \subset \mathcal{M}$ and from (iv) that $\chi(\mathcal{M}_i) \le \chi(\mathcal{M})$, so $a \le \chi(\mathcal{M})$. To prove the equality, chooose $\varepsilon > 0$ and a covering $\{\mathcal{M}_i^1, \mathcal{M}_i^2, \dots, \mathcal{M}_i^{m_i}\}$ of $\mathcal{M}_i$ with

$$\operatorname{diam} \mathcal{M}_i^j \le \chi(\mathcal{M}_i) + \varepsilon \le a + \varepsilon.$$

All of these $\mathcal{M}_i^j$ form a covering of $\mathcal{M}$, so that

$$\chi(\mathcal{M}) \le a + \varepsilon, \qquad \text{i.e., since } \varepsilon > 0 \text{ is arbitrary, } \chi(\mathcal{M}) \le a.$$

Hence, $\chi(\mathcal{M}) = a$ and (viii) is proved.

Finally, we prove (ix). It follows from $\mathcal{M} \subset \operatorname{Co} \mathcal{M}$ and (iv) that

$$\chi(\mathcal{M}) \le \chi(\operatorname{Co} \mathcal{M}).$$

Conversely, we show that $\chi(\operatorname{Co} \mathcal{M}) \le \chi(\mathcal{M})$. This will be done in three steps.

[8] $\mathcal{M} + \mathcal{N} := \{z = x + y : x \in \mathcal{M}, y \in \mathcal{N}\}$.

Step 1. We prove inequality (5.2.5) below. For every $\varepsilon > 0$ there exists a covering $\mathcal{M} \subset \bigcup_{i=1}^{N} \mathcal{M}_i$ with $\operatorname{diam} \mathcal{M}_i \le \chi(\mathcal{M}) + \varepsilon$. Since $\operatorname{diam}(\operatorname{Co} \mathcal{M}_i) = \operatorname{diam} \mathcal{M}_i$,[9] we may assume that $\mathcal{M}_i$ are all convex. Let

$$\Lambda := \left\{ \lambda = (\lambda_1, \dots, \lambda_N) \in \mathbb{R}^N : \sum_{i=1}^{N} \lambda_i = 1,\ \lambda_i \ge 0 \quad \text{for all} \quad i \right\}$$

and

$$\mathcal{A}(\lambda) := \sum_{i=1}^{N} \lambda_i \mathcal{M}_i \qquad \text{for all} \quad \lambda = (\lambda_1, \dots, \lambda_N) \in \Lambda.$$

Now it follows from (iv), (v) and (vi) that

$$\chi(\mathcal{A}(\lambda)) \le \sum_{i=1}^{N} \lambda_i \chi(\mathcal{M}_i) \le \chi(\mathcal{M}) + \varepsilon. \tag{5.2.5}$$

Step 2. We show that the union $\bigcup_{\lambda \in \Lambda} \mathcal{A}(\lambda)$ is a convex set. Indeed, let

$$x = \sum_{i=1}^{N} \lambda_i x_i, \qquad y = \sum_{i=1}^{N} \mu_i y_i, \qquad t \in [0,1] \quad \text{and} \quad z = tx + (1-t)y$$

where $\lambda, \mu \in \Lambda$ and $x_i, y_i \in \mathcal{M}_i$ for all i. The point z can be represented in the form

$$z = \sum_{i=1}^{N} \xi_i z_i \quad \text{where } \xi_i = t\lambda_i + (1-t)\mu_i,\ z_i = \varrho_i x_i + (1-\varrho_i) y_i,\ \varrho_i = \begin{cases} t\dfrac{\lambda_i}{\xi_i}, & \text{for } \xi_i > 0, \\ 0, & \text{for } \xi_i = 0. \end{cases}$$

By definition of ξ we have $0 \le \varrho_i \le 1$. The set $\mathcal{M}_i$ is convex, so $z_i \in \mathcal{M}_i$. Moreover, $\xi \in \Lambda$, by the convexity of Λ. Hence $z \in \mathcal{A}(\xi)$.

Step 3. We prove that

$$\chi(\operatorname{Co} \mathcal{M}) \le \chi(\mathcal{M}) + 3\varepsilon.$$

Since the set Λ is compact, for a given $\varepsilon > 0$ we can find finitely many points $\lambda^{(1)}, \dots,$ $\lambda^{(m)} \in \Lambda$ such that for any $x = \sum_{i=1}^{N} \lambda_i x_i \in \mathcal{A}(\lambda)$ there exists $\lambda^{(j)} = \left(\lambda_1^{(j)}, \dots, \lambda_N^{(j)}\right)$ for which

$$\left\| x - \sum_{i=1}^{N} \lambda_i^{(j)} x_i \right\| \le \max_{i=1,\dots,N} \left| \lambda_i - \lambda_i^{(j)} \right| \max_{i=1,\dots,N} |x_i| \le \frac{\varepsilon}{k} k = \varepsilon$$

where $k > 0$ is a common bound for all sets $\mathcal{M}_i$. Therefore,

$$\bigcup_{\lambda \in \Lambda} \mathcal{A}(\lambda) \subset \bigcup_{j=1}^{m} \mathcal{A}\left(\lambda^{(j)}\right) + B(o; \varepsilon).$$

So, by Step 2, we have $\operatorname{Co} \mathcal{M} \subset \bigcup_{\lambda \in \Lambda} \mathcal{A}(\lambda)$ and by the other statements and (5.2.5),

$$\chi(\operatorname{Co} \mathcal{M}) \le \chi\left(\bigcup_{\lambda \in \Lambda} \mathcal{A}(\lambda) \right) \le \chi\left(\bigcup_{j=1}^{m} \mathcal{A}\left(\lambda^{(j)}\right) + B(o; \varepsilon) \right)$$

[9] The reader is invited to prove this equality.

$$\leq \chi\left(\bigcup_{j=1}^{m} \mathcal{A}\left(\lambda^{(j)}\right)\right) + 2\varepsilon \leq \max\left\{\chi\left(\mathcal{A}\left(\lambda^{(1)}\right)\right), \dots, \chi\left(\mathcal{A}\left(\lambda^{(m)}\right)\right)\right\} + 2\varepsilon$$

$$\leq \chi(\mathcal{M}) + 3\varepsilon,$$

i.e., since $\varepsilon > 0$ is arbitrary,

$$\chi(\operatorname{Co}\mathcal{M}) \leq \chi(\mathcal{M}). \qquad \square$$

Example 5.2.15. Let $B(o;1) \subset X$ be the open unit ball in a Banach space X. If $\dim X < \infty$, then

$$\chi(B(o;1)) = \chi(\overline{B(o;1)}) = \chi(\partial B(o;1)) = 0$$

(see Proposition 1.2.3). On the other hand, if $\dim X = \infty$, then

$$\chi(B(o;1)) = \chi(\overline{B(o;1)}) = \chi(\partial B(o;1)) = 2. \tag{5.2.6}$$

The proof of this fact is not trivial. Since the diameter of $\overline{B(o;1)}$ is equal to 2, we know that $\chi(\overline{B(o;1)}) \leq 2$. In order to prove (5.2.6) we show that $\chi(\partial B(o;1)) \geq 2$. Assume the contrary. Then there exist sets $\mathcal{M}_i$ with

$$\partial B(o;1) = \bigcup_{i=1}^{n} \mathcal{M}_i$$

and the diameter of every $\mathcal{M}_i$ is strictly less than 2. We may take all $\mathcal{M}_i$s to be closed. Let $X_n \subset X$ be a subspace of X such that $\dim X_n = n$. Then we have

$$\partial B(o;1) \cap X_n = \bigcup_{i=1}^{n} (\mathcal{M}_i \cap X_n).$$

The sets $\mathcal{M}_i \cap X_n$, $i = 1, 2, \dots, n$, cover the closed unit sphere $\partial B(o;1) \cap X_n$ in X_n. By the result of Lusternik and Schnirelmann (see Exercise 5.7.40 below) there exists $\mathcal{M}_j$ such that $\mathcal{M}_j \cap X_n$ contains an antipodal pair $\{x, -x\}$. Consequently,

$$2 \leq \operatorname{diam}(\mathcal{M}_j \cap X_n) \leq \operatorname{diam}\mathcal{M}_j,$$

which is a contradiction. Finally, by (iv) and (vii) of Proposition 5.2.14 we have (5.2.6). ○

In the next definition we will consider a special class of continuous and bounded operators.

Definition 5.2.16. Let $T\colon M \subset X \to X$ be a bounded operator[10] from a Banach space X into itself. The operator T is called a *k-set contraction* if there is a number $k \geq 0$ such that

$$\chi(T(\mathcal{M})) \leq k\chi(\mathcal{M})$$

for all bounded sets $\mathcal{M}$ in M.

The bounded operator T is called *condensing* if

$$\chi(T(\mathcal{M})) < \chi(\mathcal{M})$$

for all bounded sets $\mathcal{M}$ in M with $\chi(\mathcal{M}) > 0$.

[10] The operator T is said to be *bounded* on M if $T(M \cap \mathcal{A})$ is a bounded set provided $\mathcal{A}$ is a bounded set.

Obviously, every k-set contraction for $0 \le k < 1$ is condensing. Every compact map T is a k-set contraction with $k = 0$. A typical example of a k-set contraction with $0 \le k < 1$ is the following one.

Example 5.2.17. Let $K, C \colon \mathcal{D} \subset X \to X$ be operators on a Banach space X. Let K be a k-contractive, i.e., there exists $k \in [0, 1)$ such that

$$\|K(x) - K(y)\| \le k\|x - y\| \qquad \text{for all} \quad x, y \in \mathcal{D}, \tag{5.2.7}$$

and let C be compact. Then $K + C$ is a k-set contraction. Indeed, let $\mathcal{M} \subset \mathcal{D}$ be a bounded set. By Definition 5.2.13 it follows from (5.2.7) that $\chi(K(\mathcal{M})) \le k\chi(\mathcal{M})$. By (ii) of Proposition 5.2.14 we have $\chi(C(\mathcal{M})) = 0$. Set $T := K + C$. Now (iv) and (v) of Proposition 5.2.14 imply

$$\chi(T(\mathcal{M})) \le \chi(K(\mathcal{M}) + C(\mathcal{M})) \le \chi(K(\mathcal{M})) + \chi(C(\mathcal{M})) \le k\chi(\mathcal{M}). \qquad \circ$$

The following assertion is a generalization of the Schauder Fixed Point Theorem (note that every compact operator is condensing).

Theorem 5.2.18 (Darbo). *Let us suppose that*

(i) *$\mathcal{M}$ is a nonempty, closed, bounded and convex subset of a Banach space X;*

(ii) *an operator $T \colon \mathcal{M} \subset X \to \mathcal{M}$ is condensing and continuous on $\mathcal{M}$.*

Then T has a fixed point in $\mathcal{M}$.

Proof. The idea of the proof is to find a suitable subset $\mathcal{A}$ of $\mathcal{M}$ which is mapped into itself by T in such a way that the Schauder Fixed Point Theorem can be applied to the restriction $T \colon \mathcal{A} \to \mathcal{A}$. The resulting fixed point is then trivially a fixed point of the original mapping $T \colon \mathcal{M} \to \mathcal{M}$. The set $\mathcal{A}$ is constructed in the following way. Choose a point $m \in \mathcal{M}$ and let Σ denote the system of all closed, convex subsets $\mathcal{K}$ of $\mathcal{M}$ for which $m \in \mathcal{K}$ and $T(\mathcal{K}) \subset \mathcal{K}$. Set

$$\mathcal{A} = \bigcap_{\mathcal{K} \in \Sigma} \mathcal{K} \qquad \text{and} \qquad \mathcal{C} = \overline{\operatorname{Co}\{T(\mathcal{A}) \cup \{m\}\}}.$$ [11]

Since $m \in \mathcal{A}$ and $T(\mathcal{A}) \subset \mathcal{A}$, it follows that $\mathcal{C} \subset \mathcal{A}$. This implies $T(\mathcal{C}) \subset T(\mathcal{A})$. Obviously $T(\mathcal{A}) \subset \mathcal{C}$, i.e., $T(\mathcal{C}) \subset \mathcal{C}$ which means that $\mathcal{C} \in \Sigma$. So, $\mathcal{A} \subset \mathcal{C}$. We have proved that $\mathcal{A} = \mathcal{C}$. Now, (vii), (viii) and (ix) of Proposition 5.2.14 imply that

$$\chi(\mathcal{A}) = \chi(\mathcal{C}) = \chi(T(\mathcal{A})). \tag{5.2.8}$$

Since T is condensing,

$$\chi(\mathcal{A}) = 0.$$

Since $\mathcal{A}$ is also closed, $\mathcal{A}$ is a compact set. The restriction of T to $\mathcal{A}$ is thus a compact operator. Consequently, the Schauder Fixed Point Theorem can be applied to the mapping $T \colon \mathcal{A} \to \mathcal{A}$. $\square$

Corollary 5.2.19. *Let $K, C \colon \mathcal{M} \subset X \to X$ be operators in a Banach space X such that $(K + C)(\mathcal{M}) \subset \mathcal{M}$, let $\mathcal{M}$ be a nonempty, closed, bounded and convex set in X, let K be k-contractive $(0 \le k < 1)$ and C compact. Then $K + C$ has a fixed point in $\mathcal{M}$.*

[11] Observe that $\Sigma \neq \emptyset$ because $\mathcal{M} \in \Sigma$, and $\mathcal{A} \neq \emptyset$ because $m \in \mathcal{A}$.

Proof. The proof follows immediately from Example 5.2.17 and Theorem 5.2.18. □

The following assertion generalizes the existence part of Theorem 3.1.4 and follows from the previous Corollary 5.2.19, cf. the statement with the example on page 114. Let us consider the initial value problem

$$\begin{cases} \dot{x} = f(t,x) + g(t,x), \\ x(t_0) = x_0 \end{cases} \tag{5.2.9}$$

in a Banach space Y. For fixed positive numbers a and b define

$$\mathcal{R} := [t_0 - a, t_0 + a] \times [x \in Y : \|x - x_0\| \le b\}.$$

Proposition 5.2.20. *Let us assume that*

(i) *the map* $f\colon \mathcal{R} \to Y$ *is continuous and also Lipschitz continuous with respect to the second variable, i.e., there exists* $L > 0$ *such that*

$$\|f(t,x) - f(t,y)\| \le L\|x - y\| \qquad \text{for all} \quad (t,x),(t,y) \in \mathcal{R};$$

(ii) *the map* $g\colon \mathcal{R} \to Y$ *is compact;*

(iii) *the sum* $f + g$ *is bounded, i.e., there exists* $B > 0$ *such that*

$$\|f(t,y) + g(t,y)\| \le B \qquad \text{for all} \quad (t,y) \in \mathcal{R};$$

(iv) *the number* $c > 0$ *is chosen such that*

$$cL < 1, \qquad c \le a, \qquad Bc \le b.$$

Then the problem (5.2.9) *has a solution* $x = x(t)$ *defined on* $(t_0 - c, t_0 + c)$.

Proof. It follows from Lemma 3.1.5 that the problem (5.2.9) is equivalent to the integral equation

$$x(t) = x_0 + \int_{t_0}^{t} [f(s,x(s)) + g(s,x(s))]\,\mathrm{d}s. \tag{5.2.10}$$

Let

$$X = C([t_0 - c, t_0 + c], Y) \qquad \text{and} \qquad \mathcal{M} = \{x \in X : \|x - x_0\|_X \le b\}.$$[12]

Then (5.2.10) can be regarded as the operator equation

$$x = K(x) + C(x), \qquad x \in \mathcal{M}, \tag{5.2.11}$$

where

$$K(x)(t) = x_0 + \int_{t_0}^{t} f(s,x(s))\,\mathrm{d}s, \quad C(x)(t) = \int_{t_0}^{t} g(s,x(s))\,\mathrm{d}s.$$

Similarly to the proof of Theorem 3.1.4 we obtain that

$$(K + C)(\mathcal{M}) \subset \mathcal{M}.$$

Furthermore, the operator K is k-contractive with $k = Lc$ and the operator C is compact. So, Corollary 5.2.19 yields the existence of a solution of (5.2.11), hence of (5.2.10), and thus, ultimately, of (5.2.9). □

[12] Here $x_0 \in X$ is understood to be a constant function defined on $[t_0 - c, t_0 + c]$ with value in Y.

A similar approach can be also used for functional differential equations. In the following example we describe a simple situation. For a more general treatment of evolution equations see, e.g., Milota & Petzeltová [113].

Example 5.2.21. Consider a system of ordinary functional differential equations

$$\dot{x}(t) = \boldsymbol{A}(t)x(t) + \int_0^t f(t,s,x(s))\,\mathrm{d}s, \qquad x(0) = x_0, \tag{5.2.12}$$

where $\boldsymbol{A}$ is an $N \times N$-matrix with continuous entries on the interval $[0,T]$, $f\colon M := ([0,T] \times [0,T] \times \mathbb{R}^N) \to \mathbb{R}^N$ is continuous and locally α-Hölder continuous with respect to the first variable and locally satisfies the Lipschitz condition with respect to the third variable, i.e., for any $(t_0, s_0, x_0) \in M$ there are a neighborhood $\mathcal{U}$ of this point and constants $c > 0$, $L > 0$, $\alpha \in (0,1)$ such that

$$|f(t_1,s,x_1) - f(t_2,s,x_2)| \le c|t_1 - t_2|^\alpha + L|x_1 - x_2| \qquad \text{for} \quad (t_i,s,x_i) \in \mathcal{U},\ i = 1,2.$$

Instead of (5.2.12) we consider the equivalent integral equation

$$x(t) = H(x)(t) := \boldsymbol{\Phi}(t,0)x_0 + \int_0^t \boldsymbol{\Phi}(t,s)\int_0^s f(s,\sigma,x(\sigma))\,\mathrm{d}\sigma\,\mathrm{d}s$$

where $\boldsymbol{\Phi}(t)$ is a fundamental matrix of the equation

$$\dot{x}(t) = \boldsymbol{A}(t)x(t).$$

We put

$$F(s,x) := \int_0^s f(s,\sigma,x(\sigma))\,\mathrm{d}\sigma, \qquad s \in [0,T], \quad x \in C([0,T],\mathbb{R}^N)$$

and

$$G_1(x)(t) = \int_0^t \boldsymbol{\Phi}(t,s)[F(s,x) - F(t,x)]\,\mathrm{d}s, \qquad G_2(x)(t) = \int_0^t \boldsymbol{\Phi}(t,s)\,\mathrm{d}s\,F(t,x).$$

It is not difficult to show that there are $r > 0$, $\tau > 0$ small enough such that H maps the set

$$\mathcal{Q}(r,\tau) := \left\{ y \in C([0,\tau],\mathbb{R}^N) : \sup_{t\in[0,\tau]} |y(t) - x_0| \le r \right\}$$

into itself, G_1 is a compact mapping (using the Arzelà–Ascoli Theorem) on $\mathcal{Q}(r,\tau)$ and G_2 is a contraction on $\mathcal{Q}(r,\tau)$. The local existence of a solution of (5.2.12) follows now from Corollary 5.2.19. This local solution can be continuously extended. To keep the time step τ fixed it is sufficient to assume that f satisfies the global Lipschitz condition with respect to x-variable on the whole domain M. ○

Exercise 5.2.22. Prove Proposition 5.2.14(i)–(vii).

Exercise 5.2.23. Consider the boundary value problem

$$\begin{cases} \ddot{x}(t) = f(t,x(t),\dot{x}(t)), & t \in (0,1), \\ x(0) = x(1) = 0, \end{cases} \tag{5.2.13}$$

where $f\colon \mathbb{R}^3 \to \mathbb{R}$ is a real function. Find conditions on f and apply Corollary 5.2.19 to prove the existence of a solution to (5.2.13).

Hint. Look for conditions which guarantee that the Nemytski operator given by f is a sum of a contraction and a compact operator.

Exercise 5.2.24. Let H and $\mathcal{Q}(r,\tau)$ be as in Example 5.2.21. Prove that H maps the set $\mathcal{Q}(r,\tau)$ into itself.

Exercise 5.2.25. Let G_1, G_2 and $\mathcal{Q}(r,\tau)$ be as in Example 5.2.21. Prove that G_1 is a compact mapping on $\mathcal{Q}(r,\tau)$ and G_2 is a contraction on $\mathcal{Q}(r,\tau)$.

5.3 Classical Solutions of PDEs, Functional Setting

In this section we will explain the notion of the classical solution of a semilinear problem with the Laplace operator and explain what is the "right" functional setting for it. Let Ω be an open bounded subset of $\mathbb{R}^N$ and let $u\colon \Omega \to \mathbb{R}$ be a real smooth function. We will denote by

$$\Delta u(x) := \frac{\partial^2 u(x)}{\partial x_1^2} + \frac{\partial^2 u(x)}{\partial x_2^2} + \cdots + \frac{\partial^2 u(x)}{\partial x_N^2}, \qquad x = (x_1,\dots,x_N) \in \Omega$$

the *Laplace operator* defined in Ω. Let

$$g\colon \Omega \times \mathbb{R} \to \mathbb{R}$$

be a continuous real function. We will study the Dirichlet boundary value problem

$$\begin{cases} -\Delta u(x) = g(x,u(x)) & \text{in} \quad \Omega, \\ u = 0 & \text{on} \quad \partial\Omega \end{cases} \tag{5.3.1}$$

and look for its *classical solution.* Following the definition of the classical solution for the ordinary differential equation it should be a function $u \in C^2(\Omega) \cap C(\overline{\Omega})$ such that $u(x) = 0$ for every $x \in \partial\Omega$ and the equation $-\Delta u(x) = g(x,u(x))$ is satisfied at every point $x \in \Omega$. Let us explain why *this is not a suitable definition* of the solution for partial differential equations.

In order to apply the methods of functional analysis we need an operator representation of the Laplace operator subject to the Dirichlet boundary conditions. For this purpose we need the fact that for any $f \in C(\overline{\Omega})$ the linear problem

$$\begin{cases} -\Delta u(x) = f(x) & \text{in} \quad \Omega, \\ u = 0 & \text{on} \quad \partial\Omega, \end{cases} \tag{5.3.2}$$

has a unique solution $u \in C^2(\Omega) \cap C(\overline{\Omega})$. However, not for every $f \in C(\overline{\Omega})$ such a solution does exist in general! This fact is nontrivial and can be found in the book Gilbarg & Trudinger [71, Chapter 4]. So, we have to look for a different concept of the classical solution.

To motivate the definition of the classical solution of (5.3.1) we treat first the linear problem (5.3.2). In order to simplify the notation, in this chapter, denote by $|\cdot|$ the norm in $\mathbb{R}^N$ for any $N \geq 1$. For $\gamma \in (0,1)$ let us consider the space of γ-Hölder continuous functions

$$C^{0,\gamma}(\overline{\Omega}) := \{u \in C(\overline{\Omega}) : \exists K > 0\, \forall x, y \in \Omega : |u(x) - u(y)| \leq K|x-y|^\gamma\}$$

(cf. Example 1.2.25 and Exercise 5.3.4). Let $\alpha = (\alpha_1, \dots, \alpha_N)$ be a multiindex of length

$$|\alpha| = \sum_{i=1}^N \alpha_i,$$

and for the sake of brevity denote

$$D^\alpha u = \frac{\partial^{|\alpha|} u}{\partial x_1^{\alpha_1} \partial x_2^{\alpha_2} \dots \partial x_N^{\alpha_N}} \qquad \left(= \left(\frac{\partial}{\partial x_1}\right)^{\alpha_1} \cdots \left(\frac{\partial}{\partial x_N}\right)^{\alpha_N} u\right)$$

(cf. Section 1.2). Set

$$C^{2,\gamma}(\overline{\Omega}) := \{u \in C^2(\overline{\Omega}) : \forall |\alpha| = 2,\ D^\alpha u \in C^{0,\gamma}(\overline{\Omega})\}$$

(cf. Exercise 5.3.5).

In what follows we will assume that Ω is a bounded *domain* (an open and connected set) *of the class* $C^{k,\gamma}$ ($k \in \mathbb{N} \cup \{0\}$, $\gamma \in (0,1]$), i.e., at each point $x_0 \in \partial\Omega$ there is a ball $B(x_0; \varrho)$ and a one-to-one mapping ψ from $\overline{B(x_0;\varrho)}$ onto $\overline{B(o;1)} \subset \mathbb{R}^N$ such that

(1) $\psi(\overline{B(x_0;\varrho)} \cap \Omega) \subset \mathbb{R}^N_+$; [13]

(2) $\psi(\overline{B(x_0;\varrho)} \cap \partial\Omega) \subset \partial\mathbb{R}^N_+$;

(3) $\psi \in C^{k,\gamma}(\overline{B(x_0;\varrho)})$, $\psi^{-1} \in C^{k,\gamma}(\overline{B(o;1)})$,

see Figure 5.3.1 (cf. Definition 4.3.95 and Exercise 5.3.10). For the sake of brevity we write

$$\Omega \in C^{k,\gamma}.$$

In particular, if $\Omega \in C^{0,1}$, then Ω will be called the *domain with a Lipschitz boundary.*

The following deep result is crucial for the classical setting of the problem (5.3.1).

Theorem 5.3.1. *Let $f \in C^{0,\gamma}(\overline{\Omega})$, $\gamma \in (0,1)$, $\Omega \in C^{2,\gamma}$. Then there exists a unique $u \in C^{2,\gamma}(\overline{\Omega})$ such that $u(x) = 0$, $x \in \partial\Omega$, and the equation in* (5.3.2) *is satisfied at every point $x \in \Omega$. Moreover, there exists $c > 0$ (which depends only on Ω and $\gamma \in (0,1)$) such that*

$$\|u\|_{C^{2,\gamma}(\overline{\Omega})} \leq c\|f\|_{C^{0,\gamma}(\overline{\Omega})}. \tag{5.3.3}$$

[13] $\mathbb{R}^N_+ = \{x \in \mathbb{R}^N : x = (x_1, x_2, \dots, x_N), x_N > 0\}$.

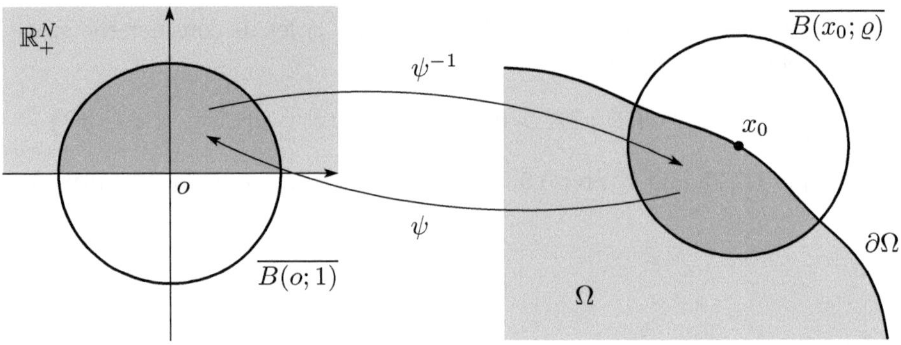

Figure 5.3.1.

Proof. The proof of this assertion can be found in Gilbarg & Trudinger [71, Chapter 6]. □

Note that (5.3.3) is a special case of more general *Schauder estimates* (see, e.g., Gilbarg & Trudinger [71]).

Set

$$X := \{u \in C^{2,\gamma}(\overline{\Omega}) : u(x) = 0, x \in \partial\Omega\} \qquad \text{and} \qquad Y := C^{0,\gamma}(\overline{\Omega}).$$

Then X and Y are Banach spaces (see Exercises 5.3.4 and 5.3.5) and the Arzelà–Ascoli Theorem (see Theorem 1.2.13) implies that the following compact embedding holds true:

$$X \subsetneq\subsetneq Y. \tag{5.3.4}$$

Let us define an operator $L\colon X \to Y$ by

$$(Lu)(x) := -\Delta u(x), \qquad u \in X. \tag{5.3.5}$$

Then L is a linear and bounded operator (Exercise 5.3.6). It follows from Theorem 5.3.1 that

$$L^{-1}\colon Y \to X$$

is a well-defined linear and bounded operator. Actually, the best constant (i.e., the least one) in (5.3.3) is nothing but $c = \|L^{-1}\|_{\mathcal{L}(Y,X)}$.

It follows from the compact embedding (5.3.4) that[14]

$$L^{-1}\colon Y \to Y$$

is a compact operator. Notice, however, that $\|L^{-1}\|_{\mathcal{L}(Y)} \neq \|L^{-1}\|_{\mathcal{L}(Y,X)}$ in general, because

$$\|L^{-1}\|_{\mathcal{L}(Y)} \le c_{\text{emb}}\|L^{-1}\|_{\mathcal{L}(Y,X)}$$

[14]To be precise, we should write $I \circ L^{-1}\colon Y \to Y$ where I is the compact embedding of X into Y. However, for the sake of brevity of notation we drop it.

where c_{emb} is the constant of embedding of X into Y, i.e., the least constant c for which

$$\|u\|_Y \le c\|u\|_X \qquad \text{holds for all} \quad u \in X.$$

This constant depends on Ω and $\gamma \in (0,1)$.

Let us simplify the situation and assume that the nonlinear function $g = g(x,u)$ is given in the special form

$$g(x,u(x)) = g(u(x)) + f(x)$$

(the "variables" u and x are "separated").

In order to find a suitable operator representation for

$$\begin{cases} -\Delta u(x) = g(u(x)) + f(x) & \text{in} \quad \Omega, \\ u = 0 & \text{on} \quad \partial\Omega, \end{cases} \tag{5.3.6}$$

we have to find conditions on g which guarantee that $G\colon Y \to Y$ defined by

$$G(u)(x) := g(u(x)) \tag{5.3.7}$$

is a correctly defined operator (see also Example 3.2.21).

The following nontrivial assertion provides a necessary and sufficient condition on g guaranteeing that the *Nemytski operator* G is continuous from Y into Y.

Lemma 5.3.2 (Drábek [47]). *The operator G defined by* (5.3.7) *maps Y continuously into Y if and only if $g \in C^1(\mathbb{R})$.*[15]

Definition 5.3.3. Let $f \in Y$ and $g \in C^1(\mathbb{R})$. Then $u \in X$ is a *classical solution* of (5.3.6) if the equation in (5.3.6) holds at every point of Ω.

Let us give an operator representation of (5.3.6). Denote

$$G_f(u)(x) := g(u(x)) + f(x).$$

Then $G_f\colon Y \to Y$ is a continuous operator by Lemma 5.3.2 and it maps bounded sets onto bounded sets (see Exercise 5.3.9). The problem (5.3.6) is then equivalent to

$$L(u) = G_f(u) \qquad \text{or} \qquad u = L^{-1}(G_f(u)).$$

The operator $T = L^{-1} \circ G_f\colon Y \to Y$ is compact (explain why!). The problem (5.3.6) thus can be written as a fixed point problem

$$u = T(u). \tag{5.3.8}$$

The equation (5.3.8) can be solved by applying some of the methods presented in the previous chapters. We will show some applications in the next section.

[15]Note that a similar assertion can be proved also for a more general Nemytski operator $G(u)(x) = g(x,u(x))$. However, the conditions on g are more complicated in that case (see, e.g., Appell & Zabreiko [9]).

Exercise 5.3.4. Prove that for $\gamma \in (0,1]$ the set $C^{0,\gamma}(\overline{\Omega})$, Ω a bounded domain in $\mathbb{R}^N$, is a Banach space equipped with the norm

$$\|u\|_{C^{0,\gamma}(\overline{\Omega})} = \sup_{x\in\overline{\Omega}} |u(x)| + \sup_{\substack{x,y\in\Omega\\ x\neq y}} \frac{|u(x)-u(y)|}{|x-y|^\gamma}.$$

Exercise 5.3.5. Prove that for $\gamma \in (0,1]$ the set $C^{2,\gamma}(\overline{\Omega})$, Ω a bounded domain in $\mathbb{R}^N$, is a Banach space equipped with the norm

$$\|u\|_{C^{2,\gamma}(\overline{\Omega})} = \|u\|_{C^2(\overline{\Omega})} + \sum_{|\alpha|=2} \sup_{\substack{x,y\in\Omega\\ x\neq y}} \frac{|D^\alpha u(x)-D^\alpha u(y)|}{|x-y|^\gamma}.$$

Exercise 5.3.6. Prove that for $\gamma \in (0,1]$ the space $C^{0,\gamma}(\overline{\Omega})$ is an algebra (see footnote 17 on page 32).

Exercise 5.3.7. Prove that for $\gamma > 1$ the following equivalence holds true:

$u \in C^{0,\gamma}(\overline{\Omega})$ *if and only if* u *is constant on* $\overline{\Omega}$.

(Cf. footnote 26 on page 38).

Exercise 5.3.8. Prove that $L\colon X \to Y$ defined by (5.3.5) is a linear and bounded operator.

Exercise 5.3.9. Prove that G_f defined above maps bounded sets in Y onto bounded sets in Y.

Exercise 5.3.10. Let $\Omega \neq \emptyset$ be an open subset of $\mathbb{R}^N$. Prove that the topological boundary $\partial\Omega$ is equal to the boundary of $\overline{\Omega}$ in the sense of Definition 4.3.95 provided Ω is of class C^1.

5.4 Classical Solutions, Applications of Fixed Point Theorems

In this section we will deal with the existence (and uniqueness) of the classical solution of the Dirichlet problem

$$\begin{cases} -\Delta u(x) = g(u(x)) + f(x) & \text{in} \quad \Omega,\\ u = 0 & \text{on} \quad \partial\Omega, \end{cases} \tag{5.4.1}$$

under the assumptions on Ω, g and f introduced in Section 5.3. Namely, we assume that Ω is a domain of the class $C^{2,\gamma}$, $\gamma \in (0,1)$, $f \in Y = C^{0,\gamma}(\overline{\Omega})$ and $g \in C^1(\mathbb{R})$.

Let us start with a direct application of the *Schauder Fixed Point Theorem.*

Theorem 5.4.1. *Let* $\sup_{s\in\mathbb{R}} |g(s)| < \infty$ *and*

$$\sup_{s\in\mathbb{R}} |g'(s)| < \frac{1}{\|L^{-1}\|_{\mathcal{L}(Y)}} \tag{5.4.2}$$

where L^{-1} *was introduced in Section* 5.3. *Then for any* $f \in Y$ *the problem* (5.4.1) *has at least one classical solution.*

Proof. We rewrite the problem (5.4.1) into the operator form (5.3.8) with T, L^{-1} and G_f introduced in Section 5.3. Due to our assumptions we find a ball $B(o;r) \subset Y$ with the property

$$T(\overline{B(o;r)}) \subset \overline{B(o;r)}.$$

The existence of at least one solution will then follow from the Schauder Fixed Point Theorem (Theorem 5.2.5).

For $f \in Y$ fixed we get

$$\begin{aligned}
\|T(u)\| &= \|L^{-1}(G_f(u))\| \le \|L^{-1}\|[\|f\| + \|G(u)\|] \\
&\le \|L^{-1}\| \left[\|f\| + \sup_{x\in\overline{\Omega}} |g(u(x))| + \sup_{\substack{x,y\in\Omega \\ x\ne y}} \frac{|g(u(x)) - g(u(y))|}{|x-y|^\gamma} \right] \\
&\le \|L^{-1}\|\|f\| + \|L^{-1}\| \sup_{s\in\mathbb{R}} |g(s)| + \|L^{-1}\| \sup_{s\in\mathbb{R}} |g'(s)| \sup_{\substack{x,y\in\Omega \\ x\ne y}} \frac{|u(x)-u(y)|}{|x-y|^\gamma} \\
&\le \|L^{-1}\|\|f\| + \|L^{-1}\| \sup_{s\in\mathbb{R}} |g(s)| + \|L^{-1}\| \sup_{s\in\mathbb{R}} |g'(s)|\|u\|
\end{aligned}$$

for any $u \in Y$. Note that the first two terms in the last sum are constants independent of $u \in Y$. So, taking $r > 0$ large enough, the assumption (5.4.2) will guarantee that

$$\|T(u)\| < r \qquad \text{for any} \quad u \in B(o;r) \subset Y.$$[16]

This completes the proof. □

Remark 5.4.2. The above theorem actually says that the problem (5.4.1) is solvable in the classical sense if the smooth nonlinearity g is uniformly bounded on $\mathbb{R}$ and, moreover, its derivative is uniformly bounded by a certain constant which depends on Ω and $\gamma \in (0,1)$.[17] This constant may be rather difficult to calculate but its estimate from above can be given for special domains Ω and exponents $\gamma \in (0,1)$ (see Gilbarg & Trudinger [71] for more details).

[16] The reader is invited to calculate the least value of such r in terms of $\|L^{-1}\|$, $\|f\|$, $\sup_{s\in\mathbb{R}} |g(s)|$ and $\sup_{s\in\mathbb{R}} |g'(s)|$!

[17] Note that $\|L^{-1}\|$ depends on Ω and $\gamma \in (0,1)$. See also Exercise 5.4.5.

It is quite natural to ask under which conditions on g the classical solution from Theorem 5.4.1 is uniquely determined. For this purpose let us consider the eigenvalue problem for the Laplace operator subject to the homogeneous Dirichlet boundary conditions

$$\begin{cases} -\Delta u(x) = \lambda u(x) & \text{in} \quad \Omega, \\ u = 0 & \text{on} \quad \partial\Omega. \end{cases} \tag{5.4.3}$$

There exists the so-called principal eigenvalue $\lambda_1 > 0$ of (5.4.3) having the property that

$$\int_\Omega |\nabla u(x)|^2 \, \mathrm{d}x \geq \lambda_1 \int_\Omega |u(x)|^2 \, \mathrm{d}x \qquad \text{for all} \quad u \in Y.^{18} \tag{5.4.4}$$

Now we can formulate the following existence and uniqueness result.

Theorem 5.4.3. *Let* $\sup\limits_{s\in\mathbb{R}} |g(s)| < \infty$ *and*

$$\sup_{s\in\mathbb{R}} |g'(s)| < \min\left\{ \frac{1}{\|L^{-1}\|_{\mathcal{L}(Y)}}, \lambda_1 \right\}. \tag{5.4.5}$$

Then for any $f \in Y$ *the problem* (5.4.1) *has exactly one classical solution.*

Proof. Let $f \in Y$ be arbitrary. The existence of at least one solution follows from Theorem 5.4.1. To prove that this solution is unique we proceed via contradiction. Let $u_1 \neq u_2$, $u_i \in X$, $i = 1, 2$, be two solutions of (5.4.1) for a given $f \in Y$. Then

$$\begin{cases} -\Delta u_i(x) = g(u_i(x)) + f(x) & \text{in} \quad \Omega, \\ u_i = 0 & \text{on} \quad \partial\Omega, \end{cases} \qquad i = 1, 2.$$

Multiply both equations by the difference $u_1 - u_2$, use the Green Formula[19] and the boundary conditions, and then subtract the first expression from the second. Thus we get

$$\int_\Omega |\nabla u_1(x) - \nabla u_2(x)|^2 \, \mathrm{d}x = \int_\Omega (g(u_1(x)) - g(u_2(x)))(u_1(x) - u_2(x)) \, \mathrm{d}x. \tag{5.4.6}$$

[18] This inequality is the *Poincaré inequality* (see Exercise 1.2.48, Example 5.6.4 and Remark 5.6.5 below).

[19] The Green Formula reads:

> *Let* Ω *be a domain with a Lipschitz boundary (see Section* 5.3*) and assume that* $w, v \in C^2(\overline{\Omega})$*. Then the relation*
>
> $$\int_\Omega \Delta w(x) v(x) \, \mathrm{d}x = -\int_\Omega (\nabla w(x), \nabla v(x)) \, \mathrm{d}x + \int_{\partial\Omega} (\nabla w(x), n(x)) v(x) \, \mathrm{d}S$$
>
> *holds where* n *is the unit vector of the outer normal to* $\partial\Omega$ *and* $\mathrm{d}S$ *indicates integration with respect to the surface measure on* $\partial\Omega$.

For more details see Appendix 4.3C, in particular, Remark 4.3.105.

Now, by virtue of (5.4.4) we have

$$\int_\Omega |\nabla u_1(x) - \nabla u_2(x)|^2 \, dx \geq \lambda_1 \int_\Omega |u_1(x) - u_2(x)|^2 \, dx, \tag{5.4.7}$$

and (5.4.5) implies that

$$\int_\Omega (g(u_1(x)) - g(u_2(x)))(u_1(x) - u_2(x)) \, dx < \lambda_1 \int_\Omega |u_1(x) - u_2(x)|^2 \, dx. \tag{5.4.8}$$

It follows from (5.4.6)–(5.4.8) that

$$\begin{aligned} \lambda_1 \int_\Omega |u_1(x) - u_2(x)|^2 \, dx &\leq \int_\Omega |\nabla u_1(x) - \nabla u_2(x)|^2 \, dx \\ &< \lambda_1 \int_\Omega |u_1(x) - u_2(x)|^2 \, dx, \end{aligned}$$

a contradiction. □

Remark 5.4.4. One can ask why not prove the existence (and uniqueness) of a classical solution to (5.4.1) applying directly the Contraction Principle. The reason consists in the fact that the key assumption in this case would be the contractivity of T. Due to the linearity of L^{-1} this is equivalent to the Lipschitz continuity of the Nemytski operator $G\colon Y \to Y$. However, according to Appell & Zabreiko [9, Theorem 7.8], G satisfies the global Lipschitz condition if and only if g is of the form

$$g(u) = a + bu$$

where $a, b \in \mathbb{R}$, i.e., g is a linear function. According to Appell & Zabreiko [9, Theorem 7.9], G satisfies the local Lipschitz condition provided g' is locally Lipschitz continuous. In other words, this means that the assumptions would be too restrictive, and so the above functional setting is not suitable for direct application of the Contraction Principle.

Exercise 5.4.5. Prove that the statement of Theorem 5.4.1 holds provided g is a uniformly Lipschitz continuous function on $\mathbb{R}$ with a constant $K < \frac{1}{\|L^{-1}\|}$. Generalize also Theorem 5.4.3.

Hint. Follow the proof of Theorem 5.4.1 and for an estimate of $g(u)$ use

$$|g(u(\cdot))| \leq |g(u(\cdot)) - g(0)| + |g(0)| \leq K|u(\cdot)| + |g(0)|.$$

5.5 Weak Solutions of PDEs, Functional Setting

As we have mentioned in the previous section the concept of the classical solution is not suitable for application of many abstract results of nonlinear analysis presented in the previous chapters. Let us list the major drawbacks connected with this fact:

- the spaces of Hölder continuous functions do not possess Hilbert structure and are not reflexive;
- to prove that the Nemytski operator between spaces of Hölder continuous functions has a certain property requires too strong assumptions about the nonlinearity.

To be more specific, the fact that $C^{k,\gamma}(\overline{\Omega})$ are not reflexive spaces prevents us from applying variational methods which strongly depend on the selection of weakly convergent sequences from bounded ones (which is usually guaranteed by reflexivity) and sometimes also on the Hilbert structure of the function space. Concerning the restrictive assumptions on nonlinearity let us recall Remark 5.4.4.

That is why it is important to look for a different concept of "solution" than that introduced in Section 5.3. Let us point out that there are also more practical reasons for the introduction of a different concept of solution. Many equations and boundary value problems are derived from globally formulated laws (coming from physics, chemistry, biology, sociology, economy, ...). To give a very simple example, let G be the primitive of g, i.e., $G'(s) = g(s)$, and let the functional

$$E(u) = \frac{1}{2}\int_\Omega |\nabla u(x)|^2 \, \mathrm{d}x - \int_\Omega G(u(x)) \, \mathrm{d}x - \int_\Omega f(x)u(x) \, \mathrm{d}x \tag{5.5.1}$$

represent the energy of a certain system. Following the well-known law a physicist will be interested in finding functions u defined on Ω, $u = 0$ on $\partial\Omega$, for which $E(u)$ attains its minimal value. The task for mathematicians is to find the corresponding formalism in the framework of which the existence of such functions is guaranteed (and they can be calculated).

Let us assume that u belongs to a normed linear space H of functions defined on Ω and let us specify the properties of H later on. Let us also understand the following calculations formally and assume that u, g, f and Ω have all properties necessary in order to perform the calculations. Assume that $u_0 \in H$ is the point of minimum of E, i.e.,

$$E(u_0) = \min_{u \in H} E(u). \tag{5.5.2}$$

Then u_0 is the point of minimum of E restricted to any direction passing through u_0 and hence

$$\delta E(u_0; v) = 0 \qquad \text{for any} \quad v \in H$$

i.e.,

$$\int_\Omega (\nabla u_0(x), \nabla v(x)) \, \mathrm{d}x = \int_\Omega g(u_0(x))v(x) \, \mathrm{d}x + \int_\Omega f(x)v(x) \, \mathrm{d}x \tag{5.5.3}$$

for any $v \in H$. Then (5.5.3) is the necessary condition for (5.5.2).

Now, if we assume that $u_0 \in C^2(\overline{\Omega})$, $v \in C^1(\overline{\Omega})$, $v = 0$ on $\partial\Omega$, and apply the Green Formula (Remark 4.3.105 and footnote 19 on page 270) to the left-hand side of (5.5.3) (using the fact that $v = 0$ on $\partial\Omega$), we arrive at

$$\int_\Omega (-\Delta u_0(x))v(x)\,\mathrm{d}x = \int_\Omega g(u_0(x))v(x)\,\mathrm{d}x + \int_\Omega f(x)v(x)\,\mathrm{d}x. \tag{5.5.4}$$

If $g \in C(\mathbb{R})$, $f \in C(\overline{\Omega})$ and H contains "enough" functions (e.g., $\{v \in C^1(\overline{\Omega}) : v = 0 \text{ on } \partial\Omega\} \subset H$), then (5.5.4) implies

$$\begin{cases} -\Delta u_0(x) = g(u_0(x)) + f(x) & \text{in} \quad \Omega, \\ u_0 = 0 & \text{on} \quad \partial\Omega. \end{cases} \tag{5.5.5}$$

On the other hand, if we find $u_0 \in H \cap C^2(\overline{\Omega})$ satisfying (5.5.5), then we can pass, using the Green Formula, easily "back" to (5.5.3).

However, looking carefully at (5.5.3), we immediately realize that all the expressions in (5.5.3) make sense under more general assumptions on u_0 (Ω, g and f) than the expressions in (5.5.5) do. One can immediately see that for (5.5.3) to hold we do not need to assume the existence of second partial derivatives of u_0 at all. On the other hand, (5.5.5) does not make any sense without them if we understand it in the classical (pointwise) sense.

It also makes a good sense from the physical point of view to consider integral identity (5.5.3) as a starting point for the definition of the "solution" u_0.[20] The advantage of the *weak solution* of (5.5.5) consists in the fact that the equation and the boundary conditions are not satisfied pointwise but in a more general sense which can correspond better to the real situation described by problems of the type (5.5.5).

Let us point out here that the notion of the weak solution is a generalization of the notion of the classical solution. We shall show later that not every weak solution is a classical solution. We will also see later that more methods of nonlinear analysis are applicable to get a weak solution instead of a classical one. On the other hand, it makes a good sense to ask whether a weak solution to some problem has some better properties (continuity, Hölder continuity, differentiability, cf. Theorem 7.1.14). Very often this is the case and it depends on Ω, g and f how much *regular* (i.e., smooth) the weak solution is. For general partial differential equations, however, this is a difficult problem and the *regularity theory* which deals with these questions is an important part of basic research in mathematics.

Let us consider now a bit more general situation when $g = g(x, u)$, and for $u \in H$ let us investigate the identity

$$\int_\Omega (\nabla u(x), \nabla v(x))\,\mathrm{d}x = \int_\Omega g(x, u(x))v(x)\,\mathrm{d}x, \qquad v \in H, \tag{5.5.6}$$

[20]The usual scheme for deriving basic equations of mathematical physics is the following: First the "global formulation" in terms of an integral identity is derived from the conservation law, balance law, etc., and then the "local formulation" is derived in terms of differential equations. The second step, however, requires some extra assumptions about the solution (e.g., smoothness, etc.).

in more detail and try to make clear under how "weak" assumptions concerning the functions u, v and g the expressions appearing in (5.5.6) make sense. Note that the Lebesgue integral is used in (5.5.6). So, in particular, the functions under the integral sign must be measurable. As for ∇u, ∇v and v this will be guaranteed by the requirement that u and v belong to a suitable space of integrable functions. The measurability of the composite function $g(x, u(x))$ will depend on the properties of the function g itself.

As was mentioned in Section 3.2 (Definition 3.2.22) the composite function $h(x) = g(x, u(x))$ is measurable provided u is measurable and g fulfils the Carathéodory conditions. For the sake of brevity we denote this fact as follows:

$$g \in \mathrm{CAR}(\Omega \times \mathbb{R}).$$

Let us return to the right-hand side of (5.5.6). Assume that there exist $r \in L^2(\Omega)$ and $c > 0$ such that for a.a. $x \in \Omega$ and for all $s \in \mathbb{R}$,

$$|g(x,s)| \le r(x) + c|s|. \tag{5.5.7}$$

Let $u, v \in L^2(\Omega)$. Then, according to Theorem 3.2.24, $g(x, u(x)) \in L^2(\Omega)$ and the Hölder inequality yields that

$$g(x, u(x))v(x) \in L^1(\Omega).$$

Now, let us take care of the left-hand side of (5.5.6). For this purpose we have to employ the Sobolev spaces introduced in Section 1.2.

Recall that the Sobolev space $W_0^{1,p}(\Omega)$, $p > 1$, was defined in Exercise 1.2.48. It follows from the Poincaré inequality (see Exercise 1.2.48(ii), (iii)) that the expression

$$\|u\| = \left(\int_\Omega |\nabla u(x)|^p \,\mathrm{d}x \right)^{\frac{1}{p}}$$

defines an equivalent norm on $W_0^{1,p}(\Omega)$. For functions $u \in W_0^{1,p}(\Omega)$, $v \in W_0^{1,p'}(\Omega)$, $\frac{1}{p} + \frac{1}{p'} = 1$, we can apply the Hölder inequality to get

$$\int_\Omega |(\nabla u(x), \nabla v(x))| \,\mathrm{d}x \le \left(\int_\Omega |\nabla u(x)|^p \,\mathrm{d}x \right)^{\frac{1}{p}} \left(\int_\Omega |\nabla v(x)|^{p'} \,\mathrm{d}x \right)^{\frac{1}{p'}}. \tag{5.5.8}$$

Let us point out that not every function from $W_0^{1,p}(\Omega)$ is continuous in general (cf. Theorem 1.2.26). This fact depends on the values of $p > 1$ and N (notice that $\Omega \subset \mathbb{R}^N$) and on the properties of the boundary of Ω.

For domains of the class $C^{0,1}$ (see page 265), i.e., domains with a Lipschitz boundary, we can define the space $L^p(\partial\Omega)$ with the norm

$$\|u\|_{L^p(\partial\Omega)} = \left(\int_{\partial\Omega} |u(x)|^p \,\mathrm{d}S \right)^{\frac{1}{p}}$$

(see Definition 4.3.87 for the meaning of integral, and, e.g., Kufner, John & Fučík [96], Nečas [117]).

The following assertion is called the *Trace Theorem* and it is the key to understanding the notion of "boundary values" of functions from $W^{1,p}(\Omega)$.

Theorem 5.5.1 (Trace Theorem). *Let $\Omega \in C^{0,1}$ be a bounded domain in $\mathbb{R}^N$. There exists one and only one continuous linear operator T which assigns to every function $u \in W^{1,p}(\Omega)$ a function $Tu \in L^p(\partial\Omega)$ and has the following property:*

$$\text{“For} \quad u \in C^\infty(\overline{\Omega}) \quad \text{we have} \qquad Tu = u|_{\partial\Omega}.\text{”}$$

The following identity holds:

$$W_0^{1,p}(\Omega) = \{u \in W^{1,p}(\Omega) : Tu = o \text{ in } L^p(\partial\Omega)\}.$$

(See, e.g., Kufner, John & Fučík [96] and Nečas [117].)

This assertion offers another way of understanding the space $W_0^{1,p}(\Omega)$. Instead of $Tu = o$ in $L^p(\partial\Omega)$ we can say

$$\text{“}u = o \quad \text{on} \quad \partial\Omega \qquad \text{in the } \textit{sense of traces}\text{”}$$

but we usually write

$$u = 0 \qquad \text{on} \quad \partial\Omega.$$

Now, if we return to the integral identity (5.5.6), we can immediately see, by applying the Hölder inequality (5.5.8) with $p = 2$, that the integral on the left-hand side is finite if $u, v \in W^{1,2}(\Omega)$. If, moreover, $u \in W_0^{1,2}(\Omega)$, then u satisfies the boundary condition in the sense of traces. These facts motivate the following definition.

Definition 5.5.2. Let $g \in \text{CAR}(\Omega \times \mathbb{R})$ and let $\Omega \in C^{0,1}$ be a bounded domain in $\mathbb{R}^N$. By a *weak solution* of the *Dirichlet problem* (with homogeneous boundary conditions)

$$\begin{cases} -\Delta u(x) = g(x, u(x)) & \text{in} \quad \Omega, \\ u = 0 & \text{on} \quad \partial\Omega, \end{cases}$$

we understand a function $u \in W_0^{1,2}(\Omega)$ such that the integral identity

$$\int_\Omega (\nabla u(x), \nabla v(x))\, \mathrm{d}x = \int_\Omega g(x, u(x)) v(x)\, \mathrm{d}x \tag{5.5.9}$$

holds for every $v \in W_0^{1,2}(\Omega)$.[21]

[21] Since $\mathcal{D}(\Omega)$ is dense in $W_0^{1,2}(\Omega)$ this is equivalent to the validity of (5.5.9) for all $v \in \mathcal{D}(\Omega)$. We use $v \in W_0^{1,2}(\Omega)$ to get the scalar product in $W_0^{1,2}(\Omega)$ on the left-hand side of (5.5.9).

In order to simplify the notation in the sequel, we write

$$\nabla u(x)\nabla v(x) := (\nabla u(x), \nabla v(x)).$$

So, (5.5.9) reads as

$$\int_\Omega \nabla u(x)\nabla v(x)\,\mathrm{d}x = \int_\Omega g(x,u(x))v(x)\,\mathrm{d}x,$$

which is a more common form used in literature.

Remark 5.5.3. The function v is called a *test function*. More general operators as well as nonhomogeneous boundary conditions are often dealt with in literature. The notion of a weak solution is then defined in a similar way. However, the definition is technically more complicated (see, e.g., Gilbarg & Trudinger [71], Fučík & Kufner [66]).

Remark 5.5.4. It was shown above that every classical solution is also a weak solution. The converse is not always true as follows, e.g., from the fact that the weak solution need not possess the second partial derivatives at all!

Roughly speaking, the weak solutions are looked for in larger spaces than the classical ones. That is why the chance to prove the existence of a weak solution is usually bigger than the chance to prove the existence of a classical solution. Hence, some "ill-posed" problems in the framework of classical solutions can appear to be "well-posed" in the framework of weak solutions.

In the forthcoming sections we will work exclusively with bounded domains $\Omega \in C^{0,1}$ without any further specification. The problems on unbounded domains are more involved and require more sophisticated technical background. For example, Theorem 1.2.28 does not hold on unbounded domains in general.

Exercise 5.5.5. A function $u \in W^{1,2}(\Omega)$ satisfying (5.5.9) for any $v \in W^{1,2}(\Omega)$ is called a weak solution of the Neumann problem

$$\begin{cases} -\Delta u(x) = g(x,u(x)) & \text{in} \quad \Omega, \\ \dfrac{\partial u}{\partial n} = 0 & \text{on} \quad \partial\Omega. \end{cases} \tag{5.5.10}$$

Here $\frac{\partial u}{\partial n}$ denotes the derivative of u in the direction of the outer normal n.

Prove that any weak solution u of (5.5.10) such that $u \in C^2(\overline{\Omega})$ satisfies

$$\frac{\partial u}{\partial n} = 0 \qquad \text{at every point of} \quad \partial\Omega.$$

Hint. Taking $v \in \mathcal{D}(\Omega)$ in (5.5.6) show that the equation (5.5.10) holds pointwise in Ω. Then take an arbitrary $v \in C^2(\overline{\Omega})$ in (5.5.6) and use the Green Formula.

5.6 Weak Solutions of PDEs, Applications of Fixed Point Theorems

In this section we give an application of the fixed point theorems to proving the existence of a weak solution to the problem

$$\begin{cases} -\Delta u(x) = g(x, u(x)) & \text{in} \quad \Omega, \\ u = 0 & \text{on} \quad \partial\Omega. \end{cases} \tag{5.6.1}$$

For this purpose we have to find suitable operator representation of (5.6.1).

For a fixed $u \in W_0^{1,2}(\Omega)$ and $g \in \mathrm{CAR}(\Omega \times \mathbb{R})$ satisfying (5.5.7) it is easy to see that

$$\hat{L}_u \colon v \mapsto \int_\Omega \nabla u(x) \nabla v(x) \, \mathrm{d}x, \qquad \hat{S}_u \colon v \mapsto \int_\Omega g(x, u(x)) v(x) \, \mathrm{d}x$$

are continuous linear functionals on the space $W_0^{1,2}(\Omega)$. Since $W_0^{1,2}(\Omega)$ is a Hilbert space, by the Riesz Representation Theorem (Theorem 1.2.40) there exist uniquely determined elements $Lu, S(u) \in W_0^{1,2}(\Omega)$ such that

$$(Lu, v)_{W_0^{1,2}(\Omega)} = \hat{L}_u(v), \qquad (S(u), v)_{W_0^{1,2}(\Omega)} = \hat{S}_u(v) \tag{5.6.2}$$

for all $v \in W_0^{1,2}(\Omega)$.

To prove that the Dirichlet problem (5.6.1) has at least one weak solution it is necessary and sufficient to prove that the operator equation

$$Lu = S(u) \tag{5.6.3}$$

has at least one solution in the space $W_0^{1,2}(\Omega)$. Let us therefore investigate the properties of the operators L and S.

There are several equivalent inner products defined on $W_0^{1,2}(\Omega)$. If we choose

$$(u, v)_{W_0^{1,2}(\Omega)} := \int_\Omega \nabla u(x) \nabla v(x) \, \mathrm{d}x,$$ [22]

then L defined by (5.6.2) is just an identity on $W_0^{1,2}(\Omega)$.

On the other hand, for any $u \in W_0^{1,2}(\Omega)$ we also have $u \in L^2(\Omega)$, and Theorem 3.2.24 then implies that $g(x, u) \in L^2(\Omega)$. Let us assume that g is Lipschitz continuous with respect to the second variable, i.e., there exists a constant $c > 0$ such that for a.a. $x \in \Omega$ and for any $s_1, s_2 \in \mathbb{R}$,

$$|g(x, s_1) - g(x, s_2)| \le c|s_1 - s_2|.$$

[22] Cf. the Poincaré inequality (5.4.4).

Then for $u_1, u_2 \in W_0^{1,2}(\Omega)$ we have

$$\begin{aligned}\|S(u_1) - S(u_2)\| &= \sup_{\|v\|\le 1} |(S(u_1) - S(u_2), v)| \\ &= \sup_{\|v\|\le 1} \left| \int_\Omega [g(x, u_1(x)) - g(x, u_2(x))]v(x)\,\mathrm{d}x \right| \\ &\le \sup_{\|v\|\le 1} \int_\Omega c|u_1(x) - u_2(x)||v(x)|\,\mathrm{d}x \qquad (5.6.4) \\ &\le \sup_{\|v\|\le 1} c\|u_1 - u_2\|_{L^2(\Omega)}\|v\|_{L^2(\Omega)} \\ &\le \sup_{\|v\|\le 1} c\,c_{\mathrm{emb}}^2\|u_1 - u_2\|\|v\| = c\,c_{\mathrm{emb}}^2\|u_1 - u_2\|.\end{aligned}$$ [23]

Hence (5.6.3) is equivalent in $W_0^{1,2}(\Omega)$ to the operator equation,

$$u = S(u) \qquad (5.6.5)$$

where S is a contraction if $c_{\mathrm{emb}}^2 c < 1$.

We can apply the Contraction Principle to get the following result.

Theorem 5.6.1. *Let $g \in \mathrm{CAR}(\Omega \times \mathbb{R})$ be Lipschitz continuous with respect to the second variable with a constant $c > 0$, $c < c_{\mathrm{emb}}^{-2}$, where c_{emb} is the constant of the embedding of $W_0^{1,2}(\Omega)$ into $L^2(\Omega)$. Then there is a unique fixed point $u \in W_0^{1,2}(\Omega)$ of the operator S, i.e., u is a unique weak solution of* (5.6.1).

Another possibility is to apply the Schauder Fixed Point Theorem. Let us assume that $g \in \mathrm{CAR}(\Omega \times \mathbb{R})$ and $|g(x, s)| \le r(x)$ for a.a. $x \in \Omega$ and all $s \in \mathbb{R}$ with a fixed $r \in L^2(\Omega)$. Then for any $u \in W_0^{1,2}(\Omega)$ we have

$$\begin{aligned}\|S(u)\| &= \sup_{\|v\|\le 1} |(S(u), v)| = \sup_{\|v\|\le 1} \left| \int_\Omega g(x, u(x))v(x)\,\mathrm{d}x \right| \\ &\le \sup_{\|v\|\le 1} \left(\int_\Omega |g(x, u(x))|^2\,\mathrm{d}x \right)^{\frac{1}{2}} \left(\int_\Omega |v(x)|^2\,\mathrm{d}x \right)^{\frac{1}{2}} \le c_{\mathrm{emb}}\|r\|_{L^2(\Omega)}.\end{aligned} \qquad (5.6.6)$$

It follows from (5.6.6) that the operator S maps the closure of the ball $B(o; R) \subset W_0^{1,2}(\Omega)$ with radius

$$R = c_{\mathrm{emb}}\|r\|_{L^2(\Omega)}$$

into itself.

[23] There exists a constant $c_{\mathrm{emb}} > 0$ such that $\|u\|_{L^2(\Omega)} \le c_{\mathrm{emb}}\|u\|$ for any $u \in W_0^{1,2}(\Omega)$ (cf. Theorem 1.2.26 and Exercise 1.2.48). It can be shown that the best value of c_{emb} is $\frac{1}{\sqrt{\lambda_1}}$ (cf. Remark 5.6.5 and also (5.4.4)).

We will show that S is compact. Indeed, let $\mathcal{M} \subset W_0^{1,2}(\Omega)$ be a bounded set and $\{w_n\}_{n=1}^\infty \subset S(\mathcal{M})$ an arbitrary sequence. Let $\{u_n\}_{n=1}^\infty \subset \mathcal{M}$ be such that

$$S(u_n) = w_n.$$

The reflexivity of $W_0^{1,2}(\Omega)$ implies that $u_n \rightharpoonup u$ in $W_0^{1,2}(\Omega)$ at least for a subsequence. It follows from Theorem 1.2.28 that $u_n \to u$ in $L^2(\Omega)$. Estimates similar to those from (5.6.4) with u_1 replaced by u_n and u_2 replaced by u yield

$$\|S(u_n) - S(u)\|_{W_0^{1,2}(\Omega)} \leq c_{\text{emb}} \|g(\cdot, u_n) - g(\cdot, u)\|_{L^2(\Omega)}. \tag{5.6.7}$$

The right-hand side approaches zero as follows from the continuity of the Nemytski operator from $L^2(\Omega)$ into $L^2(\Omega)$ (see Theorem 3.2.24), i.e.,

$$w_n \to S(u) \qquad \text{in} \quad W_0^{1,2}(\Omega)$$

(at least for a subsequence). This proves the compactness of $\overline{S(\mathcal{M})}$. Note that (5.6.4) implies the continuity of S as well, i.e., S is a compact operator.

Thus we have the following assertion.

Proposition 5.6.2. *Let $g \in \mathrm{CAR}(\Omega \times \mathbb{R})$ and let there exist $r \in L^2(\Omega)$ such that $|g(x,s)| \leq r(x)$ for all $s \in \mathbb{R}$ and a.a. $x \in \Omega$. Then there is at least one fixed point $u \in W_0^{1,2}(\Omega)$ of S (and so a weak solution of* (5.6.1)*) such that*

$$\|u\|_{W_0^{1,2}(\Omega)} \leq c_{\text{emb}} \|r\|_{L^2(\Omega)}.$$

It is not hard to see that the assumptions of Proposition 5.6.2 are unnecessarily strong. In order to find a ball $B(o; R) \subset W_0^{1,2}(\Omega)$ which is mapped by S into itself one can assume that g has *sublinear growth* with respect to the second variable. More precisely, we assume that $g \in \mathrm{CAR}(\Omega \times \mathbb{R})$ and there exist $r \in L^2(\Omega)$, $c > 0$ and $\delta \in (0,1)$ such that for all $s \in \mathbb{R}$ and a.a. $x \in \Omega$ we have

$$|g(x,s)| \leq r(x) + c|s|^\delta. \tag{5.6.8}$$

Then for any $u \in W_0^{1,2}(\Omega)$, similarly to (5.6.6), we have

$$\begin{aligned}
\|S(u)\| &\leq \sup_{\|v\| \leq 1} \left(\int_\Omega |g(x, u(x))|^2 \, \mathrm{d}x \right)^{\frac{1}{2}} \left(\int_\Omega |v(x)|^2 \, \mathrm{d}x \right)^{\frac{1}{2}} \\
&\leq c_{\text{emb}} \left(\int_\Omega \left| r(x) + c|u(x)|^\delta \right|^2 \mathrm{d}x \right)^{\frac{1}{2}} \\
&\leq c_{\text{emb}} \left[\left(\int_\Omega |r(x)|^2 \, \mathrm{d}x \right)^{\frac{1}{2}} + c \left(\int_\Omega |u(x)|^{2\delta} \, \mathrm{d}x \right)^{\frac{1}{2}} \right]
\end{aligned} \tag{5.6.9}$$

where the last estimate is due to the Minkowski inequality (1.2.4) for $p = 2$. Applying the Hölder inequality we have

$$\left(\int_\Omega |u(x)|^{2\delta} \, \mathrm{d}x \right)^{\frac{1}{2}} \leq \left(\int_\Omega |u(x)|^2 \, \mathrm{d}x \right)^{\frac{\delta}{2}} (\operatorname{meas} \Omega)^{\frac{1-\delta}{2}} \leq c_{\text{emb}}^\delta (\operatorname{meas} \Omega)^{\frac{1-\delta}{2}} \|u\|^\delta. \tag{5.6.10}$$

Now, (5.6.9) and (5.6.10) yield

$$\|S(u)\| \le \underbrace{c_{\text{emb}}\|r\|_{L^2(\Omega)}}_{C:=} + \underbrace{c\,c_{\text{emb}}^{1+\delta}(\operatorname{meas}\Omega)^{\frac{1-\delta}{2}}}_{D:=}\|u\|^{\delta}. \tag{5.6.11}$$

It follows from (5.6.11) that for any $u \in B(o;R)$ we get

$$\|S(u)\| < R \qquad \text{provided} \qquad C + DR^{\delta} < R \tag{5.6.12}$$

and hence S maps $B(o;R)$ into itself if R is large enough. The map S is compact as well (no growth restrictions on g are needed to prove the compactness of S). Therefore we have a more general result than that in Proposition 5.6.2.

Theorem 5.6.3. *Let $g \in \operatorname{CAR}(\Omega \times \mathbb{R})$ satisfy* (5.6.8) *for $\delta < 1$. Then there is at least one weak solution u of* (5.6.1).

In Remark 5.6.5 we will show that it is not possible in general to allow the linear growth of $g = g(x,s)$ with respect to s. We will need the result from the next example.

Example 5.6.4. Let

$$f(u) = \frac{\int_\Omega |\nabla u(x)|^2\,\mathrm{d}x}{\int_\Omega |u(x)|^2\,\mathrm{d}x} \quad \text{for} \quad u \in W_0^{1,2}(\Omega), \quad u \neq 0.$$

Let $\lambda_1 \in \mathbb{R}$ be defined as

$$\lambda_1 = \inf_{\substack{u\in W_0^{1,2}(\Omega)\\ u\neq o}} f(u) \tag{5.6.13}$$

or equivalently as

$$\lambda_1 = \inf\left\{\int_\Omega |\nabla u(x)|^2\,\mathrm{d}x : \int_\Omega |u(x)|^2\,\mathrm{d}x = 1,\ u \in W_0^{1,2}(\Omega)\right\}.$$

This is equivalent to the following characterization of λ_1:

$$\frac{1}{\lambda_1} = \sup_{\substack{u\in W_0^{1,2}(\Omega)\\ u\neq o}} \frac{\displaystyle\int_\Omega |u(x)|^2\,\mathrm{d}x}{\displaystyle\int_\Omega |\nabla u(x)|^2\,\mathrm{d}x}$$

or

$$\frac{1}{\lambda_1} = \sup\left\{\int_\Omega |u(x)|^2\,\mathrm{d}x : \int_\Omega |\nabla u(x)|^2\,\mathrm{d}x = 1,\ u \in W_0^{1,2}(\Omega)\right\}.$$

The operator $A\colon W_0^{1,2}(\Omega) \to W_0^{1,2}(\Omega)$ defined by

$$(A(u),v)_{W_0^{1,2}(\Omega)} = \int_\Omega u(x)v(x)\,\mathrm{d}x, \qquad u,v \in W_0^{1,2}(\Omega),$$

is positive, self-adjoint and compact.[24] We show that the supremum $\frac{1}{\lambda_1}$ is achieved and the function $u_1 \in W_0^{1,2}(\Omega)$ such that

$$\|u_1\|_{W_0^{1,2}(\Omega)} = 1, \qquad \frac{1}{\lambda_1} = \int_\Omega |u_1(x)|^2 \, dx,$$

is an eigenvector of A (corresponding to the eigenvalue $\frac{1}{\lambda_1}$). Indeed, let $\{v_n\}_{n=1}^\infty \subset W_0^{1,2}(\Omega)$ be a maximizing sequence, i.e.,

$$\int_\Omega |v_n(x)|^2 \, dx \to \frac{1}{\lambda_1} \text{ as } n \to \infty, \int_\Omega |\nabla v_n(x)|^2 \, dx = 1.$$

Compactness of the embedding $W_0^{1,2}(\Omega) \hookrightarrow\hookrightarrow L^2(\Omega)$ yields the existence of $u_1 \in W_0^{1,2}(\Omega)$ that satisfies $\int_\Omega |u_1(x)|^2 \, dx = \frac{1}{\lambda_1}$ and $\int_\Omega |\nabla u_1(x)|^2 \, dx = 1$. [25]Since u_1 minimizes the quotient in (5.6.13), we also have (cf.page 273)

$$\delta f(u_1; v) = \frac{\int_\Omega u_1(x) \nabla v(x) \, dx \int_\Omega |u_1(x)|^2 \, dx - \int_\Omega |u_1(x)|^2 \, dx \int_\Omega u_1(x) v(x) \, dx}{\left(\int_\Omega |u_1(x)|^2 \, dx\right)^2} = 0$$

for any $v \in W_0^{1,2}(\Omega)$ and the result follows. The reader is invited to compare this reasoning with Theorem 7.8.12 and Example 7.8.15. Then

$$\varphi_1 = \sqrt{\lambda_1} u_1$$

satisfies

$$\|\varphi_1\|_{L^2(\Omega)} = 1 \qquad \text{and} \qquad \lambda_1 = \int_\Omega |\nabla \varphi_1(x)|^2 \, dx.$$

Moreover, we have that

$$\int_\Omega \nabla \varphi_1(x) \nabla v(x) \, dx = \lambda_1 \int_\Omega \varphi_1(x) v(x) \, dx \qquad \text{for any} \quad v \in W_0^{1,2}(\Omega).$$ [26]

We will then call λ_1 the *principal eigenvalue* and φ_1 the corresponding *principal* (normalized) *eigenfunction* of the eigenvalue problem

$$\begin{cases} -\Delta u(x) = \lambda u(x) & \text{in} \quad \Omega, \\ u = 0 & \text{on} \quad \partial\Omega \end{cases} \tag{5.6.14}$$

(cf. page 270). It can be proved that λ_1 is actually the least eigenvalue (Exercise 7.8.19), and according to Example 6.3.42 below λ_1 is simple and the function φ_1 can be chosen positive in Ω. ○

[24]Use the compact embedding $W_0^{1,2}(\Omega) \hookrightarrow\hookrightarrow L^2(\Omega)$ (Theorem 1.2.28).

[25]Indeed $\|u\| = \left(\int_\Omega |\nabla u(x)|^2 \, dx\right)^{\frac{1}{2}}$ is a norm on $W_0^{1,2}(\Omega)$, see Excercise 1.2.48 (ii). Proposition 2.1.22 (iii) implies that $\|u_1\| \le 1$. If $\|u_1\| < 1$, we get a contradiction with the definition of $\frac{1}{\lambda_1}$.

[26]The reader is invited to prove this equality using the Lagrange Multiplier Method (Theorem 7.8.2).

Remark 5.6.5. Note that it follows from the definition of the principal eigenvalue λ_1 that,

$$\|u\|_{L^2(\Omega)} \le \frac{1}{\sqrt{\lambda_1}} \|u\|_{W_0^{1,2}(\Omega)} \qquad \text{for any} \quad u \in W_0^{1,2}(\Omega)$$

and $\frac{1}{\sqrt{\lambda_1}}$ is the least constant with this property. Hence

$$c_{\text{emb}} = \frac{1}{\sqrt{\lambda_1}}.$$

Let us now consider the problem (5.6.1) where g is of the form

$$g(x,u) = \lambda_1 u + f(x) \qquad \text{with} \quad f \in L^2(\Omega).$$

If $u \in W_0^{1,2}(\Omega)$ is a weak solution of

$$\begin{cases} -\Delta u(x) = \lambda_1 u(x) + f(x) & \text{in} \quad \Omega, \\ u = 0 & \text{on} \quad \partial\Omega, \end{cases} \tag{5.6.15}$$

then

$$\int_\Omega \nabla u(x) \nabla v(x) \, \mathrm{d}x = \lambda_1 \int_\Omega u(x) v(x) \, \mathrm{d}x + \int_\Omega f(x) v(x) \, \mathrm{d}x$$

for any $v \in W_0^{1,2}(\Omega)$. In particular, choosing $v = \varphi_1$ where φ_1 is the principal eigenfunction corresponding to λ_1, we have

$$0 = \int_\Omega \nabla u(x) \nabla \varphi_1(x) \, \mathrm{d}x - \lambda_1 \int_\Omega u(x) \varphi_1(x) \, \mathrm{d}x = \int_\Omega f(x) \varphi_1(x) \, \mathrm{d}x,$$

i.e.,

$$\int_\Omega f(x) \varphi_1(x) \, \mathrm{d}x = 0$$

is a necessary condition for (5.6.15) to have a weak solution (cf. Exercise 5.9.11). In other words, it is not possible to relax condition (5.6.8) to hold for $\delta = 1$ and to expect the existence of a weak solution of (5.6.1) at the same time.

Checking carefully estimates similar to (5.6.9) but with $\delta = 1$, we come to the conclusion that small c in (5.6.8) may compensate the linear growth and Theorem 5.6.3 will still hold true. More precisely, assuming

$$|g(x,s)| \le r(x) + c|s|, \tag{5.6.16}$$

we have for any $u \in W_0^{1,2}(\Omega)$, similarly to (5.6.9),

$$\begin{aligned}
\|S(u)\| &\le \sup_{\|v\|\le 1} \left(\int_\Omega |g(x,u(x))|^2\,\mathrm{d}x\right)^{\frac{1}{2}} \left(\int_\Omega |v(x)|^2\,\mathrm{d}x\right)^{\frac{1}{2}} \\
&\le c_{\mathrm{emb}} \left(\int_\Omega \big|r(x)+c|u(x)|\big|^2\,\mathrm{d}x\right)^{\frac{1}{2}} \\
&\le c_{\mathrm{emb}} \left[\left(\int_\Omega |r(x)|^2\,\mathrm{d}x\right)^{\frac{1}{2}} + c\left(\int_\Omega |u(x)|^2\,\mathrm{d}x\right)^{\frac{1}{2}}\right] \\
&\le c_{\mathrm{emb}}\|r\|_{L^2(\Omega)} + c\,c_{\mathrm{emb}}^2\|u\|.
\end{aligned}$$

If

$$c < \frac{1}{c_{\mathrm{emb}}^2} \quad (\le \lambda_1), \tag{5.6.17}$$

then we can choose $R > 0$ such that

$$c_{\mathrm{emb}}\|r\|_{L^2(\Omega)} + c\,c_{\mathrm{emb}}^2 R < R,$$

and for any $u \in B(o;R)$ we get

$$\|S(u)\| < R. \tag{5.6.18}$$

Since S is also a compact operator under the growth condition (5.6.16), we can formulate

Theorem 5.6.6. *Let $g \in \mathrm{CAR}(\Omega\times\mathbb{R})$ and let there exist $r \in L^2(\Omega)$ and $c > 0$ such that* (5.6.16) *and* (5.6.17) *hold true. Then problem* (5.6.1) *has at least one weak solution $u \in W_0^{1,2}(\Omega)$.*

Remark 5.6.7. It follows directly that the closer c is to the value $\frac{1}{c_{\mathrm{emb}}^2}$, the larger R has to be taken in order to get (5.6.18). At the same time one can see that the statement does not hold if $c = \lambda_1$ (see Remark 5.6.5).

Exercise 5.6.8. Formulate and prove an analogue of Theorems 5.6.1, 5.6.3 and 5.6.6 for the Neumann problem

$$\begin{cases} -\Delta u(x) = g(x,u(x)) & \text{in} \quad \Omega, \\ \dfrac{\partial u}{\partial n} = 0 & \text{on} \quad \partial\Omega. \end{cases}$$

5.7 Brouwer Topological Degree

In this section we stress the basic properties of the Brouwer degree of a continuous map in finite-dimensional spaces. We start with some elementary considerations in one dimension. The reader can find another motivation from the theory of functions of a complex variable in Appendix 5.7A. In the previous sections we have dealt with a solution of the operator equation

$$F(x) = 0.$$

Now we are asking what happens with its solution if $F\colon \mathbb{R} \to \mathbb{R}$ is slightly perturbed. Figures 5.7.1–5.7.3 show that the situation can change considerably, namely, either a solution may disappear (if a perturbation takes place in the solid arrow direction) or the number of solutions may vary (in the dashed arrow direction).

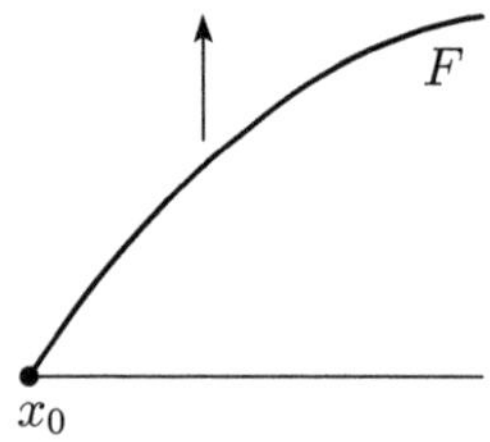

Figure 5.7.1.

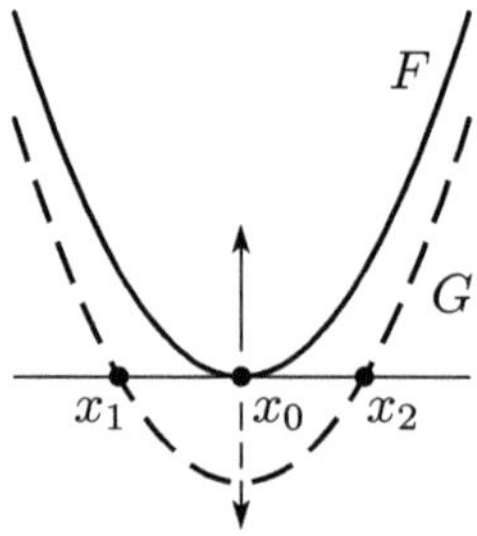

Figure 5.7.2.

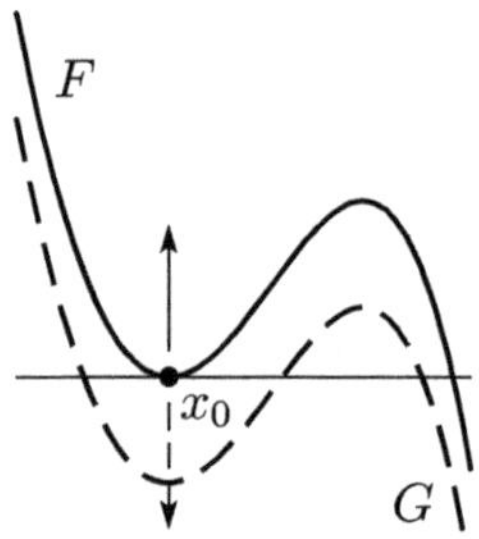

Figure 5.7.3.

A closer examination indicates that this can happen since a solution x_0 is either on the boundary (Figure 5.7.1) or the derivative F' vanishes at x_0, i.e., x_0 is a critical point of F (Figures 5.7.2 and 5.7.3). We expect that a small perturbation of F does not cause any alteration provided the just described cases do not occur.

There is another point which should be mentioned, namely the distinction between perturbations of F in the direction of one or the other arrow in Figures 5.7.2 and 5.7.3. The number of solutions changes by two, being even in Figure 5.7.2 (0 is even by definition), and being odd in Figure 5.7.3. Is there any way to describe this phenomenon? Look at the dashed curve G in Figure 5.7.2 and assume that $G \in C^1$. We have

$$G'(x_1) < 0, \qquad G'(x_2) > 0.$$

These signs remain the same in some neighborhoods $\mathcal{U}_1$, $\mathcal{U}_2$ of x_1 and x_2.

In particular, G is injective (actually a diffeomorphism) on these neighborhoods and can be regarded as a local transformation of the x-coordinate. This transformation changes the orientation at x_1 and does not do that at x_2.

The sum of signs of G' at the solutions of $G(x) = 0$ is zero (more generally even) in Figure 5.7.2 and odd for the dashed curve in Figure 5.7.3. This observation can be generalized to higher dimensions: If $A\colon \mathbb{R}^N \to \mathbb{R}^N$ is a linear transformation of coordinates (i.e., A is injective and surjective), then we say that A does not change the orientation in $\mathbb{R}^N$ provided

$$\det \boldsymbol{A} > 0$$

where $\boldsymbol{A}$ is the matrix representation of A (this does not depend on the choice of basis in which the representation is taken). This concept can be also used locally for a nonlinear C^1-transformation $G\colon \mathbb{R}^N \to \mathbb{R}^N$ by replacing $G'(a)$ for A. Then the sign of the matrix representation of $G'(a)$ is the sign of its Jacobian $J_G(a)$. This idea leads to the following preparatory definition.

Definition 5.7.1. Let Ω be an open and bounded subset of $\mathbb{R}^N$ and let $F \in C(\overline{\Omega}, \mathbb{R}^N) \cap C^1(\Omega, \mathbb{R}^N)$. Assume that $y_0 \in \mathbb{R}^N \setminus F(\partial\Omega)$ and y_0 is a regular value of F.[27] Then we define the *Brouwer degree* of F as

$$\deg(F, \Omega, y_0) = \sum_{x \in F_{-1}(y_0) \cap \Omega} \operatorname{sgn} J_{F(x)}.^{28} \tag{5.7.1}$$

We point out that the sum in (5.7.1) is finite. Indeed, otherwise the set

$$F_{-1}(y_0) \cap \Omega := \{x \in \Omega : F(x) = y_0\}$$

has an accumulation point $\tilde{x} \in \overline{\Omega}$. By the continuity of F, $F(\tilde{x}) = y_0$, and since $y_0 \notin F(\partial\Omega)$, $\tilde{x} \in \Omega$. By the Local Inverse Function Theorem (Theorem 4.1.1), F is injective in a neighborhood $\mathcal{U}$ of $\tilde{x}$. But $\mathcal{U}$ contains points of $F_{-1}(y_0)$ different from $\tilde{x}$, a contradiction. Notice that in this argument we have used all assumptions of Definition 5.7.1.

Proposition 5.7.2. *Let Ω be an open bounded subset of $\mathbb{R}^N$. The degree defined in Definition* 5.7.1 *has the following properties (I is the identity map):*

(i) $\deg(I, \Omega, y_0) = \begin{cases} 1 & \text{if} \quad y_0 \in \Omega, \\ 0 & \text{if} \quad y_0 \notin \overline{\Omega}. \end{cases}$

Suppose that $F \in C(\overline{\Omega}, \mathbb{R}^N) \cap C^1(\Omega, \mathbb{R}^N)$ and $y_0 \in \mathbb{R}^N \setminus F(\partial\Omega)$ is a regular value of F. Then

(ii) $\deg(F, \Omega, y_0) \in \mathbb{Z}$;

(iii) $\deg(F, \Omega, y_0) = \deg(F - y_0, \Omega, o)$;

(iv) *if* $\deg(F, \Omega, y_0) \neq 0$, *then the equation*

$$F(x) = y_0$$

has a solution in Ω;

(v) *if Ω_1 is an open subset of Ω and $y_0 \notin F(\overline{\Omega} \setminus \Omega_1)$, then*

$$\deg(F, \Omega, y_0) = \deg(F, \Omega_1, y_0).$$

[27] The definition of a regular value is given in Definition 4.3.6.

[28] Here $\sum\limits_{\emptyset} = 0$ as usual.

More generally, if $\Omega_1, \dots, \Omega_k$ *are pairwise disjoint open subsets of* Ω *and* $y_0 \notin F\left(\overline{\Omega} \setminus \bigcup_{j=1}^{k} \Omega_j\right)$, *then*

$$\deg(F, \Omega, y_0) = \sum_{j=1}^{k} \deg(F, \Omega_j, y_0). \tag{5.7.2}$$

(vi) *For all* $y \in \mathbb{R}^N$ *which are sufficiently close to* y_0,

$$\deg(F, \Omega, y) = \deg(F, \Omega, y_0) \qquad \textit{holds}. \tag{5.7.3}$$

(vii) *For all* $G \in C(\overline{\Omega}, \mathbb{R}^N) \cap C^1(\Omega, \mathbb{R}^N)$ *which are sufficiently close to* F *in the* C^1*-topology,*[29]

$$\deg(G, \Omega, y_0) = \deg(F, \Omega, y_0) \qquad \textit{is valid}. \tag{5.7.4}$$

Proof. The properties (i)–(v) follow immediately from Definition 5.7.1.

To prove (vi) let

$$F_{-1}(y_0) = \{x_1, \dots, x_k\}$$

and let F be a diffeomorphism of an open neighborhood $\mathcal{U}_j$ of x_j onto a neighborhood $\mathcal{V}_j$ of y_0, cf. the Local Inverse Function Theorem (Theorem 4.1.1). Denote

$$d := \inf\left\{\|F(x) - y_0\| : x \in \overline{\Omega} \setminus \bigcup_{j=1}^{k} \mathcal{U}_j\right\} > 0.$$

If $\|y - y_0\| < d$, then there is no solution of $F(x) = y$ in $\overline{\Omega} \setminus \bigcup_{j=1}^{k} \mathcal{U}_j$, and if $y \in \bigcap_{j=1}^{k} \mathcal{V}_j$ (the neighborhood of y_0), then there is exactly one $\tilde{x}_j \in \mathcal{U}_j$ such that $F(\tilde{x}_j) = y$. Moreover,

$$\operatorname{sgn} J_F(\tilde{x}_j) = \operatorname{sgn} J_F(x_j).$$

This completes the proof of (5.7.3).[30]

To prove (vii) we use the same notation as above. Let G differ a little from F in the C^1-topology, say

$$\|F - G\|_{C^1(\Omega, \mathbb{R}^N)} < \varepsilon.$$

The quantity ε will be specified later. Put

$$H(t, x) = (1 - t)F(x) + tG(x), \quad x \in \overline{\Omega}, t \in (-\delta, 1 + \delta) \text{ for } \delta > 0. \tag{5.7.5}$$

[29] I.e., there exists $\varepsilon > 0$ such that
$\|F - G\|_{C^1(\Omega, \mathbb{R}^N)} := \sup_{x \in \overline{\Omega}} \|F(x) - G(x)\|_{\mathbb{R}^N} + \sup_{x \in \Omega} \|F'(x) - G'(x)\|_{\mathcal{L}(\mathbb{R}^N)} < \varepsilon$.

[30] Notice that this proof is correct also for $F_{-1}(y_0) = \emptyset$.

Choose a fixed neighborhood $\mathcal{U}_j$ as above. Notice that we can take $\mathcal{U}_j$ so small that

$$c_1 := \sup_{x\in\mathcal{U}_j} \|F'(x)\|_{\mathcal{L}(\mathbb{R}^N)} < \infty \qquad \text{and} \qquad c_2 := \sup_{x\in\mathcal{U}_j} \|[F'(x)]^{-1}\|_{\mathcal{L}(\mathbb{R}^N)} < \infty,$$

and the determinant $\det F'(x)$ has a constant sign in $\mathcal{U}_j$. We have

$$\|H(t,x) - y_0\| \ge \|F(x) - y_0\| - |t|\|F(x) - G(x)\| \ge d - |t|\varepsilon > 0$$

for every $x \in \overline{\Omega} \setminus \bigcup_{j=1}^{k} \mathcal{U}_j$ and $t \in (-\delta, 1+\delta)$, and for sufficiently small $\varepsilon > 0$. In particular, $\deg(H(t,\cdot),\mathcal{U}_j, y_0)$ is well defined[31] and, by (v),

$$\deg(H(t,\cdot),\Omega,y_0) = \sum_{j=1}^{k} \deg(H(t,\cdot),\mathcal{U}_j,y_0).$$

We wish to prove that this degree is constant with respect to t on the interval $[0,1]$. We will study the set

$$\mathcal{M} := \{(t,x) \in [0,1]\times\mathcal{U}_j : H(t,x) = y_0\}$$

with help of the Implicit Function Theorem at the point $(0,x_j)$. This is possible since

$$\|H_2'(t,x) - F'(x)\|_{\mathcal{L}(\mathbb{R}^N)} = |t|\|F'(x) - G'(x)\|_{\mathcal{L}(\mathbb{R}^N)} \le |t|\varepsilon < \frac{1}{c_2},$$
$$(t,x) \in (-\delta, 1+\delta)\times\mathcal{U}_j, \quad \text{for small } \varepsilon > 0.$$

This estimate implies that $[H_2'(t,x)]^{-1}$ exists (Exercise 2.1.33). The Implicit Function Theorem implies that $\mathcal{M}$ contains the set

$$\{(t,\varphi(t)) : t \in [0,\beta)\}$$

in a certain neighborhood of $(0,x_j)$ $(F(x_j) = y_0)$ where $\varphi \in C^1([0,\beta),\mathbb{R}^N)$ and

$$\|\dot\varphi(t)\|_{\mathcal{L}(\mathbb{R}^N)} = \|[H_2'(t,\varphi(t))]^{-1}H_1'(t,\varphi(t))\|_{\mathcal{L}(\mathbb{R}^N)} \le \frac{c_2\varepsilon}{1-c_2\varepsilon},$$

see again Exercise 2.1.33. In particular, φ is uniformly continuous and, if necessary, it can be continued at least until $t = 1$.[32] Therefore $\mathcal{M}$ contains the graph of φ on the interval $[0,1]$, i.e.,

$$\{x \in \mathcal{U}_j : G(x) = H(1,x) = y_0\} = \{\varphi(1)\},$$

and, consequently,

$$\deg(G,\mathcal{U}_j,y_0) = \deg(F,\mathcal{U}_j,y_0). \qquad \square$$

[31] A homotopy $H(t,x)$ for which the degree is well defined is called an *admissible homotopy*.

[32] If $\beta \le 1$, then $\lim_{t\to\beta_-} \varphi(t) = \tilde{x}$ exists and $\tilde{x} \in \mathcal{U}_j$ (see Proposition 1.2.4).

One of our main goals is to show that the degree is homotopically invariant, i.e., if for H given by (5.7.5) we have

$$y_0 \neq H(t,x) \qquad \text{for all} \quad t \in [0,1], \quad x \in \partial\Omega,$$

then $\deg(H(t,\cdot),\Omega,y_0)$ is constant on $[0,1]$. In particular,

$$\deg(H(0,\cdot),\Omega,y_0) = \deg(H(1,\cdot),\Omega,y_0) \tag{5.7.6}$$

provided at least one side in (5.7.6) is defined. The problem in proving this property can be seen from Figures 5.7.2 and 5.7.3. Namely, if the dashed curve G is moving up, then it is equal to F in one instance, o is not a regular value for F, and so $\deg(F,\Omega,o)$ is not yet defined. To overcome this obstacle we approximate a critical value by a regular one. Such approximation is based on the so-called *Sard Theorem*. Its special case stated below will be sufficient for our purposes.

Theorem 5.7.3 (Sard). *Let Ω be an open subset of $\mathbb{R}^N$ and assume that $F \in C^1(\Omega,\mathbb{R}^N)$. Then the Lebesgue measure of the set of critical values of F is zero.*

Proof. Since $\mathbb{R}^N$ can be covered by a sequence of bounded open sets and a countable union of sets of measure zero has also measure zero we can suppose that Ω is bounded. Choose now an open subset $\mathcal{G} \subset \Omega$ such that $\overline{\mathcal{G}} \subset \Omega$. Let $\mathcal{S}$ be the set of critical points of F in $\mathcal{G}$. By the same argument as above, it is sufficient to show that

$$\operatorname{meas}_N F(\mathcal{S}) = 0$$

where meas_N is the Lebesgue measure in $\mathbb{R}^N$. Since $\overline{\mathcal{G}}$ is compact,

$$d := \operatorname{dist}(\overline{\mathcal{G}},\mathbb{R}^N \setminus \Omega) > 0$$

and $\mathcal{G}$ can be covered by a finite number of closed cubes $\mathcal{C}_1,\dots,\mathcal{C}_k$ with sides parallel to the coordinate hyperplanes and edges of length a. If $a < d\sqrt{N}$, then $\bigcup\limits_{i=1}^{k} \mathcal{C}_i \subset \Omega$ and

$$c := \sup_{x \in \bigcup\limits_{i=1}^{k} \mathcal{C}_i} \|F'(x)\|_{\mathcal{L}(\mathbb{R}^N)} < \infty.$$

Again it is sufficient to show that

$$\operatorname{meas}_N F(\mathcal{C}_i \cap \mathcal{S}) = 0, \qquad i = 1,\dots,k.$$

Choose one of these cubes and denote it by $\mathcal{C}$. By the Mean Value Theorem (Theorem 3.2.7),

$$\|F(y) - F(x)\| \le c\|x-y\| \quad \text{and} \quad \|F(y) - L_x y\| \le \omega(\|x-y\|)\|x-y\|, \quad x,y \in \mathcal{C}$$

where $\lim\limits_{r\to 0_+} \omega(r) = 0$ (uniform continuity of F' on the compact set $\mathcal{C}$) and

$$L_x y := F(x) + F'(x)(y-x).$$

Divide now the cube $\mathcal{C}$ into m^N small cubes with edges $\frac{a}{m}$, and consider a small cube $\tilde{\mathcal{C}}$ which contains a critical point x. Denote the diameter of $\tilde{\mathcal{C}}$ by $\tilde{r}$ ($\tilde{r} = \sqrt{N}\frac{a}{m}$). Since $L_x(\tilde{\mathcal{C}})$ lies in an $(N-1)$-dimensional hyperplane (x is a critical point!),

$$\operatorname{meas}_{N-1} L_x(\tilde{\mathcal{C}}) \le (c\tilde{r})^{N-1} \quad \text{and} \quad \operatorname{dist}(F(y), L_x(\tilde{\mathcal{C}})) \le \omega(\tilde{r})\tilde{r} \quad \text{for } y \in \tilde{\mathcal{C}},$$

we have

$$\operatorname{meas}_N F(\tilde{\mathcal{C}}) \le c^{N-1}\omega(\tilde{r})\tilde{r}^N.$$

The number of such small cubes which contain critical points is m^N at most. Therefore,

$$\operatorname{meas}_N F(\mathcal{S} \cap \mathcal{C}) \le c^{N-1}\omega(\tilde{r})\tilde{r}^N m^N \le c_1\omega(\tilde{r})$$

with a constant c_1 independent of m. Since $\omega(\tilde{r}) \to 0$ for $m \to \infty$,

$$\operatorname{meas}_N F(\mathcal{S} \cap \mathcal{C}) = 0.$$

□

Corollary 5.7.4. *Under the hypotheses of Theorem 5.7.3 the set of regular values of $F\colon \Omega \to \mathbb{R}^N$ is dense in $\mathbb{R}^N$.*

Proof. The complement of regular values, i.e., $F(\Omega \cap \mathcal{S})$, cannot have an interior point and zero Lebesgue measure simultaneously. □

Remark 5.7.5. A more general Sard Theorem concerns $F\colon \mathbb{R}^M \to \mathbb{R}^N$. It is surprising that the assertion $\operatorname{meas}_N F(\mathcal{S} \cap \mathcal{C}) = 0$ needs more smoothness of F, namely

$$F \in C^r(\Omega, \mathbb{R}^N) \qquad \text{where} \qquad r > \max\{0, M - N\}.$$

The proof is more involved (see, e.g., Hirsch [80, Chapter 3, Theorem 1.3] for the C^∞-case and comments given there, or Sard [136], or Sternberg [146, Theorem II.3.1]). If $r \le \max\{0, M - N\}$, then there exists $F \in C^r(\Omega, \mathbb{R}^N)$ such that

$$\operatorname{int} F(\Omega \cap \mathcal{S}) \ne \emptyset,$$

see Whitney [157]. The statement on the Lebesgue measure can be strengthened by considering the finer Hausdorff measure or dimension. The following result also holds:

If $F\colon \mathbb{R}^M \to \mathbb{R}$ is analytic, then $F(\Omega \cap \mathcal{S})$ is even countable.

For more detail see, e.g., Fučík et al. [68, Chapter IV and Appendix IV]. There is also a generalization for mappings $F\colon X \to Y$, X, Y Banach spaces:

If $F'(x)$ is a Fredholm operator for all $x \in \Omega$ and F is sufficiently smooth, then $F(\Omega \cap \mathcal{S})$ is nowhere dense in Y.

Sharper results can be proved for functionals (i.e., $Y = \mathbb{R}$) (see the book Fučík et al. [68] cited above).

Now we return to the degree $\deg(F, \Omega, y_0)$ where $y_0 \in \mathbb{R}^N \setminus F(\partial\Omega)$ and it is a critical value of F. According to Corollary 5.7.4 there is a sequence $\{y_n\}_{n=1}^{\infty}$ of regular values of F such that

$$\lim_{n\to\infty} y_n = y_0, \qquad y_n \notin F(\partial\Omega).$$

In particular, $\deg(F, \Omega, y_n)$ is well defined by Definition 5.7.1. Part (vi) of Proposition 5.7.2 allows us to presume that the sequence $\{\deg(F, \Omega, y_n)\}_{n=1}^{\infty}$ is eventually constant and does not depend on the choice of the sequence $\{y_n\}_{n=1}^{\infty}$ of regular values. We want to avoid a rather long explanation of this fact (see, e.g., Fučík, S., Nečas, J., Souček, J. & Souček, V. [68]). Another (geometrical) approach to the construction of the degree will be presented in Appendix 5.7A.

We thus may define

$$\deg(F, \Omega, y_0) = \lim_{n\to\infty} \deg(F, \Omega, y_n)$$

also for a critical value y_0. It is then straightforward to verify that Proposition 5.7.2 is still valid for this more general definition of the degree.

For the definition of the degree $\deg(F, \Omega, y_0)$ it is not necessary to assume that $F \in C^1(\Omega, \mathbb{R}^N)$ since any $F \in C(\overline{\Omega}, \mathbb{R}^N)$ can be approximated by smooth mappings. This is a consequence of the Stone–Weierstrass Theorem (see Theorem 1.2.14).[33]

To show that $\deg(G, \Omega, y_0)$ is the same for all

$$G \in C(\overline{\Omega}, \mathbb{R}^N) \cap C^1(\Omega, \mathbb{R}^N)$$

which are close to F in the $C(\overline{\Omega}, \mathbb{R}^N)$-norm we need the following extension of Proposition 5.7.2(vii).

Proposition 5.7.6. *Let Ω be a bounded open subset of $\mathbb{R}^N$ and let F, G be the mappings from $C(\overline{\Omega}, \mathbb{R}^N) \cap C^1(\Omega, \mathbb{R}^N)$. Put*

$$H(t, x) = (1-t)F(x) + tG(x), \qquad t \in [0,1], \quad x \in \overline{\Omega}.$$

Assume that

$$y_0 \in \mathbb{R}^N \setminus \{H(t, x) : t \in [0,1],\ x \in \partial\Omega\}.$$

Then

$$\deg(F, \Omega, y_0) = \deg(G, \Omega, y_0). \tag{5.7.7}$$

Proof. As has been stated above, Proposition 5.7.2(vii) holds for an arbitrary $y_0 \in \mathbb{R}^N \setminus F(\partial\Omega)$, in particular for $H(t, \cdot)$ and t small. Put

$$t_0 = \sup\{t \in [0,1] : \deg(H(t, \cdot), \Omega, y_0) = \deg(F, \Omega, y_0)\}.$$

[33] For the set $\mathcal{A}$ in Theorem 1.2.14 take restrictions of polynomials of N variables to $\overline{\Omega}$ and approximate separately every F_i, $i = 1, \dots, N$, where $F = (F_1, \dots, F_N)$. For another proof see Lemma 5.7.23.

By the same statement ($\deg(H(t,\cdot),\Omega,y_0)$ is defined for all $t \in [0,1]$),

$$\deg(H(t_0,\cdot),\Omega,y_0) = \deg(H(t,\cdot),\Omega,y_0) \quad \text{for } t \in (t_0-\delta, t_0+\delta)\cap[0,1],$$

i.e., $t_0 = 1$, and the equality (5.7.7) follows. □

The following result is a summarization of the previous exposition.

Theorem 5.7.7. *Let Ω be a bounded open set in $\mathbb{R}^N$. There exists a mapping*

$$\deg : C(\overline{\Omega},\mathbb{R}^N)\times\mathbb{R}^N \to \mathbb{Z}$$

defined for all $F \in C(\overline{\Omega},\mathbb{R}^N)$ and $y_0 \in \mathbb{R}^N \setminus F(\partial\Omega)$ which has the properties (i)–(vii) *from Proposition* 5.7.2 *and from Proposition* 5.7.6.[34] *If, moreover, $F \in C^1(\Omega,\mathbb{R}^N)$ and y_0 is a regular value of F, then the formula* (5.7.1) *holds.*

Remark 5.7.8. The function "deg" from the previous theorem is unique. The reader can consult, e.g., Amann & Weiss [5] or Deimling [42] to get more information.

Example 5.7.9 (Brouwer). Let F be a continuous mapping from the closed unit ball $\mathcal{B} := \overline{B(o;1)}$ into itself. Then F has a fixed point in $\mathcal{B}$.

Indeed, if there is $x \in \partial\mathcal{B}$ such that

$$x - F(x) = o,$$

then the statement is true. In the other case put

$$H(t,x) = x - tF(x).$$

Then

$$H(1,x) \neq o$$

and

$$\|H(t,x)\| \geq \|x\| - t\|F(x)\| \geq 1 - t > 0 \qquad \text{for} \quad t \in [0,1),\ \|x\| = 1.$$

By the homotopy invariance property of the degree,

$$1 = \deg(I, \operatorname{int}\mathcal{B}, o) = \deg(I - F, \operatorname{int}\mathcal{B}, o).$$

By property (iv), the equation $x - F(x) = o$ has a solution in $\operatorname{int}\mathcal{B}$. ○

Example 5.7.10. Let $\mathcal{B} := \overline{B(o;1)}$ be the closed unit ball in $\mathbb{R}^N$ and let A be a linear injective operator from $\mathbb{R}^N$ into $\mathbb{R}^N$.[35] Then

$$\deg(A, \operatorname{int}\mathcal{B}, o) = (-1)^p \qquad \text{where} \quad p = \sum_{\substack{\lambda\in\sigma(A)\\ \lambda<0}} m(\lambda)$$

[34] Mappings F, G are now supposed to be continuous only.
[35] Actually, A maps $\mathbb{R}^N$ onto $\mathbb{R}^N$ and $o \notin A(\partial\mathcal{B})$.

and $m(\lambda)$ is the multiplicity[36] of the eigenvalue λ of A.

This follows immediately from Definition 5.7.1 and Exercise 1.1.40. Notice that the same result is true when A is replaced by a C^1-mapping $F\colon \mathbb{R}^N \to \mathbb{R}^N$ which has an isolated zero at $x_0 \in \mathbb{R}^N$, $\overline{B(o;1)}$ is replaced by a sufficiently small ball $\overline{B(x_0;r)}$ (such that the equation $F(x) = o$ has a unique solution in this closed ball, namely x_0) and $F'(x_0)$ is injective. Under these hypotheses we have

$$\deg(F, B(x_0;r), o) = (-1)^p \qquad \text{where} \quad p = \sum_{\substack{\lambda\in\sigma(F'(x_0))\\ \lambda<0}} m(\lambda). \tag{5.7.8}$$

The value $\deg(F, B(x_0;r), o)$ is also called the *index of an isolated solution* x_0 of the equation $F(x) = o$. ○

Example 5.7.11. Let Ω be a bounded open subset of $\mathbb{R}^N$ and let Y be a linear subspace of $\mathbb{R}^N$. Suppose that $f \in C(\overline{\Omega}, Y)$ is such that $o \neq f(\partial\Omega)$. Denote by π a projection of $\mathbb{R}^N$ onto Y, by $\tilde{f}$ the restriction of πf onto $\overline{\Omega} \cap Y$ and $g(x) := \tilde{f}(\pi x) + (I - \pi)x$. Then

$$\deg(g, \Omega, o) = \deg(\tilde{f}, \Omega \cap Y, o). \tag{5.7.9}$$

To see this notice first that

$$\partial(\Omega \cap Y) = \partial\Omega \cap Y \qquad \text{and} \qquad g_{-1}(o) = \tilde{f}_{-1}(o).$$

This means that both sides of (5.7.9) are defined. By the construction of the degree, it suffices to prove the equality under the following additional assumptions: $f \in C^1(\Omega)$ and o is a regular value. Since

$$g'(y)(h+k) = \tilde{f}'(y)h + k \qquad \text{for} \quad y \in Y \cap \Omega, \quad h \in Y, \quad k \in \operatorname{Ker}\pi,$$

we get

$$\det g'(y) = \det \tilde{f}'(y),$$ [37]

and (5.7.9) follows. ○

Exercise 5.7.12. Modify the proof of (5.7.9) to obtain the so-called *Product Formula:*

$$\deg(g, \Omega, y_0) = \deg(f_1, \Omega_1, y_{1,0}) \deg(f_2, \Omega_2, y_{2,0})$$

where $g = (f_1, f_2)\colon \Omega \to \mathbb{R}^{N_1+N_2}$, $\Omega = \Omega_1 \times \Omega_2$, $y_0 = (y_{1,0}, y_{2,0})$, $\Omega_i \subset \mathbb{R}^{N_i}$, $f_i \in C(\overline{\Omega_i}, \mathbb{R}^{N_i})$, $y_{i,0} \notin f_i(\partial\Omega_i)$, $i = 1, 2$.

---

[36] See footnote 23 on page 85, $m(\lambda)$ is often called the algebraic multiplicity.
[37] The reader is asked to check this equality.

5.7A Brouwer Topological Degree on Manifolds

We established the main properties of the degree of a mapping $f\colon \mathbb{R}^M \to \mathbb{R}^M$ in the basic text (Section 5.7). In this appendix we will present another definition of the Brouwer degree, prove its main properties and give some of its topological applications.

This goal can be achieved in different ways but all of them are lengthy and contain intricate calculations. Here we choose the treatment based on the integration of differential forms, mainly to introduce the interested reader to a "geometric world" and to relate the notion of the degree to classical results of the theory of functions of a complex variable. Since C. F. Gauss was one of the fathers of this theory we will start with his approach to the so-called *Fundamental Theorem of Algebra.*

Theorem 5.7.13 (Fundamental Theorem of Algebra). *Let P be a polynomial with complex coefficients of a degree at least one. Then there exists $z_0 \in \mathbb{C}$ such that $P(z_0) = 0$. Equivalently, $P\colon \mathbb{C} \to \mathbb{C}$ is a surjective mapping.*

We will give two rather different proofs the ideas of which go back to Gauss. The former is purely geometric with a little analysis for introducing a differentiable structure on the sphere $\mathcal{S}^2$. The latter uses the theory of functions of a complex variable and demonstrates the above mentioned connection to the degree.

Proof (see, e.g., Milnor [112]). We will regard the sphere $\mathcal{S}^2$ as a compactification of the complex plane and endow it with the structure of a differentiable manifold by two charts (see Example 4.3.5(iii)) given by the stereographic projections π_+, π_- of $\mathcal{S}^2 \setminus \{N\}$ where N is the north pole, and $\mathcal{S}^2 \setminus \{S\}$ where S is the south pole, respectively, onto $\mathbb{R}^2$. See Figure 5.7.4.

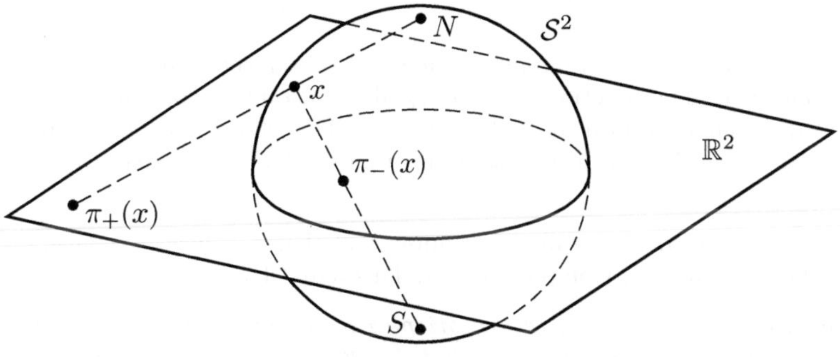

Figure 5.7.4.

Let P be a non-constant polynomial. We define f to be

$$f(x) := \begin{cases} \pi_+^{-1} \circ P \circ \pi_+(x), & x \in \mathcal{S}^2 \setminus \{N\}, \\ N, & x = N. \end{cases}$$

One can prove that $f\colon \mathcal{S}^2 \to \mathcal{S}^2$ is continuously differentiable at all points. To prove differentiability at N it is sufficient to verify another formula for f, namely

$$f = \pi_-^{-1} \circ Q \circ \pi_- \qquad \text{where} \quad Q(z) = \frac{1}{\overline{P(\overline{z}^{-1})}}, \quad Q(0) = 0$$

($\overline{z}$ is the complex conjugate to z). The calculation of f_* (see (4.3.19)) shows that $x_0 \in \mathcal{S}^2$ is a singular point of f (i.e., $f_*(T_a\mathcal{S}^2) \neq T_{f(a)}\mathcal{S}^2$) if and only if

$$P'(z_0) = o, \qquad z_0 = \pi_+(x_0).$$

Since the last equation has only a finite number of solutions, the set $\mathcal{A}$ of singular values of f is finite. This is the main point where we have used that P is a polynomial. Consider now a point $y \in \mathcal{S}^2 \setminus \mathcal{A}$, i.e., y is a regular value of f. Then the set $f_{-1}(y)$ is finite (possibly empty) since a polynomial takes any value only finitely many times. Let $\varphi(y)$ denote the number of points, i.e., the cardinality, of $f_{-1}(y)$. The main part of the proof is to show that φ is a constant function on $\mathcal{B} := \mathcal{S}^2 \setminus \mathcal{A}$. To prove that we consider two kinds of regular values,

$$\mathcal{B}_1 = \{y : f_{-1}(y) = \emptyset\}, \qquad \mathcal{B}_2 = \mathcal{B} \setminus \mathcal{B}_1.$$

Since $\mathcal{B}_1 = \mathcal{S}^2 \setminus f(\mathcal{S}^2)$ and $\mathcal{S}^2$ is compact, $\mathcal{B}_1$ is open in $\mathcal{S}^2$. For $y \in \mathcal{B}_2$ we have

$$f_{-1}(y) = \{x_1, \dots, x_k\}$$

and there are disjoint open neighborhoods $\mathcal{U}(x_1), \dots, \mathcal{U}(x_k)$ on which f is a diffeomorphism (see Local Inverse Function Theorem 4.1.1). Put

$$\mathcal{V}_i := f(\mathcal{U}(x_i)).$$

It is easy to see that φ is constant on the set

$$\bigcap_{i=1}^{k} \left[\mathcal{V}_i \setminus f\left(\mathcal{S}^2 \setminus \bigcup_{i=1}^{k} \mathcal{U}(x_i)\right)\right],$$

which is a neighborhood of the point y. Since $\mathcal{S}^2$ is connected and $\mathcal{A}$ is finite, the set $\mathcal{B}$ is also connected. The function φ, being locally constant, is constant on the whole of $\mathcal{B}$. Moreover, φ cannot vanish on $\mathcal{B}$, i.e., $\mathcal{B}_1 = \emptyset$. This shows that f actually maps $\mathcal{S}^2$ onto $\mathcal{S}^2$, i.e., $P\colon \mathbb{C} \to \mathbb{C}$ is surjective as well. In particular, there exists $z_0 \in \mathbb{C}$ such that $P(z_0) = 0$. □

The following result generalizes the above fact that f is surjective (see, e.g., Sternberg [146, Theorem 3.4.3]). Compare it with Theorem 4.1.7.

Proposition 5.7.14. *Let $\mathscr{M}_1$, $\mathscr{M}_2$ be two oriented manifolds of the same dimension, and let $\mathscr{M}_2$ be a connected space. Suppose that $f\colon \mathscr{M}_1 \to \mathscr{M}_2$ is a proper*[38] *differentiable mapping such that its realization (see Figure 4.3.15) has a nonnegative Jacobian at any point. Then either the Jacobian vanishes everywhere or $f(\mathscr{M}_1) = \mathscr{M}_2$.*

Remark 5.7.15. Write $f\colon \mathbb{C} \to \mathbb{C}$ as

$$f(z) = g(x,y) + \mathrm{i}h(x,y) \qquad \text{where} \quad z = x + \mathrm{i}y \quad \text{and} \quad g, h\colon \mathbb{R}^2 \to \mathbb{R}.$$

If f is a holomorphic function in an open set $\mathcal{G}$, then the *Cauchy–Riemann conditions*

$$\frac{\partial g}{\partial x} = \frac{\partial h}{\partial y}, \qquad \frac{\partial g}{\partial y} = -\frac{\partial h}{\partial x} \qquad \text{hold for} \quad z = x + \mathrm{i}y \in \mathcal{G},$$

[38] A mapping f is said to be *proper* if $f_{-1}(\mathcal{K})$ is compact whenever $\mathcal{K}$ is a compact set.

and the Jacobian of $(g,h)\colon \mathbb{R}^2 \to \mathbb{R}^2$ satisfies

$$J_{(g,h)}(x,y) = |f'(z)|.$$

In particular, $J_{(g,h)} \geq 0$ and if f is a polynomial, then f is proper (why?), and so Proposition 5.7.14 can be used to get the Fundamental Theorem of Algebra.

The idea of the latter proof of Theorem 5.7.13 is based on the notion of the index of a point a with respect to a curve. If γ is a closed C^1-curve in $\mathbb{C}$, $a \notin \gamma$,[39] then the *index*, $\operatorname{Ind}_\gamma a$, is defined by the formula

$$\operatorname{Ind}_\gamma a = \frac{1}{2\pi \mathrm{i}} \int_\gamma \frac{\mathrm{d}z}{z-a}.$$

We say that γ is *positively oriented* if

$$\operatorname{Ind}_\gamma a \geq 0 \qquad \text{for all} \quad a \notin \gamma.$$[40]

In particular, if

$$a = 0 \qquad \text{and} \qquad \gamma(t) = r\mathrm{e}^{\mathrm{i}t}, \qquad t \in [0, 2n\pi], \quad n \in \mathbb{Z},$$

then

$$\operatorname{Ind}_\gamma 0 = n$$

and it can be interpreted as the number of revolutions of γ and also as the increment of the argument along γ divided by 2π.

Proposition 5.7.16 (Rouché). *Let γ be a simple, closed, positively oriented C^1-curve in an open set Ω, and let*

$$\mathcal{G} := \{z \in \Omega \setminus \gamma : \operatorname{Ind}_\gamma z \neq 0\}.$$

If f is a holomorphic function in Ω for which $0 \notin f(\gamma)$, then the number $N_f(\mathcal{G})$ of solutions of the equation $f(z) = 0$ that belong to $\mathcal{G}$ is equal to

$$N_f(\mathcal{G}) = \frac{1}{2\pi \mathrm{i}} \int_\gamma \frac{f'(z)}{f(z)}\,\mathrm{d}z = \operatorname{Ind}_{f\circ\gamma} 0 \tag{5.7.10}$$[41]

provided the solutions are counted with their multiplicity.[42] *If g is another holomorphic function in Ω such that*

$$|f(z) - g(z)| < |f(z)|, \qquad z \in \gamma, \tag{5.7.11}$$

then $N_f(\mathcal{G}) = N_g(\mathcal{G})$.

Proof. The proof is based on the Residue Theorem, see, e.g., Rudin [132, Theorem 10.43]. □

---

[39] Not to complicate matters by different notation we identify the curve, i.e., a mapping $\gamma\colon$ interval $\mathcal{I} \to \mathbb{C}$, with its image.

[40] This definition, common in the theory of functions of a complex variable, coincides with our definition of an oriented manifold.

[41] This quantity is also called the *winding number* of f at 0 with respect to γ.

[42] A solution z_0 has multiplicity k if $f(z_0) = \cdots = f^{(k-1)}(z_0) = 0$, $f^{(k)}(z_0) \neq 0$. Notice that k is finite provided f is not identically zero.

We wish to point out that the condition (5.7.11) is a quantitative description of stability of the number of solutions with respect to perturbations of f.

The second proof of Theorem 5.7.13. Our second proof of the Fundamental Theorem of Algebra follows easily from the previous proposition:

Suppose that

$$P(z) = z^n + a_1 z^{n-1} + \cdots + a_n$$

and put $f(z) = z^n$. Let $R > 0$ be such that

$$|a_1 z^{n-1} + \cdots + a_n| < |z^n| = R^n \qquad \text{for} \quad |z| = R$$

(why does such an R exist?). Proposition 5.7.16 implies that

$$n = N_f(\mathcal{G}) = N_P(\mathcal{G})$$

where $\mathcal{G}$ is the open ball $B(0; R)$. □

The connection between the winding number and the degree $\deg(f, \Omega, p)$ as the latter is defined in Definition 5.7.1 for a regular value p is given by the following result. For a holomorphic function f and its regular value p the degree $\deg(f, \Omega, p)$ is defined as the number of solutions in Ω of the equation $f(z) = p$. By identification of

$$f\colon \mathbb{C} \to \mathbb{C}, \quad f(z) = g(x, y) + \mathrm{i}h(x, y) \qquad \text{with} \qquad (g, h)\colon \mathbb{R}^2 \to \mathbb{R}^2,$$

this definition coincides with Definition 5.7.1.

Lemma 5.7.17. *Let Ω be an open, bounded set whose boundary is the image of a C^1-simple, closed, positively oriented curve γ. Assume that $f\colon \mathbb{C} \to \mathbb{C}$ is holomorphic in a certain neighborhood of Ω and $p \notin f(\partial\Omega)$ is a regular value of f (i.e., $f'(z) \neq 0$ whenever $f(z) = p$). Then*

$$\deg(f, \Omega, p) = \frac{1}{2\pi\mathrm{i}} \int_\gamma \frac{f'(z)}{f(z) - p}\, \mathrm{d}z. \tag{5.7.12}$$

Proof. Denote

$$\mathcal{A} = \{z \in \Omega : f(z) = p\}.$$

If $\mathcal{A} = \emptyset$, then both sides vanish (use the Cauchy Theorem for the right-hand side). If $\mathcal{A} \neq \emptyset$, then $\mathcal{A}$ is finite since

$$\mathcal{A} \cap \partial\Omega = \emptyset,$$

and f is holomorphic. Both sides in (5.7.12) are equal to the cardinality of $\mathcal{A}$. This follows for the left-hand side from the definition of the degree and for the right-hand side by the Residue Theorem. □

We would like to point out here that the formula (5.7.12) indicates a way to remove the assumption on regularity of p. Namely, the integral exists for any holomorphic function f provided $p \notin f(\partial\Omega)$. Put

$$d := \inf_{z \in \partial\Omega} |f(z) - p|,$$

and let $\mathcal{A} = \{z_1, \dots, z_n\}$ be the same as in the proof of Lemma 5.7.17. Assume that the solution z_j is of multiplicity m_j, and denote

$$m := m_1 + \cdots + m_k.$$

According to Proposition 5.7.16, the number of solutions (including multiplicity) of the equation $f(z) = q$ is also m provided $|p-q| < d$. In this neighborhood of the point p there exists a regular value of f. This follows either from the Sard Theorem (see Theorem 5.7.3) or, in this special case more easily from the properties of holomorphic functions – see, e.g., Rudin [132, Theorem 10.32]. If the degree has the property of stability with respect to the point (Theorem 5.7.7 (vi) or Theorem 5.7.26 (viii)), the equality (5.7.12) will also hold for singular values.

In order to motivate the next result, we note that the integral in (5.7.12) can be rewritten as an integral over Ω (by the Green Theorem). We avoid such uninteresting and intricate calculation, and put aside the special case of holomorphic functions. Instead we will consider the general case of mappings $\mathbb{R}^M \to \mathbb{R}^M$. For the rest of this appendix we will suppose that

$$\text{(H)} \quad \begin{cases} \Omega \text{ is a bounded open set in } \mathbb{R}^M, \\ f\colon \overline{\Omega} \to \mathbb{R}^M \text{ is continuous on } \overline{\Omega}, \\ p \in \mathbb{R}^M \setminus f(\partial\Omega). \end{cases}$$

Proposition 5.7.18. *In addition to* (H) *let* $f \in C^1(\Omega, \mathbb{R}^M)$ *and let* p *be a regular value of* f*. Then there exist a neighborhood* $\mathcal{U}$ *of the point* p *and a smooth function* $\alpha\colon \mathbb{R}^M \to \mathbb{R}$ *supported in* $\mathcal{U}$ *with* $\int_{\mathcal{U}} \alpha(y)\,\mathrm{d}y = 1$ *such that*

$$\int_\Omega f^*\omega = \sum_{\substack{x\in\Omega \\ f(x)=p}} \operatorname{sgn} J_f(x) \qquad (=: \deg(f,\Omega,p)). \tag{5.7.13}$$

Here $f^*\omega$ *is the pull-back of the form*

$$\omega(y) = \alpha(y)\,\mathrm{d}y^1 \wedge \cdots \wedge \mathrm{d}y^M$$

(*see Remark* 4.3.57(iii))*.*

Proof.[43] By the definition of the pull-back we have

$$f^*\omega(x) = \alpha(f(x))J_f(x)\,\mathrm{d}x^1 \wedge \cdots \wedge \mathrm{d}x^M \quad \text{and} \quad \int_\Omega f^*\omega\,\mathrm{d}x = \int_\Omega \alpha(f(x))J_f(x)\,\mathrm{d}x$$

where $J_f(x)$ is the Jacobian of f at x. We consider two complementary cases:

Case 1 ($p \notin f(\overline{\Omega})$). Then there is a neighborhood $\mathcal{U}$ of p which is a subset of $\mathbb{R}^M \setminus f(\overline{\Omega})$. Choose a smooth function α on $\mathbb{R}^M$ with its support in $\mathcal{U}$ and such that $\int_\Omega \alpha(y)\,\mathrm{d}y = 1$. Then

$$\int_\Omega f^*\omega = 0 \qquad \text{since} \qquad \alpha|_{f(\Omega)} = 0.$$

The right-hand side of (5.7.13) also vanishes by definition ($\sum\limits_{\emptyset} = 0$).

[43] The proof based on a more explicit construction of the form ω is given by Mawhin [109]. There the reader can also find a different approach to the homotopy invariance property of the degree (see Lemma 5.7.19 below).

Case 2 ($p \in f(\Omega)$). Since p is a regular value, the set

$$\mathcal{A} := \{x \in \Omega : f(x) = p\}$$

is finite. Let $\mathcal{A} := \{x_1, \dots, x_k\}$. Then for any $x_i \in \mathcal{A}$ there is a neighborhood $\mathcal{V}_i \subset \Omega$ of x_i such that $f|_{\mathcal{V}_i}$ is a diffeomorphism of $\mathcal{V}_i$ onto a neighborhood $\mathcal{U}_i$ of p. These neighborhoods $\mathcal{V}_1, \dots, \mathcal{V}_k$ can be chosen mutually disjoint. Take a neighborhood $\mathcal{U} \subset \bigcap_{i=1}^{k} \mathcal{U}_i$ of p such that

$$f_{-1}(\mathcal{U}) \subset \bigcup_{i=1}^{k} \mathcal{V}_i$$

(why does such a $\mathcal{U}$ exist?). Choose a smooth function α with its support in $\mathcal{U}$ and normalize α so that $\int_{\mathcal{U}} \alpha(y)\,\mathrm{d}y = 1$. Then

$$\begin{aligned}\int_{\Omega} f^*\omega &= \sum_{i=1}^{k} \int_{\mathcal{V}_i} f^*\omega = \sum_{i=1}^{k} \int_{\mathcal{V}_i} \alpha(f(x)) J_f(x)\,\mathrm{d}x = \sum_{i=1}^{k} \operatorname{sgn} J_f(x_i) \int_{\mathcal{V}_i} \alpha(f(x))|J_f(x)|\,\mathrm{d}x \\ &= \sum_{i=1}^{k} \operatorname{sgn} J_f(x_i) \int_{f(\mathcal{V}_i)} \alpha(y)\,\mathrm{d}y = \sum_{i=1}^{k} \operatorname{sgn} J_f(x_i).\end{aligned}$$

□

We define the degree $\deg(f, \Omega, p)$ for $f \in C^1(\Omega, \mathbb{R}^M)$ and p a regular value of f by the right-hand side of (5.7.13). This definition coincides with Definition 5.7.1. We also want to point out that Case 2 of the proof contains the essence of the use of differential forms for the definition of the degree. The formula (5.7.13) can hardly be used for computation of the degree but its advantage consists in the fact that the integral exists also if p is a singular value. However, we should be careful and examine whether this integral does not depend on the choice of an M-form ω also in the case when p is a singular value.

Lemma 5.7.19. *Let ω be a smooth M-form in $\mathbb{R}^M$ with its support in a certain open cube $\mathcal{Q}$. If*

$$\int_{\mathcal{Q}} \omega = 0,$$

then there exists a smooth $(M-1)$-form η with support in $\mathcal{Q}$ such that

$$\mathrm{d}\eta = \omega,$$

i.e., ω is exact.

Proof. It is similar to that of the Poincaré Theorem (Theorem 4.3.68). □

Corollary 5.7.20. *Suppose* (H) *and $f \in C^1(\Omega, \mathbb{R}^M)$. Let ω, $\tilde{\omega}$ be two smooth M-forms with supports in a cube $\mathcal{Q} \subset \mathbb{R}^M \setminus f(\partial\Omega)$ such that $\int_{\mathcal{Q}} \omega = \int_{\mathcal{Q}} \tilde{\omega}$. Then*

$$\int_{\Omega} f^*\omega = \int_{\Omega} f^*\tilde{\omega}.$$

Proof. According to Lemma 5.7.19 there exists an $(M-1)$-form η for which $\mathrm{d}\eta = \omega - \tilde{\omega}$. Then

$$f^*(\omega - \tilde{\omega}) = f^*(\mathrm{d}\eta) = \mathrm{d}(f^*\eta)$$

(see Exercise 4.3.75(iv)). Since the support of the form $f^*\eta$ is a compact subset of Ω, the Stokes Theorem (Theorem 4.3.98) implies that

$$\int_\Omega \mathrm{d}(f^*\eta) = 0. \qquad \square$$

Let p be a singular value of f and $\mathcal{B} := B(p;r)$ be such a ball that

$$\mathcal{B} \cap f(\partial\Omega) = \emptyset.$$

According to the Sard Theorem (see Corollary 5.7.4) there is a regular value q of f in $\mathcal{B}$. We want to define

$$\deg(f,\Omega,p) := \deg(f,\Omega,q) = \int_\Omega f^*\omega.$$

To be sure that this definition does not depend on the choice of the point q we choose another regular value $\tilde{q} \in \mathcal{B}$. It is obvious that a diffeomorphism h of $\mathcal{B}$ (onto $\mathcal{B}$) can be constructed such that

$$h(q) = \tilde{q},$$

and $J_h(x) > 0$ for all $x \in \mathcal{B}$. Let $\tilde{\omega}$ be a differential form constructed in the proof of Proposition 5.7.18 supported on a cube $\tilde{\mathcal{U}}$, $\tilde{q} \in \tilde{\mathcal{U}} \subset \mathcal{B}$. Then by the definition of pull-back

$$\int_{\tilde{\mathcal{U}}} \tilde{\omega} = \int_{\mathcal{U}} h^*\tilde{\omega} = 1.$$

The form $\tilde{\omega}$ can be chosen in such a way that $h^*\tilde{\omega}$ is supported by the same cube $\mathcal{Q}$ as the form ω (explain why!). By Corollary 5.7.20,

$$\int_{\mathcal{Q}} f^*\omega = \int_{\mathcal{Q}} f^*(h^*\tilde{\omega}).$$

This shows that the assumption on regularity of p can be omitted in the definition of the degree.

Now we may drop the assumption on smoothness of f. However, it will need a certain further effort.

Lemma 5.7.21. *Let Ω be a bounded open set in $\mathbb{R}^M$ and let $H\colon [0,1] \times \overline{\Omega} \to \mathbb{R}^M$ be such that the mapping $t \in [0,1] \mapsto H(t,\cdot) \in C^1(\Omega,\mathbb{R}^M) \cap C(\overline{\Omega},\mathbb{R}^M)$ is continuous. If $p \in \mathbb{R}^M \setminus H([0,1]\times\partial\Omega)$, then*

$$\deg(H(t,\cdot),\Omega,p)$$

is constant on the interval $[0,1]$.

Proof. First note that H is continuous as a mapping considered on $[0,1]\times\overline{\Omega}$, and therefore $H([0,1]\times\partial\Omega)$ is compact and there is an open cube

$$\mathcal{Q} \subset \mathbb{R}^M \setminus H([0,1]\times\partial\Omega)$$

which is a neighborhood of p. Choose a smooth M-form ω with its support in $\mathcal{Q}$. By definition,

$$\deg(H(t,\cdot),\Omega,p) = \int_\Omega (H(t,\cdot))^*\omega.$$

This shows that the function

$$t \mapsto \deg\left(H(t,\cdot),\Omega,p\right)$$

is continuous on $[0,1]$. Taking only integer values it has to be constant. □

Corollary 5.7.22. *Suppose* (H) *and denote* $d := \operatorname{dist}(p, f(\partial\Omega))$. *If* g, h *are mappings from* $C^1(\Omega,\mathbb{R}^M)\cap C(\overline{\Omega},\mathbb{R}^M)$ *such that*

$$\|f-g\|_{C(\overline{\Omega},\mathbb{R}^M)} := \sup_{x\in\overline{\Omega}} |f(x)-g(x)| < d, \qquad \|f-h\|_{C(\overline{\Omega},\mathbb{R}^M)} < d,$$

then

$$\deg(g,\Omega,p) = \deg(h,\Omega,p).$$

Proof. Put

$$H(t,x) := (1-t)g(x) + th(x), \qquad t\in[0,1], \quad x\in\overline{\Omega}.$$

The assertion now follows from Lemma 5.7.21. □

The last step in our approach to the definition of the degree consists in the approximation of a continuous mapping by smooth ones.

Lemma 5.7.23. *Let* Ω *be a bounded open set in* $\mathbb{R}^M$ *and let* $f\colon \overline{\Omega}\to\mathbb{R}^M$ *be continuous. Then there exist mappings* $f_n \in C^1(\Omega,\mathbb{R}^M)\cap C(\overline{\Omega},\mathbb{R}^M)$ *that converge uniformly on* $\overline{\Omega}$ *to* f.

Proof. Observe first that it is sufficient to prove the statement for individual components of f. So we will assume that $f\colon\overline{\Omega}\to\mathbb{R}$ is continuous.

There are many ways how to prove the density of smooth functions in the space of continuous functions. See the discussion on page 32. We will present the approach based on convolution approximations (see Proposition 1.2.20 and the proof of Proposition 1.2.21). Extend f to a continuous bounded function g on $\mathbb{R}^M$ (such an extension exists by the Tietze Theorem[44]). Choose a nonnegative C^∞-function $\varphi\colon\mathbb{R}^M\to\mathbb{R}$ with compact support and $\displaystyle\int_{\mathbb{R}^M}\varphi(x)\,\mathrm{d}x = 1$. Put

$$\varphi_n(x) = n^M\varphi(nx).$$

Then the convolutions

$$(\varphi_n * g)(x) := \int_{\mathbb{R}^M}\varphi_n(x-y)g(y)\,\mathrm{d}y, \qquad x\in\mathbb{R}^M,$$

[44]The Tietze Theorem says:

Assume that g *is a bounded, continuous, real function on a closed non-void subset* $\mathcal{A}$ *of a metric space* X *equipped with a metric* ϱ. *Then there exists a continuous extension* $G\colon X\to\mathbb{R}$ *such that*

$$\sup_{x\in\mathcal{A}} g(x) = \sup_{x\in X} G(x), \qquad \inf_{x\in\mathcal{A}} g(x) = \inf_{x\in X} G(x).$$

This theorem permits generalization (nontrivial) to normal topological spaces (see, e.g., Dugundji [52, Section VII.5]). The proof for metric spaces is quite easy. Indeed, without loss of generality we can suppose that $1\le g\le 2$. Put $G(x) = \inf_{y\in\mathcal{A}}\frac{\varrho(x,y)g(y)}{\operatorname{dist}(x,\mathcal{A})}$ for $x\in X\setminus\mathcal{A}$, and $G(x) = g(x)$ for $x\in\mathcal{A}$. It is not difficult to show that G has all the required properties.

are C^∞-functions which converge to g locally uniformly on $\mathbb{R}^M$, in particular, uniformly to f on $\overline{\Omega}$. Proposition 1.2.20(iii) implies the smoothness of $\varphi_n * g$. To see the convergence it is convenient to visualize the form of φ_n (see Figure 5.7.5 for $M = 1$).[45]

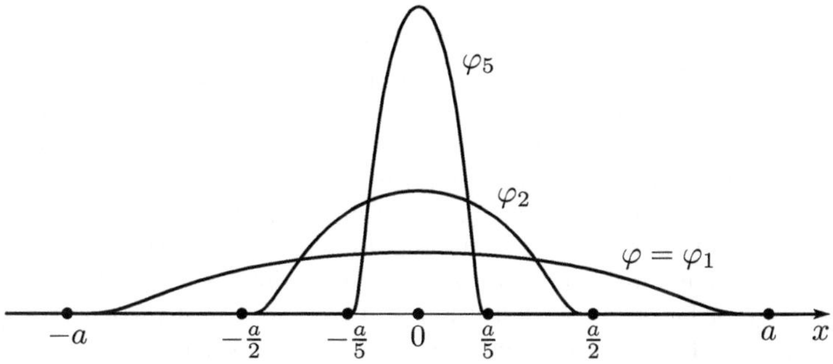

Figure 5.7.5.

The convergence is obtained as follows:

$$\begin{aligned}|(\varphi_n * g)(x) - g(x)| &\le \int_{\mathbb{R}^M} \varphi_n(x-y)|g(y)-g(x)|\,\mathrm{d}y \\ &= \int_{\mathcal{U}(x)} \varphi_n(x-y)|g(y)-g(x)|\,\mathrm{d}y \\ &\quad + \int_{\mathbb{R}^M\setminus\mathcal{U}(x)} \varphi_n(x-y)|g(y)-g(x)|\,\mathrm{d}y < \delta + 0.\end{aligned}$$

The first integral is arbitrarily small for a sufficiently small neighborhood $\mathcal{U}(x)$ of x by the continuity of g at x; the second integral is zero for sufficiently large n since $\varphi_n(x-\cdot)$ vanishes on $\mathbb{R}^M \setminus \mathcal{U}(x)$ for such n. □

Corollary 5.7.24. *Suppose* (H) *and let* $\{f_n\}_{n=1}^\infty$ *be a sequence the existence of which follows from Lemma* 5.7.23. *Then*

$$\lim_{n\to\infty} \deg(f_n, \Omega, p)$$

exists and its value does not depend on $\{f_n\}_{n=1}^\infty$ *whenever this sequence possesses the properties from Lemma* 5.7.23.

Proof. Since $\operatorname{dist}(p, f(\partial\Omega))$ is positive, we conclude that $p \notin f_n(\partial\Omega)$ for all sufficiently large n and the degrees $\deg(f_n, \Omega, p)$ are defined. Corollary 5.7.22 shows that the sequence

$$\{\deg(f_n, \Omega, p)\}_{n=1}^\infty$$

is eventually constant. This corollary also yields the independence of the limit of the sequence of degrees on the choice of $\{f_n\}_{n=1}^\infty$. □

[45] It is also said that φ_n converge to the Dirac measure (this is true in the sense of distributions) or that $\{\varphi_n\}_{n=1}^\infty$ is the so-called approximate unit (the space $L^1(\mathbb{R}^M)$ with the convolution multiplication is a Banach algebra without a unit, and the convergence $\varphi_n * g \to g$ takes place in the L^1-norm for all $g \in L^1(\mathbb{R}^M)$).

The previous corollary shows how the definition of the degree $\deg(f,\Omega,p)$ is extended to all triples (f,Ω,p) satisfying (H).

Remark 5.7.25. The approach which has just been described may be extended to a mapping

$$f\colon \mathscr{M} \to \tilde{\mathscr{M}}$$

between manifolds of the same dimension. We need to assume that both $\mathscr{M}$ and $\tilde{\mathscr{M}}$ are oriented (in order for the integral of an M-form to be defined, cf. Proposition 5.7.18 where instead of $J_f(x)$ we take the Jacobian of a realization of f). A set $\Omega \subset \mathscr{M}$ is supposed to be an open set with compact closure in the topology of $\mathscr{M}$.

We consider the degree of $f \in C(\overline{\Omega}, \tilde{\mathscr{M}})$ with respect to a point $p \in \tilde{\mathscr{M}} \setminus f(\partial\Omega)$. Here $\partial\Omega$ is again the boundary of Ω in the topology of $\mathscr{M}$. M-forms have their supports in some coordinate neighborhoods $\mathcal{V}$ of p which are disjoint with $f(\partial\Omega)$.

An analogue of Lemma 5.7.19 still holds. There are problems with an analogue of Corollary 5.7.22 – we have to use a topology on the space of mappings $\mathscr{M} \to \tilde{\mathscr{M}}$ since we have not defined any metric on a manifold. The existence of an approximating sequence similar to that of Lemma 5.7.23 is not clear, either. These obstacles can be overcome but special tools are required. Since they are beyond the scope of this book we only refer the interested reader to, e.g., the book Hirsch [80, Chapters 2 and 5].

We are now able to prove the main properties of the degree. See also Proposition 5.7.2 and Theorem 5.7.7. Notice that the following theorem is also true in the case of manifolds.

Theorem 5.7.26. *There exists a mapping* deg *which sends any triple (f,Ω,p) satisfying* (H) *into $\mathbb{Z}$ and has the following properties:*

(i) (*normalization property*) *If f is the identity map, $p \in \Omega$, then*

$$\deg(f,\Omega,p) = 1.$$

(ii) (*additivity property*) *If Ω_1, Ω_2 are disjoint open subsets of Ω and the point p is such that $p \notin f(\overline{\Omega} \setminus (\Omega_1 \cup \Omega_2))$, then*

$$\deg(f,\Omega,p) = \deg(f,\Omega_1,p) + \deg(f,\Omega_2,p).$$

(iii) (*continuity property*) *Let $\{f_n,\Omega,p\}$, $n = 0,1,\dots$, satisfy* (H). *If the sequence $\{f_n\}_{n=1}^{\infty}$ converges uniformly to f_0 on Ω, then*

$$\lim_{n\to\infty} \deg(f_n,\Omega,p) = \deg(f_0,\Omega,p).$$

(iv) (*translation invariance property*) $\deg(f,\Omega,p) = \deg(f-p,\Omega,o)$;

(v) *(solution property) If $\deg(f,\Omega,p) \neq 0$, then there exists an $x \in \Omega$ such that*

$$f(x) = p.$$

(vi) (*homotopy invariance property*) *If $H\colon [0,1] \times \overline{\Omega} \to \mathbb{R}^M$ is continuous and* (H) *is satisfied for all $(H(t,\cdot),\Omega,p)$, $t \in [0,1]$, then*

$$\deg(H(t,\cdot),\Omega,p)$$

is constant on $[0,1]$.

(vii) (*boundary values dependence property*) *If* (f,Ω,p), (g,Ω,p) *satisfy* (H) *and* f *and* g *coincide on* $\partial\Omega$, *then*

$$\deg(f,\Omega,p) = \deg(g,\Omega,p).$$

(viii) (*point dependence property*) *The mapping*

$$p \mapsto \deg(f,\Omega,p)$$

is constant on every component of $\mathbb{R}^M \setminus f(\partial\Omega)$.

(ix) (*multiplication property*) *Let* Ω *be a bounded open set in* $\mathbb{R}^M$ *and let the mapping* $f\colon \overline{\Omega} \to \mathbb{R}^M$ *be continuous. Denote by* $\mathcal{U}_1, \dots$ *all bounded components of* $\mathbb{R}^M \setminus f(\partial\Omega)$. *Suppose that the mapping* $g\colon \mathbb{R}^M \to \mathbb{R}^M$ *is continuous on* $f(\overline{\Omega})$ *and* $p \notin g(f(\partial\Omega))$. *Then*

$$\deg(g\circ f,\Omega,p) = \sum_i \deg(f,\Omega,\mathcal{U}_i)\deg(g,\mathcal{U}_i,p) \text{ [46]} \tag{5.7.14}$$

where the sum contains only a finite number of nonzero terms.

Proof. The degree is defined in Definition 5.7.1 for $f \in C(\overline{\Omega},\mathbb{R}^N) \cap C^1(\Omega,\mathbb{R}^N)$ and a regular value $p \notin f(\partial\Omega)$ and has been extended above by the procedure started with Proposition 5.7.18. It follows from this construction that it is sufficient to prove all parts of Theorem 5.7.26 for $f \in C(\overline{\Omega},\mathbb{R}^N) \cap C^1(\Omega,\mathbb{R}^N)$. Another proof is given on pages 286–287 (the proof of Proposition 5.7.2).

(i) This follows immediately from the definition.

(ii) This is a consequence of Proposition 5.7.18 since an M-form ω can be chosen in such a way that its support is disjoint with $f(\overline{\Omega} \setminus (\Omega_1 \cup \Omega_2))$. Then

$$\int_\Omega f^*\omega = \int_{\Omega_1} f^*\omega + \int_{\Omega_2} f^*\omega.$$

(iii) It is obtained directly from Corollary 5.7.24.

(iv) It follows from Proposition 5.7.18.

(v) This is slightly tricky: Suppose by contradiction that

$$f_{-1}(p) \cap \overline{\Omega} = \emptyset,$$

and choose four mutually disjoint, nonempty open subsets $\Omega_1, \dots, \Omega_4$ of Ω. Then, by the additivity property, we have

$$\deg(f,\Omega,p) = \deg(f,\Omega_1,p) + \deg(f,\Omega_2,p) = \deg(f,\Omega_3,p) + \deg(f,\Omega_4,p),$$

and also

$$\deg(f,\Omega,p) = \deg(f,\Omega_1\cup\Omega_2,p) + \deg(f,\Omega_3\cup\Omega_4,p) = \dots = 2\deg(f,\Omega,p).$$

This contradicts the inequality $\deg(f,\Omega,p) \neq 0$.

(vi) It follows from the construction – see Lemma 5.7.21.

[46] Because of property (viii), $\deg(f,\Omega,\mathcal{U}) := \deg(f,\Omega,q)$ for a $q \in \mathcal{U}$ is well defined for any component $\mathcal{U}$ of $\mathbb{R}^M \setminus f(\partial\Omega)$.

(vii) It is sufficient to apply property (vi) to

$$H(t,x) = tf(x) + (1-t)g(x).$$

(viii) Choose an M-form ω supported in an open set $\mathcal{U}$ where $\mathcal{U} \cap f(\partial\Omega) = \emptyset$. Then, by definition,

$$\deg(f,\Omega,p) = \int_\Omega f^*\omega \qquad \text{for all} \quad p \in \mathcal{U}.$$

This means that the degree is locally constant on $\mathbb{R}^M \setminus f(\partial\Omega)$, and therefore it is constant on every component of $\mathbb{R}^M \setminus f(\partial\Omega)$.

(ix) If the equation

$$g(f(x)) = p$$

has no solution in Ω, then the left-hand side of (5.7.14) vanishes (by the solution property). For the same reason all products on the right-hand side are also equal to zero. Suppose therefore that

$$f(\Omega) \cap g_{-1}(p) \neq \emptyset.$$

There is exactly one unbounded component $\mathcal{U}_0$ of $\mathbb{R}^M \setminus f(\partial\Omega)$. Regardless of whether $\mathcal{U}_0 \cap g_{-1}(p) \neq \emptyset$ or not,

$$\deg(f,\Omega,\mathcal{U}_0) = 0$$

(by (v) and (viii)). Since

$$g_{-1}(p) \cap f(\overline{\Omega})$$

is compact there is only a finite number of bounded components of $\mathbb{R}^M \setminus f(\partial\Omega)$, say $\mathcal{U}_1,\dots,\mathcal{U}_k$, which contain some points of $g_{-1}(p)$ (different components are disjoint). According to property (ii) we have

$$\deg(g\circ f,\Omega,p) = \sum_{i=1}^k \deg(g\circ f,\Omega_i,p) \tag{5.7.15}$$

where $\Omega_i = f_{-1}(\mathcal{U}_i)$. By definition, there is a neighborhood $\mathcal{V}$ of p, $\mathcal{V} \cap g(f(\partial\Omega)) = \emptyset$, and an M-form ω supported in $\mathcal{V}$ with $\int_\mathcal{V} \omega = 1$ for which

$$\deg(g\circ f,\Omega,p) = \int_\Omega (g\circ f)^*\omega.$$

First consider a component $\mathcal{U}_i$ such that

$$\deg(g,\mathcal{U}_i,p) = \int_{\mathcal{U}_i} g^*\omega \neq 0.$$

In this case put

$$\kappa_i = \frac{1}{\deg(g,\mathcal{U}_i,p)} g^*\omega|_{\mathcal{U}_i}.$$

Then κ_i is an admissible M-form for the definition of $\deg(f,\Omega_i,\mathcal{U}_i)$, and we have

$$\deg(g\circ f,\Omega_i,p) = \int_{\Omega_i} (g\circ f)^*\omega = \deg(g,\mathcal{U}_i,p)\int_{\Omega_i} f^*\kappa_i = \deg(g,\mathcal{U}_i,p)\deg(f,\Omega_i,\mathcal{U}_i). \tag{5.7.16}$$

Now consider a component $\mathcal{U}_i$ such that

$$\deg(g, \mathcal{U}_i, p) = \int_{\mathcal{U}_i} g^* \omega = 0.$$

Let

$$g^*\omega(y) = \kappa(y)\, \mathrm{d}y^1 \wedge \cdots \wedge \mathrm{d}y^M.$$

If κ does not vanish on $\mathcal{U}_i$, then there are disjoint open sets $\mathcal{U}_i^+$ and $\mathcal{U}_i^-$ which carry the positive part κ^+ and the negative part κ^-, respectively. Let $\Omega_i^{\pm} = f_{-1}(\mathcal{U}_i^{\pm})$. Then

$$\deg(f, \Omega_i, \mathcal{U}_i) = \deg(f, \Omega_i, \mathcal{U}_i^+) = \frac{\displaystyle\int_{\Omega_i^+} f^*(g^*\omega)}{\displaystyle\int_{\mathcal{U}_i^+} \kappa^+(y)\,\mathrm{d}y} = \deg(f, \Omega_i, \mathcal{U}_i^-) = -\frac{\displaystyle\int_{\Omega_i^-} f^*(g^*\omega)}{\displaystyle\int_{\mathcal{U}_i^-} \kappa^-(y)\,\mathrm{d}y}.$$

Notice that both denominators are nonzero since $\displaystyle\int_{\mathcal{U}_i} g^*\omega = 0$ and $g^*\omega$ does not vanish everywhere. This means that we have

$$\begin{aligned}\deg(g \circ f, \Omega_i, p) &= \int_{\Omega_i} f^*(g^*\omega) = \int_{\Omega_i^+} f^*(g^*\omega) + \int_{\Omega_i^-} f^*(g^*\omega)\\ &= \deg(f, \Omega_i, \mathcal{U}_i)\left[\int_{\mathcal{U}_i^+} \kappa^+(y)\,\mathrm{d}y - \int_{\mathcal{U}_i^-} \kappa^-(y)\,\mathrm{d}y\right] = \deg(f, \Omega_i, \mathcal{U}_i)\deg(g, \mathcal{U}_i, p),\end{aligned}$$

i.e., (5.7.16) holds in this case as well. Summing up and using (5.7.15) we complete the proof. □

Remark 5.7.27. It can be proved (see Amann & Weiss [5]) that the properties (i)–(iv) determine the degree uniquely.

We will conclude this appendix with some topological applications of the degree. Applications to differential equations are shown in Section 5.9. The well-known and basic Jordan Separation Theorem asserts that a Jordan curve divides the complex domain into two open components and exactly one of them is bounded (the interior domain of γ). This theorem has the following generalization.

Theorem 5.7.28 (Generalized Jordan Separation Theorem). *Let $\mathcal{K}$ be a compact set in $\mathbb{R}^M$ such that $\mathbb{R}^M \setminus \mathcal{K}$ has a finite number, say k, of components. If f is a continuous injection of $\mathcal{K}$ into $\mathbb{R}^M$, then $\mathbb{R}^M \setminus f(\mathcal{K})$ has exactly k components.*

Proof (a sketch). Notice first that f is actually a homeomorphism of $\mathcal{K}$, i.e., f^{-1} is continuous on the compact set $f(\mathcal{K})$. Applying the Tietze Theorem (see the footnote on page 300) to each coordinate f^i of $f = (f^1, \dots, f^M)$ and f^{-1} we conclude that there are continuous extensions g and h of f and f^{-1}, respectively, which are defined in $\mathbb{R}^M$. Denote by $\mathcal{G}_0, \dots, \mathcal{G}_{k-1}$ the components of $\mathbb{R}^M \setminus \mathcal{K}$ where $\mathcal{G}_0$ is the unique unbounded component. Similarly, let $\mathcal{U}_0, \dots, \mathcal{U}_m$ ($m \in \mathbb{N} \cup \{\infty\}$) denote the components of $\mathbb{R}^M \setminus f(\mathcal{K})$ where $\mathcal{U}_0$ is the unique unbounded component. The idea of the proof consists in the application of the multiplication property of the degree (Theorem 5.7.26 (ix)): Show that

$$\begin{aligned}\deg(h \circ g, \mathcal{G}_j, p) &= \delta_{ij} \quad \text{for } p \in \mathcal{G}_i,\\ \deg(g \circ h, \mathcal{U}_i, q) &= \delta_{ij} \quad \text{for } q \in \mathcal{U}_j,\end{aligned} \qquad i = 1, \dots, m,\ j = 1, \dots, k-1.$$

Define matrices

$$\boldsymbol{A} := (\deg(g, \mathcal{G}_j, \mathcal{U}_i))_{\substack{i=1,\dots,m\\ j=1,\dots,k-1}}, \qquad \boldsymbol{B} := (\deg(h, \mathcal{U}_i, \mathcal{G}_j))_{\substack{j=1,\dots,k-1\\ i=1,\dots,m}},$$

and show that

$$\boldsymbol{AB} = \boldsymbol{I}_m.$$

This means that $m \le k-1$. Similarly, $\boldsymbol{BA} = \boldsymbol{I}_{k-1}$. The equality $m = k-1$ follows. □

Corollary 5.7.29 (Invariance of domain). *Let $\mathcal{G} \subset \mathbb{R}^M$ be an open set and $f\colon \mathcal{G} \to \mathbb{R}^M$ a continuous injection. Then $f(\mathcal{G})$ is an open set.*

Proof (a sketch). Show first that it is sufficient to prove the assertion for the case when $\mathcal{G}$ is an open ball $\mathcal{B}$ and f is continuous and injective on $\overline{\mathcal{B}}$. According to Theorem 5.7.28, $\mathbb{R}^M \setminus f(\partial\mathcal{B})$ has exactly two components $\mathcal{U}_0, \mathcal{U}_1$. If $\mathcal{U}_0$ denotes the unbounded component, then

$$\mathbb{R}^M \setminus f(\overline{\mathcal{B}}) \subset \mathcal{U}_0$$

(again by Theorem 5.7.28), i.e., $\mathcal{U}_1 \subset f(\mathcal{B})$. To show the opposite inclusion recall that $f(\mathcal{B})$ is bounded and connected. □

Remark 5.7.30. If $M < N$ and $f\colon \mathbb{R}^M \to \mathbb{R}^N$ is a continuous injection, then it can be proved (not too easily) that the complement of $f(\mathbb{R}^M)$ is a dense set in $\mathbb{R}^N$. We want to recall the famous "Peano curve", i.e., a continuous (but not injective) map from the interval $[0,1]$ onto the square $[0,1] \times [0,1]$.[47] This map can be used to construct a continuous surjection of $\mathbb{R}^M$ onto $\mathbb{R}^N$ for any $M < N$. The existence of such a surjection for $M \ge N$ is trivial.

As we have mentioned in Section 4.3, our main interest in the degree theory consists in applying it to solving equations, i.e., in using the solution property (Theorem 5.7.26 (v)). This means to compute the degree, which is by no means an easy task. Fortunately, we do not need to know the exact value of the degree. It will be sufficient to show that it is not equal to zero. For this purpose the following mapping property is very important.

Definition 5.7.31.

(1) A nonempty subset $\mathcal{A}$ of a linear space X is said to be *symmetric* if

$$\text{for every} \quad x \in \mathcal{A} \qquad \text{we have} \qquad -x \in \mathcal{A}.$$

(2) A mapping $f\colon \mathcal{A} \subset X \to Y$, X, Y linear spaces, is called an *odd mapping* on a symmetric set $\mathcal{A}$ if

$$f(-x) = -f(x) \qquad \text{for each} \quad x \in \mathcal{A}.$$

Theorem 5.7.32 (Borsuk Antipodal Theorem). *Let Ω be a bounded, open, symmetric subset of $\mathbb{R}^M$ and $o \in \Omega$. Let $f\colon \overline{\Omega} \to \mathbb{R}^M$ be a continuous mapping whose restriction to $\partial\Omega$ is odd. If $o \notin f(\partial\Omega)$, then $\deg(f, \Omega, o)$ is an odd integer. In particular,*

$$\deg(f, \Omega, o) \neq 0,$$

[47]This example (for the construction see, e.g., Dugundji [52, Section IV.4]) has played an important role in developing the notion of a curve.

and there is a solution of the equation

$$f(x) = o \qquad \textit{in} \quad \Omega.$$

There are several proofs of this important topological theorem. For the proof based on algebraic machinery see, e.g., Dugundji [52, Section XVI.6]. Here we present the main steps of an "analytic" proof which is taken from Schwartz [138] (see also Nirenberg [118] or Rothe [130] or Krawcewicz & Wu [94]).

Proof of Theorem 5.7.32. The assertion is obvious for $M = 1$, therefore, we assume that $M > 1$. The main idea of this proof is quite simple. First observe that $\deg(f, \Omega, o)$ does not depend on a continuous extension of f from $\partial\Omega$ into Ω (see Theorem 5.7.26(vii)). Since $o \in \Omega$, there is a small open ball $\mathcal{B} := B(o;\delta)$ inside Ω, and there is a continuous mapping $g\colon \overline{\Omega} \to \mathbb{R}^M$ such that

$$g(x) = \begin{cases} f(x) & \text{for} \quad x \in \partial\Omega, \\ x & \text{for} \quad x \in \overline{\mathcal{B}} \end{cases}$$

(by the Tietze Theorem). Part (ii) of Theorem 5.7.26 implies that

$$\deg(g, \Omega, o) = \deg(g, \Omega \setminus \overline{\mathcal{B}}, o) + \deg(g, \mathcal{B}, o) = \deg(g, \Omega \setminus \overline{\mathcal{B}}, o) + 1.$$

If g is constructed in such a way that $g \in C^1(\Omega \setminus \overline{\mathcal{B}})$ and o is a regular value of g, then

$$\deg(g, \Omega \setminus \overline{\mathcal{B}}, o) = \sum_{x \in \mathcal{S}} \operatorname{sgn} J_g(x) \qquad \text{where} \quad \mathcal{S} = \{x \in \Omega \setminus \overline{\mathcal{B}} : g(x) = o\}.$$

If, moreover, g is odd, then $\mathcal{S}$ is either the empty set or a symmetric set and $g'(-x) = g'(x)$, $\deg(g, \Omega \setminus \overline{\mathcal{B}}, o)$ is an even integer, and the proof is completed. Unfortunately, it is not known whether such g does exist. Therefore, we will show that all the required properties of g are not actually needed. The demand for regularity of o can be replaced by the assumption that g does not vanish on a part of a hyperplane, for instance on

$$\mathcal{H} := \{(x_1, \dots, x_{M-1}, 0) \in \Omega \setminus \overline{\mathcal{B}}\}.$$

Indeed, in this case we have

$$\deg(g, \Omega \setminus \overline{\mathcal{B}}, o) = \deg(g, \mathcal{H}^+, o) + \deg(g, \mathcal{H}^-, o)$$

where

$$\mathcal{H}^{\pm} := \{(x_1, \dots, x_M) \in \Omega \setminus \overline{\mathcal{B}} : x_M > 0 \text{ or } x_M < 0\}.$$

Moreover,

$$\deg(g, \mathcal{H}^+, o) = \int_{\mathcal{H}^+} g^*\omega = \int_{\mathcal{H}^-} g^*\omega = \deg(g, \mathcal{H}^-, o)$$

since $J_g(x) = J_g(-x)$ and the mapping $x \in \mathcal{H}^+ \mapsto -x$ is a diffeomorphism. It can be shown, by smooth approximation, that the equality

$$\deg(g, \mathcal{H}^+, o) = \deg(g, \mathcal{H}^-, o)$$

holds also for continuous g. Therefore, the core of the proof is the following substantial strengthening of the Tietze theorem for odd mappings. □

Lemma 5.7.33. *Let $\mathcal{D}$ be a bounded, open, symmetric subset of $\mathbb{R}^M$ and $o \notin \mathcal{D}$. Let f be a continuous mapping from $\partial\mathcal{D}$ into $\mathbb{R}^M$ which is odd and nowhere zero on $\partial\mathcal{D}$. Then there exists a continuous odd extension $g\colon \overline{\mathcal{D}} \to \mathbb{R}^M$ such that*

$$g(x) \neq o \qquad \text{for all} \quad x \in \mathcal{H} := \{(x_1, \dots, x_M) \in \overline{\mathcal{D}} : x_M = 0\}.$$

To elucidate the problems of construction we note that the requirement of oddness of g does not cause difficulties. The crucial point is that we need g to be nowhere zero on $\mathcal{H}$. The following simple example shows that the existence of such an extension is not obvious. Let $\mathcal{D} = (-1, 1)$, $f(-1) = -1$, $f(1) = 1$. Then any continuous extension of f has a zero point in $\mathcal{D}$.

Proof of Lemma 5.7.33. We assume again $M > 1$. The notation $\mathcal{H}^{\pm}$ is used similarly as above. The key point is an odd, nowhere zero, continuous extension of f from $\partial\mathcal{D} \cap \overline{\mathcal{H}}$ to $\tilde{f}\colon \overline{\mathcal{H}} \to \mathbb{R}^M$. See the above example and notice the distinction, namely that $\dim \mathcal{H} = M - 1$ and f is a map into $\mathbb{R}^M$. Having such an extension, the rest of the proof is an application of the Tietze Theorem:

$$g(x) = \begin{cases} f(x), & x \in \partial\mathcal{D}, \\ \tilde{f}(x), & x \in \overline{\mathcal{H}}, \\ \tilde{g}(x), & x \in \mathcal{H}^+, \\ -\tilde{g}(-x), & x \in \mathcal{H}^-, \end{cases}$$

where $\tilde{g}$ is the Tietze extension of f and $\tilde{f}$.

The existence of an extension $\tilde{f}$ follows from the following slightly more general assertion:

> *Let $\mathcal{G}$ be a bounded, open, symmetric subset of $\mathbb{R}^N$, $o \notin \overline{\mathcal{G}}$, and let the mapping $f\colon \partial\mathcal{G} \to \mathbb{R}^M$ be continuous, odd and nowhere zero. If $N < M$, then there exists a continuous, odd and nowhere zero extension*
>
> $$\varphi\colon \overline{\mathcal{G}} \to \mathbb{R}^M.$$

We will prove this assertion by induction with respect to the dimension N.

If $N = 1$, then

$$\overline{\mathcal{G}} \subset [-\beta, -\alpha] \cup [\alpha, \beta] \qquad \text{for some} \quad 0 < \alpha < \beta < \infty,$$

and we need to find a continuous extension φ to $[\alpha, \beta]$ which is nowhere zero, define φ on $[-\beta, -\alpha]$ to be odd and restrict φ to $\overline{\mathcal{G}}$.

The induction step is done similarly: First use the induction hypothesis for an extension to $\overline{\mathcal{G}} \cap \mathbb{R}^{N-1}$, and then for extending it into the upper half-space. In order to show that such an extension actually exists (also for $N = 1$) we need the following key result:

> *Let $\mathcal{K}$ be a compact subset of $\mathbb{R}^N$ and let f be a continuous nowhere zero mapping from $\mathcal{K}$ into $\mathbb{R}^M$ where $N < M$. Then for any compact set $\mathcal{L} \supset \mathcal{K}$ there exists a continuous*
>
> $$\varphi\colon \mathcal{L} \to \mathbb{R}^M$$
>
> *which extends f and is nowhere zero on $\mathcal{L}$.*

Recall again the above example to see the obstacles in the proof. Denote

$$c := \min_{x \in \mathcal{K}} |f(x)|$$

and choose $\varepsilon \in \left(0, \frac{c}{2}\right)$. First we prove the existence of a smooth approximation which is defined on a neighborhood of $\mathcal{L}$. By the Tietze Theorem, there is a continuous extension $f_1 \colon \mathcal{L} \to \mathbb{R}^M$. This f_1 can be smoothly approximated on $\mathcal{L}$ (the proof of Lemma 5.7.23): Take $\varphi_1 \in C^1(\mathcal{U})$ where $\mathcal{U}$ is a neighborhood of $\mathcal{L}$ such that

$$\|f_1 - \varphi_1\|_{C(\mathcal{L})} < \frac{\varepsilon}{2}.$$

Put

$$\tilde{\varphi}_1(x_1, \dots, x_M) = \varphi_1(x_1, \dots, x_N) \qquad \text{for} \quad x = (x_1, \dots, x_N, \dots, x_M) \in \mathcal{U} \times \mathbb{R}^{M-N}.$$

In particular, this means that

$$\det \tilde{\varphi}_1'(x) = 0,$$

i.e., all points of $\varphi_1(\mathcal{U}) = \tilde{\varphi}_1(\mathcal{U} \times \mathbb{R}^{M-N})$ are critical values of $\tilde{\varphi}_1$. According to the Sard Theorem (Theorem 5.7.3)

$$\operatorname{meas} \varphi_1(\mathcal{U}) = 0$$

(meas is the Lebesgue measure in $\mathbb{R}^M$), and $\mathbb{R}^M \setminus \varphi_1(\mathcal{U})$ is dense.[48] Therefore, there is a point $y_0 \in \mathbb{R}^M \setminus \varphi_1(\mathcal{U})$ such that $\|y_0\| < \frac{\varepsilon}{2}$. Put $\varphi_2 = \varphi_1 - y_0$. Then $\varphi_2(x) \neq o$ for every $x \in \mathcal{L}$ and

$$\|f - \varphi_2\|_{C(\mathcal{K})} < \varepsilon.$$

Moreover,

$$\|\varphi_2(x)\| \geq \frac{c}{2} \qquad \text{for} \quad x \in \mathcal{K}.$$

We can assume that the last inequality holds for all $x \in \mathcal{L}$. Otherwise, we multiply φ_2 by the function $\frac{c}{2\|\varphi(\cdot)\|}$ outside the set where $\|\varphi_2(x)\| \geq \frac{c}{2}$. Since the Tietze extension

---

[48] In fact we do not need here the whole strength of the Sard Theorem. The following much weaker result is sufficient:

Let $\mathcal{G} \subset \mathbb{R}^M$ be an open set and $\psi \colon \mathcal{G} \to \mathbb{R}^M$ a C^1-mapping on $\mathcal{G}$. If $\mathcal{A} \subset \mathcal{G}$ has Lebesgue measure zero, so has $\psi(\mathcal{A})$.

Indeed, every point of $\mathcal{A}$ belongs to a ball $\mathcal{B} \subset \mathcal{G}$ on which ψ' is bounded. By the Mean Value Theorem, there is a constant K such that

$$\|\psi(x) - \psi(y)\| \leq K\|x - y\|, \qquad x, y \in \mathcal{B}. \tag{$*$}$$

Since $\mathbb{R}^M$ is separable, the set $\mathcal{A}$ can be covered by countably many balls $\{\mathcal{B}_j\}_{j \in \mathbb{N}}$, i.e., $\mathcal{A} = \bigcup_{j=1}^{\infty} (\mathcal{A} \cap \mathcal{B}_j)$, and $\psi(\mathcal{A}) = \bigcup_{j=1}^{\infty} \psi(\mathcal{A} \cap \mathcal{B}_j)$. To complete the proof we show that $\operatorname{meas}(\psi(\mathcal{A} \cap \mathcal{B}_j)) = 0$, $j \in \mathbb{N}$. So take $\eta > 0$. Since $\operatorname{meas}(\mathcal{A} \cap \mathcal{B}_j) = 0$, there are countably many cubes (or balls) $\{\mathcal{Q}_k\}_{k \in \mathbb{N}}$ $\mathcal{A} \cap \mathcal{B}_j \subset \bigcup_{k=1}^{\infty} \mathcal{Q}_k$, $\sum_{k=1}^{\infty} \operatorname{meas} \mathcal{Q}_k < \eta$, and the estimate $(*)$ holds for all $\mathcal{Q}_k$ with the same constant K. This implies $\operatorname{meas}(\psi(\mathcal{A} \cap \mathcal{B}_j)) \leq \sum_{k=1}^{\infty} \operatorname{meas} \psi(\mathcal{Q}_k) \leq c \sum_{k=1}^{\infty} \operatorname{meas} \mathcal{Q}_k < c\eta$ where the constant c depends only on the dimension M and on K. Since $\eta > 0$ is arbitrary, $\operatorname{meas}(\psi(\mathcal{A} \cap \mathcal{B}_j)) = 0$ for all $j \in \mathbb{N}$. To obtain the result required in the proof above take $\mathcal{A} = \mathcal{U} \times \underbrace{\{0, \dots, 0\}}_{(M-N)\text{-tuple}}$.

retains upper and lower bounds, we can extend $f - \varphi_2$ from the set $\mathcal{K}$ to a continuous mapping ψ on $\mathcal{L}$ for which

$$\|\psi(x)\| < \varepsilon, \qquad x \in \mathcal{L}.$$

It remains to put

$$\varphi(x) = \varphi_2(x) + \psi(x), \qquad x \in \mathcal{L}.$$

□

The above proof is cumbersome and seems to be endless. So we recommend the reader to go through the main steps once again not checking all the technicalities but concentrating on their main ideas.

Corollary 5.7.34 (Borsuk–Ulam). *Let f be a continuous mapping from the M-dimensional sphere $\mathcal{S}^M \subset \mathbb{R}^{M+1}$ into $\mathbb{R}^M$. Then there is a point $x_0 \in \mathcal{S}^M$ such that*

$$f(x_0) = f(-x_0).$$

Proof. Extend

$$\varphi(x) = f(x) - f(-x)$$

to a mapping from the unit ball $B(o;1) \subset \mathbb{R}^{M+1}$ into $\mathbb{R}^M$ which is viewed as a subset $\mathbb{R}^M \times \{0\}$ of $\mathbb{R}^{M+1}$. If $o \in \varphi(\mathcal{S}^M)$, the proof is complete. For the case $o \notin \varphi(\mathcal{S}^M)$, the application of the Borsuk Theorem yields

$$\deg(\varphi, B(o;1), o) \neq 0.$$

By Theorem 5.7.26 (viii),

$$\deg(\varphi, B(o;1), o) = \deg(\varphi, B(o;1), y) \qquad \text{where} \quad y = (0, \dots, 0, 1) \in \mathbb{R}^{M+1},$$

i.e., the equation

$$\varphi(x) = y$$

has a solution in $B(o;1)$. However, this is impossible since

$$\varphi(B(o;1)) \subset \mathbb{R}^M \times \{0\}.$$

□

For more information in this direction see Schwartz [138].

Exercise 5.7.35. Deduce the classical Jordan Separation Theorem from Theorem 5.7.28!

Hint. A Jordan curve is homeomorphic to $\mathcal{S}^1$.

Exercise 5.7.36. Show that there is no continuous injection of $\mathbb{R}^M$ into $\mathbb{R}^N$ whenever $M > N$!

Hint. Assume by contradiction that φ is such a mapping and put

$$f(x) = (\varphi(x), 0, \dots, 0).$$

Apply Corollary 5.7.29. For another proof see Exercise 5.7.39.

Exercise 5.7.37. Let Ω be a ball in $\mathbb{C}$ with sufficiently large radius and let P be a polynomial of degree $n \geq 1$. Show that

$$\deg(P, \Omega, 0) = n.$$

Hint. For $P(x) = \sum_{k=0}^{n} a_k x^k$, $a_n \neq 0$, use the homotopy

$$H(t,x) = tP(x) + (1-t)a_n x^n \qquad \text{on} \quad \overline{\Omega}.$$

Exercise 5.7.38. Let f be an odd mapping from $\mathcal{S}^M = \partial B(o;1) \subset \mathbb{R}^{M+1}$ into $\mathbb{R}^{M+1} \setminus \{o\}$. Show that there is no continuous extension of f to

$$\varphi\colon B(o;1) \to \mathbb{R}^{M+1} \setminus \{o\}.$$

(This is Borsuk's original formulation of Theorem 5.7.32.)

Exercise 5.7.39. Deduce the assertion of Exercise 5.7.36 from the Borsuk–Ulam Theorem (Corollary 5.7.34).

Exercise 5.7.40. Prove the following result due to Lusternik and Schnirelmann :

Let $\mathcal{F}_1, \dots, \mathcal{F}_{M+1}$ be closed sets which cover $\mathcal{S}^M$. Then at least one $\mathcal{F}_i$ contains a pair of antipodal points (i.e., $x, -x \in \mathcal{F}_i$).

Hint. Let

$$\varphi(x) := -x, \qquad x \in \mathcal{S}^M,$$

and suppose that

$$\varphi(\mathcal{F}_i) \cap \mathcal{F}_i = \emptyset \qquad \text{for} \quad i = 1, \dots, M.$$

There are continuous functions $f^i\colon \mathcal{S}^M \to [0,1]$ such that

$$f^i(\mathcal{F}_i) = \{0\}, \qquad f^i(\varphi(\mathcal{F}_i)) = \{1\}$$

(this consequence of the Tietze Theorem is known in a normal topological space as the Urysohn Lemma). Put $f = (f^1, \dots, f^M)$ and apply the Borsuk–Ulam Theorem to obtain a point x_0. Show that

$$x_0 \in \mathcal{F}_{M+1} \cap \varphi(\mathcal{F}_{M+1}).$$

Exercise 5.7.41. Prove the following complement of the above covering result of Lusternik and Schnirelmann:

There exist closed sets $\mathcal{F}_1, \dots, \mathcal{F}_{M+2}$ which cover $\mathcal{S}^M$, and such that no $\mathcal{F}_i$ contains a pair of antipodal points.

Hint. Proceed by induction with respect to M. The assertion is obviously true for $M = 1$. Let $M = 2$. Cover the equator of $\mathcal{S}^2$ with three closed sets $\mathcal{E}_1$, $\mathcal{E}_2$ and $\mathcal{E}_3$, with $\mathcal{E}_j \cap (-\mathcal{E}_j) = \emptyset$, $j = 1, 2, 3$. Then choose a latitude L on the southern hemisphere and extend the cover of the equator to a covering $\mathcal{A}_1$, $\mathcal{A}_2$ and $\mathcal{A}_3$ of the set of all points which lie to the north of L, including those of latitude L (see Figure 5.7.6).

Here $\mathcal{A}_j$ consists of all great circle arcs from latitude L to the north pole which contain a point of $\mathcal{E}_j$. Finally, let $\mathcal{A}_4$ consist of all points lying to the south of L, including those of latitude L . Then $\mathcal{A}_1$, $\mathcal{A}_2$, $\mathcal{A}_3$ and $\mathcal{A}_4$ is the desired covering of $\mathcal{S}^2$. Continue the argument for $M \geq 3$.

Exercise 5.7.42. Prove the following Bread–Ham–Cheese Theorem:

If $\mathcal{B}_1, \dots, \mathcal{B}_M$ are bounded measurable sets in $\mathbb{R}^M$ with $M \geq 1$, then there is an $(M-1)$-dimensional plane which divides all the sets $\mathcal{B}_j$ into two parts of the same measure.

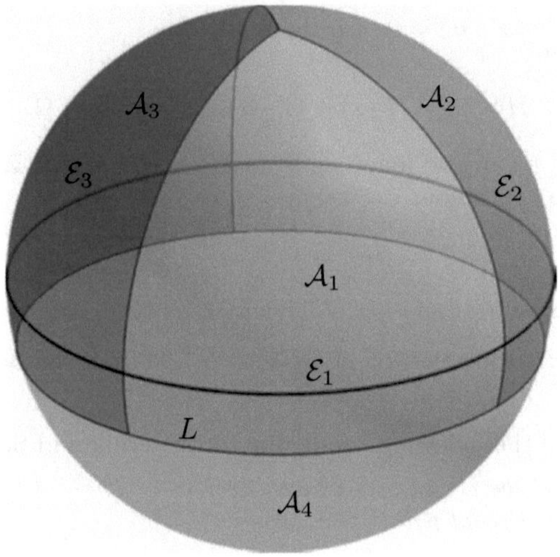

Figure 5.7.6.

This assertion can be reformulated in three dimensions as follows:

> *Suppose we have a sandwich of bread, ham and cheese with ham and cheese piled attractively but irregularly on the bread. Then the sandwich can be cut in two with one straight slash of a knife in such a way that each of two persons gets an identical share of bread, ham and cheese.*

Hint. Let $M = 2$ and let $d \in \mathcal{S}^1$ determine a direction in $\mathbb{R}^2$. Take a perpendicular line to d and move it from $-\infty$ to $+\infty$ (see Figure 5.7.7).

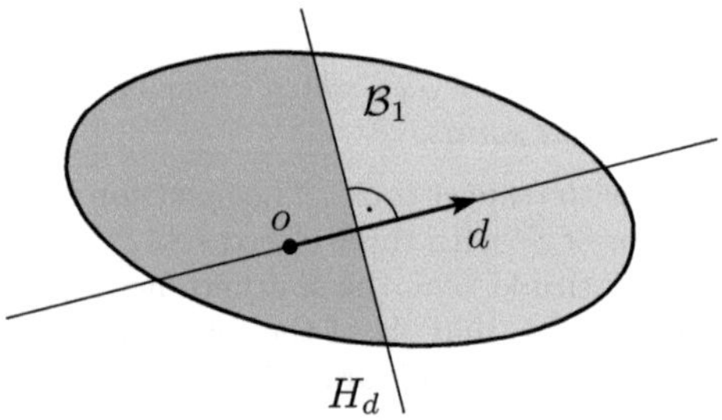

Figure 5.7.7.

Take the first and the last perpendicular (they need not be necessarily distinct) which splits the set $\mathcal{B}_1$ into two parts of the same measure. The perpendicular H_d which is half-way between these two has the equation

$$(x, d) = a(d)$$

where $a\colon \mathbb{R}^2 \to \mathbb{R}$ satisfies

$$a(-d) = -a(d).$$

In order to find d for which H_d also splits $\mathcal{B}_2$ into two parts of the same measure, we set

$$f(d) := \operatorname{meas}\{x \in \mathcal{B}_2 : (x, d) > a(d)\}.$$

Then $f\colon \mathcal{S}^1 \to \mathbb{R}$ is continuous and

$$f(d) + f(-d) = \operatorname{meas} \mathcal{B}_2.$$

By the Borsuk–Ulam Theorem, there is a point d such that $f(d) = f(-d)$. Thus the corresponding H_d divides $\mathcal{B}_2$ into two parts of the same measure.

If $M \geq 3$, then construct functions $f_2, \dots, f_M$ corresponding to $\mathcal{B}_2, \dots, \mathcal{B}_M$.

5.8 Leray-Schauder Topological Degree

Our next aim is to generalize the notion of the degree to infinite-dimensional spaces. Since the Brouwer Theorem is a corollary of the homotopy invariance property of the degree and this theorem does not hold even in an infinite-dimensional Hilbert space (Example 5.2.1) we cannot expect a meaningful generalization of the Brouwer degree which would be valid for all continuous mappings. Similarly as in the Schauder Fixed Point Theorem we restrict our attention to operators which are well approximated by finite-dimensional ones, i.e., to compact operators.

One more remark is desirable. One of the main consequences of the notion of $\deg(F, \Omega, y_0)$ is a sufficient condition for the solvability of the equation $F(x) = y_0$ in the set Ω. If F is a compact operator, then $F(\Omega)$ is rather small in an infinite-dimensional space. Therefore it is much better to solve either

$$x - F(x) - y_0 = o$$

(recall the Fredholm theory in Section 2.2) or, more generally,

$$Ax - F(x) = o \qquad \text{for a suitable} \quad A.$$

The Leray–Schauder degree concerns operators of the type

$$I - F$$

where $F\colon X \to X$ is a compact operator. Let Ω be a bounded open set in a Banach space X, $o \in X \setminus (I - F)(\partial\Omega)$. Using the compactness of F, it is easy to prove that $(I - F)(\partial\Omega)$ is closed, and hence,

$$d := \operatorname{dist}(o, (I - F)(\partial\Omega)) > 0.$$

Let $\{F_n\}_{n=1}^{\infty}$ be a sequence in $\mathscr{C}_{\mathrm{f}}(\overline{\Omega}, X)$ which converges to F uniformly on $\overline{\Omega}$ (Theorem 5.2.3). Denote

$$X_n = \operatorname{Lin} F_n(\overline{\Omega}), \qquad \Omega_n = \Omega \cap X_n,$$

and

$$G_n(x) = x - F_n(x) \qquad \text{for} \quad x \in \overline{\Omega_n}.$$

If n is sufficiently large, then $o \notin G_n(\partial\Omega_n)$ and $\deg(G_n, \Omega_n, o)$ is well defined.

Lemma 5.8.1. *Under the above stated hypotheses, the sequence of integers* $\{\deg(G_n, \Omega_n, o)\}_{n=1}^{\infty}$ *is constant for large* n, *and its limit does not depend on the choice of the approximating sequence* $\{G_n\}_{n=1}^{\infty}$.

Proof. For a given $\varepsilon > 0$ there is n_0 such that

$$\sup_{x\in\overline{\Omega}} \|F(x) - F_n(x)\| < \varepsilon \qquad \text{for} \quad n \geq n_0.$$

Choose some $n, m \geq n_0$ and put

$$\tilde{X} = X_n + X_m, \qquad \tilde{G}_k(x) = x - F_k(x), \qquad x \in \overline{\Omega} \cap \tilde{X}, \quad k = n, m.$$

Consider the homotopy

$$H(t, x) = t\tilde{G}_m(x) + (1-t)\tilde{G}_n(x), \qquad t \in [0,1],\ x \in \overline{\Omega} \cap \tilde{X}.$$

For $x \in \partial\Omega \cap \tilde{X} = \partial(\Omega \cap \tilde{X})$ we have

$$\begin{aligned} \|H(t,x)\| &= \|x - F(x) - t[F_m(x) - F(x)] - (1-t)[F_n(x) - F(x)]\| \\ &\geq \|x - F(x)\| - t\|F_m(x) - F(x)\| - (1-t)\|F_n(x) - F(x)\| \\ &\geq d - 2\varepsilon > 0 \end{aligned}$$

for small $\varepsilon > 0$. By Proposition 5.7.6 and Theorem 5.7.7,

$$\deg(\tilde{G}_m, \Omega \cap \tilde{X}, o) = \deg(\tilde{G}_n, \Omega \cap \tilde{X}, o).$$

Using Example 5.7.11 we get

$$\deg(\tilde{G}_k, \Omega \cap \tilde{X}, o) = \deg(G_k, \Omega_k, o), \qquad k = m, n,$$

i.e., $\deg(G_n, \Omega_n, o)$ is constant for $n \geq n_0$.

If Φ_n is another approximating sequence of F, then the homotopy joining the restrictions of $I - F_n$ and $I - \Phi_n$ to the span Z of $\operatorname{Im} F_n + \operatorname{Im} \Phi_n$ can be defined. The same procedure as above yields

$$\deg(x - F_n(x), \Omega \cap Z, o) = \deg(x - \Phi_n(x), \Omega \cap Z, o). \qquad \square$$

We are now able to define the *Leray–Schauder degree*

$$\deg(I - F, \Omega, y_0) := \deg(I - (F + y_0), \Omega, o)$$

as the limit of $\deg(G_n, \Omega_n, o)$ for any approximating sequence $\{G_n, \Omega_n\}_{n=1}^{\infty}$. This construction also shows that the Leray–Schauder degree inherits its properties from the Brouwer degree.

Theorem 5.8.2. *Let Ω be a bounded open subset of a Banach space X. There exists a mapping* $\deg(I - F, \Omega, y_0)$ *defined for all* $F \in \mathscr{C}(\overline{\Omega}, X)$ *and* $y_0 \in X$ *such that*

$$x - F(x) \neq y_0 \qquad \text{for all} \quad x \in \partial\Omega.$$

This mapping has the following properties:

(i) $\deg(I, \Omega, y_0) = \begin{cases} 1 & \text{if} \quad y_0 \in \Omega, \\ 0 & \text{if} \quad y_0 \notin \overline{\Omega}. \end{cases}$

(ii) $\deg(I - F, \Omega, y_0) = \deg(I - F - y_0, \Omega, o)$.

(iii) *If* $\deg(I - F, \Omega, y_0) \neq 0$, *then the equation*

$$x - F(x) = y_0$$

has a solution in Ω.

(iv) *If* $\Omega_1, \dots, \Omega_k$ *are pairwise disjoint open subsets of* Ω *and* $x - F(x) \neq y_0$ *for each* $x \in \overline{\Omega} \setminus \bigcup_{j=1}^{k} \Omega_j$, *then*

$$\deg(I - F, \Omega, y_0) = \sum_{j=1}^{k} \deg(I - F, \Omega_j, y_0).$$

(v) *If* $F, G \in \mathscr{C}(\overline{\Omega}, X)$ *and*

$$\sup_{x \in \partial\Omega} \|F(x) - G(x)\|_X < \inf_{x \in \partial\Omega} \|x - F(x) - y_0\|_X,$$

then

$$\deg(I - F, \Omega, y_0) = \deg(I - G, \Omega, y_0).$$

(vi) (*homotopy invariance property*) *If* $F, G \in \mathscr{C}(\overline{\Omega}, X)$ *and*

$$H(t, x) = (1 - t)F(x) + tG(x), \qquad t \in [0, 1], \quad x \in \overline{\Omega},$$

are such that

$$x - H(t, x) \neq y_0 \qquad \text{for every} \quad x \in \partial\Omega \quad \text{and} \quad t \in [0, 1],$$

then $\deg(I - H(t, \cdot), \Omega, y_0)$ *is constant on* $[0, 1]$. *In particular,*

$$\deg(I - F, \Omega, y_0) = \deg(I - G, \Omega, y_0).$$

Proof. Finite-dimensional approximations and the corresponding properties of the Brouwer degree are used to prove all the statements. We only give details for (iii). We can assume that $y_0 = o$ (by (ii)). Let $\{F_n\}_{n=1}^{\infty} \subset \mathscr{C}_{\mathrm{f}}(\overline{\Omega}, X)$ be a sequence of finite-dimensional approximations that converges to F uniformly on $\overline{\Omega}$. We know, by construction of the degree, that

$$\deg(I - F, \Omega, o) = \deg(I_n - F_n, \Omega_n, o) \neq 0 \qquad \text{for all } n \text{ large.}$$

By Theorem 5.7.7 there are $x_n \in \Omega_n \subset \Omega$ such that

$$F_n(x_n) = x_n.$$

Since F is compact there exists a subsequence $\{F(x_{n_k})\}_{k=1}^{\infty}$ converging to a $z \in X$. It follows from the uniform convergence of F_n that $\lim\limits_{k\to\infty} F_{n_k}(x_{n_k}) = z$, too. This means that also $\lim\limits_{k\to\infty} x_{n_k} = z \in \overline{\Omega}$ and, therefore,

$$F(z) = z.$$

But z cannot belong to $\partial\Omega$, i.e., $z \in \Omega$. □

Example 5.8.3 (Rothe's version of the Schauder Fixed Point Theorem). Assume that F is a compact operator from the closed unit ball $\overline{B(o;1)}$ of a Banach space X into X. If

$$F(\partial B(o;1)) \subset \overline{B(o;1)},$$

then F has a fixed point in $\overline{B(o;1)}$.

Indeed, suppose not and consider $H(t,x) := (1-t)F(x)$. By the homotopy invariance property (for $G = 0$),

$$\deg(I - F, B(o;1), o) = \deg(I, B(o;1), o) = 1,$$

a contradiction. ○

Example 5.8.4 (Schaefer). Let $F \in \mathscr{C}(X,X)$ and let

$$\Sigma := \{x \in X : \exists t \in [0,1] \text{ such that } x - tF(x) = o\}$$

be bounded.[49] Then F has a fixed point.

To prove this choose an $r > 0$ such that $\Sigma \subset B(o;r)$ and put $\Omega = B(o;r)$ (open ball). The homotopy invariance property of the degree can be applied to

$$H(t,x) = (1-t)x + t(x - F(x)),$$

and the result follows. ○

The next example shows that finding an a priori estimate need not be a trivial task.

[49]This assumption is often called an *a priori estimate*. Notice that it is not assumed that the equation $x - tF(x) = o$ has any solution. However, if a solution exists, then it belongs to a certain ball the radius of which is independent of t. The result given in this example is also called the *Leray–Schauder Continuation Method*.

Example 5.8.5. Consider the boundary value problem

$$\begin{cases} \ddot{x}(t) = f(t, x(t), \dot{x}(t)), & t \in (0,1), \\ x(0) = x(1) = 0, \end{cases} \tag{5.8.1}$$

where $f\colon [0,1] \times \mathbb{R}^2 \to \mathbb{R}$ is a continuous function. We know (Example 2.3.10 and also Example 5.2.8) that a solution of (5.8.1) is also a solution of the integral equation

$$F(x)(t) := \int_0^1 G(t,s) f(s, x(s), \dot{x}(s))\, ds = x(t)$$

where the Green function $G(t,s)$ is defined as follows:

$$G(t,s) = \begin{cases} s(t-1), & 0 \le s \le t \le 1, \\ t(s-1), & 0 \le t < s \le 1. \end{cases}$$

Notice the difference between this example and Example 5.2.8. Here the operator F depends also on the derivative $\dot{x}$, and it is thus defined only on a dense subset of the space $C[0,1]$. The notion of the degree cannot be used for F in this space. Therefore, we have to work either in C^1 or in the space

$$X = \{x \in C^2[0,1] : x(0) = x(1) = 0\}$$

which our solution has to belong to. In both the spaces the problem of an a priori estimate of a possible solution to (5.8.1) occurs – see Step 2 of this example. We will work in X.

Step 1. First we show that F is a compact operator on X. To prove this notice that F is the composition $\Psi \circ \Phi$ where $\Phi\colon X \to C[0,1]$ is a Nemytski type operator

$$\Phi(x)\colon t \to f(t, x(t), \dot{x}(t))$$

and $\Psi\colon C[0,1] \to X$ is the linear integral operator

$$\Psi(y)(t) = \int_0^1 G(t,s) y(s)\, ds.$$

The Nemytski operator Φ is also a composition of a compact embedding of X into C^1 (the Arzelà–Ascoli Theorem) and a continuous operator from $C^1[0,1]$ into $C[0,1]$. It is sufficient to show that Ψ is a continuous linear operator from $C[0,1]$ into X.

Indeed, since $\Psi(y)$ is a solution of the boundary value problem

$$\begin{cases} \ddot{x}(t) = y(t), & t \in (0,1), \\ x(0) = x(1) = 0 \end{cases}$$

(Example 2.3.10), we have $x = \Psi(y) \in X$. Because of the boundary conditions there is $t_0 \in (0,1)$ such that $\dot{x}(t_0) = 0$ (the classical theorem due to Rolle). This allows us to write

$$\dot{x}(t) = \int_{t_0}^{t} y(s)\,\mathrm{d}s, \qquad t \in [0,1],$$

and to get the estimate $|\dot{x}(t)| \le \|y\|_{C[0,1]}$. Since $|x(t)| = \left|\int_0^t \dot{x}(s)\,\mathrm{d}s\right| \le \|y\|_{C[0,1]}$, we have

$$\|x\|_X := \sup_{t\in[0,1]} |x(t)| + \sup_{t\in[0,1]} |\dot{x}(t)| + \sup_{t\in[0,1]} |\ddot{x}(t)| \le 3\|y\|_{C[0,1]}.$$

Step 2. In order to establish an a priori estimate we have to require estimates on the behavior of $f(t,x,y)$ for large x and y:

(H1) There is $M_0 > 0$ such that

$$xf(s,x,0) > 0 \qquad \text{for} \quad s \in [0,1] \quad \text{and} \quad |x| \ge M_0.$$

(H2) There are positive constants c_1, c_2 such that

$$|f(s,x,y)| \le c_1 y^2 + c_2 \qquad \text{for} \quad s\in[0,1], \quad |x| \le M_0, \quad y \in \mathbb{R}.$$[50]

Suppose that (H1) and (H2) hold. Let there exist $x \in X$, $x \ne o$, and $\lambda \in (0,1]$ such that

$$x = \lambda F(x).$$

First we will estimate $\|x\|_{C[0,1]}$. There is $t_0 \in (0,1)$ such that $|x(t_0)| = \|x\|_{C[0,1]}$ and we assume $x(t_0) > 0$. Then $\dot{x}(t_0) = 0$ and $\ddot{x}(t_0) \le 0$. Since

$$0 \ge x(t_0)\ddot{x}(t_0) = \lambda x(t_0) f(t_0, x(t_0), 0),$$

we have $x(t_0) \le M_0$ according to (H1). Similarly for $x(t_0) < 0$, i.e., $\|x\|_{C[0,1]} \le M_0$.

To get an estimate of $\dot{x}$ we consider the function

$$g(t) := \log[c_1(\dot{x}(t))^2 + c_2].$$

Since

$$\dot{g}(t) = 2c_1 \frac{\dot{x}(t)\ddot{x}(t)}{c_1(\dot{x}(t))^2 + c_2} = 2\lambda c_1 \frac{\dot{x}(t) f(t, x(t), \dot{x}(t))}{c_1(\dot{x}(t))^2 + c_2},$$

we obtain, by (H2),

$$|\dot{g}(t)| \le 2c_1 |\dot{x}(t)|.$$

[50] The condition (H1) is sometimes called the *sign condition* and (H2) the *Nagumo-type condition*.

Let
$$\mathcal{G} = \{t \in [0,1] : \dot{x}(t) \neq 0\}.$$
Then
$$\mathcal{G} = \bigcup \mathcal{J}_j$$
where $\overline{\mathcal{J}_j}$ is a closed interval such that $\dot{x}(t) \neq 0$ for $t \in \mathcal{J}_j$ and $\dot{x}$ vanishes at one end point of $\mathcal{J}_j$ (say τ_j) at least. Now, we have

$$0 \le \log \frac{c_1 \dot{x}^2(t) + c_2}{c_1 \cdot 0 + c_2} = g(t) - g(\tau_j) = \int_{\tau_j}^{t} \dot{g}(s)\,\mathrm{d}s \le \left| \int_{\tau_j}^{t} |\dot{g}(s)|\,\mathrm{d}s \right|$$
$$\le 2c_1 \left| \int_{\tau_j}^{t} |\dot{x}(s)|\,\mathrm{d}s \right| = 2c_1 |x(t) - x(\tau_j)|,$$

since $\dot{x}$ does not change its sign in $\mathcal{J}_j$. Hence
$$\log \frac{c_1 \dot{x}^2(t) + c_2}{c_2} \le 4c_1 M_0.$$
This inequality shows that
$$|\dot{x}(t)| \le M_1 := \sqrt{\frac{c_2}{c_1}} \mathrm{e}^{2c_1 M_0}, \qquad t \in [0,1].$$
If $M_2 := \sup\{|f(t,x,y)| : t \in [0,1], |x| \le M_0, |y| \le M_1\}$, then
$$\|\ddot{x}\|_{C[0,1]} \le M_2.$$
These estimates of $\|x\|_{C[0,1]}$, $\|\dot{x}\|_{C[0,1]}$, $\|\ddot{x}\|_{C[0,1]}$ show that the set Σ from Example 5.8.4 is bounded and therefore the proof of existence of a solution of (5.8.1) under the hypotheses (H1) and (H2) is complete.

The reader can imagine that (H1), (H2) are not the only sufficient conditions for solving (5.8.1). However, the direct use of the Schauder Fixed Point Theorem leads to more restrictive assumptions on f. A survey of results until 1980 can be found in the monograph Fučík [65]. ○

It is clear that the above stated procedure can be used for solving a more general equation
$$Ax = F(x) \qquad \text{where} \quad F \in \mathscr{C}(\overline{\Omega}, X) \tag{5.8.2}$$
and A is a linear operator with a bounded inverse. In that case (5.8.2) is equivalent to
$$x = A^{-1}F(x)$$
with a compact operator $A^{-1}F$. More interesting questions arise for a non-invertible A. Since many differential operators (both ordinary and partial) are Fredholm

operators we will suppose that A is a linear closed Fredholm operator[51] of index zero, and proceed as in Remark 4.3.14 with the exception that A is not assumed to be continuous.

We denote

$$X_1 := \operatorname{Ker} A, \qquad Y_2 := \operatorname{Im} A,$$

and choose topological complements X_2, Y_1 such that

$$X = X_1 \oplus X_2, \qquad Y = Y_1 \oplus Y_2.$$

These closed complements exist because X_1 has a finite dimension and Y_2 a finite co-dimension (Example 2.1.12 and Remark 2.1.19). By the assumption on the index of A there is also an isomorphism Λ of Y_1 onto X_1. Denote by P and Q the linear continuous projections onto X_1 and Y_1 with kernels X_2 and Y_2, respectively. Then the restriction of A to $X_2 \cap \operatorname{Dom} A$ is an injective operator with a bounded inverse B.[52]

The equation (5.8.2) is equivalent to the couple of equations

$$Ax_2 = (I-Q)F(x_1+x_2), \qquad o = QF(x_1+x_2), \quad x_1 \in X_1, \ x_2 \in X_2 \cap \operatorname{Dom} A,$$

(see Figure 5.8.1) or

$$x_2 = B(I-Q)F(x_1+x_2), \qquad x_1 = x_1 + \Lambda QF(x_1+x_2).^{53} \tag{5.8.3}$$

This couple of equations is equivalent to the equation

$$G(x) = x \qquad \text{where} \qquad G(x) := Px + \Lambda QF(x) + B(I-Q)F(x).$$

If Ω is a bounded open subset of X, $Ax - F(x) \neq o$ for each $x \in \partial\Omega$ and G is compact on $\overline{\Omega}$, then the Leray–Schauder degree

$$\deg(I-G, \Omega, o)$$

is well defined. It is called the *coincidence degree* of the couple (A, F). It can be proved that this definition does not depend on the choice of the projections P, Q and the class of Λ's which do not change the orientations in X_1 and Y_1. The coincidence degree was introduced by J. Mawhin (see, e.g., Mawhin [108] or Gaines & Mawhin [69]). J. Mawhin also proved the following theorem which generalizes the statement of Example 5.8.4.

---

[51] A linear closed operator A is said to be *Fredholm* if $\dim \operatorname{Ker} A < \infty$, $\operatorname{Im} A$ is closed and $\operatorname{codim} \operatorname{Im} A < \infty$. The index of such an operator is defined as $\operatorname{ind} A := \dim \operatorname{Ker} A - \operatorname{codim} \operatorname{Im} A$. See the definition on page 71.

[52] Indeed, B is a closed operator (as an inverse to a closed operator) defined on the Banach space Y_2. The continuity of B follows now from the Closed Graph Theorem (Corollary 2.1.10). The operator $B(I-Q)$ is called a *generalized* (or a *right*) *inverse* to A. It is characterized by the following two properties:

(i) $AB(I-Q) = I - Q$ (the reason for calling it the right inverse);

(ii) $B(I-Q)Ax = (I-P)x$ for $x \in \operatorname{Dom} A$.

[53] Since F is nonlinear, Λ need not be taken as linear. It is actually only essential that Λ maps Y_1 into X_1 and $\Lambda_{-1}(o) = \{o\}$.

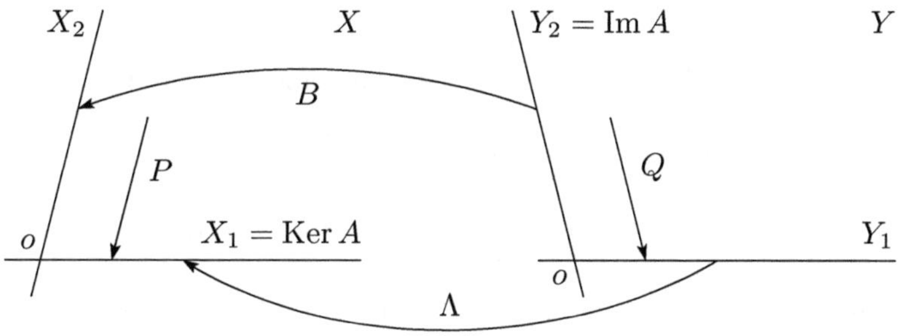

Figure 5.8.1.

Theorem 5.8.6 (J. Mawhin). *Let* $A\colon \operatorname{Dom} A \subset X \to X$ *be a Fredholm operator of index zero,* Ω *a bounded open subset of a Banach space* X*, and let* $B(I-Q)F \in \mathscr{C}(\overline{\Omega}, X)$ *where* $B(I-Q)$ *is a generalized inverse to* A*. Assume further that*

(i) $Ax - \lambda F(x) \neq o$ *for* $x \in \partial\Omega \cap \operatorname{Dom} A$, $\lambda \in (0,1)$,[54]

(ii) $\deg(\Lambda QF|_{\operatorname{Ker} A \cap \Omega}, \operatorname{Ker} A \cap \Omega, o) \neq 0$.

*Then the equation (*5.8.2*) has a solution in* $\overline{\Omega}$*.*

Proof. The proof is based on the observation that the coincidence degree can be reduced with help of the homotopy invariance property and the Product Formula to the Brouwer degree of the restriction of ΛQF to $\operatorname{Ker} A \cap \Omega$ – for details see the references given above. □

Example 5.8.7. Consider the equation

$$\dot{x} = f(t, x) \tag{5.8.4}$$

together with the periodic boundary condition

$$x(0) = x(1) \tag{5.8.5}$$

where $f \in C([0,1] \times \mathbb{R}, \mathbb{R})$. See also Example 4.3.15.

We denote $\operatorname{Dom} A = \{x \in C^1[0,1] : x(0) = x(1)\}$ and put

$$Ax = \dot{x} \qquad \text{for} \quad x \in \operatorname{Dom} A.$$

Further, let F be a Nemytski operator defined by

$$F(x)(t) = f(t, x(t)), \qquad t \in [0,1], \quad x \in X := C[0,1].$$

Then A is a Fredholm operator of index zero. We choose projections

$$Px = x(0) \quad \text{onto } \operatorname{Ker} A$$

[54]Notice that injectivity of A is actually not needed hence the assumption (i) is not assumed for $\lambda = 0$.

and

$$Qy = \int_0^1 y(s)\,\mathrm{d}s \quad \text{onto the complement of } \operatorname{Im} A.$$

Then the generalized inverse to A is given by

$$B(I-Q)y(t) = \int_0^t y(s)\,\mathrm{d}s - t\int_0^1 y(s)\,\mathrm{d}s$$

and it is an isomorphism of $\operatorname{Im} A = \operatorname{Ker} Q$ onto $X_2 \cap \operatorname{Dom} A$ where $X_2 = \operatorname{Ker} P$. Moreover, $B(I-Q)F$ is a compact operator on X. To verify conditions (i), (ii) of Theorem 5.8.6 we suppose

(H) there are functions $f_+, f_- \in C[0,1]$ such that

$$\lim_{x\to-\infty} f(t,x) = f_-(t), \qquad \lim_{x\to+\infty} f(t,x) = f_+(t)$$

uniformly with respect to $t \in [0,1]$ and

$$f_-(t) < f(t,x) < f_+(t), \qquad t \in [0,1], \quad x \in \mathbb{R}.$$[55]

Step 1. Verification of condition (i): For any solution x of (5.8.4)–(5.8.5) we have, by integration,

$$\int_0^1 f_-(t)\,\mathrm{d}t < 0 = \int_0^1 f(t,x(t))\,\mathrm{d}t < \int_0^1 f_+(t)\,\mathrm{d}t. \tag{5.8.6}$$

Take an $\varepsilon > 0$ such that

$$\int_0^1 f_-(t)\,\mathrm{d}t < -\varepsilon, \qquad \int_0^1 f_+(t)\,\mathrm{d}t > \varepsilon.$$

Then there exists $r > 0$ such that

$$0 < f(t,x) - f_-(t) < \frac{\varepsilon}{2} \qquad \text{for} \quad t \in [0,1], \quad x < -r,$$

and

$$0 < f_+(t) - f(t,x) < \frac{\varepsilon}{2} \qquad \text{for} \quad t \in [0,1], \quad x > r.$$

This implies that

$$\int_0^1 f(t,x)\,\mathrm{d}t < -\frac{\varepsilon}{2} \qquad \text{for} \quad x < -r,$$

$$\int_0^1 f(t,x)\,\mathrm{d}t > \frac{\varepsilon}{2} \qquad \text{for} \quad x > r.$$

[55] Conditions similar to (H) are called *conditions of the Landesman–Lazer type*. See, e.g., Landesman & Lazer [98], Fučík [65], Mawhin [108] or Drábek [48], and also Remark 5.9.4.

It means that for any solution x of (5.8.4)–(5.8.5) there exists $t_0 \in [0,1]$ such that $|x(t_0)| \leq r$. Since

$$x(t) = x(t_0) + \int_{t_0}^{t} f(s, x(s))\,\mathrm{d}s,$$

we get

$$\|x\|_{C[0,1]} \leq M := r + \max\{\|f_-\|_{C[0,1]}, \|f_+\|_{C[0,1]}\}.$$

If we take Ω to be a ball $B(o;R)$ of radius $R > M$, then the condition (i) from Theorem 5.8.6 is satisfied for $\lambda = 1$. The same is also true for the solution of

$$Ax = \lambda F(x), \qquad \lambda \in (0,1).$$

Step 2. Verification of condition (ii): For $x \in \operatorname{Ker} A$ we have

$$\Lambda QF(x) = \int_0^1 f(t,x)\,\mathrm{d}t$$

provided we have taken Λ as the identity map. Moreover,

$$\int_0^1 f(t,R)\,\mathrm{d}t > 0 > \int_0^1 f(t,-R)\,\mathrm{d}t. \tag{5.8.7}$$

From the construction of the Brouwer degree it follows that we can assume that

$$\Phi(x) := \int_0^1 f(t,x)\,\mathrm{d}t$$

is smooth, and 0 is a regular value of Φ. By (5.8.7), the number of zero points of Φ is odd. This means that

$$\begin{aligned}\deg(\Phi, (-R,R), 0) &= \sum_{\Phi(x)=0} \operatorname{sgn}\det \Phi'(x)\\ &= 1 = \deg(\Lambda QF|_{\operatorname{Ker} A \cap B(o;R)}, \operatorname{Ker} A \cap B(o;R), o)\end{aligned}$$

and, therefore, condition (ii) is also satisfied.

These considerations show that the problem (5.8.4)–(5.8.5) has a solution provided (H) is fulfilled. Notice that we have shown that

$$\int_0^1 f_-(t)\,\mathrm{d}t < 0 < \int_0^1 f_+(t)\,\mathrm{d}t$$

is also a necessary condition for the solvability of (5.8.4)–(5.8.5) under the assumption (H) (by (5.8.6)). ○

Remark 5.8.8. It is not necessary to consider only projections P, Q on the "small" $\operatorname{Ker} A$ and a complement of $\operatorname{Im} A$. For example, suppose that there are projections $\{Q_n\}$ converging to the identity in a certain sense, and

$$Q_n Q = Q$$

(e.g., Q_n can be the partial sums of the Fourier series of the elements of $Y = C[0,1]$ for the periodic problem). If we can take projections P_n so that

$$A(\operatorname{Im}(I - P_n)) = \operatorname{Im}(I - Q_n),$$

then there is a chance to solve the first equation in (5.8.3) for a fixed x_1 by the Contraction Principle even if F is only locally Lipschitz. This idea belongs to L. Cesari (see, e.g., his survey in Cesari [27]). Using this approach he proved the existence of a 2π-periodic solution of the equation

$$\ddot{x} + x^3 = \sin t.$$

Notice a significant difference in the sign of the nonlinear term here and in (H1) in Example 5.8.5, and the fact that the growth of the nonlinear term is faster here than in (H2).

At the end of this section we turn our attention to the bifurcations of solutions. Like in Section 4.3 we consider the equation

$$f(\lambda, x) = o$$

where $f \colon \mathbb{R} \times X \to X$ is continuous on $\mathcal{J} \times \mathcal{U}$, $\mathcal{J}$ is an open interval and $\mathcal{U}$ is a neighborhood of o in a Banach space X. We suppose that

$$f(\lambda, o) = o \qquad \text{for all} \quad \lambda \in \mathcal{J}$$

and desire to find conditions under which the point $\lambda_0 \in \mathcal{J}$ is a bifurcation point according to Definition 4.3.21. In Section 4.3 we have used a method based on the Implicit Function Theorem and now we want to employ the topological approach based on the degree theory. Notice that the definition of the index of an isolated solution (Example 5.7.10) can be literally used also in an infinite-dimensional space.

Proposition 5.8.9. *Let $h(\lambda, \cdot) \colon X \to X$ be a compact operator on the neighborhood $\mathcal{U}$ of zero in a Banach space X for all $\lambda \in \mathcal{J}$. Let o be an isolated solution of*

$$f(\lambda, x) := x - h(\lambda, x) = o \qquad \text{in} \quad \mathcal{U} \quad \text{for all} \quad \lambda \in \mathcal{J} \setminus \{\lambda_0\}.$$

Put

$$i(\lambda) = \deg(f(\lambda, \cdot), \mathcal{U}, o).$$

If

$$\lim_{\lambda \to \lambda_0 -} i(\lambda) \neq \lim_{\lambda \to \lambda_0 +} i(\lambda), \tag{5.8.8}$$

then (λ_0, o) is a bifurcation point of f.

Proof. Suppose not. Then there is a neighborhood $\mathcal{V} = \tilde{\mathcal{J}} \times \tilde{\mathcal{U}}$ of (λ_0, o) such that (λ, o) are the only solutions of

$$f(\lambda, x) = o \qquad \text{in} \quad \mathcal{V}.$$

This means that for any $\lambda \in \tilde{\mathcal{J}}$ the index

$$i(\lambda) = \deg(I - h(\lambda, \cdot), \tilde{\mathcal{U}}, o)$$

is defined and, by the general homotopy invariance property of the degree (Exercise 5.8.17), the index $i(\lambda)$ is constant, a contradiction to (5.8.8). □

The use of Proposition 5.8.9 is restricted to the problems of computing the index $i(\lambda)$. The following classical result (Theorem 5.8.12) is based on a special form of f which is often met in applications. For the proof we need two prerequisites which are of independent interest.

Proposition 5.8.10. *Let Ω be an open set in a Banach space X and let $F \in \mathscr{C}(\Omega, X)$. If the Fréchet derivative $F'(x_0)$ exists for an $x_0 \in \Omega$, then $F'(x_0)$ is a (linear) compact operator.*

Proof. If $F'(x_0)$ is not compact, then one can find $\varepsilon_0 > 0$ and a sequence $\{y_n\}_{n=1}^{\infty} \subset X$ such that $\|y_n\| \le 1$ and

$$\|F'(x_0)y_k - F'(x_0)y_l\| \ge \varepsilon_0 \qquad \text{for} \quad k \ne l.$$

By the definition of Fréchet derivative, there is $\delta > 0$ such that

$$\|F(x_0 + h) - F(x_0) - F'(x_0)h\| \le \frac{\varepsilon_0}{4}\|h\| \qquad \text{provided} \quad \|h\| < \delta \quad (\le 1).$$

Choose τ such that

$$\|\tau y_k\| < \delta \qquad \text{and} \qquad x_0 + \tau y_k \in \Omega \qquad \text{for all} \quad k \in \mathbb{N}.$$

Then

$$\begin{aligned} &\|F(x_0 + \tau y_k) - F(x_0 + \tau y_l)\| \\ &\qquad \ge \|F'(x_0)(\tau y_k - \tau y_l)\| - \|F(x_0 + \tau y_k) - F(x_0) - F'(x_0)\tau y_k)\| \\ &\qquad\qquad - \|F(x_0 + \tau y_l) - F(x_0) - F'(x_0)\tau y_l\| \ge \frac{\varepsilon_0 \tau}{2}. \end{aligned}$$

But this means that F is not compact on Ω, a contradiction. □

Proposition 5.8.11 (Leray–Schauder Index Formula). *Let Ω be an open bounded set in a Banach space X and let $F \in \mathscr{C}(\overline{\Omega}, X)$. Let $x_0 \in \Omega$ be a unique solution in $\overline{\Omega}$ of the equation*

$$x = F(x).$$

Assume that the Fréchet derivative $F'(x_0)$ *exists and* $I - F'(x_0)$ *is continuously invertible. Then*

$$\deg(I - F, \Omega, o) = (-1)^{\beta} \qquad \textit{where} \quad \beta = \sum_{\substack{\lambda \in \sigma(F'(x_0)) \cap \mathbb{R} \\ \lambda > 1}} m(\lambda) \tag{5.8.9}$$

and $m(\lambda)$ *is the multiplicity*[56] *of the eigenvalue* λ *of the operator* $F'(x_0)$.

Proof. First we recall that $F'(x_0)$ is a compact operator (Proposition 5.8.10) and, therefore, β is a finite number (Corollary 2.2.13). Choose such a small ball $B(o;\varepsilon)$ that $x_0 + B(o;\varepsilon) \subset \Omega$, and put

$$H(t,y) = \frac{F(x_0 + ty) - F(x_0)}{t}, \qquad t \in (0,1], \quad y \in B(o;\varepsilon),$$

and

$$H(0,y) = F'(x_0)y, \qquad y \in B(o;\varepsilon).$$

The ball $B(o;\varepsilon)$ can be chosen such that the equation

$$y = H(t,y)$$

has a unique solution in $\overline{B(o;\varepsilon)}$, namely $y = o$. Indeed,

$$\begin{aligned}\|H(t,y) - y\| &= \left\| \frac{1}{t}[F(x_0 + ty) - F(x_0) - F'(x_0)(ty)] + F'(x_0)y - y \right\| \\ &\geq \|F'(x_0)y - y\| - \left\| \frac{1}{t}[F(x_0 + ty) - F(x_0) - F'(x_0)(ty)\right\| \geq c\|y\| - \frac{c}{2}\|y\|\end{aligned}$$

provided $\|y\|$ is small enough (by the definition of $F'(x_0)$ and the assumption on $I - F'(x_0)$).

By the homotopy invariance property (in a more general setting – see Exercise 5.8.17), $\deg(I - H(t,\cdot), B(o;\varepsilon), o)$ is constant on the interval $[0,1]$. In particular,

$$\deg(I - F, \Omega, o) = \deg(I - F'(x_0), B(o;\varepsilon), o).$$

Put

$$X_1 := \bigcup_{\substack{\lambda \in \sigma(F'(x_0)) \\ \lambda > 1}} \bigcup_{p=1}^{\infty} \operatorname{Ker}[\lambda I - F'(x_0)]^p.$$

As we have mentioned above, $\dim X_1 = \beta < \infty$. Moreover, there exists a topological complement X_2 to X_1 in X which is $F'(x_0)$-invariant (see the decomposition (2.2.4)). This decomposition of X allows us to use the Product Formula for the

[56] For the definition of multiplicity see footnote 23 on page 85.

degree (Exercise 5.8.16) provided balls $\mathcal{B}_i \subset X_i$, $i = 1, 2$, are chosen such that $\mathcal{B}_1 \times \mathcal{B}_2 \subset B(o; \varepsilon)$. Hence we obtain

$$\deg (I - F'(x_0), B(o; \varepsilon), o) = \deg (F_1, \mathcal{B}_1, o) \deg (F_2, \mathcal{B}_2, o)$$

where F_i denotes the restriction of $I - F'(x_0)$ to X_i, $i = 1, 2$. To compute $\deg (F_2, \mathcal{B}_2, o)$ we introduce the homotopy

$$H_2(t, y) = y - tF'(x_0)y, \qquad t \in [0, 1], \quad y \in X_2.$$

Assume that $H_2(t, y) = o$ for a $y \neq o$. Then $t \in (0, 1)$ and $\frac{1}{t} \in \sigma(F'(x_0))$ and, by the definition of X_1, $y \in X_1$. Since $X_1 \cap X_2 = \{o\}$ we arrive at a contradiction. This consideration shows that we may apply the homotopy invariance property to H_2 to get

$$\deg (F_2, \mathcal{B}_2, o) = \deg (I, \mathcal{B}_2, o) = 1.$$

The degree $\deg (F_1, \mathcal{B}_1, o)$ is the Brouwer degree of the linear operator F_1 in the finite-dimensional space X_1, and it was computed in Example 5.7.10. Notice that A is here$(I - F'(x_0))|_{X_1}$, and thus

$$\{\mu \in \sigma(A) : \mu < 0\} = \{\lambda \in \sigma(F'(x_0)) : \lambda > 1\}.$$

This shows that

$$\deg (F_1, \mathcal{B}_1, o) = (-1)^{\beta}.$$

□

Theorem 5.8.12 (Krasnoselski Local Bifurcation Theorem). *Let $\mathcal{U}$ be a neighborhood of o in a Banach space X, and let*

$$f(\lambda, x) = x - \lambda Ax - G(\lambda, x), \qquad \lambda \in \mathcal{J}, \quad x \in \mathcal{U},$$

where $\mathcal{J}$ is an open interval in $\mathbb{R}$, A is a linear compact operator on X, $G(\lambda, \cdot) \colon \overline{\mathcal{U}} \to X$ is a compact operator and

$$\lim_{x \to o} \frac{G(\lambda, x)}{\|x\|} = o \qquad \text{for all} \quad \lambda \in \mathcal{J}.$$

If $\lambda_0 \in \mathcal{J}$ is such that $\frac{1}{\lambda_0}$ is an eigenvalue of A of odd multiplicity, then (λ_0, o) is a bifurcation point of f.

Proof. Suppose that (λ_0, o) is not a bifurcation point. Then there is a neighborhood $\tilde{\mathcal{J}} \times \mathcal{V}$ of (λ_0, o) such that the equation

$$f(\lambda, x) = o$$

has for every $\lambda \in \tilde{\mathcal{J}}$ a unique solution in $\mathcal{V}$, namely $x = o$. We may assume that $0 \notin \tilde{\mathcal{J}}$, $\{\lambda : \frac{1}{\lambda} \in \sigma(A)\} \cap \tilde{\mathcal{J}} = \{\lambda_0\}$, and also that $\lambda_0 > 0$ (for $\lambda_0 < 0$ consider

$\tilde{f}(\lambda, x) = f(-\lambda, x)$). The degree $\deg(f(\lambda,\cdot),\mathcal{V},o)$ is defined and it is given by the Leray–Schauder Index Formula (Proposition 5.8.11):

$$\deg(f(\lambda,\cdot),\mathcal{V},o) = (-1)^\beta \qquad \text{where} \quad \beta = \sum_{\substack{\mu\in\sigma(\lambda A)\cap\mathbb{R}\\ \mu>1}} m(\mu) = \sum_{\substack{\kappa\in\sigma(A)\cap\mathbb{R}\\ \kappa>\frac{1}{\lambda}}} m(\kappa).$$

Since $m\left(\frac{1}{\lambda_0}\right)$ is odd, the degree $\deg(f(\lambda,\cdot),\mathcal{V},o)$ changes sign at λ_0. A contradiction follows now from Proposition 5.8.9. □

The Krasnoselski Theorem 5.8.12 is of a local nature and does not say anything about the global behavior of a "branch" of nontrivial solutions of the equation

$$f(\lambda, x) = o.$$

The so-called global bifurcation theorems describe these branches (see Appendix 5.8A). The interested reader can also consult, e.g., Rabinowitz [122], Izé [82], Nirenberg [118, Chapter 3], Krasnoselski & Zabreiko [93], Krawcewicz & Wu [94]. There are also methods depending on other topological tools. See, e.g., Alexander [3] or Fitzpatrick [62] and references given there.

Remark 5.8.13 (Comparison of Theorems 4.3.22 and 5.8.12). Let $\frac{1}{\lambda_0}$ be an eigenvalue of A and

$$\dim\operatorname{Ker}\left(\frac{1}{\lambda_0}I - A\right) = 1.$$

Denote by x_0, $\|x_0\| = 1$, an eigenvector of A associated with $\frac{1}{\lambda_0}$. Let us compare the assumptions of Theorems 4.3.22 and 5.8.12. One of the essential differences consists in the smoothness assumptions: while Theorem 4.3.22 applied to

$$f(\lambda, x) = x - \lambda A x - G(\lambda, x)$$

requires G being a C^2-mapping, Theorem 5.8.12 demands G compact (and so continuous). The assumption $G(\lambda, x) = o(\|x\|)$, $x \to o$, yields

$$f_2'(\lambda_0, o) = I - \lambda_0 A, \qquad f_{1,2}''(\lambda_0, o) = -A. \tag{5.8.10}$$

Theorem 4.3.22 requires

$$\operatorname{Im}(I - \lambda_0 A) = \overline{\operatorname{Im}(I - \lambda_0 A)}, \tag{5.8.11}$$

$$\operatorname{codim}\operatorname{Im}(I - \lambda_0 A) = 1, \tag{5.8.12}$$

$$f_{1,2}'(\lambda_0, o)(1, x_0) \notin \operatorname{Im}(I - \lambda_0 A). \tag{5.8.13}$$

The compactness of A (Theorem 5.8.12) implies that $I - \lambda_0 A$ is a Fredholm operator of index 0, hence (5.8.11) holds and also

$$\operatorname{codim}\operatorname{Im}(I - \lambda_0 A) = \dim\operatorname{Ker}(I - \lambda_0 A) = 1,$$

i.e., (5.8.12) follows. The last assumption is closely connected with the multiplicity $m\left(\frac{1}{\lambda_0}\right)$ of $\frac{1}{\lambda_0}$ as follows from the assertion:

The assumption (5.8.13) *is verified if and only if*

$$m\left(\frac{1}{\lambda_0}\right) = \dim \bigcup_{k=1}^{\infty} \operatorname{Ker}(I - \lambda_0 A)^k = 1. \tag{5.8.14}$$

First, let us prove (5.8.14) $\Rightarrow$ (5.8.13). Assume the contrary:

$$f''_{1,2}(\lambda_0, o)(1, x_0) \in \operatorname{Im}(I - \lambda_0 A).$$

According to (5.8.10) it means $-Ax_0 \in \operatorname{Im}(I - \lambda_0 A)$. Since $x_0 = \lambda_0 A x_0$, we have $x_0 \in \operatorname{Im}(I - \lambda_0 A)$ as well. Then there exists $w \in X$ such that $x_0 = w - \lambda_0 A w$. But

$$x_0 \in \operatorname{Ker}(I - \lambda_0 A), \qquad \text{i.e.,} \qquad w \in \operatorname{Ker}(I - \lambda_0 A)^2.$$

Since $x \neq o$, we have $w \notin \operatorname{Ker}(I - \lambda_0 A)$, which implies

$$\dim \operatorname{Ker}(I - \lambda_0 A)^2 > \dim \operatorname{Ker}(I - \lambda_0 A), \qquad \text{i.e.,} \qquad m\left(\frac{1}{\lambda_0}\right) > 1,$$

a contradiction.

Now, let us prove (5.8.13) $\Rightarrow$ (5.8.14). Take

$$w \in \operatorname{Ker}(I - \lambda_0 A)^2 \qquad \text{and set} \qquad u = (I - \lambda_0 A) w.$$

Then $(I - \lambda_0 A) u = (I - \lambda_0 A)^2 w = o$ that implies $u \in \operatorname{Ker}(I - \lambda_0 A)$. Since $\operatorname{Ker}(I - \lambda_0 A)$ is generated by x_0, there exists $a \in \mathbb{R}$ such that $u = a x_0$. Simultaneously,

$$u = (I - \lambda_0 A) w \in \operatorname{Im}(I - \lambda_0 A).$$

For $a \neq 0$ we have

$$-Ax_0 = -\lambda_0 x_0 = -\frac{\lambda_0}{a} u \in \operatorname{Im}(I - \lambda_0 A),$$

a contradiction with (5.8.10) and (5.8.13). Hence

$$a = 0 \quad \text{and} \quad u = (I - \lambda_0 A) w = o, \qquad \text{i.e.,} \qquad w \in \operatorname{Ker} A.$$

This proves

$$\operatorname{Ker}(I - \lambda_0 A)^2 \subset \operatorname{Ker}(I - \lambda_0 A).$$

Since the opposite inclusion is evident, we have proved

$$\operatorname{Ker}(I - \lambda_0 A)^2 = \operatorname{Ker}(I - \lambda_0 A).$$

By induction by the power n we now easily prove that

$$\operatorname{Ker}(I - \lambda_0 A)^{n+1} = \operatorname{Ker}(I - \lambda_0 A)^n \qquad \text{for any} \quad n \in \mathbb{N}$$

(do it in details!).

Exercise 5.8.14. Prove the following assertion:

Let f, A and G be as in Theorem 5.8.12. Let $\lambda_0 \neq 0$ be a bifurcation point of f. Then $\frac{1}{\lambda_0}$ is an eigenvalue of A.

Hint. If λ_0 is a bifurcation point of f, there are

$$o \neq x_n \to o, \qquad \lambda_n \to \lambda_0, \qquad f(\lambda_n, x_n) = o.$$

Set $v_n := \frac{x_n}{\|x_n\|}$. Then

$$v_n = \lambda_n A v_n - \frac{G(\lambda_n, x_n)}{\|x_n\|}. \tag{5.8.15}$$

Since $\{v_n\}_{n=1}^{\infty}$ is bounded, A is compact, and passing to a subsequence if necessary we may assume that $v_n \to v$ for a $v \in X$, $v \neq o$. From (5.8.15) we obtain that

$$v = \lambda_0 A v.$$

Exercise 5.8.15. Let $F \in \mathscr{C}(\overline{\Omega}, X)$ where $\Omega \subset X$ is an open, bounded, symmetric with respect to $o \in X$, and nonempty set in a Banach space X, $F(x) \neq o$ for all $x \in \partial\Omega$. Assume that

$$F(x) = -F(-x) \qquad \text{for any} \quad x \in \partial\Omega.$$

Then the Leray–Schauder degree

$$\deg(I - F, \Omega, o)$$

is an odd number.

Hint. Use finite-dimensional approximations as in the construction of the Leray–Schauder degree (see pages 313–315) and Theorem 5.7.32.

Exercise 5.8.16. By repeating the construction of the Leray–Schauder degree show that the Product Formula (Exercise 5.7.12) and the boundary dependence of the degree also hold for the Leray–Schauder degree.

Exercise 5.8.17. Prove the following general homotopy invariance property:

Let Ω be an open bounded set in a Banach space X and assume that $h = h(t,x) \in \mathscr{C}([0,1] \times \overline{\Omega}, X)$ and

$$x - h(t,x) \neq y_0 \qquad \textit{for every} \quad x \in \partial\Omega \quad \textit{and} \quad t \in [0,1].$$

Then

$$\deg(I - h(t,\cdot), \Omega, y_0)$$

is constant with respect to $t \in [0,1]$.

The following two exercises use an idea similar to that of Example 5.8.3.

Exercise 5.8.18. Let H be a Hilbert space and F a compact operator on a bounded open set $\Omega \subset H$ into H. Assume that $o \in \Omega$ and

$$(F(x), x) \leq \|x\|^2 \qquad \text{for each} \quad x \in \partial\Omega.$$

Prove that F has a fixed point in $\overline{\Omega}$.

Hint. Suppose not and show that there is $t_0 \in [0,1]$, $x_0 \in \partial\Omega$ such that

$$x_0 = t_0 F(x_0).$$

By assumption, $t_0 = 1$.

Exercise 5.8.19. Let F be a compact operator from the closed unit ball $\overline{B(o;1)}$ of a Banach space X into X and, moreover, let

$$\|x - F(x)\|^2 \geq \|F(x)\|^2 - \|x\|^2 \qquad \text{for} \quad x \in \partial B(o;1).$$

Prove that F has a fixed point in $\overline{B(o;1)}$.

Exercise 5.8.20. Let f be continuous and satisfy the following growth conditions:

There are $K > 0$ and $0 < \gamma < 1$ such that the inequality

$$|f(t,x,y)| \leq K(1 + |x|^\gamma + |y|^\gamma) \qquad \text{holds for} \quad t \in [0,1], \ x, y \in \mathbb{R}.$$

Then the boundary value problem (5.8.1) has a solution. Prove that!

Hint. Proceed similarly to Example 5.8.5. Use the equation to estimate $x \in \Sigma$ and compute $\dot{x}$ with help of a special form of the kernel G.

Exercise 5.8.21. Apply Theorem 5.8.12 to the Dirichlet boundary value problem

$$\begin{cases} \ddot{x}(t) + \lambda x(t) + g(\lambda, t, x(t)) = 0, & t \in (0,\pi), \\ x(0) = x(\pi) = 0, \end{cases}$$

and show that every point (k^2, o), $k = 1, 2, \dots$ is a bifurcation point.

5.8A Global Bifurcation Theorem

In this appendix we study the bifurcation equation

$$f(\lambda, x) := x - \lambda A x - G(\lambda, x) = o. \tag{5.8.16}$$

The following result is due to Rabinowitz [122], Rabinowitz [123].

Theorem 5.8.22 (Rabinowitz Global Bifurcation Theorem). *Let X be a Banach space, Ω an open set in $\mathbb{R} \times X$, $(\lambda_0, o) \in \Omega$, $\lambda_0 \neq 0$. Let us assume:*

$$A \text{ is a compact linear operator from } X \text{ into } X, \tag{5.8.17}$$

$$G \text{ is a compact (nonlinear) operator from } \Omega \text{ into } X, \tag{5.8.18}$$

$$\begin{gathered} \text{for any bounded set } \mathcal{M} \subset \{v \in \mathbb{R} : (v, o) \in \Omega\} \text{ we have} \\ G(\lambda, x) = o(\|x\|), \ \|x\| \to 0, \text{ uniformly for } \lambda \in \mathcal{M}, \end{gathered} \tag{5.8.19}$$

$$\frac{1}{\lambda_0} \text{ is an eigenvalue of } A \text{ of odd multiplicity.} \tag{5.8.20}$$

Denote by $\mathcal{S}$ *the closure of all solutions of* (5.8.16) *with* $x \neq o$, *i.e.*,

$$\mathcal{S} = \overline{\{(\lambda, x) \in \Omega : x \neq o,\ f(\lambda, x) = o\}}.$$

Then $\mathcal{S}$ *contains the point* (λ_0, o).[57] *Let* $\mathcal{C}$ *be a component of* $\mathcal{S}$ *which contains* (λ_0, o). *Then at least one of the following assertions holds:*

(i) $\mathcal{C}$ *is not a compact set in* Ω.

(ii) $\mathcal{C}$ *contains an even number of points* (λ, o) *where* $\frac{1}{\lambda}$ *is an eigenvalue of* A *of odd multiplicity.*

Proof. We shall follow the proof of Izé [82]. The idea is the following. We will *assume that* $\mathcal{C}$ *is compact, and prove that it contains an even number of points described in* (ii). Since $\mathcal{C}$ is compact, it contains only a finite number of points (λ, o) where $\lambda \neq 0$ and $\frac{1}{\lambda}$ is an eigenvalue of the compact operator A (see Figure 5.8.2): We shall denote them by

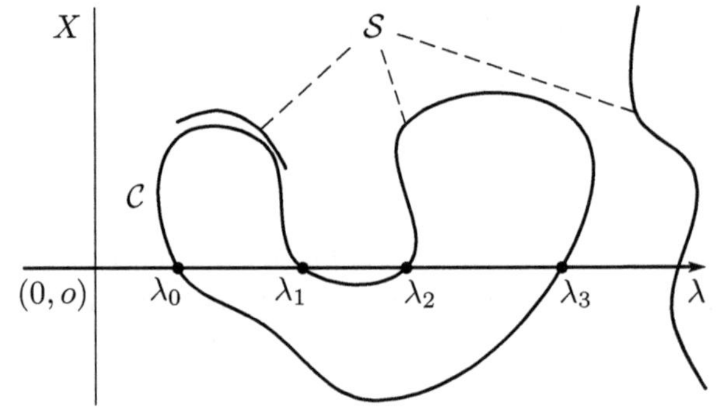

Figure 5.8.2.

$$(\lambda_0, o), \dots, (\lambda_{k-1}, o).$$

Since $\mathcal{C}$ is a component of $\mathcal{S}$ in Ω and $\mathcal{S}$ is closed, there exists an open bounded set $\tilde{\Omega}$ such that $\mathcal{C} \subset \tilde{\Omega}$ and $\mathcal{S} \cap \partial\tilde{\Omega} = \emptyset$. We prove that $\tilde{\Omega}$ can be chosen in such a way that

$$\text{“}(\lambda_j, o) \in \tilde{\Omega}, j = 0, 1, \dots, k-1, \text{ but } (\lambda, o) \notin \tilde{\Omega} \text{ for } \frac{1}{\lambda} \in \sigma(A), \lambda \neq \lambda_j, j = 0, 1, \dots, k-1\text{”}$$

(see Figure 5.8.3). Indeed, let $\mathcal{U}$ be a δ-neighborhood of $\mathcal{C}$ such that $\mathcal{U} \setminus \mathcal{C}$ does not contain any point (λ, o), $\lambda \neq 0$, $\frac{1}{\lambda} \in \sigma(A)$. The set $\mathcal{K} = \overline{\mathcal{U}} \cap \mathcal{S}$ is then compact,[58] and obviously $\mathcal{C} \cap (\partial\mathcal{U} \cap \mathcal{S}) = \emptyset$. By Deimling [42, Lemma 29.1] there exist compact disjoint sets $\mathcal{K}_1, \mathcal{K}_2 \subset \mathcal{K}$ such that

$$\mathcal{K} = \mathcal{K}_1 \cup \mathcal{K}_2, \qquad \mathcal{C} \subset \mathcal{K}_1, \qquad \partial\mathcal{U} \cap \mathcal{S} \subset \mathcal{K}_2.$$

Hence $\tilde{\Omega}$ can be chosen as an ε_0-neighborhood of $\mathcal{K}_1$ with

$$\varepsilon_0 < \min\{\operatorname{dist}(\mathcal{K}_1, \mathcal{K}_2), \operatorname{dist}(\mathcal{K}_1, \partial\mathcal{U}), \delta\}.$$

[57] I.e., (λ_0, o) is a bifurcation point in the sense of Definition 4.3.21.

[58] The reader is invited to prove it using the compactness of A and G.

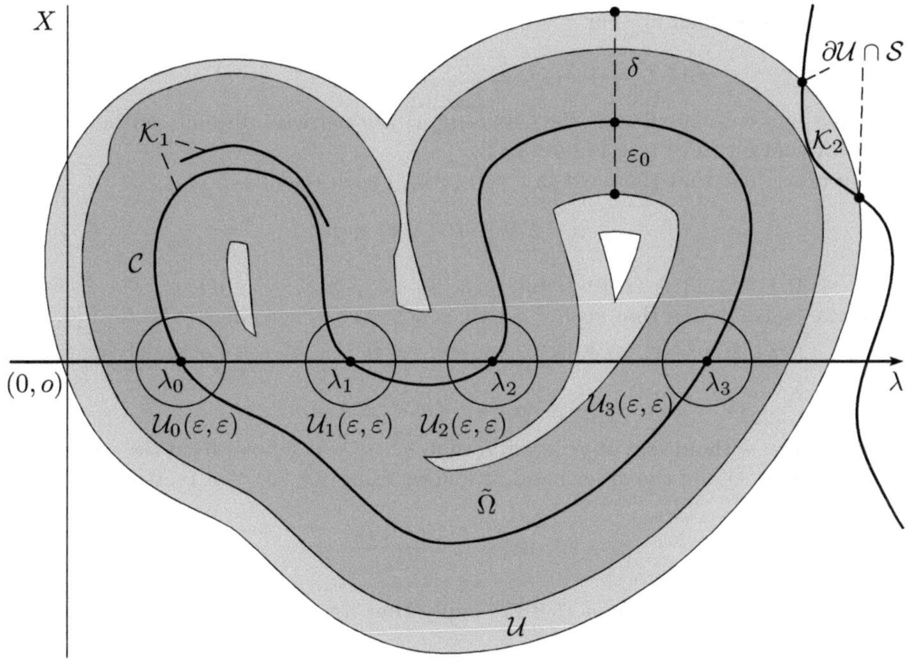

Figure 5.8.3.

For any $r > 0$ define $f_r : \tilde{\Omega} \to X \times \mathbb{R}$ as

$$f_r(\lambda, x) = (\|x\|^2 - r^2, f(\lambda, x)). \tag{5.8.21}$$

Then obviously

$$f_r(\lambda, x) = o \qquad \iff \qquad f(\lambda, x) = o \quad \text{and} \quad \|x\| = r.$$

(In other words, the function f_r "considers" the solutions of $f(\lambda, x) = o$ which belong to the sphere $\|x\| = r$.) Then thanks to the choice of $\tilde{\Omega}$ and the homotopy invariance property of the degree (Theorem 5.8.2(vi)), we conclude that

$$\deg(f_r, \tilde{\Omega}, o)$$

is well defined and independent of $r > 0$.

The rest of the proof consists in the calculation of this degree for sufficiently large r and for sufficiently small r.

Step 1 (**sufficiently large** r). The boundedness of $\tilde{\Omega}$ implies that there exists $C > 0$ such that $\|x\| < C$ for any $(\lambda, x) \in \tilde{\Omega}$. Then for $r > C$ the equation

$$f_r(\lambda, x) = o$$

has no solution in $\tilde{\Omega}$, and so, according to Theorem 5.8.2(iii), we have

$$\deg(f_r, \tilde{\Omega}, o) = 0.$$

Step 2 (**sufficiently small** r). For $j = 0, 1, \dots, k-1$ set

$$\mathcal{U}_j(\varepsilon, r) := \{(\lambda, x) : \|x\|^2 + |\lambda - \lambda_j|^2 < r^2 + \varepsilon^2\},$$

and choose $\varepsilon > 0$ so small that the sets $\overline{\mathcal{U}_j(\varepsilon,\varepsilon)}$ are pairwise disjoint, all belong to $\tilde{\Omega}$, and do not contain $(0, o)$ (see Figure 5.8.3).

We prove first that there exists $r > 0$ ($r \le \varepsilon$) such that

$$x - \lambda Ax - tG(\lambda, x) \neq o \tag{5.8.22}$$

for all $t \in [0,1]$, $(\lambda, x) \in \overline{\tilde{\Omega}}$, $0 < \|x\| \le r$, $|\lambda - \lambda_j| \ge \varepsilon$, $j = 0, 1, \dots, k-1$. Indeed, assume via contradiction that such $r > 0$ does not exist. Then there exist $t_n \in [0,1]$ and $(\lambda^n, x_n) \in \overline{\tilde{\Omega}}$, $n \in \mathbb{N}$, $o \neq x_n \to o$, $|\lambda^n - \lambda_j| \ge \varepsilon$, $j = 0, 1, \dots, k-1$, not satisfying (5.8.22), i.e.,

$$x_n - \lambda^n Ax_n - t_n G(\lambda^n, x_n) = o. \tag{5.8.23}$$

We can assume, without loss of generality, that $\lambda^n \to \tilde{\lambda}$. It follows from the construction of $\tilde{\Omega}$ that $\frac{1}{\tilde{\lambda}} \notin \sigma(A)$. On the other hand, it follows from (5.8.23) that (setting $y_n = \frac{x_n}{\|x_n\|}$)

$$y_n - \lambda^n Ay_n - t_n \frac{G(\lambda^n, x_n)}{\|x_n\|} = o. \tag{5.8.24}$$

Now, the compactness of A and (5.8.19) imply that for a $y \neq o$ ($y_{n_k} \to y$ for a subsequence) we have

$$y - \tilde{\lambda} Ay = o,$$

a contradiction.

We shall write $\mathcal{U}_j = \mathcal{U}_j(\varepsilon, r)$ for simplicity. It follows from Theorem 5.8.2(iv) that

$$\deg(f_r, \tilde{\Omega}, o) = \sum_{j=0}^{k-1} \deg(f_r, \mathcal{U}_j, o). \tag{5.8.25}$$

Let λ_j be fixed. It follows from the choice of $\varepsilon > 0$ that for $0 < |\lambda - \lambda_j| \le \varepsilon$ we have

$$\frac{1}{\lambda} \notin \sigma(A).$$

Then for any such λ the degree

$$\deg(I - \lambda A, B(o; r), o)$$

is well defined. Moreover, the homotopy invariance property of the degree implies that it is locally constant with respect to λ. Denote

$$i_-^j = \deg(I - (\lambda_j - \varepsilon)A, B(o;r), o), \qquad i_+^j = \deg(I - (\lambda_j + \varepsilon)A; B(o;r), o).$$

It follows from Lemma 5.8.23 below that

$$\deg(f_r, \mathcal{U}_j, o) = i_-^j - i_+^j. \tag{5.8.26}$$

If m_j is the multiplicity of $\frac{1}{\lambda_j}$, then Proposition 5.8.11 yields

$$i_+^j = (-1)^{m_j} i_-^j.$$

Hence for m_j even we obtain

$$\deg(f_r, \mathcal{U}_j, o) = i_-^j - i_+^j = 0, \tag{5.8.27}$$

while for m_j odd we have

$$\deg(f_r, \mathcal{U}_j, o) = 2i_-^j. \tag{5.8.28}$$

It follows from (5.8.25)–(5.8.28) that

$$\deg(f_r, \tilde{\Omega}, o) = 2 \sum_{\substack{j=0 \\ m_j \text{ odd}}}^{k-1} i_-^j.$$

Since this degree is independent of r, it must be equal to zero (see Step 1 of this proof). Hence there must be an even number of eigenvalues of odd algebraic multiplicity among $\lambda_0, \dots, \lambda_{k-1}$. □

Now we prove an analogue of the Leray–Schauder Index Formula (see Proposition 5.8.11).

Lemma 5.8.23. *Let f_r, $\mathcal{U}_j$, i_-^j, i_+^j be as above. Then*

$$\deg(f_r, \mathcal{U}_j, o) = i_-^j - i_+^j.$$

Proof. We will connect f_r with a simpler mapping using a suitable homotopy. Let us define this homotopy in the following way:

$$\forall t \in [0,1] \quad f_{t,r} \colon \mathcal{U}_j \to \mathbb{R} \times X : f_{t,r}(\lambda, x) = (\tau_t, y_t),$$

$$\tau_t = t(\|x\|^2 - r^2) + (1-t)(\varepsilon^2 - (\lambda - \lambda_j)^2), \qquad y_t = x - \lambda Ax - tG(\lambda, x).$$

We prove that for any $t \in [0,1]$

$$o \notin f_{t,r}(\partial \mathcal{U}_j).$$

Assume the contrary, i.e., there exist $t \in [0,1]$ and $(\lambda, x) \in \partial \mathcal{U}_j$ such that

$$f_{t,r}(\lambda, x) = o.$$

The fact that $(\lambda, x) \in \partial \mathcal{U}_j$ implies

$$\|x\|^2 + (\lambda - \lambda_j)^2 = r^2 + \varepsilon^2.$$

At the same time, from

$$0 = \tau_t = t(\|x\|^2 + (\lambda - \lambda_j)^2) - t(r^2 + \varepsilon^2) + \varepsilon^2 - (\lambda - \lambda_j)^2$$

we obtain $\lambda = \lambda_j \pm \varepsilon$, and so $\|x\| = r$. This together with $y_t = o$ contradicts (5.8.22). The homotopy invariance property of the degree then implies

$$\deg(f_r, \mathcal{U}_j, o) = \deg(f_{0,r}, \mathcal{U}_j, o).$$

The mapping $f_{0,r}$ is now easier to deal with. Indeed, the point o has two preimages, $(\lambda_j - \varepsilon, o)$ and $(\lambda_j + \varepsilon, o)$, with respect to the mapping

$$f_{0,r}(\lambda, x) = (\varepsilon^2 - (\lambda - \lambda_j)^2, x - \lambda Ax).$$

At both points the Fréchet differential $f'_{0,r}$ is injective:

$$f'_{0,r}(\lambda,0)(\overline{\lambda},u) = (-2(\lambda-\lambda_j)\overline{\lambda}, u-\lambda Au).$$

Let us choose sufficiently small neighborhoods of points $(\lambda_j \pm \varepsilon, o)$ in the following way: Let $\mathcal{V}^{\pm}$ be small neighborhoods of points $\lambda_j \pm \varepsilon$ and let $\mathcal{U}$ be a small neighborhood of o in X such that

$$\mathcal{U}^{\pm} := \mathcal{V}^{\pm} \times \mathcal{U} \subset \mathcal{U}_j.$$

We have

$$\deg(f_{0,r},\mathcal{U}_j,o) = \deg(f_{0,r},\mathcal{U}^-,o) + \deg(f_{0,r},\mathcal{U}^+,o)$$

and, by the Product Formula (Exercise 5.8.16) and Proposition 5.8.11,

$$\deg(f_{0,r},\mathcal{U}^-,o) = \deg(I-(\lambda_j-\varepsilon)A,\mathcal{U},o)\cdot\deg(\varepsilon^2-(\cdot-\lambda_j)^2,\mathcal{V}^-,o) = i^j_-\cdot 1.$$

Similarly, we get

$$\deg(f_{0,r},\mathcal{U}^+,o) = i^j_+\cdot(-1).$$

This completes the proof. □

Corollary 5.8.24. *If $\Omega = \mathbb{R}\times X$ in Theorem* 5.8.22, *then the first possibility* (i) *reduces to $\mathcal{C}$ is unbounded in $\mathbb{R}\times X$ and* (ii) *remains unchanged.*

Proof. Let $(\lambda,x)\in\mathcal{C}$. Then

$$x = \lambda Ax + G(\lambda,x).$$

This implies that if $\mathcal{C}$ is bounded in $\mathbb{R}\times X$, it is also relatively compact because

$$T(\lambda,x) = \lambda Ax + G(\lambda,x)$$

is a compact operator. But $\mathcal{C}$ is closed, and so it is compact. We have thus proved that if $\mathcal{C}$ is bounded, it is also compact. □

We will now discuss the special case when $\frac{1}{\lambda_0}$ is an eigenvalue of A the multiplicity of which is equal to one. If this is the case in Theorem 5.8.22 and $\mathcal{C}$ is the component containing the point (λ_0,o), then $\mathcal{C}$ consists of two connected sets $\mathcal{C}^{\pm}$ which near (λ_0,o) meet only in (λ_0,o). More precisely, the next assertion holds (see Deimling [42, Corollary 29.1]).

Corollary 5.8.25. *Under the hypotheses of Theorem* 5.8.22 *suppose, in addition, that the multiplicity of $\frac{1}{\lambda_0}$ is one. Then the component $\mathcal{C}$ containing (λ_0,o) consists of two connected sets $\mathcal{C}^+$ and $\mathcal{C}^-$, $\mathcal{C}=\mathcal{C}^+\cup\mathcal{C}^-$ such that*

$$\mathcal{C}^+\cap\mathcal{C}^-\cap B((\lambda_0,o);\varrho) = \{(\lambda_0,o)\} \quad \text{and} \quad \mathcal{C}^{\pm}\cap\partial B((\lambda_0,o);\varrho)\neq\emptyset$$

for sufficiently small $\varrho>0$.

The meaning of $\mathcal{C}^{\pm}$ is the following. Let us assume that $(\lambda_n,x_n)\in\mathcal{C}^{\pm}$, $\lambda_n\to\lambda_0$ and $x_n\to o$. Then similarly to Exercise 5.8.14 we prove that $\frac{x_n}{\|x_n\|}\to\pm v_0$ where $v_0\neq o$ is a normalized eigenvector associated with the eigenvalue λ_0. In other words, the sets $\mathcal{C}^{\pm}$ describe the "branches" of nontrivial solutions which bifurcate in the direction of the eigenvectors $\pm v_0$ (see Figure 5.8.4 for the projections of $\mathcal{C}^{\pm}$ into the space X).

The global properties of $\mathcal{C}^{\pm}$ were studied by Dancer [38]. The main result of this paper can be formulated as follows.

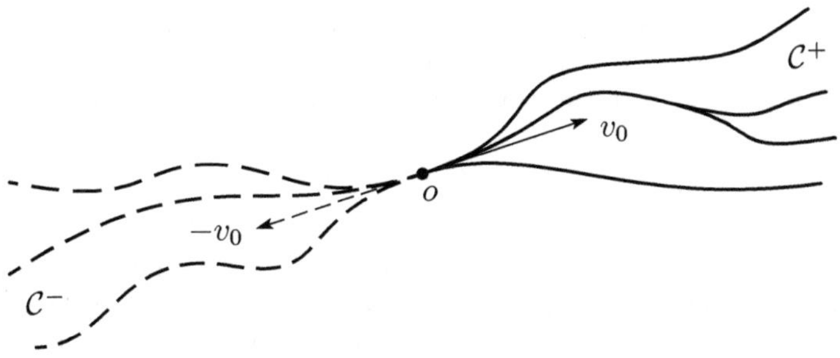

Figure 5.8.4.

Theorem 5.8.26 (Dancer Global Bifurcation Theorem). *The sets $\mathcal{C}^+$ and $\mathcal{C}^-$ are either both unbounded, or*

$$\mathcal{C}^+ \cap \mathcal{C}^- \neq \{(\lambda_0, o)\}.$$

Example 5.8.27 (Application of the Dancer Global Bifurcation Theorem). Let us consider the Dirichlet boundary value problem

$$\begin{cases} \ddot{x}(t) + \lambda x(t) = g(\lambda, t, x(t)), & t \in (0, \pi), \\ x(0) = x(\pi) = 0. \end{cases} \tag{5.8.29}$$

We assume that $g = g(\lambda, t, s)$ is a continuous function from $[0, \pi] \times \mathbb{R} \times \mathbb{R}$ into $\mathbb{R}$ and, given any bounded interval $\mathcal{I} \subset \mathbb{R}$,

$$\lim_{s \to 0} \frac{g(\lambda, t, s)}{s} = 0 \tag{5.8.30}$$

holds uniformly with respect to $t \in [0, \pi]$ and $\lambda \in \mathcal{I}$. In particular,

$$g(\lambda, t, 0) = 0, \qquad t \in [0, \pi],\ \lambda \in \mathbb{R},$$

and so (5.8.29) has a trivial solution. In this example we discuss the existence and properties of nontrivial weak solutions of (5.8.29). Let $X := W_0^{1,2}(0, \pi)$ and define operators $A, G\colon X \to X$ as follows

$$(Ax, y) = \int_0^\pi x(t)y(t)\,\mathrm{d}t, \quad (G(\lambda, x), y) = \int_0^\pi g(\lambda, t, x(t))y(t)\,\mathrm{d}t \quad \text{for any } x, y \in X.$$

The existence of a weak solution of (5.8.29) is equivalent to the existence of a solution of the operator equation (5.8.16), i.e.,

$$x - \lambda Ax - G(\lambda, x) = o,$$

cf.the beginning of Section 5.6. Moreover, (5.8.30) implies (5.8.19) (the reader is invited to check it).

Set $\lambda_0 = n^2$ where $n \in \mathbb{N}$ is fixed. Then $\frac{1}{\lambda_0} = \frac{1}{n^2}$ is an eigenvalue of A of the multiplicity 1. It follows from the above results (Theorems 5.8.22 and 5.8.26) that there is a component $\mathcal{C}$ of $\mathcal{S}$ which contains nontrivial solutions of (5.8.16), and such that

$$\mathcal{C} = \mathcal{C}^+ \cup \mathcal{C}^-, \qquad \{(n^2, o)\} \in \mathcal{C}^+ \cap \mathcal{C}^-,$$

$\mathcal{C}^\pm$ are either both unbounded, or $\mathcal{C}^+ \cap \mathcal{C}^- \neq \{(n^2, o)\}$. We show that the latter case cannot occur if $g = g(\lambda, t, s)$ is locally Lipschitz continuous with respect to the third variable s (cf. page 95). To prove this fact the properties of the initial value problem (with λ fixed)

$$\begin{cases} \ddot{x}(t) + \lambda x(t) - g(\lambda, t, x(t)) = 0, \\ x(t_0) = x_0, \ \dot{x}(t_0) = x_1, \qquad t_0 \in [0, \pi], \end{cases} \tag{5.8.31}$$

play a crucial role. In particular, we use the uniqueness of the solution to (5.8.31), which in turn implies that (5.8.31) with $x_0 = x_1 = 0$ has only the trivial solution.

The regularity result (cf. Theorem 7.1.14 and Exercise 7.1.22) for weak solutions of (5.8.29) yields that for any $(\lambda, x) \in \mathcal{C}^\pm$ we have $x \in C^2[0, \pi]$ and the above mentioned uniqueness result for (5.8.31) also implies that any such x has only a finite number of nodes in $(0, \pi)$.

Let

$$v_0(t) = \frac{1}{n}\sqrt{\frac{2}{\pi}} \sin nt$$

be a normalized eigenfunction associated with the eigenvalue $\frac{1}{n^2}$ of A. Consider $(\lambda_k, x_k) \in \mathcal{C}^+$ such that $\lambda_k \to n^2$ and $\|x_k\| \to 0$ as $k \to \infty$. Then

$$x_k - \lambda_k A x_k - G(\lambda_k, x_k) = o. \tag{5.8.32}$$

The reflexivity of X, (5.8.30) and the compactness of A imply that

$$v_k := \frac{x_k}{\|x_k\|} \to v_0 \qquad \text{in} \quad X.$$

The embedding $X = W_0^{1,2}(0, \pi) \subsetneq C[0, \pi]$ and the fact that

$$v_k = \lambda_k A v_k + \frac{G(\lambda_k, x_k)}{\|x_k\|}$$

then yield that $v_k \to v_0$ even in $C^2[0, \pi]$. In particular, it means that for large enough k, the functions x_k share the nodal properties of v_0. More precisely, let

$$\mathcal{A}^+ := \{(\lambda, x) \in \mathcal{C}^+ : x \text{ has exactly } (n-1) \text{ nodes in } (0, \pi) \text{ and } \dot{x}(0) > 0\},$$
$$\mathcal{A}^- := \{(\lambda, x) \in \mathcal{C}^- : x \text{ has exactly } (n-1) \text{ nodes in } (0, \pi) \text{ and } \dot{x}(0) < 0\}.$$

Then there exists $\varrho_0 > 0$ such that

$$\mathcal{C}^\pm \cap B((n^2, o); \varrho) = \mathcal{A}^\pm \qquad \text{for any} \quad 0 \le \varrho \le \varrho_0.$$

In particular, $\mathcal{A}^\pm \neq \emptyset$. We show that $\mathcal{A}^\pm$ is closed and open in $\mathcal{C}^\pm$. Let us consider $\mathcal{C}^+$, the case of $\mathcal{C}^-$ is similar. Recall that $\mathcal{C}^+$ is a connected set with respect to the topology induced by the topology on $\mathbb{R} \times X$. For a given $(\hat{\lambda}, \hat{x}) \in \mathcal{C}^+$ the convergence $(\lambda_k, x_k) \to (\hat{\lambda}, \hat{x})$ in this topology means that

$$\lambda_k \to \hat{\lambda} \ \text{ in } \mathbb{R} \qquad \text{and} \qquad x_k \to \hat{x} \ \text{ in } X.$$

The above mentioned regularity result and the embedding $X = W_0^{1,2}(0, \pi) \subsetneq C[0, \pi]$ then imply that

$$x_k \to \hat{x} \qquad \text{in} \quad C^2[0, \pi].$$

Let us assume that

$$(\lambda_k, x_k) \in \mathcal{A}^+, \qquad (\lambda_k, x_k) \neq (\lambda, o), \qquad (\lambda_k, x_k) \to (\hat{\lambda}, \hat{x}) \in \mathcal{C}^+.$$

The fact $x_k \to \hat{x}$ in $C^2[0,\pi]$ then yields that $(\hat{\lambda}, \hat{x}) \in \mathcal{A}^+$, i.e., $\mathcal{A}^+$ is closed in $\mathcal{C}^+$. On the other hand, if $(\hat{\lambda}, \hat{x}) \in \mathcal{A}^+$, $(\hat{\lambda}, \hat{x}) \neq (n^2, o)$, then there exists $\hat{\varrho} > 0$ such that

$$\mathcal{C}^+ \cap B((\hat{\lambda}, \hat{x}); \hat{\varrho}) \subset \mathcal{A}^+,$$

for otherwise there would be $\lambda_k \to \hat{\lambda}$ and $x_k \to \hat{x}$ in $C^2[0,\pi]$, $(\lambda_k, x_k) \notin \mathcal{A}^+$, $(\lambda_k, x_k) \in \mathcal{C}^+$, a contradiction. Hence $\mathcal{A}^+$ is open in $\mathcal{C}^+$.

We have just proved that $\mathcal{A}^\pm = \mathcal{C}^\pm$ and so the sets $\mathcal{C}^+$ and $\mathcal{C}^-$ do not have any common point besides (n^2, o). According to Theorem 5.8.26 both $\mathcal{C}^+$ and $\mathcal{C}^-$ are unbounded in $\mathbb{R} \times X$.

Let us emphasize that this means that $\mathcal{C}^\pm$ are unbounded either with respect to $\|x\|$, or with respect to λ (or with respect to both $\|x\|$ and λ!).

Some further properties of g might provide more information about the sets of $\mathcal{C}^\pm$ (e.g., boundedness with respect to x – if there are a priori estimates for all solutions – and unboundedness with respect to λ; or vice versa, boundedness with respect to λ and unboundedness with respect to $\|x\|$). ○

Exercise 5.8.28. Consider the Neumann boundary value problem

$$\begin{cases} \ddot{x}(t) + \lambda x(t) = g(\lambda, t, x(t)), & t \in (0,\pi), \\ \dot{x}(0) = \dot{x}(\pi) = 0. \end{cases} \tag{5.8.33}$$

Find conditions on $g = g(\lambda, t, s)$ and λ making it possible to apply Theorem 5.8.22.

Exercise 5.8.29. Modify assumptions from Exercise 5.8.28 on g so as to make it possible to apply Theorem 5.8.26 to (5.8.33) and to exclude the situation $\mathcal{C}^+ \cap \mathcal{C}^- \neq \{(n^2, o)\}$.

Exercise 5.8.30. Consider the periodic problem

$$\begin{cases} \ddot{x}(t) + \lambda x(t) = g(\lambda, t, x(t)), & t \in (0,2\pi), \\ x(0) = x(2\pi), \quad \dot{x}(0) = \dot{x}(2\pi). \end{cases} \tag{5.8.34}$$

Find conditions on $g = g(\lambda, t, s)$ and λ making it possible to apply Theorem 5.8.22, cf. Example 4.3.25.

5.8B Topological Degree for Generalized Monotone Operators

Let X be a reflexive real Banach space and X^* its dual. We will consider the operator

$$T\colon X \to X^*. \tag{5.8.35}$$

The purpose of this appendix is to inform the reader about a possibility how to extend the Leray–Schauder degree theory to mappings of the type (5.8.35). The following definition is the key to the theory presented in this appendix.

Definition 5.8.31. The operator $T\colon X \to X^*$ is said to satisfy the (S_+) *condition* if the assumptions

$$u_n \rightharpoonup u_0 \quad \text{(weakly) in } X \qquad \text{and} \qquad \limsup_{n\to\infty} \langle T(u_n), u_n - u_0\rangle \le 0$$ [59]

imply

$$u_n \to u_0 \qquad \text{(strongly) in} \quad X.$$

Remark 5.8.32. The topological degree for generalized monotone operators was independently introduced by Browder [23] and Skrypnik [142]. The notation (S_+) is brought from Browder [23] while the same condition is called $\alpha(X)$ in Skrypnik [142]. The (S_+) condition is a *kind of compactness condition* and plays an essential role in the construction of the degree for $T\colon X \to X^*$. This construction is based on the Brouwer degree and finite-dimensional approximations as the construction of the Leray–Schauder degree, and mappings satisfying the (S_+) condition then play a similar role as compact perturbations of the identity. The following assertion illustrates this fact. Its proof is a straightforward consequence of Definition 5.8.31.

Lemma 5.8.33. *Let $T\colon X \to X^*$ satisfy the (S_+) condition and let $K\colon X \to X^*$ be a compact operator. Then the sum $T + K\colon X \to X^*$ satisfies the (S_+) condition.*

The following assertion is an analogue of Theorem 5.8.2 and of Exercise 5.8.15.

Theorem 5.8.34 (I. V. Skrypnik [142]). *Let $T\colon X \to X^*$ be a bounded and demicontinuous*[60] *operator satisfying the (S_+) condition. Let $\mathcal{D} \subset X$ be an open, bounded and nonempty set with the boundary $\partial\mathcal{D}$ such that $T(u) \ne o$ for $u \in \partial\mathcal{D}$. Then there exists an integer*

$$\deg(T, \mathcal{D}, o)$$

(called the degree of the mapping T*) such that*

(i) $\deg(T, \mathcal{D}, o) \ne 0$ *implies that there exists an element $u_0 \in \mathcal{D}$ such that*

$$T(u_0) = o.$$

(ii) *If $\mathcal{D}$ is symmetric with respect to the origin and T satisfies $T(u) = -T(-u)$ for any $u \in \partial\mathcal{D}$, then*

$$\deg(T, \mathcal{D}, o)$$

is an odd number (and thus different from zero).

(iii) *(Homotopy invariance property) Let T_λ be a family of bounded and demicontinuous mappings which satisfy the (S_+) condition and which depend continuously on a real parameter $\lambda \in [0,1]$, and let $T_\lambda(u) \ne o$ for any $u \in \partial\mathcal{D}$ and $\lambda \in [0,1]$. Then*

$$\deg(T_\lambda, \mathcal{D}, o)$$

is constant with respect to $\lambda \in [0,1]$. In particular, we have

$$\deg(T_0, \mathcal{D}, o) = \deg(T_1, \mathcal{D}, o).$$

---

[59] Here and in the sequel we denote by $\langle f, u\rangle := f(u)$ the value of the linear form $f \in X^*$ for an element $u \in X$. If X is a Hilbert space, then according to the Riesz Representation Theorem, $\langle f, x\rangle = (x, f)$.

[60] We say that $T\colon X \to X^*$ is *demicontinuous* if T maps strongly convergent sequences in X to weakly convergent sequences in X^*.

The following assertion combined with Theorem 5.8.34(i) is a crucial tool in proving the existence of a solution or the existence of a bifurcation branch (see Section 5.9A).

Proposition 5.8.35 (I. V. Skrypnik [142]). *Let $T\colon X \to X^*$ be a bounded, demicontinuous mapping satisfying the (S_+) condition, $o \in \overline{\mathcal{D}} \setminus \partial\mathcal{D}$, $T(u) \neq o$ for $u \in \partial\mathcal{D}$, $\mathcal{D}$ being as in Theorem* 5.8.34. *Let for $u \in \partial\mathcal{D}$ the inequality*

$$\langle T(u), u\rangle \geq 0 \qquad \textit{be valid.}$$

Then

$$\deg(T, \mathcal{D}, o) = 1.$$

Let $u_0 \in X$ be an isolated solution of the equation

$$T(u) = o. \tag{5.8.36}$$

Similarly to the finite-dimensional case (and in the case of the Leray–Schauder degree) we define the *index of an isolated solution* u_0 as

$$i(T, u_0) = \lim_{r \to 0_+} \deg(T, B(u_0; r), o).$$

Then we have the following useful property of the degree.

Proposition 5.8.36 (I. V. Skrypnik [142]). *Let T and $\mathcal{D}$ be as in Theorem* 5.8.34. *Let $T(u) = o$ have only isolated solutions in $\mathcal{D}$ and let $T(u) \neq o$ for $u \in \partial\mathcal{D}$. Then there is only a finite number of solutions of* (5.8.36) *in $\mathcal{D}$, u_i, $i = 1, \dots, n$, and the equality*

$$\deg(T, \mathcal{D}, o) = \sum_{i=1}^{n} i(T, u_i) \qquad \textit{holds.}$$

The last assertion connects the properties of the functionals and the degree of their Fréchet derivatives.

Proposition 5.8.37 (I. V. Skrypnik [142]). *Assume that a real functional $F\colon X \to \mathbb{R}$ has a local minimum at $u_0 \in X$ and its Fréchet derivative $F'\colon X \to X^*$ is a bounded and demicontinuous mapping which satisfies the (S_+) condition. Let, moreover, u_0 be an isolated solution of $F'(u_0) = o$. Then*

$$i(F', u_0) = 1.$$

Example 5.8.38. Let us consider the boundary value problem

$$\begin{cases} -(|\dot{x}(t)|^{p-2}\dot{x}(t))^{\cdot} - g(x(t)) = f(t), & t \in (0,1),\\ x(0) = x(1) = 0 \end{cases} \tag{5.8.37}$$

where $p > 1$, $f \in L^{p'}(0,1)$, $p' = \frac{p}{p-1}$, and $g\colon \mathbb{R} \to \mathbb{R}$ is a continuous function. The differential operator

$$x \mapsto (|\dot{x}|^{p-2}\dot{x})^{\cdot}\ ^{61}$$

is the so-called *one-dimensional p-Laplacian* (or *half–linear differential operator* of the second order). The parameter $\lambda \in \mathbb{R}$ for which there is a nontrivial weak solution $\varphi = \varphi(t)$ (i.e., not identically equal to zero in $(0,1)$) of the problem

$$\begin{cases} -(|\dot{x}(t)|^{p-2}\dot{x}(t))^{\cdot} - \lambda|x(t)|^{p-2}x(t) = 0, & t \in (0,1), \\ x(0) = x(1) = 0 \end{cases} \tag{5.8.38}$$

is called an *eigenvalue* of the eigenvalue problem (5.8.38) and the function φ an *eigenfunction* associated with the eigenvalue λ. It is known (see, e.g., Elbert [57]) that the problem (5.8.38) has a countable set of simple eigenvalues $0 < \lambda_1 < \lambda_2 < \dots,\ \lim\limits_{n\to\infty} \lambda_n = \infty$ and the values of λ_n, $n = 1, 2, \dots$, can be explicitly calculated in terms of p and π. The eigenfunction φ_n associated with λ_n is continuously differentiable and has exactly $n-1$ zero points in $(0,1)$. In particular, we can choose $\varphi_1(t) > 0$, $t \in (0,1)$. (See Elbert [57], Došlý [46], Drábek, Krejčí & Takáč [50] and references given there.) However, the concrete values of λ_n are not important in this example. Let us assume that

$$\lim_{s\to\pm\infty} \frac{g(s)}{|s|^{p-2}s} = \lambda \qquad \text{where} \quad \lambda_n < \lambda < \lambda_{n+1} \quad \text{for an } n = 1, 2, \dots \tag{5.8.39}$$

The problem (5.8.37) is then called a *nonresonance problem* (cf. Remark 5.9.5 below). Put $X := W_0^{1,p}(0,1)$ with the norm

$$\|x\| = \left(\int_0^1 |\dot{x}(t)|^p \,\mathrm{d}t\right)^{\frac{1}{p}}.$$

Let us define operators $J, G\colon X \to X^*$ and an element $f^* \in X^*$ by

$$\langle J(x), y\rangle = \int_0^1 |\dot{x}(t)|^{p-2}\dot{x}(t)\dot{y}(t)\,\mathrm{d}t,^{62} \qquad \langle G(x), y\rangle = \int_0^1 g(x(t))y(t)\,\mathrm{d}t,$$

$$\langle f^*, y\rangle = \int_0^1 f(t)y(t)\,\mathrm{d}t \qquad \text{for any} \quad x, y \in X.$$

If we set

$$T = J - G,$$

then the operator equation

$$T(x) = f^* \tag{5.8.40}$$

is equivalent to the requirement that the integral identity

$$\int_0^1 |\dot{x}(t)|^{p-2}\dot{x}(t)\dot{y}(t)\,\mathrm{d}t - \int_0^1 g(x(t))y(t)\,\mathrm{d}t = \int_0^1 f(t)y(t)\,\mathrm{d}t \tag{5.8.41}$$

holds for all $y \in X$. The function $x \in X$ satisfying (5.8.41) is then a weak solution of (5.8.37). It follows that the existence of a weak solution of (5.8.37) is equivalent to the existence of a solution of the operator equation (5.8.40).

[61] The right-hand side is defined by $\varphi(\dot{x})$ where $\varphi(s) = |s|^{p-2}s$, $s \neq 0$, $\varphi(0) = 0$ for $p > 1$.

[62] By the Hölder inequality the integral exists and defines, for a fixed $x \in X$, a continuous linear form on X.

Our plan is to use the degree argument to prove the existence of a solution of (5.8.40). First we sketch the properties of operators J and G. The operator J satisfies

$$\langle J(x), x\rangle = \|x\|^p. \tag{5.8.42}$$

Moreover, J is an odd mapping, $(p-1)$-homogeneous,[63] it is bounded, continuous (and so demicontinuous) and satisfies the (S_+) condition. Indeed, let $x_n \rightharpoonup x_0$ in X and

$$\limsup_{n\to\infty} \langle J(x_n), x_n - x_0\rangle \le 0.$$

Then $\lim\limits_{n\to\infty} \langle J(x_0), x_n - x_0\rangle = 0$, and so

$$\begin{aligned}
0 &\ge \limsup_{n\to\infty} \langle J(x_n) - J(x_0), x_n - x_0\rangle \\
&= \limsup_{n\to\infty} \int_0^1 \left[|\dot{x}_n|^{p-2}\dot{x}_n(t) - |\dot{x}_0(t)|^{p-2}\dot{x}_0(t)\right](\dot{x}_n(t) - \dot{x}_0(t))\,\mathrm{d}t \\
&\ge \limsup_{n\to\infty} \left\{ \int_0^1 |\dot{x}_n(t)|^p\,\mathrm{d}t - \left(\int_0^1 |\dot{x}_n(t)|^p\,\mathrm{d}t\right)^{\frac{1}{p'}} \left(\int_0^1 |\dot{x}_0(t)|^p\,\mathrm{d}t\right)^{\frac{1}{p}} \right. \\
&\qquad \left. - \left(\int_0^1 |\dot{x}_0(t)|^p\,\mathrm{d}t\right)^{\frac{1}{p'}} \left(\int_0^1 |\dot{x}_n(t)|^p\,\mathrm{d}t\right)^{\frac{1}{p}} + \int_0^1 |\dot{x}_0(t)|^p\,\mathrm{d}t \right\} \\
&= \limsup_{n\to\infty} \left[\|x_n\|^{p-1} - \|x_0\|^{p-1}\right]\left[\|x_n\| - \|x_0\|\right] \ge 0
\end{aligned}$$

where the last inequality follows from the fact that $s \mapsto |s|^{p-1}$ is strictly increasing on $(0,\infty)$. Hence $\|x_n\| \to \|x_0\|$, and due to the uniform convexity of X we have $x_n \to x_0$ in X.[64] The operator J is also surjective, invertible and its inverse is continuous.[65]

The operator G is compact. This follows immediately from the compact embedding $X = W_0^{1,p}(0,1) \subsetneq\subsetneq C[0,1]$ and from the continuity of g (the reader is invited to prove it in details). Hence, due to Lemma 5.8.33 the operator T satisfies the (S_+) condition.

Let us define an operator $S\colon X \to X^*$ by

$$\langle S(x), y\rangle = \int_0^1 |x(t)|^{p-2}x(t)y(t)\,\mathrm{d}t \qquad \text{for any} \quad x, y \in X.$$

Then S is $(p-1)$-homogeneous and compact (use $X = W_0^{1,p}(0,1) \subsetneq\subsetneq L^p(0,1)$). We define a homotopy

$$T_\tau(x) = J(x) - (1-\tau)G(x) - \tau\lambda S(x) + (\tau - 1)f^* \qquad \text{for} \quad \tau \in [0,1],\ x \in X,$$

and show that there exists $R > 0$ (large enough) such that this homotopy is admissible with respect to the ball $B(o;R) \subset X$. The usual way to prove it relies on the indirect argument. Assume by contradiction that for any $k \in \mathbb{N}$ there exists $\tau_k \in [0,1]$ and $x_k \in X$, $\|x_k\| \ge k$ such that

$$T_{\tau_k}(x_k) = o,$$

[63] I.e., $J(tx) = t^{p-1}J(x)$ for any $t > 0$, $x \in X$.

[64] See Proposition 2.1.22(iv).

[65] See Exercise 5.8.40.

i.e.,

$$J(x_k) - (1-\tau_k)G(x_k) - \tau_k\lambda S(x_k) + (\tau_k - 1)f^* = o. \tag{5.8.43}$$

We divide (5.8.43) by $\|x_k\|^{p-1}$, denote $v_k := \frac{x_k}{\|x_k\|^{p-1}}$ and use that J and S are $(p-1)$-homogeneous to get

$$J(v_k) - (1-\tau_k)\frac{G(x_k)}{\|x_k\|^{p-1}} - \tau_k\lambda S(v_k) + (\tau_k - 1)\frac{f^*}{\|x_k\|^{p-1}} = o. \tag{5.8.44}$$

Due to the reflexivity of X and the compactness of the interval $[0,1]$, we may assume that

$$v_k \rightharpoonup v \quad \text{in } X \qquad \text{and} \qquad \tau_k \to \tau \in [0,1].$$

Using the compactness of the embedding $X \hookrightarrow\hookrightarrow L^p(0,1)$, the facts that G and S are continuous as operators from $L^p(0,1)$ into $L^{p'}(0,1)$, and using the assumption (5.8.39) we obtain

$$\begin{aligned}
(1-\tau_k)\frac{G(x_k)}{\|x_k\|^{p-1}} &\to (1-\tau)\lambda S(v) && \text{in } X^*,\\
\tau_k\lambda S(v_k) &\to \tau\lambda S(v) && \text{in } X^*,\\
(\tau_k - 1)\frac{f^*}{\|x_k\|^{p-1}} &\to o && \text{in } X^* \quad \text{as} \quad k\to\infty
\end{aligned}$$

(the reader is invited to justify all in detail!). Passing to the limit in (5.8.44) we thus get

$$J(v_k) \to (1-\tau)\lambda S(v) + \tau\lambda S(v) \qquad \text{in } X^* \qquad \text{as} \quad k\to\infty,$$

i.e.,

$$v_k \to J^{-1}(\lambda S(v)) \qquad \text{in } X.$$

Since at the same time $v_k \rightharpoonup v$ in X, we have

$$v_k \to v \qquad \text{in } X$$

and

$$J(v) - \lambda S(v) = o \qquad \text{in } X^*. \tag{5.8.45}$$

Since $\|v_k\| = 1$ for all $k = 1,2,\dots$, we have $\|v\| = 1$, and so (5.8.45) contradicts the fact that λ is not an eigenvalue of (5.8.38). This proves that the homotopy T_τ is admissible with respect to the ball $B(o;R)$ if R is large. Applying Theorem 5.8.34(iii) we arrive at

$$\deg(J - G - f^*, B(o;R), o) = \deg(J - \lambda S, B(o;R), o),$$

but the value of the degree on the right-hand side is an odd number according to Theorem 5.8.34(ii). Hence

$$\deg(J - G - f^*, B(o;R), o) \neq 0,$$

and the existence of at least one solution $x \in X$ of (5.8.40) which satisfies $\|x\| < R$ follows from Theorem 5.8.34(i). ○

Remark 5.8.39. It is possible to solve the problem (5.8.37) by means of the Leray–Schauder degree theory as well. In that case instead of solving the operator equation

$$J(x) - G(x) = f^*$$

one has to deal with

$$x = J^{-1}(f^* + G(x))$$

(cf. Exercise 5.8.41). Due to the properties of J^{-1} (cf. Exercise 5.8.40) this approach is more or less equivalent to that presented in Example 5.8.38. However, in more complicated applications (equations of higher order, partial differential equations, etc., see, e.g., Section 5.9A) the use of the degree presented in Theorem 5.8.34 can appear to be of essential advantage!

Exercise 5.8.40. Let J be an operator from Example 5.8.38. Prove that J is surjective and there exists an inverse operator J^{-1} which is bounded and continuous.

Hint. The strict monotonicity of $s \mapsto |s|^{p-2}s$ implies that

$$\langle J(u) - J(v), u - v\rangle > 0 \qquad \text{for} \quad u \neq v.$$

Hence J is injective. Using the Hölder inequality prove that

$$\langle J(x) - J(y), x - y\rangle \geq (\|x\|^{p-1} - \|y\|^{p-1})(\|x\| - \|y\|) \tag{5.8.46}$$

(cf. the proof of the (S_+) condition in Example 5.8.38). The boundedness of J^{-1} then follows.

To prove that J^{-1} is continuous proceed via contradiction. Suppose it is not, i.e., there exists a sequence $\{f_n\}_{n=1}^{\infty}$, $f_n \to f$ in X^* and

$$\|J^{-1}(f_n) - J^{-1}(f)\| \geq \delta \qquad \text{for a} \quad \delta > 0.$$

Let $x_n := J^{-1}(f_n)$, $x = J^{-1}(f)$. It follows that

$$\|f_n\|\|x_n\| \geq \langle f_n, x_n\rangle = \langle J(x_n), x_n\rangle = \|x_n\|^p, \qquad \text{i.e.,} \qquad \|x_n\|^{p-1} \leq \|f_n\|.$$

We may then assume $x_n \rightharpoonup \tilde{x}$ in X due to the reflexivity of X. Hence

$$\langle J(x_n) - J(\tilde{x}), x_n - \tilde{x}\rangle = \langle J(x_n) - J(x), x_n - \tilde{x}\rangle + \langle J(x) - J(\tilde{x}), x_n - \tilde{x}\rangle \to 0 \tag{5.8.47}$$

since $J(x_n) \to J(x)$ in X^*. It follows from (5.8.46) (with $x := x_n$, $y := \tilde{x}$) and (5.8.47) that $\|x_n\| \to \|\tilde{x}\|$. Hence $x_n \to \tilde{x}$ follows due to the fact that X is a uniformly convex Banach space (see page 66 or Adams [2, Theorem 3.5]). Since J is injective, $\tilde{x} = x$, a contradiction.

Exercise 5.8.41. Consider the boundary value problem (5.8.37) with g satisfying (5.8.39). Prove the existence of at least one weak solution of (5.8.37) using the Leray–Schauder degree theory.

Hint. Prove that $J^{-1} \circ G$ is a compact operator from X into itself and then use the homotopy invariance property of the Leray–Schauder degree to prove that

$$x = J^{-1}(f^* + G(x))$$

has at least one solution in X. Compare your proof with the method presented in Example 5.8.38.

Exercise 5.8.42. Consider the problem

$$\begin{cases} -\left(|\dot{x}(t)|^{p-2}\dot{x}(t)\right)^{\boldsymbol{\cdot}} = h(t, x(t), \dot{x}(t)), & t \in (0,1), \\ x(0) = x(1) = 0, \end{cases} \tag{5.8.48}$$

where $p > 1$. Formulate conditions on $h = h(t, x, s)$ which guarantee the existence of a weak solution of (5.8.48).

5.9 Weak Solutions of PDEs, Applications of Degree Theory

In this section we show how to generalize assumptions (5.6.16) and (5.6.17). We apply the degree theory and prove that more general nonlinearities $g = g(x,s)$ can be considered in (5.6.1) than those satisfying (5.6.16) and (5.6.17). At the same time we must be aware of Remark 5.6.5 and avoid the situation treated there. But first we give an example concerning the higher eigenvalues of (5.6.14).

Example 5.9.1. If for a λ there is $u \neq o$, $u \in W_0^{1,2}(\Omega)$, such that

$$\int_\Omega \nabla u(x)\nabla v(x)\,\mathrm{d}x = \lambda \int_\Omega u(x)v(x)\,\mathrm{d}x \quad \text{holds for any } v \in W_0^{1,2}(\Omega), \tag{5.9.1}$$

then λ is an eigenvalue and u is the corresponding eigenfunction of (5.6.14). Let $A\colon W_0^{1,2}(\Omega) \to W_0^{1,2}(\Omega)$ be from Example 5.6.4. Then (5.9.1) is equivalent to the eigenvalue problem

$$\mu u = Au \qquad \text{where} \quad \mu = \frac{1}{\lambda}. \tag{5.9.2}$$

It follows from the Hilbert–Schmidt Theorem – Theorem 2.2.16 that A has a countable set of eigenvalues $\{\mu_n\}_{n=1}^\infty$ of finite multiplicity such that $\mu_n \geq \mu_{n+1} \geq \cdots > 0$, $\mu_n \to 0$. Hence the eigenvalues of (5.6.14) form an increasing sequence

$$0 < \lambda_1 \leq \lambda_2 \leq \lambda_3 \leq \cdots, \qquad \lambda_n \to \infty.$$

In fact, it is also possible to prove that λ_1 has multiplicity 1 (i.e., $\lambda_1 < \lambda_2$) and the corresponding eigenspace is spanned by a function φ_1 which is positive in Ω, see Example 6.3.42 below. The system of all normalized eigenfunctions $\{\varphi_n\}_{n=1}^\infty$ forms an orthonormal basis in $W_0^{1,2}(\Omega)$ (cf. the Hilbert–Schmidt Theorem (Theorem 2.2.16)). It follows from the regularity result (which is far from being simple for $N \geq 2$, see, e.g., Gilbarg & Trudinger [71], Fučík [65], Nečas [117]) that all eigenfunctions φ_k are continuous functions in $\overline{\Omega}$. ○

Let us assume now that the nonlinear function $g = g(x,s)$ is of the form

$$g(x,s) = \lambda s + h(x,s) \qquad \text{where} \quad \lambda \in \mathbb{R}, \quad h \in \mathrm{CAR}(\Omega \times \mathbb{R})^{66} \tag{5.9.3}$$

and there exists $r \in L^2(\Omega)$ such that

$$|h(x,s)| \leq r(x) \qquad \text{for a.a.} \quad x \in \Omega \quad \text{and all} \quad s \in \mathbb{R}. \tag{5.9.4}$$

Now we can formulate the following assertion which generalizes (in the sense of the growth of g with respect to s) the results from Section 5.6.

[66] See Definition 3.2.22.

Theorem 5.9.2. *Let* $h \in \mathrm{CAR}(\Omega \times \mathbb{R})$ *satisfy* (5.9.4). *Let, moreover,* $\lambda \neq \lambda_n$, $n = 1, 2, \dots$ *Then the problem*

$$\begin{cases} -\Delta u(x) = \lambda u(x) + h(x, u(x)) & \text{in} \quad \Omega, \\ u = 0 & \text{on} \quad \partial\Omega, \end{cases} \tag{5.9.5}$$

has at least one weak solution.

Note that the assertion of Theorem 5.9.2 does not hold without the assumption $\lambda \neq \lambda_n$, $n = 1, 2, \dots$, as shown in Remark 5.6.5.

Proof of Theorem 5.9.2. We will use the Leray–Schauder degree theory to prove the result. The reader is invited to verify that the existence of a weak solution of (5.9.5) is equivalent to the existence of a solution of the operator equation

$$u = \lambda A u + S(u) \tag{5.9.6}$$

where A is as in Example 5.6.4 and $S\colon W_0^{1,2}(\Omega) \to W_0^{1,2}(\Omega)$ is defined by

$$(S(u), v) = \int_\Omega h(x, u(x)) v(x) \,\mathrm{d}x \qquad \text{for all} \quad u, v \in W_0^{1,2}(\Omega).$$

Note that S is a compact operator (see page 279).

Our plan is the following. The existence of at least one solution of (5.9.6) would follow from

$$\deg(I - \lambda A - S(\cdot), B(o; R), o) \neq 0 \tag{5.9.7}$$

if we found a ball $B(o; R)$ for which (5.9.7) is valid. To prove (5.9.7) we use the homotopy invariance property of the degree (Theorem 5.8.2(vi)) and "connect" the operator $I - \lambda A - S(\cdot)$ with the operator $I - \lambda A$ on the boundary of a ball $B(o; R)$ with a sufficiently large radius $R > 0$. Once this is done we finally use

$$\deg(I - \lambda A, B(o; R), o) \neq 0.$$[67]

So, to complete the proof, we have to find an admissible homotopy (cf.footnote 31 on page 287) connecting $I - \lambda A - S(\cdot)$ and $I - \lambda A$. Probably the simplest way to do it is the following. Define

$$H(\tau, u) = u - \lambda A(u) - \tau S(u), \qquad \tau \in [0, 1], \quad u \in W_0^{1,2}(\Omega),$$

and prove that there exists $R > 0$ such that for all $u \in W_0^{1,2}(\Omega)$, $\|u\| = R$ and $\tau \in [0, 1]$ we have

$$H(\tau, u) \neq o. \tag{5.9.8}$$

The usual way to establish (5.9.8) is an indirect proof. Assume that no such $R > 0$ exists, i.e., we can find sequences $\{u_n\}_{n=1}^\infty \subset W_0^{1,2}(\Omega)$ and $\{\tau_n\}_{n=1}^\infty \subset [0, 1]$ such that $\|u_n\| \to \infty$ and

$$u_n - \lambda A(u_n) - \tau_n S(u_n) = o. \tag{5.9.9}$$

[67] This follows from Proposition 5.8.11 or Exercise 5.8.15.

Set $v_n := \frac{u_n}{\|u_n\|}$ and divide (5.9.9) by $\|u_n\|$ to get

$$v_n - \lambda A(v_n) - \tau_n \frac{S(u_n)}{\|u_n\|} = o. \tag{5.9.10}$$

This is equivalent to

$$\int_\Omega \nabla v_n(x) \nabla w(x)\,\mathrm{d}x = \lambda \int_\Omega v_n(x) w(x)\,\mathrm{d}x + \tau_n \int_\Omega \frac{h(x, u_n(x))}{\|u_n\|} w(x)\,\mathrm{d}x$$

for any $w \in W_0^{1,2}(\Omega)$. Now, passing to suitable subsequences we can assume that $\tau_{n_k} \to \tau \in [0,1]$, $v_{n_k} \rightharpoonup v$ in $W_0^{1,2}(\Omega)$. At the same time

$$\int_\Omega \frac{|h(x, u_{n_k}(x))|}{\|u_{n_k}\|} |w(x)|\,\mathrm{d}x \le \int_\Omega \frac{r(x)}{\|u_{n_k}\|} |w(x)|\,\mathrm{d}x \to 0 \qquad \text{as} \quad k \to \infty.$$

To summarize, we have

$$\tau_{n_k} \frac{S(u_{n_k})}{\|u_{n_k}\|} \to o, \tag{5.9.11}$$

$$A(v_{n_k}) \to A(v) \tag{5.9.12}$$

(by the compactness of A, see Proposition 2.2.4(iii)). So, putting together (5.9.10)–(5.9.12) we also obtain that

$$v_{n_k} \to v^* \qquad \text{in} \quad W_0^{1,2}(\Omega).$$

But $v^* = v$ by virtue of $v_{n_k} \rightharpoonup v$. Now, passing to the limit in (5.9.10) with n replaced by n_k, we arrive at

$$v - \lambda A(v) = o, \tag{5.9.13}$$

and $v \in W_0^{1,2}(\Omega)$ satisfies $\|v\| = 1$ (it is the strong limit of elements v_{n_k} which satisfy $\|v_{n_k}\| = 1$!). However, this contradicts our assumption $\lambda \ne \lambda_n$, $n = 1, 2, \dots$ It proves that (5.9.8) holds, i.e., the homotopy H is admissible. This completes the proof. □

Following the scheme of the previous proof one can also handle nonlinearities of type (5.9.3) with $\lambda = \lambda_n$ for an $n = 1, 2, \dots$ However, the assumptions on h must be strengthened in order "to help" to prove that the corresponding homotopy H is admissible. The following assertion is a simple example of a generalization of Theorem 5.9.2 in this direction.

Theorem 5.9.3. *Let $f \in L^2(\Omega)$ and $g \colon \mathbb{R} \to \mathbb{R}$ be continuous with finite limits $\lim\limits_{s \to \pm\infty} g(s) = g(\pm\infty)$ and such that for all $s \in \mathbb{R}$ we have*

$$g(-\infty) < g(s) < g(+\infty).$$

Let $\lambda_1 > 0$ be the first eigenvalue of the Laplace operator subject to the Dirichlet boundary conditions. Then the problem

$$\begin{cases} -\Delta u(x) = \lambda_1 u(x) + g(u(x)) - f(x) & \text{in} \quad \Omega, \\ u = 0 & \text{on} \quad \partial\Omega, \end{cases} \tag{5.9.14}$$

has at least one weak solution if and only if

$$g(-\infty) < \int_\Omega f(x)\varphi_1(x)\,\mathrm{d}x < g(+\infty) \tag{5.9.15}$$

where φ_1 is the first positive eigenfunction normalized by $\int_\Omega \varphi_1(x)\,\mathrm{d}x = 1$.

Proof. We will follow a scheme similar to the proof of Theorem 5.9.2. For $\delta > 0$ so small that $\lambda_1 + \delta < \lambda_2$ we define the homotopy

$$H(\tau, u) := u - \lambda_1 A(u) - (1-\tau)\delta A(u) - \tau S(u),\text{[68]} \qquad \tau \in [0,1], \quad u \in W_0^{1,2}(\Omega).$$

Performing all steps as in the proof of Theorem 5.9.2 we arrive at an analogue of (5.9.13), namely,

$$v - [\lambda_1 + (1-\tau)\delta]A(v) = o, \qquad \|v\| = 1, \qquad \text{for a} \quad \tau \in [0,1].$$

This is a contradiction if $\tau \neq 1$ since $\lambda_1 + (1-\tau)\delta$ is not an eigenvalue and $v \neq o$. Let us assume $\tau = 1$, i.e., $\tau_{n_k} \to 1$. Now, however, we have no contradiction since λ_1 is an eigenvalue and

$$v - \lambda_1 A(v) = o$$

has a solution with $\|v\| = 1$. Another step is necessary to reach a contradiction and to prove that the homotopy H is admissible. Here the assumption(5.9.15) plays the key role. We have to revise the last step when passing to the limit in

$$v_n - \lambda_1 A(v_n) - (1-\tau_n)\delta A(v_n) - \tau_n \frac{S(u_n)}{\|u_n\|} = o$$

and employ special properties of S. Namely,

$$u_{n_k} - \lambda_1 A(u_{n_k}) - (1-\tau_{n_k})\delta A(u_{n_k}) - \tau_{n_k} S(u_{n_k}) = o$$

is equivalent to the integral identity

$$\int_\Omega \nabla u_{n_k}(x)\nabla w(x)\,\mathrm{d}x = [\lambda_1 + (1-\tau_{n_k})\delta]\int_\Omega u_{n_k}(x)w(x)\,\mathrm{d}x$$
$$+ \tau_{n_k}\int_\Omega g(u_{n_k}(x))w(x)\,\mathrm{d}x - \tau_{n_k}\int_\Omega f(x)w(x)\,\mathrm{d}x \tag{5.9.16}$$

[68]But now $(S(u), v) = \int_\Omega g(u(x))v(x)\,\mathrm{d}x + \int_\Omega f(x)v(x)\,\mathrm{d}x$, $u, v \in W_0^{1,2}(\Omega)$.

for all $w \in W_0^{1,2}(\Omega)$. Taking $w = \varphi_1$ in (5.9.16) and using the fact that

$$\int_\Omega \nabla u_{n_k}(x)\nabla\varphi_1(x)\,\mathrm{d}x = \lambda_1\int_\Omega u_{n_k}(x)\varphi_1(x)\,\mathrm{d}x,$$

we obtain

$$(1-\tau_{n_k})\delta\int_\Omega u_{n_k}(x)\varphi_1(x)\,\mathrm{d}x + \tau_{n_k}\int_\Omega g(u_{n_k}(x))\varphi_1(x)\,\mathrm{d}x = \tau_{n_k}\int_\Omega f(x)\varphi_1(x)\,\mathrm{d}x. \quad (5.9.17)$$

As above, $v_{n_k} := \frac{u_{n_k}}{\|u_{n_k}\|} \to v$ in $W_0^{1,2}(\Omega)$ and $v = \kappa\varphi_1$ with a $\kappa \neq 0$. Assume that $\kappa > 0$. Then (at least for a subsequence) $u_{n_k}(x) \to \infty$ a.e. in Ω.[69] Passing to the limit in (5.9.17) and using $\tau_{n_k} \to 1_-$ and the Lebesgue Dominated Convergence Theorem we obtain

$$\int_\Omega f(x)\varphi_1(x)\,\mathrm{d}x \geq \lim_{k\to\infty}\int_\Omega g(u_{n_k}(x))\varphi_1(x)\,\mathrm{d}x = \int_\Omega g(+\infty)\varphi_1(x)\,\mathrm{d}x = g(+\infty)$$

due to our normalization of φ_1. This contradicts the second inequality in (5.9.15). Similarly we proceed if $\kappa < 0$ to get a contradiction with the first inequality in (5.9.15). This proves that H is admissible, and so (5.9.15) is sufficient for the existence of a weak solution of (5.9.14).

To prove that (5.9.15) is also necessary we proceed as follows. Let u_0 be a weak solution of (5.9.14), i.e., for any $w \in W_0^{1,2}(\Omega)$ let us have

$$\int_\Omega \nabla u_0(x)\nabla w(x)\,\mathrm{d}x = \lambda_1\int_\Omega u_0(x)w(x)\,\mathrm{d}x + \int_\Omega g(u_0(x))w(x)\,\mathrm{d}x - \int_\Omega f(x)w(x)\,\mathrm{d}x.$$

Set $w = \varphi_1$. Then

$$\int_\Omega g(u_0(x))\varphi_1(x)\,\mathrm{d}x = \int_\Omega f(x)\varphi_1(x)\,\mathrm{d}x,$$

and the result follows from the fact that

$$g(-\infty) < \int_\Omega g(u_0(x))\varphi_1(x)\,\mathrm{d}x < g(+\infty).$$[70] □

[69] Indeed, $v_{n_k} \to \kappa\varphi_1$ in $W_0^{1,2}(\Omega)$ implies $v_{n_k} \to \kappa\varphi_1$ in $L^2(\Omega)$ (Theorem 1.2.26). Hence there exists a subsequence $\left\{v_{n_{k_l}}\right\}_{l=1}^{\infty} \subset \{v_{n_k}\}_{k=1}^{\infty}$ such that $v_{n_{k_l}} \to \kappa\varphi_1 > 0$ a.e. in Ω (Remark 1.2.18), i.e., $u_{n_{k_l}}(x) \to \infty$ a.e. in Ω.

[70] Indeed, note that $g(-\infty) < g(u_0(x)) < g(+\infty)$, $\varphi_1(x) > 0$ in Ω, and $\int_\Omega \varphi_1(x)\,\mathrm{d}x = 1$.

Remark 5.9.4. Theorem 5.9.3 holds for more general nonlinearities g (see Exercises 5.9.9 and 5.9.10). Conditions of type (5.9.15) are called the *Landesman–Lazer type conditions* according to the paper Landesman & Lazer [98] where analogous results were proved for the first time.

Remark 5.9.5. The main difference between Theorems 5.9.2 and 5.9.3 consists in the fact that λ is not an eigenvalue of (5.6.14) in Theorem 5.9.2 while λ is an eigenvalue of (5.6.14) in Theorem 5.9.3. That is why we speak about a *nonresonance problem* in the former and about a *resonance problem* in the latter case. The word "resonance" is used here because of the fact that the corresponding ordinary differential version of (5.6.14) describes the resonance in electric circuits when λ is an eigenvalue.

Remark 5.9.6. A result analogous to Theorem 5.9.3 can be proved also if λ_1 is replaced by any eigenvalue λ_n, $n \geq 2$. The Landesman–Lazer type conditions then have a different form and all linearly independent eigenfunctions associated with λ_n must be involved (see, e.g., Fučík [65]).

Remark 5.9.7. Let us mention one geometrical aspect of condition (5.9.15). It follows from the linear Fredholm alternative (see Exercise 5.9.11) that

$$\begin{cases} -\Delta u(x) = \lambda_1 u(x) - f(x) & \text{in} \quad \Omega, \\ u = 0 & \text{on} \quad \partial\Omega \end{cases} \tag{5.9.18}$$

with $f \in L^2(\Omega)$ has a weak solution if and only if f belongs to a linear subspace V of $L^2(\Omega)$ of codimension one:

$$V = \left\{ f \in L^2(\Omega) : \int_\Omega f(x)\varphi_1(x)\,\mathrm{d}x = 0 \right\}.$$

If $g(-\infty) < 0 < g(+\infty)$, then the set $\widetilde{V}$ of all f for which (5.9.14) has at least one weak solution contains V as a proper subset and is in fact much larger (see Figure 5.9.1). For example, contrary to V, the set $\widetilde{V}$ has a nonempty interior in $L^2(\Omega)$.

Exercise 5.9.8. The same method as in the proof of Theorem 5.9.2 works if we replace (5.9.4) by a more general assumption

$$|h(x,s)| \leq r(x) + |s|^\delta \quad \text{for a.a. } x \in \Omega \text{ and all } s \in \mathbb{R} \quad \text{where } \delta \in (0,1).$$

Modify the proof and get the corresponding generalization of Theorem 5.9.2.

Exercise 5.9.9. Let $h \in \mathrm{CAR}(\Omega \times \mathbb{R})$ be bounded and satisfy the following conditions:

Let there exist limits

$$h(x,+\infty) = \lim_{s\to+\infty} h(x,s), \qquad h(x,-\infty) = \lim_{s\to-\infty} h(x,s)$$
$$\text{for a.a.} \quad x \in \Omega.$$

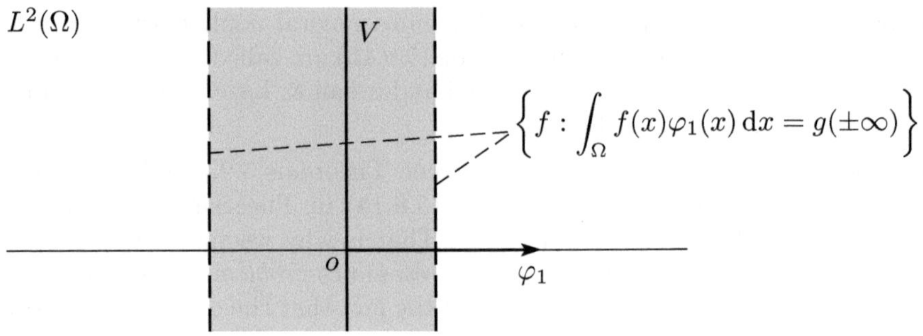

Figure 5.9.1. Problem (5.9.14) has a weak solution if and only if f belongs to the shaded area.

Assume that

$$h(x,-\infty) < h(x,+\infty), \quad (h(x,+\infty) < h(x,-\infty)) \quad \text{for a.a. } x \in \Omega.$$

Prove that

$$\begin{cases} -\Delta u(x) = \lambda_1 u(x) + h(x,u(x)) & \text{in} \quad \Omega, \\ u = 0 & \text{on} \quad \partial\Omega \end{cases} \tag{5.9.19}$$

has at least one weak solution provided

$$\begin{gathered} \int_\Omega h(x,-\infty)\varphi_1(x)\,dx < 0 < \int_\Omega h(x,+\infty)\varphi_1(x)\,dx \\ \left(\int_\Omega h(x,+\infty)\varphi_1(x)\,dx < 0 < \int_\Omega h(x,-\infty)\varphi_1(x)\,dx\right). \end{gathered} \tag{5.9.20}$$

Exercise 5.9.10. Assume that for a.a. $x \in \Omega$ and for all $s \in \mathbb{R}$ we have

$$h(x,-\infty) < h(x,s) < h(x,+\infty) \qquad (h(x,+\infty) < h(x,s) < h(x,-\infty)).$$

Prove that (5.9.20) is also a necessary condition for the existence of a weak solution of (5.9.19).

Exercise 5.9.11. Let $f \in L^2(\Omega)$. Prove that (5.9.18) has a weak solution if and only if $f \in V$, i.e.,

$$\int_\Omega f(x)\varphi_1(x)\,dx = 0.$$

(Cf. the Fredholm alternative mentioned in Theorem 2.2.9(iv).)

Hint. Let A be from Example 5.6.4. Then $I - \lambda_1 A$ is a self-adjoint operator and $\operatorname{Ker}(I - \lambda_1 A) = \operatorname{Lin}\{\varphi_1\}$. Any $f \in L^2(\Omega)$ defines a continuous linear form f^* on $W_0^{1,2}(\Omega)$ by

$$f^*(u) = \int_\Omega f(x)u(x)\,dx, \qquad u \in W_0^{1,2}(\Omega).$$

It follows from Proposition 2.1.27(iv) that given $f \in L^2(\Omega)$, $\int_\Omega f(x)\varphi_1(x)\,dx = 0$, then

$$f^* \in (\mathrm{Ker}\,(I - \lambda_1 A))_\perp = \overline{\mathrm{Im}\,(I - \lambda_1 A)}.$$

However, $\mathrm{Im}\,(I - \lambda_1 A) = \overline{\mathrm{Im}\,(I - \lambda_1 A)}$ by Theorem 2.2.9(iii). On the other hand, if (5.9.18) has a weak solution u_0 for a given $f \in L^2(\Omega)$, then taking $v = \varphi_1$ as a test function we arrive at

$$\int_\Omega f(x)\varphi_1(x)\,dx = 0.$$

Exercise 5.9.12. Prove an analogue of Theorem 5.9.2 for the Neumann problem

$$\begin{cases} -\Delta u(x) = \lambda u(x) + h(x, u(x)) & \text{in} \quad \Omega, \\ \dfrac{\partial u}{\partial n} = 0 & \text{on} \quad \partial\Omega. \end{cases}$$

Exercise 5.9.13. Prove an analogue of Theorem 5.9.3 for the Neumann problem

$$\begin{cases} -\Delta u(x) = g(u(x)) - f(x) & \text{in} \quad \Omega, \\ \dfrac{\partial u}{\partial n} = 0 & \text{on} \quad \partial\Omega. \end{cases}$$

5.9A Weak Solutions of PDEs, Application of the Degree of Generalized Monotone Mappings

In this appendix we deal with the boundary value problem

$$\begin{cases} -\Delta_p u(x) = \lambda |u(x)|^{p-2} u(x) + f(\lambda, x, u(x)) & \text{in} \quad \Omega, \\ u = 0 & \text{on} \quad \partial\Omega \end{cases} \tag{5.9.21}$$

where $p > 1$, $\Omega \in C^{0,1}$ is a bounded domain in $\mathbb{R}^N$, $f\colon \mathbb{R} \times \Omega \times \mathbb{R} \to \mathbb{R}$ is a Carathéodory function (see Remark 3.2.25) which satisfies some further conditions specified below, and

$$\Delta_p u := \operatorname{div}\left(|\nabla u|^{p-2}\nabla u\right) \text{ [71]}$$

is the *p-Laplacian*. It is well known that the problem

$$\begin{cases} -\Delta_p u(x) = \lambda |u(x)|^{p-2} u(x) & \text{in} \quad \Omega, \\ u = 0 & \text{on} \quad \partial\Omega \end{cases}$$

[71] This means $\Delta_p u = \frac{\partial}{\partial x_1}\left\{\left[\left(\frac{\partial u}{\partial x_1}\right)^2 + \cdots + \left(\frac{\partial u}{\partial x_N}\right)^2\right]^{\frac{p-2}{2}} \frac{\partial u}{\partial x_1}\right\} + \cdots + \frac{\partial}{\partial x_N}\left\{\left[\left(\frac{\partial u}{\partial x_1}\right)^2 + \cdots + \left(\frac{\partial u}{\partial x_N}\right)^2\right]^{\frac{p-2}{2}} \frac{\partial u}{\partial x_N}\right\}$, $\Delta_p u = 0$ if $\nabla u = o$.

has a principal eigenvalue (i.e., the least one) $\lambda_1 > 0$ which is simple, isolated and characterized variationally by

$$\lambda_1 = \inf \frac{\displaystyle\int_\Omega |\nabla u(x)|^p \,\mathrm{d}x}{\displaystyle\int_\Omega |u(x)|^p \,\mathrm{d}x}$$

where "inf" is taken over all $u \in W_0^{1,p}(\Omega)$, $u \neq o$; no eigenfunction associated with λ_1 changes sign in Ω and they all form a one-dimensional linear space (see, e.g., Anane [8], Lindqvist [101], or Example 7.8.5 for the case $N = 1$). Our aim in this appendix is to show that under some appropriate assumptions on f the value λ_1 is a bifurcation point of (5.9.21) in the sense that for any neighborhood of $(\lambda_1, o) \in \mathbb{R} \times W_0^{1,p}(\Omega)$ there exists at least one $(\lambda, u) \in \mathbb{R} \times W_0^{1,p}(\Omega)$, $u \neq o$, which solves (in the sense mentioned below) the boundary value problem (5.9.21).

Let us assume, for simplicity, that $f = f(\lambda, x, s)$ is uniformly bounded, i.e., there exists a constant $M > 0$ such that for any $(\lambda, s) \in \mathbb{R}^2$ and a.a. $x \in \Omega$,

$$|f(\lambda, x, s)| \leq M. \tag{5.9.22}$$

Let us define for $\lambda \in \mathbb{R}$ operators $J, S, F_\lambda \colon W_0^{1,p}(\Omega) \to (W_0^{1,p}(\Omega))^*$ by

$$\langle J(u), v\rangle = \int_\Omega |\nabla u(x)|^{p-2} \nabla u(x) \nabla v(x) \,\mathrm{d}x, \quad \langle S(u), v\rangle = \int_\Omega |u(x)|^{p-2} u(x) v(x) \,\mathrm{d}x,$$
$$\langle F_\lambda(u), v\rangle = \int_\Omega f(\lambda, x, u(x)) v(x) \,\mathrm{d}x \qquad \text{for} \quad u, v \in W_0^{1,p}(\Omega).$$

It follows from (5.9.22), Theorem 3.2.24 and Remark 3.2.26 that J, S and F_λ are well defined operators (the reader should justify it in details!). Actually, combining Theorems 1.2.26, 1.2.28, 3.2.24 and Remark 3.2.26 we have (cf. Exercise 5.9.18):[72]

(a) J and S are bounded operators in the sense that they map bounded sets onto bounded sets;

(b) F_λ is a uniformly bounded operator in the sense that there exists a constant $c > 0$ such that for any $\lambda \in \mathbb{R}$ and $u \in W_0^{1,p}(\Omega)$ we have

$$\|F_\lambda(u)\|_{(W_0^{1,p}(\Omega))^*} \leq c;$$

(c) J, S and F_λ are continuous operators;

(d) for any $u, v \in W_0^{1,p}(\Omega)$,

$$\langle J(u) - J(v), u - v\rangle \geq \left(\|u\|^{p-1} - \|v\|^{p-1}\right)(\|u\| - \|v\|);$$

(e) if $u_n \rightharpoonup u$ in $W_0^{1,p}(\Omega)$ and $\lambda_n \to \lambda$ in $\mathbb{R}$, then we have $S(u_n) \to S(u)$ and $F_{\lambda_n}(u_n) \to F_\lambda(u)$ in $(W_0^{1,p}(\Omega))^*$; in particular, S and F_λ are compact.

[72] We work with the norm $\|u\| = \left(\displaystyle\int_\Omega |\nabla u(x)|^p \,\mathrm{d}x\right)^{\frac{1}{p}}$ on $W_0^{1,p}(\Omega)$, i.e., $\langle J(u), u\rangle = \|u\|^p$.

A couple $(\lambda, u) \in \mathbb{R} \times W_0^{1,p}(\Omega)$ is a *weak solution* of (5.9.21) if

$$\langle J(u), v\rangle = \lambda\langle S(u), v\rangle + \langle F_\lambda(u), v\rangle \qquad \text{holds for any} \quad v \in W_0^{1,p}(\Omega).$$

Thus, to find a weak solution of (5.9.21) is equivalent to finding a couple $(\lambda, u) \in \mathbb{R} \times W_0^{1,p}(\Omega)$ which satisfies the operator equation

$$J(u) = \lambda S(u) + F_\lambda(u). \tag{5.9.23}$$

Lemma 5.9.14. *For any $\lambda \in \mathbb{R}$ the operator $J - \lambda S - F_\lambda$ satisfies the (S_+) condition (see Definition* 5.8.31).

Proof. We use the property (d) to prove that J satisfies the (S_+) condition (we proceed similarly to Example 5.8.38). The assertion then follows from (e) and Lemma 5.8.33. □

Assume that for any $\lambda \in \mathbb{R}$ and a.a. $x \in \Omega$ we have

$$f(\lambda, x, o) = 0. \tag{5.9.24}$$

This immediately yields

$$F_\lambda(o) = o \qquad \text{for any} \quad \lambda \in \mathbb{R}.$$

We also assume that for any bounded interval $\mathcal{I} \subset \mathbb{R}$ the limit

$$\lim_{s\to 0} \frac{f(\lambda, x, s)}{|s|^{p-2}s} = 0 \tag{5.9.25}$$

exists uniformly for a.a. $x \in \Omega$ and $\lambda \in \mathcal{I}$. This implies

$$\lim_{\|u\|\to o} \frac{F_\lambda(u)}{\|u\|^{p-1}} = o \qquad \text{uniformly for} \quad \lambda \in \mathcal{I} \tag{5.9.26}$$

(cf. Exercise 5.9.19). The fact that the eigenvalue $\lambda_1 > 0$ is isolated and the compactness of S imply that there exists $\delta > 0$ such that for any $\lambda \in (\lambda_1 - \delta, \lambda_1 + \delta)$, $\lambda \neq \lambda_1$, there exists $c = c(\lambda) > 0$ such that

$$\|J(u) - \lambda S(u)\|_{(W_0^{1,p}(\Omega))^*} \geq c(\lambda)\|u\|^{p-1} \qquad \text{for any} \quad u \in W_0^{1,p}(\Omega) \tag{5.9.27}$$

(cf. Exercise 5.9.20). Denote

$$T_\lambda := J - \lambda S - F_\lambda.$$

Then (5.9.26) and (5.9.27) imply that for $\lambda \in (\lambda_1 - \delta, \lambda_1 + \delta)$, $\lambda \neq \lambda_1$, o is an isolated solution of $T_\lambda(u) = o$ and its index $i(T_\lambda, o)$ [73] is well defined.

Proposition 5.9.15. *Let us assume*

$$\lim_{\lambda\to\lambda_{1-}} i(T_\lambda, o) \neq \lim_{\lambda\to\lambda_{1+}} i(T_\lambda, o).$$

Then (λ_1, o) is a bifurcation point of T_λ, i.e., there exist $\lambda_n \to \lambda_1$, $u_n \to o$, $u_n \neq o$, such that

$$T_{\lambda_n}(u_n) = o.$$

[73] Here we mean the index from Appendix 5.8B, page 341.

Proof. To prove this assertion we can follow the proof of Proposition 5.8.9, but working with the index defined on page 341. □

Combining (5.9.26) and (5.9.27) we obtain that

$$i(T_\lambda, o) = i(J - \lambda S, o) \qquad \text{for} \quad \lambda \in (\lambda_1 - \delta, \lambda_1 + \delta), \quad \lambda \neq \lambda_1 \tag{5.9.28}$$

(cf. Exercise 5.9.21). It then follows that to prove that (λ_1, o) is a bifurcation point of T_λ it suffices to prove that

$$\lim_{\lambda \to \lambda_{1-}} i(J - \lambda S, o) \neq \lim_{\lambda \to \lambda_{1+}} i(J - \lambda S, o). \tag{5.9.29}$$

We are now ready to prove the following assertion.

Theorem 5.9.16. *Let f satisfy all the assumptions stated above. Then (λ_1, o) is a bifurcation point of T_λ, i.e., of (5.9.21).*

Proof. Let $\delta > 0$ be such that (5.9.28) holds. Taking into account the above discussion it is enough to prove

$$i(J - \lambda S, o) = 1 \qquad \text{for} \quad \lambda \in (\lambda_1 - \delta, \lambda_1), \tag{5.9.30}$$

$$i(J - \lambda S, o) = -1 \qquad \text{for} \quad \lambda \in (\lambda_1, \lambda_1 + \delta). \tag{5.9.31}$$

It follows from the definition of $\lambda_1 > 0$ on page 354 that

$$\langle J(u) - \lambda S(u), u \rangle > 0 \qquad \text{for} \quad u \neq o \qquad \text{if} \qquad \lambda \in (\lambda_1 - \delta, \lambda_1).$$

Applying Proposition 5.8.35 we immediately get (5.9.30).

To prove (5.9.31) we proceed in the following way. There exists $K > 0$ large enough so that we can define a function $\psi \colon \mathbb{R} \to \mathbb{R}$ by

$$\psi(t) = \begin{cases} 0 & \text{for} \quad t \leq K, \\ \dfrac{2\delta}{\lambda_1}(t - 2K) & \text{for} \quad t \geq 3K, \end{cases}$$

and such that $\psi(t)$ is continuously differentiable in $\mathbb{R}$, positive and strictly convex in $(K, 3K)$ (see Figure 5.9.2, the reader is invited to write an explicit formula for such ψ).

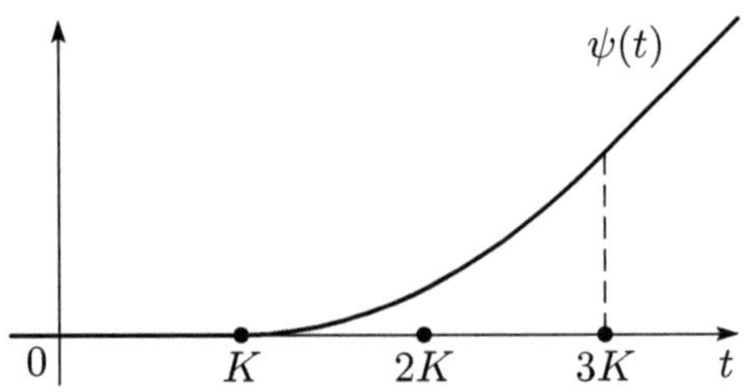

Figure 5.9.2.

We define a functional

$$\Psi_\lambda(u) := \frac{1}{p}\|u\|^p - \frac{\lambda}{p}\|u\|^p_{L^p(\Omega)} + \psi\left(\frac{1}{p}\|u\|^p\right), \qquad u \in W_0^{1,p}(\Omega).$$

Then Ψ_λ is continuously Fréchet differentiable and its critical point $u_0 \in W_0^{1,p}(\Omega)$ corresponds to a solution of the equation

$$J(u_0) - \frac{\lambda}{1+\psi'\left(\frac{1}{p}\|u_0\|^p\right)} S(u_0) = o.$$

However, since $\lambda \in (\lambda_1, \lambda_1 + \delta)$ and there is only one eigenvalue below $\lambda_1 + \delta$, a nonzero critical point u_0 of Ψ_λ has to satisfy

$$\frac{\lambda}{1+\psi'\left(\frac{1}{p}\|u_0\|^p\right)} = \lambda_1, \qquad \text{i.e.,} \qquad \psi'\left(\frac{1}{p}\|u_0\|^p\right) = \frac{\lambda}{\lambda_1} - 1. \tag{5.9.32}$$

Due to the properties of ψ we necessarily have

$$\frac{1}{p}\|u_0\|^p \in (K, 3K)$$

and due to (5.9.32) and the simplicity of λ_1, either $u_0 = -u_1$ or $u_0 = u_1$ for a $u_1 > 0$ which is an eigenfunction associated with λ_1. So, we conclude that there are precisely three isolated critical points of Ψ_λ:

$$-u_1, \quad o, \quad u_1.$$

The functional Ψ_λ is weakly sequentially lower semicontinuous. Indeed, assume that $u_n \rightharpoonup v_0$ in $W_0^{1,p}(\Omega)$. Then

$$\|u_n\|^p_{L^p(\Omega)} \to \|v_0\|^p_{L^p(\Omega)} \tag{5.9.33}$$

due to the compactness of $W_0^{1,p}(\Omega) \subsetneq\subsetneq L^p(\Omega)$, and then

$$\liminf_{n\to\infty} \Psi_\lambda(u_n) \geq \Psi_\lambda(v_0)$$

by the fact that $\liminf\limits_{n\to\infty} \|u_n\| \geq \|v_0\|$, (5.9.33) holds, and ψ is increasing. Observe that Ψ_λ is weakly coercive, i.e.,

$$\lim_{\|u\|\to\infty} \Psi_\lambda(u) = \infty. \tag{5.9.34}$$

Indeed, we have

$$\Psi_\lambda(u) = \frac{1}{p}\|u\|^p - \frac{\lambda_1}{p}\|u\|^p_{L^p(\Omega)} + \frac{\lambda_1-\lambda}{p}\|u\|^p_{L^p(\Omega)} + \psi\left(\frac{1}{p}\|u\|^p\right). \tag{5.9.35}$$

Since

$$\|u\|^p - \lambda_1\|u\|^p_{L^p(\Omega)} \geq 0 \qquad \text{for any} \quad u \in W_0^{1,p}(\Omega),$$

we also have ($\lambda > \lambda_1$)

$$\begin{aligned} \frac{\lambda_1-\lambda}{p}\|u\|^p_{L^p(\Omega)} + \psi\left(\frac{1}{p}\|u\|^p\right) &\geq \frac{\lambda_1-\lambda}{p\lambda_1}\|u\|^p + \psi\left(\frac{1}{p}\|u\|^p\right) \\ &\geq -\frac{\delta}{p\lambda_1}\|u\|^p + \frac{2\delta}{\lambda_1}\left(\frac{1}{p}\|u\|^p - 2K\right) \to \infty \end{aligned} \tag{5.9.36}$$

for $\|u\| \to \infty$, and (5.9.34) follows. Since Ψ_λ is an even functional, there are precisely two its critical points at which the global minimum is achieved: $-u_1$ and u_1. The third critical point o is obviously an isolated critical point of "saddle type".[74] In virtue of Proposition 5.8.37 we have

$$i(\Psi'_\lambda, -u_1) = i(\Psi'_\lambda, u_1) = 1.^{75} \tag{5.9.37}$$

At the same time, due to the definition of ψ, we have

$$\langle \Psi'_\lambda(u), u\rangle > 0 \qquad \text{for any} \quad u \in W_0^{1,p}(\Omega), \quad \|u\| = \kappa \tag{5.9.38}$$

with $\kappa > 0$ large enough (cf. Exercise 5.9.22). Hence, Proposition 5.8.35 implies that

$$\deg(\Psi'_\lambda, B(o;\kappa), o) = 1. \tag{5.9.39}$$

Take $\kappa > 0$ so large that $\pm u_1 \in B(o;\kappa)$. By Proposition 5.8.36 (the additivity property), (5.9.37) and (5.9.39), we deduce

$$i(\Psi'_\lambda, o) = -1. \tag{5.9.40}$$

Since ψ vanishes in a small neighborhood of 0, we also have

$$i(J - \lambda S, o) = i(\Psi'_\lambda, o). \tag{5.9.41}$$

Hence (5.9.31) follows from (5.9.40) and (5.9.41). □

Remark 5.9.17. Note that a global bifurcation result in the sense of Appendix 5.8A can be proved from (5.9.30) and (5.9.31) (see, e.g., Drábek, Kufner & Nicolosi [51]). We do not treat it here because other technical details have to be involved.

Exercise 5.9.18. Prove rigorously properties (a)–(e) from the beginning of this section.

Hint. Use the Hölder inequality and (5.9.22) for the proofs of (a)–(c). For the proof of (d) adopt the estimate similar to the one from Example 5.8.38. For the proof of (e) use the Rellich–Kondrachov Theorem (Theorem 1.2.28) and Theorem 3.2.24.

Exercise 5.9.19. Prove that (5.9.25) implies (5.9.26).

Hint. Apply the Lebesgue Dominated Convergence Theorem.

Exercise 5.9.20. Prove (5.9.27).

Hint. The inequality (5.9.27) is equivalent to

$$\|J(u) - \lambda S(u)\|_{(W_0^{1,p}(\Omega))^*} \ge c(\lambda) \qquad \text{for all} \quad \|u\| = 1.$$

Proceed via contradiction and use the compactness of S and the fact that λ is not an eigenvalue.

Exercise 5.9.21. Prove (5.9.28).

Hint. Use (5.9.26) and (5.9.27) and the homotopy invariance property of the degree on the ball with a sufficiently small radius.

---

[74] Consider Ψ_λ in the direction of u_1 to get a local maximum at o due to the variational characterization of λ_1, and, on the other hand, in the direction of $u \neq o$ satisfying $\displaystyle\int_\Omega |\nabla u(x)|^p \,\mathrm{d}x - \lambda \int_\Omega |u(x)|^p \,\mathrm{d}x > 0$ to get $\Psi_\lambda(tu) > 0$, $t \neq 0$.

[75] The reader is invited to check that the assumptions of Proposition 5.8.37 are satisfied.

Exercise 5.9.22. Prove (5.9.38).

Hint.

$$\begin{aligned}\langle \Psi'_\lambda(u), u\rangle &= \|u\|^p - \lambda\|u\|^p_{L^p(\Omega)} + \psi'\left(\frac{1}{p}\|u\|^p\right)\|u\|^p\\ &= \|u\|^p - \lambda_1\|u\|^p_{L^p(\Omega)} + \psi'\left(\frac{1}{p}\|u\|^p\right)\left[\|u\|^p - \frac{\lambda-\lambda_1}{\psi'\left(\frac{1}{p}\|u\|^p\right)}\|u\|^p_{L^p(\Omega)}\right]\\ &\geq \frac{2\delta}{\lambda_1}\left[\|u\|^p - \frac{\lambda_1}{2}\|u\|^p_{L^p(\Omega)}\right] \to \infty \qquad \text{for} \quad \|u\| \to \infty,\end{aligned}$$

by the variational characterization of λ_1.

Exercise 5.9.23. Let $\lambda < \lambda_1$ and let $f \in L^{p'}(\Omega)$. Prove that there exists a weak solution $u \in W_0^{1,p}(\Omega)$ of the problem

$$\begin{cases} -\Delta_p u(x) = \lambda|u(x)|^{p-2}u(x) + f(x) & \text{in} \quad \Omega,\\ u = 0 & \text{on} \quad \partial\Omega\end{cases}$$

where $p > 1$, $p' = \frac{p}{p-1}$.

Exercise 5.9.24. Let $\lambda < \lambda_1$ and let $f\colon \Omega \times \mathbb{R} \to \mathbb{R}$ be a bounded Carathéodory function. Prove that there exists a weak solution $u \in W_0^{1,p}(\Omega)$ of the problem

$$\begin{cases} -\Delta_p u(x) = \lambda|u(x)|^{p-2}u(x) + f(x, u(x)) & \text{in} \quad \Omega,\\ u = 0 & \text{on} \quad \partial\Omega\end{cases}$$

where $p > 1$.

Chapter 6

Monotonicity Methods

6.1 Theory of Monotone Operators

The motivation for the methods presented in this section can be described by the following simple example of a real function of one real variable $f\colon \mathbb{R} \to \mathbb{R}$. We would like to find conditions on f which guarantee that for any $y \in \mathbb{R}$ the equation $f(x) = y$ has a (unique) solution x. One possible way to solve this first semester calculus problem is to consider f which is continuous, (strictly) monotone and $\lim\limits_{|x|\to\infty} |f(x)| = \infty$ (see Figure 6.1.1).

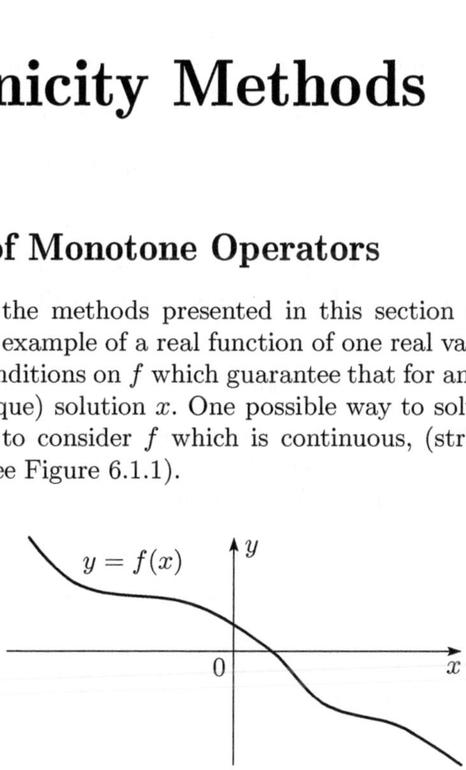

Figure 6.1.1.

If f is replaced by an operator

$$T\colon H \to H$$

from a real Hilbert space H (with a scalar product $(\cdot,\cdot)$ and the induced norm $\|\cdot\|$) into itself and the same question is posed, then similar conditions appear to be appropriate to prove that for any $h \in H$ the equation

$$T(u) = h$$

has a unique solution $u \in H$. It is clear how to reformulate the first condition on f in the case of a general operator T. The third condition motivates the following definition.

Definition 6.1.1. Let H be a real Hilbert space. An operator $T\colon H \to H$ satisfying

$$\lim_{\|u\|_H \to \infty} \|T(u)\| = \infty$$

is called *weakly coercive.*

In order to reformulate the second condition we should first note that a real function of one real variable is increasing (decreasing) if and only if

$$(f(x) - f(y))(x - y) \ge 0 \qquad (\le 0) \qquad \text{for any} \quad x, y \in \mathbb{R}.$$

Definition 6.1.2. Let H be a real Hilbert space. An operator $T\colon H \to H$ satisfying

$$(T(u) - T(v), u - v) \ge 0 \qquad \text{for any} \quad u, v \in H \tag{6.1.1}$$

is called a *monotone operator.* An operator T is called *strictly monotone* if for $u \neq v$ the strict inequality holds in (6.1.1). An operator T is called *strongly monotone* if there exists $c > 0$ such that

$$(T(u) - T(v), u - v) \ge c\|u - v\|^2 \qquad \text{for any} \quad u, v \in H.$$

Remark 6.1.3. It is clear that a strongly monotone operator is strictly monotone and, therefore, monotone. Also, every strongly monotone operator is weakly coercive. Indeed, T being strongly monotone implies

$$(T(u) - T(o), u) \ge c\|u\|^2. \tag{6.1.2}$$

The Schwartz inequality (see Proposition 1.2.30(i)) yields

$$(T(u) - T(o), u) \le [\|T(u)\| + \|T(o)\|]\|u\|. \tag{6.1.3}$$

Putting (6.1.2) and (6.1.3) together we get

$$\|T(u)\| \ge c\|u\| - \|T(o)\|,$$

and the weak coercivity follows.

The following theorem is the basic assertion of this section.

Theorem 6.1.4. *Let H be a real Hilbert space and let $T\colon H \to H$ be continuous, monotone and weakly coercive. Then*

$$T(H) = H.$$

If, moreover, T is strictly monotone, then for any $h \in H$ the equation

$$T(u) = h \tag{6.1.4}$$

has a unique solution.

Proof. The uniqueness of the solution is a direct consequence of the strict monotonicity of T. The existence of a solution to (6.1.4) for any $h \in H$ is proved in two steps:

Step 1. Assume for a while that the assertion of the theorem holds if T is continuous and strongly monotone. We prove this fact later, in Proposition 6.1.5. Since $T_n\colon H \to H$, $n \in \mathbb{N}$, defined by

$$T_n\colon u \mapsto \frac{1}{n}u + T(u)$$

is strongly monotone (prove it!) for any $n \in \mathbb{N}$, we claim that given $h \in H$ there exists $u_n \in H$ such that

$$T_n(u_n) = h. \tag{6.1.5}$$

Step 2. Let us prove that $\{u_n\}_{n=1}^{\infty}$ is a bounded sequence in H. Assume the contrary, i.e., there exists a subsequence which will be denoted by $\{u_n\}_{n=1}^{\infty}$ again such that

$$\lim_{n\to\infty} \|u_n\| = \infty.$$

It follows from the monotonicity of T that

$$\begin{aligned}\|h\| &\geq \left(h, \frac{u_n}{\|u_n\|}\right) = \frac{1}{n}\|u_n\| + \frac{1}{\|u_n\|}(T(u_n) - T(o), u_n) + \frac{1}{\|u_n\|}(T(o), u_n)\\ &\geq \frac{1}{n}\|u_n\| - \|T(o)\|,\end{aligned}$$

i.e., $\left\{\frac{1}{n}u_n\right\}_{n=1}^{\infty}$ is a bounded sequence (and therefore weakly sequentially compact – see Theorem 2.1.25 and note that H is reflexive). Hence there exists a subsequence $\left\{\frac{1}{n_k}u_{n_k}\right\}_{k=1}^{\infty} \subset \left\{\frac{1}{n}u_n\right\}_{n=1}^{\infty}$ which is weakly convergent, i.e.,

$$\frac{1}{n_k}u_{n_k} \rightharpoonup w.$$

According to (6.1.5),

$$T(u_{n_k}) \rightharpoonup h - w.$$

This implies that $\{T(u_{n_k})\}_{k=1}^{\infty}$ is a bounded sequence (Proposition 2.1.22(iii)), which contradicts the weak coercivity of T. This proves the boundedness of $\{u_n\}_{n=1}^{\infty}$. In particular, $\frac{1}{n}u_n \to o$ and $T(u_n) \to h$.

By Theorem 2.1.25, there is a subsequence $\{u_{m_k}\}_{k=1}^{\infty} \subset \{u_n\}_{n=1}^{\infty}$ such that

$$u_{m_k} \rightharpoonup u_0.$$

We prove that $T(u_0) = h$. Indeed, for any $v \in H$ and $k \in \mathbb{N}$ we have

$$(T(u_{m_k}) - T(v), u_{m_k} - v) \geq 0.$$

Passing to the limit for $k \to \infty$ we obtain

$$(h - T(v), u_0 - v) \geq 0 \qquad \text{for any} \quad v \in H.^1$$

Set $v = u_0 + \lambda w$, $\lambda > 0$, $w \in H$. Then

$$(h - T(u_0 + \lambda w), w) \leq 0 \qquad \text{holds for any} \quad \lambda > 0 \text{ and } w \in H. \tag{6.1.6}$$

Passing to the limit for $\lambda \to 0_+$ in (6.1.6) and using the continuity of T and of the scalar product in H, we get

$$(h - T(u_0), w) \leq 0 \qquad \text{for any} \quad w \in H. \tag{6.1.7}$$

Since (6.1.7) holds simultaneously for any w and $-w$, we actually have

$$(h - T(u_0), w) = 0 \qquad \text{for any} \quad w \in H, \qquad \text{i.e.,} \qquad T(u_0) = h. \qquad \square$$

Now, it remains to justify the assumption made in Step 1. For this purpose we prove the following assertion.

Proposition 6.1.5. *Let H be a real Hilbert space and $S\colon H \to H$ a continuous and strongly monotone operator. Then*

$$S(H) = H.$$

Proof. The idea of the proof is easy. Since H is a connected metric space, it is enough to prove (see Lemmas 6.1.6 and 6.1.8) that $S(H)$ is both open and closed in H. Then $S(H) = H$ because the only nonempty subset of H which is both *open* and *closed* is the entire space H. $\square$

First we prove that $S(H)$ is closed.

Lemma 6.1.6. *Let $\mathcal{D}$ be a closed set in H, let $S\colon \mathcal{D} \to H$ be a continuous and strongly monotone operator. Then $S(\mathcal{D})$ is a closed set in H.*

Proof. Let $\{u_n\}_{n=1}^{\infty} \subset \mathcal{D}$ be such that

$$S(u_n) \to h.$$

Since S is strongly monotone, we have

$$(S(u_n) - S(u_m), u_n - u_m) \geq c\|u_n - u_m\|^2,$$

and using the Schwartz inequality we obtain

$$\frac{1}{c}\|S(u_n) - S(u_m)\| \geq \|u_n - u_m\|.$$

[1]Here we use that $x_n \rightharpoonup x$ and $y_n \to y$ imply $(x_n, y_n) \to (x, y)$. See Exercise 2.1.37.

Hence $\{u_n\}_{n=1}^{\infty}$ is a Cauchy sequence, and there exists $u_0 \in \mathcal{D}$ such that

$$u_n \to u_0.$$

The continuity of S implies that

$$S(u_n) \to S(u_0), \qquad \text{i.e.,} \qquad S(u_0) = h. \qquad \square$$

To prove that $S(H)$ is an open set is more tricky. For this purpose we need an auxiliary assertion about an extension of Lipschitz continuous operators.

Lemma 6.1.7. *Let $\mathcal{D}$ be a subset of a real Hilbert space H, let $V\colon \mathcal{D} \to H$ be an operator satisfying*

$$\|V(u) - V(v)\| \le \|u - v\| \qquad \text{for any} \quad u, v \in \mathcal{D}.$$

Then there exists an operator $W\colon H \to H$ such that

$$\|W(u) - W(v)\| \le \|u - v\| \qquad \text{for any} \quad u, v \in H \tag{6.1.8}$$

and, moreover,

$$W(u) = V(u) \qquad \text{for any} \quad u \in \mathcal{D}.$$

Proof. It follows from Zorn's Lemma (see Theorem 1.1.4) that there exists a maximal extension W of the operator V, the domain of which satisfies

$$\operatorname{Dom} W \subset H, \qquad \mathcal{D} \subset \operatorname{Dom} W,$$

and for any $u, v \in \operatorname{Dom} W$ the inequality (6.1.8) holds. Our aim is to prove

$$\operatorname{Dom} W = H.$$

Assume the contrary, i.e., there exists $u_0 \in H \setminus \operatorname{Dom} W$. In order to reach a contradiction it is enough to prove the existence of $v_0 \in H$ such that

$$\|v_0 - W(u)\| \le \|u_0 - u\| \qquad \text{for any} \quad u \in \operatorname{Dom} W. \tag{6.1.9}$$

Indeed, setting

$$\tilde{W}\colon u \mapsto \begin{cases} v_0, & u = u_0, \\ W(u), & u \in \operatorname{Dom} W, \end{cases}$$

we obtain an operator

$$\tilde{W}\colon \operatorname{Dom} W \cup \{u_0\} \to H$$

satisfying (6.1.8) for any $u, v \in \operatorname{Dom} W \cup \{u_0\}$. This will be a contradiction with the maximality of the extension W. So, in the rest of the proof we concentrate on the existence of v_0 satisfying (6.1.9).

Let $\mathcal{B}$ be a finite subset of $\operatorname{Dom} W$. Denote by $\mathcal{A}_{\mathcal{B}}$ the set of all $v_0 \in H$ satisfying (6.1.9) for any $u \in \mathcal{B}$. Let $\mathcal{A}$ denote the set of all $v_0 \in H$ satisfying

(6.1.9) for all $u \in \operatorname{Dom} W$. Let $\mathscr{B}_n$ be the system of all finite subsets $\mathcal{B}$ of $\operatorname{Dom} W$ which belong to the closed ball $\{u \in H : \|u\| \le n\}$, $n \in \mathbb{N}$. Set

$$\mathcal{A}_n = \bigcap_{\mathcal{B} \in \mathscr{B}_n} \mathcal{A}_{\mathcal{B}}.$$

Clearly, we have

$$\mathcal{A} = \bigcap_{n=1}^{\infty} \mathcal{A}_n, \qquad \mathcal{A}_{n+1} \subset \mathcal{A}_n \subset \mathcal{A}_1.$$

We wish to prove that $\mathcal{A} \neq \emptyset$.

Observe first that $\mathcal{A}_{\mathcal{B}}$ and $\mathcal{A}_n$ are weakly compact sets (they are bounded and weakly closed[2]). If $\mathcal{A}_{\mathcal{B}} \neq \emptyset$ for any finite subset $\mathcal{B} \subset \operatorname{Dom} W$, then $\mathcal{A}_n \neq \emptyset$ for any $n \in \mathbb{N}$ by Exercise 1.2.44. Applying this procedure again we finally obtain $\mathcal{A} \neq \emptyset$ and the proof will be complete.

Assume now that there exists $\mathcal{B} = \{u_1, \dots, u_m\} \subset \operatorname{Dom} W$ such that $\mathcal{A}_{\mathcal{B}} = \emptyset$. We want to reach a contradiction which will complete the proof. Denote

$$H_{\mathrm{f}} = \operatorname{Lin}\{u_1 - u_0, \dots, u_m - u_0, W(u_1), \dots, W(u_m)\}.$$

Then H_{f} is a subspace of H and

$$\dim H_{\mathrm{f}} \le 2m.$$

For any $w \in H_{\mathrm{f}}$ set

$$h(w) = \max_{1 \le j \le m} \frac{\|w - W(u_j)\|}{\|u_0 - u_j\|}.$$

If there exists $v_0 \in H_{\mathrm{f}}$ such that $h(v_0) \le 1$, then $v_0 \in \mathcal{A}_{\mathcal{B}}$, a contradiction.

So, assume that $h(w) > 1$ for any $w \in H_{\mathrm{f}}$. Note that the real function h is continuous on H_{f} and

$$\lim_{\substack{\|w\| \to \infty \\ w \in H_{\mathrm{f}}}} h(w) = \infty.$$

Hence, there exists $w_0 \in H_{\mathrm{f}}$[3] such that

$$h(w_0) = \min_{w \in H_{\mathrm{f}}} h(w) = \lambda > 1.$$

Let us re-enumerate $u_1, \dots, u_m$ in such a way that

$$\begin{aligned} \frac{\|w_0 - W(u_j)\|}{\|u_0 - u_j\|} &= \lambda > 1, & 1 \le j \le k, \\ \frac{\|w_0 - W(u_j)\|}{\|u_0 - u_j\|} &< \lambda, & k+1 \le j \le m. \end{aligned} \tag{6.1.10}$$

[2]Since the weak topology is not metrizable the fact that $\mathcal{A}_{\mathcal{B}}$ is weakly closed has to be shown with help of weak neighborhoods (Remark 2.1.23). But this is simple due to the Dual Characterization of the Norm (Corollary 2.1.16).

[3]Recall that bounded sets in a finite-dimensional space H_{f} are relatively compact.

We prove that w_0 belongs to the convex hull $\mathcal{M} = \mathrm{Co}\{W(u_1), \dots, W(u_k)\}$ of $\{W(u_1), \dots, W(u_k)\}$.[4] Let us assume the contrary. Then we can find $w_1 \in H_{\mathrm{f}}$ such that

$$\frac{\|w_1 - W(u_j)\|}{\|u_0 - u_j\|} < \frac{\|w_0 - W(u_j)\|}{\|u_0 - u_j\|} = \lambda, \qquad 1 \le j \le k,$$
$$\frac{\|w_1 - W(u_j)\|}{\|u_0 - u_j\|} < \lambda, \qquad k+1 \le j \le m,$$

(see Figure 6.1.2 for $m = 5$ and $k = 2$).[5] Hence $h(w_1) < h(w_0)$, a contradiction.

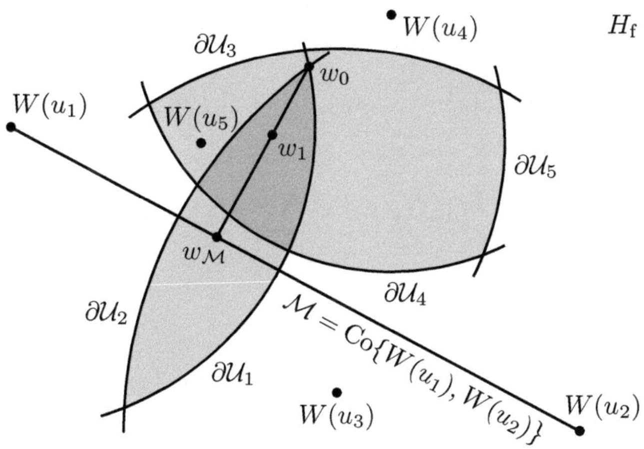

Figure 6.1.2. $\mathcal{U}_j = \left\{w \in H_{\mathrm{f}} : \frac{\|w - W(u_j)\|}{\|u_0 - u_j\|} < \lambda\right\} = B(W(u_j); \lambda\|u_0 - u_j\|) \cap H_{\mathrm{f}}$

Consequently, there are $c_1, \dots, c_k$ such that

$$w_0 = \sum_{j=1}^{k} c_j W(u_j), \qquad c_j \ge 0, \qquad \sum_{j=1}^{k} c_j = 1.$$

---

[4]The convex hull of the set $\mathcal{A}$ is the least convex set containing $\mathcal{A}$.

[5]In other words, $\bigcap_{1 \le j \le m} \mathcal{U}_j \ne \emptyset$ where $\mathcal{U}_j = \left\{w \in H_{\mathrm{f}} : \frac{\|w - W(u_j)\|}{\|u_0 - u_j\|} < \lambda\right\}$ (see Figure 6.1.2). Indeed, $\bigcap_{k+1 \le j \le m} \mathcal{U}_j$ is a nonempty open subset of H_{f} and the latter $(m-k)$ inequalities in (6.1.10) imply $w_0 \in \bigcap_{k+1 \le j \le m} \mathcal{U}_j$. Using the convexity and compactness of $\mathcal{M}$, the reader is invited to show that $\bigcap_{1 \le j \le k} \mathcal{U}_j$ contains the segment $\{t w_0 + (1-t) w_{\mathcal{M}} : 0 < t < 1\}$ where $\|w_0 - w_{\mathcal{M}}\| = \mathrm{dist}\,(w_0, \mathcal{M})$. Consequently, $\bigcap_{1 \le j \le k} \mathcal{U}_j$ is a nonempty set, too, and w_0 belongs to its boundary.

Set $z_j = w_0 - W(u_j)$, $\hat{z}_j = u_0 - u_j$, $1 \le j \le k$. Then

$$\sum_{j=1}^{k} c_j z_j = o, \qquad \|\hat{z}_j\|^2 < \|z_j\|^2, \qquad 1 \le j \le k. \tag{6.1.11}$$

For $1 \le j, n \le k$ we also have

$$\|z_j - z_n\|^2 = \|W(u_n) - W(u_j)\|^2 \le \|u_n - u_j\|^2 = \|\hat{z}_j - \hat{z}_n\|^2,$$

i.e.,

$$\|z_j\|^2 + \|z_n\|^2 - 2(z_j, z_n) \le \|\hat{z}_j\|^2 + \|\hat{z}_n\|^2 - 2(\hat{z}_j, \hat{z}_n). \tag{6.1.12}$$

We conclude from (6.1.11) and (6.1.12) that

$$(\hat{z}_j, \hat{z}_n) < (z_j, z_n), \qquad 1 \le j, n \le k,$$

and thus

$$\sum_{j,n=1}^{k} c_j c_n (\hat{z}_j, \hat{z}_n) < \sum_{j,n=1}^{k} c_j c_n (z_j, z_n).$$

However,

$$\sum_{j,n=1}^{k} c_j c_n (z_j, z_n) = \left\| \sum_{j=1}^{k} c_j z_j \right\|^2 = 0, \qquad \sum_{j,n=1}^{k} c_j c_n (\hat{z}_j, \hat{z}_n) = \left\| \sum_{j=1}^{k} c_j \hat{z}_j \right\|^2,$$

i.e.,

$$\left\| \sum_{j=1}^{k} c_j \hat{z}_j \right\|^2 < 0,$$

a contradiction. This proves that there exists $v_0 \in H_{\mathrm{f}}$ such that $h(v_0) \le 1$, i.e., $\mathcal{A}_{\mathcal{B}} \ne \emptyset$ for any finite set $\mathcal{B} \subset \operatorname{Dom} W$, and the proof is complete. □

Now, we are ready to prove that $S(H)$ is also an open set in H.

Lemma 6.1.8. *Let $\mathcal{D} \subset H$ be an open set, let $S\colon \mathcal{D} \to H$ be continuous and strongly monotone. Then $S(\mathcal{D})$ is an open subset of H.*

Proof. It is enough to prove this lemma for S satisfying the strong monotonicity assumption with $c = 1$ (explain why!). Let us denote $\mathcal{R} := S(\mathcal{D})$. We are going to construct a continuous mapping $Z\colon H \to H$, $\operatorname{Dom} Z = H$, such that $Z_{-1}(\mathcal{D}) = \mathcal{R}$ and this will imply that $\mathcal{R}$ is open. The operator S is injective in $\mathcal{D}$ and S^{-1} is continuous on $\mathcal{R}$ (see Exercise 6.1.11). So we intend to construct Z as an extension of S^{-1}. For this purpose set

$$F(u) = S(u) - u.$$

Then for $u, u_1 \in \mathcal{D}$ we have

$$(F(u) - F(u_1), u - u_1) \geq 0,$$

i.e., F is monotone. For $v \in \mathcal{R}$ set

$$K(v) = S^{-1}(v) - F(S^{-1}(v)).$$

Let $v, v_1 \in \mathcal{R}$ be such that $v = S(u)$, $v_1 = S(u_1)$. Then

$$\|K(v) - K(v_1)\|^2 = \|u - u_1\|^2 + \|F(u) - F(u_1)\|^2 - 2(F(u) - F(u_1), u - u_1),$$
$$\|v - v_1\|^2 = \|F(u) - F(u_1)\|^2 + \|u - u_1\|^2 + 2(F(u) - F(u_1), u - u_1).$$

The monotonicity of F implies that for any $v, v_1 \in \mathcal{R}$,

$$\|K(v) - K(v_1)\| \leq \|v - v_1\|.$$

It follows from Lemma 6.1.7 that there exists a continuous extension K_1 of K which is defined on the whole H and for any $v, v_1 \in H$ we have

$$\|K_1(v) - K_1(v_1)\| \leq \|v - v_1\|.$$

For $v \in H$ set

$$Z(v) = \frac{1}{2}(v + K_1(v)).$$

If $v \in \mathcal{R}$ and $v = S(u)$, then

$$v + K_1(v) = v + K(v) = 2u, \qquad \text{i.e.,} \qquad Z|_{\mathcal{R}} = S^{-1} \quad \text{and} \quad \mathcal{R} \subset Z_{-1}(\mathcal{D}).$$

The inclusion

$$Z_{-1}(\mathcal{D}) \subset \mathcal{R} \tag{6.1.13}$$

will imply that

$$Z_{-1}(\mathcal{D}) = \mathcal{R}$$

and, by the continuity of Z, the set $\mathcal{R} = S(\mathcal{D})$ is open.

To prove (6.1.13) it is enough to show that for any $v \in Z_{-1}(\mathcal{D})$ we have $v = S(Z(v))$. Assume by contradiction that there is $v \in Z_{-1}(\mathcal{D})$ such that for $u = Z(v)$ we have

$$\|v - S(u)\| > 0. \tag{6.1.14}$$

The continuity of S implies the existence of $d > 0$ such that $B(u; d) \subset \mathcal{D}$ and for $u_1 \in B(u; d)$ we have

$$\|S(u) - u - S(u_1) + u_1\| \leq \frac{1}{2}\|v - S(u)\|.$$

Let us choose $t > 0$ so small that

$$\|t(v - S(u))\| < d.$$

Set

$$u_1 = u + t(v - S(u)), \qquad v_1 = S(u_1).$$

Then $\|u - u_1\| < d$, and so

$$\begin{aligned}(S(u_1) - u_1 - v + u, t(v - S(u))) &= (v_1 - Z(v_1) - v + Z(v), Z(v_1) - Z(v)) \\ &= \frac{1}{4}(v_1 - K_1(v_1) - v + K_1(v), v_1 + K_1(v_1) - v - K_1(v)) \\ &= \frac{1}{4}(\|v - v_1\|^2 - \|K_1(v) - K_1(v_1)\|^2) \geq 0.\end{aligned}$$

Furthermore,

$$\begin{aligned}&(S(u_1) - u_1 - S(u) + u, u - u_1) \\ &\quad = (S(u_1) - u_1 - v + u, -t(v - S(u))) + (v - S(u), -t(v - S(u))) \\ &\quad \leq (v - S(u), -t(v - S(u))) = -t\|v - S(u)\|^2,\end{aligned}$$

and so

$$\begin{aligned}t\|v - S(u)\|^2 &\leq (S(u_1) - u_1 - S(u) + u, t(v - S(u))) \\ &\leq t\|S(u_1) - u_1 - S(u) + u\|\|v - S(u)\| \leq \frac{1}{2}t\|v - S(u)\|^2.\end{aligned}$$

Since $t > 0$ this contradicts (6.1.14). This proves (6.1.13) and the proof is complete. □

Let us point out that for operator equations with strongly monotone operators we obtain the continuous dependence of the solution on the right-hand side.

Corollary 6.1.9. *Let H be a real Hilbert space and $T\colon H \to H$ a continuous and strongly monotone operator. Then for any $h \in H$ the equation*

$$T(u) = h$$

has a unique solution. Let $T(u_1) = h_1$ and $T(u_2) = h_2$. Then

$$\|u_1 - u_2\| \leq \frac{1}{c}\|h_1 - h_2\|$$

with $c > 0$ from Definition 6.1.2, i.e., T^{-1} is Lipschitz continuous.

Proof. The existence part follows from Proposition 6.1.5. Uniqueness is obvious. For $T(u_1) = h_1$ and $T(u_2) = h_2$ we have (using the Schwartz inequality)

$$c\|u_1 - u_2\|^2 \leq (T(u_1) - T(u_2), u_1 - u_2) \leq \|u_1 - u_2\|\|h_1 - h_2\|,$$

which completes the proof. □

Example 6.1.10. Let us consider the boundary value problem

$$\begin{cases} -\ddot{x}(t) + g(x(t)) = f(t), & t \in (0,1), \\ x(0) = x(1) = 0 \end{cases} \tag{6.1.15}$$

where $g\colon \mathbb{R} \to \mathbb{R}$ is a continuous function and $f \in L^2(0,1)$ is a given function. Reformulate (6.1.15) as an operator equation. Put $H := W_0^{1,2}(0,1)$ and define operators $J, G\colon H \to H$ and an element $f^* \in H$ by

$$(J(x), y) = \int_0^1 \dot{x}(t)\dot{y}(t)\,\mathrm{d}t, \qquad (G(x), y) = \int_0^1 g(x(t))y(t)\,\mathrm{d}t,$$
$$(f^*, y) = \int_0^1 f(t)y(t)\,\mathrm{d}t \qquad \text{for any} \quad x, y \in H.$$

We will work with the scalar product

$$(x, y) = \int_0^1 \dot{x}(t)\dot{y}(t)\,\mathrm{d}t, \qquad x, y \in H,$$

and with the norm

$$\|x\| = \left(\int_0^1 |\dot{x}(t)|^2\,\mathrm{d}t\right)^{\frac{1}{2}},$$

cf. Exercise 1.2.48. The reader is invited to prove that the operators J and G as well as the element f^* are well defined and that J is linear. Set

$$S = J + G.$$

Then the operator equation

$$S(x) = f^*$$

is equivalent to the requirement that the integral identity

$$\int_0^1 \dot{x}(t)\dot{y}(t)\,\mathrm{d}t + \int_0^1 g(x(t))y(t)\,\mathrm{d}t = \int_0^1 f(t)y(t)\,\mathrm{d}t \tag{6.1.16}$$

holds for all $y \in H$. This is the weak formulation of (6.1.15), and $x \in H$ satisfying (6.1.16) for any $y \in H$ is a weak solution of (6.1.15).

Let us prove that S is a continuous operator. This fact follows from the continuity of J and G. By the definition of J and of the scalar product in H, J is the identity on H, and so it is a continuous operator. Assume now that $x_n \to x$. The embedding of $H = W_0^{1,2}(0,1)$ into $C[0,1]$ (see Theorem 1.2.26) implies that $x_n \rightrightarrows x$ uniformly in $[0,1]$. It follows then from the continuity of g on $\mathbb{R}$ that $g \circ x_n \rightrightarrows g \circ x$ uniformly in $[0,1]$ (justify this statement carefully!). Then using the

Dual Characterization of the Norm and the Hölder inequality, we conclude that

$$\begin{aligned}\|G(x_n) - G(x)\| &= \sup_{\|w\|\le 1} |(G(x_n) - G(x), w)| \\ &= \sup_{\|w\|\le 1} \left| \int_0^1 [g(x_n(t)) - g(x(t))]w(t)\,\mathrm{d}t \right| \\ &\le \left(\int_0^1 |g(x_n(t)) - g(x(t))|^2\,\mathrm{d}t \right)^{\frac{1}{2}} \sup_{\|w\|\le 1} \|w\|_{L^2(0,1)} \\ &\le c \left(\int_0^1 |g(x_n(t)) - g(x(t))|^2\,\mathrm{d}t \right)^{\frac{1}{2}} \to 0 \quad \text{as } n \to \infty.^6\end{aligned}$$

Hence G is also a continuous operator.

Next we prove that S is a strongly monotone operator provided g is an increasing function. Indeed, for any $x, y \in H$ we have

$$\begin{aligned}&(S(x) - S(y), x - y) \\ &\quad = \int_0^1 |\dot{x}(t) - \dot{y}(t)|^2\,\mathrm{d}t + \int_0^1 [g(x(t)) - g(y(t))][x(t) - y(t)]\,\mathrm{d}t \ge \|x - y\|^2.^7\end{aligned}$$

It follows from Corollary 6.1.9 that the problem (6.1.15) has a unique weak solution for any $f \in L^2(0,1)$.

If $f_n \to f$ in $L^2(0,1)$, then $f_n^* \to f^*$ strongly in H (prove it in details!), and according to Corollary 6.1.9 the corresponding weak solutions $x_n \in H$ satisfy

$$\|x_n - x\| \to 0.$$

In particular, this means that a weak solution of (6.1.15) depends continuously on the right-hand side $f \in L^2(0,1)$. ○

Exercise 6.1.11. Let H be a real Hilbert space and let $S\colon H \to H$ be a strongly monotone operator. Prove that S is injective and S^{-1} is Lipschitz continuous.

Exercise 6.1.12. Let $\varepsilon > 0$ and $T\colon B(o; R+\varepsilon) \subset \mathbb{R}^N \to \mathbb{R}^N$ be a monotone operator. Prove that $T(\overline{B(o;R)})$ is a bounded set.

Exercise 6.1.13. Let $\overline{B(o;1)} \subset \mathbb{R}^N$, $N \ge 2$. Prove that there exists a strongly monotone operator $T\colon \overline{B(o;1)} \to \mathbb{R}^N$ for which $T(\overline{B(o;1)})$ is an unbounded set.

Hint. Let $\{x_n\}_{n=1}^\infty \subset \mathbb{R}^N$, $\|x_n\| = 1$, $x_n \ne x_m$ for $n \ne m$, $x_n \to x_0$. Set

$$T\colon x \mapsto \begin{cases} x & \text{for} \quad x \in \overline{B(o;1)}),\ x \ne x_m, \quad m = 1, 2, \dots, \\ x_n + n x_n & \text{for} \quad x = x_n, \quad n = 1, 2, \dots \end{cases}$$

Prove that $T(\overline{B(o;1)})$ is unbounded and T is strongly monotone.

---

[6]Provide optimal value of c! Cf. Exercise 7.4.12 below.

[7]Here we use that $(g(r) - g(s))(r - s) \ge 0$.

Exercise 6.1.14. Define

$$f_n\colon t \mapsto \begin{cases} 0 & \text{for} \quad t \le \frac{1}{2}, \\ nt - \dfrac{n}{2} & \text{for} \quad t > \frac{1}{2}. \end{cases}$$

For $x = (x_1, x_2, \dots) \in l^2$ set

$$Tx = (f_1(x_1), f_2(x_2), \dots) + (x_1, x_2, \dots).$$

Prove that

$$(T(x) - T(y), x - y)_{l^2} \ge \|x - y\|^2_{l^2} \qquad \text{for any} \quad x, y \in l^2$$

and $T(\overline{B(o;1)})$ is an unbounded set.

Exercise 6.1.15. Let $T\colon \mathbb{R}^N \to \mathbb{R}^N$ be a monotone operator and $T(\mathbb{R}^N) = \mathbb{R}^N$. Prove that T is weakly coercive.

Hint. Assume the contrary, i.e., there exist $M > 0$ and a sequence $\{u_n\}_{n=1}^{\infty}$ such that

$$\|u_n\| \to \infty \quad \text{as} \quad n \to \infty \qquad \text{and} \qquad \|T(u_n)\| \le M.$$

Choose a subsequence of $\left\{\frac{u_n}{\|u_n\|}\right\}_{n=1}^{\infty}$ which is convergent to w. Since $T(\mathbb{R}^N) = \mathbb{R}^N$ there is $u \in \mathbb{R}^N$ such that $T(u) = (M+1)w$. By the monotonicity of T we have

$$\left(T(u_n), \frac{u_n - u}{\|u_n\|}\right) \ge \left(T(u), \frac{u_n - u}{\|u_n\|}\right).$$

Taking lim sup on both sides we obtain a contradiction.

Exercise 6.1.16. Let $f\colon \mathbb{R} \to \mathbb{R}$ be defined as follows:

$$f\colon x \mapsto \begin{cases} x & \text{for} \quad x < 0, \\ x + 1 & \text{for} \quad x \ge 0. \end{cases}$$

For any $(x, y) \in \mathbb{R}^2$ set

$$T(x, y) = (y + f(x), -x).$$

Prove that T is an injective monotone operator, $T(\mathbb{R}^2) = \mathbb{R}^2$ and T is not continuous. Can there exist an injective, monotone function $T\colon \mathbb{R} \to \mathbb{R}$ which is not continuous and $T(\mathbb{R}) = \mathbb{R}$?

Exercise 6.1.17. Let H be a real Hilbert space and $T\colon H \to H$ a strongly monotone and Lipschitz continuous operator, i.e., there exist numbers $m > 0$, $M > 0$, $M > m$ such that

$$(T(u) - T(v), u - v) \ge m\|u - v\|^2, \qquad \|T(u) - T(v)\| \le M\|u - v\|$$

hold for all $u, v \in H$. Prove that the equation

$$T(u) = h$$

has precisely one solution for every $h \in H$, and it is possible to construct this solution by iterations.

Hint. Let $h \in H$, $\varepsilon > 0$. Define an operator

$$A_\varepsilon(u) = u - \varepsilon(T(u) - h).$$

Prove that for $u, v \in H$,

$$\|A_\varepsilon(u) - A_\varepsilon(v)\|^2 \le (1 - 2\varepsilon m + \varepsilon^2 M^2)\|u - v\|^2,$$

and show that for $\varepsilon < \frac{2m}{M^2}$ the operator A_ε is contractive. Apply the Contraction Principle (Theorem 2.3.1).

Exercise 6.1.18. Let H be a Hilbert space and $T\colon H \to H$ a contraction. Prove that $I - T$ is a monotone operator.

Exercise 6.1.19. A function $x \in W^{1,2}(0,1)$ satisfying

$$\int_0^1 \dot{x}(t)\dot{y}(t)dt = \int_0^1 h\left(t, x(t), \dot{x}(t)\right) y(t)dt$$

for any $y \in W^{1,2}(0,1)$ is called a weak solution of the Neumann problem

$$\begin{cases} -\ddot{x}(t) = h(t, x(t), \dot{x}(t)), & t \in (0,1), \\ \dot{x}(0) = \dot{x}(1) = 0. \end{cases} \tag{6.1.17}$$

Prove that any weak solution x of (6.1.17) such that $x \in C^2[0,1]$ satisfies the equation in (6.1.17) and $\dot{x}(0) = \dot{x}(1) = 0$, i.e., it is a classical solution of (6.1.17).

Hint. Use the argument similar to that in Exercise 5.5.5.

Exercise 6.1.20. Find conditions on λ, $g = g(t,x)$ and f for the Neumann problem

$$\begin{cases} -\ddot{x}(t) + \lambda x(t) + g(t, x(t)) = f(t), & t \in (0,1), \\ \dot{x}(0) = \dot{x}(1) = 0, \end{cases}$$

to have a unique weak solution.

Hint. Use Corollary 6.1.9.

6.1A Browder and Leray–Lions Theorem

In this appendix we will discuss generalizations of the previous assertions from the basic text. We will present two assertions: one attributed to F. E. Browder, the other named after J. Leray and J.-L. Lions.

Theorem 6.1.21 (Browder). *Let X be a reflexive real Banach space. Moreover, let $T\colon X \to X^*$ be an operator satisfying the conditions*

(i) *T is bounded;*

(ii) *T is demicontinuous;*

(iii) *T is coercive, i.e.,*

$$\lim_{\|u\|\to\infty} \frac{\langle T(u), u\rangle}{\|u\|} = +\infty,$$

cf. Definition 7.2.17 in the Hilbert space setting;

(iv) *T is monotone on the space X, i.e., for all $u, v \in X$ we have*

$$\langle T(u) - T(v), u - v\rangle \geq 0, \tag{6.1.18}$$

cf. Definition 6.1.2 in the Hilbert space setting.

Then the equation

$$T(u) = f^* \tag{6.1.19}$$

has at least one solution $u \in X$ for every $f^* \in X^*$. If, moreover, the inequality (6.1.18) is strict for all $u, v \in X$, $u \neq v$, then the equation (6.1.19) has precisely one solution $u \in X$ for every $f^* \in X^*$.

The second assertion is more general since the monotonicity condition (iv) is replaced by a set of weaker conditions.

Theorem 6.1.22 (Leray–Lions). *Let X be a reflexive real Banach space. Let $T\colon X \to X^*$ be an operator satisfying the conditions*

(i) *T is bounded;*

(ii) *T is demicontinuous;*

(iii) *T is coercive.*

Moreover, let there exist a bounded mapping $\Phi\colon X \times X \to X^*$ such that

(iv) *$\Phi(u, u) = T(u)$ for every $u \in X$;*

(v) *for all $u, w, h \in X$ and any sequence $\{t_n\}_{n=1}^{\infty}$ of real numbers such that $t_n \to 0$, we have*

$$\Phi(u + t_n h, w) \rightharpoonup \Phi(u, w);$$

(vi) *for all $u, w \in X$ we have*

$$\langle \Phi(u, u) - \Phi(w, u), u - w\rangle \geq 0$$

(the so-called condition of monotonicity in the principal part*);*

(vii) *if* $u_n \rightharpoonup u$ *and*

$$\lim_{n\to\infty} \langle \Phi(u_n, u_n) - \Phi(u, u_n), u_n - u\rangle = 0,$$

then we have

$$\Phi(w, u_n) \rightharpoonup \Phi(w, u) \qquad \textit{for arbitrary} \quad w \in X;$$

(viii) *if* $w \in X$, $u_n \rightharpoonup u$, $\Phi(w, u_n) \rightharpoonup z$, *then*

$$\lim_{n\to\infty} \langle \Phi(w, u_n), u_n\rangle = \langle z, u\rangle.$$

Then the equation

$$T(u) = f^*$$

has at least one solution $u \in X$ *for every* $f^* \in X^*$.

The conditions (iv)–(viii) of Theorem 6.1.22 are somewhat unintuitive at the first glance. We will try to clarify these conditions in Section 6.2A where an application to boundary value problems for partial differential equations is given.

Next we will discuss the main steps of the proof of Theorem 6.1.21. The proof of Theorem 6.1.22 is similar, nonetheless it is technically more demanding (see, e.g., Leray & Lions [100]).

Proof of Theorem 6.1.21. We divide the proof into eight steps.

Step 1. Observe that the operator

$$T_{f^*}(u) := T(u) - f^*$$

also satisfies all the conditions of Theorem 6.1.21. Hence it suffices to prove that the equation

$$T(u) = o \tag{6.1.20}$$

has at least one solution.

Step 2. We construct an "approximation of the infinite-dimensional equation (6.1.20) by an equation in a space of finite dimension". More precisely: Let Λ be the family of all subspaces of finite dimension in the space X. If $F \in \Lambda$, define the operator $j_F\colon F \to X$ by

$$j_F(u) = u.$$

Obviously j_F is linear and continuous on F. Let j_F^* be the adjoint operator to j_F (see Section 2.1). Then

$$j_F^*\colon X^* \to F^*$$

and for $u \in F$, put

$$T_F(u) := j_F^*(T(u)).$$

This defines a mapping T_F from the space F into the space F^* (see Figure 6.1.3).

Step 3. Since a continuous linear operator maps a weakly convergent sequence to a weakly convergent one (see Proposition 2.1.27(i)) and the weak convergence coincides with the strong convergence on the subspace F of finite dimension (cf. Remark 2.1.23), it follows from (ii) that T_F is continuous. Put

$$c(r) := \inf_{\substack{u\in X\\ \|u\|=r}} \frac{\langle T(u), u\rangle}{\|u\|},$$

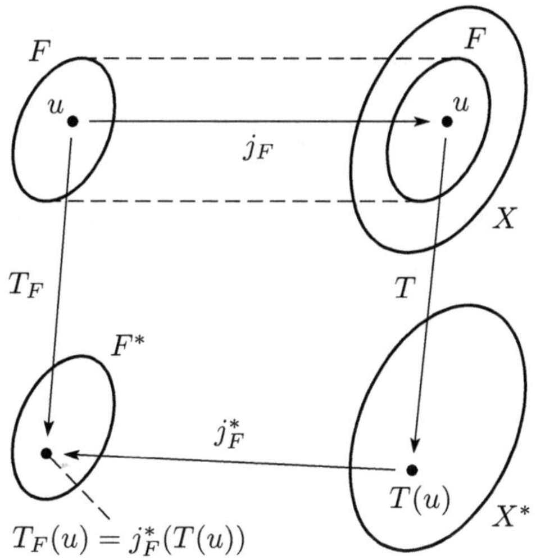

Figure 6.1.3.

By condition (iii), we have

$$\lim_{r\to\infty} c(r) = \infty,$$

i.e.,

$$\langle T_F(u), u\rangle = \langle T(u), j_F(u)\rangle = \langle T(u), u\rangle \geq c(\|u\|)\|u\| \tag{6.1.21}$$

holds for all $u \in F$.

Step 4. In Exercise 6.1.24 it is proved that the equation

$$T_F(u) = o_{F^*} \tag{6.1.22}$$

has at least one solution $u_F \in F$.

Step 5 **(a priori estimate)**. There exists an $r_0 > 0$ such that

$$\|u_F\| \leq r_0$$

holds for arbitrary $F \in \Lambda$ and for every solution $u_F \in F$ of the equation (6.1.22). Indeed, if such an r_0 did not exist, there would be a sequence $\{u_n\}_{n=1}^{\infty}$ of solutions of the equation (6.1.22) with $F = F_n$, $n = 1, 2, \dots$, such that

$$\lim_{n\to\infty} \|u_n\| = \infty, \qquad \lim_{n\to\infty} c(\|u_n\|) = \infty.$$

This would lead to a contradiction in view of the inequality (6.1.21)

$$c(\|u_n\|)\|u_n\| \leq 0 = \langle T_{F_n}(u_n), u_n\rangle.$$

Step 6. Let $F_0 \in \Lambda$ and

$$\Lambda_{F_0} = \{F \in \Lambda : F_0 \subset F\}.$$

We denote by $\mathcal{U}_{F_0}$ the set of all elements $u \in X$ which are solutions of the equation (6.1.22) for a $F \in \Lambda_{F_0}$. Furthermore, let $\overline{\mathcal{U}_{F_0}}^{w}$ be the weak closure of the set $\mathcal{U}_{F_0}$.[8] Note

[8]In other words, $\overline{\mathcal{U}_{F_0}}^{w}$ is the least weakly closed set which contains $\mathcal{U}_{F_0}$.

that $\overline{\mathcal{U}_{F_0}}^{w} \subset \overline{B(o;r_0)}$ for any $F_0 \in \Lambda$ (cf. Exercise 2.1.40 and the fact that $\mathcal{U}_{F_0} \subset \overline{B(o;r_0)}$ for all $F_0 \in \Lambda$). Let $\Lambda_{\mathrm{f}} \subset \Lambda$ be any finite subset of Λ. Then

$$\bigcap_{F_0 \in \Lambda_{\mathrm{f}}} \overline{\mathcal{U}_{F_0}}^{w} \neq \emptyset.$$

Indeed, let $\Lambda_{\mathrm{f}} = \{F_0^i \in \Lambda : \dim F_0^i < \infty,\ i = 1, \dots, n\}$. Then each of the sets $\mathcal{U}_{F_0^i}$ contains all solutions $u \in X$ of the equation (6.1.22) in $\bigcup_{i=1}^{n} F_0^i$.

Since $\overline{B(o;r_0)}$ is a compact topological space with respect to the weak topology (notice that X is reflexive), it follows from the result of Exercise 1.2.44 that

$$\bigcap_{F_0 \in \Lambda} \overline{\mathcal{U}_{F_0}}^{w} \neq \emptyset.$$

Hence there exists

$$u_0 \in \bigcap_{F_0 \in \Lambda} \overline{\mathcal{U}_{F_0}}^{w}.$$

In the next two steps we prove that u_0 is the desired solution of (6.1.20).

Step 7. Let $v \in X$. Choose $F_0 \in \Lambda$ such that $v \in F_0$, and let $F \in \Lambda_{F_0}$. If $u_F \in F$ is a solution of the equation (6.1.22), then condition (iv) implies that

$$\begin{aligned} 0 &\le \langle T(v) - T(u_F), v - u_F\rangle = \langle T(v), v - u_F\rangle - \langle T(u_F), v - u_F\rangle \\ &= \langle T(v), v - u_F\rangle - \langle T(u_F), j_F(v - u_F)\rangle = \langle T(v), v - u_F\rangle - \langle T_F(u_F), v - u_F\rangle \\ &= \langle T(v), v - u_F\rangle. \end{aligned}$$

Thus

$$\langle T(v), v - u\rangle \ge 0 \qquad \text{holds for arbitrary} \quad u \in \mathcal{U}_{F_0}. \tag{6.1.23}$$

By the definition of weak topology (see Remark 2.1.23), (6.1.23) is valid even for arbitrary $u \in \overline{\mathcal{U}_{F_0}}^{w}$. In particular, we then have

$$\langle T(v), v - u_0\rangle \ge 0. \tag{6.1.24}$$

Step 8 **(the Minty trick)**. Choose $w \in X$, $t > 0$, and put $v = u_0 + tw$ in the inequality (6.1.24). Then

$$0 \le \langle T(u_0 + tw), tw\rangle = t\langle (T(u_0 + tw), w\rangle, \qquad \text{i.e.,} \qquad 0 \le \langle T(u_0 + tw), w\rangle.$$

By passing to the limit as $t \to 0_+$, we obtain (applying condition (ii) – demicontinuity of T) the inequality

$$\langle T(u_0), w\rangle \ge 0 \tag{6.1.25}$$

which is valid for all $w \in X$. Replacing the element w in (6.1.25) by the element $-w$, we have

$$0 \le \langle T(u_0), -w\rangle = -\langle T(u_0), w\rangle, \tag{6.1.26}$$

and thus

$$\langle T(u_0), w\rangle = 0 \qquad \text{for every} \quad w \in X, \qquad \text{i.e.,} \qquad T(u_0) = o. \qquad \square$$

Example 6.1.23. Let us consider the boundary value problem

$$\begin{cases} -(|\dot{x}(t)|^{p-2}\dot{x}(t))^{\cdot} + g(x(t)) = f(t), & t \in (0,1), \\ x(0) = x(1) = 0 \end{cases} \tag{6.1.27}$$

where $p > 1$, $f \in L^{p'}(0,1)$ and g is as in Example 6.1.10 (continuous and increasing). Put $X := W_0^{1,p}(0,1)$ with the norm

$$\|x\| = \left(\int_0^1 |\dot{x}(t)|^p \,\mathrm{d}t\right)^{\frac{1}{p}}$$

and define operators $J, G\colon X \to X^*$ and an element $f^* \in X^*$ as in Example 5.8.38. Set

$$T = J + G.$$

Then the operator equation

$$T(x) = f^* \tag{6.1.28}$$

is equivalent to the requirement that the integral identity

$$\int_0^1 |\dot{x}(t)|^{p-2}\dot{x}(t)\dot{y}(t)\,\mathrm{d}t + \int_0^1 g(x(t))y(t)\,\mathrm{d}t = \int_0^1 f(t)y(t)\,\mathrm{d}t \tag{6.1.29}$$

holds for all $y \in X$. So, as in Example 5.8.38, to find a weak solution x of (6.1.27) (i.e., x satisfying (6.1.29)) is equivalent to finding a solution of (6.1.28). We have, by the Hölder inequality,

$$\begin{aligned} \|J(x_n) - J(x)\| &= \sup_{\|y\|\le 1} |\langle J(x_n) - J(x), y\rangle| \\ &= \sup_{\|y\|\le 1} \left| \int_0^1 \left[|\dot{x}_n(t)|^{p-2}\dot{x}_n(t) - |\dot{x}(t)|^{p-2}\dot{x}(t)\right] \dot{y}(t)\,\mathrm{d}t \right| \\ &\le \left(\int_0^1 \left||\dot{x}_n(t)|^{p-2}\dot{x}_n(t) - |\dot{x}(t)|^{p-2}\dot{x}(t)\right|^{p'} \mathrm{d}t\right)^{\frac{1}{p'}} \sup_{\|y\|\le 1} \left(\int_0^1 |\dot{y}(t)|^p \,\mathrm{d}t\right)^{\frac{1}{p}} \\ &\le \left(\int_0^1 \left||\dot{x}_n(t)|^{p-2}\dot{x}_n(t) - |\dot{x}(t)|^{p-2}\dot{x}(t)\right|^{p'} \mathrm{d}t\right)^{\frac{1}{p'}}. \end{aligned} \tag{6.1.30}$$

The last integral tends to zero as $\|x_n - x\| \to 0$ due to the continuity of the Nemytski operator $\Phi(\dot{x})(t) = \varphi(\dot{x}(t))$ from $L^p(0,1)$ into $L^{p'}(0,1)$ with $\varphi(s) = |s|^{p-2}s$, $s \neq 0$, $\varphi(0) = 0$, $p > 1$ (see Theorem 3.2.24).

Using the Hölder inequality we also have

$$\begin{aligned} \|G(x_n) - G(x)\| &= \sup_{\|y\|\le 1} |\langle G(x_n) - G(x), y\rangle| = \sup_{\|y\|\le 1} \left| \int_0^1 (g(x_n(t)) - g(x(t)))y(t)\,\mathrm{d}t \right| \\ &\le \sup_{\|y\|\le 1} \left(\int_0^1 |g(x_n(t)) - g(x(t))|^{p'} \mathrm{d}t\right)^{\frac{1}{p'}} \left(\int_0^1 |y(t)|^p \,\mathrm{d}t\right)^{\frac{1}{p}} \\ &\le c\left(\int_0^1 |g(x_n(t)) - g(x(t))|^{p'} \mathrm{d}t\right)^{\frac{1}{p'}} \to 0 \end{aligned} \tag{6.1.31}$$

as $\|x_n - x\| \to 0$ (cf. Example 6.1.10 and the continuous embedding $W_0^{1,p}(0,1) \subsetneq C[0,1]$ for $p > 1$). It follows from (6.1.30) and (6.1.31) that T is continuous and hence demicontinuous. The boundedness of T follows from estimates similar to (6.1.30), (6.1.31) (the reader is invited to do it in details!). We also have

$$\begin{aligned}\langle T(x) - T(y), x - y\rangle &= \int_0^1 \left[|\dot{x}(t)|^{p-2}\dot{x}(t) - |\dot{y}(t)|^{p-2}\dot{y}(t)\right](\dot{x}(t) - \dot{y}(t))\,\mathrm{d}t \\ &\quad + \int_0^1 [g(x(t)) - g(y(t))](x(t) - y(t))\,\mathrm{d}t \\ &\geq \int_0^1 |\dot{x}(t)|^p\,\mathrm{d}t - \int_0^1 |\dot{y}(t)|^{p-2}\dot{y}(t)\dot{x}(t)\,\mathrm{d}t \\ &\quad - \int_0^1 |\dot{x}(t)|^{p-2}\dot{x}(t)\dot{y}(t)\,\mathrm{d}t + \int_0^1 |\dot{y}(t)|^p\,\mathrm{d}t \\ &\geq \int_0^1 |\dot{x}(t)|^p\,\mathrm{d}t - \left(\int_0^1 |\dot{y}(t)|^p\,\mathrm{d}t\right)^{\frac{1}{p'}}\left(\int_0^1 |\dot{x}(t)|^p\,\mathrm{d}t\right)^{\frac{1}{p}} \\ &\quad - \left(\int_0^1 |\dot{x}(t)|^p\,\mathrm{d}t\right)^{\frac{1}{p'}}\left(\int_0^1 |\dot{y}(t)|^p\,\mathrm{d}t\right)^{\frac{1}{p}} + \int_0^1 |\dot{y}(t)|^p\,\mathrm{d}t \\ &= [\|x\|^{p-1} - \|y\|^{p-1}][\|x\| - \|y\|] \geq 0\end{aligned}$$

with strict inequality for $x \neq y$, since $s \mapsto |s|^{p-1}$ is a strictly increasing function on $(0,\infty)$. Hence the monotonicity of T follows. Finally,

$$\begin{aligned}\langle T(x), x\rangle &= \int_0^1 |\dot{x}(t)|^p\,\mathrm{d}t + \int_0^1 g(x(t))x(t)\,\mathrm{d}t \\ &= \|x\|^p + \int_0^1 [g(x(t)) - g(0)](x(t) - 0)\,\mathrm{d}t + \int_0^1 g(0)x(t)\,\mathrm{d}t \geq \|x\|^p - |g(0)|\|x\|,^{9}\end{aligned}$$

i.e., T is coercive. It follows then from Theorem 6.1.21 that there is a unique solution of (6.1.28) (which in turn is a unique weak solution of (6.1.27)). ○

The advantage of the Browder Theorem is more transparent in the case of partial differential equations when the embedding $W_0^{1,p}(\Omega) \subsetneq C(\overline{\Omega})$ does not hold in general, and so only the demicontinuity of T can be proved.

An application of the more general Theorem 6.1.22 is postponed to the last chapter, Section 6.2A.

Exercise 6.1.24. Prove the following assertion:

> *Let T be a continuous mapping defined on a Banach space X of finite dimension with values in X^*. Assume that there exists a real function $c = c(r)$, defined on the interval $(0,\infty)$, such that*
>
> $$\lim_{r\to\infty} c(r) = \infty \text{ and that } \langle T(u), u\rangle \geq c(\|u\|)\|u\| \text{ holds for all } u \in X.$$
>
> *Then $T(X) = X^*$, i.e., the equation*
>
> $$T(u) = f^*$$

---

[9] We have used $\|x\|_{L^1(0,1)} \leq \|x\|_{W_0^{1,p}(0,1)}$. Prove it!

has at least one solution in the space X for arbitrary $f^* \in X^*$.

Hint. Let $f^* \in X^*$. In the case when $X = X^* = \mathbb{R}^N$ and $\langle u, v\rangle = (u, v)_{\mathbb{R}^N}$ is the scalar product in $\mathbb{R}^N$, there exists $r > 0$ such that the operator $F\colon \mathbb{R}^N \to \mathbb{R}^N$ defined by the relation

$$F(u) = T(u) - f^*$$

satisfies the assumption

$$(F(u), u)_{\mathbb{R}^N} > 0 \qquad \text{for} \quad u \in \partial B(o; R) \quad \text{with } R > 0 \text{ large enough.} \tag{6.1.32}$$

Apply the homotopy invariance property of the Brouwer degree and show that (6.1.32) implies that there exists $u_0 \in B(o; R)$ such that

$$F(u_0) = o, \qquad \text{i.e.,} \qquad T(u_0) = f^*.$$

In the general case Remark 1.1.12(i) must be employed.

Exercise 6.1.25. Consider the problem

$$\begin{cases} -\left(|\dot{x}(t)|^{p-2}\dot{x}(t)\right)^{\cdot} = h(t, x(t), \dot{x}(t)), & t \in (0,1), \\ x(0) = x(1) = 0, \end{cases} \tag{6.1.33}$$

where $p > 1$. Formulate conditions on $h = h(t, x, s)$ which guarantee the existence of a weak solution of (6.1.33).

Hint. Apply Theorem 6.1.21.

Exercise 6.1.26. How do the conditions on h change if we replace the homogeneous Dirichlet boundary conditions in (6.1.33) by the Neumann ones?

6.2 Weak Solutions of PDEs, Application of Monotone Operators

Before we apply Corollary 6.1.9 we revise our growth conditions on $g = g(x, s)$ with respect to the second variable which were considered in Section 5.6. According to (5.5.7) and Theorem 3.2.24 we have $g(x, u(x)) \in L^2(\Omega)$ and the corresponding Nemytski operator is continuous from $L^2(\Omega)$ into $L^2(\Omega)$. The operator $S\colon W_0^{1,2}(\Omega) \to W_0^{1,2}(\Omega)$ defined at the beginning of Section 5.6 is then continuous as follows from the estimate (5.6.4). Our goal is to show that a growth condition more general than (5.5.7) can be considered in order to get analogous results. For this purpose, however, we have to substitute the embedding

$$W_0^{1,2}(\Omega) \hookrightarrow\hookrightarrow L^2(\Omega) \tag{6.2.1}$$

by a more general one. Recall that $\Omega \in C^{0,1}$ is a bounded domain. Namely, we have (see Theorem 1.2.26 or Kufner, John & Fučík [96])

(i) $N = 1 \implies \|u\|_{C(\overline{\Omega})} \le c_1 \|u\|_{W_0^{1,2}(\Omega)}$, $u \in W_0^{1,2}(\Omega)$;

(ii) $N = 2 \implies \|u\|_{L^q(\Omega)} \le c_{2,q}\|u\|_{W_0^{1,2}(\Omega)}$, $u \in W_0^{1,2}(\Omega)$, where $q \ge 1$ is arbitrary;

(iii) $N \ge 3 \implies \|u\|_{L^q(\Omega)} \le c_{N,q}\|u\|_{W_0^{1,2}(\Omega)}$, $u \in W_0^{1,2}(\Omega)$, where $1 \le q \le \frac{2N}{N-2}$.

The estimate (5.6.4) can then be modified as follows:

$$\|S(u_1) - S(u_2)\|_{W_0^{1,2}(\Omega)} \le \|G(u_1) - G(u_2)\|_X \qquad \text{where} \quad G(u) = g(\cdot, u(\cdot))$$

and

(i) for $N = 1$, $X = L^1(\Omega)$;

(ii) for $N = 2$, $q \ge 1$ arbitrary, $X = L^{q'}(\Omega)$ where $q' = \frac{q}{q-1}$;

(iii) for $N \ge 3$, $1 \le q \le \frac{2N}{N-2}$, $X = L^{q'}(\Omega)$ where $q' = \frac{q}{q-1}$.

The operator $S\colon W_0^{1,2}(\Omega) \to W_0^{1,2}(\Omega)$ is continuous provided the Nemytski operator G is continuous

(i) from $L^\infty(\Omega)$ to $L^1(\Omega)$ for $N = 1$;

(ii) from $L^q(\Omega)$ to $L^{q'}(\Omega)$ for $N = 2$ where $q \ge 1$ is arbitrary;

(iii) from $L^q(\Omega)$ to $L^{q'}(\Omega)$ for $N \ge 3$ where $1 \le q \le \frac{2N}{N-2}$.

It follows from Theorem 3.2.24 that the following *growth conditions* guarantee the desired continuity of G:

(i) for $N = 1$:

$$|g(x,s)| \le r(x) + C(|s|) \qquad \text{where} \quad r \in L^1(\Omega) \tag{6.2.2}$$

and $C(t)$ is a nonnegative continuous function of the variable $t \ge 0$;

(ii) for $N = 2$:

$$|g(x,s)| \le r(x) + c|s|^{q-1} \tag{6.2.3}$$

where $r \in L^{\frac{q}{q-1}}(\Omega)$, $c > 0$, $q \ge 1$ is arbitrary;

(iii) for $N \ge 3$:

$$|g(x,s)| \le r(x) + c|s|^{\frac{N+2}{N-2}} \qquad \text{where} \quad r \in L^{\frac{2N}{N+2}}(\Omega),\ c > 0. \tag{6.2.4}$$

The reader should verify in details that the growth conditions (6.2.2)–(6.2.4) generalize the condition

$$|g(x,s)| \le r(x) + c|s|, \qquad r \in L^2(\Omega),\ c > 0. \tag{6.2.5}$$

It is clear that the larger q we choose in (6.2.3) the more general growth condition for g we obtain. It is also clear that all conditions (6.2.2)–(6.2.4) generalize (6.2.5) in the sense that more nonlinearities g can be taken into account in the nonlinear problem

$$\begin{cases} -\Delta u(x) = g(x, u(x)) & \text{in} \quad \Omega, \\ u = 0 & \text{on} \quad \partial\Omega, \end{cases} \tag{6.2.6}$$

and the definition of weak solution still makes sense. Moreover, the operator $S\colon W_0^{1,2}(\Omega) \to W_0^{1,2}(\Omega)$ defined by

$$(S(u), v) = \int_\Omega g(x, u(x))v(x)\,\mathrm{d}x \qquad \text{for all} \quad u, v \in W_0^{1,2}(\Omega)$$

is also a well defined continuous operator.

Warning. *However, S is not compact in general!*

Remark 6.2.1. In order to prove the compactness of S, we need to employ some compact embeddings of $W_0^{1,2}(\Omega)$. Namely, we use the following ones (see Theorem 1.2.28):

(i) $N = 1 \implies W_0^{1,2}(\Omega) \hookrightarrow\hookrightarrow C(\overline{\Omega})$;
(ii) $N = 2 \implies W_0^{1,2}(\Omega) \hookrightarrow\hookrightarrow L^q(\Omega)$ where $q \geq 1$ is arbitrary;
(iii) $N \geq 3 \implies W_0^{1,2}(\Omega) \hookrightarrow\hookrightarrow L^q(\Omega)$ where $1 \leq q < \frac{2N}{N-2}$.

So, S is compact if $N = 1$ and $N = 2$ provided (6.2.2) and (6.2.3), respectively, hold.[10] To get compactness also for $N \geq 3$ we need to modify the growth condition (6.2.4) as follows:

> there exists $\varepsilon > 0$ (arbitrarily small) such that for a.a. $x \in \Omega$ and all $s \in \mathbb{R}$ we have

$$|g(x, s)| \leq r(x) + c|s|^{\frac{N+2}{N-2}(1-\varepsilon)} \qquad \text{where} \quad r \in L^{\frac{2N}{N+2}}(\Omega),\ c > 0. \tag{6.2.7}$$

Indeed, let $u_n \rightharpoonup u$ in $W_0^{1,2}(\Omega)$. Then $u_n \to u$ in $L^q(\Omega)$ for q arbitrarily close to $\frac{2N}{N-2}$, e.g., $q = \frac{2N}{N-2}(1 - \varepsilon)$. Then (6.2.7) and Theorem 3.2.24 imply that

$$g(\cdot, u_n) \to g(\cdot, u) \qquad \text{in} \quad L^{\frac{2N}{N+2}}(\Omega).$$

As a consequence we obtain

$$S(u_n) \to S(u) \qquad \text{in} \quad W_0^{1,2}(\Omega).$$ [11]

Let us give an application of Corollary 6.1.9.

Theorem 6.2.2. *Let $g \in \mathrm{CAR}(\Omega \times \mathbb{R})$ satisfy one of the growth conditions* (6.2.2)–(6.2.4) *depending on N. Moreover, let $g(x, \cdot)$ be a decreasing function for a.a. $x \in \Omega$. Then* (6.2.6) *has a unique weak solution.*

---

[10] The reader is invited to repeat the argument from the beginning of Section 5.6 to prove the compactness of $S\colon W_0^{1,2}(\Omega) \to W_0^{1,2}(\Omega)$ for $N = 1, 2$.
[11] The reader is invited to perform these steps in details.

Proof. Set $T = L - S$, i.e.[12],

$$(T(u), v) = \int_\Omega \nabla u(x) \nabla v(x) \, \mathrm{d}x - \int_\Omega g(x, u(x)) v(x) \, \mathrm{d}x \text{ for any } u, v \in W_0^{1,2}(\Omega).$$

Then T is a continuous operator from $W_0^{1,2}(\Omega)$ into itself. Moreover, for $u_1, u_2 \in W_0^{1,2}(\Omega)$ we have

$$\begin{aligned}(T(u_1) - T(u_2), u_1 - u_2) &= \int_\Omega \nabla(u_1(x) - u_2(x)) \nabla(u_1(x) - u_2(x)) \, \mathrm{d}x \\ &\quad - \int_\Omega [g(x, u_1(x)) - g(x, u_2(x))](u_1(x) - u_2(x)) \, \mathrm{d}x \\ &\geq \int_\Omega |\nabla(u_1(x) - u_2(x))|^2 \, \mathrm{d}x = \|u_1 - u_2\|^2.\end{aligned}$$

Hence T is a strongly monotone operator. The operator equation

$$T(u) = o$$

has a unique solution according to Corollary 6.1.9, i.e., (6.2.6) has a unique weak solution. □

Reading carefully the proof of Theorem 6.2.2 one can easily see that the assumptions on monotonicity of g could be relaxed. However, we have to pay the prize for this modification (see the below conditions (6.2.10) and (6.2.13)). Actually, strict monotonicity of T would be enough to prove the same assertion provided we apply Theorem 6.1.4. For instance, the following assumption on $g = g(x, s)$ guarantees the strict monotonicity of T:

$$\int_\Omega [g(x, u_1(x)) - g(x, u_2(x))](u_1(x) - u_2(x)) \, \mathrm{d}x < \int_\Omega |\nabla(u_1(x) - u_2(x))|^2 \, \mathrm{d}x \tag{6.2.8}$$

for any $u_1, u_2 \in W_0^{1,2}(\Omega)$, $u_1 \neq u_2$. Since

$$\int_\Omega |\nabla(u_1(x) - u_2(x))|^2 \, \mathrm{d}x \geq \lambda_1 \int_\Omega |u_1(x) - u_2(x)|^2 \, \mathrm{d}x$$

by the definition of the first eigenvalue λ_1 (see Example 5.6.4), the inequality (6.2.8) follows from

$$\int_\Omega [g(x, u_1(x)) - g(x, u_2(x))](u_1(x) - u_2(x)) \, \mathrm{d}x < \lambda_1 \int_\Omega |u_1(x) - u_2(x)|^2 \, \mathrm{d}x. \tag{6.2.9}$$

[12] For the definition of $L\colon W_0^{1,2}(\Omega) \to W_0^{1,2}(\Omega)$ see (5.6.2) and the definition of $S\colon W_0^{1,2}(\Omega) \to W_0^{1,2}(\Omega)$ see above in this section.

A sufficient condition for (6.2.9) to hold is the Lipschitz continuity of g with respect to the second variable, i.e.,

$$|g(x,s_1) - g(x,s_2)| < \lambda_1 |s_1 - s_2| \tag{6.2.10}$$

for a.a. $x \in \Omega$ and any $s_1, s_2 \in \mathbb{R}$, $s_1 \neq s_2$. Then (6.2.8) is satisfied and T is then a strictly monotone operator (the reader should verify it in detail!). In order to apply Theorem 6.1.4 we also need T to be weakly coercive, i.e.,

$$\lim_{\|u\|\to\infty} \|T(u)\| = \infty. \tag{6.2.11}$$

We have

$$\begin{aligned} \|T(u)\| &= \sup_{\|v\|\le 1} |(T(u),v)| = \sup_{\|v\|\le 1} \left| \int_\Omega \nabla u(x)\nabla v(x)\,\mathrm{d}x - \int_\Omega g(x,u(x))v(x)\,\mathrm{d}x \right| \\ &\ge \|u\| - \sup_{\|v\|\le 1} \left| \int_\Omega g(x,u(x))v(x)\,\mathrm{d}x \right| \\ &\ge \|u\| - \frac{1}{\sqrt{\lambda_1}} \left(\int_\Omega |g(x,u(x))|^2\,\mathrm{d}x \right)^{\frac{1}{2}} \\ &\ge \|u\| - \frac{1}{\sqrt{\lambda_1}} \left[\|r\|_{L^2(\Omega)} + (\lambda_1 - \varepsilon)\|u\|_{L^2(\Omega)} \right]\ ^{13} \end{aligned} \tag{6.2.12}$$

provided g satisfies the growth condition

$$|g(x,s)| \le r(x) + (\lambda_1 - \varepsilon)|s| \qquad \text{for a.a.} \quad x \in \Omega \quad \text{and all} \quad s \in \mathbb{R}. \tag{6.2.13}$$

Since

$$\frac{1}{\sqrt{\lambda_1}}(\lambda_1 - \varepsilon)\|u\|_{L^2(\Omega)} \le \frac{\lambda_1 - \varepsilon}{\lambda_1}\|u\|,$$

we get from (6.2.12) that

$$\|T(u)\| \ge \frac{\varepsilon}{\lambda_1}\|u\| - \frac{1}{\sqrt{\lambda_1}}\|r\|_{L^2(\Omega)},$$

and so (6.2.11) follows. We have just proved the following assertion.

Theorem 6.2.3. *Let $g \in \mathrm{CAR}(\Omega\times\mathbb{R})$ satisfy* (6.2.10) *and* (6.2.13). *Then the boundary value problem* (6.2.6) *has a unique weak solution.*

Exercise 6.2.4. Let g be as in Theorem 6.2.2 and let $\lambda < \lambda_1$ (here $\lambda_1 > 0$ is the principal eigenvalue of the Laplacian, see Example 5.6.4). Prove that

$$\begin{cases} -\Delta u(x) = \lambda u(x) + g(x,u(x)) & \text{in} \quad \Omega, \\ u = 0 & \text{on} \quad \partial\Omega \end{cases}$$

has a unique weak solution $u \in W_0^{1,2}(\Omega)$.

[13]Note that $\frac{1}{\sqrt{\lambda_1}}$ is the best embedding constant as it folows from Example 5.6.4.

Exercise 6.2.5. Let g be as in Theorem 6.2.2 and let $\lambda < 0$. Prove that

$$\begin{cases} -\Delta u(x) = \lambda u(x) + g(x, u(x)) & \text{in} \quad \Omega, \\ \dfrac{\partial u}{\partial n} = 0 & \text{on} \quad \partial\Omega \end{cases}$$

has a unique weak solution $u \in W^{1,2}(\Omega)$.

Exercise 6.2.6. Consider the problem

$$\begin{cases} -\Delta u(x) = h(x, u(x), \nabla u(x)) & \text{in} \quad \Omega, \\ u = 0 & \text{on} \quad \partial\Omega. \end{cases} \tag{6.2.14}$$

Formulate conditions on $h = h(x, u, \xi)$ which guarantee that (6.2.14) has a unique weak solution.

Exercise 6.2.7. Replace in (6.2.14) the homogeneous Dirichlet condition by the Neumann one.

6.2A Weak Solutions of PDEs, Application of Leray–Lions Theorem

Let $p > 1$, let $\Omega \in C^{0,1}$ be a bounded domain in $\mathbb{R}^N$ and $g \colon \Omega \times \mathbb{R}^{N+1} \to \mathbb{R}$ a Carathéodory function (see Remark 3.2.25). We shall consider the Dirichlet problem

$$\begin{cases} -\Delta_p u(x) + g(x, u(x), \nabla u(x)) = f(x) & \text{in} \quad \Omega, \\ u = 0 & \text{on} \quad \partial\Omega. \end{cases} \tag{6.2.15}$$

Here $\Delta_p u$ is the p-Laplacian (see Section 5.9A). Assume that $1 < p < N$[14] and that $g = g(x, s, t_1, \dots, t_N)$ satisfies the following *growth condition*:

> there exist (possibly small) $\varepsilon > 0$, a constant (possibly large) $c > 0$ and a function $g_0 \in L^{(p^*)'}(\Omega)$ such that
>
> $$|g(x, s, t_1, \dots, t_N)| \le c\left(g_0(x) + |s|^{q(s)-\varepsilon} + \sum_{i=1}^{N} |t_i|^{q(t)-\varepsilon} \right) \tag{6.2.16}$$
>
> for a.a. $x \in \Omega$ and for all $(s, t_1, \dots, t_N) \in \mathbb{R}^{N+1}$ where
>
> $$q(s) = \frac{p^*}{(p^*)'} = p^* - 1, \qquad q(t) = \frac{p}{(p^*)'}.$$[15]

We consider the Sobolev space $W_0^{1,p}(\Omega)$ with the norm

$$\|u\| = \left(\int_\Omega |\nabla u(x)|^p \,\mathrm{d}x \right)^{\frac{1}{p}}.$$

[14] This is a technical assumption. The case $p \ge N$ is easier since stronger embeddings are available (see Theorems 1.2.26 and 1.2.28). However, a slightly different technique must be employed.

[15] Recall that p^* is the critical Sobolev exponent (see Theorem 1.2.26), $(p^*)' = \frac{p^*}{p^*-1} = \frac{pN}{pN-N+p}$, i.e., $q(s) = \frac{pN-N+p}{N-p}$, $q(t) = p - 1 + \frac{p}{N}$ for $1 < p < N$.

Let us recall the continuous embedding

$$W_0^{1,p}(\Omega) \subsetneq L^{p^*}(\Omega) \tag{6.2.17}$$

(see Theorem 1.2.26(i) and Remark 1.2.27), and the compact embedding

$$W_0^{1,p}(\Omega) \subsetneq\subsetneq L^q(\Omega) \qquad \text{where} \quad q \in [1, p^*) \tag{6.2.18}$$

(see Theorem 1.2.28(i)). Similarly to Appendix 5.9A we define (nonlinear) operators $J, G\colon W_0^{1,p}(\Omega) \to (W_0^{1,p}(\Omega))^*$ and an element $f^* \in (W_0^{1,p}(\Omega))^*$ by

$$\begin{aligned}
\langle J(u), v\rangle &= \int_\Omega |\nabla u(x)|^{p-2} \nabla u(x) \nabla v(x)\, \mathrm{d}x,\\
\langle G(u), v\rangle &= \int_\Omega g(x, u(x), \nabla u(x)) v(x)\, \mathrm{d}x,\\
\langle f^*, v\rangle &= \int_\Omega f(x) v(x)\, \mathrm{d}x \qquad \text{for all} \quad u, v \in W_0^{1,p}(\Omega).
\end{aligned}$$

It follows from (6.2.16) and Remark 3.2.26 that G is well defined (the reader is invited to justify this statement!). We have the following properties of J and G (cf. Section 5.9A):

(a) J and G are bounded operators;

(b) J and G are continuous operators;[16]

(c) for any $u, v \in W_0^{1,p}(\Omega)$,

$$\langle J(u) - J(v), u - v\rangle \ge \left(\|u\|^{p-1} - \|v\|^{p-1}\right)\left(\|u\| - \|v\|\right);$$

(d) if $u_n \rightharpoonup u$ in $W_0^{1,p}(\Omega)$, then

$$G(u_n) \to G(u) \qquad \text{in} \quad (W_0^{1,p}(\Omega))^*.$$

Theorem 6.2.8. *Let $1 < p < N$ and assume that $g\colon \Omega \times \mathbb{R}^{N+1} \to \mathbb{R}$ is a Carathéodory function satisfying (6.2.16) and that for all $(s, t_1, \dots, t_N) \in \mathbb{R}^{N+1}$ and almost all $x \in \Omega$ we have*

$$s g(x, s, t_1, \dots, t_N) \ge 0. \tag{6.2.19}$$

Then (6.2.15) has at least one weak solution.

Proof. Set $T := J + G$. Then the operator equation

$$T(u) = f^* \tag{6.2.20}$$

is equivalent to the validity of the integral identity

$$\int_\Omega |\nabla u(x)|^{p-2} \nabla u(x) \nabla v(x)\, \mathrm{d}x + \int_\Omega g(x, u(x), \nabla u(x)) v(x)\, \mathrm{d}x = \int_\Omega f(x) v(x)\, \mathrm{d}x$$

for all $v \in W_0^{1,p}(\Omega)$. This fact shows that the solutions of (6.2.20) correspond one-to-one to the weak solutions of (6.2.15). Next we verify the assumptions (i)–(viii) of Theorem 6.1.22 to prove that there is a solution of (6.2.20).

---

[16] To prove boundedness and continuity of G we have to use the embedding $W_0^{1,p}(\Omega) \subsetneq L^{p^*}(\Omega)$ and the continuity of the Nemytski operator given by g from $L^p(\Omega)$ into the dual space of $L^{p^*}(\Omega)$.

Assumptions (i) and (ii) follow directly from (a) and (b). The assumption (iii), i.e., the coercivity of T, is a direct consequence of (6.2.19):

$$\begin{aligned}\lim_{\|u\|\to\infty}\frac{1}{\|u\|}\langle T(u),u\rangle &= \lim_{\|u\|\to\infty}\frac{1}{\|u\|}\left[\int_\Omega |\nabla u(x)|^p\,dx+\int_\Omega g(x,u(x),\nabla u(x))u(x)\,dx\right]\\ &\ge \lim_{\|u\|\to\infty}\|u\|^{p-1}=\infty.\end{aligned}$$

Let us define an operator $\Phi\colon W_0^{1,p}(\Omega)\times W_0^{1,p}(\Omega)\to (W_0^{1,p}(\Omega))^*$ by

$$\langle \Phi(u,w),v\rangle := \langle J(u),v\rangle + \langle G(w),v\rangle \qquad \text{for all}\quad u,w,v\in W_0^{1,p}(\Omega).$$

It is straightforward to verify the assumption (iv).

In order to verify the assumption (v), let $u,w,h\in W_0^{1,p}(\Omega)$ and $t_n\to 0$. Then

$$\Phi(u+t_nh,w)=J(u+t_nh)+G(w)\to J(u)+G(w)=\Phi(u,w)$$

by continuity of J (see (b)). The validity of the assumption (vi) follows directly from (c).

In order to verify the assumption (vii) let us assume that $u_n\rightharpoonup u$ in $W_0^{1,p}(\Omega)$ and

$$\lim_{n\to\infty}\langle \Phi(u_n,u_n)-\Phi(u,u_n),u_n-u\rangle=0,$$

i.e.,

$$\lim_{n\to\infty}\langle J(u_n)-J(u),u_n-u\rangle=0. \tag{6.2.21}$$

But (6.2.21) together with (c) implies that

$$\|u_n\|\to\|u\|.$$

The last fact and the fact that $W_0^{1,p}(\Omega)$ is a uniformly convex Banach space (see footnote 10 on page 66 and check details) together with the weak convergence imply

$$u_n\to u \qquad \text{in}\quad W_0^{1,p}(\Omega).$$

(see Proposition 2.1.22(iv)). Now,

$$\Phi(w,u_n)=J(w)+G(u_n)\to J(w)+G(u)=\Phi(w,u)\quad \text{for arbitrary } w\in W_0^{1,p}(\Omega).$$

Finally, to verify the assumption (viii), let $w\in W_0^{1,p}(\Omega)$ and $u_n\rightharpoonup u$ in $W_0^{1,p}(\Omega)$. Then $G(u_n)\to G(u)$ in $(W_0^{1,p}(\Omega))^*$ by (d), and so

$$\langle\Phi(w,u_n),u_n\rangle=\langle J(w)+G(u_n),u_n\rangle\to\langle J(w),u\rangle+\langle G(u),u\rangle=\langle\Phi(w,u),u\rangle.$$

Since also $u_n\rightharpoonup u$ in $W_0^{1,p}(\Omega)$ implies that $\Phi(w,u_n)\to J(w)+G(u)$, the last assumption of Theorem 6.1.22 is verified. □

The advantage of the Leray–Lions Theorem becomes more transparent when one deals with partial differential equations of higher order. The reader is asked to see, e.g., Zeidler [163] for more advanced but also technically more involved problems.

Exercise 6.2.9. Modify the assumptions on g in such a way that Theorem 6.1.21 could be applied to get at least one weak solution of (6.2.15).

Exercise 6.2.10. Prove the implication (d) from page 387.

Hint. Use Theorem 1.2.28(i) and Remark 3.2.26.

Exercise 6.2.11. Prove the following assertion:

Let $p \geq 2$, then for all $x_1, x_2 \in \mathbb{R}^N$,

$$|x_2|^p \geq |x_1|^p + p|x_1|^{p-2}x_1(x_2 - x_1) + \frac{|x_2 - x_1|^p}{2^{p-1} - 1}.^{17} \tag{6.2.22}$$

Hint ((Lindqvist [101])). The strict convexity of $x \mapsto |x|^p$ implies that for any $x_1, x_2 \in \mathbb{R}^N$, $p > 1$,

$$|x_2|^p > |x_1|^p + p|x_1|^{p-2}x_1(x_2 - x_1). \tag{6.2.23}$$

Then writing $\frac{x_2+x_1}{2}$ instead of x_2 in (6.2.23) we get

$$\left|\frac{x_1 + x_2}{2}\right|^p \geq |x_1|^p + \frac{1}{2}p|x_1|^{p-2}x_1(x_2 - x_1).$$

Using the Clarkson inequality (see, e.g., Adams [2, Theorem 2.28]) for $p \geq 2$,

$$|x_1|^p + |x_2|^p \geq 2\left|\frac{x_2 + x_2}{2}\right|^p + 2\left|\frac{x_1 - x_2}{2}\right|^p, \tag{6.2.24}$$

we arrive at

$$|x_2|^p \geq |x_1|^p + p|x_1|^{p-2}x_1(x_2 - x_1) + 2\left|\frac{x_1 - x_2}{2}\right|^p. \tag{6.2.25}$$

This is actually (6.2.22) but with 2^{1-p} in place of $\frac{1}{2^{p-1}-1}$. Repeating this procedure, starting again with (6.2.24) but now using (6.2.25) instead of (6.2.23), we get the constant improved to $2^{1-p} + 4^{1-p}$. By iteration one obtains the constant

$$2^{1-p} + 4^{1-p} + 8^{1-p} + \cdots = \frac{1}{2^{p-1} - 1}$$

in (6.2.22).

Exercise 6.2.12. Prove the following assertion

Let $1 < p < 2$, then for all $x_1, x_2 \in \mathbb{R}^N$,

$$|x_2|^p \geq |x_1|^p + p|x_1|^{p-2}x_1(x_2 - x_1) + c(p)\frac{|x_1 - x_2|^2}{(|x_1| + |x_2|)^{2-p}}. \tag{6.2.26}$$

Hint ((Lindqvist [101])). Fix x_1, x_2 and expand the real function

$$f(t) = |x_1 + t(x_2 - x_1)|^p$$

using the Taylor formula

$$f(1) = f(0) + f'(0) + \int_0^1 (1 - t)f''(t)\, dt.$$

Then, provided $f(t) \neq 0$ for all $0 \leq t \leq 1$,

$$|x_2|^p = |x_1|^p + p|x_1|^{p-2}x_1(x_2 - x_1) + \int_0^1 (1 - t)f''(t)\, dt. \tag{6.2.27}$$

---

[17] Recall that $xy := (x, y)_{\mathbb{R}^N}$.

(In the case when there exists t, $0 \le t \le 1$, such that $|x_1 + t(x_2 - x_1)| = 0$ it is easily checked that (6.2.26) holds!) At the same time

$$f''(t) = p(p-2)|x_1 + t(x_2 - x_1)|^{p-4}[(x_1 + t(x_2 - x_1))(x_2 - x_1)]^2 + p|x_1 + t(x_2 - x_1)|^{p-2}|x_2 - x_1|^2,$$

and the Schwartz inequality yields

$$f''(t) \ge p(p-1)|x_1 + t(x_2 - x_1)|^{p-2}|x_2 - x_1|^2. \qquad (6.2.28)$$

Returning to (6.2.27) we estimate

$$\int_0^1 (1-t)f''(t)\,\mathrm{d}t \ge \frac{3}{4}\int_0^{\frac{1}{4}} f''(t)\,\mathrm{d}t \qquad (6.2.29)$$

and since $|x_1 + t(x_2 - x_1)| \le |x_1| + |x_2|$, we use (6.2.28), (6.2.29) and arrive at (6.2.26) with $c(p) = \frac{3}{16}p(p-1)$.

Exercise 6.2.13. Prove that the operator J defined on page 387 is strictly monotone[18] for $1 < p < 2$ and strongly monotone[19] for $p \ge 2$.

Hint. For $u, v \in W_0^{1,p}(\Omega)$ we have

$$\begin{aligned}\langle J(u)-J(v), u-v\rangle &= \int_\Omega \big(|\nabla u(x)|^{p-2}\nabla u(x) - |\nabla v(x)|^{p-2}\nabla v(x)\big)(\nabla u(x) - \nabla v(x))\,\mathrm{d}x \\ &= \int_\Omega |\nabla u(x)|^{p-2}\nabla u(x)(\nabla u(x) - \nabla v(x))\,\mathrm{d}x \\ &\quad - \int_\Omega |\nabla v(x)|^{p-2}\nabla v(x)\,(\nabla u(x) - \nabla v(x))\,\mathrm{d}x = I_1 + I_2.\end{aligned}$$

For $p \ge 2$, it follows from Exercise 6.2.11 that

$$I_1 + I_2 \ge \frac{2}{p\,(2^{p-1}-1)}\int_\Omega |\nabla u(x) - \nabla v(x)|^p\,\mathrm{d}x = c\|u - v\|^p.$$

For $1 < p < 2$, it follows from Exercise 6.2.12 that

$$I_1 + I_2 \ge \frac{2c(p)}{p}\int_\Omega \frac{|\nabla u(x) - \nabla v(x)|^2}{(|\nabla u(x)| + |\nabla v(x)|)^{2-p}}\,\mathrm{d}x > 0 \text{ provided } u, v \in W_0^{1,p}(\Omega),\ u \ne v.$$

Exercise 6.2.14. Prove that the weak solution from Exercise 5.9.23 is unique provided $\lambda \le 0$.

Exercise 6.2.15. Let $\lambda < 0$ and let $f\colon \Omega \times \mathbb{R} \to \mathbb{R}$ be a Carathéodory function which is decreasing with respect to the second variable, i.e.,

$$f(x, s_1) \ge f(x, s_2) \qquad \text{for a.a.} \quad x \in \Omega \quad \text{and} \quad s_1, s_2 \in \mathbb{R}, \quad s_1 \le s_2.$$

[18] I.e., for any $u, v \in W_0^{1,p}(\Omega)$, $u \ne v$: $\langle J(u) - J(v), u - v\rangle > 0$, cf. Definition 6.1.2.

[19] I.e., there exists $c > 0$ such that for any $u, v \in W_0^{1,p}(\Omega)$: $\langle J(u) - J(v), u - v\rangle \ge c\|u - v\|^p$, cf. Definition 6.1.2.

Assume, moreover, that there exist $f_0 \in L^{p'}(\Omega)$, $p' = \frac{p}{p-1}$, $p > 1$ and $c > 0$ such that

$$|f(x,s)| \le f_0(x) + c|s|^{p-1}.$$

Prove that there is a unique weak solution $u \in W_0^{1,p}(\Omega)$ of the problem

$$\begin{cases} -\Delta_p u(x) = \lambda |u(x)|^{p-2}u(x) + f(x,u(x)) & \text{in} \quad \Omega, \\ u = 0 & \text{on} \quad \partial\Omega. \end{cases}$$

6.3 Supersolutions, Subsolutions, Monotone Iterations

In this section we deal with another possibility how to extend the notion of a monotone function to operators between Banach spaces of infinite dimension. Instead of characterizing an increasing function $f\colon \mathbb{R} \to \mathbb{R}$ in terms of the inequality

$$(f(x) - f(y))(x - y) \ge 0 \qquad \text{for any} \quad x, y \in \mathbb{R}$$

(cf. Section 6.1), we use the usual "first semester calculus" definition

$$\text{for any } x, y \in \mathbb{R} \text{ satisfying} \quad x \le y \qquad \text{we have} \qquad f(x) \le f(y). \tag{6.3.1}$$

However, to generalize the implication (6.3.1) to the case of general operators we have to introduce an inequality relation for Banach spaces which can be used analogously to the inequality relation for the set of real numbers.

Definition 6.3.1. Let X be a real Banach space and let $\mathcal{K}$ be a subset of X. Then $\mathcal{K}$ is called an *order cone* if

(1) $\mathcal{K}$ is closed, nonempty, and $\mathcal{K} \ne \{o\}$;

(2) $a, b \in \mathbb{R}$, $a, b \ge 0$, $x, y \in \mathcal{K}$ implies $ax + by \in \mathcal{K}$;

(3) $x \in \mathcal{K}$ and $-x \in \mathcal{K}$ implies $x = o$.

On this basis we define

$$\begin{aligned} x &\le y && \text{provided} && y - x \in \mathcal{K}, \\ x &< y && \text{provided} && x \le y \quad \text{and} \quad x \ne y, \\ x &\ll y && \text{provided} && y - x \in \operatorname{int} \mathcal{K}, \end{aligned} \tag{6.3.2}$$

The set

$$[x, y] = \{z \in X : x \le z \le y\}$$

is called an *order interval* in X. Note that

$$x \ge y \qquad \text{means} \qquad x - y \in \mathcal{K}$$

and similarly for ">" and "≫".

Remark 6.3.2. Condition (2) is equivalent to saying that $\mathcal{K}$ is convex, and if $x \in \mathcal{K}$ and $a \geq 0$, then $ax \in \mathcal{K}$.

Definition 6.3.3. By an *ordered Banach space* we mean a real Banach space together with an order cone.

Remark 6.3.4. The reader should notice the difference between order cones and cones. A subset $\mathcal{C}$ of the Banach space X is called a *cone* if $x \in \mathcal{C}$ and $a > 0$ implies $ax \in \mathcal{C}$. So, every order cone is a cone, but the converse is not true in general.

Example 6.3.5. Let $X = \mathbb{R}^N$. We set

$$\mathbb{R}^{N,+} = \{(\xi_1, \dots, \xi_N) \in \mathbb{R}^N : \xi_i \geq 0 \text{ for all } i = 1, \dots, N\}.$$

Then $\mathcal{K} = \mathbb{R}^{N,+}$ is an order cone (see Figure 6.3.1). We have

$$(\xi_1, \dots, \xi_N) \leq (\eta_1, \dots, \eta_N) \quad \text{if and only if} \quad \xi_i \leq \eta_i \text{ for all } i = 1, \dots, N,$$

$$(\xi_1, \dots, \xi_N) \ll (\eta_1, \dots, \eta_N) \quad \text{if and only if} \quad \xi_i < \eta_i \text{ for all } i = 1, \dots, N.$$

○

Example 6.3.6. The set $\mathcal{C}$ in Figure 6.3.2 is a cone in $\mathbb{R}^2$ but it is not an order cone.[20] ○

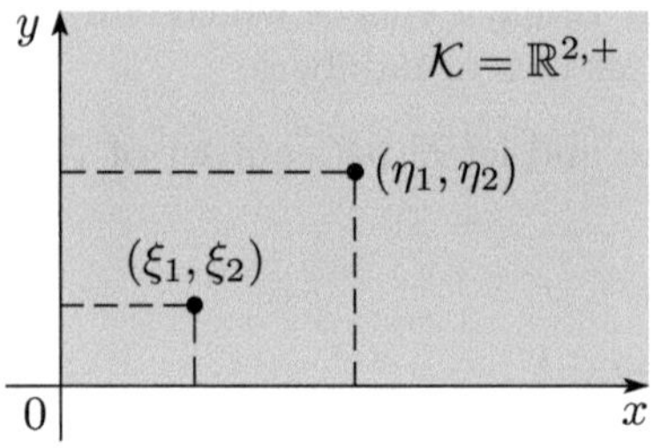

Figure 6.3.1.

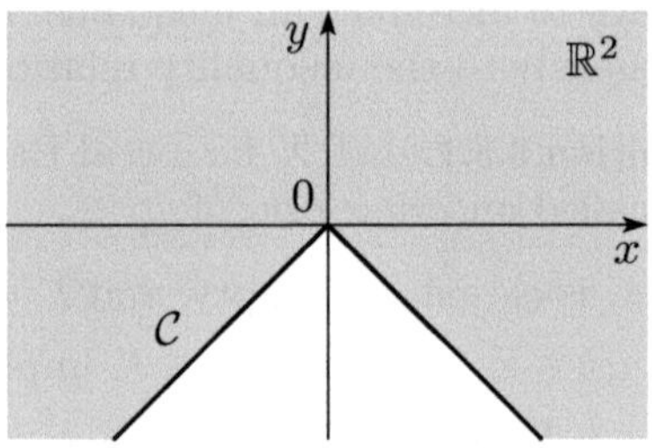

Figure 6.3.2.

Example 6.3.7. Let $X = C(\overline{\Omega})$ for a bounded set $\Omega \subset \mathbb{R}^N$. We set

$$C^+(\overline{\Omega}) = \{f \in C(\overline{\Omega}) : f(x) \geq 0 \text{ for every } x \in \overline{\Omega}\}.$$

Then $\mathcal{K} = C^+(\overline{\Omega})$ is an order cone in X, and we have

$$f \leq g \quad \text{if and only if} \quad f(x) \leq g(x) \quad \text{for all} \quad x \in \overline{\Omega},$$

$$f \ll g \quad \text{if and only if} \quad f(x) < g(x) \quad \text{for all} \quad x \in \overline{\Omega}.$$

○

[20] Explain why!

The following assertion summarizes the basic properties of ordering in a Banach space X.

Proposition 6.3.8. *For all $u, x, x_n, y, y_n, z \in X$ and all $a, b \in \mathbb{R}$, we have*

$$\begin{aligned} & x \leq x, \\ x \leq y \quad \text{and} \quad y \leq x \quad & \text{imply} \quad x = y, \\ x \leq y \quad \text{and} \quad y \leq z \quad & \text{imply} \quad x \leq z. \end{aligned}$$

Furthermore, we have

$$\begin{aligned} x \leq y \quad \text{and} \quad 0 \leq a \leq b \quad & \text{imply} \quad ax \leq by, \\ x \leq y \quad \text{and} \quad u \leq z \quad & \text{imply} \quad x + u \leq y + z \quad \text{and} \\ x_n \leq y_n \quad \text{for all} \quad n \quad & \text{implies} \quad \lim_{n\to\infty} x_n \leq \lim_{n\to\infty} y_n \end{aligned}$$

provided the limits exist [21]. *For the symbol "$\ll$", the following implications hold:*

$$\begin{aligned} x \ll y \quad \text{and} \quad y \leq z \quad & \text{imply} \quad x \ll z, \\ x \leq y \quad \text{and} \quad y \ll z \quad & \text{imply} \quad x \ll z, \\ x \ll y \quad \text{and} \quad a > 0 \quad & \text{imply} \quad ax \ll ay. \end{aligned}$$

Proof. Use (6.3.2) and the properties of $\mathcal{K}$. For example, if $x_n \leq y_n$ for all n, then $y_n - x_n \in \mathcal{K}$. Since $\mathcal{K}$ is closed and limits of $\{x_n\}_{n=1}^{\infty}$, $\{y_n\}_{n=1}^{\infty}$ exist, we conclude that

$$y - x \in \mathcal{K}, \qquad \text{i.e.,} \qquad x \leq y. \qquad \square$$

Definition 6.3.9. The order cone $\mathcal{K}$ is called *normal* if there is a number $c > 0$ such that for all $x, y \in X$, $o \leq x \leq y$ we have

$$\|x\| \leq c\|y\|.$$

Example 6.3.10. For $X = \mathbb{R}^N$, $\mathcal{K} = \mathbb{R}^{N,+}$ is a normal order cone in $\mathbb{R}^N$. Similarly, $C^+(\overline{\Omega})$ is a normal order cone in $C(\overline{\Omega})$. ○

Lemma 6.3.11. *If an order cone is normal, then every order interval $[x, y]$ is bounded in the norm.*

Proof. If $x \leq w \leq y$, then $o \leq w - x \leq y - x$, and hence

$$\|w\| \leq \|w - x\| + \|x\| \leq c\|x - y\| + \|x\|. \qquad \square$$

Now we can introduce the definition of a monotone increasing operator between ordered Banach spaces.

[21] The limits are understood in the norm topology of X.

Definition 6.3.12. Let X and Y be ordered Banach spaces. An operator $T\colon \operatorname{Dom} T \subset X \to Y$ is said to be *monotone increasing* if

$$x < y \qquad \text{implies} \qquad T(x) \le T(y) \qquad \text{for any} \quad x, y \in \operatorname{Dom} T.$$

An operator T is said to be *strictly* or *strongly monotone increasing* if the symbol "$\le$" is replaced by "$<$" or "$\ll$", respectively. Similarly we define a (*strictly, strongly*) *monotone decreasing* operator.

The operator T is said to be *positive* if $T(o) \ge o$ and for all $x \in \operatorname{Dom} T$,

$$x > o \qquad \text{implies} \qquad T(x) \ge o.$$

As above, the operator is *strictly* or *strongly positive* if the symbol "$\ge$" is replaced by "$>$" or "$\gg$", respectively.

Example 6.3.13. Let $X = Y = \mathbb{R}$, $\mathcal{K} = \mathbb{R}^+$. Then for a real function $f\colon \operatorname{Dom} f \subset \mathbb{R} \to \mathbb{R}$ the concepts of (strictly) monotone increasing (or decreasing) above coincide with the usual definitions. Because of the equivalence of $x \ll y$ and $x < y$ on $\mathbb{R}$ there is no difference here between strongly monotone increasing (decreasing) and strictly monotone increasing (decreasing). ○

Example 6.3.14. For a linear operator T, the concepts of (strictly, strongly) positive are the same as those of (strictly, strongly) monotone increasing. Indeed, let T be positive, for example. Then we have the following sequence of implications:

$$\begin{aligned} x < y \quad &\Longrightarrow \quad o < y - x \quad \Longrightarrow \quad o \le T(y - x) \\ &\Longrightarrow \quad o \le T(y) - T(x) \quad \Longrightarrow \quad T(x) \le T(y), \end{aligned}$$

i.e., T is monotone increasing. The other proofs are similar. ○

Let $T\colon X \to X$ be an operator on a Banach space X. We consider the operator equation

$$u = T(u) \tag{6.3.3}$$

and apply an iterative method to solve it. For this purpose we consider the iterations (successive approximations)

$$u_{n+1} = T(u_n) \qquad \text{and} \qquad v_{n+1} = T(v_n), \qquad n = 0, 1, 2, \dots \tag{6.3.4}$$

We illustrate the idea of approximations in Figure 6.3.3.

The next definition is a basic definition of the existence theory for operator equations in ordered Banach spaces.

Definition 6.3.15. A point $\overline{u}$ is called a *supersolution* of (6.3.3) (or of the operator T) if

$$T(\overline{u}) \le \overline{u}.$$

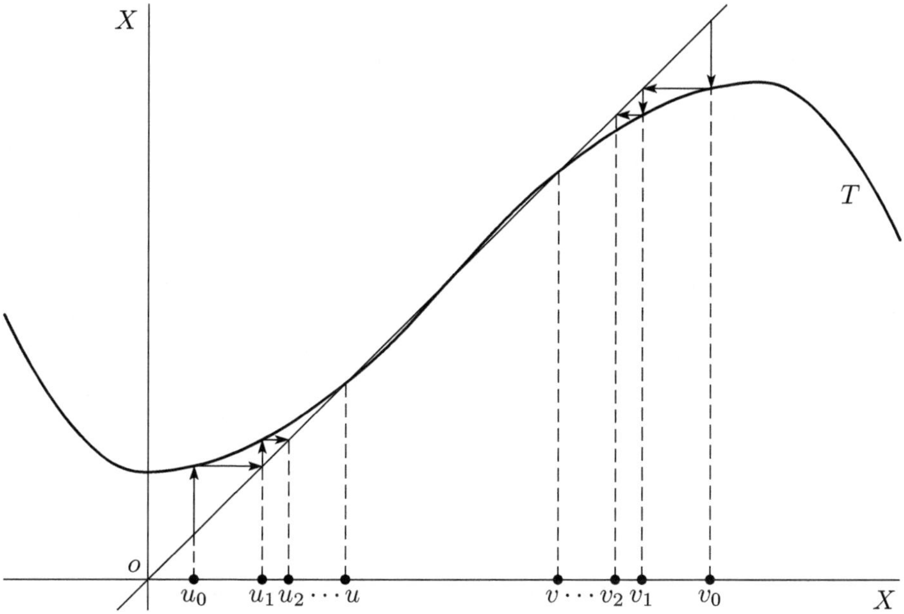

Figure 6.3.3. Fixed points u, v of T.

The prefix "super" is replaced by "sub" when the respective inequalities are reversed. For example, $\underline{u} \in X$ satisfying $\underline{u} \le T(\underline{u})$ is a *subsolution* of (6.3.3).[22]

The following assertions justify the general principle of super- and subsolutions. This principle can be formulated as follows:

> *If there is a super- and subsolution, then a solution can be obtained by the convergent iterative method* (6.3.4)*.*

Namely, we have the following results.

Theorem 6.3.16. *Let $T\colon X \to X$ be a compact monotone increasing operator on a real Banach space X with a normal order cone X^+ and u_0 be a subsolution (v_0 a supersolution) of* (6.3.3)*. Then $\{u_n\}_{n=1}^{\infty}$ ($\{v_n\}_{n=1}^{\infty}$) in* (6.3.4) *converges if and only if this sequence is bounded above (below).*[23] *In the case of convergence, the limit point u is the smallest fixed point u of T with $u_0 \le u$ (v is the greatest fixed point v of T with $v \le v_0$).*[24]

[22]The terminology is not fixed. Instead of super- and subsolution the notions *upper* and *lower solutions* have been also used.

[23]The set $\mathcal{M} \subset X$ is bounded above if there is $m \in X$ such that $u \le m$ for any $u \in \mathcal{M}$.

[24]Note that the concepts of "smallest" and "greatest" are used in the usual sense, e.g., a smallest fixed point u in X is characterized by $u \le w$ for all other fixed points $w \ge u_0$.

Proof. We will consider the case of a subsolution. The case of a supersolution is very similar. Since T is monotone increasing, we have the sequence of implications

$$u_0 \le T(u_0) = u_1 \implies u_1 \le u_2 \implies \cdots, \quad \text{i.e.,} \quad u_0 \le u_1 \le u_2 \le \cdots$$

Now $u_n \to u$ implies that $u_n \le u$ for all n. Consequently, the sequence $\{u_n\}_{n=1}^{\infty}$ is bounded above, if it is convergent.

Conversely, the sequence $\{u_n\}_{n=1}^{\infty}$ is convergent if it is bounded above. Indeed, suppose $u_n \le v$ for all n. By Lemma 6.3.11, the sequence $\{u_n\}_{n=1}^{\infty}$ is bounded in the norm. Since $u_n = T(u_{n-1})$ and since T is compact, the set of all u_n is relatively compact. Thus there exist a convergent subsequence $\{u_{n_k}\}_{k=1}^{\infty}$ and u such that $u_{n_k} \to u$. Since the sequence $\{u_n\}_{n=1}^{\infty}$ is monotone, all convergent subsequences have the same limit point. Therefore, the whole sequence $\{u_n\}_{n=1}^{\infty}$ converges to u as well. Since $u_{n+1} = T(u_n)$, letting $n \to \infty$ shows that

$$u = T(u).$$

Let w be a solution of (6.3.3) with $u_0 \le w$. Then $u_1 = T(u_0) \le T(w) = w$, etc., so that $u_n \le w$ for all n. Hence $u \le w$. □

The intuitive meaning of the following assertion is demonstrated in Figure 6.3.3.

Corollary 6.3.17 (Monotone Iterative Method). *Let X be a real Banach space with a normal order cone and $T\colon X \to X$. Assume that u_0 and v_0 is a subsolution and a supersolution of* (6.3.3)*, respectively, and $u_0 \le v_0$. If T is a compact monotone increasing operator on the order interval $[u_0, v_0]$, then both the iterative sequences $\{u_n\}_{n=1}^{\infty}$ and $\{v_n\}_{n=1}^{\infty}$ from* (6.3.4) *are defined, converge, and*

$$u = \lim_{n\to\infty} u_n \qquad \text{and} \qquad v = \lim_{n\to\infty} v_n$$

is the smallest fixed point and the largest fixed point, respectively, of T in $[u_0, v_0]$.

Furthermore, we have the error estimates

$$u_n \le u \le v \le v_n \qquad \text{for all} \quad n = 0, 1, \dots$$

(Figure 6.3.3*).*

Proof. Since $u_0 \le T(u_0)$, $T(v_0) \le v_0$ and $u_0 \le v_0$ together imply that $u_0 \le u_1 \le v_0$ and similarly that

$$u_n \le v_0 \qquad \text{for all} \quad n,$$

it follows that $\{u_n\}_{n=1}^{\infty}$ is bounded above. The proof then follows from Theorem 6.3.16. □

Example 6.3.18 (Integral Equation). Let Ω be a bounded domain in $\mathbb{R}^N$, $G\colon \overline{\Omega} \times \overline{\Omega} \to \mathbb{R}$ be a continuous and nonnegative function and $f\colon \overline{\Omega} \times \mathbb{R} \to \mathbb{R}$ be a continuous and increasing function in the second variable. Consider the integral equation

$$u(x) = \int_{\Omega} G(x,y) f(y, u(y))\, \mathrm{d}y \tag{6.3.5}$$

in the space $C(\overline{\Omega})$. We write this equation in the form

$$u = T(u), \qquad u \in C(\overline{\Omega}).$$

Let us consider the normal order cone $C^+(\overline{\Omega})$ from Example 6.3.7. Then the operator $T\colon C(\overline{\Omega}) \to C(\overline{\Omega})$ is compact and monotone increasing.[25]

Considering subsolutions and supersolutions now means that we replace "$=$" by "$\le$" and "$\ge$", respectively, in the integral equation (6.3.5). Corollary 6.3.17 implies that if $u_0 \in C(\overline{\Omega})$ is a subsolution and $v_0 \in C(\overline{\Omega})$ is a supersolution with $u_0 \le v_0$ on $\overline{\Omega}$, then for $n \to \infty$ the iterative method

$$u_{n+1}(x) = \int_\Omega G(x,y) f(y, u_n(y))\,\mathrm{d}y, \qquad n = 0, 1, 2, \dots,$$

converges uniformly on $\overline{\Omega}$ to a solution $u \in C(\overline{\Omega})$ of the integral equation with $u_0 \le u \le v_0$ on $\overline{\Omega}$. Here u is the smallest solution with this property. If, instead, the iterative method starts with v_0, then we obtain the greatest solution v with $u_0 \le v \le v_0$. The most difficult task in solving (6.3.5) is to find at least one subsolution u_0 and/or supersolution. ○

Example 6.3.19 (Differential Equation). Let us consider the Dirichlet boundary value problem

$$\begin{cases} -\ddot{x}(t) = f(t, x(t)), & t \in (0,1), \\ x(0) = x(1) = 0 \end{cases} \tag{6.3.6}$$

where $f\colon [0,1] \times \mathbb{R} \to \mathbb{R}$ is a function continuous in the first variable and continuously differentiable in the second one. Suppose $u_0, v_0 \in C^2[0,1]$ are such that

$$\begin{cases} -\ddot{u}_0(t) \le f(t, u_0(t)), & t \in (0,1), \\ u_0(0) \le 0, \quad u_0(1) \le 0, \end{cases} \qquad \begin{cases} -\ddot{v}_0(t) \ge f(t, v_0(t)), & t \in (0,1), \\ v_0(0) \ge 0, \quad v_0(1) \ge 0.^{26} \end{cases} \tag{6.3.7}$$

We will show that u_0, v_0 is a subsolution and a supersolution, respectively, for the operator $w = T(z)$ defined by a solution of the problem

$$\begin{cases} -\ddot{w}(t) + cw(t) = f(t, z(t)) + cz(t) =: F(t, z), & t \in (0,1), \\ w(0) = w(1) = 0, \end{cases}$$

where $c > 0$ is chosen in such a way that

$$\frac{\partial f}{\partial s}(t,s) + c > 0 \quad \text{for } t \in [0,1] \quad \text{and} \quad s \in \mathcal{I}_0 := \left[\min_{t\in[0,1]}\{u_0(t)\}, \max_{t\in[0,1]}\{v_0(t)\}\right].$$

[25]The compactness of T has been proved in Example 5.2.8; $f = f(y,u)$ increasing in u immediately implies that T is monotone increasing.

[26]The functions u_0, v_0 are called a subsolution and supersolution of the boundary value problem (6.3.6), respectively.

Notice that $0 \in \mathcal{I}_0$. The map T is correctly defined because the Dirichlet problem

$$\begin{cases} -\ddot{w}(t) + cw(t) = g(t), & t \in (0,1), \\ w(0) = w(1) = 0 \end{cases} \tag{6.3.8}$$

has a unique solution for any fixed $g \in C[0,1]$. Then $T\colon C[0,1] \to C[0,1]$ is a compact operator. This follows from the fact that T is composed from the Nemytski operator

$$N\colon z(t) \mapsto f(t, z(t)) + cz(t)$$

which is continuous, and a compact linear operator A^{-1} where

$$A(w(t)) = -\ddot{w}(t) + cw(t), \qquad \operatorname{Dom} A = \{w \in C^2[0,1] : w(0) = w(1) = 0\}$$

(cf. Example 2.2.17), i.e.,

$$T = A^{-1} \circ N.$$

We will prove that $T\colon C[0,1] \to C[0,1]$ is a monotone increasing operator. Indeed, let $z_1, z_2 \in C[0,1]$, $z_1 \le z_2$. By definition,

$$\begin{cases} -(T(z_i))^{\cdot\cdot}(t) + cT(z_i)(t) = f(t, z_i(t)) + cz_i(t), & t \in (0,1), \\ (T(z_i))(0) = (T(z_i))(1) = 0 \end{cases} \quad \text{for } i = 1, 2.$$

Putting

$$w = T(z_2) - T(z_1)$$

we get

$$\begin{cases} -\ddot{w}(t) + cw(t) = f(t, z_2(t)) - f(t, z_1(t)) + c(z_2(t) - z_1(t)), & t \in (0,1), \\ w(0) = w(1) = 0. \end{cases}$$

However, the function

$$F\colon (t, s) \mapsto f(t, s) + cs$$

is increasing in s on the interval $\mathcal{I}_0$ by the choice of c. Hence for $z_1 \le z_2$, $z_1(t), z_2(t) \in \mathcal{I}_0$ for every $t \in [0,1]$, we have

$$0 \le F(t, z_2) - F(t, z_1) = f(t, z_2(t)) - f(t, z_1(t)) + c(z_2(t) - z_1(t)) = -\ddot{w}(t) + cw(t).$$

Therefore

$$\begin{cases} -\ddot{w}(t) + cw(t) \ge 0, \\ w(0) = w(1) = 0. \end{cases} \tag{6.3.9}$$

Assume that there is $t \in (0,1)$ such that $w(t) < 0$. Then there is $t_0 \in (0,1)$ such that

$$0 > w(t_0) = \min_{t \in [0,1]} w(t).$$

But then $\ddot{w}(t_0) \geq 0$, a contradiction with the inequality (6.3.9). Hence $w(t) \geq 0$ in $(0,1)$, i.e., $T(z_1) \leq T(z_2)$.[27]

We now prove that $v_0 \geq T(v_0)$, i.e., v_0 is a supersolution of T. Set $v_1 = T(v_0)$. We get

$$\begin{cases} -\ddot{v}_1(t) + cv_1(t) = f(t, v_0(t)) + cv_0(t), & t \in (0,1), \\ v_1(0) = v_1(1) = 0, \end{cases}$$

therefore

$$-(v_1(t) - v_0(t))^{\cdot\cdot} + c(v_1(t) - v_0(t)) = f(t, v_0(t)) + cv_0(t) + \ddot{v}_0(t) - cv_0(t) \leq 0$$

for $t \in (0,1)$. The same argument as above yields that $v_1(t) \leq v_0(t)$, $t \in [0,1]$. Analogously we prove that $u_0 \leq T(u_0)$, i.e., u_0 is a subsolution of T. If, moreover, $u_0 \leq v_0$, then Corollary 6.3.17 can be used (cf. Exercise 6.3.22). ○

Exercise 6.3.20. Let $T\colon \mathbb{R}^N \to \mathbb{R}^N$. Then the equation $x = T(x)$, $x \in \mathbb{R}^N$, in (6.3.3) describes a system of nonlinear equations

$$x_i = T_i(x_1, \dots, x_N), \qquad i = 1, \dots, N.$$

Consider the order cone $\mathbb{R}^{N,+}$ from Example 6.3.5. Translate all the assumptions and conclusions of Theorem 6.3.16 and Corollary 6.3.17 to this system.

Exercise 6.3.21. Formulate conditions on a nonlinear function $f\colon \overline{\Omega} \times \mathbb{R} \to \mathbb{R}$ which guarantee that there exist a subsolution $u_0 \in C(\overline{\Omega})$ and a supersolution $v_0 \in C(\overline{\Omega})$ of the operator T from Example 6.3.18 such that $u_0 \leq v_0$ on $\overline{\Omega}$. Then formulate the corresponding existence result for the integral equation (6.3.5).

Exercise 6.3.22. Formulate conditions on a function $f\colon [0,1] \times \mathbb{R} \to \mathbb{R}$ which guarantee that there exist a subsolution $u_0 \in C^2[0,1]$ and a supersolution $v_0 \in C^2[0,1]$ of the operator T from Example 6.3.19 such that $u_0 \leq v_0$ in $[0,1]$. Then formulate the corresponding existence result for the boundary value problem (6.3.6).

Exercise 6.3.23. Let u be a solution of the problem (5.3.2) where Ω is a domain of the class $C^{2,\gamma}$, $\gamma \in (0,1)$ and $f \in C^{0,\gamma}(\overline{\Omega})$. The maximum principle (see, e.g., Section 6.5 or Protter & Weinberger [121]) states that

$$f \geq 0 \quad \text{in } \Omega \qquad \text{implies} \qquad u \geq 0 \quad \text{in } \Omega.$$

Use this fact to generalize the result of Example 6.3.19 to the problem

$$\begin{cases} -\Delta u(x) = f(x, u(x)) & \text{in} \quad \Omega, \\ u = 0 & \text{on} \quad \partial\Omega. \end{cases} \tag{6.3.10}$$

[27]The argument used to prove $w(t) \geq 0$ in $(0,1)$ is a special version of the more general Maximum Principle (see Section 6.4, or e.g., Protter & Weinberger [121]). The monotonicity of T can be also shown by proving that the Green function corresponding to the operator A (Example 2.2.17) is nonnegative.

Exercise 6.3.24. Formulate conditions on $f = f(x,u)$ which guarantee that there is a pair of a subsolution u_0 and a supersolution v_0 of (6.3.10) satisfying

$$u_0 \le v_0 \qquad \text{in} \quad \Omega.$$

Hint. Look for u_0 and v_0 constant on Ω, cf. Exercise 7.2.38.

Exercise 6.3.25. Replace in (6.3.6) the homogeneous Dirichlet boundary conditions by the Neumann ones. Modify the definitions of a subsolution and a supersolution in such a way that Corollary 6.3.17 could be applied. Formulate conditions on $f = f(t,x)$ which guarantee the existence of a solution of the corresponding Neumann problem.

Hint. Use Corollary 6.3.17.

Exercise 6.3.26. Consider the Dirichlet boundary value problem

$$\begin{cases} -\ddot{x}(t) = h(t, x(t), \dot{x}(t)), & t \in (0,1), \\ x(0) = x(1) = 0. \end{cases} \tag{6.3.11}$$

Formulate conditions on $h = h(t,x,s)$ which guarantee the existence of a solution of (6.3.11).

Hint. Use Corollary 6.3.17.

Exercise 6.3.27. How do the conditions on h change if we replace the homogeneous Dirichlet boundary conditions in (6.3.11) by the Neumann ones?

6.3A Minorant Principle and Krein–Rutman Theorem

In this appendix we study the eigenvalue problem

$$T(x) = \lambda x, \tag{6.3.12}$$

and the corresponding inhomogeneous equation

$$\lambda x - T(x) = y, \qquad y > o, \tag{6.3.13}$$

on a real Banach space X with an order cone $\mathcal{K} := X^+$.

Definition 6.3.28. By a *positive solution* (x, λ) of (6.3.12), we mean a solution of $T(x) = \lambda x$ with $x > o$ and $\lambda > 0$. If we replace "=" by "$\ge$", then we speak about a *positive subsolution*.

Although we present mainly statements about linear problems, the following results play a central role in the investigation of nonlinear problems, for example, in the bifurcation theory, variational principles, etc. The essential tools for investigating (6.3.12) are the *Minorant Principle* (see below) and the *Separation Theorem* for convex sets (see Corollary 2.1.18).

Set

$$\mathcal{K}^r = \{x \in \mathcal{K} : \|x\| \le r\} \qquad \text{for a fixed, given} \quad r > 0.$$

The key is to find a suitable *minorant* M for T, so that

$$T(x) \geq M(x) \qquad \text{for all} \quad x \in \mathcal{K}^r, \tag{6.3.14}$$

and which satisfies appropriate conditions. Furthermore, it is important to know a *subsolution* x_0, i.e.,

$$M(x_0) \geq cx_0, \qquad c > 0, \quad x_0 > o. \tag{6.3.15}$$

The general *Minorant Principle*:

> *If we know a subsolution of* (6.3.12), *then we can obtain a positive eigenvalue with a positive eigenvector of* (6.3.12),

is formulated precisely in the following two theorems.

Theorem 6.3.29 (Krasnoselski). *Suppose that*

(i) X *is a real Banach space with an order cone* $\mathcal{K}$;

(ii) *an operator* $T\colon \mathcal{K}^r \subset X \to X$ *is compact and* (6.3.14) *holds;*

(iii) *a linear operator* $M\colon \mathcal{K} \to X$ *is positive, and there are an* $x_0 > o$ *and a positive real number* c *such that* (6.3.15) *holds.*

Then for every ϱ *with* $0 < \varrho < r$ *the problem* (6.3.12) *has a positive solution* (x, λ) *satisfying*

$$\|x\| = \varrho.$$

Theorem 6.3.30 (Zeidler [162]). *Let us set*

$$\alpha(x) = \sup\{t \geq 0 : x \geq tx_0\} \qquad \textit{for fixed} \quad x_0 > o \quad \textit{and all} \quad x \in \mathcal{K}.$$

The conclusion of Theorem 6.3.29 *still holds if we replace* (iii) *by the following condition:*

(iii′) *suppose that* $M\colon \mathcal{K} \subset X \to \mathcal{K}$ *is an operator, not necessarily linear, for which there is an* $x_0 > o$ *and there are real numbers* s *with* $0 < s \leq 1$ *and* $c > 0$ *such that*

$$M(x) \geq (\alpha(x))^s cx_0 \qquad \textit{for all} \quad x \in \mathcal{K}^r. \tag{6.3.16}$$

Theorem 6.3.29 is a special case of Theorem 6.3.30. Indeed, since $x \geq \alpha(x)x_0$ for $x \in \mathcal{K}^r$, we have

$$M(x) \geq \alpha(x)M(x_0) \geq \alpha(x)cx_0 \qquad \text{for all} \quad x \in \mathcal{K}^r.$$

Thus, (6.3.15) implies (6.3.16) with $s = 1$.

Proof of Theorem 6.3.30. We will use a regularization method and the Schauder Fixed Point Theorem (Theorem 5.2.5).

Let us first solve an auxiliary problem

$$\lambda_n x_n = T_n(x_n), \qquad \lambda_n > 0, \quad x_n > o, \quad \|x_n\| = \varrho \tag{6.3.17}$$

where

$$T_n(x) := T(x) + \frac{x_0}{n}, n = 1, 2, \dots, 0 < \varrho \leq r.$$

Let n be fixed. Set

$$z(x) = \varrho \frac{x}{\|x\|} \qquad \text{for} \quad x \neq o \qquad \text{and} \qquad z(o) = o.$$

For $x \in \mathcal{K}^\varrho$ we set

$$S(x) = \|x\|T_n(z(x)) + \frac{(\varrho - \|x\|)x_0}{n}.$$

Then S is compact on $\mathcal{K}^\varrho$ (explain why!), and by (6.3.14), (6.3.17)

$$S(x) \geq \frac{\|x\|x_0}{n} + \frac{(\varrho - \|x\|)x_0}{n} = \frac{\varrho x_0}{n} > o \qquad \text{for all} \quad x \in \mathcal{K}^\varrho.$$

So there is an $a_n > 0$ such that

$$\|S(x)\| \geq a_n \qquad \text{for all} \quad x \in \mathcal{K}^\varrho.$$

It follows from the boundedness of $S(\mathcal{K}^\varrho)$ that there exists a number $b_n > 0$ such that

$$0 < a_n \leq \|S(x)\| \leq b_n \qquad \text{for all} \quad x \in \mathcal{K}^\varrho. \tag{6.3.18}$$

By (6.3.18),

$$V(x) = \varrho \frac{S(x)}{\|S(x)\|}$$

is well defined on $\mathcal{K}^\varrho$. Furthermore, the operator $V\colon \mathcal{K}^\varrho \to \mathcal{K}^\varrho$ is compact on the closed, bounded, and convex set $\mathcal{K}^\varrho$ (why?). By Theorem 5.2.5 (the Schauder Fixed Point Theorem) there is an $x_n \in \mathcal{K}^\varrho$ such that

$$x_n = V(x_n).$$

In particular, $\|x_n\| = \|V(x_n)\| = \varrho$, so $z(x_n) = x_n$. Therefore $x_n = \varrho^2 \frac{T_n(x_n)}{\|S(x_n)\|}$, which means that x_n is a solution of (6.3.17) with $\lambda_n = \frac{\|S(x_n)\|}{\varrho^2}$.

Before we pass to the limit for $n \to \infty$, we estimate the value of λ_n. Namely, we will show that there exist numbers $a, b > 0$ such that

$$0 < a \leq \lambda_n \leq b \qquad \text{for all} \quad n \in \mathbb{N}. \tag{6.3.19}$$

It follows from (6.3.17) that

$$\lambda_n \varrho \leq \|T(x_n)\| + \|x_0\|, \qquad \text{so} \qquad b := \sup_{n\in\mathbb{N}} \lambda_n < \infty.$$

On the other hand, $x_n \geq \alpha(x_n)x_0$ implies that there exists γ such that

$$\gamma := \sup_{n\in\mathbb{N}} \alpha(x_n) < \infty.$$

Indeed, otherwise there would be a subsequence, again denoted by $\{x_n\}_{n=1}^\infty$, with $\alpha(x_n) \to \infty$ as $n \to \infty$, contradicting $o < x_0 \leq \frac{x_n}{\alpha(x_n)} \to o$ as $n \to \infty$. Now (6.3.17), (6.3.14) and (6.3.16) imply that

$$\lambda_n x_n = T(x_n) + \frac{x_0}{n} \geq M(x_n) + \frac{x_0}{n} \geq \frac{x_0}{n}.$$

Therefore, $\alpha(x_n) > 0$, and furthermore

$$\lambda_n x_n \geq M(x_n) \geq (\alpha(x_n))^s c x_0,$$

i.e., the definition of $\alpha(x_n)$ implies

$$\alpha(x_n) \geq \frac{(\alpha(x_n))^s c}{\lambda_n}, \qquad \text{so} \qquad \lambda_n \geq \alpha(x_n)^{s-1} c \geq \gamma^{s-1} c =: a.$$

This proves (6.3.19).

Now, we pass to the limit $n \to \infty$ in (6.3.17). Using (6.3.19) and $\|x_n\| = \varrho$, we can find convergent subsequences, again denoted by $\{\lambda_n\}_{n=1}^{\infty}$ and $\{T(x_n)\}_{n=1}^{\infty}$, with $\lambda_n \to \lambda$ and $T(x_n) \to y$ strongly in X. By (6.3.19), $\lambda > 0$. Then we have also strong convergence in X for

$$x_n = \lambda_n^{-1} \left(T(x_n) + \frac{x_0}{n} \right) \to x.$$

Hence

$$\lambda x = T(x) \qquad \text{and} \qquad x \in \mathcal{K}, \quad \|x\| = \varrho. \qquad \square$$

Example 6.3.31. We will consider the nonlinear system of equations

$$\lambda \xi_i = f_i(\xi_1, \ldots, \xi_N), \qquad i = 1, \ldots, N, \tag{6.3.20}$$

with $x = (\xi_1, \ldots, \xi_N)$ and $x \in \mathcal{K} := \mathbb{R}^{N,+}$, $\|x\| = \varrho$, $\lambda > 0$. The following assertion (the *Generalized Perron Theorem*) is a consequence of Theorem 6.3.29:

Suppose that $f_i \colon \mathcal{K} \to (0, \infty)$ is continuous for $i = 1, \ldots, N$ and that there is a fixed $r > 0$ for which

$$f_i(\xi_1, \ldots, \xi_N) \geq \sum_{j=1}^{N} \mu_{ij} \xi_j \qquad \textit{holds for all} \quad x \in \mathcal{K} \quad \textit{with} \quad \|x\| \leq r, \tag{6.3.21}$$

and $i = 1, \ldots, N$. Assume that all the real numbers μ_{ij} are nonnegative, and that

$$\min_{1 \leq i \leq N} \sum_{j=1}^{N} \mu_{ij} > 0.$$

Then (6.3.20) *has a positive solution for every ϱ with $0 < \varrho \leq r$.*

Indeed, we can write (6.3.20) as $\lambda x = T(x)$ and apply Theorem 6.3.29 with $X = \mathbb{R}^N$, $X^+ = \mathbb{R}^{N,+}$, $x_0 = (1, \ldots, 1)$, and

$$M(x) := (\eta_1, \ldots, \eta_N) \qquad \text{where} \quad \eta_i = \sum_{j=1}^{N} \mu_{ij} \xi_j. \qquad \circ$$

Example 6.3.32. We will consider the nonlinear integral equation

$$\lambda x(t) = \int_a^b A(t,s) f(s, x(s)) \, \mathrm{d}s \tag{6.3.22}$$

on a finite interval $[a, b]$ with $\lambda > 0$. This time, for fixed $\mu > 0$ and $r > 0$, the key condition (substituting (6.3.21)) is

$$f(s, x) \geq \mu x \qquad \text{for all} \quad (s, x) \in [a, b] \times [0, r]. \tag{6.3.23}$$

Applying Theorem 6.3.29 we have the following assertion (the *Generalized Jentzsch Theorem*):

Suppose $A\colon [a,b]\times[a,b]\to\mathbb{R}$ is continuous, nonnegative, and

$$\min_{t\in[a,b]}\int_a^b A(t,s)\,\mathrm{d}s>0.$$

Let $f\colon [a,b]\times\mathbb{R}^+\to\mathbb{R}$ be continuous and let (6.3.23) *be satisfied. Then for every ϱ with $0<\varrho\le r$,* (6.3.22) *has a positive solution $x\in C[a,b]$ with $\|x\|=\varrho$.*

Indeed, we write (6.3.22) as $\lambda x=T(x)$ and apply Theorem 6.3.29 with $X=C[a,b]$, $X^+=C^+[a,b]$, $x_0(t)\equiv 1$ and

$$M(x)(t):=\mu\int_a^b A(t,s)x(s)\,\mathrm{d}s. \qquad \circ$$

Proposition 6.3.33. *Let $T\colon X\to X$ be a compact linear positive operator on a real ordered Banach space X. Then there exists a positive solution of* (6.3.12) *if and only if* (6.3.12) *has a positive subsolution.*

Proof. This assertion is an immediate consequence of the Minorant Principle (Theorem 6.3.29 with $M=T$). □

Our goal is to sharpen this result. Let $T\colon X\to X$ be a linear operator and let $r(T)$ denote the spectral radius of the complexification of T.[28]

We call λ a *simple eigenvalue* of T if its multiplicity $m(\lambda)$ is equal to 1.[29] Recall that this means

$$\dim\operatorname{Ker}(\lambda I-T)=1 \qquad\text{and}\qquad \operatorname{Ker}(\lambda I-T)^2=\operatorname{Ker}(\lambda I-T).$$

Let $\mathcal{K}^*:=X^{*,+}$ denote the set of all *positive functionals* $x^*\in\mathcal{K}^*$, i.e.,

$$\langle x^*,x\rangle\ge 0 \qquad\text{for all}\quad x\in\mathcal{K}:=X^+.$$

We write $x^*\ge o$ if x^* is *positive*. Furthermore, $x^*>o$ means

$$x^*\ge o \qquad\text{and}\qquad \langle x^*,x\rangle>0 \qquad\text{for a certain}\quad x\in\mathcal{K}.$$

We call x^* *strictly positive* if

$$x>o \qquad\text{always implies}\qquad \langle x^*,x\rangle>0.$$

A cone $\mathcal{K}:=X^+\subset X$ is called *total* if $\operatorname{Lin}(\mathcal{K})$ is dense in X. Then $\mathcal{K}$ being total implies that $\mathcal{K}^*$ is an order cone (cf. Exercise 6.3.45). In this case we call $\mathcal{K}^*$ the *dual order cone* of $\mathcal{K}$. In particular, $\mathcal{K}$ is total if

$$\operatorname{int}\mathcal{K}\ne\emptyset$$

(cf. Exercise 6.3.43). For $X=\mathbb{R}^N$, $\mathcal{K}=\mathbb{R}^{N,+}$ we have $X^*=X$, $\mathcal{K}^*=\mathcal{K}$ (explain why!).

---

[28] If X is a real Banach space, then by the *complexification* of T we mean the operator $\boldsymbol{T}\colon X_{\mathbb{C}}\to X_{\mathbb{C}}$ defined by $\boldsymbol{T}(x+\mathrm{i}y)=T(x)+\mathrm{i}T(y)$, $x,y\in X$, where $X_{\mathbb{C}}$ is the complexification of X in the sense of Example 1.1.6(iii).

[29] The significance of simple eigenvalues, roughly speaking, is that their behavior is very stable under perturbations of the operator (cf. Example 4.2.4 and in more details Kato [87]). For this reason, simple eigenvalues play a special role also in the bifurcation theory.

Proposition 6.3.34 (Krein–Rutman). *Let X be a real Banach space with a total order cone $\mathcal{K}$. Suppose that $T\colon X \to X$ is linear, compact, and positive, with spectral radius $r(T) > 0$. Then $r(T)$ is an eigenvalue of both T and T^* with eigenvectors in $\mathcal{K}$ and $\mathcal{K}^*$, respectively.*

If T is strongly positive, we get a sharper version of the previous assertion.

Theorem 6.3.35 (Krein–Rutman). *Let X be a real Banach space with an order cone $\mathcal{K}$ having nonempty interior. Then any linear, compact, and strongly positive operator $T\colon X \to X$ has the following properties:*

(i) *T has exactly one eigenvector with $x > o$ and $\|x\| = 1$. The corresponding eigenvalue is $r(T)$ and it is algebraically simple. Furthermore, $x \gg o$.*

(ii) *If $\lambda \in \sigma(T)$, $\lambda \neq r(T)$, then $|\lambda| < r(T)$.*

(iii) *The dual operator T^* has $r(T)$ as an algebraically simple eigenvalue with a strictly positive eigenvector x^*.*

Remark 6.3.36. Recall that by the Riesz–Schauder theory (see Theorem 2.2.9), the spectrum of T consists of at most countably many nonzero eigenvalues of finite multiplicity which can accumulate only at the origin, and $o \in \sigma(T)$ whenever $\dim X = \infty$. The spectra of T and T^* coincide (X is a real space).

Now, we will give proofs of Proposition 6.3.34 and Theorem 6.3.35.

Proof of Proposition 6.3.34. Let us consider T on the complexification $X_{\mathbb{C}} = X + \mathrm{i}X$. By the Riesz–Schauder theory (see Theorem 2.2.9), all of the nonzero points of the spectrum of T consist of eigenvalues of finite multiplicity. The same holds for T^*. Note that

$$\sigma(T) \cap \{\lambda : |\lambda| = r(T)\} \neq \emptyset.$$

We consider the eigenvalues λ of T satisfying $|\lambda| = r(T)$, and distinguish three cases.

Case 1 ($\lambda_0 = r(T)$ **is an eigenvalue**). Our goal is to construct an $x > o$ and an $x^* > o$ such that

$$T(x) = \lambda_0 x \qquad \text{and} \qquad T^*(x^*) = \lambda_0 x^*.$$

From footnote 3 on page 57 we have

$$(\lambda I - T)^{-1} u = \sum_{j=0}^{\infty} \frac{T^j u}{\lambda^{j+1}}, \qquad \lambda > r(T),$$

and, therefore,

$$(\lambda I - T)^{-1} u \geq o \qquad \text{for} \quad \lambda > \lambda_0 \quad \text{and} \quad u \geq o.$$

Since T is compact, $\lambda_0 \neq 0$ is an eigenvalue of finite multiplicity (Remark 2.2.10) and in the Laurent series

$$(\lambda I - T)^{-1} = \sum_{n=-\infty}^{+\infty} (\lambda - \lambda_0)^n A_n, \tag{6.3.24}$$

there is an index k such that

$$A_{-n} = O \qquad \text{for all} \quad k < n \qquad \text{and} \qquad A_{-k} \neq O$$

(Proposition 3.1.20). So, there is $u > o$ such that

$$x := A_{-k}u \neq o$$

(otherwise $A_{-k} = O$ since $\mathcal{K}$ is total). It follows from Proposition 3.1.20 that

$$Tx = \lambda_0 x.$$

Moreover, by (6.3.24) and its proof (cf. page 122)

$$x = A_{-k}u = \lim_{\lambda \to \lambda_0+} (\lambda - \lambda_0)^k (\lambda I - T)^{-1} u \geq o, \qquad \text{i.e.,} \qquad x > o.$$

Let us construct the element x^*. By the previous step, $A_{-n}(\mathcal{K}) \subset \mathcal{K}$. We choose a $u^* \in \mathcal{K}^*$ with $\langle u^*, x\rangle > 0$. This is possible by Exercise 6.3.44. We set $x^* = A^*_{-n}u^*$. Then $v \geq o$ implies

$$\langle x^*, v\rangle = \langle u^*, A_{-n}v\rangle \geq 0 \qquad \text{and} \qquad \langle x^*, u\rangle = \langle u^*, x\rangle > 0.$$

Thus $x^* > o$. Passing to the dual operator in (6.3.24), we obtain

$$\lambda_0 x^* = T^*(x^*)$$

analogously as above.

Case 2 (**there is an eigenvalue $\lambda_0 \in \mathbb{C}$ of T with $|\lambda_0| = r(T)$ and $\lambda_0^n > 0$ for an $n \in \mathbb{N}$**). Now T^n has a positive eigenvalue λ_0^n which lies on the spectral circle of T^n, so by Case 1 there exists a $u > o$ with

$$T^n(u) = \lambda_0^n u.$$

If we set

$$x = |\lambda_0|^{n-1}u + |\lambda_0|^{n-2}T(u) + \cdots + T^{n-1}(u),$$

then $x > o$ and $T(x) = |\lambda_0|x$. Analogously we construct an x^* for T^*.

Case 3 (**none of the eigenvalues of T with $|\lambda| = r(T)$ has the property from Case 2**). We show that this is impossible. So, let λ_0 be an eigenvalue of T with $|\lambda_0| = r(T)$ and with the greatest possible real part. We set

$$T_\varepsilon = T + \varepsilon T^2 \qquad \text{for} \quad \varepsilon > 0.$$

By the Spectral Mapping Theorem (see Proposition 3.1.19(v)), all eigenvalues of T_ε are of the form $\lambda + \varepsilon\lambda^2$ where λ is an eigenvalue of T. One can check that $\lambda_1 = \lambda_0 + \varepsilon\lambda_0^2$ and $\overline{\lambda}_1$ are the eigenvalues of T_ε of greatest absolute value (the reader is asked to justify it!). There is a sequence $\{\varepsilon_k\}_{k=1}^\infty$, $\varepsilon_k \to 0$, such that $\operatorname{Arg}\lambda_{1,k}$ is a rational multiple of 2π where $\lambda_{1,k} = \lambda_0 + \varepsilon_k\lambda_0^2$ (explain why!). According to Case 2, there is $n \in \mathbb{N}$ such that $\lambda_{1,k}^n > 0$. Since

$$\lim_{k\to\infty} \lambda_{1,k}^n = \lambda_0^n > 0,$$

we get a contradiction. □

Before we prove Theorem 6.3.35 we need the following result.

Lemma 6.3.37. *Let X be a real Banach space with an order cone $\mathcal{K} := X^+$ containing an interior point. Let $u \gg o$. Then for every $v \notin \mathcal{K}$ there is a uniquely determined number $\alpha_u(v) > 0$ such that*

(i) $0 \le \alpha \le \alpha_u(v)$ *implies* $u + \alpha v \ge o$;

(ii) $\alpha > \alpha_u(v)$ *implies* $u + \alpha v \notin \mathcal{K}$.

In particular,

$$u + \alpha v \gg o \quad \textit{and} \quad \alpha > 0 \qquad \textit{imply} \qquad \alpha < \alpha_u(v). \tag{6.3.25}$$

Proof. Consider the ray

$$\varrho = \{u + \alpha v : \alpha \ge 0\}.$$

For small $\alpha \ge 0$ we have $u + \alpha v \in \operatorname{int} \mathcal{K}$, and for large $\alpha \ge \alpha_0$ we have $u + \alpha v \notin \mathcal{K}$. Otherwise $u + nv \in \mathcal{K}$ for large $n \in \mathbb{N}$, and $\frac{u}{n} + v \in \mathcal{K}$. Passing to the limit for $n \to \infty$, we obtain a contradiction $v \in \mathcal{K}$. Set

$$\alpha_u(v) := \sup\{\alpha > 0 : u + \alpha v \in \operatorname{int} \mathcal{K}\}.$$

It is easy to show that $\alpha_u(v)$ has the desired properties. □

Proof of Theorem 6.3.35. We proceed in six steps.

Step 1 **(existence of a positive solution)**. We choose an $x > o$. Since T is strongly positive, $T(x) \gg o$, so $T(x) \in \operatorname{int} \mathcal{K}$. Thus $T(x) - \gamma x \in \mathcal{K}$ for small $\gamma > 0$, so $T(x) \ge \gamma x$. By Proposition 6.3.33, there exists a positive solution (e, λ_0):

$$T(e) = \lambda_0 e \qquad \text{with} \quad e > o \quad \text{and} \quad \lambda_0 > 0.$$

Since $T(e) \gg o$, we have also $e \gg o$.

Step 2. We show:

If $T(x) = \lambda x$, $x > o$ *and* $\lambda \in \mathbb{R}$, *then* $x = \gamma e$ *for a positive* γ *and* $\lambda = \lambda_0$.

To begin with, $T(x) \gg o$, so $\lambda > 0$ and $x \gg o$. We consider two identities

$$T(e - \beta x) = \lambda_0(e - \beta\lambda_0^{-1}\lambda x), \tag{6.3.26}$$

$$T(x - \gamma e) = \lambda(x - \gamma\lambda^{-1}\lambda_0 e), \tag{6.3.27}$$

and choose

$$\beta = \alpha_e(-x) \qquad \text{and} \qquad \gamma = \alpha_x(-e).$$

Then $x = \gamma e$. Otherwise, $x - \gamma e > o$. This implies $T(x - \gamma e) \gg o$, and hence $\lambda^{-1}\lambda_0 < 1$ by (6.3.27) and (6.3.25). On the other hand, $e - \beta x \ge o$ immediately implies $T(e - \beta x) \ge o$, and (6.3.26) yields the contradiction $\lambda_0^{-1}\lambda \le 1$.

Step 3. We show:

If $T(x) = \lambda x$ *and* $x \ne o$, $\lambda \in \mathbb{R} \setminus \{0\}$ *as well as* $x \ne \alpha e$ *for all* $\alpha \in \mathbb{R}$, *then* $|\lambda| < \lambda_0$.

By Proposition 6.3.34 $\lambda_0 = r(T)$ now follows, and with respect to Step 2,

$$\dim \operatorname{Ker}(\lambda_0 I - T) = 1.$$

By Step 2, $\pm x \notin \mathcal{K}$. We consider

$$T(e \pm \beta_\pm x) = \lambda_0(e \pm \beta_\pm \lambda_0^{-1}\lambda x) \tag{6.3.28}$$

and set

$$\beta_\pm = \alpha_e(\pm x).$$

Since $e \pm \beta_{\pm} x \neq o$, we have $e \pm \beta_{\pm} x > o$, so

$$T(e \pm \beta_{\pm} x) \gg o.$$

Then (6.3.28) and (6.3.25) immediately imply $\lambda_0^{-1}|\lambda| < 1$.

Step 4. We consider the complexification $X_{\mathbb{C}} = X + \mathrm{i}X$ and $\boldsymbol{T}: X_{\mathbb{C}} \to X_{\mathbb{C}}$ (see footnote 28 on page 404). In this step we show:

If λ is a complex eigenvalue of $\boldsymbol{T}$, then $|\lambda| < \lambda_0$.

Let $\lambda = \sigma + \mathrm{i}\tau$, $\sigma, \tau \in \mathbb{R}$, be an eigenvalue of $\boldsymbol{T}$ and $z = x + \mathrm{i}y$, $x, y \in X$, the corresponding eigenvector, i.e., according to the definition of $\boldsymbol{T}$, we have

$$\boldsymbol{T}(x + \mathrm{i}y) = (\sigma + \mathrm{i}\tau)(x + \mathrm{i}y),$$

which is equivalent to

$$T(x) = \sigma x - \tau y, \qquad T(y) = \tau x + \sigma y. \tag{6.3.29}$$

Our goal is to show that (6.3.29) implies

$$|\lambda| = \sqrt{\sigma^2 + \tau^2} < \lambda_0.$$

The reader is invited to prove that if λ is not real and (6.3.29) holds, then x and y are linearly independent elements of X (cf. Remark 1.1.35(ii)). In particular, $x \neq o$ and $y \neq o$.

Let $\mathcal{P}$ be a two-dimensional plane in X which consists of elements $\xi x + \eta y$, $\xi, \eta \in \mathbb{R}$. Then $\mathcal{P}$ is an invariant subspace of the operator T, i.e.,

$$T(\mathcal{P}) \subset \mathcal{P}.$$

Let $\tilde{T}$ be the restriction of T onto $\mathcal{P}$. Since also $T(\mathcal{K}) \subset \mathcal{K}$, the cone $\tilde{\mathcal{K}} = \mathcal{K} \cap \mathcal{P}$ is invariant with respect to $\tilde{T}$, i.e.,

$$\tilde{T}(\tilde{\mathcal{K}}) \subset \tilde{\mathcal{K}}.$$

We want to prove that

$$\tilde{\mathcal{K}} = \{o\}. \tag{6.3.30}$$

Assume the opposite, then $\tilde{\mathcal{K}}$ is an order cone in $\mathcal{P}$ and $\tilde{T}: \mathcal{P} \to \mathcal{P}$ is strongly positive since T is strongly positive. According to Step 1, there exists a positive eigenvector $\tilde{e} \in \mathcal{P}$ of $\tilde{T}$ (and hence also of T) such that $\tilde{e} \in \tilde{\mathcal{K}}$. According to Step 2 we necessarily have $\tilde{e} = \gamma e$ for a certain $\gamma \neq 0$, $\gamma \in \mathbb{R}$. But this fact combined with (6.3.29) implies that x and y are linearly dependent, which is a contradiction, i.e., (6.3.30) is proved.

It now follows from (6.3.30) that no elements $\xi x + \eta y$ with $|\xi| + |\eta| > 0$ belong to $\mathcal{K}$. In particular, $x \notin \mathcal{K}$. Since $\operatorname{int} \mathcal{K} \neq \emptyset$ implies that $\mathcal{K}$ is total, there exist nonzero elements $x' \in \operatorname{int} \mathcal{K}$ and $x'' \in \operatorname{int}(\mathcal{K})$ such that

$$x = x' - x''.$$[30]

There exists $\beta > 0$ such that

$$T(x'') \leq \beta e.$$

[30] Indeed, if $v_0 \in \operatorname{int} \mathcal{K}$, $v_0 \neq o$ and $\varrho > 0$ is small enough, then $u = v_0 + \varrho x \in \operatorname{int} \mathcal{K}$, $u \neq o$. Hence $x = \frac{u}{\varrho} - \frac{v_0}{\varrho}$.

Indeed, since $e \in \operatorname{int} \mathcal{K}$ we find $\beta > 0$ large enough to satisfy $e - \frac{1}{\beta}T(x'') \in \mathcal{K}$.

So, we have

$$T(x) = T(x') - T(x'') \geq -T(x'') \geq -\beta e, \qquad \text{i.e.,} \qquad e + \frac{1}{\beta}T(x) \in \mathcal{K}.$$

It follows from (6.3.29) that $\psi = e + \frac{1}{\beta}T(x)$ can be written in the form

$$\psi = \xi x + \eta y + e, \qquad |\xi| + |\eta| > 0. \tag{6.3.31}$$

Let $\mathcal{A}$ be the set of all elements of the form (6.3.31) which belong to $\mathcal{K}$. We have just shown that $\mathcal{A} \neq \emptyset$. Let us consider a continuous function of two variables $f\colon \mathcal{A} \to \mathbb{R}$ which with every $\psi \in \mathcal{A}$ associates the number $\xi^2 + \eta^2$. Since $x \notin \mathcal{K}$, $y \notin \mathcal{K}$, the function f must be bounded. It follows from the Extreme Value Theorem ($\mathcal{K}$ is closed) that there is

$$\psi_0 = \xi_0 x + \eta_0 y + e \in \mathcal{A} \qquad \text{such that} \qquad f(\psi_0) = \max_{\psi \in \mathcal{A}} f(\psi) =: M.$$

It follows from the strict positivity of T that there exists $\delta > 0$ such that

$$T(\psi_0) \geq \delta e.$$

Indeed, $\psi_0 \in \mathcal{K}$, $\psi_0 \neq o$, implies $T(\psi_0) \in \operatorname{int} \mathcal{K}$. We then can find $\delta > 0$ small enough to satisfy

$$T(\psi_0) - \delta e \in \mathcal{K}.$$

Let us assume without loss of generality that $\frac{\delta}{\lambda_0} < 1$. Let us rewrite $T(\psi_0) \geq \delta e$ as

$$\left(1 - \frac{\delta}{\lambda_0}\right)\lambda_0 e + (\xi_1 x + \eta_1 y) \geq o \qquad \text{where} \quad \xi_1 x + \eta_1 y = T(\xi_0 x + \eta_0 y). \tag{6.3.32}$$

Using (6.3.29), we have

$$T(\xi_0 x + \eta_0 y) = (\xi_0 \sigma + \eta_0 \tau)x + (-\xi_0 \tau + \eta_0 \sigma)y$$

and hence

$$\xi_1 = \xi_0 \sigma + \eta_0 \tau, \qquad \eta_1 = -\xi_0 \tau + \eta_0 \sigma. \tag{6.3.33}$$

Then

$$\xi_1^2 + \eta_1^2 = (\xi_0^2 + \eta_0^2)(\sigma^2 + \tau^2) = M|\lambda|^2.$$

It follows from (6.3.32) that

$$\psi_1 = e + \frac{\xi_1}{\left(1 - \frac{\delta}{\lambda_0}\right)\lambda_0}x + \frac{\eta_1}{\left(1 - \frac{\delta}{\lambda_0}\right)\lambda_0}y$$

is an element of the form (6.3.31). Hence

$$M \geq \left(\frac{\xi_1}{\lambda_0 - \delta}\right)^2 + \left(\frac{\eta_1}{\lambda_0 - \delta}\right)^2 = \frac{M|\lambda|^2}{(\lambda_0 - \delta)^2},$$

which implies $|\lambda| < \lambda_0$.

Step 5. We show that λ_0 *is simple.* Since $\dim \operatorname{Ker}(\lambda_0 I - T) = 1$ (Step 3), it is enough to prove

$$\operatorname{Ker}(\lambda_0 I - T)^2 = \operatorname{Ker}(\lambda_0 I - T).$$

Let

$$(\lambda_0 I - T)^2(x) = o.$$

By Step 2, this implies

$$(\lambda_0 I - T)x = \gamma e.$$

We want to show that $\gamma = 0$. Suppose $\gamma \neq 0$. We may assume that $\gamma > 0$, for otherwise we pass to $-x$. Set $\mu_0 = \lambda_0^{-1}$. Now $x = \mu_0 T(x + \gamma e)$ implies

$$x + \gamma e = \mu_0 T(x + 2\gamma e) \qquad \text{and} \qquad x = \mu_0^2 T^2(x + 2\gamma e).$$

It follows by induction that $x = \mu_0^n T^n(x + n\gamma e)$, so

$$\frac{x}{n} = \mu_0^n T^n \left(\gamma e + \frac{x}{n}\right) \qquad \text{for all} \quad n \in \mathbb{N}. \tag{6.3.34}$$

Since $e \in \operatorname{int} \mathcal{K}$, we have $\gamma e + \frac{x}{n} \geq o$ for large n. By (6.3.34) and the positivity of T, we have $\frac{x}{n} \geq o$. Furthermore, from (6.3.34) and $\mu_0^n T^n(e) = e$ we immediately conclude

$$\frac{x}{n} - \gamma e = \mu_0^n T^n \left(\frac{x}{n}\right) \geq o.$$

Passing to the limit for $n \to \infty$ we get $-\gamma e \geq o$, so $\gamma = 0$, contradicting $\gamma > 0$.

Step 6 **(properties of T^*)**. By Proposition 6.3.34 there exists $e^* > o$ such that

$$T^*(e^*) = \lambda_0 e^*.$$

We show that

$$\langle e^*, x\rangle > 0 \qquad \text{provided} \qquad x > o, \tag{6.3.35}$$

i.e., e^* is strictly positive. Indeed, let $x > o$. Then $T(x) \gg o$ and by Exercise 6.3.46, $\langle e^*, T(x)\rangle > 0$. So

$$\lambda_0 \langle e^*, x\rangle = \langle T^*(e^*), x\rangle = \langle e^*, T(x)\rangle > 0.$$

According to the Riesz–Schauder Theory (see Theorem 2.2.9),

$$\dim \operatorname{Ker}(\lambda_0 I^* - T^*) = \dim \operatorname{Ker}(\lambda_0 I - T)$$

which is equal to 1 by Steps 2 and 3. To prove that λ_0 is an algebraically simple eigenvalue of T^* choose $x^* \in \operatorname{Ker}(\lambda_0 I^* - T^*)^2$. Let $y^* = \lambda_0 x^* - T^* x^*$. Then $y^* = \alpha e^*$ for an $\alpha \in \mathbb{R}$. For any $x > o$ we have

$$\alpha \langle e^*, x\rangle = \langle y^*, x\rangle = \langle x^*, \lambda_0 x - Tx\rangle.$$

In particular, taking $x = e$ we obtain $\alpha = 0$, i.e., $y^* = o$ and $x^* \in \operatorname{Ker}(\lambda_0 I^* - T^*)$. This proves

$$\operatorname{Ker}(\lambda_0 I^* - T^*)^2 = \operatorname{Ker}(\lambda_0 I^* - T^*).$$

This completes the proof of Theorem 6.3.35. □

The authors want to point out that another proof of the Krein–Rutman Theorem can be found in, e.g., Takáč [148].

Corollary 6.3.38. *Let X and T be as in Theorem* 6.3.35. *For every $y > o$,* (6.3.13) *has exactly one solution $x > o$ if $\lambda > r(T)$, and no such solution if $\lambda \le r(T)$. Moreover,*

$$\lambda x - T(x) = \mu y \quad \text{and} \quad x > o,\ y > o \qquad \text{imply} \qquad \operatorname{sgn}(\mu) = \operatorname{sgn}(\lambda - r(T)).$$

Here λ and μ are real numbers.

Proof. The resolvent R_λ exists for $\lambda > r(T)$ and thus the equation

$$\lambda x - T(x) = y \tag{6.3.36}$$

has a unique solution for any $y \in X$. Since $R_\lambda \colon \mathcal{K} \to \mathcal{K}$ by the proof of Proposition 6.3.34, hence $y > o$ implies $x > o$. On the other hand, if $\lambda \le r(T)$ and there is a positive solution x of (6.3.36) for $y > o$, then choosing $e^* \in X^*$ as in Step 6 of the proof of Theorem 6.3.35 we arrive at

$$(\lambda - r(T))\langle e^*, x\rangle = \langle e^*, \lambda x - T(x)\rangle = \langle e^*, y\rangle > 0,$$

a contradiction. Finally, let $x > o$, $y > o$ and

$$\lambda x - T(x) = \mu y \qquad \text{for a certain} \quad \mu \in \mathbb{R}.$$

Then

$$(\lambda - r(T))\langle e^*, x\rangle = \langle e^*, \lambda x - T(x)\rangle = \mu\langle e^*, y\rangle, \qquad \text{i.e.,} \quad \operatorname{sgn}(\lambda - r(T)) = \operatorname{sgn}\mu. \qquad \square$$

Corollary 6.3.39 (Comparison Principle). *Let X and T be as in Theorem* 6.3.35. *If $S\colon X \to X$ is a compact linear operator with*

$$S(x) \ge T(x) \qquad \textit{for all} \quad x \ge o,$$

then

$$r(S) \ge r(T).$$

If $S(x) > T(x)$ for all $x > o$, then $r(S) > r(T)$.

Proof. Let

$$S(x) \ge T(x) \qquad \text{for all} \quad x \ge o.$$

Choose $e > o$ such that $T(e) = r(T)e$. Then

$$S(e) \ge T(e) = r(T)e.$$

By Proposition 6.3.33, $r(S) \in \sigma(S)$ and therefore $r(S) \ge r(T)$.

In order to prove the second part of the statement we choose $x > o$ with $S(x) = r(S)x$ (see Proposition 6.3.34). We now set

$$A := S - T$$

and choose e^* as in Step 6 of the proof of Theorem 6.3.35. Then

$$\begin{aligned} r(S)\langle e^*, x\rangle &= \langle e^*, A(x)\rangle + \langle e^*, T(x)\rangle = \langle e^*, A(x)\rangle + \langle T^*(e^*), x\rangle \\ &= \langle e^*, A(x)\rangle + r(T)\langle e^*, x\rangle. \end{aligned}$$

By (6.3.35), we have $\langle e^*, x\rangle > 0$ and also $\langle e^*, A(x)\rangle > 0$, and thus

$$r(S) > r(T). \qquad \square$$

Example 6.3.40. Let $X = \mathbb{R}^N$ and $X^+ = \mathbb{R}^{N,+}$. Further, let $\boldsymbol{T}$ be a real $(N \times N)$ matrix of positive elements only. Then $T\colon X \to X$ is linear, compact, and strongly positive. The conclusions of Theorem 6.3.35 coincide with those of the *Classical Perron Theorem.* ○

Example 6.3.41. Let Ω be a bounded domain in $\mathbb{R}^N$. We set $X = C(\overline{\Omega})$, $X^+ = C^+(\overline{\Omega})$ (cf. Example 6.3.7) and consider the integral equation

$$\lambda x(t) = \int_\Omega A(t,s)x(s)\,\mathrm{d}s \qquad \text{for all} \quad t \in \overline{\Omega}, \tag{6.3.37}$$

with a positive continuous kernel $A\colon \overline{\Omega} \times \overline{\Omega} \to \mathbb{R}$. If we write (6.3.37) in the form

$$\lambda x = T(x), \qquad x \in X,$$

then Theorem 6.3.35 is the *Classical Jentzsch Theorem.* ○

Example 6.3.42. Let us consider the eigenvalue problem for the Laplace operator subject to the homogeneous Dirichlet boundary conditions

$$\begin{cases} -\Delta u(x) = \mu u(x) & \text{in} \quad x \in \Omega, \\ u(x) = 0 & \text{on} \quad x \in \partial\Omega, \end{cases} \tag{6.3.38}$$

where Ω is a bounded domain in $\mathbb{R}^N$ and $\partial\Omega$ is its boundary (cf. Remark 5.4.2). Then (6.3.38) can be written in the form (6.3.37) with $\lambda = \frac{1}{\mu}$ where $A = A(t,s)$ is the Green function associated with the Laplace equation with the homogeneous Dirichlet boundary conditions. Since A is a positive continuous function $A\colon \overline{\Omega} \times \overline{\Omega} \to \mathbb{R}$ (see, e.g., Gilbarg & Trudinger [71]), we can apply the result of Example 6.3.41.

Multiplying the equation in (6.3.38) by $u = u(x)$ (u is a real function) and using the Green Formula (cf. footnote 19 on page 270), we find

$$\int_\Omega \|\nabla u(x)\|^2\,\mathrm{d}x = \mu \int_\Omega u^2(x)\,\mathrm{d}x,$$

which shows that (6.3.38) has only positive real eigenvalues. It then follows from Example 6.3.41 (and hence from the Krein–Rutman Theorem) that (6.3.38) has the least eigenvalue $\mu_1 > 0$ which is simple and which is the only eigenvalue of (6.3.38) having a positive eigenfunction $\varphi_1(x) > 0$, $x \in \Omega$. ○

Exercise 6.3.43. Show that if $\operatorname{int}\mathcal{K} \neq \emptyset$, then $\mathcal{K}$ is a total cone and construct an example of a cone which is not total.

Hint. If $y \in \operatorname{int}\mathcal{K}$, then $y \pm \alpha x \in \mathcal{K}$ for every $x \in X$ with $\alpha > 0$ sufficiently small. Thus $X = \mathcal{K} - \mathcal{K}$ because

$$x = \frac{(y + \alpha x) - (y - \alpha x)}{2\alpha}.$$

Exercise 6.3.44. Show that for every $x \in \mathcal{K} \setminus \{o\}$, there exists an $x^* \in X^*$ such that $\langle x^*, x\rangle > 0$.

Hint. Since $-x \notin \mathcal{K}$ and $\mathcal{K}$ is closed, $-x$ is an exterior point of $\mathcal{K}$. Consequently, there is an open convex neighborhood $\mathcal{U}$ of $-x$ which is disjoint from $\mathcal{K}$. By the Separation Theorem for convex sets,[31] there is an $x^* \in X^*$ with $x^*(\mathcal{K}) \geq 0$ and $x^*(\mathcal{U}) < 0$. Hence $\langle x^*, x\rangle > 0$.

[31]This is a minor supplement of Corollary 2.1.18.

Exercise 6.3.45. Show that if $\mathcal{K}$ is total, then $\mathcal{K}^*$ is an order cone on X^*.

Hint. $\mathcal{K} \neq \{o\}$ implies $\mathcal{K}^* \neq \{o\}$ by Exercise 6.3.44. Suppose $\pm x^* \in \mathcal{K}^*$. We have to show that $x^* = o$. Indeed, $\langle x^*, \pm x\rangle \geq 0$ for all $x \in \mathcal{K}$ implies $\langle x^*, x\rangle \geq 0$ for all $x \in X$, because $\mathcal{K}$ is total. Hence $x^* = o$.

Exercise 6.3.46. Let $x^* \in X^*$. Show that if $x^* > o$ (i.e., $x^* \geq o$ and $\langle x^*, y\rangle > 0$ for a $y > o$), then $\langle x^*, x\rangle > 0$ for all $x \in \operatorname{int}\mathcal{K}$.

Hint. Suppose $\langle x^*, x\rangle = 0$ for an $x \in \operatorname{int}\mathcal{K}$. Then $x \pm \alpha y \in \mathcal{K}$ for sufficiently small $\alpha > 0$. Hence $\langle x^*, x \pm \alpha y\rangle \geq 0$, so $\langle x^*, y\rangle = 0$. This is a contradiction.

Exercise 6.3.47. Prove that the functional $v \mapsto \alpha_u(v)$ from Lemma 6.3.37 is continuous.

Exercise 6.3.48. Apply the Krein–Rutman Theorem to the problems from Examples 2.1.32 and 2.2.17.

6.3B Supersolutions, Subsolutions and Topological Degree

In this appendix we show the connection between the supersolution and subsolution on the one hand and the topological degree on the other. We consider the quasilinear boundary value problem

$$\begin{cases} -(|\dot{x}(t)|^{p-2}\dot{x}(t))^{\cdot} = f(t, x(t)), & t \in (0,1), \\ x(0) = x(1) = 0, \end{cases} \tag{6.3.39}$$

as a model example. A special case of it was studied in Examples 5.8.38 and 6.1.23. However, in this appendix we work in different function spaces. Here $p > 1$ is a real number and $f\colon [0,1] \times \mathbb{R} \to \mathbb{R}$ is a function the properties of which will be specified later.

By a *solution* of (6.3.39) we understand a function $x \in C^1[0,1]$ with $x(0) = x(1) = 0$ such that $|\dot{x}|^{p-2}\dot{x}$ is absolutely continuous and the equation in (6.3.39) holds a.e. in $(0,1)$.

Clearly, the problem (6.3.39) formally coincides with (6.3.6) if $p = 2$.

Definition 6.3.49. A function $u_0 \in C^1[0,1]$ with $|\dot{u}_0|^{p-2}\dot{u}_0$ absolutely continuous is called a *subsolution* of (6.3.39) if

$$u_0(0) \leq 0, \qquad u_0(1) \leq 0$$

and

$$-(|\dot{u}_0(t)|^{p-2}\dot{u}_0(t))^{\cdot} \leq f(t, u_0(t)) \qquad \text{for a.e.} \quad t \in (0,1).$$

In the analogous way we define a *supersolution* v_0 of (6.3.39).

We write $x \ll y$ if and only if

$$x(t) < y(t), \qquad t \in (0,1),$$

and

$$\text{either} \quad x(0) < y(0) \quad \text{or} \quad x(0) = y(0) \ \text{ and } \ \dot{x}(0) < \dot{y}(0),$$

and the same alternatives hold at 1.[32]

Definition 6.3.50. A *subsolution* u_0 of (6.3.39) is said to be *strict* if every possible solution x of (6.3.39) such that $u_0 \leq x$ on $[0,1]$ satisfies $u_0 \ll x$.

In the analogous way we define a *strict supersolution* of (6.3.39).

[32] Here $\dot{x}(0)$ and $\dot{x}(1)$ mean the derivative from the right and from the left, respectively.

Let us formulate (6.3.39) as a "fixed point" operator equation. Assume that for any

$$y \in C_0^1[0,1] := \{x \in C^1[0,1] : x(0) = x(1) = 0\}$$

we have

$$f(t, y(t)) \in L^\infty(0,1).$$

We denote by $T\colon C_0^1[0,1] \to C_0^1[0,1]$ the *solution operator* of

$$\begin{cases} -(|\dot{x}(t)|^{p-2}\dot{x}(t))^\cdot = f(t, y(t)), & t \in (0,1), \\ x(0) = x(1) = 0, \end{cases} \tag{6.3.40}$$

i.e., for $x, y \in C_0^1[0,1]$,

$$x = T(y)$$

if and only if the equation in (6.3.40) holds a.e. in $(0,1)$. For any fixed $y \in C_0^1[0,1]$ it follows by integration of (6.3.40) and the injectivity of $\varphi(s) = |s|^{p-2}s$ that the operator T is well defined.

Clearly, the problem (6.3.39) has a solution x if and only if

$$x = T(x),$$

i.e., x is a fixed point of T.

Let f be a Carathéodory function and for any $r > 0$ let there exist a constant $h_r > 0$ such that for a.e. $t \in (0,1)$ and for all $s \in (-r, r)$,

$$|f(t,s)| < h_r. \tag{6.3.41}$$

This condition is satisfied if, e.g., $f(t, x(t)) = h(t) - g(x(t))$ where $h \in L^\infty(0,1)$ and $g\colon \mathbb{R} \to \mathbb{R}$ is a continuous function (cf. Examples 5.8.38 and 6.1.23).

We prove that the operator T is compact. For that purpose we express T in the integral form. By the Rolle Theorem for any $x = T(y)$ there exists $t_x \in [0,1]$ such that $\dot{x}(t_x) = 0$, i.e.,

$$\dot{x}(t) = \left|\int_t^{t_x} f(\tau, y(\tau))\,\mathrm{d}\tau\right|^{p'-2} \int_t^{t_x} f(\tau, y(\tau))\,\mathrm{d}\tau \tag{6.3.42}$$

and

$$x(t) = \int_0^t \left(\left|\int_\sigma^{t_x} f(\tau, y(\tau))\,\mathrm{d}\tau\right|^{p'-2} \int_\sigma^{t_x} f(\tau, y(\tau))\,\mathrm{d}\tau\right) \mathrm{d}\sigma \tag{6.3.43}$$

where $p' = \frac{p}{p-1}$. If $y_n \to y_0$ in $C_0^1[0,1]$, then the continuity of the Nemytski operator

$$y \mapsto f(\cdot, y) \tag{6.3.44}$$

from $C[0,1]$ into $C[0,1]$, and (6.3.42), (6.3.43) imply that $x_n \to x_0$ in $C_0^1[0,1]$ where $x_n = T(y_n)$, $x_0 = T(y_0)$, i.e., T is continuous. Let $\mathcal{M} \subset C_0^1[0,1]$ be a bounded set. To prove the compactness of T we have to show that $T(\mathcal{M})$ is relatively compact. Let $\{x_n\}_{n=1}^\infty \subset T(\mathcal{M})$ be an arbitrary sequence, $x_n = T(y_n)$, $y_n \in \mathcal{M}$. It follows from the compact embedding $C_0^1[0,1] \hookrightarrow\hookrightarrow C[0,1]$ (see Theorem 1.2.13) that there exists a subsequence $\{y_{n_k}\}_{k=1}^\infty \subset \{y_n\}_{n=1}^\infty$ which converges uniformly on $[0,1]$. But the continuity of the Nemytski operator (6.3.44) and (6.3.42), (6.3.43) imply that $\{x_{n_k}\}_{k=1}^\infty$ converges in $C_0^1[0,1]$, i.e., $T(\mathcal{M})$ is relatively compact. Hence the compactness of T is proved.

The following assertion is referred to as a *well-ordered case* of supersolution and subsolution.

Theorem 6.3.51 (well-ordered case). *Let f be a Carathéodory function satisfying* (6.3.41). *Assume that u_0 and v_0 are a subsolution and a supersolution of* (6.3.39), *respectively, with $u_0 \le v_0$ (see Figure* 6.3.4*). Then the problem* (6.3.39) *has at least one solution x satisfying*

$$u_0 \le x \le v_0 \qquad \text{in} \quad [0,1].$$

If, moreover, u_0 and v_0 are strict and satisfy $u_0 \ll v_0$, then there exists $R_0 > 0$ such that for, all $R > R_0$,

$$\deg(I - T, \Omega_1, o) = 1 \qquad \text{where} \quad \Omega_1 := \{x \in C_0^1[0,1] : u_0 \ll x \ll v_0\} \cap B(o; R),$$

is an open set in $C_0^1[0,1]$ (cf. Exercise 6.3.55*).*

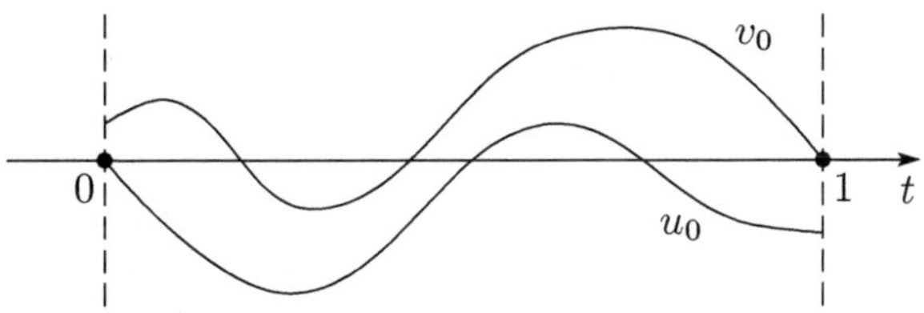

Figure 6.3.4. Well-ordered case

Proof. Set

$$\tilde{f}(t,y) := \begin{cases} f(t,y) & \text{if} \quad u_0(t) \le y \le v_0(t), \\ f(t,u_0(t)) & \text{if} \quad y \le u_0(t), \\ f(t,v_0(t)) & \text{if} \quad y \ge v_0(t). \end{cases}$$

Every solution of

$$\begin{cases} -(|\dot{x}(t)|^{p-2}\dot{x}(t))^{\cdot} = \tilde{f}(t,x(t)), & t \in (0,1), \\ x(0) = x(1) = 0, \end{cases} \tag{6.3.45}$$

is a solution of (6.3.39). Indeed, assume that x solves (6.3.45) and $x > v_0$ in an interval $\mathcal{I}_+ \subset (0,1)$ and $x = v_0$ on $\partial\mathcal{I}_+$. Then

$$\int_0^1 \left|\frac{\mathrm{d}x(t)}{\mathrm{d}t}\right|^{p-2} \frac{\mathrm{d}x(t)}{\mathrm{d}t}\frac{\mathrm{d}}{\mathrm{d}t}(x(t) - v_0(t))^* \,\mathrm{d}t = \int_0^1 f(t,v_0(t))(x(t) - v_0(t))^* \,\mathrm{d}t \tag{6.3.46}$$

where

$$(x(t) - v_0(t))^* = \begin{cases} x(t) - v_0(t) & \text{on} \quad \mathcal{I}_+ \\ 0 & \text{on} \quad [0,1] \setminus \mathcal{I}_+. \end{cases}$$

Since v_0 is a supersolution, we have

$$\int_0^1 \left|\frac{\mathrm{d}v_0(t)}{\mathrm{d}t}\right|^{p-2} \frac{\mathrm{d}v_0(t)}{\mathrm{d}t}\frac{\mathrm{d}}{\mathrm{d}t}(x(t) - v_0(t))^* \,\mathrm{d}t \ge \int_0^1 f(t,v_0(t))(x(t) - v_0(t))^* \,\mathrm{d}t. \tag{6.3.47}$$

Hence, combining (6.3.46) and (6.3.47), we obtain

$$\int_{\mathcal{I}_+} (|\dot{x}(t)|^{p-2}\dot{x}(t) - |\dot{v}_0(t)|^{p-2}\dot{v}_0(t))(\dot{x}(t) - \dot{v}_0(t))\,\mathrm{d}t \le 0.$$

This is a contradiction,[33] which proves that

$$x(t) \le v_0(t), \qquad t \in (0,1).$$

The same argument shows that

$$x(t) \ge u_0(t), \qquad t \in (0,1).$$

Now, denote by $\tilde{T}(y)$ a solution of the boundary value problem

$$\begin{cases} -(|\dot{x}(t)|^{p-2}\dot{x}(t))^{\cdot} = \tilde{f}(t, y(t)), & t \in (0,1), \\ x(0) = x(1) = 0 \end{cases}$$

for $y \in C_0^1[0,1]$. Then $\tilde{T}\colon C_0^1[0,1] \to C_0^1[0,1]$ is compact[34] and the solutions of (6.3.45) are in a one-to-one correspondence with the fixed points of $\tilde{T}$. The definition of $\tilde{f}$ ensures that there exists a constant $R_0 > 0$ such that for any $y \in C_0^1[0,1]$ we have

$$\|\tilde{T}(y)\|_{C_0^1[0,1]} < R_0 \tag{6.3.48}$$

(see (6.3.42), (6.3.43)). By the Schauder Fixed Point Theorem $\tilde{T}$ has a fixed point x in $B(o; R_0)$, i.e., x is a solution of (6.3.45). It follows from the above considerations that $u_0 \le x \le v_0$, and so x is also a desired solution of (6.3.39).

The proof of the second part follows from the fact that due to (6.3.48), we can construct an admissible homotopy

$$H(\tau, \cdot) := I - \tau\tilde{T}, \qquad \tau \in [0,1],$$

which shows that

$$\deg(I - \tilde{T}, B(o; R_0), o) = \deg(I, B(o; R_0), o) = 1.$$

Since u_0 and v_0 are strict and there is no solution x of the equation

$$x - \tilde{T}(x) = o$$

for which either $x(t) < u_0(t)$ or $x(t) > v_0(t)$ for a $t \in (0,1)$, it follows from Theorem 5.8.2(iv) that

$$\deg(I - \tilde{T}, \Omega_1, o) = \deg(I - \tilde{T}, B(o; R_0), o) = 1.$$

The assertion now follows from the fact that T and $\tilde{T}$ coincide in Ω_1. □

The next assertion is referred to as a *non-well-ordered case* of a supersolution and a subsolution.

[33] Note that $s \mapsto |s|^{p-2}s$ is a strictly increasing function!

[34] The proof of this fact is the same as that for T.

Theorem 6.3.52 (non-well-ordered case). *Let f be a Carathéodory function which satisfies the following assumption:*

there are $c_i > 0$, $i = 1, 2$, such that

$$|f(t,s)| \le c_1 + c_2|s|^{p-1} \qquad \text{for a.e.} \quad t \in (0,1) \quad \text{and for all} \quad s \in \mathbb{R} \tag{6.3.49}$$

and, moreover,

$$\lim_{|s|\to\infty} \frac{f(t,s)}{|s|^{p-2}s} = \lambda_1.^{35} \tag{6.3.50}$$

Assume that u_0 and v_0 are a subsolution and a supersolution of (6.3.39)*, respectively, and there exists t_0 such that $u_0(t_0) > v_0(t_0)$ (see Figure* 6.3.5*).*

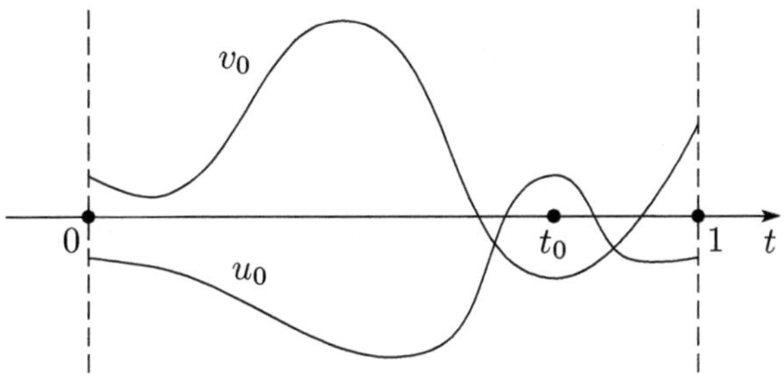

Figure 6.3.5. Non-well-ordered case.

Then (6.3.39) *has at least one solution in the closure (with respect to the C^1-norm) of the set*

$$\mathcal{S} := \{x \in C_0^1[0,1] : \exists t_1, t_2 \in (0,1),\ x(t_1) < u_0(t_1),\ x(t_2) > v_0(t_2)\}.$$

Set $\Omega_2 := \mathcal{S} \cap B(o; R)$ and assume that there is no solution of (6.3.39) *on $\partial\Omega_2$. Then there exists $R_0 > 0$ such that for all $R > R_0$,*

$$\deg(I - T, \Omega_2, o) = -1.$$

Proof. If (6.3.39) has a solution on $\partial\mathcal{S}$, we are done. Let us assume in the sequel that (6.3.39) does not have a solution on $\partial\mathcal{S}$. For $r > 0$ let us define

$$f_r(t,y) = \begin{cases} f(t,y) & \text{if} \quad |y| < r, \\ (1 + r - |y|)f(t,y) & \text{if} \quad r < |y| < r+1, \\ 0 & \text{if} \quad |y| > r+1. \end{cases}$$

Next we show that there is $K > 0$ such that for any $r > 0$ and for any possible solution of

$$\begin{cases} -(|\dot{x}(t)|^{p-2}\dot{x}(t))^\cdot = f_r(t, x(t)), & t \in (0,1), \\ x(0) = x(1) = 0, \end{cases} \tag{6.3.51}$$

[35] Here λ_1 is the first eigenvalue of (5.8.38), see Example 5.8.38.

the following a priori estimate holds:

$$\|x\|_{C_0^1[0,1]} \le K. \tag{6.3.52}$$

To prove this fact we argue by contradiction, and thus we assume that for any $k \in \mathbb{N}$ there are $r_k > 0$, $x_k \in \mathcal{S}$ solving

$$\begin{cases} -(|\dot{x}_k(t)|^{p-2}\dot{x}_k(t))^{\cdot} = f_{r_k}(t, x_k(t)), & t \in (0,1), \\ x_k(0) = x_k(1) = 0, \end{cases} \tag{6.3.53}$$

and satisfying $\|x_k\| \ge k$. Set $y_k := \frac{x_k}{\|x_k\|}$ and divide (6.3.51) by $\|x_k\|^{p-1}$ to obtain

$$\begin{cases} -(|\dot{y}_k(t)|^{p-2}\dot{y}_k(t))^{\cdot} = \dfrac{f_{r_k}(t, x_k(t))}{\|x_k\|^{p-1}}, & t \in (0,1), \\ y_k(0) = y_k(1) = 0. \end{cases}$$

By integration we find that $\{y_k\}_{k=1}^{\infty}$ equivalently satisfies

$$\dot{y}_k(t) = \varphi_{p'}\left(\varphi_p(\dot{y}_k(0)) + \int_0^t \frac{f_{r_k}(\sigma, x_k(\sigma))}{\|x_k\|^{p-1}}\,\mathrm{d}\sigma\right) \tag{6.3.54}$$

and

$$y_k(t) = \int_0^t \varphi_{p'}\left(\varphi_p(\dot{y}_k(0)) + \int_0^\tau \frac{f_{r_k}(\sigma, x_k(\sigma))}{\|x_k\|^{p-1}}\,\mathrm{d}\sigma\right)\mathrm{d}\tau, \qquad t \in [0,1], \tag{6.3.55}$$

where for $s > 1$ we set $\varphi_s(\xi) = |\xi|^{s-2}\xi$ if $\xi \ne 0$ and $\varphi_s(0) = 0$.

Now, since $\|y_k\| = 1$, by passing to a subsequence if necessary, we have

$$y_k \to y \quad \text{in} \quad C_0[0,1] := \{x \in C[0,1] : x(0) = x(1) = 0\} \quad \text{for a} \quad y \in C_0[0,1].$$[36]

But then (6.3.54) yields

$$y_k \to y \quad \text{in} \quad C_0^1[0,1]$$

(note that without loss of generality we may also assume that $\{\dot{y}_k(0)\}_{k=1}^{\infty}$ forms a convergent sequence!). It follows from (6.3.55), (6.3.49), (6.3.50) and the Lebesgue Dominated Convergence Theorem that y solves the problem

$$\begin{cases} -(|\dot{y}(t)|^{p-2}\dot{y}(t))^{\cdot} = \lambda_1 |y(t)|^{p-2}y(t), & t \in (0,1), \\ y(0) = y(1) = 0. \end{cases}$$

Since $\|y\| = 1$, it follows that y is a nonzero multiple of the first eigenfunction $\varphi_1(t) > 0$ in $(0,1)$ (see Example 5.8.38). If $y > 0$ in $(0,1)$, then we find that $x_k(t) \to \infty$ for any $t \in (0,1)$, which contradicts $x_k \in \mathcal{S}$. Also $y < 0$ in $(0,1)$ leads to a contradiction. Hence the a priori estimate (6.3.52) is proved.

Now choose

$$R > R_0 = \max\{K, \|u_0\|_{C[0,1]}, \|v_0\|_{C[0,1]}\} + 1$$

[36] This is a consequence of the Arzelà–Ascoli Theorem.

and consider (6.3.51) with $r = R$, i.e.,

$$\begin{cases} -(|\dot{x}(t)|^{p-2}\dot{x}(t))^{\cdot} = f_R(t, x(t)), & t \in (0,1), \\ x(0) = x(1) = 0. \end{cases} \tag{6.3.56}$$

Obvious modifications of the definition of a strict subsolution and supersolution of (6.3.39) lead to the same notions associated with (6.3.56). Then $\alpha = -R-2$ and $\beta = R+2$ are a subsolution and a supersolution, respectively, associated with (6.3.56). Both are actually strict. Indeed, assume, e.g., that x is a solution of (6.3.56), $x(t) \geq -R-2$ and $x(t_0) = -R-2$ for a certain $t_0 \in (0,1)$. Then $x(t_0) = \min\limits_{\tau \in (0,1)} x(\tau)$, i.e., $\dot{x}(t_0) = 0$ and there exists $\eta > 0$ such that $x(t) < -R-1$ for $t \in [t_0, t_0 + \eta)$. But $f_R(t, x(t)) = 0$ by definition, so $x(t) \equiv -R-2$ in $(t_0, t_0+\eta]$. Since this implies that $x(t) \equiv -R-2$ in $(t_0, 1]$, we obtain a contradiction. The same argument applies to $R+2$. Notice also that $\alpha \ll v_0$ and $u_0 \ll \beta$.

Now, let us define $T_R \colon C_0^1[0,1] \to C_0^1[0,1]$ by

$$x := T_R(y)$$

where x is a solution of the problem

$$\begin{cases} -(|\dot{x}(t)|^{p-2}\dot{x}(t))^{\cdot} = f_R(t, y(t)), & t \in (0,1), \\ x(0) = x(1) = 0, \end{cases}$$

and define the sets

$$\mathcal{S}_{\alpha\beta} := \{x \in C_0^1[0,1] : \alpha \ll x \ll \beta\},$$

$$\mathcal{S}_{u_0\beta} := \{x \in C_0^1[0,1] : u_0 \ll x \ll \beta\} \quad \text{and} \quad \mathcal{S}_{\alpha v_0} := \{x \in C_0^1[0,1] : \alpha \ll x \ll v_0\}$$

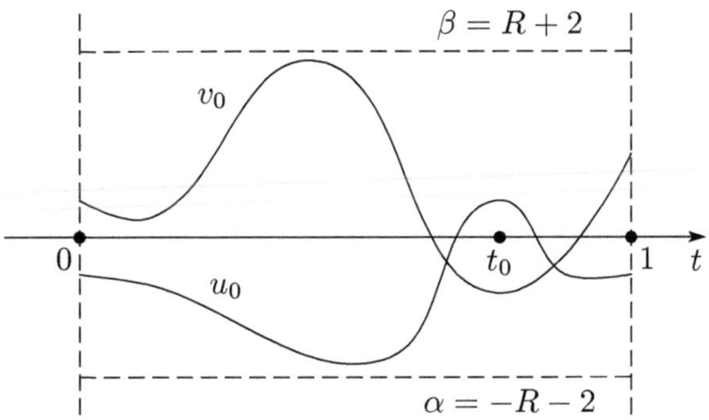

Figure 6.3.6.

(see Figure 6.3.6). By definition, T_R and T coincide in the ball $B(o;R)$. Applying Theorem 6.3.51 and Theorem 5.8.2(iv) we obtain

$$\begin{aligned} 1 &= \deg(I - T_R, B(o;R) \cap \mathcal{S}_{\alpha\beta}, o) \\ &= \deg(I - T_R, B(o;R) \cap \mathcal{S}_{\alpha v_0}, o) + \deg(I - T_R, B(o;R) \cap \mathcal{S}_{u_0\beta}, o) \\ &\quad + \deg(I - T_R, \Omega_2, o) = 2 + \deg(I - T_R, \Omega_2, o), \end{aligned}$$

which completes the proof. □

Remark 6.3.53. There are several applications of Theorems 6.3.51 and 6.3.52. Also generalizations of these results to the case of partial differential equations can be found in literature, see, e.g., Drábek, Girg & Manásevich [49].

In the next assertion we present one application of Theorems 6.3.51 and 6.3.52 which under suitable assumptions on f yields the multiplicity of solutions of (6.3.39).

Theorem 6.3.54. *Let f be as in Theorem 6.3.52 and let u_0^i and v_0^i, $i = 1, 2$, be subsolutions and supersolutions of (6.3.39), respectively, which satisfy*

$$u_0^1 \ll v_0^1, \qquad u_0^2 \ll v_0^2,$$

and let there exist $t_0 \in (0,1)$ such that

$$u_0^2(t_0) > v_0^1(t_0)$$

(see Figure 6.3.7).

Then the problem (6.3.39) has at least three distinct solutions.

Proof. It follows from Theorem 6.3.51 that there are solutions $x_i = x_i(t)$, $i = 1, 2$, of (6.3.39) which satisfy

$$u_0^1 \ll x_1 \ll v_0^1, \qquad u_0^2 \ll x_2 \ll v_0^2.$$

Now, let us apply Theorem 6.3.52 with a subsolution u_0^2 and a supersolution v_0^1. We get another solution $x_3 = x_3(t)$ of (6.3.39). Clearly, all x_i, $i = 1, 2, 3$, are mutually different. □

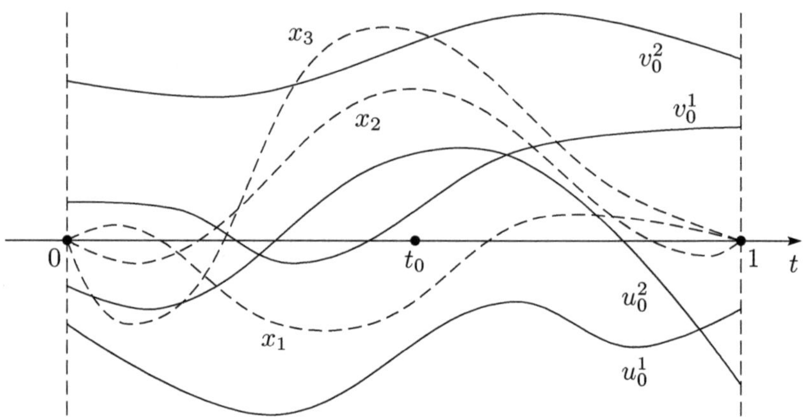

Figure 6.3.7.

Exercise 6.3.55. Prove that Ω_1 from Theorem 6.3.51 is an open set in $C_0^1[0,1]$.

Exercise 6.3.56. Formulate conditions on $f = f(t,x)$ which guarantee that the problem (6.3.39) has a pair of well-ordered supersolution and subsolution.

Exercise 6.3.57. Formulate conditions on $f = f(t, x)$ which guarantee that the problem (6.3.39) has a pair of non-well-ordered supersolution and subsolution.

Exercise 6.3.58. Formulate conditions on $f = f(t, x)$ which guarantee that the problem (6.3.39) has two pairs of supersolutions and subsolutions which satisfy the assumptions from Theorem 6.3.51.

6.4 Maximum Principle for ODEs

Let x be a continuous function in $[a, b]$ with the second derivative in (a, b). If x attains its maximum over $[a, b]$ at a point $c \in (a, b)$, then

$$\dot{x}(c) = 0 \quad \text{and} \quad \ddot{x}(c) \leq 0. \tag{6.4.1}$$

Let g be a bounded function on (a, b) and suppose that a function x is known to satisfy inequality

$$L(x)(t) := \ddot{x}(t) + g(t)\dot{x}(t) > 0 \tag{6.4.2}$$

in (a, b). It follows that (6.4.1) cannot be satisfied at any point $c \in (a, b)$. In other words, the maximal value of a function x cannot be attained at any interior point of the interval $[a, b]$. Hence, it must be attained either at a or at b. This is a very simple form of the *maximum principle*.

The strict inequality in (6.4.2) is essential in the argument above but may be restrictive in many applications. That is why we are going to prove the maximum principle with the nonstrict inequality in (6.4.2), i.e.,

$$L(x)(t) \geq 0 \quad \text{in} \quad (a, b). \tag{6.4.3}$$

We show that a constant solution of (6.4.3) is the only one which attains its maximum over $[a, b]$ at an interior point.

Lemma 6.4.1. *Let x be continuous on $[a, b]$ and satisfy (6.4.3) in (a, b), g be bounded in (a, b). If $M = \max_{t \in [a,b]} x(t)$ is attained at an interior point $c \in (a, b)$, then $x \equiv M$ on $[a, b]$.*

Proof. Let $c \in (a, b)$, $x(c) = M$ and there exists $d \in (a, b)$ such that $x(d) < M$. We derive a contradiction.

Assume first that $d > c$ and consider the function

$$z(t) = e^{\alpha(t-c)} - 1,$$

with $\alpha > 0$ to be specified later. Then $z(c) = 0$, $z(t) < 0$ for $a < t < c$ and $z(t) > 0$ for $c < t < b$. Elementary calculation yields

$$L(z)(t) = \ddot{z}(t) + g(t)\dot{z}(t) = \alpha\,(\alpha + g(t))\,e^{\alpha(t-c)}.$$

Since g is bounded, we can select our α so large that $\alpha > -g(t)$, $t \in (a,b)$. With this α we have $L(z) > 0$ in (a,b). Define

$$y(t) := x(t) + \varepsilon z(t), \tag{6.4.4}$$

where $\varepsilon > 0$ is chosen to satisfy

$$\varepsilon < \frac{M - x(d)}{z(d)}. \tag{6.4.5}$$

It follows from $z(t) < 0$, $t \in (a,c)$, (6.4.4) and (6.4.5) that

$$y(t) < M, \quad t \in (a,c).$$

Moreover,

$$y(d) = x(d) + \varepsilon z(d) < x(d) + M - x(d) = M$$

and

$$y(c) = x(c) + \varepsilon z(c) = M.$$

Hence

$$\widetilde{M} := \max_{t\in[a,d]} y(t) \geq M$$

and $\widetilde{M}$ is attained at some point $\xi \in (a,d)$. But due to the choice of α, for $t \in (a,b)$ we have

$$L(y)(t) = L(x)(t) + \varepsilon L(z)(t) > 0$$

and so the discussion prior to this lemma applies on y, i.e., we have a contradiction.

If $d < c$, then the auxiliary function

$$z(t) = e^{-\alpha(t-c)} - 1$$

with a positive $\alpha > g(t)$ allows to reach a contradiction in a similar way. $\square$

Remark 6.4.2. If Lemma 6.4.1 is applied to $-x$ then we have the *minimum principle*. It asserts that a nonconstant function x satisfying the differential inequality $L(x) \leq 0$ cannot attain its minimum at an interior point.

Remark 6.4.3. The boundedness condition on g may be relaxed. See Exercise 6.4.10.

Below we present a stronger version of Lemma 6.4.1 which provides an information about one-sided derivative of nonconstant solutions of (6.4.3) at the point of maximum (attained at the boundary point).

Lemma 6.4.4. *Let x be a nonconstant function which satisfies* (6.4.3) *in (a,b) and has one-sided derivatives at a and b. Suppose that g is bounded on every closed subinterval of (a,b). If the maximum of x occurs at $t = a$ and g is bounded below at the right neighborhood of a, then $\dot{x}_+(a) < 0$. If the maximum occurs at $t = b$ and g is bounded above in the left neighborhood of b, then $\dot{x}_-(b) > 0$.*

Proof. Let $x(a) = M$, $x(t) \le M$, $t \in [a,b]$, and for some $d \in (a,b)$ we have $x(d) < M$.

Let us define an auxiliary function

$$z(t) = e^{\alpha(t-a)} - 1$$

with positive $\alpha > -g(t)$, $t \in [a,d]$. Then

$$L(z)(t) > 0, \quad t \in (a,d).$$

Define

$$y(t) := x(t) + \varepsilon z(t)$$

with ε chosen again so that

$$0 < \varepsilon < \frac{M - x(d)}{z(d)}.$$

Similarly as in the proof of Lemma 6.4.1, we have $L(y)(t) > 0$ in $[a,d]$ and hence the maximum of y in the interval $[a,d]$ must occur at one of the end points. Since

$$y(a) = M > y(d),$$

we have $\max\limits_{t\in[a,d]} y(t) = y(a)$.

Hence

$$0 \ge \dot{y}_+(a) = \dot{x}_+(a) + \varepsilon \dot{z}_+(a).$$

Since $\dot{z}_+(a) = \alpha > 0$, we get $\dot{x}_+(a) < 0$.

The proof is similar in the other case. □

The following example shows that the assumption of the boundedness of g on every closed subinterval of (a,b) is crucial.

Example 6.4.5. (see [121]). The equation

$$\ddot{x}(t) + g(t)\dot{x}(t) = 0,$$

where

$$g(t) = \begin{cases} -\frac{3}{t} & \text{for} \quad t \ne 0, \\ 0 & \text{for} \quad t = 0. \end{cases}$$

has the solution

$$x(t) = 1 - t^4.$$

Since x is not a constant and has its maximum at $t = 0$, Lemma 6.4.1 is violated on $[a,b] = [-1,1]$. Lemma 6.4.4 is then violated on $[a,b] = [0,1]$ as $\dot{x}(0) = 0$. ○

We focus now on more general differential inequality

$$(L+h)(x)(t) := \ddot{x}(t) + g(t)\dot{x}(t) + h(t)x(t) \geq 0. \tag{6.4.6}$$

It is easy to see that if

$$(L+h)(x)(t) > 0 \text{ and } h(t) \leq 0, \ t \in (a,b) \tag{6.4.7}$$

then x cannot have a *nonnegative maximum* in the interior point of $[a,b]$. This observation unables us to extend Lemma 6.4.1 and Lemma 6.4.4 choosing α in their proofs so large that $(L+h)(z) > 0$. In fact, we have

Theorem 6.4.6. *Let x be continuous on $[a,b]$ and satisfy the differential inequality* (6.4.6) *in an interval (a,b), g and h be bounded on every closed subinterval of (a,b) and $h(t) \leq 0$, $t \in (a,b)$. If x assumes a nonnegative maximum value M at an interior point $c \in (a,b)$, then $x \equiv M$ on $[a,b]$.*

Proof. Let c, d, α and z be as in the proof of Lemma 6.4.1, $[a',b'] \subset (a,b)$ be such that $c, d \in (a',b')$ and let $\alpha > 0$ be chosen so that

$$\alpha^2 - \alpha|g(t)| + h(t) > 0$$

for all $t \in [a',b']$. Then

$$\alpha^2 + \alpha g(t) + h(t)\left[1 - e^{-\alpha(t-c)}\right] > 0$$

$$\left(\text{or} \quad \alpha^2 - \alpha g(t) + h(t)\left[1 - e^{\alpha(t-c)}\right] > 0\right)$$

if $d > c$ (or $d < c$) and hence

$$(L+h)(z) > 0 \text{ in } (a',b').$$

The proof now follows as that of Lemma 6.4.1 on any subinterval $[a',b'] \subset (a,b)$ containing the points c and d in its interior (cf. Exercise 6.4.10). □

Corollary 6.4.7. *Let x satisfy* (6.4.6) *in (a,b), $h(t) \leq 0$, $t \in (a,b)$, x be continuous on $[a,b]$, g and h be bounded on every closed subinterval of (a,b). Let, moreover, $x(a) \leq 0$, $x(b) \leq 0$. Then either $x \equiv 0$ in $[a,b]$ or else $x(t) < 0$, $t \in (a,b)$.*

Example 6.4.8. Let us consider the boundary value problem (6.3.6) from Example 6.3.19. The key point here is the proof of monotonicity of $T : C[0,1] \to C[0,1]$. Recall that for $z_1 \leq z_2$ in $[0,1]$ we put $w = T(z_2) - T(z_1)$ and w is a solution of

$$\begin{cases} \ddot{w}(t) - cw(t) = f(t, z_2(t)) - f(t, z_1(t)) + c(z_2(t) - z_1(t)), \\ w(0) = w(1) = 0. \end{cases}$$

The choice of $c > 0$ guarantees that

$$f(t, z_2(t)) - f(t, z_1(t)) + c(z_2(t) - z_1(t)) \geq 0,$$

i.e., w is a solution to

$$\begin{cases} \ddot{w}(t) - cw(t) \geq 0, & t \in (0,1), \\ w(0) = w(1) = 0. \end{cases}$$

The maximum principle applied to $(-w)$ then yields $w \geq 0$ in $[0,1]$, i.e., $T(z_1) \leq T(z_2)$. ○

Example 6.4.9. Let us consider the boundary value problem

$$\begin{cases} \ddot{x}(t) + g(t)\dot{x}(t) + h(t)x(t) = f(t), & t \in (a,b), \\ x(a) = A, x(b) = B. \end{cases} \tag{6.4.8}$$

If x is a solution of (6.4.8), $h(t) \leq 0$, $t \in (a,b)$ and g, h are bounded on closed subintervals of (a,b), then x is *unique solution*. Indeed, let x_1 and x_2 be two solutions of (6.4.8) and set $x = x_1 - x_2$. Then x solves the problem

$$\begin{cases} \ddot{x}(t) + g(t)\dot{x}(t) + h(t)x(t) = 0, & t \in (a,b), \\ x(a) = x(b) = 0. \end{cases} \tag{6.4.9}$$

Theorem 6.4.6 implies that $x(t) \leq 0$, $t \in (a,b)$. But since $-x$ solves also (6.4.9) we may apply Theorem 6.4.6 on $-x$ as well. Therefore $x \equiv 0$ in $[a,b]$ and the uniqueness follows. ○

Simple equation $\ddot{x}(t) + x(t) = 0$ has two different solutions $x_1(t) = \sin t$ and $x_2(t) = 0$ in $(a,b) = (0,\pi)$. This fact illustrates the importance of the assumption $h(t) \leq 0$ in (a,b).

More general variants of the maximum principle can be found, e.g., in the book Protter & Weinberger [121].

Next we relate previous results to the notions of subsolution and supersolution introduced in Section 6.3. For this purpose consider again the Dirichlet problem (6.4.8). Assume that f, g and h are bounded functions and $h \leq 0$ in (a,b). We use Theorem 6.4.6 to obtain an upper (lower) bound for a solution x of (6.4.8) without any actual knowledge of x itself. We thus get an *a priori estimate*, which, combined with the method of sub- and supersolutions, may lead to the proof of the existence of a solution.

Let us assume that we can find a continuous function v_0 defined on $[a,b]$ which satisfies

$$\begin{cases} (L+h)(v_0)(t) \leq f(t), & t \in (a,b), \\ v_0(a) \geq A, \ v_0(b) \geq B. \end{cases} \tag{6.4.10}$$

i.e., v_0 is a *supersolution* of (6.4.8). Then the function $y_1 = x - v_0$ satisfies

$$\begin{cases} (L+h)(y_1)(t) \geq 0, & t \in (a,b), \\ y_1(a) \leq 0, \ y_1(b) \leq 0. \end{cases}$$

Theorem 6.4.6 applied to y_1 yields $y_1(t) \le 0$, $t \in [a,b]$. In other words $x(t) \le v_0(t)$ for $t \in [a,b]$. Then the supersolution v_0 of (6.4.8) is an *upper bound* for all possible solutions x of (6.4.8).

Similarly, a *lower bound* for x is obtained by finding a *subsolution* u_0 with the properties

$$\begin{cases} (L+h)(u_0)(t) \ge f(t), & t \in (a,b), \\ u_0(a) \le A, \ u_0(b) \le B. \end{cases} \tag{6.4.11}$$

Theorem 6.4.6 applied to $y_2 = u_0 - x$ then shows that $x(t) \ge u_0(t)$, $t \in [a,b]$.

In order to determine v_0 we may set

$$v_0(t) = C\left[2 - e^{-\alpha(t-a)}\right] \tag{6.4.12}$$

with C and α chosen suitably as follows. Let α be so large that

$$(L+h)\left(e^{-\alpha(t-a)}\right) = \left(\alpha^2 - \alpha g(t) + h(t)\right) e^{-\alpha(t-a)} > 0$$

for $a \le t \le b$. Set

$$k := \min_{t\in[a,b]} \left(\alpha^2 - \alpha g(t) + h(t)\right) e^{-\alpha(t-a)}$$

and choose C so that

$$C := \max\left\{A, B, \frac{1}{k}\max_{t\in[a,b]} (-f(t)), 0\right\}.$$

Show that with these constants C and α the function v_0 defined in (6.4.12) satisfies the differential inequality (6.4.10).

Similarly, we may set

$$u_0(t) = D\left[2 - e^{-\alpha(t-a)}\right],$$

where α is chosen as above and

$$D := \min\left\{A, B, \frac{1}{k}\min_{t\in[a,b]} (-f(t)), 0\right\}.$$

Verify that u_0 satisfies (6.4.11).

It follows from our construction that

$$D\left[2 - e^{-\alpha(t-a)}\right] \le x(t) \le C\left[2 - e^{-\alpha(t-a)}\right]$$

for $t \in [a,b]$. In particular, we get

$$|x(t)| \le 2\max\left\{|A|, |B|, \frac{1}{k}\max_{t\in[a,b]} |f(t)|\right\} \tag{6.4.13}$$

for $t \in [a, b]$.

In fact, this approach yields a *continuous dependence* of a solution of (6.4.8) *on the right-hand side* f and *on the boundary values* A and B. Indeed, if x is a solution of (6.4.8) and $\overline{x}$ is a solution of

$$\begin{cases} \ddot{x}(t) + g(t)\dot{x}(t) + h(t)x(t) = \overline{f(t)}, & t \in (a, b), \\ x(a) = \overline{A},\ x(b) = \overline{B} \end{cases}$$

then the difference $y = x - \overline{x}$ satisfies

$$\begin{cases} \ddot{y}(t) + g(t)\dot{y}(t) + h(t)y(t) = f(t) - \overline{f(t)}, & t \in (a, b), \\ y(a) = A - \overline{A},\ y(b) = B - \overline{B}. \end{cases}$$

Inequality (6.4.13) shows that

$$\|x - \overline{x}\|_{C[a,b]} \leq 2 \max \left\{ |A - \overline{A}|, |B - \overline{B}|, \frac{1}{k} \|f - \overline{f}\|_{C[a,b]} \right\}, \ t \in [a, b].$$

Similar estimates as above can be obtained for more general boundary conditions and without the assumption $h(t) \leq 0$, $t \in [a, b]$. See Protter & Weinberger [121] for details.

Exercise 6.4.10. Prove Lemma 6.4.1 under the assumption that g is a continuous function on (a, b) (and so, possibly unbounded).

Hint. Notice that g is bounded on every closed interval $[a', b'] \subset (a, b)$. Repeat the argument from the proof of Lemma 6.4.1 on any subinterval $[a', b']$ containing the points c and d in its interior.

Exercise 6.4.11. The function $x(t) = \cos t$ solves the equation $\ddot{x}(t) + x(t) = 0$ and yet x has a maximum at $t = 0$. Is this a contradiction with Theorem 6.4.6?

Exercise 6.4.12. Consider the boundary value problem

$$\begin{cases} \ddot{x}(t) + e^t \dot{x}(t) = -1, & t \in (0, 1), \\ x(0) = x(1) = 1. \end{cases}$$

Prove that a solution x has no minimum in $(0, 1)$ and show that $\dot{x}_+(0) > 0$, $\dot{x}_-(1) < 0$.

Exercise 6.4.13. The equation $\ddot{x}(t) - x(t) = 0$ has the solution $x(t) = -\left(e^t + e^{-t}\right)$ which attains its maximum value at $t = 0$. Does it contradict Theorem 6.4.6?

Exercise 6.4.14. Prove that if $\ddot{x}(t) + e^{3x(t)} = -t$, $t \in (0, 1)$, then x cannot attain a minimum in $(0, 1)$.

6.5 Maximum Principle for PDEs

Let u be a twice continuously differentiable function defined in a domain $\Omega \subset \mathbb{R}^N$. If the equation $\Delta u = 0$ is satisfied at each point of Ω, then u is called a *harmonic function* in Ω.

Suppose that u has a local maximum in the interior point of Ω. Then $\frac{\partial u}{\partial x_i} = 0$, $\frac{\partial^2 u}{\partial x_i^2} \le 0$, $i = 1, \dots, N$, at this point. In particular, we have $\Delta u \le 0$ there. In other words, if a function u satisfies the strict inequality $\Delta u > 0$ in Ω, then u cannot attain its maximum at any interior point of Ω. As in the previous section we shall prove the maximum principle for functions satisfying non-strict inequality $\Delta u \ge 0$ in Ω and show that a constant function is the only one which solves this inequality and attains its maximum over $\overline{\Omega}$ at an interior point, i.e., in Ω. For simplicity, we work with function u of two variables $u = u(x, y)$ which satisfies

$$\Delta u = \frac{\partial^2 u}{\partial x^2} + \frac{\partial^2 u}{\partial y^2} \ge 0 \tag{6.5.1}$$

in a planar domain $\Omega \subset \mathbb{R}^2$. A function u satisfying (6.5.1) is called *subharmonic* in Ω. If $(-u)$ is subharmonic, i.e., $\Delta u \le 0$ in Ω, then u is called *superharmonic* in Ω.

Let $B_r := B(x_0, y_0; r)$ be a disk centered at $(x_0, y_0) \in \Omega$ with radius $r > 0$ which entirely lies in Ω, i.e., $B_r \subset \Omega$. The *Divergence Theorem* (see Exercise 4.3.115) applied to the gradient of u in the disk B_r yields

$$\iint\limits_{B_r} \Delta u \, dxdy = \int\limits_{\partial B_r} \frac{\partial u}{\partial r} ds.\text{[37]} \tag{6.5.2}$$

In polar coordinates $x = x_0 + r\cos\theta$, $y = y_0 + r\sin\theta$, $ds = rd\theta$ the identity (6.5.2) reads as

$$\iint\limits_{B_r} \Delta u \, dxdy = r \int\limits_0^{2\pi} \frac{\partial u}{\partial r} d\theta.$$

Then $\Delta u \ge 0$ in Ω forces

$$\int\limits_0^{2\pi} \frac{\partial u}{\partial r} d\theta \ge 0. \tag{6.5.3}$$

Let $R > 0$ be such that $B_R \subset \Omega$ and let $r \in (0, R)$. Integrating (6.5.3) from 0 to R with respect to r and using the Fubini Theorem, we obtain

$$0 \le \int\limits_0^R \int\limits_0^{2\pi} \frac{\partial u}{\partial r} d\theta dr = \int\limits_0^{2\pi} u(R, \theta) d\theta - 2\pi u(x_0, y_0),$$

[37] For the definition of the integral see Definition 4.3.87.

i.e.,

$$u(x_0, y_0) \le \frac{1}{2\pi R} \int_0^{2\pi} u(R,\theta) R d\theta = \frac{1}{2\pi R} \int_{\partial B_R} u ds. \tag{6.5.4}$$

Since $(x_0, y_0) \in \Omega$ is arbitrary, the value of u at any point of Ω is bounded above by its mean value over any circle in Ω having that point as a center. If $\Delta u = 0$ in Ω, this inequality is true for both u and $-u$ and hence

$$u(x_0, y_0) = \frac{1}{2\pi R} \int_{\partial B_R} u ds.$$

(The Mean Value Theorem for a harmonic function in Ω.)

Proposition 6.5.1. *Let $\Delta u \ge 0$ in a domain Ω. If u attains its maximum M at a point of Ω then $u \equiv M$ in Ω.*

Proof. Let $(x_0, y_0) \in \Omega$, $u(x_0, y_0) = M$. Since $u \le M$ in Ω, we conclude from (6.5.4) that u must be identically equal to M on every circle centered at (x_0, y_0) and situated in Ω. Let $(x_1, y_1) \in \Omega$ and $u(x_1, y_1) < M$. Then the same is true in a neighborhood of (x_1, y_1). We connect (x_0, y_0) with (x_1, y_1) by a curve in Ω (Proposition 1.2.7) and let (x_2, y_2) be the last point on this curve where $u = M$ (see Figure 6.5.1). Then u is not identically equal to M on any sufficiently small

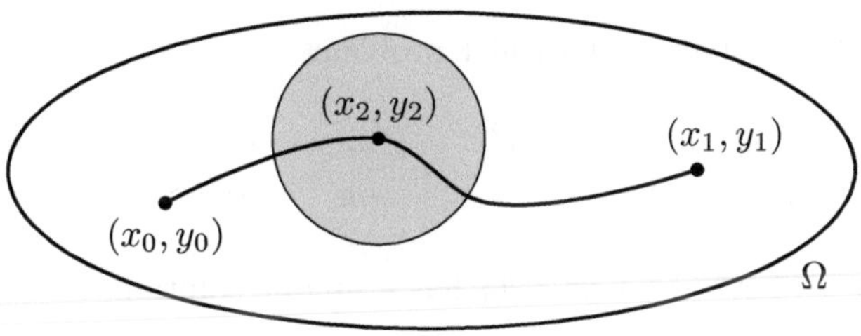

Figure 6.5.1.

circle centered at (x_2, y_2) which is a contradiction. □

Remark 6.5.2. Let Ω be a *bounded* domain. If u is subharmonic in Ω, continuous on $\overline{\Omega}$ and is not constant, then the values of u in Ω are below the maximum of u on $\partial\Omega$. If Ω is an *unbounded* domain, these values are either below the maximum of u on $\partial\Omega$ or below the lim sup of u as $(x^2 + y^2) \to \infty$.

Since a corresponding minimum principle holds for functions satisfying $\Delta u \le 0$, we have the following assertion.

Theorem 6.5.3. *A nonconstant harmonic function in Ω can attain neither its maximum nor its minimum at any interior point of Ω.*

Remark 6.5.4. Let u be subharmonic and v harmonic in Ω. If u and v coincide on $\partial\Omega$, then $w = u - v$ is subharmonic in Ω and vanishes on $\partial\Omega$. By the maximum principle $w \le 0$, i.e., $u \le v$ in Ω. This is the property where the notion *sub*harmonic comes from.

Remark 6.5.5. Proposition 6.5.1 and Theorem 6.5.3 have extensions in higher dimensions. Indeed, if $B_R := B(x_0; r)$ be a ball in $\mathbb{R}^N$ centered at $x_0 \in \Omega$ with radius $r > 0$ then

$$\text{meas}\, \partial B_r = \omega_N r^{N-1},$$

where ω_N is an absolute constant depending only on N (the measure of a unit sphere in $\mathbb{R}^N$). An analogue of (6.5.4) now reads as

$$u(x_0) \le \frac{1}{\omega_N R^{N-1}} \int_{\partial B_R} u dS. \tag{6.5.5}$$

The Mean Value Theorem for harmonic functions as well as the Maximum Principle for N-dimensional subharmonic functions are then direct consequences of (6.5.5).

The Maximum Principle holds also for more general elliptic second order partial differential operators, see, e.g., Protter & Weinberger [121].

Exercise 6.5.6. Prove in detail the mean value inequality (6.5.5).

Exercise 6.5.7. Show that the Dirichlet problem

$$\begin{cases} \Delta u = u^3 & \text{in} \quad D, \\ u = 0 & \text{on} \quad \partial D \end{cases}$$

where $D = \{(x, y) \in \mathbb{R}^2 : x^2 + y^2 < 1\}$ has a unique solution $u \equiv 0$.

Exercise 6.5.8. Prove that the Dirichlet problem

$$\begin{cases} \Delta u = f & \text{in} \quad \Omega, \\ u = g & \text{on} \quad \partial D \end{cases}$$

has at most one solution $u \in C^2(\Omega) \cap C(\overline{\Omega})$.

The heat flow in a finite bar of length $l > 0$, whose ends are held at time $t > 0$ at temperature $g(t)$ and $h(t)$, with the initial distribution of temperature at time $t = 0$ given by a continuous function $\varphi(x)$ is expressed by the initial boundary value problem

$$\begin{cases} k\frac{\partial^2 u}{\partial x^2} - \frac{\partial u}{\partial t} = f(x, t), \ x \in (0, l), \ t \ge 0, \\ u(x, 0) = \varphi(x), \\ u(0, t) = g(t), \ u(l, t) = h(t), \end{cases} \tag{6.5.6}$$

where unknown function $u = u(x,t)$ is the thermodynamic temperature at a point x and time t, the constant k depends on the material properties of the bar and the function f is the rate of heat exchange between the bar and a surrounding environment.

The so-called space-time cylinder $\mathcal{R}$ (parabolic cylinder) is defined as a closed rectangle in the xt-plane with one vertex placed at the origin $(0,0)$, and two sides lying on coordinate axes, see Figure 6.5.2.

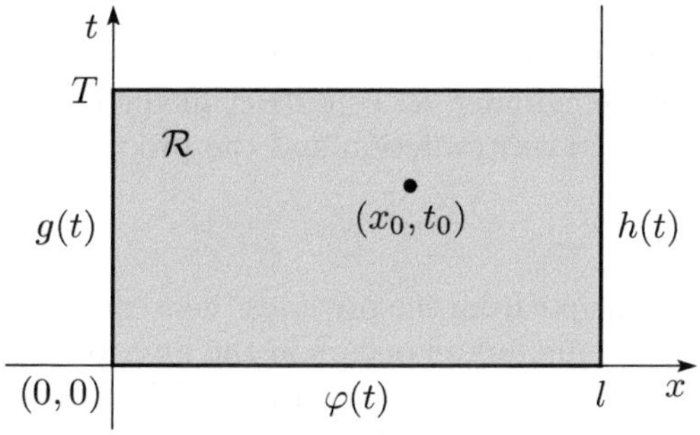

Figure 6.5.2.

The length of the side on the x-axis is l, the length of the side on the t-axis is T. By the bottom of the cylinder $\mathcal{R}$ we understand the horizontal side lying on the x-axis, by the cylinder jacket we understand both lateral sides. The upper horizontal line is then called the top of the cylinder $\mathcal{R}$.

Note that the notion of "cylinder" becomes for $\mathcal{R}$ more realistic when the heat equation is considered in more spatial variables.

Using a suitable change of variables we can assume that $k = 1$ in the equation (6.5.6).

Theorem 6.5.9. *Let $u = u(x,t)$ be continuous on $\mathcal{R}$ and satisfy differential inequality*

$$L(u) := \frac{\partial^2 u}{\partial x^2} - \frac{\partial u}{\partial t} \geq 0 \tag{6.5.7}$$

in the interior of the space-time cylinder $\mathcal{R}$. Then the maximum of u over $\mathcal{R}$ must occur either at the bottom or at the jacket of $\mathcal{R}$.

Proof. Let M be the maximal value of the function u on the sides $t = 0$, $x = 0$ and $x = l$. We assume that there is a point $(x_0, t_0) \in \operatorname{int} \mathcal{R}$ such that

$$M_1 = u(x_0, t_0) > M$$

and derive a contradiction. Set

$$w(x) := \frac{M_1 - M}{2l^2}(x - x_0)^2.$$

Due to $u \le M$ on the sides $t = 0$, $x = 0$ and $x = l$, we obtain

$$v(x,t) := u(x,t) + w(x) \le M + \frac{M_1 - M}{2} < M_1 \tag{6.5.8}$$

at the bottom and at the jacket of cylinder $\mathcal{R}$. Moreover, $v(x_0,t_0) = u(x_0,t_0) = M_1$ and

$$L(v) = L(u) + L(w) = L(u) + \frac{M_1 - M}{l^2} > 0 \tag{6.5.9}$$

in int $\mathcal{R}$. It follows from (6.5.8) that the maximum of v occurs either at an interior point or at the top of the cylinder $\mathcal{R}$. But strict inequality (6.5.9) rules out both options. This is the desired contradiction and the proof is complete. □

Remark 6.5.10.

(i) The Maximum Principle from the previous theorem is not one of the strongest forms. In fact, if the maximum occurs in the interior or at the top of $\mathcal{R}$ then the solution of (6.5.7) must be constant in $\mathcal{R}$. Theorem 6.5.9 is then a special case of this result (cf. Protter & Weinberger [121]).

(ii) For solutions of $L(u) \le 0$ we obtain an associated *minimum principle* when we replace u by $-u$.

(iii) Contrary to the elliptic differential inequalities, where the maximum of a solution could occur anywhere at the boundary, in case of the heat equation we have a stronger result. In fact, the maximum can occur only on a specified portion of the boundary (unless u is a constant). This fact is true also for more general partial differential equations of the parabolic type (the heat equation is a prototype) and for more general domains than $\mathcal{R}$. In particular, it is true also for solutions u of the differential inequality

$$\Delta u - \frac{\partial u}{\partial t} \ge 0,$$

where $u = (x_1, \dots, x_N, t)$ is a function of $N + 1$ independent variables (N spatial variables and time). See [121].

(iv) There is also an extension of the Maximum Principle to the case when the bottom of the cylinder $\mathcal{R}$ is an unbounded domain in $\mathbb{R}^N$. In particular, if u satisfies (6.5.7) for all $(x,t) \in (-\infty,\infty) \times (0,T)$, $u \le 0$ for $t = 0$, and moreover, there exists a constant $c > 0$ such that

$$\liminf_{R\to\infty} e^{-cR^2} \max_{\substack{|x|=R\\ t\in[0,T]}} u(x,t) \le 0,$$

then $u \le 0$ in the entire $(-\infty,\infty) \times (0,T)$.

The proof of this and even more general versions of this assertion can be found in [121].

Exercise 6.5.11. Prove that the initial boundary value problem (6.5.6) has at most one classical solution.

Exercise 6.5.12. Let u_1, u_2 be two classical solutions of the initial boundary value problem (6.5.6) corresponding to two initial conditions φ_1, φ_2. Prove that

$$\max_{x\in[0,l]} |u_1(x,t) - u_2(x,t)| \le \max_{x\in[0,l]} |\varphi_1(x) - \varphi_2(x)|,\ t \ge 0.$$

Exercise 6.5.13. Let u_1, u_2, φ_1 and φ_2 be as above. Prove that

$$\|u_1(\cdot,t) - u_2(\cdot,t)\|_{L^2(0,l)} \le \|\varphi_1 - \varphi_2\|_{L^2(0,l)}$$

holds for any $t > 0$.

Hint. Set $w = u_1 - u_2$. Then $\frac{\partial w}{\partial t} = k\frac{\partial^2 w}{\partial x^2}$ and $w(0,t) = w(l,t) = 0$. Hence

$$0 = \left(\frac{\partial w}{\partial t} - k\frac{\partial^2 w}{\partial x^2}\right) w = \frac{1}{2}\frac{\partial}{\partial t}\left(w^2\right) - k\frac{\partial}{\partial x}\left(\frac{\partial w}{\partial x} w\right) + k\left(\frac{\partial w}{\partial x}\right)^2.$$

Integrate and derive

$$\frac{d}{dt}\int_0^l \frac{1}{2} w^2(x,t)dx = -k\int_0^l \left(\frac{\partial w}{\partial t}(x,t)\right)^2 dx \le 0,$$

$$\int_0^l w^2(x,t)dx \le \int_0^l w^2(x,0)dx$$

for all $t > 0$.

Chapter 7

Variational Methods

7.1 Local Extrema

In this section we present necessary and/or sufficient conditions for local extrema of real functionals. The most famous ones are the Euler and Lagrange necessary conditions and the Lagrange sufficient condition. We also present the brachistochrone problem, one of the oldest problems in the calculus of variations. We also discuss regularity of the point of a local extremum. The methods presented in this section are motivated by the equation

$$f(x) = 0 \tag{7.1.1}$$

where f is a continuous real function defined in $\mathbb{R}$. The solution of this equation can be transformed to the problem of finding a local extremum of the integral F of f (i.e., $F'(x) = f(x)$, $x \in \mathbb{R}$). Indeed, if there exists a point $x_0 \in \mathbb{R}$ at which the function F has its local extremum, then the derivative $F'(x_0)$ necessarily vanishes due to a familiar theorem of the first-semester calculus. The problem of finding solutions of (7.1.1) can be thus transformed to the problem of finding local extrema of the function F. On the other hand, one should keep in mind that the equation (7.1.1) may have a solution which is not a local extremum of F.

In what follows we will deal with real functionals

$$F \colon X \to \mathbb{R}$$

where X is a normed linear space with the norm $\|\cdot\|$.

Definition 7.1.1. We say that F has a *local minimum* (*maximum*) at a point $a \in X$ if there exists a neighborhood $\mathcal{U}$ of a such that for all $x \in \mathcal{U} \setminus \{a\}$ we have

$$F(x) \geq F(a) \qquad (F(x) \leq F(a)).$$

If the inequalities are strict, we speak about a *strict local minimum* (*strict local maximum*). If the functional F has a (strict) local minimum or (strict) local maximum at a, we say that it has a (*strict*) *local extremum* at a.

In Figure 7.1.1 the critical point a is not a point of extremum of F.

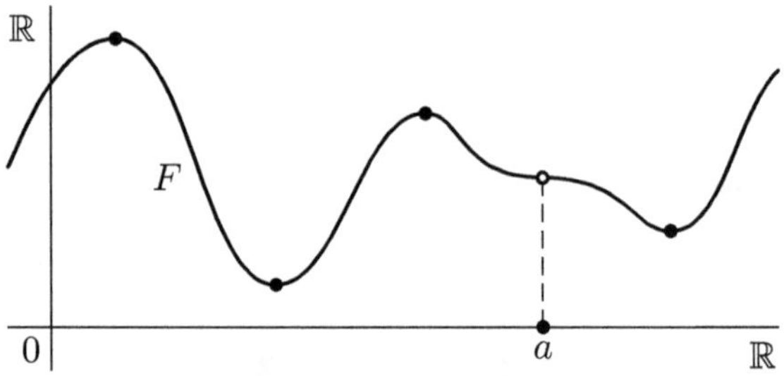

Figure 7.1.1.

The fundamental assertion is the following *Euler* (or *Fermat*) *Necessary Condition.*

Proposition 7.1.2 (Euler Necessary Condition). *Let $F\colon X \to \mathbb{R}$ have a local extremum at $a \in X$. If for $v \in X$ the derivative $\delta F(a;v)$ exists, then*

$$\delta F(a;v) = 0.$$

Proof. Set

$$g(t) = F(a+tv), \qquad t \in \mathbb{R}.$$

Then g attains a local minimum at $t=0$, thus

$$0 = g'(0) = \delta F(a;v). \qquad \square$$

Definition 7.1.3. If

$$\delta F(a;v) = 0 \qquad \text{for all} \quad v \in X,$$

then a is called a *critical point* of the functional F.[1]

The more precise *Lagrange Necessary Condition* distinguishes between local minima and maxima, but requires the existence of the second derivative in the given direction.

Proposition 7.1.4 (Lagrange Necessary Condition). *Let $F\colon X \to \mathbb{R}$ have a local minimum (maximum) at $a \in X$. If for $v \in X$ the second derivative $\delta^2 F(a;v,v)$ exists, then*

$$\delta^2 F(a;v,v) \geq 0 \qquad (\delta^2 F(a;v,v) \leq 0).$$

Proof. Let g be as in the proof of Proposition 7.1.2. Then

$$g''(0) = \delta^2 F(a;v,v).$$

Now we can apply the Lagrange necessary condition for local extrema of the real function g of one real variable to get the conclusion. $\square$

[1] Cf. Definition 4.3.6.

Contrary to Propositions 7.1.2 and 7.1.4, the *Lagrange Sufficient Condition* provides the information when a critical point of F is a point of its local minimum or local maximum.

Theorem 7.1.5 (Lagrange Sufficient Condition). *Let $a \in X$ be a critical point of $F\colon X \to \mathbb{R}$. Let there exist a neighborhood $\mathcal{U}$ of a such that the mapping $x \mapsto D^2F(x)$ is continuous in $\mathcal{U}$. If there exists $\alpha > 0$ such that*

$$D^2F(a)(v,v) \ge \alpha\|v\|^2 \qquad (D^2F(a)(v,v) \le -\alpha\|v\|^2) \qquad \textit{for any} \quad v \in X,$$

then F has a strict local minimum (maximum) at a.

Proof. Let $v \in X$ be such that $a + v \in \mathcal{U}$. Then according to Proposition 3.2.27 we have

$$F(a+v) - F(a) = \int_0^1 (1-t)D^2F(a+tv)(v,v)\,\mathrm{d}t.^2 \tag{7.1.2}$$

On the other hand,

$$\begin{aligned} D^2F(a+tv)(v,v) &\ge D^2F(a)(v,v) - |D^2F(a+tv)(v,v) - D^2F(a)(v,v)| \\ &\ge \left[\alpha - \|D^2F(a+tv) - D^2F(a)\|_{\mathcal{B}_2(X,\mathbb{R})}\right] \|v\|^2. \end{aligned}$$

The continuity of $D^2F(x)$ in $\mathcal{U}$ implies that there is $\delta > 0$ so small that for $\|v\| < \delta$, $t \in [0,1]$,

$$\|D^2F(a+tv) - D^2F(a)\|_{\mathcal{B}_2(X,\mathbb{R})} < \alpha, \tag{7.1.3}$$

i.e., for $0 < \|v\| < \delta$ we have (due to (7.1.2) and (7.1.3))

$$F(a+v) > F(a).$$

The proof for a strict local maximum is similar. □

Let us illustrate the general statements at first on a function of several real variables

$$F\colon \mathbb{R}^N \to \mathbb{R}.$$

Example 7.1.6. Let $F\colon \mathbb{R}^N \to \mathbb{R}$ have all partial derivatives of the first order at a point $a \in \mathbb{R}^N$ and, moreover, let the function F have a local extremum at a. Then Proposition 7.1.2 states that

$$\frac{\partial F}{\partial x_1}(a) = \frac{\partial F}{\partial x_2}(a) = \cdots = \frac{\partial F}{\partial x_N}(a) = 0. \tag{7.1.4}$$

On the other hand, it is well known that (7.1.4) does not imply that F has a local extremum at the point a. To check that this is the case we can apply Theorem 7.1.5.

[2]We can assume that $\mathcal{U}$ is convex. Then $D^2F(a+tv)$ exists and is continuous for all $t \in [0,1]$.

If F has continuous second partial derivatives in a neighborhood of a, then we should investigate the quadratic form

$$D^2F(a)(v,v) = \sum_{i,j=1}^{N} \frac{\partial^2 F}{\partial x_i \partial x_j}(a) v_i v_j. \tag{7.1.5}$$

To prove that F has, e.g., a local minimum at a, it is enough to show that there exists $\alpha > 0$ such that for any $v \in \mathbb{R}^N$, $\|v\| = 1$,

$$D^2F(a)(v,v) \geq \alpha. \tag{7.1.6}$$

(Here we have used the fact that the quadratic form is homogeneous.) Since we are in finite dimension, the unit sphere in $\mathbb{R}^N$ is a compact set. Then (7.1.6) holds with an $\alpha > 0$ whenever

$$D^2F(a)(v,v) > 0 \qquad \text{for all} \quad \|v\| = 1.^3 \tag{7.1.7}$$

The reader is invited to justify that (7.1.7) implies (7.1.6) and to explain why this is not the case when $\mathbb{R}^N$ is replaced by a space of infinite dimension.

It follows from linear algebra[4] that for any quadratic form on $\mathbb{R}^N$ there exists a basis $\{u_1, \dots, u_N\}$ of $\mathbb{R}^N$ and numbers $\lambda_1, \dots, \lambda_N$ such that for any v of the form

$$v = \sum_{i=1}^{N} \xi_i u_i$$

we have

$$D^2F(a)(v,v) = \sum_{i=1}^{N} \lambda_i \xi_i^2.$$

The inequality (7.1.7) holds if and only if all λ_i, $i = 1, \dots, N$, are positive, and so according to Theorem 7.1.5 the function F has a strict local minimum at a. If there is at least one positive and at least one negative number among λ_i, $i = 1, \dots, N$, then according to Proposition 7.1.4 the function F does not have a local extremum at a. ○

Before we give an application in an infinite-dimensional space, we prove the following assertion for convex functionals.

Definition 7.1.7. Let $\mathcal{M} \subset X$ be a convex set. A functional $F\colon X \to \mathbb{R}$ is said to be *convex in* $\mathcal{M}$ if for any $u, v \in \mathcal{M}$ and $t \in [0,1]$ we have

$$F(tu + (1-t)v) \leq tF(u) + (1-t)F(v).$$

[3] Here we use the fact that a positive continuous function on a compact set achieves its minimal value which has to be positive.

[4] See also Corollary 7.8.9. (Remember that $\frac{\partial^2 F}{\partial x_i \partial x_j}(a) = \frac{\partial^2 F}{\partial x_j \partial x_i}(a)$.)

The functional F is said to be *strictly convex in* $\mathcal{M}$ if for any $u, v \in \mathcal{M}$, $u \neq v$ and $t \in (0,1)$ we have

$$F(tu + (1-t)v) < tF(u) + (1-t)F(v).$$

Proposition 7.1.8. *Let $F\colon X \to \mathbb{R}$ be a convex functional on a normed linear space X. Then every critical point of F in X is a point of minimum of F over X.*

Proof. Without loss of generality we can assume that

$$F(o) = 0 \qquad \text{and} \qquad \delta F(o; v) = 0 \qquad \text{for any} \quad v \in X$$

(i.e., $o \in X$ is a critical point). Assume that F does not achieve the minimum value over X at $o \in X$. Then there exists $u \in X$ for which $F(u) = \alpha < 0$. The convexity of F implies that

$$F(tu + (1-t)o) \le t\alpha \qquad \text{for any} \quad t \in (0,1),$$

i.e.,

$$\frac{F(tu) - F(o)}{t} \le \alpha < 0. \tag{7.1.8}$$

But (7.1.8) implies

$$\delta F(o; u) \le \alpha < 0,$$

which is a contradiction. □

The following result will be needed several times in the further text.

Lemma 7.1.9 (Fundamental Lemma in Calculus of Variations). *Let $\mathcal{I}$ be an open interval and $f \in L^1_{\mathrm{loc}}(\mathcal{I})$. If*

$$\int_{\mathcal{I}} f(x)\varphi'(x)\,\mathrm{d}x = 0 \qquad \textit{for any} \quad \varphi \in \mathcal{D}(\mathcal{I}),^5 \tag{7.1.9}$$

then $f = $ const. a.e. in $\mathcal{I}$.

Proof. Let $\mathcal{J}$ be a compact subinterval of $\mathcal{I}$ and φ a mollifier, $\varphi \in \mathcal{D}(\mathbb{R})$, $\operatorname{supp}\varphi \subset [-1,1]$ (see Proposition 1.2.20(iv)). For

$$g(x) = \begin{cases} f(x), & x \in \mathcal{J}, \\ 0, & x \in \mathbb{R} \setminus \mathcal{J}, \end{cases}$$

we have $g \in L^1(\mathbb{R})$, and thus

$$\lim_{n\to\infty} g * \varphi_n = g \qquad \text{in the } L^1(\mathbb{R})\text{-norm}^6$$

[5] See page 36 for the definition of $\mathcal{D}(\mathcal{I})$.

and (passing to a subsequence – cf. Remark 1.2.18) also a.e. in $\mathbb{R}$. Since

$$(g * \varphi_n)'(y) = \int_{\mathbb{R}} g(x)\varphi_n'(y-x)\,\mathrm{d}x = \int_{\mathcal{I}} f(x)\varphi_n'(y-x)\,\mathrm{d}x$$

whenever $\left(y - \frac{1}{n}, y + \frac{1}{n}\right) \subset \mathcal{J}$, by the assumption (7.1.9), $(g * \varphi_n)(y)$ is constant for all such y. The convergence of $g * \varphi_n$ to g means that g is constant a.e. in $\mathcal{J}$, i.e., $f = $ const. a.e. in $\mathcal{I}$. For a slightly different proof see Lemma 3.1.16. □

One of the oldest problems in the calculus of variations is studied in detail in the following example.

Example 7.1.10 (Brachistochrone Problem). The problem is formulated as follows:

> "For given two points A and B in a vertical plane find a curve connecting A and B which is optimal among all other such curves in the following sense. The point P of unit mass which starts from A with zero velocity and moves along this curve only due to the gravitational force will reach the point B in a minimal time."[7]

In order to find a suitable mathematical model we shall assume that the points $A = (0,0)$ and $B = (a,b)$, $b \geq 0$, are situated in a vertical plane with the coordinate system chosen as in Figure 7.1.2. The reader is invited to verify that such a position of A and B can be considered without loss of generality. We shall concentrate first only on curves which are graphs of nonnegative functions $y = u(x)$ which belong to the space $C^1[0,a]$.

The point P moves according to the second Newton Law. The resulting force is a composition of the gravitational force and the reaction force of the constraint (the point P moves along the given curve). The resulting direction is given by the tangent line of the curve, see Figure 7.1.2.

The Second Newton Law says that for the velocity v of the point P the following identity holds:

$$m\dot{v} = F = mg\cos\alpha$$

(see Figure 7.1.2). Multiplying this identity by v and taking into account that $\dot{x} = v\cos\alpha$, we obtain

$$\left(\frac{1}{2}v^2\right)^{\cdot} = gv\cos\alpha = g\dot{x},$$

i.e.,

$$\frac{1}{2}v^2 = gx \tag{7.1.10}$$

(the Principle of Conservation of Energy).

[6] φ_n is defined in Proposition 1.2.20(iv).

[7] This problem was posed by Johann Bernoulli (see Berkovitz [15]).

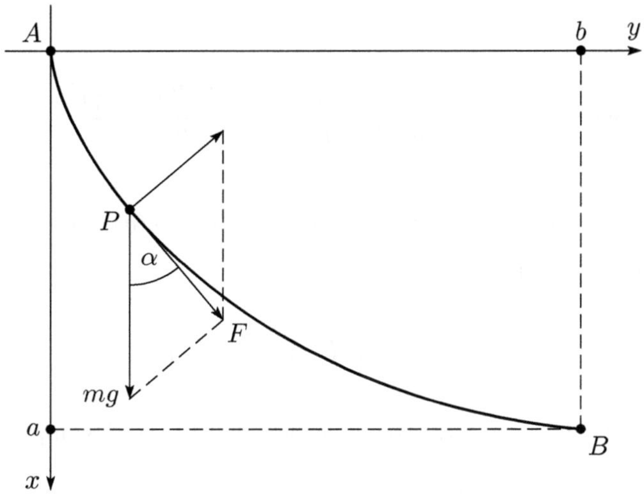

Figure 7.1.2. The x-axis is oriented in the (downward) direction of the gravitational force.

Since the point P moves along the graph of $u = u(x)$, its trajectory $s = s(t)$ is given by

$$s(t) = \int_0^{x(t)} \sqrt{1 + (u'(x))^2}\,\mathrm{d}x.^8 \tag{7.1.11}$$

Hence

$$v(t) = \frac{\mathrm{d}s(t)}{\mathrm{d}t} = \frac{\mathrm{d}s(t)}{\mathrm{d}x}\frac{\mathrm{d}x}{\mathrm{d}t} = \sqrt{1 + (u'(x(t)))^2}\dot{x}(t).$$

Using (7.1.10) and the strict monotonicity of x we have

$$\frac{\mathrm{d}t}{\mathrm{d}x} = \frac{\sqrt{1 + (u'(x))^2}}{\sqrt{2gx}}.$$

Therefore the time needed to get from A to B is given by

$$\tilde{F}(u) = \int_0^a \frac{\sqrt{1 + (u'(x))^2}}{\sqrt{2gx}}\,\mathrm{d}x. \tag{7.1.12}$$

We wish to apply Proposition 7.1.2 to the functional $\tilde{F}$. However, $\tilde{F}$ is not defined on a linear space ($u(a) = b \neq 0$). To avoid this obstacle we change the variable u for this moment by a substitution

$$w(x) = u(x) - \frac{b}{a}x.$$

[8] We use the formula for the length of a curve given by the graph of $u = u(x)$: $s = \int_0^{x_0} \sqrt{1 + (u'(x))^2}\,\mathrm{d}x$.

So, we can write (7.1.12) as

$$F(w) = \tilde{F}\left(w + \frac{b}{a}x\right) = \int_0^a \frac{\sqrt{1 + \left(w'(x) + \frac{b}{a}\right)^2}}{\sqrt{2gx}}\,\mathrm{d}x$$

where

$$w \in C_0^1[0,a] := \{w \in C^1[0,a] : w(0) = w(a) = 0\}.$$

We equip $C_0^1[0,a]$ with the norm

$$\|u\|_{C_0^1[0,a]} = \left(\int_0^a |u'(x)|^2\,\mathrm{d}x\right)^{\frac{1}{2}}.$$

For a given $h \in C_0^1[0,a]$ we have (see Corollary 3.2.14 and Example 3.2.21)

$$\delta F(w;h) = \int_0^a \frac{w'(x) + \frac{b}{a}}{\sqrt{2gx\left[1 + \left(w'(x) + \frac{b}{a}\right)^2\right]}} h'(x)\,\mathrm{d}x.$$

The Euler Necessary Condition (Proposition 7.1.2) for the original variable u reads

$$\int_0^a \frac{u'(x)}{\sqrt{2gx[1 + (u'(x))^2]}} h'(x)\,\mathrm{d}x = 0 \qquad \text{for all} \quad h \in C_0^1[0,a]. \tag{7.1.13}$$

Let us denote

$$M(x) = \frac{u'(x)}{\sqrt{2gx[1 + (u'(x))^2]}}, \qquad x \in (0,a).$$

Applying Lemma 7.1.9 we obtain that there is a constant $K \in \mathbb{R}$ such that $M(x) = K$ a.e. in $(0,a)$. However, the continuity of M actually implies that

$$\frac{u'(x)}{\sqrt{2gx[1 + (u'(x))^2]}} = K \qquad \text{for all} \quad x \in (0,a). \tag{7.1.14}$$

We will find a solution of the *Euler equation* (7.1.14). Note that $K = 0$ implies $b = 0$, and so in this case $u = 0$ is a unique solution of (7.1.14). Assume now that $b > 0$, and write K as $\pm\frac{1}{\sqrt{4gc}}$ with a $c > 0$. The equation (7.1.14) then implies

$$\left(1 - \frac{x}{2c}\right)(u'(x))^2 = \frac{x}{2c}, \qquad x \in [0,a]. \tag{7.1.15}$$

Hence $0 \le \frac{x}{2c} < 1$. After the change of variables $x = c(1 - \cos\tau)$, $\tau \in [0,\tau_0]$ (here $\tau_0 < \pi$ is such that $a = c(1 - \cos\tau_0)$) we obtain

$$\frac{\mathrm{d}u}{\mathrm{d}\tau} = \frac{\mathrm{d}u}{\mathrm{d}x} c\sin\tau$$

and (7.1.15) is transformed into

$$\left(\frac{\mathrm{d}u}{\mathrm{d}\tau}\right)^2 = c^2(1-\cos\tau)^2.$$

Hence

$$u(\tau) = \pm c(\tau - \sin\tau), \qquad \tau \in [0,\tau_0].$$

(Notice that the integration constant is zero since $u(0)=0$, and only the sign plus corresponds to our problem.) Hence the parametric equations of the graph of u are given by

$$x = c(1-\cos\tau), \quad y = c(\tau - \sin\tau), \qquad \tau \in [0,\tau_0].$$

This is a part of the cycloid, and we have to determine parameters c and τ_0 so that B is the end point of this curve. This means

$$\frac{b}{a} = \frac{\tau_0 - \sin\tau_0}{1-\cos\tau_0}, \qquad \tau_0 \in (0,\pi). \tag{7.1.16}$$

Since the function

$$\tau \mapsto \frac{\tau - \sin\tau}{1-\cos\tau}, \qquad \tau \in (0,\pi),$$

is strictly increasing with the supremum (over $(0,\pi)$) equal to $\frac{\pi}{2}$, we have that for $0 \le \frac{b}{a} < \frac{\pi}{2}$ the functional F has a unique critical point $v \in C_0^1[0,a]$ such that the graph of the function $u(x) = v(x) + \frac{b}{a}x$ has parametric equations

$$x = a\frac{1-\cos\tau}{1-\cos\tau_0}, \quad y = a\frac{\tau-\sin\tau}{1-\cos\tau_0}, \qquad \tau\in[0,\tau_0], \tag{7.1.17}$$

where τ_0 is given by (7.1.16).

On the other hand, for $\frac{b}{a} \ge \frac{\pi}{2}$ the functional F does not have critical points in $C_0^1[0,a]$. However, this does not mean that the original problem has no solution at all! The restriction we made during the formulation of the mathematical model (considering only curves which are graphs of functions $y = u(x)$) does not fit with the real situation if $\frac{b}{a} \ge \frac{\pi}{2}$! In this case one has to parametrize the curves $x = x(\tau)$, $y = y(\tau)$ and to investigate the functional

$$\tilde{F}(x,y) = \int_0^{\tau_0} \frac{\sqrt{\left(\frac{\mathrm{d}x}{\mathrm{d}\tau}(\tau)\right)^2 + \left(\frac{\mathrm{d}y}{\mathrm{d}\tau}(\tau)\right)^2}}{\sqrt{2gx(\tau)}}\,\mathrm{d}\tau.$$

An analogous procedure leads to the solution of two differential equations for x and y and one can prove the existence of a unique critical point.[9]

[9]The reader is invited to prove it in details as an exercise.

Let us return to the case $\frac{b}{a} < \frac{\pi}{2}$. It still remains to show that the solution (7.1.17) is a global minimum of F over $C_0^1[0,a]$. This follows from Proposition 7.1.8. Indeed, the function

$$z \mapsto \sqrt{1+z^2}$$

is convex in $\mathbb{R}$. This immediately implies that the functional F is convex on $C_0^1[0,a]$ (the reader is invited to prove both facts in details). Hence the unique critical point of F in $C_0^1[0,a]$ must be the point of its global minimum. ○

Let us now consider a more general situation. Namely, let

$$\mathcal{M} = \{u \in C^1[a,b] : u(a) = u_1,\ u(b) = u_2\},$$

and let us introduce the functional

$$F(u) = \int_a^b f(x, u(x), u'(x))\,\mathrm{d}x, \qquad u \in \mathcal{M},$$

where $f = f(x,y,z)$ is a function defined on $[a,b] \times \mathbb{R}^2$ with continuous second partial derivatives with respect to all its variables. This assumption will hold throughout the rest of this section. Applying the Euler Necessary Condition (Proposition 7.1.2) we get the following assertion.

Proposition 7.1.11. *Let $u_0 \in \mathcal{M}$ be a local extremum of F with respect to $\mathcal{M}$. Then the function*

$$x \mapsto \frac{\partial f}{\partial z}(x, u_0(x), u_0'(x)) \tag{7.1.18}$$

is continuously differentiable on $[a,b]$ and

$$\frac{\partial f}{\partial y}(x, u_0(x), u_0'(x)) - \frac{\mathrm{d}}{\mathrm{d}x}\left[\frac{\partial f}{\partial z}(x, u_0(x), u_0'(x))\right] = 0 \tag{7.1.19}$$

for all $x \in [a,b]$.

Proof. Let us first assume $u_1 = u_2 = 0$. Let $w \in C_0^1[a,b]$. Since

$$0 = \delta F(u_0; w) = \int_a^b \left[\frac{\partial f}{\partial y}(x, u_0(x), u_0'(x))w(x) + \frac{\partial f}{\partial z}(x, u_0(x), u_0'(x))w'(x)\right] \mathrm{d}x,$$

we get, by integrating by parts,

$$\int_a^b \left[\frac{\partial f}{\partial z}(x, u_0(x), u_0'(x)) - \int_a^x \frac{\partial f}{\partial y}(\xi, u_0(\xi), u_0'(\xi))\,\mathrm{d}\xi\right] w'(x)\,\mathrm{d}x = 0. \tag{7.1.20}$$

Using Lemma 7.1.9 we get from (7.1.20) that there is $c \in \mathbb{R}$ such that

$$\frac{\partial f}{\partial z}(x, u_0(x), u_0'(x)) - \int_a^x \frac{\partial f}{\partial y}(\xi, u_0(\xi), u_0'(\xi))\,\mathrm{d}\xi = c \tag{7.1.21}$$

for all $x \in [a,b]$. This equality shows that the function (7.1.18) is continuously differentiable and (7.1.19) holds for all $x \in [a,b]$.

In a general case we can consider $u - \frac{u_2-u_1}{b-a}(x-a) - u_1$ instead of u and apply the previous result on the transformed functional. □

Remark 7.1.12. Equation (7.1.19) is the *Euler Equation* of the functional F. Taking the "formal" derivative of the second term in (7.1.19) we obtain

$$\begin{aligned}\frac{\mathrm{d}}{\mathrm{d}x}\left[\frac{\partial f}{\partial z}(x,u_0(x),u_0'(x))\right] &= \frac{\partial^2 f}{\partial x \partial z}(x,u_0(x),u_0'(x)) \\ &\quad + \frac{\partial^2 f}{\partial y \partial z}(x,u_0(x),u_0'(x))u_0'(x) + \frac{\partial^2 f}{\partial z^2}(x,u_0(x),u_0'(x))u_0''(x).\end{aligned}$$

Hence (7.1.19) indicates that $u_0''(x)$ should exist. This motivates the following assertion.

Theorem 7.1.13 (Regularity of the "classical solution"). *Let $u_0 \in \mathcal{M}$ be a local extremum of F with respect to $\mathcal{M}$, and let $x_0 \in (a,b)$ be such that*

$$\frac{\partial^2 f}{\partial z^2}(x_0,u_0(x_0),u_0'(x_0)) \neq 0.$$

Then there exists $\delta > 0$ such that $u_0 \in C^2(x_0-\delta, x_0+\delta)$.

Proof. For $x \in [a,b]$ and $z \in \mathbb{R}$ define a function φ by

$$\varphi(x,z) = \frac{\partial f}{\partial z}(x,u_0(x),z) - \int_a^x \frac{\partial f}{\partial y}(\xi,u_0(\xi),u_0'(\xi))\,\mathrm{d}\xi - c$$

where c is the constant from the proof of Proposition 7.1.11. The Implicit Function Theorem (see Theorem 4.2.1) implies that there exist $\delta_1 > 0$, $\hat{\delta} > 0$ with the following properties: for any $x \in (x_0-\delta_1, x_0+\delta_1)$ there exists a unique $z(x) \in (u_0'(x_0)-\hat{\delta}, u_0'(x_0)+\hat{\delta})$ such that

$$\varphi(x,z(x)) = 0.$$

Moreover, $z \in C^1(x_0-\delta_1, x_0+\delta_1)$. The continuity of u_0' and the uniqueness of z imply the existence of $\delta \in (0,\delta_1)$ such that

$$u_0'(x) = z(x) \qquad \text{for} \quad x \in (x_0-\delta, x_0+\delta).$$ □

It is more convenient to look for critical points of F on "greater" sets than $\mathcal{M}$ in several situations. As we will see later (Section 7.2) this is mainly connected with the fact that the space of continuously differentiable functions $C^1[a,b]$ is not reflexive and it does not possess a Hilbert structure, either. For this purpose it is more convenient to work in the Sobolev space $W^{1,2}(a,b)$ and to look for extrema of F on the set

$$\mathcal{N} = \{u \in W^{1,2}(a,b) : u(a) = u_1,\ u(b) = u_2\}.$$

Notice that it is not obvious whether the functional F is well defined on the set $\mathcal{N}$. We have to assume that f satisfies certain growth conditions (see Theorem 3.2.24 and Remark 3.2.25; the Carathéodory property is guaranteed by the continuity of f and its derivatives).

In this case we have

Theorem 7.1.14 (Regularity of the "weak solution"). *Let $h \in L^2(a,b)$, $c_1 \ge 0$ be such that for a.a. $x \in [a,b]$ and for all $(y,z) \in \mathbb{R}^2$,*

$$|f(x,y,z)| \le h(x) + c_1(y^2+z^2), \tag{7.1.22}$$

$$\left|\frac{\partial f}{\partial y}(x,y,z)\right| \le h(x) + c_1(|y|+|z|), \tag{7.1.23}$$

$$\left|\frac{\partial f}{\partial z}(x,y,z)\right| \le h(x) + c_1(|y|+|z|). \tag{7.1.24}$$

Let $u_0 \in W^{1,2}(a,b)$ be a local minimum of F on $\mathcal{N}$. For $x \in [a,b]$ and $z \in \mathbb{R}$ set

$$\psi(x,z) = \frac{\partial f}{\partial z}(x,u_0(x),z).$$

Assume that $\frac{\partial \psi}{\partial z} > 0$ on $[a,b]\times\mathbb{R}$ and that for every fixed $x \in [a,b]$ the function $z \mapsto \psi(x,z)$ maps $\mathbb{R}$ onto $\mathbb{R}$.

Then $u_0 \in C^2[a,b]$.

Proof. First, let us assume that $u_1 = u_2 = 0$. Conditions (7.1.22)–(7.1.24) guarantee that F is well defined on $W_0^{1,2}(a,b)$ and that $\delta F(u_0;v)$ exists for any $v \in W^{1,2}(a,b)$.[10] It follows from Proposition 7.1.2 that for any $w \in W_0^{1,2}(a,b)$,

$$\delta F(u_0;w) = \int_a^b \left[\frac{\partial f}{\partial z}(x,u_0(x),u_0'(x))w'(x) + \frac{\partial f}{\partial y}(x,u_0(x),u_0'(x))w(x)\right] \mathrm{d}x = 0.$$

If we proceed literally as in the proof of Proposition 7.1.11 we arrive at (7.1.21) which now holds for a.a. $x \in [a,b]$. Since the function

$$g(x,z) = \psi(x,z) - c - \int_a^x \frac{\partial f}{\partial y}(\xi,u_0(\xi),u_0'(\xi))\,\mathrm{d}\xi$$

is continuous on $[a,b]\times\mathbb{R}$, hence by the assumptions on the function ψ, for any $x \in [a,b]$ the equation

$$g(x,z) = 0$$

has a unique solution $z = z(x)$. Moreover, by the Implicit Function Theorem (see Remark 4.2.3, not Theorem 4.2.1!), the function $x \mapsto z(x)$ is continuous on (a,b).

[10] The reader is invited to check these facts in details, see Remark 3.2.25.

It can be shown (Exercise 7.1.21) that it is continuous also at the end points a, b. So, it follows from (7.1.21) that

$$u_0'(x) = z(x) \qquad \text{for a.a.} \quad x \in [a,b], \qquad \text{i.e.,} \qquad u_0(x) = \int_a^x z(y)\,\mathrm{d}y.$$

Hence $u_0 \in C_0^1[a,b]$ and it is a local minimum of F in the space $C_0^1[a,b]$. The assertion now follows from Theorem 7.1.13.

In the general case, we consider again

$$u - \frac{u_2 - u_1}{b-a}(x-a) - u_1$$

instead of u and apply the previous result on the transformed functional. □

Exercise 7.1.15. Consider a function of two real variables

$$F(x,y) = \sin x + \sin y - \sin(x+y)$$

on the set

$$\mathcal{M} = \left(-\frac{\pi}{2}, \frac{3\pi}{2}\right) \times \left(-\frac{\pi}{2}, \frac{3\pi}{2}\right).$$

Prove that F has a local maximum at $\left(\frac{2\pi}{3}, \frac{2\pi}{3}\right)$, a local minimum at $\left(\frac{4\pi}{3}, \frac{4\pi}{3}\right)$, and there is no extremum at the critical point $(0,0)$. For the graph of F see Figure 7.1.3.

Exercise 7.1.16. Find local and global extrema of the functional

$$F\colon C[0,1] \to \mathbb{R}: \qquad F(u) = \int_0^1 [|u(t)|^2 + u(t)v(t) + w(t)]\,\mathrm{d}t$$

where $v, w \in C[0,1]$ are given functions.[11]

Exercise 7.1.17. Use Theorem 7.1.5 to prove that the solution of the Euler equation (7.1.14) is a local minimum of F from Example 7.1.10.

Hint. Show that

$$\delta^2 F(v; h, h) \geq \frac{(2c-a)^{\frac{3}{2}}}{4c\sqrt{gca}} \int_0^a |h'(x)|^2\,\mathrm{d}x.$$

Exercise 7.1.18. Prove that the functional

$$F(u) = \int_0^\pi |u(x)|^2[1 - |u'(x)|^2]\,\mathrm{d}x$$

has in $C_0^1[0,\pi]$ a unique local minimum at $u = 0$.

[11]The functional $F\colon X \to \mathbb{R}$ reaches its *global minimum* over $\mathcal{M} \subset X$ if there exists $u_0 \in \mathcal{M}$ such that $F(u) \geq F(u_0)$ for all $u \in \mathcal{M}$. *Global maximum* is defined similarly. See Section 7.2 for more detail on the existence of global extrema.

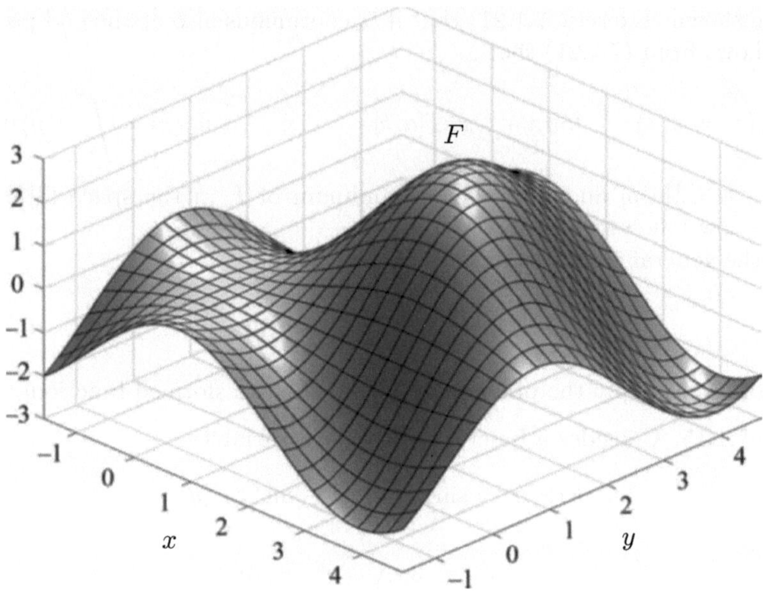

Figure 7.1.3. Graph of F.

Exercise 7.1.19 (Weierstrass Example). Prove that the functional

$$F(u) = \int_{-1}^{1} x^2 |u'(x)|^2 \, dx$$

does not have its global minimum over the set

$$\mathcal{M} = \{u \in C^1[-1,1] : u(-1) = -1, \ u(1) = 1\}.$$

Hint. Set $u_n(x) = \frac{\arctan nx}{\arctan n}$ and prove that $\lim_{n\to\infty} F(u_n) = 0$.

Exercise 7.1.20. Prove that the functional

$$F(u) = \int_{-1}^{1} x^{\frac{2}{5}} |u'(x)|^2 \, dx$$

does not have its global minimum over the set $\mathcal{M}$ from Exercise 7.1.19.

Hint. The corresponding Euler equation has no solution.

Exercise 7.1.21. Prove the following statement:

Let $g\colon [a,b]\times\mathbb{R} \to \mathbb{R}$ be a function and assume that for any $x \in [a,b]$ the equation $g(x,z) = 0$ has a solution denoted by $z = z(x)$ (not necessarily unique). If

$$\frac{\partial g}{\partial z}(x,z) > 0 \qquad on \quad [a,b]\times\mathbb{R},$$

then this solution is unique. If, moreover, g and $\frac{\partial g}{\partial z}$ are continuous on $[a, b] \times \mathbb{R}$, then $z = z(x)$ is continuous on $[a, b]$ as well.

Hint. For the continuity of $z = z(x)$ use the Implicit Function Theorem in the form of Remark 4.2.3 and notice that usage of the Contraction Principle is also possible at the end points a, b.

Exercise 7.1.22. Prove that the unique weak solution $x = x(t)$ of (6.1.27) belongs to $C^1[0, 1]$, $|\dot{x}|^{p-2}\dot{x}$ is absolutely continuous and the equation

$$-(|\dot{x}|^{p-2}\dot{x})^{\cdot} + g(x(t)) = f(t) \qquad \text{holds a.e. in} \quad (0, 1).$$

Hint. Integrating by parts in (6.1.29), we obtain that

$$\int_0^1 \left[|\dot{x}(t)|^{p-2}\dot{x}(t) + \int_0^t (g(x(\tau)) - f(\tau))\,\mathrm{d}\tau \right] \dot{y}(t)\,\mathrm{d}t = 0 \tag{7.1.25}$$

for every $y \in \mathcal{D}(0, 1)$. Set

$$M(t) = |\dot{x}(t)|^{p-2}\dot{x}(t) + \int_0^t (g(x(\tau)) - f(\tau))\,\mathrm{d}\tau.$$

It follows from Lemma 7.1.9 that

$$M(t) = c \qquad \text{a.e. in} \quad (0, 1). \tag{7.1.26}$$

The assertion now follows from (7.1.26) as in the proof of Theorem 7.1.14.

7.2 Global Extrema

In contrast with the previous section we focus now on points of global extrema. The key assertions deal with weakly coercive and weakly sequentially lower semicontinuous functionals. Let us consider a differentiable function of one real variable,

$$F\colon \mathbb{R} \to \mathbb{R}.$$

It is not difficult to give an example which shows that local extrema of F need not be its global extrema – see Figures 7.2.1, 7.2.2.

It is quite natural to ask:

> What properties of F guarantee the existence of the point of global extremum of F?

First of all let us note that we can look for global minima only because global maxima of F are global minima of $-F$ and vice versa.

Let us consider the following very simple model example of a function $F\colon \mathbb{R} \to \mathbb{R}$ which is continuous in a bounded interval $[a, b]$. Then there exists a point $x_0 \in [a, b]$ such that

$$F(x_0) = \min_{x \in [a,b]} F(x),$$

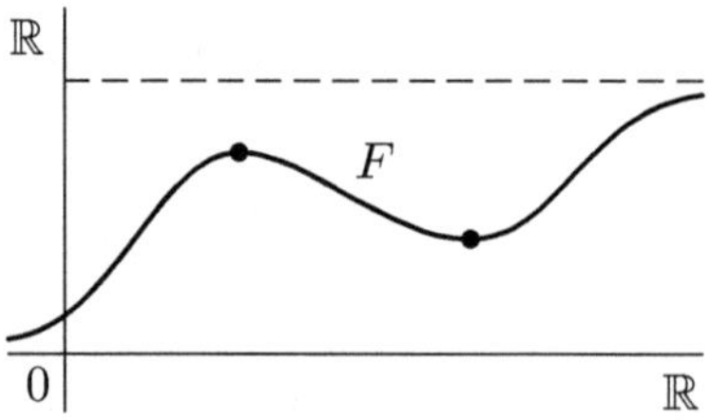

Figure 7.2.1. F attains neither its maximum nor its minimum on $\mathbb{R}$.

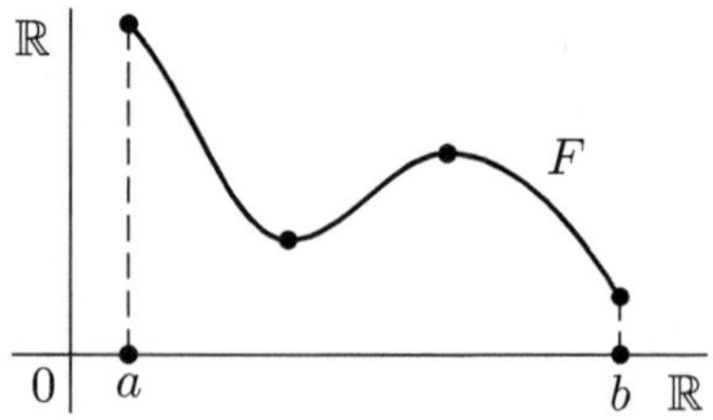

Figure 7.2.2. F attains its extrema on $[a, b]$ at a and b, respectively.

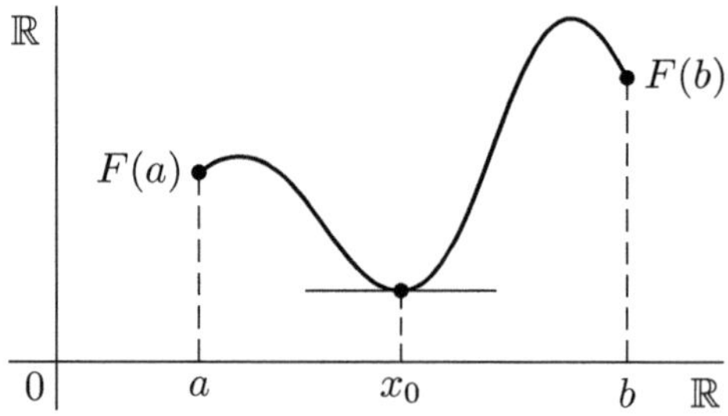

Figure 7.2.3.

i.e., the minimum of F over $[a, b]$ is at the point x_0 (see Figure 7.2.3). The proof of this fact is typical for this section. Assume that $\{x_n\}_{n=1}^{\infty}$ is a *minimizing sequence* for F on $[a, b]$, i.e.,

$$F(x_n) \searrow \inf_{x \in [a,b]} F(x).^{12} \tag{7.2.1}$$

The compactness of $[a, b]$ implies that there is a subsequence $\{x_{n_k}\}_{k=1}^{\infty} \subset \{x_n\}_{n=1}^{\infty}$ and a point $x_0 \in [a, b]$ such that

$$x_{n_k} \to x_0.$$

The continuity of F then implies that

$$F(x_0) = \inf_{x \in [a,b]} F(x).$$

The reader should notice that a property weaker than the continuity of F is sufficient to get this conclusion, namely

$$F(x_0) \leq \liminf_{k \to \infty} F(x_{n_k}) \tag{7.2.2}$$

[12] Note that, for a general $\mathcal{M}$, we set $\inf \mathcal{M} = -\infty$ if $\mathcal{M}$ is not bounded below.

(cf. Definition 7.2.1 below). It follows now from (7.2.1) and (7.2.2) that

$$F(x_0) = \inf_{x\in[a,b]} F(x)$$

(see Figure 7.2.4). If, moreover,

$$F(a) > \inf_{x\in(a,b)} F(x), \qquad F(b) > \inf_{x\in(a,b)} F(x), \tag{7.2.3}$$

then x_0 is also a local minimum of F (see Figures 7.2.3 and 7.2.4).

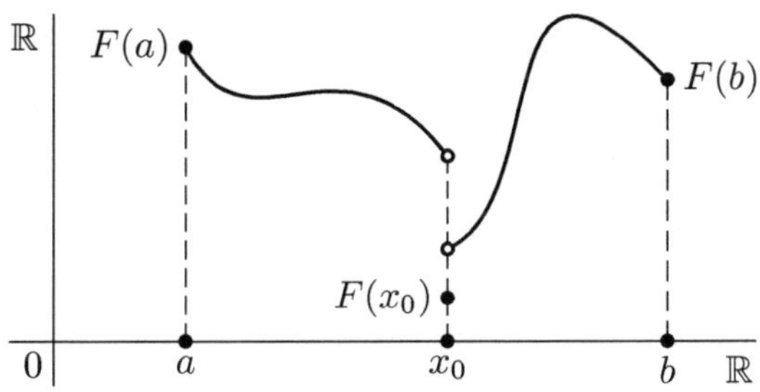

Figure 7.2.4.

Assume in the sequel in this section that

$$F\colon X \to \mathbb{R}$$

is a functional on a (infinite-dimensional) Banach space X. It is quite natural to ask if a similar result as above holds if $[a,b]$ is replaced by a *closed* and *bounded* set $\mathcal{D} \subset X$ and (7.2.3) is substituted by

$$\inf_{u\in\partial\mathcal{D}} F(u) > \inf_{u\in\operatorname{int}\mathcal{D}} F(u).$$

Unfortunately, the answer is *no* in general (see Exercise 7.2.24). The reason lies in the fact that the compactness of the bounded and closed interval $[a,b]$ is the crucial property which plays the essential role in the proof. In fact, one can imitate the proof above to get the following result:

> *Let F be a lower semi-continuous real functional on a compact set $\mathcal{K} \subset X$. Then F has a minimum in $\mathcal{K}$.*

However, this assertion has practically no applications because compact subsets of the infinite-dimensional Banach space X are "too thin" (see Proposition 1.2.15). For instance, for any compact set $\mathcal{K} \subset X$ we have

$$\operatorname{int}\mathcal{K} = \emptyset.$$

Because of this fact we have to look for a different (weaker – why?) topology on X than that induced by the norm. We would like to find a new topology on X with respect to which any bounded (in the norm) set $\mathcal{D} \subset X$ is relatively compact. The lower semi-continuity of a functional F with respect to this topology will then allow us to prove the above assertion with $\mathcal{K}$ substituted by a bounded and closed set $\mathcal{D}$ with respect to this new topology. These problems gave an impulse for the study of *weak convergence* introduced in Definition 2.1.21. The reader should notice that we will discuss weak sequential continuity of functionals instead of weak continuity (these are different concepts since weak topology is not metrizable in general). The reason is quite practical: weak sequential (semi-) continuity is easier to prove for a concrete (e.g., integral) functional. In order to make the exposition in this section as clear as possible we will restrict our attention to *real Hilbert spaces* H. The reader should have in mind that the following notions can also be defined in any Banach space.

Definition 7.2.1. Let $F\colon H \to \mathbb{R}$ be a functional, $\mathcal{M} \subset H$. Then F is said to be *weakly sequentially lower semi-continuous at a point* $u_0 \in \mathcal{M}$ *relative to* $\mathcal{M}$ if for any sequence $\{u_n\}_{n=1}^{\infty} \subset \mathcal{M}$ such that $u_n \rightharpoonup u_0$ we have

$$F(u_0) \le \liminf_{n\to\infty} F(u_n).$$

We say that F is *weakly sequentially lower semi-continuous in* $\mathcal{M} \subset H$ if it is weakly sequentially lower semi-continuous at every point $u \in \mathcal{M}$ relative to $\mathcal{M}$.

Example 7.2.2. The norm $\|\cdot\|$ on H is a weakly sequentially lower semi-continuous functional in H as follows immediately from Proposition 2.1.22(iii). ○

Example 7.2.3. Let $L\colon H \to \mathbb{R}$ be a continuous linear form. Then L is weakly sequentially lower semi-continuous in H. Indeed, it follows from the Riesz Representation Theorem (Theorem 1.2.40) that there is $v \in H$ such that

$$L(u) = (u, v) \qquad \text{for all} \quad u \in H.$$

Hence

$$u_n \rightharpoonup u_0 \qquad \text{implies} \qquad L(u_n) \to L(u_0),$$ [13]

in particular,

$$L(u_0) \le \liminf_{n\to\infty} L(u_n).$$ ○

The following assertion is a counterpart of Proposition 1.2.2 which is known as the *Extreme Value Theorem for* $H = \mathbb{R}$.

Theorem 7.2.4 (Extreme Value Theorem). *Let $\mathcal{M}$ be a weakly sequentially compact nonempty subset of H and let F be a weakly sequentially lower semi-continuous*

[13] If $u_n \rightharpoonup u_0$ implies $F(u_n) \to F(u_0)$, then the functional F is called *weakly sequentially continuous at* u_0.

functional in $\mathcal{M}$. Then F is bounded below in $\mathcal{M}$, and there exists $u_0 \in \mathcal{M}$ such that

$$F(u_0) = \min_{u \in \mathcal{M}} F(u).$$

Proof. Let $\{u_n\}_{n=1}^{\infty}$ be a minimizing sequence for F relative to $\mathcal{M}$, i.e.,

$$\{u_n\}_{n=1}^{\infty} \subset \mathcal{M} \qquad \text{and} \qquad F(u_n) \searrow \inf_{u \in \mathcal{M}} F(u).$$

Since $\mathcal{M}$ is weakly sequentially compact there exist $u_0 \in \mathcal{M}$ and a subsequence $\{u_{n_k}\}_{k=1}^{\infty} \subset \{u_n\}_{n=1}^{\infty}$ such that

$$u_{n_k} \rightharpoonup u_0.$$

The assumption on F implies

$$\inf_{u \in \mathcal{M}} F(u) \le F(u_0) \le \liminf_{k \to \infty} F(u_{n_k}) = \lim_{n \to \infty} F(u_n) = \inf_{u \in \mathcal{M}} F(u),$$

i.e.,

$$F(u_0) = \inf_{u \in \mathcal{M}} F(u) > -\infty. \qquad \square$$

Corollary 7.2.5. *Let $\mathcal{M} \subset H$, $F\colon H \to \mathbb{R}$, and u_0 be as in Theorem* 7.2.4. *Assume, moreover, that $u_0 \in \operatorname{int} \mathcal{M}$. If $\delta F(u_0; v)$ exists for a $v \in H$, then*

$$\delta F(u_0; v) = 0.$$

Proof. The assumption $u_0 \in \operatorname{int} \mathcal{M}$ implies that F attains also its local minimum at u_0. The assertion now follows from Proposition 7.1.2. $\square$

Example 7.2.6. Let us consider the boundary value problem for the second order ordinary differential equation

$$\begin{cases} -\ddot{x}(t) + x^3(t) = f(t), & t \in (0,1), \\ x(0) = x(1) = 0, \end{cases} \tag{7.2.4}$$

where $f \in L^2(0,1)$ is a given function. Put $H := W_0^{1,2}(0,1)$ with the norm

$$\|x\| = \left(\int_0^1 |\dot{x}(t)|^2 \, \mathrm{d}t \right)^{\frac{1}{2}}.$$

A weak solution of (7.2.4) is a function $x \in H$ for which the integral identity

$$\int_0^1 \dot{x}(t)\dot{y}(t) \, \mathrm{d}t + \int_0^1 x^3(t) y(t) \, \mathrm{d}t = \int_0^1 f(t) y(t) \, \mathrm{d}t$$

holds for any function $y \in H$.

Let us define a functional $F\colon H \to \mathbb{R}$ by

$$F(x) = \frac{1}{2} \int_0^1 |\dot{x}(t)|^2 \, \mathrm{d}t + \frac{1}{4} \int_0^1 |x(t)|^4 \, \mathrm{d}t - \int_0^1 f(t) x(t) \, \mathrm{d}t, \qquad x \in H.$$ [14]

Then for $x, y \in H$ we have

$$\delta F(x;y) = \int_0^1 \dot{x}(t)\dot{y}(t)\,\mathrm{d}t + \int_0^1 x^3(t)y(t)\,\mathrm{d}t - \int_0^1 f(t)y(t)\,\mathrm{d}t,$$

and any critical point of F, i.e., $x \in H$ satisfying

$$\delta F(x;y) = 0 \qquad \text{for an arbitrary} \quad y \in H,$$

is a weak solution of (7.2.4) and vice versa.

We will show that Corollary 7.2.5 applies to F and a suitably chosen set $\mathcal{M} \subset H$. First let us prove that F is a weakly sequentially lower semi-continuous functional on H. Consider an arbitrary $z \in H$ and $\{x_n\}_{n=1}^\infty \subset H$ such that $x_n \rightharpoonup z$ in H. Due to the compact embedding (Theorem 1.2.28(iii)) $H = W_0^{1,2}(0,1) \subsetneq\subsetneq C[0,1]$, we have that

$$x_n \to z \qquad \text{in} \quad C[0,1]$$

(Proposition 2.2.4(iii)). This implies

$$\int_0^1 |x_n(t)|^4\,\mathrm{d}t \to \int_0^1 |z(t)|^4\,\mathrm{d}t, \qquad \int_0^1 f(t)x_n(t)\,\mathrm{d}t \to \int_0^1 f(t)z(t)\,\mathrm{d}t. \tag{7.2.5}$$

The weak sequential lower semi-continuity of the Hilbert norm $\|\cdot\|$ (Example 7.2.2) implies

$$\liminf_{n\to\infty} \|x_n\|^2 \ge \|z\|^2. \tag{7.2.6}$$

We obtain from (7.2.5) and (7.2.6) that

$$\liminf_{n\to\infty} F(x_n) \ge F(z).$$

To find a suitable set $\mathcal{M}$ we first note that

$$\|x\|_{L^2(0,1)} \le \|x\|_{W_0^{1,2}(0,1)}.^{15} \tag{7.2.7}$$

Due to this fact we can estimate F using the Hölder inequality as follows:

$$F(x) \ge \frac{1}{2}\|x\|^2 - \|f\|_{L^2(0,1)}\|x\|_{L^2(0,1)} \ge \frac{1}{2}\|x\|\left(\|x\| - 2\|f\|_{L^2(0,1)}\right). \tag{7.2.8}$$

It is clear that for $\|x\| > 2\|f\|_{L^2(0,1)}$ we have

$$F(x) > 0,$$

[14] This functional can represent the energy of a certain system. For this reason it is often called the *energy functional.*

[15] This follows by a direct calculation using the Hölder inequality for $x(t) = \int_0^t \dot{x}(s)\,\mathrm{d}s$.

and at the same time

$$F(o) = 0.$$

So, taking

$$\mathcal{M} = \{x \in H : \|x\| \le 2\|f\|_{L^2(0,1)} + 1\},$$

the assumptions of Corollary 7.2.5 are fulfilled, since a closed ball in a Hilbert space is weakly sequentially compact (see Theorem 2.1.25 and Proposition 2.1.22(iii)). We then conclude that there exists at least one weak solution $x_0 \in H$ of the boundary value problem (7.2.4). ○

From (7.2.8) it is easy to see that the functional F from the previous example satisfies

$$\lim_{\|x\|\to\infty} F(x) = \infty.$$

This motivates the following general definition.

Definition 7.2.7. A functional $F\colon H \to \mathbb{R}$ is said to be *weakly coercive* on H if

$$\lim_{\|u\|\to\infty} F(u) = \infty.$$

This notion together with Corollary 7.2.5 leads to the following global result.

Theorem 7.2.8. *Let $F\colon H \to \mathbb{R}$ be a weakly sequentially lower semi-continuous and weakly coercive functional. Then F is bounded below on H, and there exists $u_0 \in H$ such that*

$$F(u_0) = \min_{u\in H} F(u).$$

Moreover, if $\delta F(u_0; v)$ exists for a $v \in H$, then

$$\delta F(u_0; v) = 0.$$

Proof. Let $d > \inf\limits_{u\in H} F(u)$. There exists $R > 0$ such that for $u \in H$, $\|u\| \ge R$, we have

$$F(u) > d.$$

Hence

$$\inf_{\|u\|\le R} F(u) = \inf_{u\in H} F(u).$$

Now, we apply Theorem 7.2.4 with

$$\mathcal{M} = \{u \in H : \|u\| \le R\}.$$

The assertion on a directional derivative follows from Corollary 7.2.5. □

From the point of view of applications, it is convenient to have sufficient conditions "in the language of the topology on H induced by the norm" which guarantee that

the set $\mathcal{M}$ is weakly sequentially compact;
the functional F is weakly sequentially lower semi-continuous in $\mathcal{M}$.[16]

We recall the results from Chapter 2 which state that

every closed, convex and bounded set $\mathcal{M} \subset H$ is weakly sequentially compact

(see Exercise 2.1.40, Theorem 2.1.25 and Remark 2.1.24). Concerning the desired property of F we need the following auxiliary assertion.

Lemma 7.2.9. *Let $\mathcal{M} \subset H$. Then $F\colon H \to \mathbb{R}$ is weakly sequentially lower semi-continuous in $\mathcal{M}$ if and only if for every $a \in \mathbb{R}$ the set*

$$\mathcal{E}(a) = \{u \in \mathcal{M} : F(u) \le a\}$$

is weakly sequentially closed in $\mathcal{M}$.[17]

Proof. Let F be a weakly sequentially lower semi-continuous functional in $\mathcal{M}$, $a \in \mathbb{R}$, $\{u_n\}_{n=1}^{\infty} \subset \mathcal{E}(a)$, $u_n \rightharpoonup u_0$, $u_0 \in \mathcal{M}$. Then

$$F(u_0) \le \liminf_{n\to\infty} F(u_n) \le a, \qquad \text{i.e.,} \qquad u_0 \in \mathcal{E}(a).$$

Hence $\mathcal{E}(a)$ is weakly sequentially closed in $\mathcal{M}$.

On the other hand, assume that for every $a \in \mathbb{R}$ the set $\mathcal{E}(a)$ is weakly sequentially closed in $\mathcal{M}$. Let $\{u_n\}_{n=1}^{\infty} \subset \mathcal{M}$, $u_n \rightharpoonup u_0 \in \mathcal{M}$ and denote

$$\gamma = \liminf_{n\to\infty} F(u_n)$$

Then there is a subsequence $\{u_{n_k}\}_{k=1}^{\infty}$ such that

$$F(u_{n_k}) \to \gamma.$$

We have $u_{n_k} \in \mathcal{E}(\widetilde{\gamma})$ for any $\widetilde{\gamma} > \gamma$ and k sufficiently large. Since $\mathcal{E}$ is weakly sequentially closed in $\mathcal{M}$, $u_0 \in \mathcal{E}(\widetilde{\gamma})$. Hence $u_0 \in \mathcal{E}(\gamma)$, i.e.,

$$F(u_0) \le \liminf_{n\to\infty} F(u_n).$$ □

Proposition 7.2.10. *Let F be a convex and continuous functional defined in a convex set $\mathcal{M} \subset H$. Then F is weakly sequentially lower semi-continuous in $\mathcal{M}$.*

Proof. It follows from the convexity of F that the set

$$\mathcal{E}(a) = \{u \in \mathcal{M} : F(u) \le a\}$$

is convex. The continuity of F implies that $\mathcal{E}(a)$ is closed in $\mathcal{M}$. It follows from Exercise 2.1.40 and Remark 2.1.24 that it is also weakly sequentially closed in $\mathcal{M}$. The result now follows from Lemma 7.2.9. □

[16] Not every continuous functional is weakly sequentially lower semi-continuous (cf. Exercise 7.2.32).

[17] The set $\mathcal{E} \subset \mathcal{M}$ is called *weakly sequentially closed* in $\mathcal{M}$ if for any $\{x_n\}_{n=1}^{\infty} \subset \mathcal{E}$, $x_n \rightharpoonup x \in \mathcal{M}$, we have $x \in \mathcal{E}$.

These results combined with Theorems 7.2.4 and 7.2.8 allow us to formulate the following assertions, very often used in applications.

Theorem 7.2.11. *Let $\mathcal{M}$ be a closed, convex, bounded and nonempty subset of H. Let $F\colon H \to \mathbb{R}$ be a convex and continuous functional on $\mathcal{M}$. Then F is bounded below on $\mathcal{M}$ and there exists $u_0 \in \mathcal{M}$ such that*

$$F(u_0) = \inf_{u\in\mathcal{M}} F(u).$$

If, moreover, F is strictly convex, then u_0 is the unique point with this property.[18]

Theorem 7.2.12. *Let $F\colon H \to \mathbb{R}$ be continuous, convex and weakly coercive on H. Then F is bounded below on H, and there exists $u_0 \in H$ such that*

$$F(u_0) = \inf_{u\in H} F(u).$$

If $\delta F(u_0; v)$ exists for a $v \in H$, then

$$\delta F(u_0; v) = 0.$$

If, moreover, F is strictly convex, then u_0 is uniquely determined.

Example 7.2.13. For any real continuous linear form $L\colon H \to \mathbb{R}$ there exists $u \in H$ such that $\|u\| = 1$ and

$$\|L\| = L(u).$$

Indeed, the set $\mathcal{M} = \{u \in H : \|u\| \le 1\}$ and the functional $F = -L$ satisfy the assumptions of Theorem 7.2.11. Hence there exists $u_0 \in \mathcal{M}$ such that

$$-L(u_0) = \inf_{u\in\mathcal{M}} (-L(u)).$$

By the linearity of L and the symmetry of $\mathcal{M}$ we have

$$-\inf_{u\in\mathcal{M}} (-L(u)) = \sup_{u\in\mathcal{M}} |L(u)|, \qquad \text{i.e.,} \qquad L(u_0) = \sup_{u\in\mathcal{M}} |L(u)| = \|L\|.$$

Assume that $L \neq 0$ and $\|u_0\| < 1$. Then there exists $t > 1$ such that $\|tu_0\| = 1$, i.e., $tu_0 \in \mathcal{M}$, and

$$L(tu_0) = tL(u_0) = t\|L\| > \sup_{u\in\mathcal{M}} |L(u)|,$$

a contradiction.

Note that this assertion can be proved directly using the Riesz Representation Theorem (Theorem 1.2.40). ○

[18]The reader is invited to prove the uniqueness of u_0!

Example 7.2.14. Let us consider the boundary value problem (7.2.4) and the energy functional

$$F(x) = \frac{1}{2}\int_0^1 |\dot{x}(t)|^2\,\mathrm{d}t + \frac{1}{4}\int_0^1 |x(t)|^4\,\mathrm{d}t - \int_0^1 f(t)x(t)\,\mathrm{d}t, \quad x \in H := W_0^{1,2}(0,1)$$

associated with (7.2.4). We have actually proved in Example 7.2.6 that F is weakly coercive on H. The continuity of F on H follows from the continuity of the norm in H, the continuity of the embedding $H = W_0^{1,2}(0,1) \hookrightarrow L^4(0,1)$ and from the continuity of the linear form

$$x \mapsto \int_0^1 f(t)x(t)\,\mathrm{d}t \qquad \text{on} \quad H$$

under the assumption $f \in L^2(0,1)$. The strict convexity of F follows from the strict convexity of the real functions

$$t \mapsto t^2, \qquad t \mapsto t^4,$$

and the convexity of the linear form. We conclude (see Theorem 7.2.12) that there exists a unique $x_0 \in H$ such that

$$F(x_0) = \min_{x \in H} F(x).$$

It follows then from Proposition 7.1.8 that x_0 is the unique weak solution of (7.2.4). ○

Remark 7.2.15. The reader should compare Examples 7.2.6 and 7.2.14. In the latter one we have used Theorem 7.2.12 which enables us to avoid verifying the assumption of the weak sequential lower semi-continuity of F. This might be a difficult task in general (it can not be always done so easily by means of the compact embedding as in Example 7.2.6).

The reader should also notice that the continuity of F without any additional assumptions does not imply the weak sequential lower semi-continuity of F (see Exercise 7.2.32).

In the last part of this section we show another possibility how to find critical points of F under the assumption that F is differentiable. First we need two auxiliary assertions.

Lemma 7.2.16. *Let F be a functional defined on H and ∇F its gradient.*[19] *Let $\nabla F\colon H \to H$ be a monotone operator. Then F is weakly sequentially lower semi-continuous on H.*

[19] Remember that according to the Riesz Representation Theorem (Theorem 1.2.40), Gâteaux derivative $DF(u)$ is identified with an element of H which is denoted by $\nabla F(u)$ and called a gradient of F at u. Remember also that ∇F is a mapping from H into itself. (Cf. Example 3.2.4.)

Proof. Let $u, v \in H$. According to the Mean Value Theorem applied to the real function $\varphi\colon s \mapsto F(v + s(u - v))$, $s \in [0, 1]$, there exists $t \in (0, 1)$ such that

$$\begin{aligned} F(u) - F(v) &= (\nabla F(v + t(u - v)), u - v) \\ &= (\nabla F(v), u - v) + (\nabla F(v + t(u - v)) - \nabla F(v), u - v) \\ &\geq (\nabla F(v), u) - (\nabla F(v), v).^{20} \end{aligned} \tag{7.2.9}$$

Let $\{v_n\}_{n=1}^{\infty}$ be a sequence in H such that $v_n \rightharpoonup v$ in H, i.e.,

$$(\nabla F(v), v_n) \to (\nabla F(v), v).$$

It follows from (7.2.9) that

$$\liminf_{n\to\infty} F(v_n) \geq F(v) - (\nabla F(v), v) + \lim_{n\to\infty} (\nabla F(v), v_n) = F(v). \qquad \square$$

Definition 7.2.17. Let $T\colon H \to H$ be an operator from H into itself. We say that T is *coercive* if

$$\lim_{\|u\|\to\infty} \frac{(T(u), u)}{\|u\|} = \infty.$$

Lemma 7.2.18. *Let $F\colon H \to \mathbb{R}$ be a functional and $\nabla F\colon H \to H$ its gradient. Let ∇F be a coercive and bounded operator. Then F is weakly coercive.*

Proof. Since

$$\frac{\mathrm{d}}{\mathrm{d}t} F(tu) = (\nabla F(tu), u),$$

we obtain by integration

$$F(u) = F(o) + \int_0^1 (\nabla F(tu), tu) \frac{\mathrm{d}t}{t} = F(o) + \int_0^{\|u\|} \left(\nabla F\left(\tau \frac{u}{\|u\|}\right), \tau \frac{u}{\|u\|} \right) \frac{\mathrm{d}\tau}{\tau}$$

for any $u \in H$, $u \neq o$. The coercivity of ∇F implies that there exists $r \geq 0$ such that

$$\frac{1}{\tau} \left(\nabla F\left(\tau \frac{u}{\|u\|}\right), \tau \frac{u}{\|u\|} \right) \geq 1 \qquad \text{for any} \quad \tau \geq r \quad \text{and} \quad u \in H, \quad u \neq o.$$

The boundedness of ∇F implies

$$m := \sup_{\substack{\tau \in [0,r] \\ u \in H,\, u \neq o}} \left\| \nabla F\left(\tau \frac{u}{\|u\|}\right) \right\| < \infty.$$

[20]Since the monotonicity of ∇F implies that φ' is increasing, φ is a convex function, i.e., F is convex.

Consequently, we obtain

$$\begin{aligned} F(u) &= F(o) + \int_0^r \left(\nabla F\left(\tau\frac{u}{\|u\|}\right), \tau\frac{u}{\|u\|}\right)\frac{\mathrm{d}\tau}{\tau} \\ &\quad + \int_r^{\|u\|} \left(\nabla F\left(\tau\frac{u}{\|u\|}\right), \tau\frac{u}{\|u\|}\right)\frac{\mathrm{d}\tau}{\tau} \\ &\geq F(o) - rm + \|u\| - r \qquad \text{for any} \quad u \in H, \quad \|u\| > r. \end{aligned}$$

The last inequality yields the weak coercivity of F. □

Remark 7.2.19. Let $F\colon H \to \mathbb{R}$, $h \in H$. Assume that the gradient $\nabla F(u)$ of F exists at any point $u \in H$. Then the following equivalence obviously holds true:

> *There exists $u_0 \in H$ such that $\nabla F(u_0) = h$ if and only if there exists $u_0 \in H$ such that $\nabla G(u_0) = o$ where*

$$G\colon u \mapsto F(u) - (u, h). \tag{7.2.10}$$

Theorem 7.2.20. *Let $F\colon H \to \mathbb{R}$ and let $\nabla F\colon H \to H$ be the gradient of F. Let ∇F be a monotone, coercive and bounded operator. Then*

$$\nabla F(H) = H.^{21}$$

Proof. It follows from Remark 7.2.19 that it is enough to prove that for any $h \in H$, the functional G defined by (7.2.10) has a critical point. But Lemmas 7.2.16 and 7.2.18 yield that G is weakly sequentially lower semi-continuous and weakly coercive. The existence of a critical point of G follows from Theorem 7.2.8. □

Example 7.2.21. Let us consider again the boundary value problem (7.2.4) and the associated energy functional

$$F(x) = \frac{1}{2}\int_0^1 |\dot{x}(t)|^2\,\mathrm{d}t + \frac{1}{4}\int_0^1 |x(t)|^4\,\mathrm{d}t - \int_0^1 f(t)x(t)\,\mathrm{d}t.$$

Then

$$(\nabla F(x), y) = \int_0^1 \dot{x}(t)\dot{y}(t)\,\mathrm{d}t + \int_0^1 x^3(t)y(t)\,\mathrm{d}t - \int_0^1 f(t)y(t)\,\mathrm{d}t,$$

here ∇F is the gradient of F in $H := W_0^{1,2}(0,1)$. We verify the assumption of Theorem 7.2.20. Using the continuous embedding $H = W_0^{1,2}(0,1) \subsetneq C[0,1]$

[21] Compare this result with Theorem 6.1.4.

(Theorem 1.2.26) we prove the boundedness (and even continuity!) of ∇F in the space H. Since $s \mapsto s^3$ is monotone we have

$$(\nabla F(x_1) - \nabla F(x_2), x_1 - x_2) = \int_0^1 |\dot{x}_1(t) - \dot{x}_2(t)|^2 \,\mathrm{d}t + \int_0^1 (x_1^3(t) - x_2^3(t))(x_1(t) - x_2(t)) \,\mathrm{d}t \geq \|x_1 - x_2\|^2 \quad (7.2.11)$$

for $x_1, x_2 \in H$ and the monotonicity of ∇F follows.

Finally, we have

$$\frac{(\nabla F(x), x)}{\|x\|} = \|x\| + \frac{\|x\|^4_{L^4(0,1)}}{\|x\|} - \frac{1}{\|x\|}\int_0^1 f(x)x(t)\,\mathrm{d}t \geq \|x\| - \|f\|_{L^2(0,1)}\frac{\|x\|_{L^2(0,1)}}{\|x\|}.$$

Using the inequality (7.2.7) we get

$$\frac{(\nabla F(x), x)}{\|x\|} \geq \|x\| - \|f\|_{L^2(0,1)},$$

i.e., ∇F is coercive. We conclude from Theorem 7.2.20 that

$$\nabla F(H) = H,$$

in particular, there exists $x_0 \in H$ such that

$$\nabla F(x_0) = o.$$

Hence x_0 is a weak solution of (7.2.4). The estimate (7.2.11) then implies the uniqueness of x_0.[22] ○

Remark 7.2.22. Most of the previous results hold true when the Hilbert space H is replaced by a real reflexive Banach space X and the scalar product $(\cdot,\cdot)$ is replaced by the duality pairing $\langle\cdot,\cdot,\rangle$ between X^* and X, i.e., for $f \in X^*$ and $x \in X$ we write

$$\langle f, x\rangle := f(x).$$

However, the proofs are technically more involved and the gradient ∇F has to be replaced by the Gâteaux derivative DF.

Exercise 7.2.23. Prove the Fundamental Theorem of Algebra (Theorem 5.7.13) using the methods of Section 7.2.

[22] The reader is invited to apply Theorem 6.1.4 to get the same result.

Hint. Let $P(z) = a_0 + a_1 z + \cdots + a_n z^n (n \geq 1)$, $a_n \neq 0$. We have $\lim_{|z|\to\infty} |P(z)| = \infty$ and the continuity of P yields the existence of $z_0 \in \mathbb{C}$ such that $|P(z_0)| = \min_{z\in\mathbb{C}} |P(z)|$. We show that $P(z_0) = 0$. Without loss of generality we assume $z_0 = 0$ (for otherwise we set $z := z - z_0$).

Hence

$$|P(z)| \geq |P(0)|$$

for any $z \in \mathbb{C}$. Let a_k be the first nonzero coefficient, i.e.,

$$P(z) = P(0) + a_k z^k + \cdots + a_n z^n (1 \leq k \leq n).$$

Then

$$\begin{aligned} |P(z)|^2 = P(z)\overline{P(z)} &= P(0)\overline{P(0)} + P(0)\overline{a_k z^k} \\ &+ \overline{P(0)} a_k z^k + \text{“higher order terms”}. \end{aligned}$$

Hence

$$0 \leq |P(z)|^2 - |P(0)|^2 = 2 \operatorname{Re} [\overline{P(0)} a_k z^k] + \text{“higher order terms”}.$$

Set $z = te^{i\omega}$, $t > 0$, $\omega \in [0, 2\pi)$. Then

$$\begin{aligned} 0 \leq |P(te^{i\omega})|^2 - |P(0)|^2 &= 2 \operatorname{Re} [\overline{P(0)} a_k t^k e^{i\omega k}] \\ &+ \text{“higher order terms”}. \end{aligned}$$

For $t \to 0_+$ we get

$$0 \leq \frac{|P(te^{i\omega})|^2 - |P(0)|^2}{t^k} = 2 \operatorname{Re} [\overline{P(0)} a_k e^{i\omega k}] + o(1)$$

for any $\omega \in [0, 2\pi) \Rightarrow \overline{P(0)} a_k = 0 \Rightarrow P(0) = 0$.

Exercise 7.2.24. Let $\{e_n\}_{n=1}^{\infty}$ be an orthonormal basis in a Hilbert space H. Put

$$\mathcal{D}_n = \left\{ x \in H : \|x - e_n\| \leq \frac{1}{2} \right\}$$

and define a functional

$$f(x) = \begin{cases} \|x\| & \text{for } \ x \notin \bigcup_{n=1}^{\infty} \mathcal{D}_n, \\ \|x\| + \dfrac{2(n-1)}{n} \left(\dfrac{1}{2} - \|x - e_n\| \right) & \text{for } \ x \in \mathcal{D}_n. \end{cases}$$

Show that f is continuous on H,

$$\sup_{\|x\| \leq \frac{3}{2}} f(x) = 2,$$

but f does not have maximum on the ball

$$\left\{x \in H : \|x\| \leq \frac{3}{2}\right\}.$$

Exercise 7.2.25. The mapping $U\colon \mathbb{R}^2 \to \mathbb{R}^2$ is defined by

$$U\colon (x,y) \mapsto (y,-x).$$

Prove that U is monotone and satisfies

$$\lim_{\|(x,y)\|\to\infty} \|U(x,y)\| = \infty$$

(i.e., is weakly coercive) but is not coercive.

Exercise 7.2.26. Prove that any coercive map $F\colon H \to H$ is weakly coercive.

Exercise 7.2.27. Prove that the same conclusion as in Example 7.2.14 holds true also if $f \in L^1(0,1)$.
Hint. Use the embedding $W_0^{1,2}(0,1) \subsetneq L^\infty(0,1)$.

Exercise 7.2.28. Prove that the norm on H and linear forms on H are convex functionals.

Exercise 7.2.29. Prove that in Theorem 7.2.12 the weak coercivity of F can be substituted by a weaker assumption:

> *For any $u \in H$ there exists $r > 0$ such that for all $v \in H$, $\|v\| \geq r$, we have $F(v) > F(u)$.*

Exercise 7.2.30. Let $\mathcal{M}$ be an open convex subset of a real Hilbert space H, let $F\colon H \to \mathbb{R}$ be a functional such that for any $u \in \mathcal{M}$ there exists the second Gâteaux derivative $D^2F(u)$. Prove that

$$\text{(a)} \implies \text{(b)} \implies \text{(c)}$$

where

(a) $D^2F(u)(h,h) \geq 0$ for $u \in \mathcal{M}$, $h \in H$;
(b) $(\nabla F(u) - \nabla F(v), u - v) \geq 0$ for $u, v \in \mathcal{M}$;
(c) F is convex on $\mathcal{M}$.

Hint. Use the Mean Value Theorem (see Theorem 3.2.6) as for real functions.

Exercise 7.2.31. Prove that for any $n \in \mathbb{N}$ and $f \in L^2(0,1)$ the boundary value problem

$$\begin{cases} -\ddot{x}(t) + x^{2n+1}(t) = f(t), & t \in (0,1), \\ x(0) = x(1) = 0 \end{cases}$$

has a unique weak solution.

Exercise 7.2.32. Let f be the functional from Exercise 7.2.24. Prove that f is not weakly sequentially upper semi-continuous (i.e., $-f$ is not weakly sequentially lower semi-continuous).

Hint. Remember that $e_n \rightharpoonup o$.

7.2A Supersolutions, Subsolutions and Global Extrema

In this appendix we show the connection between the supersolutions and subsolutions (see Section 6.3) on the one hand and the existence of global minima (see Section 7.2) on the other. We will illustrate it on the Dirichlet boundary value problem

$$\begin{cases} \ddot{x}(t) = f(t, x(t)), & t \in (0,1), \\ x(0) = x(1) = 0, \end{cases} \tag{7.2.12}$$

where f is a continuous function on $[0,1] \times \mathbb{R}$ (cf. Example 6.3.19). Put $H := W_0^{1,2}(0,1)$. The functional

$$\psi(x) := \int_0^1 \int_0^{x(t)} f(t,s)\,\mathrm{d}s\,\mathrm{d}t$$

defined on H is of the class $C^1(H,\mathbb{R})$ and

$$\psi'(x)(h) = \int_0^1 f(t,x(t))h(t)\,\mathrm{d}t, \qquad x, h \in H.^{23} \tag{7.2.13}$$

Then

$$F(x) = \int_0^1 \left[\frac{1}{2}|\dot{x}(t)|^2 + \int_0^{x(t)} f(t,s)\,\mathrm{d}s\right] \mathrm{d}t$$

is of the class $C^1(H,\mathbb{R})$ and its critical points correspond to weak solutions of (7.2.12). A regularity argument applied to (7.2.12) (similar to that from Theorem 7.1.13) implies that every weak solution is a classical solution in the sense that

$$x \in C_0^2[0,1] := \{x \in C^2[0,1] : x(0) = x(1) = 0\}$$

and the equation in (7.2.12) holds at every point $t \in (0,1)$.

The *link* between the method of supersolutions and subsolutions on the one side and the method of finding the global minimizer on the other side is that

> *the existence of a well-ordered pair of a subsolution and supersolution u_0 and v_0, respectively, implies that the functional F has a minimum on the convex but noncompact set*
>
> $$\mathcal{M} = \{x \in H : u_0(t) \le x(t) \le v_0(t) \text{ for all } t \in [0,1]\}.$$

This minimum then solves (7.2.12). Namely, we have the following assertion.

[23] Cf. Section 3.2 in order to prove this fact.

Theorem 7.2.33. *Let u_0 and v_0 be a subsolution and supersolution of* (7.2.12) *such that* $u_0(t) \le v_0(t)$, $t \in [0,1]$,

$$\mathcal{E} := \{(t,x) \in [0,1] \times \mathbb{R} : u_0(t) \le x \le v_0(t)\},$$

let $f\colon \mathcal{E} \to \mathbb{R}$ be a continuous function. Then the functional F has a global minimum on $\mathcal{M}$, i.e., there exists $x_0 \in \mathcal{M}$ such that

$$F(x_0) = \min_{\substack{x \in H \\ u_0 \le x \le v_0}} F(x).$$

Moreover, x_0 is a solution of (7.2.12).

Proof. Let

$$\gamma(t,x) := \max\{u_0(t), \min\{x, v_0(t)\}\}$$

and consider the modified problem

$$\begin{cases} \ddot{x}(t) = f(t, \gamma(t, x(t))), & t \in (0,1), \\ x(0) = x(1) = 0. \end{cases} \tag{7.2.14}$$

Define the energy functional associated with this modified problem by

$$\tilde{F}(x) = \int_0^1 \left[\frac{1}{2}|\dot{x}(t)|^2 + \int_0^{x(t)} f(t, \gamma(t,s))\,\mathrm{d}s\right] \mathrm{d}t.$$

Then $\tilde{F} \in C^1(H, \mathbb{R})$ and its critical points correspond to the solutions of (7.2.14). It is easy to prove (the reader should do it as an exercise) that $\tilde{F}$ is weakly sequentially lower semicontinuous and weakly coercive. It then follows from Theorem 7.2.8 that $\tilde{F}$ has a global minimum on H at $x_0 \in H$, $\tilde{F}'(x_0) = o$. This x_0 is a weak solution of (7.2.14) and it is regular, i.e., $x_0 \in C^2[0,1]$, by Theorem 7.1.14.

We shall show that $u_0(t) \le x_0(t) \le v_0(t)$. Indeed, assume by contradiction that

$$\min_{t \in [0,1]} (x_0(t) - u_0(t)) < 0$$

and define

$$t_0 := \max\left\{t \in [0,1] : x_0(t) - u_0(t) = \min_{s \in [0,1]} (x_0(s) - u_0(s))\right\}.$$

From the definition of a subsolution u_0 and of γ we obtain that $0 < t_0 < 1$, and for $t \ge t_0$, t close to t_0, we have

$$\dot{x}_0(t) - \dot{u}_0(t) = \int_{t_0}^t [\ddot{x}_0(s) - \ddot{u}_0(s)]\,\mathrm{d}s = \int_{t_0}^t [f(s, u_0(s)) - \ddot{u}_0(s)]\,\mathrm{d}s \le 0.$$

This contradicts the definition of t_0. Hence $x_0(t) \ge u_0(t)$, $t \in [0,1]$. Similarly we prove $x_0(t) \le v_0(t)$, $t \in [0,1]$.

Notice that if x is such that $u_0(t) \le x(t) \le v_0(t)$, then $\gamma(t, x(t)) = x(t)$, i.e., x_0 is a minimizer for F on $\mathcal{M}$ and $F'(x_0) = o$. □

Example 7.2.34. Consider the problem

$$\begin{cases} \ddot{x}(t) = \lambda f(t, x(t)), & t \in (0,1), \\ x(0) = x(1) = 0, \end{cases} \tag{7.2.15}$$

where f is continuous on $[0,1] \times \mathbb{R}$, $f(t,0) = 0$, $f(t,R) \geq 0$ for an $R > 0$ and there exists $w \in H := W_0^{1,2}(0,1)$, $0 \leq w(t) \leq R$, $t \in [0,1]$, such that

$$\int_0^1 \int_0^{w(t)} f(t,s)\,\mathrm{d}s\,\mathrm{d}t < 0.$$

Then there exists $\Lambda \geq 0$ such that for all $\lambda \geq \Lambda$, (7.2.15) has, besides the trivial solution, at least one nontrivial nonnegative solution.

Indeed, $u_0 \equiv 0$ is a subsolution and $v_0 \equiv R$ is a supersolution for $\lambda \geq 0$, and according to Theorem 7.2.33 there exists

$$x_0 \in \mathcal{M} := \{x \in H : 0 \leq x(t) \leq R\}$$

which solves (7.2.15) and minimizes the energy functional F on $\mathcal{M}$. Moreover, taking λ large enough, we have

$$F(w) = \int_0^1 \left[\frac{1}{2} |\dot{w}(t)|^2 + \lambda \int_0^{w(t)} f(t,s)\,\mathrm{d}s \right] \mathrm{d}t < 0,$$

and so

$$F(x_0) = \min_{x \in \mathcal{M}} F(x) \leq F(w) < 0 = F(o). \qquad \bigcirc$$

Remark 7.2.35. The same results as in Theorem 7.2.33 and Example 7.2.34 hold if the continuity of f is relaxed to $f \in \mathrm{CAR}([0,1] \times \mathbb{R})$ and for all $r > 0$ there exists $h \in L^1(0,1)$ such that for a.e. $t \in (0,1)$ and all $s \in \mathbb{R}$, $|s| \leq r$, we have $|f(t,s)| \leq h(t)$. The reader is invited to verify all the previous steps as an exercise.

The reader who wants to learn more is referred to De Coster & Habets [41] where also the relation between non-well ordered supersolutions and subsolutions on the one hand and the minimax method on the other is discussed.

Exercise 7.2.36. How does the proof of Theorem 7.2.33 change if the homogeneous Dirichlet boundary conditions in (7.2.12) are replaced by the Neumann ones?

Exercise 7.2.37. Consider the problem

$$\begin{cases} \left(|\dot{x}(t)|^{p-2}\dot{x}(t)\right)^{\cdot} = f(t, x(t)), & t \in (0,1), \\ x(0) = x(1) = 0, \end{cases} \tag{7.2.16}$$

where $p > 1$ and

$$F(x) = \int_0^1 \left[\frac{1}{p} |\dot{x}(t)|^p + \int_0^{x(t)} f(t,s)\,\mathrm{d}s \right] \mathrm{d}t.$$

Prove the analogue of Theorem 7.2.33 for (7.2.16).

Exercise 7.2.38. Find conditions on a continuous function $f\colon [0,1] \times \mathbb{R} \to \mathbb{R}$ which guarantee that the problem (7.2.16) has a subsolution u_0 and a supersolution v_0 satisfying

$$u_0(t) \leq v_0(t) \qquad \text{for all} \quad t \in [0,1].$$

Hint. Look for u_0 and v_0 constant on $[0,1]$.

7.2B Ritz Method

In this part of the text we want to address one fundamental numerical approach to finding the global minimum of a real functional on a real Banach space. In applications such a minimum corresponds to a solution of a certain boundary value problem and the general method we will discuss below is a starting point for many numerical methods. Let us mention the Galerkin Method, the Finite Elements Method, the Katchanov–Galerkin Method, etc., which are powerful tools in the numerical solution of differential equations.

Let X be a real Banach space and F a real functional defined on X. An element $u_0 \in X$ satisfying

$$F(u_0) = \inf_{u \in X} F(u) \tag{7.2.17}$$

will be called a solution of the variational problem (7.2.17). We will discuss the *Ritz Method* which actually yields directly an algorithm for finding a solution of the variational problem. The *basic idea* of the Ritz Method is rather simple:

> *Instead of looking for the minimum of the functional F on the entire space X, we look for its minimum on suitable subspaces of the space X in which we know how to solve the variational problem.*

Let us now formulate this idea precisely:

> *To every $n \in \mathbb{N}$, let a closed subspace X_n of the space X be assigned. The problem of finding an element $u_n \in X_n$ such that*
>
> $$F(u_n) = \inf_{u \in X_n} F(u) \tag{7.2.18}$$
>
> *holds is called the* Ritz approximation of the problem (7.2.17) *and the element $u_n \in X_n$ is called a solution of the problem* (7.2.18).

The following two fundamental problems immediately present themselves:

(a) the problem of the existence and uniqueness of a solution of the problem (7.2.18);

(b) the relation between the solutions of the problems (7.2.17) and (7.2.18).

Problem (a) has already been solved by Theorem 7.2.12 in the framework of Hilbert spaces. It follows from Remark 7.2.22 that the same assertion can be proved in a reflexive Banach space X. Since a closed subspace X_n of a reflexive Banach space X is also a reflexive Banach space (see, e.g., [36, § III.11]), we have the following assertion which follows directly from Theorem 7.2.12 and Remark 7.2.22.

Proposition 7.2.39. *Let X be a reflexive Banach space, and let a functional F defined on the space X be continuous, strictly convex and weakly coercive on X. Then each of the problems* (7.2.17) *and* (7.2.18) *has precisely one solution u_0 and u_n, respectively.*

We now focus our effort on problem (b). We investigate under what condition

$$\lim_{n \to \infty} \|u_0 - u_n\| = 0 \tag{7.2.19}$$

is true. If (7.2.19) is valid, then we say that the Ritz Method converges for the problem (7.2.17) and the solutions u_n of the problems (7.2.18) approximate the solution of the problem (7.2.17) in the sense of the norm of the space X.

Proposition 7.2.40. *Let F be a continuous linear functional on a normed linear space X, let $\{X_n\}_{n=1}^{\infty}$ be a sequence of closed subspaces of X such that for every $v \in X$ there exist elements $v_n \in X_n$, $n \in \mathbb{N}$, such that*

$$\lim_{n\to\infty} \|v - v_n\| = 0. \tag{7.2.20}$$

Let u_n be such an element of X_n that (7.2.18) holds. Then $\{u_n\}_{n=1}^{\infty}$ is a minimizing sequence for the functional F on X, i.e.,

$$\lim_{n\to\infty} F(u_n) = \inf_{u\in X} F(u). \tag{7.2.21}$$

Proof. Let $\{\alpha_k\}_{k=1}^{\infty}$ be a sequence such that

$$\alpha_k \searrow \inf_{u\in X} F(u).$$

Then there exist elements $v^{(k)} \in X$ for which

$$F\left(v^{(k)}\right) < \alpha_k.$$

By the assumption (7.2.20) we can find $w_n^{(k)} \in X_n$ satisfying $w_n^{(k)} \to v^{(k)}$ for $n \to \infty$. Hence

$$\inf_{u\in X} F(u) \le F(u_n) \le F\left(w_n^{(k)}\right).$$

By the continuity of F we get

$$\limsup_{n\to\infty} F(u_n) \le \lim_{n\to\infty} F\left(w_n^{(k)}\right) = F\left(v^{(k)}\right) < \alpha_k.$$

This implies that

$$\lim_{n\to\infty} F(u_n) = \inf_{u\in X} F(u). \qquad \square$$

The assertion on the convergence of the Ritz Method for the problem (7.2.17) is the following proposition.

Proposition 7.2.41 (Ritz Method). *Let H be a real Hilbert space,*[24] *and let F be a continuous functional on the space H which has the second Gâteaux derivative $D^2F(u) \in \mathcal{B}_2(H,\mathbb{R})$.*[25]

Assume, further, that there exists a constant $c > 0$ such that for all $u, v \in H$ we have

$$D^2F(u)(v,v) \ge c\|v\|^2. \tag{7.2.22}$$

Let subspaces H_n of the space H satisfy condition (7.2.20). Then

(i) *there exists precisely one solution $u_0 \in H$ of problem (7.2.17);*

(ii) *for every $n \in \mathbb{N}$ there exists precisely one solution $u_n \in H_n$ of problem (7.2.18);*

[24] We will state and prove Proposition 7.2.41 in the Hilbert space setting. The generalization to the Banach space setting can be obtained in lines of Remark 7.2.22. The reader can find details in specialized literature (see, e.g., Saaty [135]).

[25] See Section 3.2.

(iii) *the Ritz Method converges for problem* (7.2.17), *i.e.*,

$$\lim_{n\to\infty} \|u_0 - u_n\| = 0.$$

Proof. It follows from the Taylor Formula (Proposition 3.2.27) that

$$F(u+v) = F(u) + DF(u)(v) + \int_0^1 (1-t)D^2F(u+tv)(v,v)\,\mathrm{d}t. \tag{7.2.23}$$

Choosing $u = o$, we have due to (7.2.22)

$$\begin{aligned} F(v) &= F(o) + DF(o)(v) + \int_0^1 (1-t)D^2F(tv)(v,v)\,\mathrm{d}t \\ &\geq F(o) + \frac{c}{2}\|v\|^2 - \|DF(o)\|_{H^*}\|v\| = F(o) + \|v\| \left[\frac{c}{2}\|v\| - \|DF(o)\|_{H^*}\right] \end{aligned}$$

for $v \in H$. This implies that F is weakly coercive on H (and also on H_n for arbitrary n).

Choosing now $w := u + v$ in (7.2.23), then for $w \neq u$ (i.e., for $v \neq o$) we conclude from (7.2.23) that

$$F(w) - F(u) - DF(u)(w-u) > 0.$$

In particular, for $u = tw_1 + (1-t)w_2$, $w_1 \neq w_2$, $t \in (0,1)$, $w = w_1$ and $w = w_2$, we have

$$F(w_1) - F(u) > (1-t)DF(u)(w_1 - w_2) \qquad \text{and} \qquad F(w_2) - F(u) > -tDF(u)(w_1 - w_2),$$

respectively. Multiplying the first inequality by t, the second by $(1-t)$ and adding both of them, we obtain that F is strictly convex on H (and also on H_n for arbitrary n).

The assertions (i) and (ii) now follow from Theorem 7.2.12.

It remains to prove assertion (iii). Let u_0 and u_n be a solution of (7.2.17) and (7.2.18), respectively. Set $u := u_0$ and $v := u_n - u_0$ in (7.2.23). From (7.2.22) and (7.2.23) we obtain

$$F(u_n) \geq F(u_0) + DF(u_0)(u_n - u_0) + \frac{c}{2}\|u_n - u_0\|^2.$$

Since $u_0 \in H$ is the point of minimum for F on H, it follows from Theorem 7.2.12 that

$$DF(u_0)(u_n - u_0) = o, \qquad \text{i.e.,} \qquad F(u_n) \geq F(u_0) + \frac{c}{2}\|u_n - u_0\|^2 \tag{7.2.24}$$

holds for arbitrary $n \in \mathbb{N}$. On the other hand, due to Proposition 7.2.40, the elements u_n, $n \in \mathbb{N}$, constitute a minimizing sequence for F on H, i.e.,

$$\lim_{n\to\infty} F(u_n) = \inf_{u\in H} F(u) = F(u_0). \tag{7.2.25}$$

It follows from (7.2.24) and (7.2.25) that

$$\lim_{n\to\infty} \|u_0 - u_n\| = 0$$

and the proof is complete. □

So far, we have answered theoretically problems (a) and (b) formulated at the beginning of this appendix. However, from the point of view of practical (numerical) calculations the most interesting problems start right now. The most frequent and most important case arises in practice when the spaces H_n are of finite dimension, e.g., $\dim H_n = N$. If $e_1, \dots, e_N$ is a basis of H_n and

$$F_n(c_1, \dots, c_N) := F\left(\sum_{i=1}^{N} c_i e_i\right),$$

then the problem (7.2.18) means to find $\tilde{c} = (\tilde{c}_1, \dots, \tilde{c}_N) \in \mathbb{R}^N$ such that

$$F_n(\tilde{c}_1, \dots, \tilde{c}_N) = \inf_{(c_1, \dots, c_N) \in \mathbb{R}^N} F_n(c_1, \dots, c_N). \tag{7.2.26}$$

If the assumptions of Proposition 7.2.41 are satisfied, then the function F_n is continuous, strictly convex on the space $\mathbb{R}^N$, satisfies

$$\lim_{\|c\| \to \infty} F_n(c) = \infty,$$

and then the vector $\tilde{c}$ is a solution of problem (7.2.26) if and only if all partial derivatives of the first order of the function F_n vanish at $\tilde{c}$ (cf. Theorem 7.2.12). Thus the problem of finding a solution of problem (7.2.26) is equivalent to the problem of finding a solution of the system

$$\begin{aligned} \frac{\partial F_n(c_1, \dots, c_N)}{\partial c_1} &= 0, \\ &\vdots \\ \frac{\partial F_n(c_1, \dots, c_N)}{\partial c_N} &= 0. \end{aligned} \tag{7.2.27}$$

The system (7.2.27) is a system of N algebraic equations which are generally nonlinear. However, note that if the functional F is *quadratic*, then the system (7.2.27) is a system of linear algebraic equations.

Remark 7.2.42. We have not been concerned with the question which is fundamental from the practical point of view: "How to solve system (7.2.27) numerically?" A vast literature dedicated to numerical methods deals with this problem. Just for an illustration we mention one *minimization method.* Choose arbitrarily a vector $c^0 = (c_1^0, \dots, c_N^0) \in \mathbb{R}^N$. Let us present an algorithm for the construction of a sequence $\{c^m\}_{m=1}^{\infty}$ which converges under appropriate assumptions on f to the solution of system (7.2.27). If we know the vector $c^m = (c_1^m, \dots, c_N^m) \in \mathbb{R}^N$, we calculate the components of the vector $c^{m+1} = (c_1^{m+1}, \dots, c_N^{m+1}) \in \mathbb{R}^N$ as follows: Let the function

$$F_n(c_1^{m+1}, \dots, c_{i-1}^{m+1}, \xi, c_{i+1}^m, \dots, c_N^m)$$

of the variable ξ on $\mathbb{R}$ assume its minimum at the point $\tilde{c}_i^{m+1}$. Put, then,

$$c_i^{m+1} := c_i^m + \omega(\tilde{c}_i^{m+1} - c_i^m) \qquad \text{where} \quad 0 < \omega \le 2.$$

Here ω is the so-called *relaxation parameter.* If we choose $\omega = 1$ and if F is a quadratic functional, we obtain the so-called *Gauss–Seidel Iterative Method* (see, e.g., Stoer & Bulirsch [147]). Nowadays there are plenty of packages available in MATHEMATICA, MAPLE, MATLAB, etc. and offering different solvers of system (7.2.27).

From the practical point of view it is important that the system (7.2.27) be as simple as possible. The form of the system (7.2.27) depends in an essential way on the actual choice of the subspaces H_n. One special choice depends on the notion and the properties of the Schauder basis.

Let $\{e_i\}_{i=1}^{\infty}$ be a Schauder basis (see Section 1.2) of a Hilbert space H (not necessarily orthonormal) and define the subspace H_n as the set of all elements $u \in H$ which are of the form

$$u = c_1 e_1 + \cdots + c_n e_n.$$

It follows from the definition of the Schauder basis that $\{H_n\}_{n=1}^{\infty}$ satisfies condition (7.2.20).

Example 7.2.43. Let $H := W_0^{1,2}(0,1)$, $f \in L^1(0,1)$ and

$$F(x) := \frac{1}{2}\int_0^1 |\dot{x}(t)|^2 \,\mathrm{d}t + \frac{1}{4}\int_0^1 |x(t)|^4 \,\mathrm{d}t - \int_0^1 f(t)x(t)\,\mathrm{d}t, \qquad x \in H. \tag{7.2.28}$$

Then F is the energy functional associated with the Dirichlet problem

$$\begin{cases} -\ddot{x}(t) + x^3(t) = f(t), & t \in (0,1), \\ x(0) = x(1) = 0 \end{cases} \tag{7.2.29}$$

(cf. Example 7.2.6). We have

$$DF(x)(y) = \int_0^1 \dot{x}(t)\dot{y}(t)\,\mathrm{d}t + \int_0^1 x^3(t)y(t)\,\mathrm{d}t - \int_0^1 f(t)y(t)\,\mathrm{d}t,$$
$$D^2F(x)(y,y) = \int_0^1 |\dot{y}(t)|^2\,\mathrm{d}t + 3\int_0^1 |x(t)|^2|y(t)|^2\,\mathrm{d}t,$$

and the assumptions of Proposition 7.2.41 are satisfied.[26] The sequence of functions e_i, $i = 1, 2, \dots$, which are defined by

$$e_i(t) := t^i(1-t),$$

constitutes a Schauder basis of the space H (see, e.g., Michlin [111]). Thus, if we construct the subspaces H_n as above, the condition (7.2.20) will be satisfied. If we rewrite the system (7.2.27) for this particular case, we obtain the system of nonlinear equations for unknowns $c_1, \dots, c_n$

$$\sum_{k=1}^{n} c_k \int_0^1 \frac{\mathrm{d}}{\mathrm{d}t}\left[t^k(1-t)\right] \frac{\mathrm{d}}{\mathrm{d}t}\left[t^j(1-t)\right] \mathrm{d}t + \int_0^1 \left[\sum_{k=1}^{n} c_k t^k(1-t)\right]^3 t^j(1-t)\,\mathrm{d}t$$
$$= \int_0^1 f(t)t^j(1-t)\,\mathrm{d}t, \tag{7.2.30}$$

$j = 1, \dots, n$. In each of the equations of system (7.2.30), all unknowns $c_1, \dots, c_n$ appear – this fact is rather unpleasant from the computational point of view!

---

[26] Note that we consider $\|x\| = \left(\int_0^1 (|\dot{x}(t)|^2\,\mathrm{d}t\right)^{\frac{1}{2}}$ as the norm on H.

The question then arises whether it is possible to choose the spaces H_n so that each of the equations of the system (7.2.27) depend on only a “small number” of unknowns. This is one of the fundamental questions of numerical mathematics. Such a choice of H_n is possible, there are different ways how to do it and each of them leads to a particular numerical method. Below we indicate one possible choice of H_n which is different from the previous one and which meets the above mentioned requirements.

Let $n \in \mathbb{N}$, and put $t_i = \frac{i}{n}$ for $i = 0, 1, \dots, n$ and $\mathcal{I}_j = [t_j, t_{j+1}]$ for $j = 0, 1, \dots, n-1$. We define the spaces H_n as follows:

> H_n is the set of functions $x = x(t)$ continuous on the interval $[0,1]$ which are linear on every interval $[t_i, t_{i+1}]$ and for which $x(0) = x(1) = 0$.

Let $e_i \in H_n$, $i = 1, \dots, n-1$, be functions such that

$$e_i(t_j) = \begin{cases} 1 & \text{for} \quad i = j, \\ 0 & \text{for} \quad i \neq j, \end{cases} \qquad j = 0, \dots, n.$$

It is easily established that the set $\{e_i\}_{i=1}^{n-1}$ constitutes a basis of the space H_n and that for all $y \in H_n$ we have

$$y(t) = \sum_{j=1}^{n-1} y(t_j) e_j(t), \qquad t \in [0,1].$$

The system (7.2.27) constructed for this basis will now be itself a system for the unknown values $x_n(t_j)$ of the solution of problem (7.2.18). The crucial point in this construction is the fact that the functions $e_i(t)$ vanish outside the interval $\mathcal{I}_{i-1} \cup \mathcal{I}_i = \left[\frac{i-1}{n}, \frac{i+1}{n}\right]$. We then have

$$e_i(t)e_j(t) = \dot{e}_i(t)\dot{e}_j(t) = 0$$

for $i, j = 1, \dots, n-1$, $|i - j| > 1$ at every point $t \in [0,1]$ (with the obvious exception, for derivatives, of the points $t_1, \dots, t_{n-1}$, which constitute a set of measure zero). Therefore, in each of the equations

$$\sum_{i=1}^{n-1} c_i \int_0^1 \dot{e}_i(t)\dot{e}_j(t)\,\mathrm{d}t + \int_0^1 \left[\sum_{i=1}^{n-1} c_i e_i(t)\right]^3 e_j(t)\,\mathrm{d}t = \int_0^1 f(t)e_j(t)\,\mathrm{d}t, \tag{7.2.31}$$

$j = 1, \dots, n-1$, of system (7.2.27) only the unknowns c_{j-1}, c_{j+1} appear apart from c_j. If we compute a solution $c_1, \dots, c_{n-1}$ from these equations, and if we put

$$u_n(t) = c_1 e_1(t) + \dots + c_{n-1} e_{n-1}(t), \qquad t \in [0,1],$$

we obtain a solution of problem (7.2.18). Now, we wish to know whether

$$\lim_{n \to \infty} \|u_n - u\| = 0.$$

By Proposition 7.2.41, it suffices to show that the spaces H_n satisfy condition (7.2.20). Let $y \in H$ and $\varepsilon > 0$. We shall show that there exist $n \in \mathbb{N}$ and $y_n \in H_n$ such that

$$\|y - y_n\| < \varepsilon. \tag{7.2.32}$$

Indeed, the set $\mathcal{D}(0,1)$ is dense in H (see Exercise 1.2.48). Hence there exists $w \in \mathcal{D}(0,1)$ such that

$$\|y - w\| < \frac{\varepsilon}{2}. \tag{7.2.33}$$

Let $n \in \mathbb{N}$ be arbitrary, and let us construct a function $y_n \in H_n$ such that

$$y_n(t_i) = w(t_i) \qquad \text{for all} \quad i = 0, \dots, n.$$

Then we have (due to the Mean Value Theorem):

$$\begin{aligned}\|w - y_n\| &= \sum_{i=0}^{n-1} \int_{t_i}^{t_{i+1}} |\dot{w}(t) - \dot{y}_n(t)|^2 \,\mathrm{d}t \le \sum_{i=0}^{n-1} \max_{t\in[0,1]} \frac{1}{n^2} |\ddot{w}(t)|^2 (t_{i+1} - t_i) \\ &= \frac{1}{n^2} \max_{t\in[0,1]} |\ddot{w}(t)|.\end{aligned}$$

This implies that for sufficiently large $n \in \mathbb{N}$ we have

$$\|w - y_n\| < \frac{\varepsilon}{2}. \tag{7.2.34}$$

The desired inequality (7.2.32) now follows from (7.2.33) and (7.2.34). ○

Remark 7.2.44.

(i) Let us point out that to get system (7.2.30) it was not essential that an equidistant division of the interval $[0,1]$ has been selected. Nonetheless, the norm of the division (i.e., the maximal distance between two consecutive points) must approach zero.

(ii) The spaces H_n are the simplest which could be chosen for the given example. It is also possible to choose spaces of C^1-functions which are polynomials of higher degree on every interval $\mathcal{I}_i$. For instance, one can choose

$$\begin{aligned}H_n = \{&y \in C^1[0,1] : y(0) = y(1) = 0,\\ &y|_{\mathcal{I}_i} \text{ is a polynomial of the third degree for all } i = 0, \dots, n-1\}.\end{aligned}$$[27]

There exists a basis of this space whose dimension is $2n$ which consists of the functions $e_1, \dots, e_{n-1}, \psi_0, \dots, \psi_n$ such that

$$e_i(t_j) = \begin{cases} 1 & \text{for} \quad i = j, \\ 0 & \text{for} \quad i \ne j, \end{cases} \qquad \dot{e}_i(t_j) = 0, \qquad i = 1, \dots, n-1, \ j = 0, \dots, n;$$

$$\psi_i(t_j) = 0, \qquad \dot{\psi}_i(t_j) = \begin{cases} 1 & \text{for} \quad i = j, \\ 0 & \text{for} \quad i \ne j, \end{cases} \qquad i, j, = 0, \dots, n,$$

see Figure 7.2.5. Every function $y \in H_n$ can be written in the form

$$y(t) = \sum_{j=1}^{n-1} y(t_j) e_j(t) + \sum_{j=0}^{n} \dot{y}(t_j) \psi_j(t), \qquad t \in [0,1].$$

[27] These functions are called *cubic splines* (see, e.g., de Boor [40]).

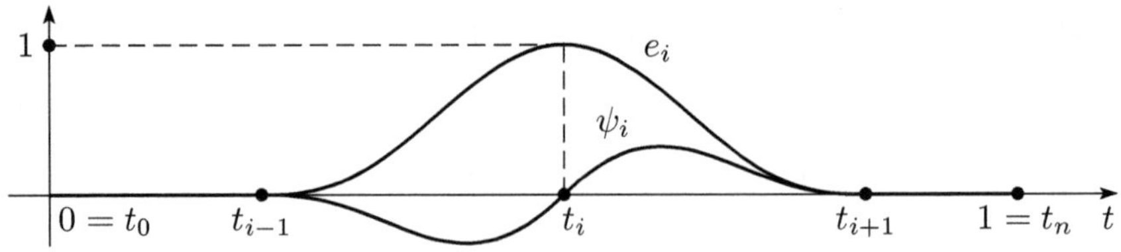

Figure 7.2.5.

(iii) From the computational point of view the question of how rapidly the solutions u_n of problem (7.2.18) converge to a solution of problem (7.2.17) is very important. This question is closely related to the *regularity of solutions* of equations. If, e.g., $f \in C^0[0,1]$, then $u_0 \in C^2[0,1]$ (cf. Proposition 7.1.11 and Theorem 7.1.13) and using this it can be proved that there exists a constant $c > 0$ such that for all $n \in \mathbb{N}$ we have

$$\|u_0 - u_n\| \le \frac{c}{n}.$$

If, e.g., $u_0 \in C^4[0,1]$, then we even have

$$\|u_0 - u_n\| \le \frac{c}{n^3}.$$

Remark 7.2.45 (Finite Elements Method). Similarly to Example 7.2.43 we could proceed even in the case $H := W_0^{k,2}(\Omega)$, $\Omega \subset \mathbb{R}^N$, $\Omega \in C^{0,1}$. The situation then corresponds to the boundary value problem for partial differential equations. Suppose that we can divide the set Ω into a finite number of open subsets Ω_i, $i = 1,\dots,k$, such that their diameter

$$\operatorname{diam}\Omega_i = \sup_{x,y\in\Omega_i} \|x - y\| < \frac{1}{n} \quad \text{and such that} \quad \overline{\Omega} = \bigcup_{i=1}^{k} \overline{\Omega_i},\ \Omega_i \cap \Omega_j = \emptyset \text{ for } i \ne j.$$

Each of the sets $\overline{\Omega_i}$ is called a *finite element*. The space H_n will consist of functions whose restrictions to Ω_i are smooth functions, for instance polynomials in N variables, and satisfy certain conditions on the common boundary of the sets Ω_i and Ω_j $(i \ne j)$. For simplicity and greater intuitive appeal we will consider Ω to be a polygon in $\mathbb{R}^2$ and for every $n \in \mathbb{N}$ we perform a *triangulation* $\mathscr{T}_n$ of the set $\overline{\Omega}$, i.e., we put

$$\overline{\Omega} = \bigcup_{i=1}^{k} \overline{\mathcal{K}_i} \quad \text{where } \mathcal{K}_i \text{ are open triangles such that} \quad \operatorname{diam}\mathcal{K}_i \le \frac{1}{n},\ i = 1,\dots,k,$$

see Figure 7.2.6. Assume that precisely one of the following situations arises for the mutual position of triangles $\mathcal{K}_i, \mathcal{K}_j \in \mathscr{T}_n$ $(i \ne j)$:

(a) the closures of two distinct triangles have no common point;

(b) the closures of two distinct triangles have only one vertex in common;

(c) the closures of two distinct triangles have an entire side in common.

The spaces H_n will be sets of continuous functions whose restrictions to $\mathcal{K}_i$ are polynomials of the kth order. Below, we give examples of the spaces H_n for the case $k = 1$ and

$k = 3$. The continuity of a function $v \in H_n$ is ensured on the set $\overline{\Omega}$ by choosing the values of parameters (used for the construction of the function) to be equal at the common vertices. The reader will find more details in specialized literature on the Finite Elements Method (see, e.g., Brenner & Scott [19], Křížek & Neitaanmäki [95], Rektorys [126]).

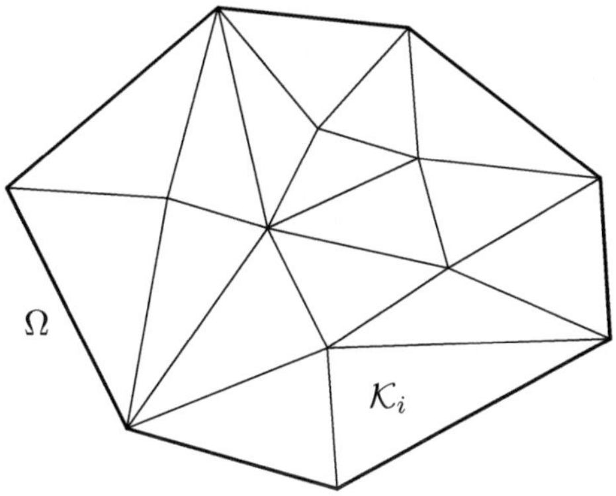

Figure 7.2.6.

Example 7.2.46. We illustrate above approach for $k = 1$. Let $\mathcal{K}$ be an open triangle with vertices Q_1, Q_2, Q_3. Let $\mathcal{P}_1(\mathcal{K})$ be the set of all polynomials of the first degree defined on $\mathcal{K}$, i.e., $P \in \mathcal{P}_1(\mathcal{K})$ if

$$P(x, y) = \alpha_0 + \alpha_1 x + \alpha_2 y, \qquad (x, y) \in \overline{\mathcal{K}}.$$

It is easily shown that any function $P(x, y) \in \mathcal{P}_1(\mathcal{K})$ is uniquely determined by its values at the vertices Q_1, Q_2, Q_3. The values $P(Q_1)$, $P(Q_2)$, $P(Q_3)$ serve as parameters by means of which the function $P(x, y)$ is constructed.

The function $P \in \mathcal{P}_1(\mathcal{K})$ for which

$$P(Q_i) = v(Q_i), \qquad i = 1, 2, 3,$$

is called the *Lagrange interpolation of the function* $v \in C(\overline{\mathcal{K}})$. The function $P(x, y)$ constructed in this way is denoted by $\Pi_{\mathcal{K}} v$. Clearly, $\Pi_{\mathcal{K}}$ is a linear operator from the space $C(\overline{\mathcal{K}})$ into $\mathcal{P}_1(\mathcal{K})$ and

$$\|v - \Pi_{\mathcal{K}} v\|_{W^{1,2}(\mathcal{K})} \le c h_{\mathcal{K}} \|v\|_{W^{2,2}(\mathcal{K})} \tag{7.2.35}$$

holds for arbitrary functions $v \in W^{2,2}(\mathcal{K})$ (here $h_{\mathcal{K}} = \operatorname{diam} \mathcal{K}$ and $c > 0$ is a constant independent of v and $h_{\mathcal{K}}$).[28] Define the space H_n as follows:

$$H_n := \{v \in C(\overline{\Omega}) : v|_{\overline{\mathcal{K}_i}} \in \mathcal{P}_1(\mathcal{K}_i) \text{ for all } \mathcal{K}_i \in \mathscr{T}_n\}.$$

Obviously,

$$H_n \subset H := W^{1,2}(\Omega).$$

[28]The reader is invited to prove it in details!

Let $v \in W^{2,2}(\Omega)$. Construct a function $v_n \in H_n$ in following way:

$$v_n|_{\overline{\mathcal{K}_i}} = \Pi_{\mathcal{K}_i} v.$$

Applying inequality (7.2.35), we obtain

$$\|v - v_n\| \le \frac{c}{n}\|v\|_{W^{2,2}(\Omega)}.$$

Thus, the function v_n is arbitrarily close to the function v provided n is a sufficiently large nonnegative integer. Hence, making use of the fact that the space $W^{2,2}(\Omega)$ is dense in the space H (explain why!), we conclude that the spaces H_n, $n \in \mathbb{N}$, satisfy condition (7.2.20).

We can construct the basis functions $e_1, \dots, e_k$ of H_n just as in Example 7.2.43. If $\{Q_i\}_{i=1}^m$ are all vertices of all triangles of the triangulation $\mathscr{T}_n$, then

$$e_i(Q_j) = \begin{cases} 1 & \text{for} \quad i = j, \\ 0 & \text{for} \quad i \neq j, \end{cases} \qquad j = 1, \dots, m.$$ ○

Example 7.2.47. Now we illustrate the approach from Remark 7.2.45 for $k = 3$. Let $\mathcal{K}$ be an open triangle with vertices Q_1, Q_2, Q_3 and with the center of gravity Q_0. Let $\mathcal{P}_3(\mathcal{K})$ be the set of polynomials of the third degree defined on $\overline{\mathcal{K}}$, i.e., $P \in \mathcal{P}_3(\mathcal{K})$ if

$$P(x, y) = \alpha_0 + \alpha_1 x + \alpha_2 x^2 + \alpha_3 x^3 + \alpha_4 xy + \alpha_5 xy^2 + \alpha_6 x^2 y + \alpha_7 y + \alpha_8 y^2 + \alpha_9 y^3,$$

$(x, y) \in \overline{\mathcal{K}}$. A function $P(x, y) \in \mathcal{P}_3(\mathcal{K})$ is uniquely determined by its values at the vertices and at the center of gravity and by the values of the first partial derivatives at the vertices of the triangle $\mathcal{K}$. A function $\Pi_{\mathcal{K}} v \in \mathcal{P}_3(\mathcal{K})$ for which

$$\Pi_{\mathcal{K}} v(Q_i) = v(Q_i), \qquad i = 0, 1, 2, 3;$$
$$\frac{\partial \Pi_{\mathcal{K}} v(Q_i)}{\partial x} = \frac{\partial v(Q_i)}{\partial x}, \qquad \frac{\partial \Pi_{\mathcal{K}} v(Q_i)}{\partial y} = \frac{\partial v(Q_i)}{\partial y}, \qquad i = 1, 2, 3,$$

is called the *Hermite interpolation of the function* $v \in C^1(\overline{\mathcal{K}})$. Just as in the preceding example, the inequality

$$\|v - \Pi_{\mathcal{K}} v\|_{W^{3,2}(\mathcal{K})} \le c h_{\mathcal{K}} \|v\|_{W^{4,2}(\mathcal{K})} \qquad \text{holds for all} \quad v \in W^{4,2}(\mathcal{K}).$$

If we put

$$H_n := \{v \in C^1(\overline{\mathcal{K}}) : v|_{\mathcal{K}_i} \in \mathcal{P}_3(\mathcal{K}_i) \text{ for every triangle } \mathcal{K}_i \in \mathscr{T}_n\},$$

then

$$H_n \subset H := W^{3,2}(\Omega)$$

and the spaces H_n, $n \in \mathbb{N}$, again satisfy condition (7.2.20) since the set $W^{4,2}(\Omega)$ is dense in the space H. ○

Exercise 7.2.48. Apply the spaces H_n described in Remark 7.2.44(ii) to Example 7.2.43.

7.3 Weak Solutions of PDEs, Applications of Global Minimizers

Let us illustrate an application of Theorem 7.2.8 (the existence of a global minimizer) to the energy functional associated with the boundary value problem

$$\begin{cases} -\Delta u(x) = g(u(x)) + f(x) & \text{in} \quad \Omega, \\ u = 0 & \text{on} \quad \partial\Omega, \end{cases} \tag{7.3.1}$$

where $\Omega \in C^{0,1}$ is a bounded domain in $\mathbb{R}^N$. We will assume that $g = g(s)$ is a continuous function and, moreover,

(i) $f \in L^1(\Omega)$ if $N = 1$;

(ii) $f \in L^{\frac{q}{q-1}}(\Omega)$ and there exists $c > 0$ such that

$$|g(s)| \le c|s|^{q-1} \qquad \text{where} \quad \begin{cases} q \in [1, \infty) \quad \text{is arbitrary if} & N = 2, \\ q \in \left[1, \dfrac{2N}{N-2}\right) \quad \text{if} & N \ge 3. \end{cases}$$

The energy functional[29] associated with (7.3.1) is defined as follows:

$$E(u) := \frac{1}{2}\int_\Omega |\nabla u(x)|^2\,\mathrm{d}x - \int_\Omega G(u(x))\,\mathrm{d}x - \int_\Omega f(x)u(x)\,\mathrm{d}x,^{30} \quad u \in W_0^{1,2}(\Omega),$$

where

$$G(t) = \int_0^t g(s)\,\mathrm{d}s.$$

We will assume that g satisfies the *sign condition*:

$$sg(s) \le 0 \qquad \text{for any} \quad s \in \mathbb{R}. \tag{7.3.2}$$

Then E is a weakly coercive functional on $W_0^{1,2}(\Omega)$. Indeed, the condition (7.3.2) immediately implies $G \le 0$, and thus

$$\int_\Omega G(u(x))\,\mathrm{d}x \le 0 \qquad \text{for any} \quad u \in W_0^{1,2}(\Omega),$$

and so

$$E(u) \ge \frac{1}{2}\|u\|^2_{W_0^{1,2}(\Omega)} - \|f\|_X\|u\|_{X'},$$

---

[29] Here f represents the influence of external forces, g a nonlinear damping or restoring force, and $\frac{1}{2}|\nabla u|^2$ the kinetic energy, respectively. Hence $E(u)$ corresponds to the total energy of the system (cf. Examples 7.2.6 and 7.2.14) – that is where the expression "energy functional" comes from. For more details cf. Hlaváček & Nečas [81].

[30] The reader is invited to check that if $u_0 \in W_0^{1,2}(\Omega)$ satisfies $\delta E(u_0; v) = 0$ for any $v \in W_0^{1,2}(\Omega)$, then it is a weak solution of (7.3.1).

where $X = L^1(\Omega)$, $X' = C(\overline{\Omega})$ if $N = 1$, $X = L^{\frac{q}{q-1}}(\Omega)$, $X' = L^q(\Omega)$ if $N \geq 2$. Hence there is a constant $c_1 > 0$ such that

$$E(u) \geq \frac{1}{2}\|u\|^2_{W_0^{1,2}(\Omega)} - c_1\|u\|_{W_0^{1,2}(\Omega)}.$$

Hence

$$\lim_{\|u\|\to\infty} E(u) = \infty.$$

Next we prove that E is weakly sequentially lower semicontinuous. Clearly, for any $u_n \rightharpoonup u_0$ in $W_0^{1,2}(\Omega)$ we have

$$\frac{1}{2}\int_\Omega |\nabla u_0(x)|^2\,\mathrm{d}x \leq \liminf_{n\to\infty} \frac{1}{2}\int_\Omega |\nabla u_n(x)|^2\,\mathrm{d}x, \tag{7.3.3}$$

$$\int_\Omega f(x)u_0(x)\,\mathrm{d}x = \lim_{n\to\infty}\int_\Omega f(x)u_n(x)\,\mathrm{d}x. \tag{7.3.4}$$

Our assumptions imply that for $N \geq 2$ the function $G(s)$ satisfies the estimate

$$|G(s)| \leq \frac{c}{q}|s|^q.$$

So, the Nemytski operator defined by

$$N_G(u)(x) = G(u(x))$$

is continuous from $L^q(\Omega)$ into $L^1(\Omega)$ (see Theorem 3.2.24). Then this fact and the compact embedding $W_0^{1,2}(\Omega) \hookrightarrow\hookrightarrow L^q(\Omega)$ imply that $u_n \to u_0$ in $L^q(\Omega)$, and thus

$$G(u_n(\cdot)) \to G(u(\cdot)) \qquad \text{in} \quad L^1(\Omega). \tag{7.3.5}$$

Since $u_n \to u_0$ in $C(\overline{\Omega})$ for $N = 1$ we obtain (7.3.5) easily in this case, too.

Summarizing (7.3.3)–(7.3.5) we obtain

$$E(u_0) \leq \liminf_{n\to\infty} E(u_n).$$

We have proved the following assertion.

Theorem 7.3.1. *Let g be a continuous function satisfying* (7.3.2) *and* (ii) *(for $N \geq 2$) , let f satisfy* (i) *(for $N = 1$) or* (ii) *(for $N \geq 2$). Then the boundary value problem* (7.3.1) *has at least one weak solution $u_0 \in W_0^{1,2}(\Omega)$.*

The growth assumptions on g stated above for $N \geq 3$ are not optimal and can be relaxed if we assume the monotonicity of g and apply Theorem 7.2.12. Indeed, let $G(s) = \int_0^s g(\tau)\,\mathrm{d}\tau$ be a concave function.[31] Then the energy functional E

[31] In particular, if g is decreasing, then G is concave.

is a strictly convex functional. This functional is continuous even if the growth assumption (ii) for $N \geq 3$ is relaxed to

$$q \in \left[1, \frac{2N}{N-2}\right],$$

i.e., the value $q = \frac{2N}{N-2}$ is admissible, too.[32] Hence we have the following assertion.

Theorem 7.3.2. *Let f and g be as in Theorem* 7.3.1. *If g is decreasing, then* (7.3.1) *has a unique weak solution. The assertion remains true even if g satisfies* (ii) *with $q = \frac{2N}{N-2}$ for $N \geq 3$.*

7.4 Mountain Pass Theorem

One of the most efficient tools to prove that a given functional having a local extremum at a point possesses another critical point is the Mountain Pass Theorem. In order to motivate the main ideas of this section we will consider a real function of two real independent variables

$$F \colon \mathbb{R} \times \mathbb{R} \to \mathbb{R}$$

which is continuously differentiable and satisfies the following condition:

There exist $r > 0$, $e \in \mathbb{R}^2$, $\|e\| > r$ such that

$$\inf_{\|x\|=r} F(x) > F(o) \geq F(e). \tag{7.4.1}$$

The graph of such a function is sketched in Figure 7.4.1.

The Extreme Value Theorem and the first inequality in (7.4.1) immediately imply that F has a local minimum and thus a critical point in the set

$$\{x \in \mathbb{R}^2 : \|x\| < r\}.$$

Hiker's experience suggests the idea that F should have another critical point different from that local minimum. Indeed, if the values of F are interpreted as mountains on the plastic map, then the valley (containing the origin) is surrounded by mountains. At the same time the altitude of every place the distance of which from the origin is equal to r is greater than that of the origin itself. So, there should be an "optimal pass" through the mountain range. Practical experience even suggests how to find such a critical point. Let us consider all continuous finite paths which lie on the graph of F and which connect the points $(o, F(o))$ and $(e, F(e))$. On every curve we have at least one "highest" point. It seems that if we select the "highest" point with the "lowest" altitude, we have found a critical point of F.

[32] Use Theorem 3.2.24 and explain why!

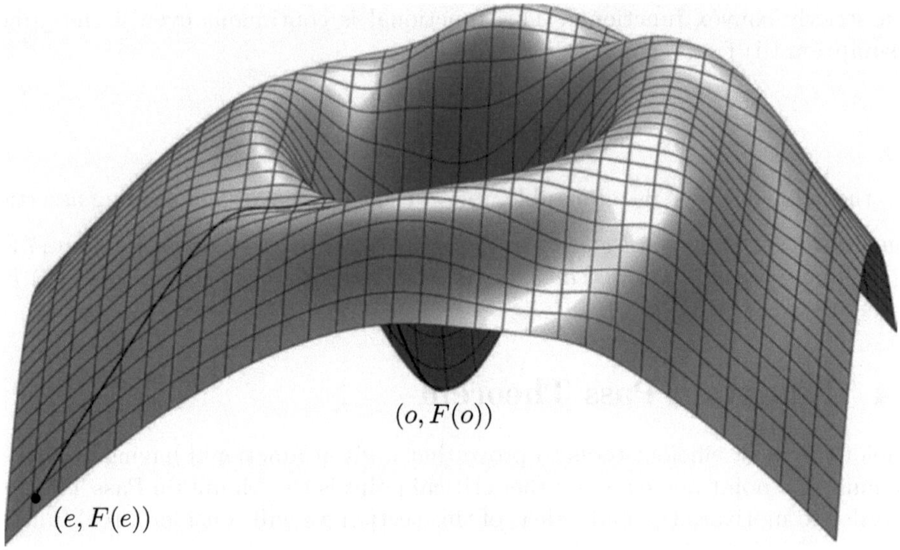

Figure 7.4.1.

If we formulate precisely the considerations made above, then the "lowest" altitude of the "highest" points corresponds to the value

$$c := \inf_{\gamma\in\Gamma} \max_{t\in[0,1]} F(\gamma(t)) \tag{7.4.2}$$

where

$$\Gamma = \{\gamma \in C([0,1],\mathbb{R}^2) : \gamma(0) = o, \gamma(1) = e\}.$$

If c is a critical value of F, then there exists $x_c \in \mathbb{R}^2$ such that

$$F(x_c) = c \qquad \text{and} \qquad F'(x_c) = o.$$

However, the value c defined above need not be a critical value of F! An example which illustrates this phenomenon is rather elementary.

Example 7.4.1 (Brézis–Nirenberg). Let

$$F(x,y) = x^2 + (1-x)^3y^2 \qquad \text{and} \qquad r = \frac{1}{2}, \qquad e = (2,2).$$

Then

$$F(o) = F(e) = 0, \qquad \inf_{\|(x,y)\|=r} F(x,y) = \min_{\|(x,y)\|=r} F(x,y) > 0,$$

and so the value c defined by (7.4.2) is positive. Since

$$\frac{\partial F}{\partial x}(x,y) = 2x - 3(1-x)^2y^2, \qquad \frac{\partial F}{\partial y}(x,y) = 2(1-x)^3y,$$

the origin is the only critical point of F and obviously $F(o) < c$. The reader is invited to sketch the level sets of the function F. ○

It is natural to ask why this happens. Such a situation corresponds, roughly speaking, to the fact that the altitude of the "highest" points approaches the value of c but the distance of these points from the origin diverges to infinity.

More precisely, if $x_n \in \mathbb{R}^2$ are such that

$$F(x_n) = \max_{t\in[0,1]} F(\gamma_n(t)) \qquad \text{for} \quad \gamma_n \in \Gamma$$

and

$$F(x_n) \to c, \tag{7.4.3}$$

then

$$\|x_n\| \to \infty. \tag{7.4.4}$$

It follows that the existence of $r > 0$ and $e \in \mathbb{R}^2$ satisfying (7.4.1) is not sufficient to guarantee the existence of a critical point which is different from the local minimum in $\{x \in \mathbb{R}^2 : \|x\| \le r\}$. On the other hand, we will prove later that (7.4.1) guarantees the existence of a sequence $\{x_n\}_{n=1}^{\infty} \subset \mathbb{R}^2$ such that

$$F(x_n) \to c \qquad \text{and} \qquad \nabla F(x_n) \to o. \tag{7.4.5}$$

Now, let us assume for a moment that F satisfies the following condition:

($\widehat{\text{PS}}$) Let $\{x_n\}_{n=1}^{\infty} \subset \mathbb{R}^2$ be such that $\{F(x_n)\}_{n=1}^{\infty}$ is a bounded sequence in $\mathbb{R}$ and $\nabla F(x_n) \to o$. Then $\{x_n\}_{n=1}^{\infty}$ is a bounded sequence in $\mathbb{R}^2$.

Then the situation described in (7.4.3) and (7.4.4) cannot occur. Moreover, (7.4.5) together with ($\widehat{\text{PS}}$) already implies that c is a critical value of F. Indeed, let $\{x_n\}_{n=1}^{\infty}$ satisfy (7.4.5). According to ($\widehat{\text{PS}}$) there exists a subsequence $\{x_{n_k}\}_{k=1}^{\infty} \subset \{x_n\}_{n=1}^{\infty}$ such that $x_{n_k} \to x_c$. The continuous differentiability of F and (7.4.5) imply that

$$\nabla F(x_c) = o.$$

Let us consider a more general situation, namely

$$F\colon H \to \mathbb{R}$$

where H is a real Hilbert space with a scalar product $(\cdot,\cdot)$ and the induced norm $\|\cdot\|$. In order to simplify the proofs we will also require more smoothness of F, i.e., let $F \in C^2(H,\mathbb{R})$. For the sake of brevity for given $d \in \mathbb{R}$ we will denote

$$\mathcal{F}^d := F_{-1}((-\infty, d])$$

sublevel set of F. The key assertion of this section is the following *Quantitative Deformation Lemma.* The reader should have in mind that it can be proved under more general assumptions (H a Banach space, $F \in C^1(H,\mathbb{R})$) – see also footnotes 35, 36 and 37 on pages 482–485.

Lemma 7.4.2 (Quantitative Deformation Lemma). *Let H be a real Hilbert space and F a C^2-functional, $c \in \mathbb{R}$, $\varepsilon > 0$. Assume that*

$$\|\nabla F(u)\| \geq 2\varepsilon \qquad \textit{for any} \quad u \in F_{-1}([c-2\varepsilon, c+2\varepsilon]).$$

Then there exists $\eta \in C(H, H)$ such that

(i) $\eta(u) = u$ *for any* $u \notin F_{-1}([c-2\varepsilon, c+2\varepsilon])$,

(ii) $\eta(\mathcal{F}^{c+\varepsilon}) \subset \mathcal{F}^{c-\varepsilon}$.

Proof. Let us introduce closed sets

$$\mathcal{A} = F_{-1}([c-2\varepsilon, c+2\varepsilon]), \qquad \mathcal{B} = F_{-1}([c-\varepsilon, c+\varepsilon])$$

(see Figure 7.4.2) and a functional

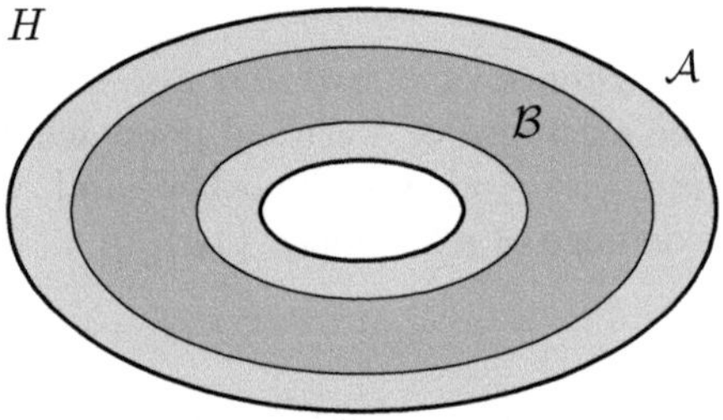

Figure 7.4.2.

$$\psi(u) = \frac{\operatorname{dist}(u, H \setminus \mathcal{A})}{\operatorname{dist}(u, H \setminus \mathcal{A}) + \operatorname{dist}(u, \mathcal{B})}.$$ [33]

Then

$$\psi = \begin{cases} 1 & \text{on} \quad \mathcal{B}, \\ 0 & \text{on} \quad H \setminus \mathcal{A}, \end{cases} \qquad \psi \text{ is locally Lipschitz continuous,}^{34} \qquad 0 \leq \psi \leq 1.$$

Let us define a vector field

$$f(u) = \begin{cases} -\psi(u)\dfrac{\nabla F(u)}{\|\nabla F(u)\|^2} & \text{for} \quad u \in \mathcal{A}, \\ o & \text{for} \quad u \in H \setminus \mathcal{A}. \end{cases}$$

Then f is also locally Lipschitz continuous[35] and for any $u \in H$ we have

$$\|f(u)\| \leq \frac{1}{2\varepsilon}.$$

[33] Recall that $\operatorname{dist}(u, \mathcal{C}) := \inf\{\|u - v\|_X : v \in \mathcal{C}\}$ for $u \in X$, $\mathcal{C} \subset X$.

[34] Cf. Exercise 7.4.8.

[35] Here the assumption $F \in C^2(H, \mathbb{R})$ is essentially used (cf. Exercise 7.4.9).

Indeed, for $u \in H \setminus \mathcal{A}$ we have $f(u) = 0$ and for $u \in \mathcal{A}$ we have

$$\|f(u)\| \le |\psi(u)| \frac{\|\nabla F(u)\|}{\|\nabla F(u)\|^2} \le \frac{1}{\|\nabla F(u)\|} \le \frac{1}{2\varepsilon}.$$

Consider the Cauchy problem

$$\begin{cases} \dot\sigma = f(\sigma), \\ \sigma(0) = u. \end{cases} \tag{7.4.6}$$

It follows from Corollary 3.1.6 that (7.4.6) has a unique solution, denoted by $\sigma(\cdot, u)$, which is defined on $\mathbb{R}$ for any $u \in H$, and for any $t > 0$, $\sigma(t, \cdot)\colon H \to H$ is continuous (continuous dependence on the initial data – see Remark 3.1.7). Let us define

$$\eta(u) = \sigma(2\varepsilon, u), \qquad u \in H.$$

We will prove that η satisfies (i) and (ii). Property (i) follows from the fact that $f(u) = 0$ for $u \in H \setminus \mathcal{A}$. Let us prove that (ii) is also satisfied. Since

$$\begin{aligned} \frac{\mathrm{d}}{\mathrm{d}t} F(\sigma(t,u)) &= \left(\nabla F(\sigma(t,u)), \frac{\mathrm{d}}{\mathrm{d}t}\sigma(t,u)\right) \\ &= (\nabla F(\sigma(t,u)), f(\sigma(t,u))) = -\psi(\sigma(t,u)), \end{aligned} \tag{7.4.7}$$

the function $t \mapsto F(\sigma(t,u))$ is decreasing. Let $u \in \mathcal{F}^{c+\varepsilon}$, i.e., $F(u) \le c + \varepsilon$. We have to show that $F(\sigma(2\varepsilon, u)) \le c - \varepsilon$. If there is $t \in [0, 2\varepsilon]$ such that

$$F(\sigma(t,u)) \le c - \varepsilon,$$

then also $F(\sigma(2\varepsilon, u)) \le c - \varepsilon$ and (ii) is satisfied. If, on the other hand,

$$c - \varepsilon < F(\sigma(t,u)) \le c + \varepsilon \qquad \text{for all} \quad t \in [0, 2\varepsilon], \qquad \text{i.e.,} \qquad \psi(\sigma(t,u)) = 1,$$

then we obtain from (7.4.7)

$$\begin{aligned} F(\sigma(2\varepsilon, u)) &= F(u) + \int_0^{2\varepsilon} \frac{\mathrm{d}}{\mathrm{d}t} F(\sigma(t,u))\,\mathrm{d}t = F(u) - \int_0^{2\varepsilon} \psi(\sigma(t,u))\,\mathrm{d}t \\ &\le c + \varepsilon - 2\varepsilon = c - \varepsilon, \end{aligned}$$

a contradiction, and so (ii) is also satisfied. □

The Quantitative Deformation Lemma provides a tool for proving the existence of "almost critical" points of functionals which have the so-called *mountain pass type geometry* (see (7.4.8) below).

Proposition 7.4.3. *Let $F \in C^2(H, \mathbb{R})$, $e \in H$ and $r > 0$ be such that $\|e\| > r$ and*

$$b := \inf_{\|u\|=r} F(u) > F(o) \ge F(e). \tag{7.4.8}$$

Let

$$c := \inf_{\gamma\in\Gamma} \max_{t\in[0,1]} F(\gamma(t)) \qquad \text{and} \qquad \Gamma := \{\gamma \in C([0,1], H) : \gamma(0) = o, \gamma(1) = e\}.$$

Then for each $\varepsilon > 0$ *there exists* $u \in H$ *such that*

(i) $c - 2\varepsilon \le F(u) \le c + 2\varepsilon$,

(ii) $\|\nabla F(u)\| < 2\varepsilon$.

Proof. Let $\gamma \in \Gamma$ be arbitrary. Then (7.4.8) implies

$$b \le \max_{t\in[0,1]} F(\gamma(t)), \qquad \text{and so} \qquad b \le c.$$

Without loss of generality, we can restrict ourselves to ε small, satisfying

$$c - 2\varepsilon > F(o) \ge F(e). \tag{7.4.9}$$

Suppose that the conclusion of the proposition is not satisfied for an $\varepsilon > 0$, i.e., for each $u \in H$ satisfying (i) condition (ii) is violated. We apply Lemma 7.4.2 to get a contradiction. By the definition of c, there exists $\gamma \in \Gamma$ such that

$$\max_{t\in[0,1]} F(\gamma(t)) < c + \varepsilon. \tag{7.4.10}$$

Consider $\beta(t) = \eta(\gamma(t))$ where η is from Lemma 7.4.2. Using Lemma 7.4.2(i), $\gamma(0) = o$, $\gamma(1) = e$, and (7.4.9) we conclude

$$\beta(0) = \eta(o) = o \qquad \text{and} \qquad \beta(1) = \eta(e) = e.$$

Hence $\beta \in \Gamma$, i.e., $c \le \max\limits_{t\in[0,1]} F(\beta(t))$. It follows from Lemma 7.4.2(ii) and (7.4.10) that

$$\max_{t\in[0,1]} F(\beta(t)) \le c - \varepsilon,$$

a contradiction. □

In order to prove that c is a critical value, our functional F has to satisfy a "compactness" condition of type $(\widehat{\text{PS}})$. However, in infinite dimensions we have to strengthen its formulation.

Definition 7.4.4. Let $F \in C^2(H,\mathbb{R})$ and $c \in \mathbb{R}$. The functional F satisfies the *Palais–Smale condition on the level* c (shortly $(\text{PS})_c$) if any sequence $\{u_n\}_{n=1}^{\infty} \subset H$ such that

$$F(u_n) \to c, \qquad \nabla F(u_n) \to o \tag{7.4.11}$$

has a convergent (in the norm of H) subsequence.[36]

[36]Definition 7.4.4 in a more general setting with H replaced by a Banach space X and $F \in C^1(X,\mathbb{R})$ is due to Brézis, Corón, Nirenberg. Cf. Corollary 7.4.20.

Now we are ready to formulate the *Mountain Pass Theorem* which is the simplest and one of the most useful "variational" theorems. It is one of the most efficient tools to prove *the existence of at least two critical points of a given functional* (see, e.g., Example 7.4.7).

Theorem 7.4.5 (Mountain Pass Theorem). *Let the assumptions of Proposition 7.4.3 be satisfied. Let F satisfy* $(\mathrm{PS})_c$. *Then c is a critical value of F.*[37]

Proof. It follows from Proposition 7.4.3 that there is a sequence $\{u_n\}_{n=1}^{\infty} \subset H$ satisfying (7.4.11). By $(\mathrm{PS})_c$ there exist $\{u_{n_k}\}_{k=1}^{\infty} \subset \{u_n\}_{n=1}^{\infty}$ and $u_0 \in H$ such that $u_{n_k} \to u_0$. But $F \in C^1(H,\mathbb{R})$ implies that

$$F(u_0) = c \qquad \text{and} \qquad \nabla F(u_0) = o. \qquad \square$$

Remark 7.4.6. Theorem 7.4.5 actually states that there exists a critical point $u_0 \neq o$ of F since

$$c \geq \inf_{\|u\|=r} F(u) > F(o).$$

Example 7.4.7. Let us consider the boundary value problem

$$\begin{cases} -\ddot{x}(t) + \lambda x(t) = |x(t)|^{p-2}x(t), & t \in (0,\pi), \\ x(0) = x(\pi) = 0, \end{cases} \tag{7.4.12}$$

where $p > 2$ is a given real number and $\lambda \in \mathbb{R}$ is a parameter. Notice that the function identically equal to zero is a solution. We will prove that problem (7.4.12) has also a positive C^2-solution on $(0,\pi)$ if and only if

$$\lambda > -1. \tag{7.4.13}$$

Let us prove that (7.4.13) is a *necessary condition.* Let $x \in C^2[0,\pi]$ be a positive solution of (7.4.12). Multiply the equation in (7.4.12) by $\sin t$ and integrate by parts:

$$\begin{aligned} \lambda \int_0^{\pi} x(t)\sin t \,\mathrm{d}t &= \int_0^{\pi} |x(t)|^{p-2}x(t)\sin t\,\mathrm{d}t + \int_0^{\pi} \ddot{x}(t)\sin t\,\mathrm{d}t \\ &> \int_0^{\pi} \ddot{x}(t)\sin t\,\mathrm{d}t = -\int_0^{\pi} x(t)\sin t\,\mathrm{d}t. \end{aligned}$$

Hence (7.4.13) follows.

Next we show that (7.4.13) is also a *sufficient condition.* Let us define the following two functions from $\mathbb{R}$ into $\mathbb{R}$:

$$g(s) = \begin{cases} 0 & \text{for} \quad s \leq 0, \\ s^{p-1} & \text{for} \quad s > 0, \end{cases} \qquad G(s) = \begin{cases} 0 & \text{for} \quad s \leq 0, \\ \dfrac{1}{p}s^p & \text{for} \quad s > 0. \end{cases}$$

[37] The assertion of Theorem 7.4.5 where H is replaced by a Banach space X and $F \in C^1(X,\mathbb{R})$ is due to Ambrosetti, Rabinowitz, see Theorem 7.4.25.

Then $G \in C^2(\mathbb{R})$ and $G'(s) = g(s)$ for all $s \in \mathbb{R}$ (remember that $p > 2$). Put $H := W_0^{1,2}(0,\pi)$ and define

$$F(x) := \frac{1}{2}\int_0^\pi |\dot{x}(t)|^2 \,\mathrm{d}t + \frac{\lambda}{2}\int_0^\pi |x(t)|^2 \,\mathrm{d}t - \int_0^\pi G(|x(t)|)\,\mathrm{d}t, \qquad x \in H.$$

Then $F \in C^2(H,\mathbb{R})$ (see Exercise 7.4.10). We shall verify the assumptions of Theorem 7.4.5.

Note that for $\lambda > -1$ the expression

$$|||x||| := \left(\int_0^\pi |\dot{x}(t)|^2 \,\mathrm{d}t + \lambda \int_0^\pi |x(t)|^2 \,\mathrm{d}t\right)^{\frac{1}{2}}$$

satisfies

$$c_1\|x\| \le |||x||| \le c_2\|x\| \qquad \text{for any} \quad x \in H \tag{7.4.14}$$

where $c_i > 0$, $i = 1,2$, are constants independent of x and

$$\|x\| = \left(\int_0^\pi |\dot{x}(t)|^2 \,\mathrm{d}t\right)^{\frac{1}{2}}$$

(cf. Exercise 7.8.18).

Let us show that the functional F has a mountain pass type geometry. It follows from the Sobolev Embedding Theorem (Theorem 1.2.26) that

$$\left(\int_0^\pi |x(t)|^p \,\mathrm{d}t\right)^{\frac{1}{p}} \le c_p \left(\int_0^\pi |\dot{x}(t)|^2 \,\mathrm{d}t\right)^{\frac{1}{2}}. \tag{7.4.15}$$

Hence combining (7.4.14) and (7.4.15) we obtain

$$\begin{aligned} F(x) &= \frac{1}{2}\int_0^\pi |\dot{x}(t)|^2 \,\mathrm{d}t + \frac{\lambda}{2}\int_0^\pi |x(t)|^2 \,\mathrm{d}t - \frac{1}{p}\int_0^\pi |x(t)|^p \,\mathrm{d}t \\ &\ge \frac{1}{2}|||x|||^2 - \frac{1}{p}\left(\frac{c_p}{c_1}\right)^p |||x|||^p = |||x|||^2\left[\frac{1}{2} - \frac{1}{p}\left(\frac{c_p}{c_1}\right)^p |||x|||^{p-2}\right]. \end{aligned}$$

So, because $p > 2$, due to (7.4.14) there exists $r > 0$ (small enough) such that

$$b = \inf_{\|x\|=r} F(x) > 0 = F(o).$$

Let $x \in H$, $x > 0$ in $(0,\pi)$. Then for $s \ge 0$ we have

$$\frac{1}{s^p}F(sx) = \frac{s^{2-p}}{2}\left(\int_0^\pi |\dot{x}(t)|^2 \,\mathrm{d}t + \lambda\int_0^\pi |x(t)|^2 \,\mathrm{d}t\right) - \frac{1}{p}\int_0^\pi |x(t)|^p \,\mathrm{d}t.$$

For $s > 0$ set $e = sx$. Then for s large we obtain

$$\|e\| > r \qquad \text{and} \qquad F(e) \le 0.$$

It remains to verify that F satisfies the $(\mathrm{PS})_c$ condition. Actually, we will verify that F satisfies even a stronger version of $(\mathrm{PS})_c$. Namely, we will prove that any sequence $\{x_n\}_{n=1}^\infty \subset H$ satisfying

$$d := \sup_n F(x_n) < \infty, \qquad \nabla F(x_n) \to o, \tag{7.4.16}$$

contains a convergent subsequence.[38] A typical scheme of the proof is the following: In Step 1 we prove that $\{x_n\}_{n=1}^\infty$ is a *bounded sequence*. In Step 2 we pass to a weakly convergent subsequence and show that it *converges strongly* as well.

Step 1. For n large enough, we have by (7.4.16)[39]

$$\begin{aligned} d + \|x_n\| &\geq F(x_n) - \frac{1}{p}(\nabla F(x_n), x_n) \\ &= \left(\frac{1}{2} - \frac{1}{p}\right)\left(\int_0^\pi |\dot{x}_n(t)|^2\,\mathrm{d}t + \lambda \int_0^\pi |x_n(t)|^2\,\mathrm{d}t\right) \geq c_1^2\left(\frac{1}{2} - \frac{1}{p}\right)\|x_n\|^2. \end{aligned}$$

It follows from this quadratic inequality that $\|x_n\|$ is bounded.

Step 2. Passing to a subsequence if necessary, we can assume that

$$x_n \rightharpoonup x \quad \text{in} \quad H.$$

By the compact embedding $H = W_0^{1,2}(0,\pi) \hookrightarrow\hookrightarrow C[0,\pi]$ (see Theorem 1.2.13) we have

$$x_n \to x \quad \text{in} \quad C[0,\pi], \qquad \text{and so} \qquad g(x_n) \to g(x) \quad \text{in} \quad C[0,\pi].$$

Observe that we have

$$\begin{aligned} |||x_n - x|||^2 &= (\nabla F(x_n) - \nabla F(x), x_n - x) \\ &\quad + \int_0^\pi (g(x_n(t)) - g(x(t)))(x_n(t) - x(t))\,\mathrm{d}t. \end{aligned} \tag{7.4.17}$$

It is clear that

$$(\nabla F(x_n) - \nabla F(x), x_n - x) \to 0 \qquad \text{as} \quad n \to \infty$$

(cf. (7.4.16)). The uniform convergence of $\{x_n\}_{n=1}^\infty$ and $\{g(x_n)\}_{n=1}^\infty$ implies that also

$$\int_0^\pi (g(x_n(t)) - g(x(t)))(x_n(t) - x(t))\,\mathrm{d}t \to 0 \qquad \text{as} \quad n \to \infty.$$

Thus it follows from (7.4.17) that

$$|||x_n - x||| \to 0 \qquad \text{as} \quad n \to \infty, \qquad \text{i.e.,} \quad x_n \to x \quad \text{in } H$$

due to (7.4.14).

[38] The reader should justify that $(\mathrm{PS})_c$ in the sense of Definition 7.4.4 is satisfied as well.

[39] Due to (7.4.16) we can actually assume that $\left|\frac{1}{p}(\nabla F(x_n), x_n)\right| \leq \|x_n\|$.

It follows from Theorem 7.4.5 that there exists a critical point $x_0 \in H$ of F (and hence a weak solution of (7.4.12)) with $F(x_0) = c \geq b > 0$. In particular, $x_0 \neq o$. We prove that $x_0 > 0$ in $(0, \pi)$. Indeed,

$$\int_0^\pi \dot{x}_0(t)\dot{y}(t)\,\mathrm{d}t + \lambda \int_0^\pi x_0(t)y(t)\,\mathrm{d}t = \int_0^\pi g(x_0(t))y(t)\,\mathrm{d}t \text{ holds for any } y \in H.$$

Taking $y = x_0^-$,[40] we get

$$\int_0^\pi \left| \frac{\mathrm{d}x_0^-(t)}{\mathrm{d}t} \right|^2 \mathrm{d}t + \lambda \int_0^\pi |x_0^-(t)|^2\,\mathrm{d}t = 0. \tag{7.4.18}$$

Hence $|||x_0^-||| = 0$, i.e., $x_0(t) \geq 0$ for all $t \in [0, \pi]$ due to (7.4.14).

A similar argument to that used in Section 7.1 yields that $x_0 \in C^2[0, \pi]$ (cf. Exercise 7.4.11). Now, if there were $t_0 \in (0, \pi)$ such that $x_0(t_0) = 0$, then, due to $x_0^- \equiv 0$, also $\dot{x}_0(t_0) = 0$ would hold. However, the uniqueness theorem for the second order initial value problem

$$\begin{cases} -\ddot{x}(t) = -\lambda x(t) + |x(t)|^{p-2}x(t), \\ x(t_0) = \dot{x}(t_0) = 0 \end{cases}$$

implies that $x_0 \equiv 0$, i.e., a contradiction to $x_0 \neq o$. Hence $x_0 > 0$ in $(0, \pi)$. ○

Exercise 7.4.8. Prove that ψ defined in the proof of Lemma 7.4.2 is a locally Lipschitz continuous functional on H.

Hint. For u_1, u_2 from a bounded set we have

$$\operatorname{dist}(u_i, H \setminus \mathcal{A}) + \operatorname{dist}(u_i, \mathcal{B}) \geq \delta, \qquad i = 1, 2, \qquad \text{with a} \quad \delta \geq 0.$$

Using the triangle inequality prove that

$$\begin{aligned} |\psi(u_2) - \psi(u_1)| \leq {} & \frac{\operatorname{dist}(u_2, H \setminus \mathcal{A})}{\delta^2} |\operatorname{dist}(u_1, \mathcal{B}) - \operatorname{dist}(u_2, \mathcal{B})| \\ & + \frac{\operatorname{dist}(u_2, \mathcal{B})}{\delta^2} |\operatorname{dist}(u_1, H \setminus \mathcal{A}) - \operatorname{dist}(u_2, H \setminus \mathcal{A})|, \end{aligned}$$

and then apply Exercise 1.2.47.

Exercise 7.4.9. Prove that f defined in the proof of Lemma 7.4.2 is a locally Lipschitz continuous map from H into itself.

Hint. Use the facts that ψ is locally Lipschitz continuous (Exercise 7.4.8) and $F \in C^2(H, \mathbb{R})$.

Exercise 7.4.10. Prove that the functional F from Example 7.4.7 satisfies $F \in C^2(H, \mathbb{R})$.

---

[40] Here $x^- = \max\{0, -x\}$. One can prove that for $x \in W_0^{1,2}(0, \pi)$ we have $x^- \in W_0^{1,2}(0, \pi)$ (cf. Exercise 1.2.49 and the embedding of $W_0^{1,2}(0, \pi)$ into $C[0, \pi]$).

Hint. Use the fact that G defined in Example 7.4.7 belongs to $C^2(\mathbb{R})$ if $p > 2$.

Exercise 7.4.11. Prove that $x_0 \in C^2[0, \pi]$ for any weak solution x_0 of (7.4.12).

Hint. Look at the proof of Proposition 7.1.11.

Exercise 7.4.12. Prove that for all $x \in W_0^{1,2}(0, \pi)$ the inequality

$$\int_0^\pi |x(t)|^2 \,\mathrm{d}t \le \int_0^\pi |\dot{x}(t)|^2 \,\mathrm{d}t \qquad \text{holds true.}$$

Exercise 7.4.13. Consider the boundary value problem

$$\begin{cases} -\ddot{x}(t) - \lambda x(t) = g(t, x(t)), & t \in (0, \pi), \\ x(0) = x(\pi) = 0. \end{cases} \tag{7.4.19}$$

Formulate conditions on λ and g which guarantee that the energy functional associated with (7.4.19) has a geometry corresponding to the Mountain Pass Theorem.

Exercise 7.4.14. Consider the Neumann boundary value problem

$$\begin{cases} -\ddot{x}(t) = h(t, x(t)), & t \in (0, \pi), \\ \dot{x}(0) = \dot{x}(\pi) = 0. \end{cases} \tag{7.4.20}$$

Formulate conditions on $h = h(t, x)$ which guarantee the existence of a weak solution of (7.4.20).

Exercise 7.4.15. Consider the Dirichlet boundary value problem for the fourth order equation

$$\begin{cases} \dfrac{\mathrm{d}^4 x(t)}{\mathrm{d}t^4} = h(t, x(t)), & t \in (0, \pi), \\ x(0) = \dot{x}(0) = x(\pi) = \dot{x}(\pi) = 0. \end{cases} \tag{7.4.21}$$

Formulate conditions on $h = h(t, x)$ which guarantee the existence of a weak solution of (7.4.21).

7.4A Pseudogradient Vector Fields in Banach Spaces

The aim of this appendix is to show how to extend the Quantitative Deformation Lemma (Lemma 7.4.2) to continuously differentiable functionals defined on a Banach space. For this purpose the notion of the *pseudogradient* introduced by Palais is crucial.

Definition 7.4.16. Let M be a separable metric space, X a normed linear space and $h\colon M \to X^* \setminus \{o\}$ a continuous mapping. A *pseudogradient vector field* for h on M is a locally Lipschitz continuous map $g\colon M \to X$ such that for every $u \in M$,

$$\|g(u)\|_X \le 2\|h(u)\|_{X^*}, \qquad \langle h(u), g(u)\rangle_X \ge \|h(u)\|_{X^*}^2.$$

Lemma 7.4.17. *For any h as above there exists a pseudogradient vector field for h on M.*

Proof. For every $v \in M$ there exists $x \in X$ such that

$$\|x\| = 1 \qquad \text{and} \qquad \langle h(v), x\rangle > \frac{2}{3}\|h(v)\|.^{41}$$

Define $y := \frac{3}{2}\|h(v)\|x$. Then

$$\|y\| < 2\|h(v)\| \qquad \text{and} \qquad \langle h(v), y\rangle > \|h(v)\|^2.$$

Since h is continuous, there exists an open neighborhood $\mathcal{U}(v) \subset M$ such that

$$\|y\| \le 2\|h(u)\| \qquad \text{and} \qquad \langle h(u), y\rangle \ge \|h(u)\|^2 \qquad \text{for every} \quad u \in \mathcal{U}(v). \tag{7.4.22}$$

The family

$$\mathscr{U} := \{\mathcal{U}(v) : v \in M\}$$

is an open covering of M. Since M is a separable metric space, there exists a locally finite open covering $\mathscr{M} := \{\mathcal{M}_i : i \in \mathbb{N}\}$ of M which is subordinate to $\mathscr{U}$ (cf. Lemma 4.3.81), i.e., for each $i \in \mathbb{N}$ there exists $v \in M$ such that $\mathcal{M}_i \subset \mathcal{U}(v)$.[42] Hence there exists $y = y_i$ such that (7.4.22) is satisfied for every $u \in \mathcal{M}_i$.

Define, on M, $\varrho_i(u) := \operatorname{dist}(u, M \setminus \mathcal{M}_i)$ and

$$g(u) := \sum_{i\in\mathbb{N}} \frac{\varrho_i(u)}{\sum\limits_{j\in\mathbb{N}} \varrho_j(u)} y_i.^{43}$$

It is now straightforward to verify that g is the desired pseudogradient vector field for h on M. □

The following generalization of Lemma 7.4.2 was proved by Willem [159].

Lemma 7.4.18 (Quantitative Deformation Lemma). *Let X be a Banach space, and let $F\colon X \to \mathbb{R}$, $F \in C^1(X,\mathbb{R})$, $\mathcal{S} \subset X$, $\mathcal{S} \neq \emptyset$, $c \in \mathbb{R}$, $\varepsilon, \delta > 0$ be such that for any $u \in F_{-1}([c-2\varepsilon, c+2\varepsilon]) \cap \mathcal{S}_{2\delta}$ we have*[44]

$$\|F'(u)\|_{X^*} \ge \frac{8\varepsilon}{\delta}. \tag{7.4.23}$$

Then there exists $\eta \in C([0,1]\times X, X)$ such that

(i) $\eta(t,u) = u$ *if* $t = 0$ *or* $u \notin F_{-1}([c-2\varepsilon, c+2\varepsilon]) \cap \mathcal{S}_{2\delta}$,

(ii) $\eta(1, \mathcal{F}^{c+\varepsilon} \cap \mathcal{S}) \subset \mathcal{F}^{c-\varepsilon}$,[45]

(iii) *for any* $t \in [0,1]$, $\eta(t,\cdot)$ *is a homeomorphism of* X,

(iv) *for any* $u \in X$ *and any* $t \in [0,1]$, $\|\eta(t,u) - u\|_X \le \delta$,

(v) *for any* $u \in X$, $F(\eta(\cdot,u))$ *is decreasing,*

(vi) *for any* $u \in \mathcal{F}^c \cap \mathcal{S}_\delta$ *and any* $t \in (0,1]$, $F(\eta(t,u)) < c$.

---

[41] Note that for any $v \in M$ we have $h(v) \neq o \in X^*$!

[42] Separability of M is actually not necessary. In the case of a general metric space M its paracompactness is used instead of separability (see Dugundji [52], Zeidler [163]).

[43] Note that the sums contain only a finite number of nonzero terms.

[44] Here $\mathcal{S}_{2\delta} := \{u \in X : \operatorname{dist}(u, \mathcal{S}) \le 2\delta\}$.

[45] Recall that $\mathcal{F}^{c\pm\varepsilon} := F_{-1}((-\infty, c\pm\varepsilon])$.

Proof.[46] By Lemma 7.4.17 there exists a pseudogradient vector field g for F' on

$$\mathcal{M} := \{u \in X : F'(u) \neq o\}.$$

Let us define sets

$$\mathcal{A} := F_{-1}([c-2\varepsilon, c+2\varepsilon]) \cap \mathcal{S}_{2\delta}, \qquad \mathcal{B} := F_{-1}([c-\varepsilon, c+\varepsilon]) \cap \mathcal{S}_{\delta}$$

and a functional

$$\psi(u) := \frac{\operatorname{dist}(u, X \setminus \mathcal{A})}{\operatorname{dist}(u, X \setminus \mathcal{A}) + \operatorname{dist}(u, \mathcal{B})}.$$

Then ψ is locally Lipschitz continuous (see Exercise 7.4.8) and

$$\psi = \begin{cases} 1 & \text{on} \quad \mathcal{B}, \\ 0 & \text{on} \quad X \setminus \mathcal{A}. \end{cases}$$

Let us define a vector field

$$f(u) = \begin{cases} -\psi(u)\dfrac{g(u)}{\|g(u)\|^2} & \text{for} \quad u \in \mathcal{A}, \\ o & \text{for} \quad u \in X \setminus \mathcal{A}. \end{cases}$$

Then f is also locally Lipschitz continuous (cf. Exercise 7.4.9) and by assumption (7.4.23) and by Definition 7.4.16, for any $u \in X$ we have

$$\|f(u)\| \leq \frac{\delta}{8\varepsilon}. \tag{7.4.24}$$

Consider the Cauchy problem

$$\begin{cases} \dot{\sigma} = f(\sigma), \\ \sigma(o) = u. \end{cases} \tag{7.4.25}$$

It follows from Corollary 3.1.6 and Remark 3.1.7 that (7.4.25) has a unique solution $\sigma(\cdot, u)$ which is defined on the whole $\mathbb{R}$ and σ is continuous on $\mathbb{R} \times X$. Let us define $\eta\colon [0,1] \times X \to X$ by

$$\eta(t, u) := \sigma(8\varepsilon t, u).$$

It follows from Definition 7.4.16, assumption (7.4.23) and from (7.4.24) that for $t \geq 0$ the inequalities

$$\|\sigma(t,u) - u\| = \left\| \int_0^t f(\sigma(\tau, u))\, \mathrm{d}\tau \right\| \leq \int_0^t \|f(\sigma(\tau, u))\|\, \mathrm{d}\tau \leq \frac{\delta t}{8\varepsilon} \tag{7.4.26}$$

and

$$\frac{\mathrm{d}}{\mathrm{d}t} F(\sigma(t,u)) = \left\langle F'(\sigma(t,u)), \frac{\mathrm{d}}{\mathrm{d}t}\sigma(t,u) \right\rangle = \langle F'(\sigma(t,u)), f(\sigma(t,u)) \rangle \leq -\frac{1}{4}\psi(\sigma(t,u)) \leq 0 \tag{7.4.27}$$

hold. To verify (i), (iii), (iv), (v) and (vi) is a matter of straightforward calculation (cf. Exercise 7.4.26).

[46] Cf. the proof of Lemma 7.4.2.

In oder to verify (ii), let $u \in \mathcal{F}^{c+\varepsilon} \cap \mathcal{S}$. If there is $t \in [0, 8\varepsilon]$ such that $F(\sigma(t,u)) < c - \varepsilon$, then $F(\sigma(8\varepsilon, u)) < c - \varepsilon$ by (7.4.27) and (ii) is satisfied. If, on the other hand, we have

$$\sigma(t,u) \in F_{-1}([c-\varepsilon, c+\varepsilon]) \qquad \text{for any} \quad t \in [0, 8\varepsilon],$$

we obtain from (7.4.26) that $\sigma(t,u) \in \mathcal{B}$ and hence (7.4.27) yields

$$\begin{aligned} F(\sigma(8\varepsilon, u)) &= F(u) + \int_0^{8\varepsilon} \frac{\mathrm{d}}{\mathrm{d}t} F(\sigma(t,u))\,\mathrm{d}t \le F(u) - \frac{1}{4}\int_0^{8\varepsilon} \psi(\sigma(t,u))\,\mathrm{d}t \\ &\le c + \varepsilon - 2\varepsilon = c - \varepsilon \end{aligned}$$

and (ii) is also satisfied. □

A special case of the *Ekeland Variational Principle* is considered to be the first application of Lemma 7.4.18. For a more general form of this principle and its applications see, e.g., Ekeland [56].

Theorem 7.4.19 (Ekeland Variational Principle). *Let X be a Banach space, let $F \in C^1(X, \mathbb{R})$ be bounded below, and let $\varepsilon, \delta > 0$ be arbitrary. If*

$$F(v) \le \inf_{u \in X} F(u) + \varepsilon \qquad \textit{for a} \quad v \in X,$$

then there exists $u_0 \in X$ such that

$$F(u_0) \le \inf_{u\in X} F(u) + 2\varepsilon, \qquad \|u_0 - v\|_X \le 2\delta, \qquad \textit{and} \qquad \|F'(u_0)\|_{X^*} < \frac{8\varepsilon}{\delta}.$$

Proof. We apply Lemma 7.4.18 with

$$\mathcal{S} := \{v\} \qquad \text{and} \qquad c := \inf_{u\in X} F(u).$$

We proceed via contradiction. Assume that there exist ε and δ such that $\|F'(u)\| \ge \frac{8\varepsilon}{\delta}$ for every $u \in F_{-1}[c, c+2\varepsilon] \cap \mathcal{S}_{2\delta}$. Then

$$\eta(1, v) \in \mathcal{F}^{c-\varepsilon}$$

by (ii) of Lemma 7.4.18. However, the definition of c implies $\mathcal{F}^{c-\varepsilon} = \emptyset$, a contradiction. □

Corollary 7.4.20. *Let $F \in C^1(X, \mathbb{R})$ be bounded below. If F satisfies the $(\mathrm{PS})_c$ condition with $c := \inf_{u\in X} F(u)$, then every minimizing sequence for F contains a converging subsequence. In particular, there exists $u_0 \in X$ such that*

$$F(u_0) = \min_{u \in X} F(u).$$

Proof. Let $\{v_n\}_{n=1}^{\infty} \subset X$ be a minimizing sequence for F. We apply Theorem 7.4.19 with

$$\varepsilon_n := \max\left\{\frac{1}{n}, F(v_n) - c\right\} \qquad \text{and} \qquad \delta_n := \sqrt{\varepsilon_n}.$$

Then there exists a sequence $\{u_n\}_{n=1}^{\infty} \subset X$ such that

$$F(u_n) \to c, \qquad F'(u_n) \to o \qquad \text{and} \qquad \|u_n - v_n\| \to 0.$$

The assertion follows now from $(\mathrm{PS})_c$. □

Another application of Lemma 7.4.18 is the following result.

Theorem 7.4.21 (Brézis and Nirenberg). *Let* $F \in C^1(X, \mathbb{R})$*. If*

$$c := \liminf_{\|u\| \to \infty} F(u) \in \mathbb{R},$$

then for every $\varepsilon, \delta > 0$*,* $R > 2\delta$*, there exists* $u \in X$ *such that*

(i) $c - 2\varepsilon \le F(u) \le c + 2\varepsilon$,

(ii) $\|u\| > R - 2\delta$,

(iii) $\|F'(u)\| < \frac{8\varepsilon}{\delta}$.

Proof. We proceed by contradiction similarly to the proof of Theorem 7.4.19. Suppose that the assertion does not hold. Then there exist ε, δ and R such that for any $u \in X$ satisfying (i) and (ii) the inequality in (iii) is false. Hence we can apply Lemma 7.4.18 with $\mathcal{S} := X \setminus B(o; R)$. By the definition of c, $\mathcal{F}^{c+\varepsilon} \cap \mathcal{S}$ is an unbounded set and $\mathcal{F}^{c-\varepsilon} \subset B(o; r)$ for $r > 0$ large enough. By (ii) and (iv) of Lemma 7.4.18,

$$\eta(1, \mathcal{F}^{c+\varepsilon} \cap \mathcal{S}) \subset \mathcal{F}^{c-\varepsilon} \qquad \text{and} \qquad \mathcal{F}^{c+\varepsilon} \cap \mathcal{S} \subset B(o; r + \delta),$$

a contradiction. □

Corollary 7.4.22 (Shujie Le). *Let* $F \in C^1(X, \mathbb{R})$ *be bounded below. If for arbitrary* $d \in \mathbb{R}$ *every sequence* $\{u_n\}_{n=1}^{\infty} \subset X$ *such that*

$$F(u_n) \to d, \qquad F'(u_n) \to o$$

is bounded, then

$$\lim_{\|u\| \to \infty} F(u) = \infty.$$

Proof. We proceed again by contradiction. Assume the assertion does not hold. Then

$$c := \liminf_{\|u\| \to \infty} F(u) \in \mathbb{R}.$$

By Theorem 7.4.21 there exists a sequence $\{u_n\}_{n=1}^{\infty} \subset X$ such that

$$F(u_n) \to c, \qquad F'(u_n) \to o \qquad \text{and} \qquad \|u_n\| \to \infty,$$

a contradiction. □

Let us present now the most important application of Lemma 7.4.18, the *General Minimax Principle.*

Theorem 7.4.23. *Let* X *be a Banach space. Let* $\mathcal{M}_0$ *be a subset of a metric space* M *and* $\Gamma_0 \subset C(\mathcal{M}_0, X)$*. Define*

$$\Gamma := \{\gamma \in C(M, X) : \gamma|_{\mathcal{M}_0} \in \Gamma_0\}.$$

If $F \in C^1(X, \mathbb{R})$ *satisfies*

$$a := \sup_{\gamma_0 \in \Gamma_0} \sup_{u \in \mathcal{M}_0} F(\gamma_0(u)) < c := \inf_{\gamma \in \Gamma} \sup_{u \in M} F(\gamma(u)) < \infty, \tag{7.4.28}$$

then for every $\varepsilon \in \left(0, \frac{c-a}{2}\right)$*,* $\delta > 0$ *and* $\gamma \in \Gamma$ *such that*

$$\sup_{u \in M} F(\gamma(u)) \le c + \varepsilon, \tag{7.4.29}$$

there exists $u_0 \in X$ *such that*

(i) $c - 2\varepsilon \le F(u_0) \le c + 2\varepsilon$,

(ii) $\operatorname{dist}(u_0, \gamma(M)) \le 2\delta$,

(iii) $\|F'(u_0)\| < \frac{8\varepsilon}{\delta}$.

Proof. Suppose, by contradiction, that the assertion is false. Then there exist $0 < \varepsilon < \frac{c-a}{2}$, $\delta > 0$ and $\gamma \in \Gamma$ such that (7.4.29) holds and for any $u \in X$ satisfying (i) and (ii), the inequality in (iii) is false. Hence Lemma 7.4.18 can be applied with $\mathcal{S} = \gamma(M)$. Define $\beta(u) := \eta(1, \gamma(u))$. Since $c - 2\varepsilon > a$, we obtain from (7.4.28) that

$$\beta(u) = \eta(1, \gamma(u)) = \gamma(u) \qquad \text{for every} \quad u \in \mathcal{M}_0 \qquad \text{so that} \quad \beta \in \Gamma.$$

It follows from (7.4.29) and Lemma 7.4.18 that

$$\sup_{u \in M} F(\beta(u)) = \sup_{u \in M} F(\eta(1, \gamma(u))) \le c - \varepsilon,$$

contradicting the definition of c. □

We now have the following consequence.

Corollary 7.4.24. *Let the assumptions of Theorem 7.4.23 be fulfilled. Then there exists a sequence* $\{u_n\}_{n=1}^{\infty} \subset X$ *satisfying*

$$F(u_n) \to c, \qquad F'(u_n) \to o.$$

In particular, if F satisfies the $(\mathrm{PS})_c$ *condition, then c is a critical value of F.*

The special choices of M, $\mathcal{M}_0$, Γ and Γ_0 in Theorem 7.4.23 can yield the Mountain Pass Theorem (see below) and the Saddle Point Theorem (see Appendix 7.6A) under more general assumptions than in Sections 7.4 and 7.6, respectively. It is possible to slightly relax the $(\mathrm{PS})_c$ condition and also the Fréchet differentiability of F, see [11, § 5.5].

Theorem 7.4.25 (Mountain Pass Theorem, Ambrosetti & Rabinowitz [7]). *Let X be a Banach space and let $F \in C^1(X, \mathbb{R})$, $e \in X$ and $r > 0$ be such that $\|e\| > r$ and*

$$\inf_{\substack{u \in X \\ \|u\| = r}} F(u) > F(o) \ge F(e).$$

If F satisfies the $(\mathrm{PS})_c$ *condition with*

$$c := \inf_{\gamma \in \Gamma} \max_{t \in [0,1]} F(\gamma(t)) \qquad \text{where} \quad \Gamma := \{\gamma \in C([0,1], X) : \gamma(0) = o, \ \gamma(1) = e\},$$

then c is a critical value of F.

Proof. It suffices to apply Corollary 7.4.24 with $M = [0,1]$, $\mathcal{M}_0 = \{0, 1\}$, $\Gamma_0 = \{\gamma_0\}$ where $\gamma_0(0) = o$ and $\gamma_0(1) = e$. □

Exercise 7.4.26. Verify that (i), (iii), (iv), (v) and (vi) of Lemma 7.4.18 hold true. Let g and f be from the proof of Lemma 7.4.18. Explain why locally Lipschitz continuity of g implies that f is locally Lipschitz continuous. Compare this with the proof of Lemma 7.4.2.

Hint. Use (7.4.26) and (7.4.27).

Exercise 7.4.27. Consider the boundary value problem

$$\begin{cases} -(|\dot{x}(t)|^{p-2}\dot{x}(t))^{\cdot} + \lambda x(t) = |x(t)|^{r-2}x(t), & t \in (0,1), \\ x(0) = x(1) = 0, \end{cases} \tag{7.4.30}$$

where $p > 1$ and $r > p$. Let λ_1 be the first eigenvalue of (5.8.38). Prove that the problem (7.4.30) has a positive solution on $(0,1)$ provided $\lambda > -\lambda_1$.

Hint. Follow the idea of proving the "sufficient condition" in Example 7.4.7 and apply Theorem 7.4.25.

Exercise 7.4.28. Consider the problem

$$\begin{cases} -\left(|\dot{x}(t)|^{p-2}\dot{x}(t)\right)^{\cdot} - \lambda|x(t)|^{p-2}x(t) = g(t, x(t)), & t \in (0,1), \\ x(0) = x(1) = 0, \end{cases} \tag{7.4.31}$$

where $p > 1$. Formulate conditions on λ and g which guarantee that the energy functional associated with (7.4.31) has a geometry corresponding to the Mountain Pass Theorem (Theorem 7.4.25).

Exercise 7.4.29. Consider the Dirichlet boundary value problem

$$\begin{cases} \left(|\dot{x}(t)|^{p-2}\dot{x}(t)\right)^{\cdot} = h(t, x(t)), & t \in (0,1), \\ x(0) = x(1) = 0, \end{cases} \tag{7.4.32}$$

where $p > 1$. Formulate conditions on $h = h(t, x)$ which guarantee that (7.4.32) has a weak solution.

7.5 Weak Solutions of PDEs, Applications of Mountain Pass Theorem

Let us now consider the Dirichlet problem

$$\begin{cases} -\Delta u(x) + \lambda u(x) = |u(x)|^{p-2}u(x) & \text{in} \quad \Omega, \\ u = 0 & \text{on} \quad \partial\Omega, \end{cases} \tag{7.5.1}$$

and look for nonnegative solutions $u \geq 0$, $u \not\equiv 0$ a.e. in Ω (cf. Example 7.4.7 for $N = 1$). For future purposes we denote by 2^* an arbitrary value greater or equal to 1 for $N = 2$ and

$$2^* = \frac{2N}{N-2} \qquad \text{for} \quad N \geq 3.$$

The real numbers λ and p in (7.5.1) are parameters. We will apply the Mountain Pass Theorem to prove the following assertion about nonnegative solutions of (7.5.1)

Theorem 7.5.1 (Willem [159]). *Let $N \geq 2$, $2 < p < 2^*$. Then* (7.5.1) *has at least one nonnegative nontrivial weak solution if and only if $\lambda > -\lambda_1$ where $\lambda_1 > 0$ is the first eigenvalue of $-\Delta$ subject to the homogeneous Dirichlet boundary conditions on $\partial\Omega$ (see Example* 5.6.4).

Proof. The necessity part is simple (cf. Example 7.4.7). Indeed, let $u \in W_0^{1,2}(\Omega)$ be a weak solution of (7.5.1), $u \geq 0$, $u \not\equiv 0$ a.e. in Ω. Then taking $v = \varphi_1$ (see Example 5.6.4) as a test function in

$$\int_\Omega \nabla u(x) \nabla v(x)\, \mathrm{d}x + \lambda \int_\Omega u(x)v(x)\, \mathrm{d}x = \int_\Omega |u(x)|^{p-2} u(x) v(x)\, \mathrm{d}x$$

we obtain

$$(\lambda + \lambda_1) \int_\Omega u(x)\varphi_1(x)\, \mathrm{d}x = \int_\Omega |u(x)|^{p-2} u(x)\varphi_1(x)\, \mathrm{d}x \tag{7.5.2}$$

due to

$$\int_\Omega \nabla u(x) \nabla \varphi_1(x)\, \mathrm{d}x - \lambda_1 \int_\Omega u(x)\varphi_1(x)\, \mathrm{d}x = 0.$$

Since $\varphi_1 > 0$ in Ω, we get from (7.5.2) that

$$\lambda + \lambda_1 > 0, \qquad \text{i.e.,} \qquad \lambda > -\lambda_1.$$

The sufficiency part is more involved and we will apply the Mountain Pass Theorem (Theorem 7.4.5) to prove it. So, we assume $\lambda > -\lambda_1$. Let us start with the observation that the expression

$$|||u||| = \left(\int_\Omega |\nabla u(x)|^2\, \mathrm{d}x + \lambda \int_\Omega |u(x)|^2\, \mathrm{d}x \right)^{\frac{1}{2}} \qquad \text{for} \quad u \in W_0^{1,2}(\Omega)$$

satisfies

$$c_1 \|u\| \leq |||u||| \leq c_2 \|u\|$$

with constants $c_i > 0$, $i = 1, 2$, independent of $u \in W_0^{1,2}(\Omega)$ where

$$\|u\| = \left(\int_\Omega |\nabla u(x)|^2\, \mathrm{d}x \right)^{\frac{1}{2}}.$$

Indeed, (5.4.4) yields

$$\int_\Omega |\nabla u(x)|^2\, \mathrm{d}x + \lambda \int_\Omega |u(x)|^2\, \mathrm{d}x \geq d \int_\Omega |\nabla u(x)|^2\, \mathrm{d}x \tag{7.5.3}$$

for any $u \in W_0^{1,2}(\Omega)$ where $d = 1 + \min\left\{0, \frac{\lambda}{\lambda_1}\right\}$.[47] Then

$$d^{\frac{1}{2}} \|u\| \leq |||u||| \leq \left(1 + \frac{|\lambda|}{\lambda_1} \right)^{\frac{1}{2}} \|u\| \qquad \text{for any} \quad u \in W_0^{1,2}(\Omega) \tag{7.5.4}$$

by (5.4.4) and (7.5.3).

[47] The reader should prove (7.5.3) in details!

Let us define $F(u) = \frac{|u^+|^p}{p}$. Then the functional

$$E(u) = \int_\Omega \left[\frac{|\nabla u(x)|^2}{2} + \lambda\frac{|u(x)|^2}{2} - F(u(x))\right] \mathrm{d}x, \qquad u \in W_0^{1,2}(\Omega),$$

is the energy functional associated with the problem

$$\begin{cases} -\Delta u(x) + \lambda u(x) = |u^+(x)|^{p-1} & \text{in} \quad \Omega, \\ u = 0 & \text{on} \quad \partial\Omega. \end{cases} \tag{7.5.5}$$

To prove the existence of a weak solution of (7.5.5) (which possibly changes sign in Ω) we apply the Mountain Pass Theorem (see Theorem 7.4.5). For this purpose we verify that E

(i) has the mountain pass type geometry (see Proposition 7.4.3),

(ii) satisfies the $(\mathrm{PS})_c$ condition (see Definition 7.4.4).

We have

$$E(o) = 0$$

and

$$\begin{aligned} E(u) &\geq \frac{1}{2}|||u|||^2 - \frac{1}{p}\int_\Omega |u(x)|^p \,\mathrm{d}x \geq \frac{1}{2}c_1^2\|u\|^2 - \frac{c_{\mathrm{emb}}^p}{p}\|u\|^p \\ &= \|u\|^2\left[\frac{c_1^2}{2} - \frac{c_{\mathrm{emb}}^p}{p}\|u\|^{p-2}\right].\text{[48]} \end{aligned}$$

Hence there exists $r > 0$ small enough and such that

$$b := \inf_{\|u\|=r} E(u) > 0 = E(o).$$

On the other hand, taking $w_0 > 0$ in Ω fixed, $w_0 \in W_0^{1,2}(\Omega)$, then

$$E(tw_0) \leq \frac{c_2^2 t^2}{2}\|w_0\|^2 - \frac{t^p}{p}\int_\Omega |w_0(x)|^p \,\mathrm{d}x \qquad \text{for} \quad t \geq 0.$$

So, there exists $t > 0$ (large enough) that for $e = tw_0 \in W_0^{1,2}(\Omega)$ we have both

$$\|e\| > r \qquad \text{and} \qquad E(e) < 0.$$

(Remember that $p > 2$.)

In order to verify the $(\mathrm{PS})_c$ condition we proceed as follows. Let $\{u_n\}_{n=1}^\infty \subset W_0^{1,2}(\Omega)$ be a sequence satisfying

$$E(u_n) \to c, \qquad \nabla E(u_n) \to o \qquad \text{with a} \quad c \in \mathbb{R}.$$

[48] Note that by the Sobolev Embedding Theorem and (7.5.4) there exists a constant $c_{\mathrm{emb}} > 0$ such that $\|u\|_{L^p(\Omega)} \leq c_{\mathrm{emb}}\|u\|_{W_0^{1,2}(\Omega)}$ for any $u \in W_0^{1,2}(\Omega)$.

(For ∇E see Exercise 7.7.2.) For n large enough, we also have

$$\|\nabla E(u_n)\| \le 1. \tag{7.5.6}$$

Since

$$(\nabla E(u), v) = \int_\Omega \nabla u(x)\nabla v(x)\,\mathrm{d}x + \lambda \int_\Omega u(x)v(x)\,\mathrm{d}x - \int_\Omega f(u(x))v(x)\,\mathrm{d}x$$

where $f(u) = (u^+)^{p-1}$, we have

$$(\nabla E(u), u) = |||u|||^2 - p\int_\Omega F(u(x))\,\mathrm{d}x.$$

Since also

$$E(u) = \frac{1}{2}|||u|||^2 - \int_\Omega F(u(x))\,\mathrm{d}x,$$

we get due to (7.5.6),

$$pE(u) = \left(\frac{p}{2} - 1\right)|||u|||^2 + (\nabla E(u), u) \ge \left(\frac{p}{2} - 1\right)c_1^2\|u\|^2 - \|u\|.$$

Put $u := u_n$ to see that $\{u_n\}_{n=1}^\infty$ is a bounded sequence.

Now, passing to a subsequence if necessary, we can assume that $u_n \rightharpoonup u$ in $W_0^{1,2}(\Omega)$ for a $u \in W_0^{1,2}(\Omega)$. By the Rellich–Kondrachov Theorem we have $u_n \to u$ in $L^p(\Omega)$. It follows from the continuity of the Nemytski operator (see Theorem 3.2.24) that

$$f(u_n) \to f(u) \qquad \text{in} \quad L^{p'}(\Omega), \quad p' = \frac{p}{p-1}.$$

Observe that

$$\begin{aligned} |||u_n - u|||^2 &= (\nabla E(u_n) - \nabla E(u), u_n - u) \\ &\quad + \int_\Omega (f(u_n(x)) - f(u(x)))(u_n(x) - u(x))\,\mathrm{d}x. \end{aligned} \tag{7.5.7}$$

By the assumption $\nabla E(u_n) \to o$ and by $u_n \rightharpoonup u$ in $W_0^{1,2}(\Omega)$ we have

$$(\nabla E(u_n) - \nabla E(u), u_n - u) \to 0 \qquad \text{as} \quad n \to \infty,$$

and by the Hölder inequality we conclude that

$$\begin{aligned} &\left|\int_\Omega (f(u_n(x)) - f(u(x)))(u_n(x) - u(x))\,\mathrm{d}x\right| \\ &\qquad \le \left(\int_\Omega |f(u_n(x)) - f(u(x))|^{p'}\,\mathrm{d}x\right)^{\frac{1}{p'}} \left(\int_\Omega |u_n(x) - u(x)|^p\,\mathrm{d}x\right)^{\frac{1}{p}} \to 0 \end{aligned}$$

as $n \to \infty$. Hence (7.5.7) implies that

$$|||u_n - u||| \to 0 \quad \text{as} \quad n \to \infty, \qquad \text{i.e.,} \qquad u_n \to u \quad \text{in} \quad W_0^{1,2}(\Omega).$$

Consequently, E satisfies the $(\mathrm{PS})_c$ condition.

It follows from Theorem 7.4.5 and Remark 7.4.6 that E has a critical point $u_0 \in W_0^{1,2}(\Omega)$, $u_0 \neq o$. Since u_0 is also a weak solution of (7.5.5) we have

$$\int_\Omega \nabla u_0(x)\nabla v(x)\,\mathrm{d}x + \lambda\int_\Omega u_0(x)v(x)\,\mathrm{d}x = \int_\Omega |u_0^+(x)|^{p-1}v(x)\,\mathrm{d}x \tag{7.5.8}$$

for any $v \in W_0^{1,2}(\Omega)$. Taking $v = u_0^-$ in (7.5.8) we arrive at

$$|||u_0^-||| = \int_\Omega |\nabla u_0^-(x)|^2\,\mathrm{d}x + \lambda\int_\Omega |u_0^-(x)|^2\,\mathrm{d}x = 0,\,^{49}$$

hence $u_0^- = 0$ a.e. in Ω. This proves that u_0 is a nonnegative weak solution of (7.5.5). Since $u_0 \geq 0$ in Ω, $u_0 \not\equiv 0$ in Ω, we have $u_0^+ = u_0$ in Ω, and so u_0 is a nonnegative nontrivial weak solution of (7.5.1). □

Remark 7.5.2. The nonlinearity $u \mapsto |u|^{p-2}u$ in (7.5.1) has the so-called *subcritical growth* due to the inequality $p < 2^*$. The reader should notice that some existence results for (7.5.1) are known also in the case of the *critical growth* $N \geq 3$, $p = 2^*$, (see, e.g., Willem [159]). The proofs are based on the *Concentration Compactness Principle* which is attributed to Lions (see Lions [102], Lions [103], Lions [104]). These techniques go beyond the limits of this book and the reader can consult, e.g., the book of Flucher [63] to get more information in this direction.

7.6 Saddle Point Theorem

The main assertion in this section, the Saddle Point Theorem, is a useful tool to prove existence of a critical point which is neither a local minimum nor a local maximum of a given functional. Let us start again by considering a real function of two independent variables

$$F\colon \mathbb{R}\times\mathbb{R} \to \mathbb{R}$$

which is continuously differentiable and satisfies the following condition:

[49] The implication $u \in W_0^{1,2}(\Omega) \Rightarrow u^- \in W_0^{1,2}(\Omega)$ is nontrivial if $\Omega \subset \mathbb{R}^N$, $N \geq 2$, and it is not true in general if we replace $W_0^{1,2}(\Omega)$ by $W_0^{k,2}(\Omega)$ with $k \geq 2$! For the case $N = 1$ see Exercise 1.2.49. It follows from Gilbarg & Trudinger [71, Section 7.4] (or Leinfelder & Simander [99, Appendix], Ziemer [164, Corollary 2.1.8 and Theorem 2.1.11]) that

$$\nabla u^+ = \begin{cases} \nabla u & \text{if } u > 0,\\ 0 & \text{if } u \leq 0,\end{cases} \quad \nabla u^- = \begin{cases} 0 & \text{if } u \geq 0,\\ \nabla u & \text{if } u < 0,\end{cases} \quad \text{and} \quad \nabla|u| = \begin{cases} \nabla u & \text{if } u > 0,\\ 0 & \text{if } u = 0,\\ -\nabla u & \text{if } u < 0,\end{cases} \quad \text{for } u \in$$

$W_0^{1,2}(\Omega)$.

There exists $\varrho > 0$ such that

$$\inf_{y\in\mathbb{R}} F(0,y) > \max\{F(-\varrho,0), F(\varrho,0)\}. \tag{7.6.1}$$

The graph of such a function is sketched in Figure 7.6.1.The impression one can

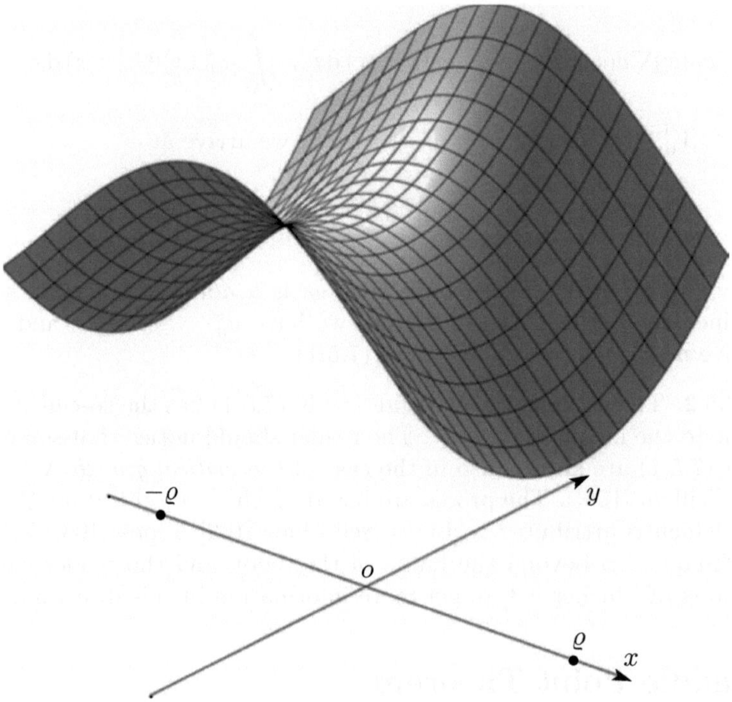

Figure 7.6.1.

get from the graph is that

$$c = \inf_{\gamma\in\Gamma} \max_{t\in[-\varrho,\varrho]} F(\gamma(t))$$

$$\text{where} \quad \Gamma = \{\gamma \in C([-\varrho,\varrho],\mathbb{R}^2) : \gamma(-\varrho) = (-\varrho,0), \gamma(\varrho) = (\varrho,0)\}$$

is a critical value of the functional F. The following example, however, shows that this is not the case in general.

Example 7.6.1. Let

$$F(x,y) = 2\mathrm{e}^{-x^2} + \mathrm{e}^y$$

(see Figure 7.6.2). Set $\varrho = 1$. Then we have

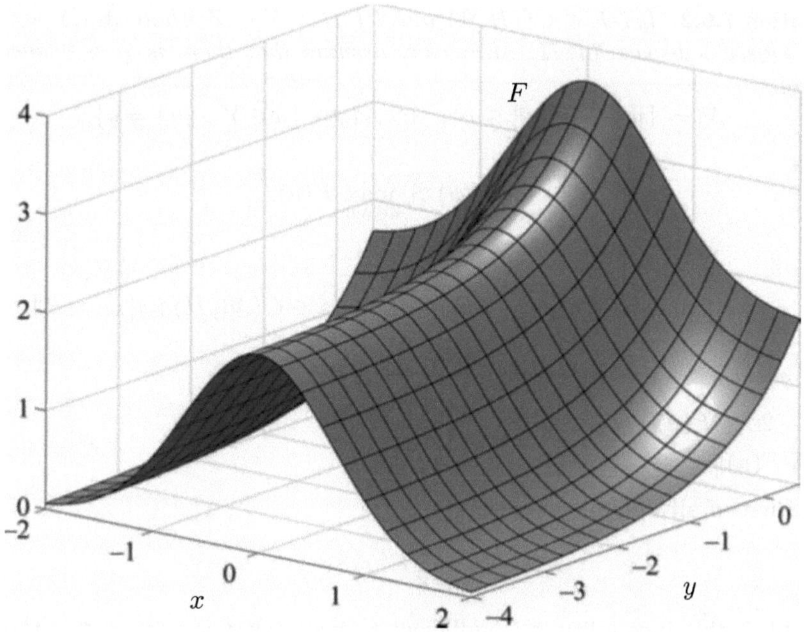

Figure 7.6.2.

$$\inf_{y\in\mathbb{R}} F(0,y) = 2 > \max\{F(-1,0), F(1,0)\} = \frac{2}{\mathrm{e}} + 1,$$

i.e., (7.6.1) is satisfied.

On the other hand, since

$$\frac{\partial F}{\partial x}(x,y) = -4x\mathrm{e}^{-x^2}, \qquad \frac{\partial F}{\partial y}(x,y) = \mathrm{e}^{y},$$

there is no critical point of F. ○

The reason why the geometric condition (7.6.1) is not sufficient to guarantee the existence of a critical point of F is the same as in the previous section. If we introduce the assumption $(\widehat{\mathrm{PS}})$ (see page 481), then the value c is a critical value of F provided F satisfies $(\widehat{\mathrm{PS}})$.

Let us consider a more general situation

$$F\colon H \to \mathbb{R}$$

where H is a real Hilbert space. We will use the Quantitative Deformation Lemma (Lemma 7.4.2) and prove the following analogue of Proposition 7.4.3.

Proposition 7.6.2. *Let $F \in C^2(H, \mathbb{R})$ and let $H = Y \oplus Z$ where $\dim Y < \infty$ and Z is a closed subspace of H. Moreover, assume that there is $\varrho > 0$ such that, denoting*

$$\mathcal{M} = \{u \in Y : \|u\| \le \varrho\}, \qquad \mathcal{M}_0 = \{u \in Y : \|u\| = \varrho\},$$

we have

$$\inf_{u \in Z} F(u) > \max_{u \in \mathcal{M}_0} F(u). \tag{7.6.2}$$

Let

$$c := \inf_{\gamma \in \Gamma} \max_{u \in \mathcal{M}} F(\gamma(u)) \qquad \text{where} \quad \Gamma := \{\gamma \in C(\mathcal{M}, H) : \gamma|_{\mathcal{M}_0} = I\}.$$

Then for each $\varepsilon > 0$ there exists $u \in H$ such that

(i) $c - 2\varepsilon \le F(u) \le c + 2\varepsilon$,

(ii) $\|\nabla F(u)\| < 2\varepsilon$.

Proof. First of all we will show that

$$\tilde{c} := \inf_{u \in Z} F(u) \le c.$$

To establish this inequality it is sufficient to prove that for any $\gamma \in \Gamma$ there is a point $\tilde{u} \in \mathcal{M}$ for which

$$\gamma(\tilde{u}) \in Z.$$

Let P be a continuous projection of H into Y such that

$$\operatorname{Ker} P = Z.$$

With this P we wish to find a solution in $\mathcal{M}$ of the equation

$$P\gamma(u) = o. \tag{7.6.3}$$

To do that we will use the Brouwer degree. Since $P\gamma|_{\mathcal{M}_0} = I$, the homotopy invariance property (Proposition 5.7.6 and Theorem 5.7.7) yields that

$$\deg(P\gamma, \operatorname{int} \mathcal{M}, o) = \deg(I, \operatorname{int} \mathcal{M}, o) = 1.$$

Therefore (7.6.3) has a solution in $\mathcal{M}$ (again Theorem 5.7.7).

Suppose that the conclusion of this proposition does not hold, i.e., assume that $\varepsilon > 0$ is so small that

$$\max_{u \in \mathcal{M}_0} F(u) < c - 2\varepsilon,$$

and for all $u \in H$ satisfying (i) the condition (ii) is violated. By the definition of c there exists $\gamma \in \Gamma$ such that

$$\max_{u \in \mathcal{M}} F(\gamma(u)) \le c + \varepsilon.^{50} \tag{7.6.4}$$

Consider $\beta(u) = \eta(\gamma(u))$ where η is from Lemma 7.4.2. Using Lemma 7.4.2(i) we conclude that for $u \in \mathcal{M}_0$ we have

$$\beta(u) = \eta(\gamma(u)) = \eta(u) = u, \qquad \text{i.e.,} \qquad \beta \in \Gamma, \qquad \text{i.e.,} \qquad c \le \max_{u\in\mathcal{M}} F(\beta(u)).$$

On the other hand, it follows from Lemma 7.4.2(ii) and (7.6.4) that

$$\max_{u\in\mathcal{M}} F(\beta(u)) \le c - \varepsilon,$$

a contradiction. □

Similarly to the previous section, employing the $(\mathrm{PS})_c$ condition, we have the following assertion called the *Saddle Point Theorem.*

Theorem 7.6.3 (Saddle Point Theorem). *Let the assumptions of Proposition* 7.6.2 *be satisfied. Let F satisfy* $(\mathrm{PS})_c$. *Then c is a critical value of F.*

Remark 7.6.4. The reader should have in mind that the Saddle Point Theorem was also proved under more general assumptions when H is a Banach space and $F \in C^1(H,\mathbb{R})$. In this more general form it is attributed to Rabinowitz (see Theorem 7.6.12).

Example 7.6.5. Let us consider the boundary value problem

$$\begin{cases} -\ddot{x}(t) - x(t) = f(t) + g(x(t)), & t \in (0,\pi), \\ x(0) = x(\pi) = 0, \end{cases} \tag{7.6.5}$$

where $f \in L^2(0,\pi)$ is a given function and $g\colon \mathbb{R} \to \mathbb{R}$ is a continuous function having finite limits $\lim\limits_{s\to\pm\infty} g(s) = g(\pm\infty)$ and such that

$$g(-\infty) < g(s) < g(+\infty) \qquad \text{for all} \quad s \in \mathbb{R}.$$

We will prove that the problem (7.6.5) has a weak solution if and only if

$$-g(+\infty) < \frac{1}{2}\int_0^\pi f(t)\sin t\,\mathrm{d}t < -g(-\infty). \tag{7.6.6}$$

First let us prove that (7.6.6) is a *necessary condition* for the solvability of (7.6.5). Assume that $x \in H := W_0^{1,2}(0,\pi)$ [51] is a weak solution of (7.6.5), i.e.,

$$\int_0^\pi (\dot{x}(t)\dot{y}(t) - x(t)y(t))\,\mathrm{d}t = \int_0^\pi f(t)y(t)\,\mathrm{d}t + \int_0^\pi g(x(t))y(t)\,\mathrm{d}t$$

[50] Note that the “max” exists due to the assumption $\dim Y < \infty$.

[51] We consider the norm $\|x\| = \left(\int_0^\pi |\dot{x}(t)|^2\,\mathrm{d}t\right)^{\frac{1}{2}}$ on H.

for any $y \in H$. Take $y = \sin t$, then

$$\int_0^\pi f(t)\sin t\,\mathrm{d}t = -\int_0^\pi g(x(t))\sin t\,\mathrm{d}t.$$

However, an easy calculation yields

$$2g(-\infty) < \int_0^\pi g(x(t))\sin t\,\mathrm{d}t < 2g(+\infty).$$

To prove that (7.6.6) is also a *sufficient condition* we apply Theorem 7.6.3 and Remark 7.6.4. Define

$$F(x) = \frac{1}{2}\int_0^\pi |\dot{x}(t)|^2\,\mathrm{d}t - \frac{1}{2}\int_0^\pi |x(t)|^2\,\mathrm{d}t - \int_0^\pi\int_0^{x(t)} g(s)\,\mathrm{d}s\,\mathrm{d}t - \int_0^\pi f(t)x(t)\,\mathrm{d}t,$$

$x \in H$.

Let us verify that F has a suitable geometry which corresponds to the Saddle Point Theorem. Let $Y = \mathrm{Lin}\{\sin t\}$, $Z = Y^\perp$.[52] For $z \in Z$ we have

$$\int_0^\pi |z(t)|^2\,\mathrm{d}t \le \frac{1}{4}\int_0^\pi |\dot{z}(t)|^2\,\mathrm{d}t \tag{7.6.7}$$

(cf. Exercise 7.6.6). From (7.6.7) we get that F is weakly coercive on Z. Namely, we have

$$\begin{aligned} F(z) &\ge \frac{3}{8}\int_0^\pi |\dot{z}(t)|^2\,\mathrm{d}t - \int_0^\pi \left|\int_0^{z(t)} g(s)\,\mathrm{d}s\right|\mathrm{d}t - \|f\|_{L^2(0,\pi)}\|z\|_{L^2(0,\pi)} \\ &\ge \frac{3}{8}\|z\|^2 - c\|z\|_{L^2(0,\pi)} \ge \|z\|\left[\frac{3}{8}\|z\| - \frac{c}{2}\right], \end{aligned}$$

and so $F(z) \to \infty$ for $\|z\| \to \infty$, $z \in Z$. The functional F is weakly sequentially lower semi-continuous on Z (the argument is the same as that used in Example 7.2.6 and the reader should check it carefully). Then it follows from Theorem 7.2.8 that there exists $z_0 \in Z$ such that

$$-\infty < F(z_0) = \min_{z\in Z} F(z).$$

On the other hand,

$$\begin{aligned} F(\varrho\sin t) = {} & \underbrace{\frac{\varrho^2}{2}\int_0^\pi \cos^2 t\,\mathrm{d}t - \frac{\varrho^2}{2}\int_0^\pi \sin^2 t\,\mathrm{d}t}_{=0} \\ & - \int_0^\pi\int_0^{\varrho\sin t} g(s)\,\mathrm{d}s\,\mathrm{d}t - \varrho\int_0^\pi f(t)\sin t\,\mathrm{d}t. \end{aligned}$$

[52] $H = Y \oplus Z$. Notice that Y and Z are orthogonal to each other also in the L^2-scalar product.

Denote

$$\int_0^\sigma g(s)\,\mathrm{d}s = G(\sigma).$$

Then

$$F(\varrho \sin t) = -\varrho \left[\int_0^\pi \frac{G(\varrho \sin t)}{\varrho}\,\mathrm{d}t + \int_0^\pi f(t)\sin t\,\mathrm{d}t \right].$$

Since, by the l'Hospital Rule,

$$\lim_{\varrho\to\pm\infty} \frac{G(\varrho \sin t)}{\varrho} = \lim_{\varrho\to\pm\infty} g(\varrho \sin t)\sin t = g(\pm\infty)\sin t,$$

the Lebesgue Dominated Convergence Theorem and (7.6.6) yield

$$\lim_{|\varrho|\to\infty} F(\varrho \sin t) = -\infty. \tag{7.6.8}$$

Taking ϱ_0 large enough we then have

$$F(\pm\varrho_0 \sin t) < F(z_0),$$

i.e., the assumptions of Theorem 7.6.3 are satisfied with

$$\mathcal{M} = \{\varrho \sin t : \varrho \in [-\varrho_0, \varrho_0]\}, \qquad \mathcal{M}_0 = \{-\varrho_0 \sin t, \varrho_0 \sin t\}.$$

It remains to prove that F satisfies $(\mathrm{PS})_c$. Similarly to Example 7.4.7 we will prove that F satisfies even a stronger version of $(\mathrm{PS})_c$:

Any sequence $\{x_n\}_{n=1}^\infty \subset H$ such that $\{F(x_n)\}_{n=1}^\infty$ is bounded in $\mathbb{R}$ and $\nabla F(x_n) \to o$ contains a convergent subsequence.

To prove it we follow the usual scheme.

Step 1. We will first show that $\{x_n\}_{n=1}^\infty$ is bounded in H. To do that decompose

$$x_n = y_n + z_n$$

where $y_n \in Y$ (i.e., $y_n(t) = \varrho_n \sin t$) and $z_n \in Z = Y^\perp$. First we prove that $\{z_n\}_{n=1}^\infty$ is bounded in H. To see this consider $(\nabla F(x_n), z_n)_H$. We have

$$\begin{aligned}
(\nabla F(x_n), z_n) &= \int_0^\pi \dot{x}_n(t)\dot{z}_n(t)\,\mathrm{d}t - \int_0^\pi x_n(t)z_n(t)\,\mathrm{d}t \\
&\quad - \int_0^\pi z_n(t)g(x_n(t))\,\mathrm{d}t - \int_0^\pi f(t)z_n(t)\,\mathrm{d}t \\
&= {}^{53}\|z_n\|^2 - \|z_n\|^2_{L^2(0,\pi)} - \int_0^\pi z_n(t)g(x_n(t))\,\mathrm{d}t - \int_0^\pi f(t)z_n(t)\,\mathrm{d}t \\
&\ge \frac{3}{4}\|z_n\|^2 - k\|z_n\|_{L^2(0,\pi)} \ge \frac{3}{4}\|z_n\|^2 - \frac{k}{2}\|z_n\|
\end{aligned}$$

with a positive constant k (for the last two inequalities we have used (7.6.7)). Since we have assumed that $\nabla F(x_n) \to o$ we know that

$$\|\nabla F(x_n)\| \leq \text{const.} \qquad \text{for all sufficiently large } n.$$

In particular, this means that for all these n the inequalities

$$(\nabla F(x_n), z_n) \leq \|z_n\| \qquad \text{hold.}$$

Hence

$$\frac{3}{4}\|z_n\|^2 - \frac{k}{2}\|z_n\| \leq \|z_n\|,$$

and the boundedness of $\{z_n\}_{n=1}^{\infty}$ is shown.

For the investigation of y_n we will use the boundedness of $F(x_n)$. We have

$$\begin{aligned}
F(x_n) = F(y_n + z_n) &= \frac{1}{2}\|y_n\|^2 + \frac{1}{2}\|z_n\|^2 - \frac{1}{2}\|y_n\|^2_{L^2(0,\pi)} - \frac{1}{2}\|z_n\|^2_{L^2(0,\pi)} \\
&\quad - \int_0^\pi \int_0^{y_n(t)} g(s)\,\mathrm{d}s\,\mathrm{d}t - \int_0^\pi \int_{y_n(t)}^{y_n(t)+z_n(t)} g(s)\,\mathrm{d}s\,\mathrm{d}t \\
&\quad - \int_0^\pi f(t)y_n(t)\,\mathrm{d}t - \int_0^\pi f(t)z_n(t)\,\mathrm{d}t \\
&= F(y_n) + \frac{1}{2}\|z_n\|^2 - \frac{1}{2}\|z_n\|^2_{L^2(0,\pi)} \\
&\quad - \int_0^\pi \int_{y_n(t)}^{y_n(t)+z_n(t)} g(s)\,\mathrm{d}s\,\mathrm{d}t - \int_0^\pi f(t)z_n(t)\,\mathrm{d}t.
\end{aligned}$$

By the boundedness of $\{z_n\}_{n=1}^{\infty}$, g and $\{F(x_n)\}_{n=1}^{\infty}$, we obtain that $\{F(y_n)\}_{n=1}^{\infty}$ is bounded. Since $y_n(t) = \varrho_n \sin t$ and $\lim\limits_{|\varrho|\to\infty} F(\varrho \sin t) = -\infty$ (see (7.6.8)), $\{\varrho_n\}_{n=1}^{\infty}$ and also $\|y_n\|$ have to be bounded.

Step 2. Passing to a subsequence if necessary we may assume that $x_n \rightharpoonup x_0$ in H and $x_n \to x_0$ in $C[0,\pi]$. Since $\nabla F(x_n) \to o$ we get

$$(\nabla F(x_n) - \nabla F(x_0), x_n - x_0) \to 0 \qquad \text{as} \quad n \to \infty.$$

This means that

$$\int_0^\pi |\dot{x}_n(t) - \dot{x}_0(t)|^2\,\mathrm{d}t - \int_0^\pi |x_n(t) - x_0(t)|^2\,\mathrm{d}t \\
- \int_0^\pi (g(x_n(t)) - g(x_0(t)))(x_n(t) - x_0(t))\,\mathrm{d}t \to 0.$$

However, the last two integrals tend to zero, i.e.,

$$\|x_n - x_0\| \to 0 \qquad \text{as} \quad n \to \infty.$$

Hence $x_n \to x_0$ in H. ○

[53] See footnote 52 on page 504.

Exercise 7.6.6. Prove the inequality (7.6.7) for $z \in Z$.

Hint. Remember that $\left\{\sqrt{\frac{2}{\pi}}\sin nt\right\}_{n=1}^{\infty}$, $\left\{\sqrt{\frac{2}{\pi}}\cos nt\right\}_{n=0}^{\infty}$ form an orthonormal basis in $L^2(0,\pi)$. For $z \in Z$ one has

$$z(t) = \sum_{k=2}^{\infty} a_k \sin kt \qquad \text{and (the Parseval equality)} \qquad \|z\|^2_{L^2(0,\pi)} = \frac{\pi}{2}\sum_{k=2}^{\infty} a_k^2.$$

A similar argument with $\dot{z}$ and integration by parts leads to (7.6.7).

Exercise 7.6.7. Find conditions on $\Phi\colon [0,1]\times\mathbb{R}\to\mathbb{R}$ such that the procedure for proving the existence of a solution of (7.6.5) could be used for the boundary value problem

$$\begin{cases} -\ddot{x}(t) = \dfrac{\partial \Phi}{\partial x}(t,x(t)), & t \in (0,1),\\ x(0) = x(1) = 0. \end{cases}$$

Exercise 7.6.8. Consider the boundary value problem

$$\begin{cases} -\ddot{x}(t) - \lambda x(t) = g(t,x(t)), & t \in (0,\pi),\\ x(0) = x(\pi) = 0. \end{cases} \tag{7.6.9}$$

Formulate conditions on λ and g which guarantee that the energy functional associated with (7.6.9) has a geometry corresponding to the Saddle Point Theorem.

Exercise 7.6.9. Consider the Dirichlet boundary value problem

$$\begin{cases} -\ddot{x}(t) = h(t,x(t)), & t \in (0,\pi),\\ x(0) = x(\pi) = 0. \end{cases} \tag{7.6.10}$$

Formulate conditions on $h = h(t,x)$ which guarantee that (7.6.10) has a weak solution.

Exercise 7.6.10. Consider the Neumann boundary value problem

$$\begin{cases} -\ddot{x} = h(t,x(t)), & t \in (0,\pi),\\ \dot{x}(0) = \dot{x}(\pi) = 0. \end{cases} \tag{7.6.11}$$

Formulate conditions on $h = h(t,x)$ which guarantee that (7.6.11) has a weak solution.

Exercise 7.6.11. Consider the Dirichlet boundary value problem

$$\begin{cases} -\ddot{x}(t) - n^2 x(t) = f(t) + g(x(t)), & t \in (0,\pi),\\ x(0) = x(\pi) = 0, \end{cases} \tag{7.6.12}$$

where $n \in \mathbb{N}$, $f \in L^2(0,\pi)$, $g\colon \mathbb{R} \to \mathbb{R}$ is a continuous function having finite limits

$$\lim_{s\to\pm\infty} g(s) = g(\pm\infty) \quad \text{and such that} \quad g(-\infty) < g(s) < g(+\infty) \quad \text{for all } s.$$

Prove that (7.6.12) has a weak solution if and only if

$$g(-\infty)\int_0^\pi (\sin nt)^- \,\mathrm{d}t - g(+\infty)\int_0^\pi (\sin nt)^+ \,\mathrm{d}t < \int_0^\pi f(t)\sin nt \,\mathrm{d}t$$
$$< g(+\infty)\int_0^\pi (\sin nt)^- \,\mathrm{d}t - g(-\infty)\int_0^\pi (\sin nt)^+ \,\mathrm{d}t$$

where $(\sin nt)^+$ and $(\sin nt)^-$ are the positive and the negative part of $\sin nt$, respectively.

Hint. Modify the estimates from Example 7.6.5.

7.6A Linking Theorem

The aim of this appendix is to apply the General Minimax Principle (see Theorem 7.4.23) from Appendix 7.4A and to generalize the assertion of Theorem 7.6.3. Namely, we start with the *Saddle Point Theorem* (cf. Remark 7.6.4). For more information we recommend [124].

Theorem 7.6.12 (Saddle Point Theorem, Rabinowitz). *Let $X = Y \oplus Z$ be a Banach space with Z closed in X and $0 < \dim Y < \infty$. For $\varrho > 0$ define*

$$\mathcal{M} := \{u \in Y : \|u\| \le \varrho\}, \qquad \mathcal{M}_0 := \{u \in Y : \|u\| = \varrho\}.$$

Let $F \in C^1(X,\mathbb{R})$ be such that

$$b := \inf_{u\in Z} F(u) > a := \max_{u\in\mathcal{M}_0} F(u).$$

If F satisfies the $(\mathrm{PS})_c$ condition with

$$c := \inf_{\gamma\in\Gamma}\max_{u\in\mathcal{M}} F(\gamma(u)) \qquad \textit{where} \quad \Gamma := \{\gamma \in C(\mathcal{M},X) : \gamma|_{\mathcal{M}_0} = I\},$$

then c is a critical value of F.

Proof. We set $\Gamma_0 = \{I\}$ and apply Theorem 7.4.23 and Corollary 7.4.24. For this purpose it is enough to verify that $c \ge b$.

Let us prove that $\gamma(\mathcal{M}) \cap Z \ne \emptyset$ for every $\gamma \in \Gamma$. Denote by P the projection onto Y such that $PZ = \{o\}$. If $\gamma(\mathcal{M}) \cap Z = \emptyset$, then the map

$$u \mapsto \varrho\frac{P\gamma(u)}{\|P\gamma(u)\|}$$

is a *retraction*[54] of the ball $\mathcal{M}$ onto its boundary $\mathcal{M}_0$ in the space Y. But this is impossible since $\dim Y < \infty$. Hence, for every $\gamma \in \Gamma$,

$$\max_{u\in\mathcal{M}} F(\gamma(u)) \ge \inf_{u\in Z} F(u), \qquad \text{i.e.,} \qquad c \ge b.$$

This completes the proof. □

[54] For the notion of the retraction see Exercise 7.6.16.

We postpone the application of Theorem 7.6.12 to Appendix 7.7A. We prove now the *Linking Theorem.*

Theorem 7.6.13 (Linking Theorem, Rabinowitz). *Let $X = Y \oplus Z$ be a Banach space with Z closed in X and $0 < \dim Y < \infty$. Let $\varrho > r > 0$ and let $z \in Z$ be such that $\|z\| = r$. Define*

$$\mathcal{M} := \{u = y + \lambda z : \|u\| \le \varrho,\ \lambda \ge 0,\ y \in Y\}, \qquad \mathcal{N} := \{u \in Z : \|u\| = r\},$$
$$\mathcal{M}_0 := \{u = y + \lambda z : y \in Y,\ \|u\| = \varrho \text{ and } \lambda \ge 0,\ \text{or } \|u\| \le \varrho \text{ and } \lambda = 0\}.$$

Let $F \in C^1(X, \mathbb{R})$ be such that

$$b := \inf_{u \in \mathcal{N}} F(u) > a := \max_{u \in \mathcal{M}_0} F(u).$$

If F satisfies the $(\mathrm{PS})_c$ condition with

$$c := \inf_{\gamma \in \Gamma} \max_{u \in \mathcal{M}} F(\gamma(u)) \qquad \text{where} \quad \Gamma := \{\gamma \in C(\mathcal{M}, X) : \gamma|_{\mathcal{M}_0} = I\},$$

then c is a critical value of F.

Proof. We set $\Gamma_0 = \{I\}$ and apply Theorem 7.4.23 and Corollary 7.4.24. As in the previous proof it is sufficient to verify that $c \ge b$.

Now, we prove that $\gamma(\mathcal{M}) \cap \mathcal{N} \neq \emptyset$ for every $\gamma \in \Gamma$. Denote by P the projection onto Y such that $PZ = \{o\}$, and by R the retraction of $(Y \oplus \mathbb{R}_z) \setminus \{z\}$ to $\mathcal{M}_0$,[55] see Figure 7.6.3. If $\gamma(\mathcal{M}) \cap \mathcal{N} = \emptyset$, then the map

$$u \mapsto R\left(P\gamma(u) + \frac{1}{r}\|(I - P)\gamma(u)\| z\right)$$

is well defined on $\mathcal{M}$ and hence it is a retraction of $\mathcal{M}$ to its boundary $\mathcal{M}_0$. This is impossible since $\mathcal{M}$ is homeomorphic to a finite-dimensional ball (see Exercise 7.6.16). Hence for every $\gamma \in \Gamma$ we obtain

$$\max_{u \in \mathcal{M}} F(\gamma(u)) \ge \inf_{u \in \mathcal{N}} F(u), \qquad \text{i.e.,} \qquad c \ge b. \qquad \square$$

As an application we give the following example.

Example 7.6.14. Let us consider the boundary value problem

$$\begin{cases} -\ddot{x}(t) + a(t)x(t) = f(t, x(t)), & t \in (0,1), \\ x(0) = x(1) = 0, \end{cases} \tag{7.6.13}$$

where $a = a(t)$ is a continuous function on $[0,1]$ and $f = f(t,s)$ is a continuous function on $[0,1] \times \mathbb{R}$ satisfying some additional hypotheses formulated below.

It follows from the Sturm–Liouville theory for linear ordinary differential equations of the second order (see Example 2.2.17 and, e.g., Walter [156]) that the eigenvalues $\{\lambda_n\}_{n=1}^{\infty}$ of

$$\begin{cases} -\ddot{x}(t) + a(t)x(t) = \lambda x(t), & t \in (0,1), \\ x(0) = x(1) = 0 \end{cases}$$

[55] By $\mathbb{R}_z$ we denote the set $\{tz : t \in \mathbb{R}\} \subset Z$.

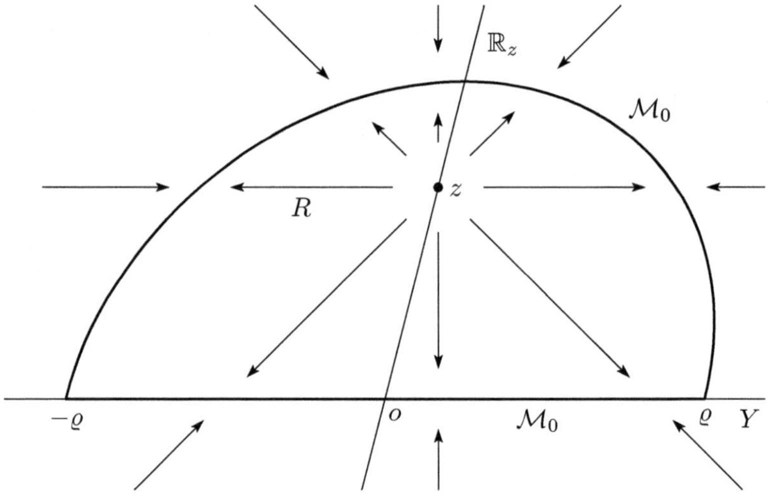

Figure 7.6.3.

form a strictly increasing sequence where each eigenvalue is simple and $\lim_{n\to\infty} \lambda_n = \infty$. If we denote by $e_1, e_2, \dots, e_n, \dots$ the corresponding eigenfunctions, then they are mutually orthogonal in $L^2(0,1)$. Suppose that there exists $k \in \mathbb{N}$ such that

$$\lambda_1 < \lambda_2 < \cdots < \lambda_k < 0 < \lambda_{k+1} < \lambda_{k+2} < \cdots$$

Let us assume that f satisfies the following assumptions:

(f1) there exist $p \in (1,2)$ and $c > 0$ such that

$$|f(t,s)| \le c\left(1 + |s|^{p-1}\right) \qquad \text{for any} \quad t \in [0,1], \quad s \in \mathbb{R};$$

(f2) there exist $\alpha > 2$ and $R > 0$ such that for $t \in [0,1]$ and $|s| > R$ we have

$$0 < \alpha \int_0^s f(t,\sigma)\,\mathrm{d}\sigma \le s f(t,s);$$

(f3) $f(t,s) = o(|s|)$ as $|s| \to 0$ uniformly on $[0,1]$;

(f4) $\lambda_k \dfrac{s^2}{2} \le \displaystyle\int_0^s f(t,\sigma)\,\mathrm{d}\sigma$ for all $s \in \mathbb{R}$ and $t \in [0,1]$.

We now consider the functional

$$F(x) := \int_0^1 \left(\frac{1}{2}|\dot{x}(t)|^2 + \frac{1}{2}a(t)|x(t)|^2 - \int_0^{x(t)} f(t,s)\,\mathrm{d}s \right) \mathrm{d}t$$

on $H := W_0^{1,2}(0,1)$.[56] The critical points of F correspond to weak solutions of (7.6.13). Our plan is to apply the Linking Theorem (Theorem 7.6.13) so as to prove that F has

[56] Note that we use the norm $\|x\| = \|\dot{x}\|_{L^2(0,1)}$.

a critical point. Then the existence of a solution of (7.6.13) will follow from a regularity argument similar to that from Theorem 7.1.13.

Denote by

$$\psi(x) := \int_0^1 \int_0^{x(t)} f(t,s)\,\mathrm{d}s\,\mathrm{d}t$$

the functional defined on the Sobolev space H. Then ψ is of the class $C^1(H,\mathbb{R})$, and

$$(\psi'(x),h) = \int_0^1 f(t,x(t))h(t)\,\mathrm{d}t \tag{7.6.14}$$

(cf. Section 3.2). The fact $\psi \in C^1(H,\mathbb{R})$ implies immediately that $F \in C^1(H,\mathbb{R})$ as well. Let us define

$$Y := \operatorname{Lin}\{e_1,\dots,e_k\}, \qquad Z := \left\{x \in H : \int_0^1 x(t)y(t)\,\mathrm{d}t = 0,\ y \in Y\right\}.$$

Then

$$\delta := \inf_{\substack{x\in Z\\ \|x\|=1}} \int_0^1 (|\dot{x}(t)|^2 + a(t)|x(t)|^2)\,\mathrm{d}t > 0. \tag{7.6.15}$$

Indeed, by definition, on Z we have (see Remark 7.8.13)

$$\int_0^1 (|\dot{x}(t)|^2 + a(t)|x(t)|^2)\,\mathrm{d}t \ge \lambda_{k+1} \int_0^1 |x(t)|^2\,\mathrm{d}t.$$

Since $\inf\limits_{\|x\|=1} \|x\|_{L^2(0,1)} = 0$ we need some more work in order to establish (7.6.15). Consider a minimizing sequence $\{x_n\}_{n=1}^\infty \subset Z$:

$$\|x_n\| = 1, \qquad \int_0^1 (|\dot{x}_n(t)|^2 + a(t)|x_n(t)|^2)\,\mathrm{d}t \to \delta.$$

Going to a subsequence if necessary, we may assume $x_n \rightharpoonup x$ in H, i.e., $x_n \to x$ in $L^2(0,1)$ by the compact embedding $H = W_0^{1,2}(0,1) \subsetneq\subsetneq L^2(0,1)$ (cf. Theorem 1.2.28). The continuity of $a = a(t)$ in $[0,1]$ then implies that

$$\int_0^1 a(t)|x_n(t)|^2\,\mathrm{d}t \to \int_0^1 a(t)|x(t)|^2\,\mathrm{d}t.$$

Since Z is weakly closed and the norm on H is weakly lower semicontinuous, we obtain

$$\delta = 1 + \int_0^1 a(t)|x(t)|^2\,\mathrm{d}t \ge \int_0^1 |\dot{x}(t)|^2\,\mathrm{d}t + \int_0^1 a(t)|x(t)|^2\,\mathrm{d}t \ge \lambda_{k+1} \int_0^1 |x(t)|^2\,\mathrm{d}t.$$

If $x = o$, we have $\delta = 1$, and if $x \ne o$, we have

$$\delta \ge \lambda_{k+1} \int_0^1 |x(t)|^2\,\mathrm{d}t > 0$$

and so (7.6.15) is proved. Using (f1) and (f3), we obtain that for any $\varepsilon > 0$ there exists $c_\varepsilon > 0$ such that

$$\left|\int_0^s f(t,\sigma)\,\mathrm{d}\sigma\right| \le \varepsilon|s|^2 + c_\varepsilon|s|^p. \tag{7.6.16}$$

It follows from (7.6.15) and (7.6.16) that on Z we have

$$F(x) \geq \frac{\delta}{2}\|x\|^2 - \int_0^1 (\varepsilon|x(t)|^2 + c_\varepsilon|x(t)|^p)\,\mathrm{d}t = \frac{\delta}{2}\|x\|^2 - \varepsilon\|x\|^2_{L^2(0,1)} - c_\varepsilon\|x\|^p_{L^p(0,1)}.$$

For $\varepsilon > 0$ small enough, by virtue of the inequality $1 < p < 2$ and the embedding $H = W_0^{1,2}(0,1) \subsetneq L^q(0,1)$ for any $q > 1$, there exists $r > 0$ such that

$$b := \inf_{\substack{\|x\|=r\\ x\in Z}} F(x) > 0.$$

By (f4) we have

$$F(x) \leq \int_0^1 \left[\lambda_k \frac{|x(t)|^2}{2} - \int_0^{x(t)} f(t,s)\,\mathrm{d}s\right] \mathrm{d}t \leq 0 \qquad \text{for} \quad x \in Y. \tag{7.6.17}$$

It follows from the first inequality in (f2) that there exist $c_1, c_2 > 0$ such that

$$\int_0^s f(t,\sigma)\,\mathrm{d}\sigma \geq c_1|s|^\alpha - c_2 \qquad \text{for all} \quad s \in \mathbb{R},\ t \in [0,1] \tag{7.6.18}$$

(cf. Exercise 7.6.17). Hence for $x \in H$ we have

$$F(x) \leq \frac{1}{2}\|x\|^2 + \|a\|_{C[0,1]}\|x\|^2_{L^2(0,1)} - c_1\|x\|^\alpha_{L^\alpha(0,1)} + c_2. \tag{7.6.19}$$

Set $z := r\frac{e_{k+1}}{\|e_{k+1}\|}$ with $r > 0$ given above. All norms are equivalent on the finite-dimensional space $Y \oplus \mathbb{R}_z$. In particular, there is a constant $c > 0$ such that

$$\|x\| \leq c\|x\|_{L^\alpha(0,1)} \qquad \text{for any} \quad x \in Y \oplus \mathbb{R}_z.$$

Since $\alpha > 2$ we obtain from (7.6.19) that

$$\lim_{\substack{\|x\|\to\infty\\ x\in Y\oplus\mathbb{R}_z}} F(x) = -\infty. \tag{7.6.20}$$

Define

$$\mathcal{M} := \{x := y + \lambda z : y \in Y,\ \lambda \geq 0,\ \|x\| \leq \varrho\},$$
$$\mathcal{M}_0 := \{x := y + \lambda z : y \in Y,\ \|x\| = \varrho \text{ and } \lambda \geq 0, \text{ or } \|x\| \leq \varrho \text{ and } \lambda = 0\}.$$

Since $F(z) \geq b > 0$, (7.6.17) and (7.6.20) imply that there is $\varrho > r$ such that

$$a := \max_{x\in\mathcal{M}_0} F(x) \leq 0.$$

It remains to verify that F satisfies the $(\mathrm{PS})_c$ condition. This will be the case if we show that any sequence $\{x_n\}_{n=1}^\infty \subset H$ such that

$$d := \sup_n F(x_n) < \infty, \qquad F'(x_n) \to o, \tag{7.6.21}$$

contains a converging subsequence. We will prove it in two steps.

Step 1. First we prove that $\{x_n\}_{n=1}^{\infty}$ is bounded in H. Let $\beta \in \left(\frac{1}{\alpha}, \frac{1}{2}\right)$. For n large enough we have for some $c_3, c_4 > 0$, by using (f2),

$$\begin{aligned} d + \|x_n\| \geq F(x_n) - \beta(F'(x_n), x_n) &= \int_0^1 \left[\left(\frac{1}{2} - \beta\right)(|\dot{x}_n(t)|^2 + a(t)|x_n(t)|^2)\right. \\ &\quad \left. + \beta f(t, x_n(t))x_n(t) - \int_0^{x_n(t)} f(t,s)\,\mathrm{d}s\right] \mathrm{d}t \\ &\geq \left(\frac{1}{2} - \beta\right)\left(\delta\|z_n\|^2 + \lambda_1\|y_n\|^2_{L^2(0,1)}\right) + (\alpha\beta - 1)\int_0^1\int_0^{x_n(t)} f(t,s)\,\mathrm{d}s\,\mathrm{d}t - c_3 \\ &\geq \left(\frac{1}{2} - \beta\right)\left(\delta\|z_n\|^2 + \lambda_1\|y_n\|^2_{L^2(0,1)}\right) + c_1(\alpha\beta - 1)\|x_n\|^{\alpha}_{L^{\alpha}(0,1)} - c_4 \end{aligned} \tag{7.6.22}$$

where $x_n = y_n + z_n$, $y_n \in Y$, $z_n \in Z$, δ is from (7.6.15), and we have also used

$$\|y_n\|^2 + \int_0^1 a(t)|y_n(t)|^2\,\mathrm{d}t \geq \lambda_1\|y_n\|^2_{L^2(0,1)}$$

and the fact that

$$\int_0^1 [\dot{y}_n(t)\dot{z}_n(t) + a(t)y_n(t)z_n(t)]\,\mathrm{d}t = 0.$$ [57]

Since $\dim Y < \infty$, the norms $\|\cdot\|$ and $\|\cdot\|_{L^2(0,1)}$ are equivalent on Y, and (7.6.22) implies that $\{x_n\}_{n=1}^{\infty}$ is bounded (cf. Exercise 7.6.18).

Step 2. In the second step we prove that $\{x_n\}_{n=1}^{\infty}$ contains a convergent subsequence. Going to a subsequence if necessary, we can assume that $x_n \rightharpoonup x$ in H. By the Rellich–Kondrachov Theorem (Theorem 1.2.28), $x_n \to x$ in $C[0,1]$. Observe that

$$\begin{aligned} \|x_n - x\|^2 &= (F'(x_n) - F'(x), x_n - x) \\ &\quad + \int_0^1 \left[(f(t, x_n(t)) - f(t, x(t)))(x_n(t) - x(t)) - a(t)|x_n(t) - x(t)|^2\right] \mathrm{d}t. \end{aligned}$$

The boundedness of $\{x_n\}_{n=1}^{\infty}$ and (7.6.21) imply

$$(F'(x_n) - F'(x), x_n - x) \to 0, \qquad x_n \to x \quad \text{in } C[0,1]$$

implies

$$\int_0^1 a(t)(|x_n(t)|^2 - |x(t)|^2)\,\mathrm{d}t \to 0,$$

and the continuity of f implies

$$\begin{aligned} &\int_0^1 (f(t, x_n(t)) - f(t, x(t)))(x_n(t) - x(t))\,\mathrm{d}t \\ &\qquad \leq \|f(\cdot, x_n(\cdot)) - f(\cdot, x(\cdot))\|_{C[0,1]}\|x_n - x\|_{C[0,1]} \to 0 \qquad \text{as} \quad n \to \infty. \end{aligned}$$

Thus we have proved that

$$\|x_n - x\| \to 0, \qquad n \to \infty.$$ ○

---

[57] Note that $\int_0^1 [\dot{e}_j(t)\dot{z}(t) + a(t)e_j(t)z(t)]\,\mathrm{d}t = \lambda_j \int_0^1 e_j(t)z(t)\,\mathrm{d}t = 0$ for e_j, $j = 1, \dots, k$, $z \in Z$.

Remark 7.6.15. If $\lambda_1 > 0$ (this is the case if, e.g., $a(t) \geq 0$ in $[0,1]$), it suffices to use the Mountain Pass Theorem instead of the Linking Theorem. The interested reader is invited to carry out the proof in detail as an exercise.

Exercise 7.6.16. A *retraction* of a topological space X to a subspace Y is a continuous map $r\colon X \to Y$ such that

$$r(y) = y \qquad \text{for every} \quad y \in Y.$$

Prove that there is no retraction of $\overline{B(o;1)} \subset \mathbb{R}^N$ to $\mathcal{S}^{N-1} = \partial B(o;1)$.

Hint. Assume, by contradiction, that there is a retraction $r\colon \overline{B(o;1)} \to \mathcal{S}^{N-1}$. Using the homotopy

$$H(t,u) := (1-t)u + tr(u)$$

we obtain (see Theorem 5.7.7)

$$\deg(r, B(o;1), o) = \deg(I, B(o;1), o) = 1,$$

i.e., there is $u_0 \in B(o;1)$ such that $r(u_0) = o$. This contradicts $r(u) \in \mathcal{S}^{N-1}$ for any $u \in B(o;1)$!

Exercise 7.6.17. Prove that the condition (f2) implies (7.6.18).

Hint. Let $s > R$. It follows from (f2) that

$$\frac{\alpha}{s} \leq \frac{f(t,s)}{\displaystyle\int_0^s f(t,\sigma)\,\mathrm{d}\sigma}$$

and integrating over $[R,s]$ yields

$$\alpha \log s - \alpha \log R \leq \log \int_0^s f(t,\sigma)\,\mathrm{d}\sigma - \log \int_0^R f(t,\sigma)\,\mathrm{d}\sigma.$$

Taking the exponential of both sides, we obtain

$$\left(\frac{s}{R}\right)^{\alpha} \leq \frac{\displaystyle\int_0^s f(t,\sigma)\,\mathrm{d}\sigma}{\displaystyle\int_0^R f(t,\sigma)\,\mathrm{d}\sigma}, \qquad \text{i.e.,} \qquad \frac{\displaystyle\int_0^R f(t,\sigma)\,\mathrm{d}\sigma}{R^{\alpha}} s^{\alpha} \leq \int_0^s f(t,\sigma)\,\mathrm{d}\sigma.$$

Since f is continuous, we have $\displaystyle\int_0^s f(t,\sigma)\,\mathrm{d}\sigma \leq \text{const.}$ for $t \in [0,1]$, $s \in [0,R]$. Hence

$$\int_0^s f(t,\sigma)\,\mathrm{d}\sigma \geq c_1 s^{\alpha} - c_2.$$

Similarly for $s < 0$.

Exercise 7.6.18. Prove the boundedness of the sequence $\{x_n\}_{n=1}^{\infty}$ from Step 1 on page 513.

Hint. Write $x_n = y_n + z_n$, $y_n \in Y$, $z_n \in Z$. Since Y and Z are L^2-orthogonal, we have

$$\|y_n\|^2_{L^2(0,1)} = \|x_n\|^2_{L^2(0,1)} - \|z_n\|^2_{L^2(0,1)}.$$

Write (7.6.22) in the equivalent form

$$\begin{aligned} d + \|x_n\| &\geq \left(\frac{1}{2} - \beta\right)\left(\delta\|z_n\|^2 - \lambda_1\|z_n\|^2_{L^2(0,1)}\right) \\ &\quad + c_1(\alpha\beta - 1)\|x_n\|^\alpha_{L^\alpha(0,1)} + \lambda_1\left(\frac{1}{2} - \beta\right)\|x_n\|^2_{L^2(0,1)} - c_4 \\ &\geq \left(\frac{1}{2} - \beta\right)\left(\delta\|z_n\|^2 - \lambda_1\|z_n\|^2_{L^2(0,1)}\right) \\ &\quad + \|x_n\|^2_{L^\alpha(0,1)}\left[c_1(\alpha\beta - 1)\|x_n\|^{\alpha-2}_{L^\alpha(0,1)} + c_5\lambda_1\left(\frac{1}{2} - \beta\right)\right] - c_4 \end{aligned}$$

(where the inequality $\|x_n\|_{L^2(0,1)} \leq c_5\|x_n\|_{L^\alpha(0,1)}$ is used) and get the boundedness of $\{x_n\}_{n=1}^\infty$.

Exercise 7.6.19. Consider the Dirichlet boundary value problem

$$\begin{cases} -\left(|\dot{x}(t)|^{p-2}\dot{x}(t)\right)^{\cdot} - \lambda|x(t)|^{p-2}x(t) = g(t, x(t)), & t \in (0,1), \\ x(0) = x(1) = 0, \end{cases} \tag{7.6.23}$$

where $p > 1$. Formulate conditions on λ and $g = g(t,x)$ which guarantee that (7.6.23) has a geometry corresponding to

(i) the Saddle Point Theorem,

(ii) the Linking Theorem.

Exercise 7.6.20. How do the conditions on λ and g change if the homogeneous Dirichlet conditions in (7.6.23) are replaced by the Neumann ones?

7.7 Weak Solutions of PDEs, Applications of Saddle Point Theorem

Now we show an application of the Saddle Point Theorem. Let us consider the Dirichlet boundary value problem

$$\begin{cases} -\Delta u(x) = \lambda u(x) + h(x, u(x)) & \text{in} \quad \Omega, \\ u = 0 & \text{on} \quad \partial\Omega. \end{cases} \tag{7.7.1}$$

If h is bounded and λ is not an eigenvalue of (5.6.14), the existence of a solution of (7.7.1) follows from Theorem 5.9.2.

We prove the following assertion.

Theorem 7.7.1. *Let λ be an eigenvalue of* (5.6.14) *and h, $\frac{\partial h}{\partial s}$ be bounded and continuous. If, moreover,*

$$H(x,s) = \int_0^s h(x,\tau)\,\mathrm{d}\tau \rightrightarrows \infty \qquad \text{as } |s| \to \infty \text{ uniformly for } x \in \Omega, \tag{7.7.2}$$

then (7.7.1) *possesses a weak solution.*[58]

Proof. Let $\lambda = \lambda_k < \lambda_{k+1}$ for a $k \in \mathbb{N}$ where λ_k, $k \in \mathbb{N}$, are the eigenvalues of (5.6.14). The energy functional associated with (7.7.1),

$$E(u) = \frac{1}{2}\int_\Omega |\nabla u(x)|^2\,\mathrm{d}x - \frac{\lambda_k}{2}\int_\Omega |u(x)|^2\,\mathrm{d}x - \int_\Omega H(x,u(x))\,\mathrm{d}x, \tag{7.7.3}$$

$u \in W_0^{1,2}(\Omega)$, has the property $E \in C^2(W_0^{1,2}(\Omega),\mathbb{R})$ due to the assumptions on h, $\frac{\partial h}{\partial s}$ (Exercise 7.7.4). Let $\{\varphi_j\}_{j=1}^\infty$ be an orthonormal basis of $W_0^{1,2}(\Omega)$ consisting of the eigenfunctions associated with the eigenvalues $\{\lambda_j\}_{j=1}^\infty$, $0 < \lambda_1 \le \lambda_2 \le \cdots$ (see Example 5.9.1). In particular,

$$\lambda_j \int_\Omega |\varphi_j(x)|^2\,\mathrm{d}x = \int_\Omega |\nabla \varphi_j(x)|^2\,\mathrm{d}x = 1 \qquad \text{holds for all} \quad j \in \mathbb{N}. \tag{7.7.4}$$

Let

$$Y := \mathrm{Lin}\{\varphi_1,\varphi_2,\dots,\varphi_k\},\ Z = \left\{u \in W_0^{1,2}(\Omega) : \int_\Omega u(x)v(x)\,\mathrm{d}x = 0, v \in Y\right\},$$

i.e.,

$$W_0^{1,2}(\Omega) = Y \oplus Z, \qquad \dim Y < \infty$$

and $\{\varphi_j\}_{j\ge k+1}^\infty$ forms an orthonormal basis of Z.

Step 1. We prove that E has a geometry of the Saddle Point Theorem. If $u \in Z$, then $u = \sum\limits_{j=k+1}^\infty a_j\varphi_j$ and (see (7.7.4))

$$\int_\Omega \left(|\nabla u(x)|^2 - \lambda_k|u(x)|^2\right)\,\mathrm{d}x = \sum_{j=k+1}^\infty a_j^2\left(1 - \frac{\lambda_k}{\lambda_j}\right) \ge \left(1 - \frac{\lambda_k}{\lambda_{k+1}}\right)\|u\|^2. \tag{7.7.5}$$

Let $M := \sup\limits_{\substack{x\in\Omega\\ s\in\mathbb{R}}} |h(x,s)|$. Then

$$\left|\int_\Omega H(x,u(x))\,\mathrm{d}x\right| \le M\int_\Omega |u(x)|\,\mathrm{d}x \le M_1\|u\| \tag{7.7.6}$$

[58] The assertion can be proved under weaker assumptions on h, cf. Rabinowitz [124] and Remark 7.6.4.

for all $u \in W_0^{1,2}(\Omega)$, by the Hölder and Poincaré inequalities. Combining (7.7.5) and (7.7.6) shows that E is bounded below on Z, i.e.,

$$\inf_{u\in Z} E(u) > -\infty. \tag{7.7.7}$$

Next, if $u \in Y$, then $u = u^0 + \hat{u}$ where

$$u^0 \in Y^0 := \operatorname{Lin}\{\varphi_j : \lambda_j = \lambda_k\}^{59} \qquad \text{and} \qquad \hat{u} \in \hat{Y} := \operatorname{Lin}\{\varphi_j : \lambda_j < \lambda_k\}.$$

Then for $u \in Y$, $u = \sum\limits_{j=1}^{k} a_j\varphi_j$,

$$\begin{aligned} E(u) &= \frac{1}{2}\sum_{j:\lambda_j<\lambda_k} a_j^2\left(1-\frac{\lambda_k}{\lambda_j}\right) - \int_\Omega H(x,u^0(x))\,\mathrm{d}x \\ &\quad - \int_\Omega \left(H(x,u^0(x)+\hat{u}(x)) - H(x,u^0(x))\right)\,\mathrm{d}x. \end{aligned} \tag{7.7.8}$$

There is a constant $M_2 > 0$ such that

$$\frac{1}{2}\sum_{j:\lambda_j<\lambda_k} a_j^2\left(1-\frac{\lambda_k}{\lambda_j}\right) \le -M_2\|\hat{u}\|^2, \tag{7.7.9}$$

and

$$\begin{aligned} &\left|\int_\Omega [H(x,u^0(x)+\hat{u}(x)) - H(x,u^0(x))]\,\mathrm{d}x\right| \\ &= \left|\int_\Omega \left[\int_0^{u^0(x)+\hat{u}(x)} h(x,s)\,\mathrm{d}s - \int_0^{u^0(x)} h(x,s)\,\mathrm{d}s\right]\mathrm{d}x\right| \\ &= \left|\int_\Omega \int_{u^0(x)}^{u^0(x)+\hat{u}(x)} h(x,s)\,\mathrm{d}s\,\mathrm{d}x\right| \le M\int_\Omega |\hat{u}(x)|\,\mathrm{d}x \le M_1\|\hat{u}\|. \end{aligned} \tag{7.7.10}$$

It follows from (7.7.8)–(7.7.10) that

$$E(u) \le -M_2\|\hat{u}\|^2 - \int_\Omega H(x,u^0(x))\,\mathrm{d}x + M_1\|\hat{u}\|.$$

This implies that

$$E(u) \to -\infty \qquad \text{as} \quad \|u\| \to \infty, \quad u \in Y. \tag{7.7.11}$$

Indeed, either $\|\hat{u}\| \to \infty$ or $\|u^0\| \to \infty$, and then (7.7.11) follows from the assumption (7.7.2). It follows from (7.7.7) and (7.7.11) that E verifies the hypotheses of Proposition 7.6.2, i.e., E has the geometry corresponding to the Saddle Point Theorem.

[59] Note that the multiplicity of λ_k need not be equal to one in general but it is finite.

Step 2. Now, we prove that E satisfies the $(\mathrm{PS})_c$ condition. Let us assume that $|E(u_m)| \le K$ for some $K > 0$ and $\nabla E(u_m) \to o$.[60] Let us write

$$u_m = u_m^0 + \hat{u}_m + \tilde{u}_m \qquad \text{where} \quad u_m^0 \in Y^0,\ \hat{u}_m \in \hat{Y},\ \tilde{u}_m \in Z.$$

For large m, we have

$$\begin{aligned}\|\tilde{u}_m\| &\ge |(\nabla E(u_m), \tilde{u}_m)| \\ &= \left| \int_\Omega [\nabla u_m(x) \nabla \tilde{u}_m(x) - \lambda_k u_m(x) \tilde{u}_m(x) - h(x, u_m(x)) \tilde{u}_m(x)]\, \mathrm{d}x \right|,\end{aligned} \tag{7.7.12}$$

and the same for $\hat{u}_m$. On the other hand, since $Z = Y^\perp$, by (7.7.12), (7.7.5) and the boundedness of h we obtain

$$\begin{aligned}&\left| \int_\Omega \nabla u_m(x) \nabla \tilde{u}_m(x) - \lambda_k u_m(x) \tilde{u}_m(x) - h(x, u_m(x)) \tilde{u}_m(x)]\, \mathrm{d}x \right| \\ &\qquad\qquad \ge \left(1 - \frac{\lambda_k}{\lambda_{k+1}}\right) \|\tilde{u}_m\|^2 - M_1 \|\tilde{u}_m\|.\end{aligned} \tag{7.7.13}$$

From (7.7.12) and (7.7.13) we obtain

$$\|\tilde{u}_m\| \ge \left(1 - \frac{\lambda_k}{\lambda_{k+1}}\right) \|\tilde{u}_m\|^2 - M_1 \|\tilde{u}_m\|,$$

which shows that $\{\|\tilde{u}_m\|\}_{m=1}^\infty$ is bounded. Similarly, we prove that $\{\|\hat{u}_m\|\}_{m=1}^\infty$ is bounded, too. Finally, we claim that $\left\{\|u_m^0\|\right\}_{m=1}^\infty$ is bounded. To verify the claim, observe that

$$\begin{aligned}K \ge |E(u_m)| = \Bigg| \int_\Omega \Big\{ &\frac{1}{2} \left[|\nabla \tilde{u}_m(x)|^2 + |\nabla \hat{u}_m(x)|^2 - \lambda_k (|\tilde{u}_m(x)|^2 + |\hat{u}_m(x)|^2) \right] \\ &- \left(H(x, u_m(x)) - H(x, u_m^0(x)) \right) \Big\}\, \mathrm{d}x - \int_\Omega H(x, u_m^0(x))\, \mathrm{d}x \Bigg|.\end{aligned}$$

By what has already been shown, the first integral on the right-hand side is bounded independently of m.[61] Therefore $\displaystyle\int_\Omega H(x, u_m^0(x))\, \mathrm{d}x$ is bounded. In order to show that $\left\{u_m^0\right\}_{m=1}^\infty$ is bounded it is sufficient to prove that

$$\int_\Omega H(x, v(x))\, \mathrm{d}x \to \infty \qquad \text{as} \quad \|v\| \to \infty \quad \text{for } v \in Y^0. \tag{7.7.14}$$

By (7.7.2) for any $l > 0$, there is d_l such that

$$H(x, s) \ge l \qquad \text{if} \quad |s| \ge d_l \qquad \text{for all} \quad x \in \Omega.$$

---

[60] $E(u_m) \to c$, $c \in \mathbb{R}$, implies that there exists $K > 0$ such that $|E(u_m)| \le K$.

[61] The reader should justify it using an estimate similar to (7.7.10).

Let $v \in Y^0$, $v \neq o$, and write

$$v = t\varphi \qquad \text{where} \quad \varphi \in \partial B(o;1) := \{w \in Y^0 : \|w\| = 1\}.$$

Then

$$\int_\Omega H(x, t\varphi(x))\, dx \geq \int_{\Omega_l^t(\varphi)} H(x, t\varphi(x))\, dx - M_0 \tag{7.7.15}$$

where

$$\Omega_l^t(\varphi) = \{x \in \Omega : |t\varphi(x)| \geq d_l\} \qquad \text{and} \qquad M_0 \geq (\operatorname{meas}\Omega) \left| \inf_{\substack{x\in\Omega \\ s\in\mathbb{R}}} H(x,s) \right|.$$

For any $\psi \in \partial B(o;1)$ we find an open neighborhood $\mathcal{U}(\psi) \subset \partial B(o;1)$ of ψ, $x = x(\psi) \in \Omega$ and $r = r(\psi) > 0$ with the following property:

for an arbitrary $l \in \mathbb{N}$ there exists $t_l(\psi)$ such that

$$B(x(\psi); r(\psi)) := \{x \in \Omega : |x - x(\psi)| < r(\psi)\} \subset \Omega_l^t(\varphi)$$ [62]

for any $t \geq t_l(\psi)$ and $\varphi \in \mathcal{U}(\psi)$.

Then (7.7.15) implies that for any $l \in \mathbb{N}$ we have

$$\int_\Omega H(x, t\varphi(x))\, dx \geq l \operatorname{meas} B(x(\psi); r(\psi)) - M_0 \tag{7.7.16}$$

for any $t \geq t_l(\psi)$, $\varphi \in \mathcal{U}(\psi)$. The system $\{\mathcal{U}(\psi) : \psi \in \partial B(o;1)\}$ is an open covering of $\partial B(o;1)$. The compactness of $\partial B(o;1)$ implies that there exists a finite subcovering $\{\mathcal{U}(\psi_i) : \psi_i \in \partial B(o;1)\}$, $i = 1, \dots, n$, of $\partial B(o;1)$. Let

$$c := \min_{i=1,\dots,n} \{\operatorname{meas} B(x(\psi_i); r(\psi_i))\}, \quad t_l := \max_{i=1,\dots,n} \{t_l(\psi_i)\} \quad \text{for any } l \in \mathbb{N}.$$

Then from (7.7.16) we obtain that

$$\int_\Omega H(x, t\varphi(x))\, dx \geq cl - M_0 \qquad \text{for any} \quad \varphi \in \partial B(o;1) \text{ and } t \geq t_l,$$

i.e., (7.7.14) holds for $\|v\| \to \infty$, $v \in Y^0$.

So, we have proved that $\{u_m\}_{m=1}^\infty \subset W_0^{1,2}(\Omega)$ is a bounded sequence. Passing to a subsequence if necessary, we may assume that $u_m \rightharpoonup u$ in $W_0^{1,2}(\Omega)$ and $u_m \to u$ in $L^2(\Omega)$. Then we have

$$(\nabla E(u_m) - \nabla E(u), u_m - u) \to 0. \tag{7.7.17}$$

[62] Here we use the fact that the eigenfunctions of (7.5.8) are continuous in $\overline{\Omega}$ (cf. Example 5.9.1).

But we also have

$$\int_\Omega |u_m(x) - u(x)|^2 \, dx \to 0,$$

$$\int_\Omega [h(x, u_m(x)) - h(x, u(x))](u_m(x) - u(x)) \, dx \to 0.$$

These facts together with (7.7.17) imply

$$\int_\Omega |\nabla u_m(x) - \nabla u(x)|^2 \, dx \to 0, \qquad \text{i.e.,} \qquad u_m \to u \quad \text{in} \quad W_0^{1,2}(\Omega).$$

This proves that E verifies the $(\mathrm{PS})_c$ condition, and the proof of Theorem 7.7.1 is complete. □

Exercise 7.7.2. Let

$$E(u) = \frac{1}{2}\int_\Omega |\nabla u(x)|^2 \, dx + \frac{\lambda}{2}\int_\Omega |u(x)|^2 \, dx - \frac{1}{p}\int_\Omega |u^+(x)|^p \, dx, \quad \lambda \in \mathbb{R}, \ p > 2.$$

Prove that $E \in C^2(W_0^{1,2}(\Omega), \mathbb{R})$ and

$$(\nabla E(u), h) = \int_\Omega \nabla u(x) \nabla h(x) \, dx + \lambda \int_\Omega u(x) h(x) \, dx - \int_\Omega |u^+(x)|^{p-1} h(x) \, dx.$$

Exercise 7.7.3. Compare the assertion and the proof of Theorem 7.5.1 with Example 7.4.7. Point out the differences between the one-dimensional and higher-dimensional cases.

Exercise 7.7.4. Prove that the functional $E(u)$ from (7.7.3) has the property

$$E \in C^2(W_0^{1,2}(\Omega), \mathbb{R}).$$

Hint. First prove that the second Gâteaux derivative is given by

$$D^2E(u)(w, z) = \int_\Omega \nabla w(x) \nabla z(x) \, dx - \lambda_k \int_\Omega w(x) z(x) \, dx - \int_\Omega \frac{\partial h}{\partial s}(x, u(x)) w(x) z(x) \, dx.$$

Then show that $D^2E(u)$ is continuous in u (Remark 3.2.29). To prove continuity of the third term in $D^2E(u)$ use the boundedness of $\frac{\partial h}{\partial s}$, the embedding $W_0^{1,2}(\Omega) \hookrightarrow\hookrightarrow L^r(\Omega)$, $r > 2$ (Remark 1.2.24) and the continuity of the Nemytski operator from $L^r(\Omega)$ into $L^{\frac{r}{r-2}}(\Omega)$ (Theorem 3.2.24(ii)).

Exercise 7.7.5. Let $X \subset W_0^{1,2}(\Omega)$, $\dim X < \infty$. Why the norms

$$\|u\| = \left(\int_\Omega |\nabla u(x)|^2 \, dx\right)^{\frac{1}{2}}, \qquad \|u\|_{L^p(\Omega)} = \left(\int_\Omega |u(x)|^p \, dx\right)^{\frac{1}{p}},$$

$$\|u\|_{L^{2^*}(\Omega)} = \left(\int_\Omega |u(x)|^{2^*} \, dx\right)^{\frac{1}{2^*}}, \qquad 1 < p < 2^*,$$

are equivalent on X? Why these norms are not equivalent on the whole space $X = W_0^{1,2}(\Omega)$?

Hint. Cf. Corollary 1.2.11.

Exercise 7.7.6. Replace (7.7.2) by the assumption

$$H(x,s) \to -\infty \qquad \text{as} \quad |s| \to \infty$$

and prove the assertion of Theorem 7.7.1.

Exercise 7.7.7. Consider the boundary value problem

$$\begin{cases} -\Delta u(x) + \lambda u(x) = g(x, u(x)) & \text{in} \quad \Omega, \\ u = 0 & \text{on} \quad \partial\Omega. \end{cases} \tag{7.7.18}$$

Formulate conditions on λ and $g = g(x,s)$ which guarantee that the energy functional associated with (7.7.18)

(i) is weakly coercive,

(ii) has a geometry corresponding to the Mountain Pass Theorem,

(iii) has a geometry corresponding to the Saddle Point Theorem.

7.7A Weak Solutions of PDEs, Applications of General Saddle Point Theorem

In this appendix we will give another application of Theorem 7.6.12. Consider the existence of weak solutions of the boundary value problem

$$\begin{cases} -\Delta_p u(x) = \lambda_1 |u(x)|^{p-2} u(x) + f(x, u(x)) - h(x) & \text{in} \quad \Omega, \\ u = 0 & \text{on} \quad \partial\Omega, \end{cases} \tag{7.7.19}$$

where $p > 1$, $\Omega \in C^{0,1}$ is a bounded domain in $\mathbb{R}^N$, $f\colon \Omega \times \mathbb{R} \to \mathbb{R}$ is a bounded Carathéodory function and $h \in L^{p'}(\Omega)$, $p' = \frac{p}{p-1}$. As in Appendix 5.9A, let $\lambda_1 > 0$ be the principal eigenvalue of $-\Delta_p$ on Ω with zero Dirichlet boundary conditions, and let us denote by φ_1 the positive (in Ω) eigenfunction associated with λ_1 normalized by $\|\varphi_1\| = \left(\int_\Omega |\nabla \varphi_1(x)|^p \,\mathrm{d}x \right)^{\frac{1}{p}} = 1.$

We will suppose that f satisfies the following condition:

for a.a. $x \in \Omega$ there exist limits

$$\lim_{s \to -\infty} f(x,s) = f_{-\infty}(x), \qquad \lim_{s \to +\infty} f(x,s) = f_{+\infty}(x).$$

It is well known that under this condition the problem (7.7.19) need not have solutions (cf. Exercise 7.7.9).

The following result extends the classical result of Landesman & Lazer [98].

Theorem 7.7.8. *Suppose that either*

$$\int_\Omega f_{+\infty}(x)\varphi_1(x)\,\mathrm{d}x < \int_\Omega h(x)\varphi_1(x)\,\mathrm{d}x < \int_\Omega f_{-\infty}(x)\varphi_1(x)\,\mathrm{d}x \tag{7.7.20}$$

or else

$$\int_\Omega f_{-\infty}(x)\varphi_1(x)\,\mathrm{d}x < \int_\Omega h(x)\varphi_1(x)\,\mathrm{d}x < \int_\Omega f_{+\infty}(x)\varphi_1(x)\,\mathrm{d}x. \tag{7.7.21}$$

Then there exists at least one weak solution $u \in W_0^{1,p}(\Omega)$ *of the problem* (7.7.19).

Proof. We follow the proof from Arcoya & Orsina [10]. Let us introduce the energy functional $E\colon W_0^{1,p}(\Omega) \to \mathbb{R}$ associated with (7.7.19):

$$E(u) := \frac{1}{p}\int_\Omega |\nabla u(x)|^p\,\mathrm{d}x - \frac{\lambda_1}{p}\int_\Omega |u(x)|^p\,\mathrm{d}x - \int_\Omega F(x,u(x))\,\mathrm{d}x + \int_\Omega h(x)u(x)\,\mathrm{d}x \tag{7.7.22}$$

where

$$F(x,s) = \int_0^s f(x,t)\,\mathrm{d}t \qquad \text{for a.a.} \quad x \in \Omega \quad \text{and} \quad s \in \mathbb{R}.$$

Then $E \in C^1(W_0^{1,p}(\Omega),\mathbb{R})$ (cf. Exercise 7.7.10) and its critical points correspond to the weak solutions of (7.7.19).

We proceed in three steps.

Step 1. Let $\{u_n\}_{n=1}^\infty \subset W_0^{1,p}(\Omega)$ be such that there exists $c > 0$ such that

$$|E(u_n)| \le c \qquad \text{for any} \quad n \in \mathbb{N} \tag{7.7.23}$$

and there exists a strictly decreasing sequence $\{\varepsilon_n\}_{n=1}^\infty$, $\lim\limits_{n\to\infty} \varepsilon_n = 0$, such that

$$\big|\langle E'(u_n), v\rangle\big| \le \varepsilon_n\|v\| \qquad \text{for any} \quad n \in \mathbb{N} \quad \text{and any} \quad v \in W_0^{1,p}(\Omega).^{63} \tag{7.7.24}$$

Then we will prove that $\{u_n\}_{n=1}^\infty$ contains a subsequence which converges strongly in $W_0^{1,p}(\Omega)$.

Let us begin by proving that the sequence $\{u_n\}_{n=1}^\infty$ is bounded in $W_0^{1,p}(\Omega)$. Suppose, by contradiction, that $\|u_n\| \to \infty$, and define $v_n = \frac{u_n}{\|u_n\|}$. Thus $\{v_n\}_{n=1}^\infty$ is bounded in $W_0^{1,p}(\Omega)$ and hence, at least its subsequence, converges to a function v_0 weakly in $W_0^{1,p}(\Omega)$ and strongly in $L^p(\Omega)$. Dividing (7.7.22) with $u = u_n$ by $\|u_n\|^p$, we get, due to (7.7.23),

$$\limsup_{n\to\infty}\left[\frac{1}{p} - \frac{\lambda_1}{p}\int_\Omega |v_n(x)|^p\,\mathrm{d}x - \int_\Omega \frac{F(x,u_n(x))}{\|u_n\|^p}\,\mathrm{d}x + \int_\Omega h(x)\frac{u_n(x)}{\|u_n\|^p}\,\mathrm{d}x\right] \le 0.$$

Since

$$\lim_{n\to\infty}\left[\int_\Omega \frac{F(x,u_n(x))}{\|u_n\|^p}\,\mathrm{d}x + \int_\Omega h(x)\frac{u_n(x)}{\|u_n\|^p}\,\mathrm{d}x\right] = 0$$

by the hypotheses on f, h, and $\{u_n\}_{n=1}^\infty$ while

$$\lim_{n\to\infty}\int_\Omega |v_n(x)|^p\,\mathrm{d}x = \int_\Omega |v_0(x)|^p\,\mathrm{d}x,$$

[63] It will be convenient to express the assumption $E'(u_n) \to 0$ in this form.

we have

$$\lambda_1 \int_\Omega |v_0(x)|^p \, dx \geq 1.$$

Using the weak lower semicontinuity of the norm and the variational characterization of λ_1 (see Appendix 5.8B), we get

$$1 \leq \lambda_1 \int_\Omega |v_0(x)|^p \, dx \leq \int_\Omega |\nabla v_0(x)|^p \, dx \leq \liminf_{n\to\infty} \int_\Omega |\nabla v_n(x)|^p \, dx = 1.$$

Thus

$$\|v_0\| = 1 \qquad \text{and} \qquad \int_\Omega |\nabla v_0(x)|^p \, dx = \lambda_1 \int_\Omega |v_0(x)|^p \, dx.$$

This implies, by the definition of φ_1, that $v_0 = \pm\varphi_1$.[64]

Now we write (7.7.23) and (7.7.24) with $v = u_n$ in the equivalent form

$$\begin{aligned} -cp \leq & \int_\Omega |\nabla u_n(x)|^p \, dx - \lambda_1 \int_\Omega |u_n(x)|^p \, dx \\ & - p \int_\Omega F(x, u_n(x)) \, dx + p \int_\Omega h(x) u_n(x) \, dx \leq cp, \\ -\varepsilon_n \|u_n\| \leq & - \int_\Omega |\nabla u_n(x)|^p \, dx + \lambda_1 \int_\Omega |u_n(x)|^p \, dx \\ & + \int_\Omega f(x, u_n(x)) u_n(x) \, dx - \int_\Omega h(x) u_n(x) \, dx \leq \varepsilon_n \|u_n\|. \end{aligned}$$

Summing up and dividing by $\|u_n\|$, we obtain

$$\left| \int_\Omega [f(x, u_n(x)) v_n(x) - p g(x, u_n(x)) v_n(x) + (p-1) h(x) v_n(x)] \, dx \right| \leq \frac{cp}{\|u_n\|} + \varepsilon_n$$

where

$$g(x, s) = \begin{cases} \dfrac{F(x,s)}{s} & \text{if } s \neq 0, \\ f(x, 0) & \text{if } s = 0. \end{cases} \tag{7.7.25}$$

Letting n tend to infinity and supposing that v_n converge to $+\varphi_1$ (for example), we obtain

$$\lim_{n\to\infty} \int_\Omega [f(x, u_n(x)) v_n(x) - p g(x, u_n(x)) v_n(x)] \, dx = (1-p) \int_\Omega h(x) \varphi_1(x) \, dx.$$

Since v_n converge to φ_1, we have $\lim_{n\to\infty} u_n(x) = \infty$ for a.a. $x \in \Omega$, and so

$$\begin{aligned} f(x, u_n(x)) &\to f_{+\infty}(x) \qquad \text{for a.a.} \quad x \in \Omega, \\ g(x, u_n(x)) &\to f_{+\infty}(x) \qquad \text{for a.a.} \quad x \in \Omega. \end{aligned}$$

The properties of f and F and the Lebesgue Theorem then imply

$$\lim_{n\to\infty} \int_\Omega [f(x, u_n(x)) v_n(x) - p g(x, u_n(x)) v_n(x)] \, dx = (1-p) \int_\Omega f_{+\infty}(x) \varphi_1(x) \, dx,$$

[64] Note that we have proved that $\|v_n\| \to \|v_0\|$ and so by the uniform convexity of $W_0^{1,p}(\Omega)$, $v_n \to \pm\varphi_1$, too.

and so, since $p > 1$,

$$\int_\Omega f_{+\infty}(x)\varphi_1(x)\,\mathrm{d}x = \int_\Omega h(x)\varphi_1(x)\,\mathrm{d}x,$$

which contradicts both (7.7.20) and (7.7.21).

Thus $\{u_n\}_{n=1}^\infty$ is bounded. This implies that there exists $u \in W_0^{1,p}(\Omega)$ such that, at least its subsequence, u_n converge to u weakly in $W_0^{1,p}(\Omega)$ and strongly in $L^p(\Omega)$.

Choosing $v = u_n - u$ in (7.7.24), we obtain

$$\bigg|\int_\Omega |\nabla u_n(x)|^{p-2}\nabla u_n(x)(\nabla u_n(x) - \nabla u(x))\,\mathrm{d}x - \lambda_1 \int_\Omega |u_n(x)|^{p-2}u_n(x)(u_n(x) - u(x))\,\mathrm{d}x$$
$$- \int_\Omega f(x, u_n(x))(u_n(x) - u(x))\,\mathrm{d}x + \int_\Omega h(x)(u_n(x) - u(x))\,\mathrm{d}x\bigg| \le \varepsilon_n \|u_n - u\|.$$

Since $u_n \to u$ in $L^p(\Omega)$ and, by the hypotheses on f and h,

$$\lim_{n\to\infty} \int_\Omega |u_n(x)|^{p-2}u_n(x)(u_n(x) - u(x))\,\mathrm{d}x = 0,$$
$$\lim_{n\to\infty} \int_\Omega f(x, u_n(x))(u_n(x) - u(x))\,\mathrm{d}x = 0,$$
$$\lim_{n\to\infty} \int_\Omega h(x)(u_n(x) - u(x))\,\mathrm{d}x = 0,$$

we have

$$\lim_{n\to\infty} \int_\Omega |\nabla u_n(x)|^{p-2}\nabla u_n(x)(\nabla u_n(x) - \nabla u(x))\,\mathrm{d}x = 0.$$

Subtracting

$$\int_\Omega |\nabla u(x)|^{p-2}\nabla u(x)(\nabla u_n(x) - \nabla u(x))\,\mathrm{d}x$$

(which converges to zero as n tends to infinity since u belongs to $W_0^{1,p}(\Omega)$), we conclude that

$$0 = \lim_{n\to\infty} \int_\Omega (|\nabla u_n(x)|^{p-2}\nabla u_n(x) - |\nabla u(x)|^{p-2}\nabla u(x))(\nabla u_n(x) - \nabla u(x))\,\mathrm{d}x$$
$$\ge \lim_{n\to\infty} (\|u_n\|^{p-1} - \|u\|^{p-1})(\|u_n\| - \|u\|) \ge 0,$$ [65]

which implies $\|u_n\| \to \|u\|$. The uniform convexity of $W_0^{1,p}(\Omega)$ yields that u_n converge strongly to u in $W_0^{1,p}(\Omega)$.

This completes the proof of Step 1. Note that it follows from Step 1 that E satisfies $(\mathrm{PS})_c$ on any level $c \in \mathbb{R}$.

Step 2. Note also that, in the proof of the Palais–Smale condition, we have proved that if $\{E(u_n)\}_{n=1}^\infty$ is a sequence bounded above with $\|u_n\| \to \infty$, then (at least its subsequence) $v_n = \frac{u_n}{\|u_n\|} \to \pm\varphi_1$ in $W_0^{1,p}(\Omega)$ (see footnote 64 on page 523). Using this fact, it is easy to prove that E is weakly coercive provided (7.7.20) holds. Otherwise, it is possible to choose a sequence $\{u_n\}_{n=1}^\infty$ such that

$$\|u_n\| \to \infty, \qquad E(u_n) \le c \qquad \text{and} \qquad v_n = \frac{u_n}{\|u_n\|} \to \pm\varphi \qquad \text{in} \quad W_0^{1,p}(\Omega).$$

[65] Cf. computation on page 380.

Assume (for example) that $v_n \to \varphi_1$; arguing as in the previous proof we get

$$\begin{aligned}
&\int_\Omega h(x)\varphi_1(x)\,\mathrm{d}x - \int_\Omega f_{+\infty}(x)\varphi_1(x)\,\mathrm{d}x \\
&\quad = \lim_{n\to\infty}\left[\int_\Omega h(x)v_n(x)\,\mathrm{d}x - \int_\Omega \frac{F(x,u_n(x))}{\|u_n\|}\,\mathrm{d}x\right] \le \limsup_{n\to\infty}\frac{E(u_n)}{\|u_n\|} \le \lim_{n\to\infty}\frac{c}{\|u_n\|} = 0,
\end{aligned}$$

which contradicts (7.7.20). The weak coerciveness of E and the weak sequential lower semicontinuity (cf. Exercise 7.7.11) are enough in order to prove that E attains its infimum (see Theorem 7.2.8 and Remark 7.2.22), so that (7.7.19) has at least one weak solution.

Step 3. If (7.7.21) holds, then E has the geometry of the Saddle Point Theorem. Indeed, splitting $W_0^{1,p}(\Omega)$ as the direct sum of

$$Y = \mathrm{Lin}\{\varphi_1\} \qquad \text{and} \qquad Z := \left\{u \in W_0^{1,p}(\Omega) : \int_\Omega u(x)\varphi_1(x)\,\mathrm{d}x = 0\right\},$$

we see that there exists $\bar\lambda > \lambda_1$ such that

$$\int_\Omega |\nabla u(x)|^p\,\mathrm{d}x \ge \bar\lambda \int_\Omega |u(x)|^p\,\mathrm{d}x \qquad \text{for all} \quad u \in Z$$

(cf. Exercise 7.7.12). Thus, by the Hölder inequality and by the properties of F, there exists $c > 0$ such that for every u in Z,

$$\begin{aligned}
E(u) &\ge \frac{1}{p}\left(1 - \frac{\lambda_1}{\bar\lambda}\right)\int_\Omega |\nabla u(x)|^p\,\mathrm{d}x - \int_\Omega F(x,u(x))\,\mathrm{d}x + \int_\Omega h(x)u(x)\,\mathrm{d}x \\
&\ge \frac{1}{p}\left(1 - \frac{\lambda_1}{\bar\lambda}\right)\int_\Omega |\nabla u(x)|^p\,\mathrm{d}x \\
&\quad - \frac{c}{\bar\lambda^{\frac{1}{p}}}\left((\mathrm{meas}\,\Omega)^{\frac{1}{p'}} + \|h\|_{L^{p'}(\Omega)}\right)\left(\int_\Omega |\nabla u(x)|^p\,\mathrm{d}x\right)^{\frac{1}{p}}.
\end{aligned}$$

Hence, E is weakly coercive on Z, so that

$$B_Z = \min_{u\in Z} E(u) > -\infty.$$

Observe that we have not yet used the fact that (7.7.21) holds. On the other hand, for every $t \in \mathbb{R}$ we have

$$\int_\Omega |\nabla(t\varphi_1)(x)|^p\,\mathrm{d}x - \lambda_1 \int_\Omega |t\varphi_1(x)|^p\,\mathrm{d}x = 0$$

as follows from the definition of λ_1 and φ_1. Thus,

$$\begin{aligned}
E(t\varphi_1) &= t\int_\Omega h(x)\varphi_1(x)\,\mathrm{d}x - \int_\Omega F(x,t\varphi_1(x))\,\mathrm{d}x \\
&= t\left[\int_\Omega h(x)\varphi_1(x)\,\mathrm{d}x - \int_\Omega g(x,t\varphi_1(x))\varphi_1(x)\,\mathrm{d}x\right]
\end{aligned}$$

where g has been defined by (7.7.25). Using the positivity of φ_1 and the hypotheses on f, it is easy to see that

$$\lim_{t\to+\infty} g(x, t\varphi_1(x))\varphi_1(x) = f_{+\infty}(x)\varphi_1(x) \qquad \text{for a.a.} \quad x \in \Omega.$$

Furthermore, there exists $c > 0$ such that

$$|g(x, t\varphi_1(x))\varphi_1(x)| \le c\varphi_1(x) \in L^1(\Omega),$$

so that the Lebesgue Theorem implies

$$\lim_{t\to+\infty} \left[\int_\Omega h(x)\varphi_1(x)\,\mathrm{d}x - \int_\Omega g(x, t\varphi_1(x))\varphi_1(x)\,\mathrm{d}x\right] = \int_\Omega [h(x) - f_{+\infty}(x)]\varphi_1(x)\,\mathrm{d}x,$$

and the limit is negative by (7.7.21). Analogously, if t tends to $-\infty$, we have the same result with $f_{+\infty}$ replaced by $f_{-\infty}$, so that the limit is positive by (7.7.21). In both the cases we have

$$\lim_{t\to\pm\infty} E(t\varphi_1) = -\infty.$$

Thus, there exists $R > 0$ such that if $|t| = R$, we have

$$E(t\varphi_1) < B_Z \le E(u) \qquad \text{for all} \quad u \in Z.$$

Hence E satisfies the hypotheses of Theorem 7.6.12, and so there exists a critical point for E, that is, a weak solution of (7.7.19). □

Exercise 7.7.9. Let $p = 2$, let $f(x, \cdot)$ be strictly increasing (decreasing). Then a necessary condition for the existence of a solution of (7.7.19) is that

$$\int_\Omega f_{-\infty}(x)\varphi_1(x)\,\mathrm{d}x < \int_\Omega h(x)\varphi_1(x)\,\mathrm{d}x < \int_\Omega f_{+\infty}(x)\varphi_1(x)\,\mathrm{d}x$$

$$\Big(\int_\Omega f_{+\infty}(x)\varphi_1(x)\,\mathrm{d}x < \int_\Omega h(x)\varphi_1(x)\,\mathrm{d}x < \int_\Omega f_{-\infty}(x)\varphi_1(x)\,\mathrm{d}x\Big).$$

Hint. Assume that u is a solution of (7.7.19), i.e.,

$$\int_\Omega \nabla u(x)\nabla v(x)\,\mathrm{d}x = \lambda_1 \int_\Omega u(x)v(x)\,\mathrm{d}x + \int_\Omega f(x, u(x))v(x)\,\mathrm{d}x - \int_\Omega h(x)v(x)\,\mathrm{d}x$$

for any $v \in W_0^{1,2}(\Omega)$. Choose $v = \varphi_1$ and use the fact that

$$\int_\Omega \nabla u(x)\nabla\varphi_1(x)\,\mathrm{d}x = \lambda_1 \int_\Omega u(x)\varphi_1(x)\,\mathrm{d}x.$$

Exercise 7.7.10. Prove that the functional $E(u)$ defined by (7.7.22) belongs to the space $C^1(W_0^{1,p}(\Omega), \mathbb{R})$.

Hint. Use an approach similar to that in Exercise 7.7.4.

Exercise 7.7.11. Prove that E is a weakly sequentially lower semicontinuous functional on $W_0^{1,p}(\Omega)$.

Hint. Use the weak sequential lower semicontinuity of the norm in $W_0^{1,p}(\Omega)$, the compact embedding $W_0^{1,p}(\Omega) \hookrightarrow\hookrightarrow L^p(\Omega)$, and the continuity of the Nemytski operator $u \mapsto F(\cdot, u)$ from $L^p(\Omega)$ to $L^1(\Omega)$.

Exercise 7.7.12. Prove that there exists $\bar{\lambda} > \lambda_1$ such that

$$\int_\Omega |\nabla u(x)|^p \, dx \geq \bar{\lambda} \int_\Omega |u(x)|^p \, dx$$

$$\text{for all} \quad u \in Z := \left\{ u \in W_0^{1,p}(\Omega) : \int_\Omega u(x)\varphi_1(x) \, dx = 0 \right\}.$$

Hint. Assume by contradiction that there exist $\varepsilon_n \to 0$ and $u_n \in Z$, $\|u_n\| = 1$ such that

$$1 = (\lambda_1 + \varepsilon_n) \int_\Omega |u_n(x)|^p \, dx.$$

Pass to a subsequence $u_n \rightharpoonup u$ in $W_0^{1,p}(\Omega)$, $u_n \to u$ in $L^p(\Omega)$ and show $u \neq o$,

$$\int_\Omega |\nabla u(x)|^p \, dx \leq \lambda_1 \int_\Omega |u(x)|^p \, dx.$$

This contradicts $u \in Z$ due to the simplicity of λ_1.

Exercise 7.7.13. Consider the boundary value problem

$$\begin{cases} -\Delta_p u(x) + \lambda |u(x)|^{p-2} u(x) = g(x, u(x)) & \text{in} \quad \Omega, \\ u = 0 & \text{on} \quad \partial\Omega, \end{cases} \tag{7.7.26}$$

where $p > 1$. Formulate conditions on λ and $g = g(x,s)$ which guarantee that the energy functional associated with (7.7.26)

(i) is weakly coercive,

(ii) has a geometry corresponding to the Mountain Pass Theorem,

(iii) has a geometry corresponding to the Saddle Point Theorem.

7.8 Relative Extrema and Lagrange Multipliers

In this section we will investigate the local minima or maxima of a real function f on a smooth manifold $\mathscr{M}$ (in particular, on a surface in $\mathbb{R}^3$). Such a manifold is often determined by various constraints which are given by certain equations like

$$\Phi(x) = o$$

(cf. Remark 4.3.9). The key assertions of this section are the Lagrange Multiplier Method, the Courant–Fischer and Courant–Weinstein Variational Principles.

Definition 7.8.1. Let X be a metric (or, more generally, topological) space, $\mathscr{M} \subset X$. We say that a function $f \colon \mathscr{M} \to \mathbb{R}$ has a *local minimum (maximum)* at a point $a \in \mathscr{M}$ with respect to $\mathscr{M}$ (or a *constrained minimum* on $\mathscr{M}$) if there is a neighborhood $\mathcal{U}$ of a such that

$$f(x) \geq f(a) \qquad (f(x) \leq f(a)) \qquad \text{for all} \quad x \in \mathscr{M} \cap \mathcal{U}.$$

We will suppose that $\mathscr{M}$ is given as the zero set of a map $\Phi\colon X \to Y$, i.e.,

$$\mathscr{M} = \{x \in X : \Phi(x) = o\}.$$

The way of investigating the behavior of f in a relative neighborhood $\mathcal{U} \cap \mathscr{M}$ of a point $a \in \mathscr{M}$ is simple and transparent. It consists in expressing $\mathscr{M} \cap \mathcal{U}$ as the graph of a map $\varphi\colon Z \to X$ and subsequently studying $f \circ \varphi$. This is always possible if $\mathscr{M}$ is a differentiable manifold in $X = \mathbb{R}^N$ (Definition 4.3.4) or if $\mathscr{M}$ is given by Φ as above and Φ satisfies certain regularity conditions (Proposition 4.3.8 and Remark 4.3.9(i)).

Theorem 7.8.2 (Lagrange Multiplier Method). *Let X be a Banach space, $f\colon X \to \mathbb{R}$, $\Phi = (\Phi_1, \dots, \Phi_N)\colon X \to \mathbb{R}^N$. Let f have a local minimum or maximum with respect to*

$$\mathscr{M} = \{x \in X : \Phi(x) = o\} \qquad \text{at a point} \quad a \in \mathscr{M}.$$

Let there be a neighborhood $\mathcal{U}$ of a in X such that $f, \Phi \in C^1(\mathcal{U})$ and let a be a regular point of Φ (i.e., $\Phi'(a)$ is a surjective map onto $\mathbb{R}^N$). Then there exist numbers $\lambda_1, \dots, \lambda_N$ [66] *such that*

$$\left(f - \sum_{i=1}^{N} \lambda_i \Phi_i\right)'(a) = o. \tag{7.8.1}$$

Proof. Proposition 4.3.8 and Remark 4.3.9(i) yield a diffeomorphism φ of a neighborhood $\mathcal{U}$ of $o \in X$ onto a neighborhood $\mathcal{V}$ of a such that

$$\varphi(\mathcal{U} \cap \operatorname{Ker} \Phi'(a)) = \mathscr{M} \cap \mathcal{V}, \qquad \varphi(o) = a.$$

If φ_1 denotes the restriction of φ to $\mathcal{U} \cap \operatorname{Ker} \Phi'(a)$, then $f \circ \varphi_1$ has a local minimum (or maximum) at o and therefore

$$(f \circ \varphi_1)'(o) = o.$$

Since $\varphi_1'(o)h = h$ for any $h \in \operatorname{Ker} \Phi'(a)$ (see the proof of Proposition 4.3.8), it follows that

$$\operatorname{Ker} \Phi'(a) \subset \operatorname{Ker} f'(a).$$

The use of Proposition 1.1.19 completes the proof. □

Remark 7.8.3.

(i) The main significance of Theorem 7.8.2 consists in reducing a (difficult) problem of finding the constrained extremal points to an easier task of finding the local ones for a function

$$f - \sum_{i=1}^{N} \lambda_i \Phi_i$$

[66] The numbers $\lambda_1, \dots, \lambda_N$ are called *Lagrange multipliers.*

with unknown coefficients $\lambda_1, \dots, \lambda_N$ (they have to be determined in the course of calculation – see Example 7.8.4).

(ii) For an infinite number of constraints (i.e., $\Phi\colon X \to Y$, Y is a Banach space of infinite dimension) the proof of Theorem 7.8.2 still holds provided there exists a continuous projection of X onto $\operatorname{Ker} \Phi'(a)$. It is interesting that the statement (now $(f - F \circ \Phi)'(a) = 0$ for a certain $F \in Y^*$) is true without the assumption on existence of a projection (the so-called *Lusternik Theorem*), but the proof is more difficult (see Lusternik & Sobolev [107]).

Example 7.8.4. Find the minimal and maximal values of

$$f(x, y, z) = x^2y + xy^2 + z^2 \quad \text{on the set } \mathscr{M} = \{(x, y, z) \in \mathbb{R}^3 : x^2 + y^2 + z^2 = 1\}.$$

Notice first that all points of $\mathscr{M}$ are regular. The necessary condition (7.8.1) given by Theorem 7.8.2 for extremal points requires solving the following four equations:

$$2xy + y^2 - 2\lambda x = 0, \tag{7.8.2}$$

$$x^2 + 2xy - 2\lambda y = 0, \tag{7.8.3}$$

$$2z - 2\lambda z = 0, \tag{7.8.4}$$

$$x^2 + y^2 + z^2 = 1. \tag{7.8.5}$$

We have either $z = 0$ or $\lambda = 1$ from the third equation. Adding x^2 and y^2 to (7.8.2) and (7.8.3) we obtain

$$x^2 + 2\lambda x = y^2 + 2\lambda y, \qquad \text{i.e.,} \qquad (x - y)(x + y + 2\lambda) = 0.$$

Case 1 ($z = 0$). If $x = y$, then

$$x = y = \pm\frac{\sqrt{2}}{2} \qquad \text{and} \qquad f\left(\pm\frac{\sqrt{2}}{2}, \pm\frac{\sqrt{2}}{2}, 0\right) = \pm\frac{\sqrt{2}}{2}.$$

If $x + y = -2\lambda$, then (7.8.2) and (7.8.5) imply $xy = -\frac{1}{3}$ and from equation (7.8.5) we find

$$x + y = \pm\frac{\sqrt{3}}{3} \qquad \text{and hence} \qquad f(x, y, 0) = xy(x + y) = \mp\frac{\sqrt{3}}{9}.$$

Case 2 ($\lambda = 1$). Again we have either $x = y$ or $x + y = -2$. Putting $x = y$ into (7.8.2) we find

$$x = y = 0, \quad z = \pm 1 \qquad \text{or} \qquad x = y = \frac{2}{3}, \quad z = \pm\frac{1}{3}$$

and

$$f(0, 0, \pm 1) = 1, \qquad f\left(\frac{2}{3}, \frac{2}{3}, \pm\frac{1}{3}\right) = \frac{19}{27}.$$

If $x + y = -2$, then (from (7.8.2) and (7.8.3))

$$x^2 + 2x - 4 = y^2 + 2y - 4 = 0.$$

Summing these equations we get

$$0 = x^2 + y^2 - 4 - 8,$$

i.e., there cannot exist z such that

$$x^2 + y^2 + z^2 = 1.$$

We have found several points in $\mathscr{M}$ for which the necessary condition is satisfied. Since $\mathscr{M}$ is a compact set in $\mathbb{R}^3$ and f is continuous, the maximum and the minimum of f on $\mathscr{M}$ have to exist. Comparing the values of f at points at which the necessary condition is satisfied we find that

$$\max_{\mathscr{M}} f = f(0,0,\pm 1) = 1, \qquad \min_{\mathscr{M}} f = f\left(-\frac{\sqrt{2}}{2}, -\frac{\sqrt{2}}{2}, 0\right) = -\frac{\sqrt{2}}{2}.$$

If we were interested in local minima/maxima of f with respect to $\mathscr{M}$, we would need some sufficient conditions. Since we are able to reduce the problem of constrained minima/maxima to that of local ones (see the proof of Theorem 7.8.2), we might employ the sufficient condition which uses the second differential (Theorem 7.1.5). Cf. Exercise 7.8.17. ○

Example 7.8.5 (Existence of the principal eigenvalue). Let $p > 1$ be a real number, $X := W_0^{1,p}(0,1)$.[67] Consider the eigenvalue problem

$$\begin{cases} -(|\dot{x}(t)|^{p-2}\dot{x}(t))^{\cdot} = \lambda |x(t)|^{p-2}x(t), & t \in (0,1), \\ x(0) = x(1) = 0 \end{cases} \tag{7.8.6}$$

with a real parameter λ. This problem is linear for $p = 2$ and nonlinear for $p \neq 2$. We say that $\lambda \in \mathbb{R}$ is an *eigenvalue* of (7.8.6) if there is a weak solution $x \in X$, $x \neq o$, of (7.8.6), i.e.,

$$\int_0^1 |\dot{x}(t)|^{p-2}\dot{x}(t)\dot{y}(t)\,\mathrm{d}t = \lambda \int_0^1 |x(t)|^{p-2}x(t)y(t)\,\mathrm{d}t \tag{7.8.7}$$

holds for every $y \in X$. The corresponding x is then called an *eigenfunction* associated with the eigenvalue λ.[68]

[67] We will work with the norm $\|x\| = \left(\int_0^1 |\dot{x}(t)|^p\,\mathrm{d}t\right)^{\frac{1}{p}}$.

[68] To see the analogue to the eigenvalue of a linear operator, the identity (7.8.7) can be interpreted (for $p = 2$) as the operator equation $x = \lambda A x$ where A is defined by the equality $(Ax, y)_{W_0^{1,2}(0,1)} = (x, y)_{L^2(0,1)}$. The eigenvalues of the boundary value problem (7.8.6) are then reciprocal values of the eigenvalues of the linear operator A.

Since (7.8.7) must also hold for $y = x$, we obtain

$$\lambda = \frac{\int_0^1 |\dot{x}(t)|^p \, dt}{\int_0^1 |x(t)|^p \, dt},$$

which implies that $\lambda > 0$ for any eigenvalue λ.

We will prove that the value

$$\lambda_1 = \inf_{\substack{x \in X \\ x \neq o}} \frac{\int_0^1 |\dot{x}(t)|^p \, dt}{\int_0^1 |x(t)|^p \, dt}, \tag{7.8.8}$$

i.e.,

$$\lambda_1 = \inf_{x \in X} \left\{ \int_0^1 |\dot{x}(t)|^p \, dt : \int_0^1 |x(t)|^p \, dt = 1 \right\}$$

is attained and use the Lagrange Multiplier Method to show that λ_1 is the least eigenvalue (*principal eigenvalue*) of (7.8.6). Let us prove that the infimum in (7.8.8) is achieved at an $x_1 \in X$ with

$$\int_0^1 |x_1(t)|^p \, dt = 1.$$

Indeed, there exists a minimizing sequence $\{x_n\}_{n=1}^\infty \subset X$ such that

$$\int_0^1 |x_n(t)|^p \, dt = 1 \qquad \text{and} \qquad \int_0^1 |\dot{x}_n(t)|^p \, dt \to \lambda_1.$$

In particular, this means that the sequence $\{x_n\}_{n=1}^\infty$ is bounded in X. By the reflexivity of X and the compact embedding $X = W_0^{1,p}(0,1) \subsetneq\subsetneq L^p(0,1)$ (see Theorem 1.2.28 and Exercise 1.2.48(i)) there exists a subsequence $\{x_{n_k}\}_{k=1}^\infty \subset \{x_n\}_{n=1}^\infty$ and a function $x_1 \in X$ such that

$$x_{n_k} \rightharpoonup x_1 \quad \text{in} \quad X, \qquad x_{n_k} \to x_1 \quad \text{in} \quad L^p(0,1).$$

Hence

$$\int_0^1 |x_1(t)|^p \, dt = 1 \qquad \text{and} \qquad \|x_1\|^p \le \liminf_{n\to\infty} \|x_n\|^p = \lambda_1,$$

i.e.,

$$\int_0^1 |\dot{x}_1(t)|^p \, dt = \lambda_1.$$

Now we apply Theorem 7.8.2 with

$$f(x) = \int_0^1 |\dot{x}(t)|^p \,\mathrm{d}t \qquad \text{and} \qquad g(x) = \int_0^1 |x(t)|^p \,\mathrm{d}t - 1.$$

The Fréchet derivatives of f and g at x_1 (in the space X) are given by

$$\left.\begin{aligned} f'(x_1)y &= p\int_0^1 |\dot{x}_1(t)|^{p-2}\dot{x}_1(t)\dot{y}(t)\,\mathrm{d}t,\\ g'(x_1)y &= p\int_0^1 |x_1(t)|^{p-2}x_1(t)y(t)\,\mathrm{d}t \end{aligned}\right\} \quad \text{for any} \quad y \in X$$

(cf. Exercise 3.2.35). Since $x_1 \neq o$, we also have $g'(x_1) \neq o$, and so the assumptions of Theorem 7.8.2 are fulfilled. Hence there exists $\lambda \in \mathbb{R}$ such that

$$f'(x_1) = \lambda g'(x_1),$$

which is equivalent to

$$\int_0^1 |\dot{x}_1(t)|^{p-2}\dot{x}_1(t)\dot{y}(t)\,\mathrm{d}t = \lambda\int_0^1 |x_1(t)|^{p-2}x_1(t)y(t)\,\mathrm{d}t \tag{7.8.9}$$

for any $y \in X$. Setting $y = x_1$ in (7.8.9) we get

$$\lambda = \lambda_1.$$

Now it follows from (7.8.7) and (7.8.8) that λ_1 is the least eigenvalue of (7.8.6). ○

Remark 7.8.6. Let us emphasize that Theorem 7.8.2 provides a necessary condition only. It means that not every point $a \in \mathscr{M}$ for which

$$f'(a) - \sum_{i=1}^{N} \lambda_i \Phi_i'(a) = o \qquad \text{with some} \quad \lambda_i \in \mathbb{R}, \quad i = 1, \dots, N,$$

need be a point of local extremum of f relative to $\mathscr{M}$! On the other hand, to find all local extrema of f relative to $\mathscr{M}$ one has to start with finding all $\lambda_i \in \mathbb{R}$, $i = 1, \dots, N$, such that the functional $f - \sum\limits_{i=1}^{N} \lambda_i \Phi_i$ has a critical point $a \in \mathscr{M}$. It is a well-known fact from the calculus of several real variables (when $X = \mathbb{R}^N$) that the set of all such a's is "almost always" finite (see, e.g., Example 7.8.4). Hence a very natural and deep question arises: "How many points a do we have if $\dim X = \infty$?"

Remark 7.8.7. Let us denote by $\Lambda \subset \mathbb{R}$ the set of all $\lambda \in \mathbb{R}$ such that $f - \lambda g$ has a critical point $a \in \mathscr{M}$. If X is a Hilbert space of infinite dimension, then in

Krasnoselski [92, Chapter 6] the reader can find the proof of the assertion that the set Λ contains a sequence of nonzero numbers $\lambda_n \neq 0$ such that $\lambda_n \to 0$. The same assertion for a Banach space X can be found in Citlanadze [34], Browder [22], Fučík & Nečas [67]. Actually, the whole Chapter 7 of the lecture notes by Fučík et al. [68] is devoted to this problem. As for more recent references the reader can confer Zeidler [163] and the bibliography therein.

Let us emphasize that in all above results the authors prove that the cardinality of the set Λ is equal to infinity. The question: "When is Λ a countable set?" is much more involved. Some partial results in this direction can be found in Fučík et al. [68]. The proofs are based on a stronger version of the Morse Theorem and go beyond the scope of this book.

Proposition 7.8.8. *Let H be a N-dimensional Hilbert space and let A be a self-adjoint operator in H. Then A has N real eigenvalues $\lambda_1, \dots, \lambda_N$ (if they are counted with their multiplicities), and the corresponding eigenvectors $e_1, \dots, e_N$ form an orthonormal basis in H.*

Proof. Consider two functions $f, \varphi_1 \colon H \to \mathbb{R}$ defined by

$$f(x) = (Ax, x), \qquad \varphi_1(x) = (x, x) - 1, \qquad x \in H.$$

Then the set

$$\mathscr{M}_1 = \{x \in H : \varphi_1(x) = 0\}$$

(the unit sphere in H) is a compact subset of H and the continuous function f assumes its maximum in $\mathscr{M}_1$ at a point $e_1 \in \mathscr{M}_1$. By Theorem 7.8.2, there is a $\lambda_1 \in \mathbb{R}$ such that

$$f'(e_1) - \lambda_1 \varphi_1'(e_1) = o.$$

A simple calculation shows that $f'(e_1)h = 2(Ae_1, h)$, $\varphi_1'(e_1)h = 2(e_1, h)$. Therefore

$$(Ae_1 - \lambda_1 e_1, h) = 0 \qquad \text{for all} \quad h \in H, \qquad \text{i.e.,} \qquad Ae_1 = \lambda_1 e_1.$$

Taking $h = e_1$ we also get

$$\lambda_1 = (Ae_1, e_1) = \max_{x \in \mathscr{M}_1} (Ax, x).$$

In particular, λ_1 is the largest (equivalently, first) eigenvalue of A. To find the second eigenvalue we add another constraint

$$\varphi_2(x) := (x, e_1) = 0$$

(remember that eigenvectors of a symmetric matrix are pairwise orthogonal). The function f has again a maximum with respect to

$$\mathscr{M}_2 = \{x \in H : \varphi_1(x) = \varphi_2(x) = 0\}$$

and thus there are $e_2 \in \mathscr{M}_2$, $\lambda_2, \tilde{\lambda}_2 \in \mathbb{R}$ such that

$$f'(e_2)h - \lambda_2\varphi_1'(e_2)h - \tilde{\lambda}_2\varphi_2'(e_2)h = (2Ae_2 - 2\lambda_2 e_2 - \tilde{\lambda}_2 e_1, h) = 0 \qquad (7.8.10)$$

for all $h \in H$. In particular, for $h = e_1$ we get

$$0 = (2Ae_2, e_1) - \tilde{\lambda}_2\|e_1\|^2 = 2(e_2, Ae_1) - \tilde{\lambda}_2 = 2\lambda_1(e_2, e_1) - \tilde{\lambda}_2,$$

and consequently $\tilde{\lambda}_2 = 0$. The equality (7.8.10) hence yields

$$Ae_2 = \lambda_2 e_2$$

and, similarly as above,

$$\lambda_2 = \max_{\substack{\|x\|=1\\(x,e_1)=0}} (Ax, x).$$

It is obvious that we can proceed by induction to obtain all eigenvalues $\lambda_1, \dots \lambda_N$ and to show that the corresponding eigenvectors $e_1, \dots, e_N$ are orthonormal and form a basis of H. □

Corollary 7.8.9. *Let $\boldsymbol{A} = (a_{ij})_{i,j=1,\dots,N}$ be a symmetric matrix ($a_{ij} = a_{ji}$ for $i, j = 1, \dots, N$). Then there exist real numbers $\lambda_1, \dots, \lambda_N$ and a basis $e_1, \dots, e_N$ of $\mathbb{R}^N$ such that*

$$\sum_{i,j=1}^{N} a_{ij}x_i x_j = \sum_{i=1}^{N} \lambda_i \xi_i^2, \qquad \textit{where} \quad x = (x_1, \dots, x_N), \quad x = \sum_{i=1}^{N} \xi_i e_i.$$

Remark 7.8.10. The procedure explored in the proof of Proposition 7.8.8 has a disadvantage, namely, to find the kth eigenvalue λ_k it is necessary to know the first $k-1$ eigenvectors $e_1, \dots, e_{k-1}$. Because of that it can be convenient to have another expression for λ_k. We will now prove that

$$\lambda_k = \min_{y_1,\dots,y_{k-1}} \max \left\{ \frac{(Ax, x)}{\|x\|^2} : (x, y_1) = \dots = (x, y_{k-1}) = 0 \text{ and } x \neq o \right\} \quad (7.8.11)$$

provided $\dim H \geq k$. Expression (7.8.11) is called the *Minimax Principle*.

Let $e_1, \dots, e_k$ be eigenvectors corresponding to the first k eigenvalues $\lambda_1 \geq \dots \geq \lambda_k$. Take $y_1, \dots, y_{k-1} \in H$ and let

$$\mathscr{N} = \{x \neq o : (x, y_1) = \dots = (x, y_{k-1}) = 0\}.$$

There is an $\tilde{x} \in \mathscr{N} \cap \operatorname{Lin}\{e_1, \dots, e_k\}$, say $\tilde{x} = \sum_{i=1}^{k} \alpha_i e_i$. A simple argument to see this consists in the observation that the linear operator $\Phi \colon \mathbb{R}^k \to \mathbb{R}^{k-1}$ (or $\mathbb{C}^k \to \mathbb{C}^{k-1}$) given by

$$\Phi\alpha = \left(\sum_{i=1}^{k} \alpha_i (e_i, y_j) \right)_{j=1,\dots,k-1}$$

must have a nontrivial kernel. For such an $\tilde{x}$ we have

$$(A\tilde{x}, \tilde{x}) = \left(\sum_{i=1}^{k} \alpha_i \lambda_i e_i, \sum_{j=1}^{k} \alpha_j e_j \right) = \sum_{i=1}^{k} \lambda_i |\alpha_i|^2 \geq \lambda_k \sum_{i=1}^{k} |\alpha_i|^2 = \lambda_k \|\tilde{x}\|^2.$$

This shows that the maximum in (7.8.11) (denoted by $m(y_1, \dots, y_{k-1})$) is not less than λ_k and therefore

$$\inf_{y_1, \dots, y_{k-1}} m(y_1, \dots, y_{k-1}) \geq \lambda_k,$$

too. But the above calculation yields that

$$m(e_1, \dots, e_{k-1}) = \lambda_k.$$

Remark 7.8.11. This method of finding eigenvalues of a self-adjoint continuous operator A cannot be extended to infinite-dimensional Hilbert spaces. The reason is rather simple: such an operator need not have any eigenvector (Example: $Ax(t) = tx(t)$, $x \in L^2(0,1)$). On the other hand, if we assume that A is, in addition to self-adjointness, also compact, then similar result holds.

Theorem 7.8.12 (Courant–Fischer Principle). *Let $A\colon H \to H$ be a compact, self-adjoint and positive*[69] *linear operator from an (infinite-dimensional) separable real Hilbert space H into itself. Then all eigenvalues of A are positive reals and there exists an orthonormal basis of H which consists of eigenvectors of A. Moreover, if*

$$\lambda_1 \geq \lambda_2 \geq \lambda_3 \geq \cdots > 0, \qquad \lambda_n \to 0 \qquad (n \to \infty),$$

denote the eigenvalues of A, then

$$\lambda_1 = \max\{(Au, u) : \|u\| = 1\}$$

and

$$\lambda_{k+1} = \min_{v_1, \dots, v_k} \max\{(Au, u) : \|u\| = 1, (u, v_1) = \cdots = (u, v_k) = 0\},$$

$k = 1, 2, \dots$[70]

Proof. Set

$$F(u) = (Au, u), \qquad \varphi_1(u) = \|u\|^2 - 1 \qquad \text{for} \quad u \in H,$$

and

$$\mathscr{M}_1 = \{u \in H : \varphi_1(u) = 0\}.$$

[69] A linear self-adjoint operator A is said to be *positive* if $(Au, u) > 0$ for all $u \neq o$.

[70] The reader should compare this assertion and its proof with the Hilbert–Schmidt Theorem (Theorem 2.2.16).

Let $\{u_n\}_{n=1}^{\infty}$ be a maximizing sequence for F subject to $\mathscr{M}_1$, i.e., $\|u_n\| = 1$, $n = 1, \dots,$ and

$$\lim_{n\to\infty} F(u_n) = \sup\{F(u) : u \in \mathscr{M}_1\}.$$

The boundedness of $\mathscr{M}_1$ and the compactness of A imply (Proposition 2.2.4(iii)) that we can pass to a subsequence (denoted again as $\{u_n\}_{n=1}^{\infty}$) for which

$$u_n \rightharpoonup e_1 \qquad \text{and} \qquad Au_n \to Ae_1 \qquad \text{in} \quad H \qquad \text{with an} \quad e_1 \in H.$$

Then

$$|(Au_n, u_n) - (Ae_1, e_1)| \le |(Au_n - Ae_1, u_n)| + |(Ae_1, u_n - e_1)| \to 0$$

since both terms on the right-hand side approach zero. So

$$F(e_1) = \sup\{F(u) : u \in \mathscr{M}_1\}.$$

In particular, we have

$$F(e_1) > 0 \qquad \text{and} \qquad e_1 \neq o.$$

Let us prove that $\|e_1\| = 1$. Indeed, we have

$$\|e_1\| \le \liminf_{n\to\infty} \|u_n\| = 1.$$

Assume that $\|e_1\| < 1$. Then there exists $t > 1$ such that for $\tilde{e}_1 = te_1$ we have $\|\tilde{e}_1\| = 1$, i.e., $\tilde{e}_1 \in \mathscr{M}_1$. Also

$$F(\tilde{e}_1) = (A(te_1), te_1) = t^2(Ae_1, e_1) = t^2 F(e_1) > \sup\{F(u) : u \in \mathscr{M}_1\},$$

a contradiction. Hence

$$\lambda_1 = F(e_1) = \max\{F(u) : u \in \mathscr{M}_1\}.$$

Applying Theorem 7.8.2 we prove exactly as in Proposition 7.8.8 that λ_1 is an eigenvalue of A and e_1 is the corresponding eigenvector. Now, we proceed by induction using

$$\mathscr{M}_n = \{u \in H : \|u\| = 1 \text{ and } (u, e_1) = \cdots = (u, e_{n-1}) = 0\}$$

as above to get the sequence of eigenvalues

$$\lambda_1 \ge \lambda_2 \ge \cdots > 0 \tag{7.8.12}$$

and the sequence of the corresponding eigenvectors

$$e_1, e_2, \dots$$ [71]

which are pairwise orthogonal. The infinite dimension of H causes that the above sequences are infinite in general.

Suppose now that there is $w \in H$ such that

$$\|w\| = 1 \qquad \text{and} \qquad (w, e_n) = 0 \quad \text{for all} \quad n \in \mathbb{N}.$$

Then

$$w \in \bigcap_{n=1}^{\infty} \mathscr{M}_n, \qquad \text{and thus} \qquad (Aw, w) \le \lambda_n \quad \text{for} \quad n = 1, 2, \dots$$

Since $\lambda_n \to 0$ (Corollary 2.2.13), we have $(Aw, w) = 0$. The assumption on the positivity of A implies $w = o$, a contradiction. This result shows that $\{e_n\}_{n=1}^{\infty}$ is an orthonormal basis of H (Corollary 1.2.36). Moreover, the sequence (7.8.12) contains all eigenvalues of A. Indeed, if

$$Aw = \lambda w \qquad \text{for} \quad w = \sum_{n=1}^{\infty} \alpha_n e_n \neq 0,$$

then

$$\lambda_n \alpha_n = \lambda \alpha_n \qquad \text{for} \quad n = 1, 2, \dots$$

Therefore $\alpha_n = 0$ provided $\lambda_n \neq \lambda$.

The "min max" characterization of λ_n's follows as in the finite-dimensional case (Remark 7.8.10). □

Remark 7.8.13. It is remarkable that the Minimax Principle holds even without the assumption on the continuity of A in the sense that

$$\inf_{y_1,\dots,y_{k-1}} \sup \{(Ax, x) : x \in \operatorname{Dom} A, \|x\| = 1, (x, y_1) = \dots = (x, y_{k-1}) = 0\}$$

yields either the kth eigenvalue or an upper bound of the essential spectrum of a linear self-adjoint operator A provided A is bounded above. For details see, e.g., Reed & Simon [125].

There is also a dual characterization of the eigenvalues of A called the *Courant–Weinstein Variational Principle.*

Theorem 7.8.14 (Courant–Weinstein Variational Principle). *Let H be a real separable Hilbert space, $A\colon H \to H$ a positive compact self-adjoint linear operator. Assume that the eigenvalues λ_n of A form a decreasing sequence*

$$\lambda_1 \ge \lambda_2 \ge \lambda_3 \ge \dots \ge \lambda_n \ge \dots > 0, \qquad \lambda_n \to 0 \qquad (n \to \infty)$$

[71] The reader should perform this part of the proof in details.

(cf. Theorem 7.8.12), and the multiplicity of an eigenvalue λ indicates how many times this λ repeats in the above sequence. Then for any $n \in \mathbb{N}$,

$$\lambda_n = \sup_{\substack{X \subset H \\ \dim X = n}} \inf_{\substack{u \in X \\ \|u\|=1}} (Au, u).$$

(Here X is an arbitrary linear subspace of H of dimension equal to n.)

Proof. Keeping the notation from Theorem 7.8.12, in particular, $Ae_n = \lambda_n e_n$, we denote for $n \in \mathbb{N}$ fixed

$$\tilde{\lambda}_n = \sup_{\substack{X \subset H \\ \dim X = n}} \inf_{\substack{u \in X \\ \|u\|=1}} (Au, u).$$

Our aim is to prove $\tilde{\lambda}_n = \lambda_n$.

Step 1. We prove that $\tilde{\lambda}_n \geq \lambda_n$. Set

$$X_0 = \operatorname{Lin}\{e_1, \dots, e_n\}.$$

Then X_0 is a linear subspace of H, $\dim X_0 = n$, and clearly

$$\tilde{\lambda}_n \geq \min_{\substack{u \in X_0 \\ \|u\|=1}} (Au, u).$$

However, we can estimate the minimum of the quadratic form on the right-hand side in terms of λ_n. For $u \in X_0$, $\|u\| = 1$ we have

$$u = \sum_{i=1}^{n} x_i e_i, \qquad \sum_{i=1}^{n} x_i^2 = 1.$$

Then

$$(Au, u) = \left(\sum_{i=1}^{n} x_i \lambda_i e_i, \sum_{j=1}^{n} x_j e_j \right) = \sum_{i=1}^{n} \lambda_i x_i^2 \geq \lambda_n, \qquad \text{i.e.,} \qquad \tilde{\lambda}_n \geq \lambda_n.$$

Step 2. We prove $\tilde{\lambda}_n \leq \lambda_n$. Set

$$Y = \overline{\operatorname{Lin}\{e_i\}_{i=n}^{\infty}}.$$

Then $\operatorname{codim} Y = n - 1$. Let X be an arbitrary linear subspace of H, $\dim X = n$. Then necessarily

$$\dim (X \cap Y) > 0,$$

and the space $X \cap Y$ must contain an element $w \neq o$. We can assume $\|w\| = 1$. Since $w \in Y$, we have

$$w = \sum_{i=n}^{\infty} x_i e_i, \qquad \sum_{i=n}^{\infty} x_i^2 = 1.$$

The estimate of the quadratic form (Au, u) on the unit sphere in X yields

$$\min_{\substack{u\in X\\ \|u\|=1}} (Au, u) \le (Aw, w) = \sum_{i=n}^{\infty} \lambda_i x_i^2 \le \lambda_n \sum_{i=n}^{\infty} x_i^2 = \lambda_n.$$

Since X is arbitrary, the equality $\tilde{\lambda}_n \le \lambda_n$ follows. □

Example 7.8.15 (Higher eigenvalues of a boundary value problem). Let $p = 2$ in (7.8.6), i.e., let us consider the eigenvalue problem

$$\begin{cases} \ddot{x}(t) + \lambda x(t) = 0, & t \in (0,1), \\ x(0) = x(1) = 0. \end{cases} \tag{7.8.13}$$

The eigenvalues of the linear problem (7.8.13) can be calculated in an elementary way. On the other hand, if we set $H := W_0^{1,2}(0,1)$ and define a positive and compact operator $A\colon H \to H$ by

$$(Ax, y)_{W_0^{1,2}(0,1)} = \int_0^1 x(t)y(t)\,\mathrm{d}t,\text{[72]}$$

then $\mu \ne 0$ is an eigenvalue of A if and only if $\lambda = \frac{1}{\mu}$ is an eigenvalue of (7.8.13) (cf. footnote 68 on page 530). It follows from Theorem 7.8.14 that

$$\frac{1}{\lambda_n} = \sup_{\substack{X\subset H\\ \dim X = n}} \min_{\substack{\|x\|=1\\ x\in X}} \int_0^1 |x(t)|^2\,\mathrm{d}t.$$

An eigenvalue λ of (7.8.13) can be also interpreted in the following way. Set

$$B := -\ddot{x} \quad \mathrm{Dom}B := \{x \in W^{2,2}(0,1) : x(0) = x(1) = 0\}.$$

Then eigenvalue λ of (7.8.13) is an eigenvalue of the operator B in the sense of Definition 1.1.27. ○

The following two exercises show the relation between the local (global) extremum subject to a constraint and the local (global) extremum of the functional depending on a parameter (without the constraint).

Exercise 7.8.16. Prove the following assertion:

Let f, Φ be two real functionals defined on a real Hilbert space H. Let the functional $f - \lambda\Phi$ ($\lambda \in \mathbb{R}$) have a local (global) extremum at a point $x_0 \in H$. Then the functional f has a local (global) extremum subject to the constraint $\{x \in H : \Phi(x) = \Phi(x_0)\}$ at the point x_0.

---

[72]By Example 2.2.17 the operator A is also defined as $(Ax)(t) = \int_0^1 G(t,s)x(s)\,\mathrm{d}s$, and the compactness of A follows.

Exercise 7.8.17. Prove the following assertion:

Let $f, \Phi\colon X \to \mathbb{R}$ satisfy the assumptions of Theorem 7.8.2 and let $x_0 \in X$, $\lambda \in \mathbb{R}$ be such that

$$f'(x_0) - \lambda\Phi'(x_0) = 0.$$

Assume, moreover, that there exist $D^2 f(x_0; h, h)$, $D^2\Phi(x_0; h, h)$. Then x_0 is a local minimum of $f - \lambda\Phi$ (without the constraint) provided the quadratic form

$$h \mapsto D^2 f(x_0; h, h) - \lambda D^2\Phi(x_0; h, h), \qquad h \in X,$$

is positive definite in X.

Exercise 7.8.18. Show that the first eigenvalue of

$$\begin{cases} \ddot{x}(t) + \lambda x(t) = 0, & t \in (0, \pi), \\ x(0) = x(\pi) = 0 \end{cases}$$

is simple and equal to 1, and that given $\lambda > -1$ there exists $c = c(\lambda) > 0$ such that for any $x \in W_0^{1,2}(0, \pi)$,

$$\int_0^\pi |\dot{x}(t)|^2 \,\mathrm{d}t + \lambda \int_0^\pi |x(t)|^2 \,\mathrm{d}t \ge c \int_0^\pi |\dot{x}(t)|^2 \,\mathrm{d}t.$$

Exercise 7.8.19. Prove that if λ is an eigenvalue of (5.6.14), then $\lambda \ge \lambda_1$.
Hint. Apply Theorem 7.8.12 similarly to Example 7.8.15.

7.8A Lusternik–Schnirelmann Method

The purpose of this appendix is to extend the results presented in Section 7.8, namely, we concentrate on the Lusternik–Schnirelmann Method which generalizes the Courant–Fischer and Courant–Weinstein Principles. In order to motivate the topic let us consider the unit circle $\mathcal{S}^1$ in the plane and a continuous function φ defined on it. The Extreme Value Theorem implies that this function has to attain its maximum and minimum values. If the function φ is the restriction of a non-vanishing linear function of two independent variables to $\mathcal{S}^1$, then φ has exactly one maximum at M and one minimum at m as in Figure 7.8.1. The set $\mathcal{S}^1$ can be covered by two closed sets which are contractible to a point in $\mathcal{S}^1$ (see Figure 7.8.2). As another example consider a two-dimensional torus $\mathcal{T}^2$ and identify it with the quotient space $\mathbb{R}^2|_{\mathbb{Z}^2}$ (see Figure 7.8.3).

Let us consider a function $\varphi \in C^1(\mathcal{T}^2, \mathbb{R})$ [73] having a maximum at M and a minimum at m. We also assume that the level sets of φ are the curves indicated in Figure 7.8.3. The function φ has three critical points on the torus: the maximum at M, the minimum at m and a saddle point at S. In Figure 7.8.4 we give a covering of the torus $\mathcal{T}^2$ by three closed sets which are contractible to a point in $\mathcal{T}^2$ (see Appendix 5.1A for the notion of a contractible set).

[73] For the notion of the differentiability of functions defined on manifolds, like $\mathcal{S}^1$, $\mathcal{T}^2$, see Definition 4.3.36.

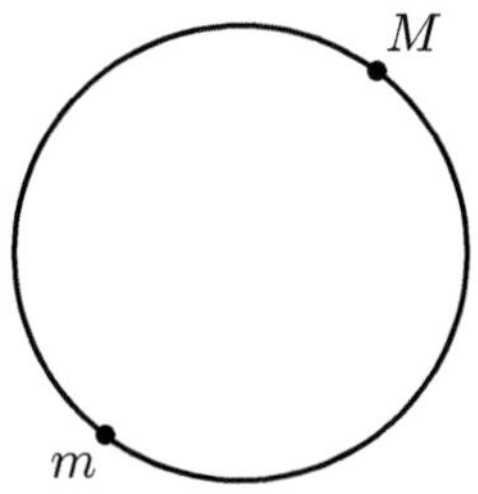

Figure 7.8.1.

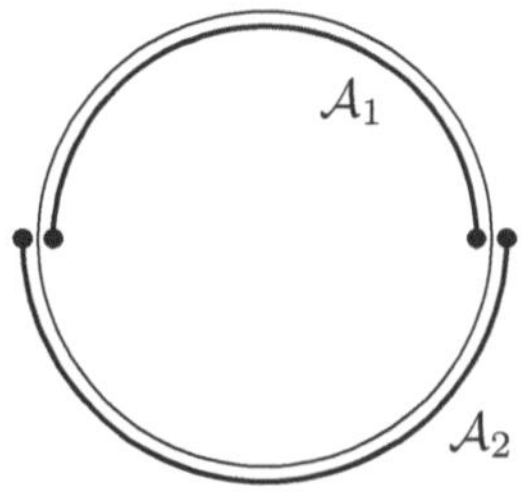

Figure 7.8.2.

Definition 7.8.20. We define the *Lusternik–Schnirelmann category* $\mathrm{cat}_Y(\mathcal{A})$ of a closed nonempty subset $\mathcal{A}$ of a topological space Y as the least integer n such that there exists a covering of $\mathcal{A}$ by n closed sets contractible to a point in Y.[74]

The essential idea of the *Lusternik–Schnirelmann method* is the following one: *The number of critical points of a C^1-functional φ defined on a compact manifold Y is greater than or equal to $\mathrm{cat}_Y(Y)$. The corresponding critical values are given by*

$$c_k := \inf_{\mathcal{A}\in\mathscr{A}_k} \sup_{u\in\mathcal{A}} \varphi(u) \quad \textit{where } \mathscr{A}_k := \{\mathcal{A}\subset Y : \mathcal{A} \textit{ closed}, \ \mathrm{cat}_Y(\mathcal{A}) \geq k\}.$$

Let us prove some elementary properties of the Lusternik–Schnirelmann category.

Lemma 7.8.21. *Let $\mathcal{A}$ and $\mathcal{B}$ be closed subsets of Y. Then we have*

(i) *(normalization)* $\mathrm{cat}_Y(\emptyset) = 0$,

(ii) *(subadditivity)* $\mathrm{cat}_Y(\mathcal{A}\cup\mathcal{B}) \leq \mathrm{cat}_Y(\mathcal{A}) + \mathrm{cat}_Y(\mathcal{B})$,

(iii) *(monotonicity) if $\mathcal{A} \prec \mathcal{B}$,*[75] *then* $\mathrm{cat}_Y(\mathcal{A}) \leq \mathrm{cat}_Y(\mathcal{B})$.

Proof. Properties (i) and (ii) follow directly from Definition 7.8.20. Let us prove (iii). Assume that $\mathcal{A} \prec \mathcal{B}$ by means of homotopy h, and let $\{\mathcal{B}_1,\dots,\mathcal{B}_n\}$ be a covering of $\mathcal{B}$ corresponding to $n := \mathrm{cat}_Y(\mathcal{B})$ according to Definition 7.8.20. Define sets

$$\mathcal{A}_j := \{u\in\mathcal{A} : h(1,u)\in\mathcal{B}_j\}, \qquad j=1,\dots,n.$$

Then

$$\mathcal{A} = \bigcup_{j=1}^{n} \mathcal{A}_j, \qquad \mathcal{A}_j \prec \mathcal{B}_j, \quad j=1,\dots,n.$$

According to Lemma 5.1.9, $\mathrm{cat}_Y(\mathcal{A}) \leq n$. □

Definition 7.8.22. A metric space Y is an *absolute neighborhood extensor* if for every metric space E, every closed subset $\mathcal{F}\subset E$ and every continuous mapping $f\colon \mathcal{F}\to Y$ there exists a continuous extension of f defined on a neighborhood of $\mathcal{F}$ in E.[76]

---

[74] See Definition 5.1.8.

[75] See Definition 5.1.8 for the relation $\prec$.

[76] The terminology is not fixed in the literature. We follow here the books Willem [159] and Zeidler [163]. On the other hand, the same objects are called "absolute neighborhood retract" in Deimling [42] and Dugundji [52].

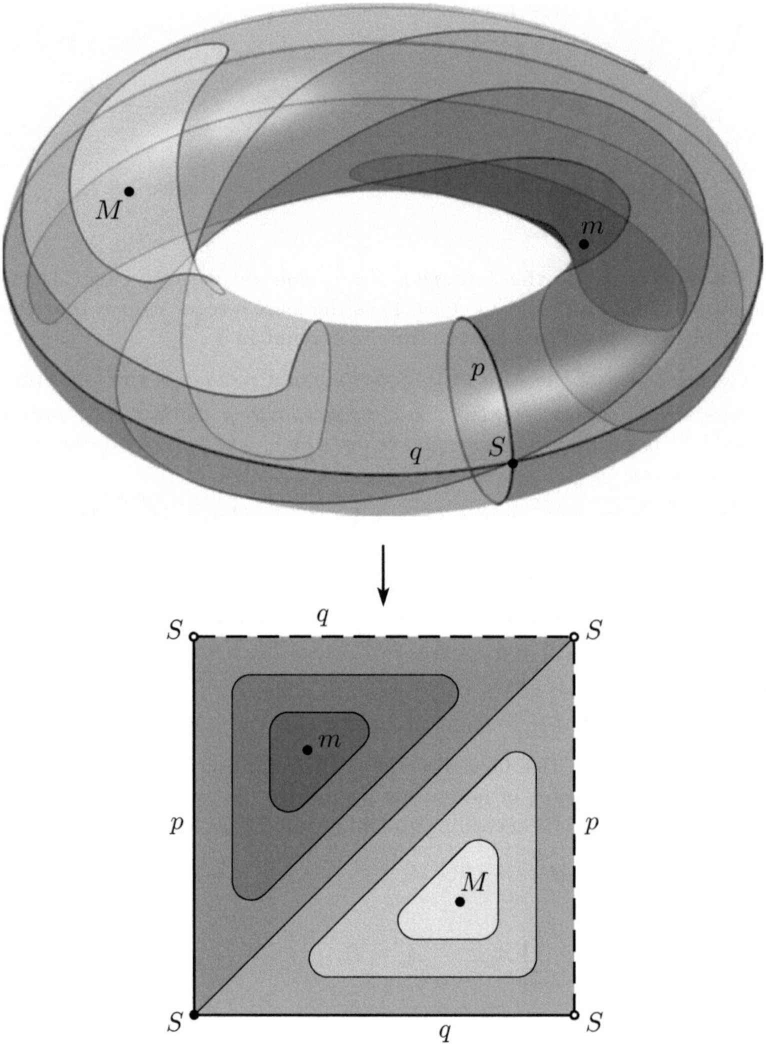

Figure 7.8.3.

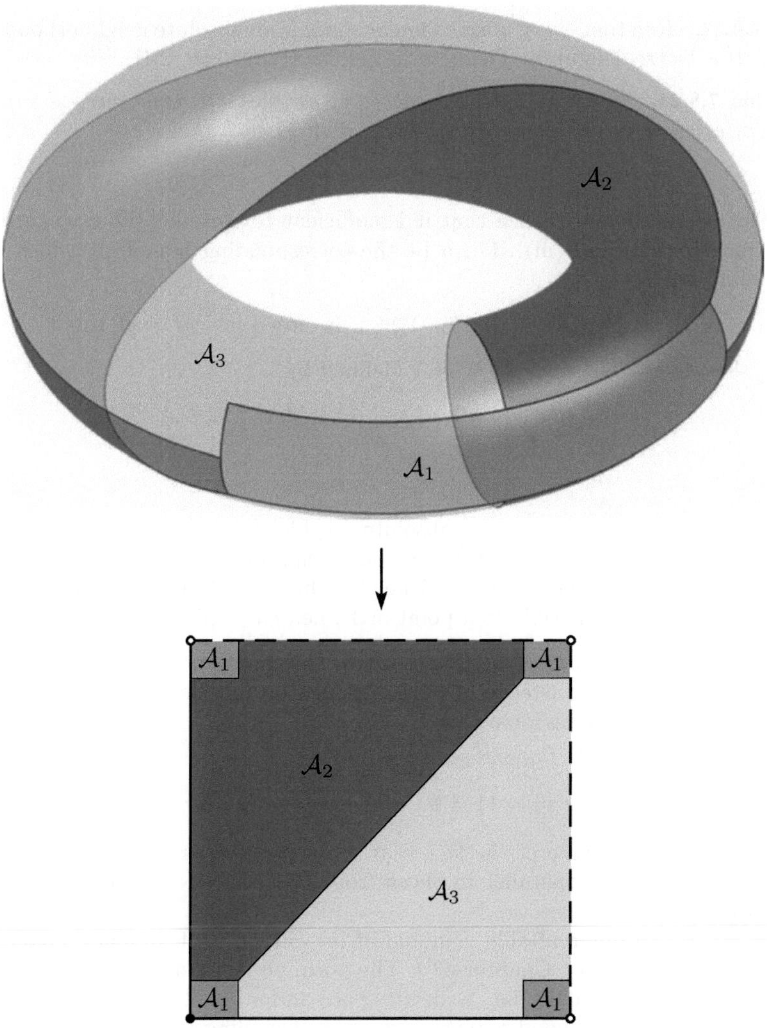

Figure 7.8.4.

Remark 7.8.23. Note that every normed linear space is an absolute neighborhood extensor (see, e.g., the Tietze–Dugundji Theorem in Zeidler [163, Prop. 2.1]).

Proposition 7.8.24. *Let $\mathcal{A}$ be a closed subset of an absolute neighborhood extensor Y. Then there exists a closed neighborhood $\mathcal{B}$ of $\mathcal{A}$ in Y such that*

$$\operatorname{cat}_Y(\mathcal{B}) = \operatorname{cat}_Y(\mathcal{A}).$$

Proof. The reader should realize that it is sufficient to consider the case $\operatorname{cat}_Y(\mathcal{A}) = 1$ (cf. Lemma 7.8.21(ii) and (iii)). Let h be the corresponding homotopy which contracts $\mathcal{A}$ to a point. The set

$$\mathcal{N} := ([0,1] \times \mathcal{A}) \cup (\{0,1\} \times Y) \qquad \text{is closed in} \quad M := [0,1] \times Y.$$

Let $u_0 \in \mathcal{A}$ be fixed. The map $f\colon \mathcal{N} \to Y$ defined by

$$f(t,u) := \begin{cases} h(t,u), & t \in [0,1], \quad u \in \mathcal{A}, \\ u, & t = 0, \quad u \in Y, \\ h(1,u_0), & t = 1, \quad u \in Y, \end{cases}$$

is continuous. The fact that Y is an absolute neighborhood extensor implies that there exists a continuous extension g of f defined on a neighborhood $\mathcal{U}$ of $\mathcal{N}$. The compactness of $[0,1]$ implies the existence of a closed neighborhood $\mathcal{B}$ of $\mathcal{A}$ such that $[0,1] \times \mathcal{B} \subset \mathcal{U}$. However, then $\mathcal{B}$ is contractible to a point in Y, i.e., $\operatorname{cat}_Y(\mathcal{B}) = 1$. □

Our aim is now to prove the Quantitative Deformation Lemma which will be the key tool for proving the existence of critical points on manifolds. In the following considerations we will always assume that

X is a real separable Banach space, $\psi \in C^2(X,\mathbb{R})$,

$$\mathscr{V} := \{v \in X : \psi(v) = 1\} \neq \emptyset \quad \text{and} \quad \psi'(v) \neq 0 \quad \text{for every} \quad v \in \mathscr{V}.$$

The reader should be aware of the fact that some of these assumptions can be relaxed and more general results parallel to those from this section can be proved (see, e.g., Ghoussoub [70]).

The set $\mathscr{V}$ is a differentiable manifold of the class C^2 (cf. Remark 4.3.39 or Deimling [42, § 27], Zeidler [163, Chapter 43]). The norm on X induces a metric on $\mathscr{V}$ and so $\mathscr{V}$ becomes a metric manifold, i.e., a metric space and a manifold. It can be proved that $\mathscr{V}$ is an absolute neighborhood extensor (see, e.g., Deimling [42, Proposition 27.6]). We denote by $T_v\mathscr{V}$ its tangent space at v (see Remark 4.3.40), i.e.,

$$T_v\mathscr{V} := \{y \in X : \langle \psi'(v), y \rangle_X = 0\}.$$

Let $\varphi \in C^1(X,\mathbb{R})$ be given. The norm of the restriction of the derivative $\varphi'(v)$ to $T_v\mathscr{V}$ is given by

$$\|\varphi'(v)\|_* := \sup_{\substack{y \in T_v\mathscr{V} \\ \|y\|_X = 1}} \langle \varphi'(v), y \rangle_X,$$

where $\langle \cdot, \cdot \rangle$ is the duality pairing between X^* and X (cf. Remark 7.2.22). The point v is a *critical point of the restriction* of φ to $\mathscr{V}$ if the restriction of $\varphi'(v)$ to $T_v\mathscr{V}$ is equal to o. We define

$$\varphi^d := \{v \in \mathscr{V} : \varphi(v) \leq d\}.$$

We will use the following *Duality Lemma*.

Lemma 7.8.25 (Duality Lemma). *If $f, g \in X^*$, then*

$$\sup_{\substack{\langle g,y\rangle=0\\ \|y\|=1}} \langle f, y\rangle = \min_{\lambda\in\mathbb{R}} \|f - \lambda g\|.$$

Proof. For every $\lambda \in \mathbb{R}$ we have

$$\sup_{\substack{\langle g,y\rangle=0\\ \|y\|=1}} \langle f, y\rangle = \sup_{\substack{\langle g,y\rangle=0\\ \|y\|=1}} \langle f - \lambda g, y\rangle \le \sup_{\|y\|=1} \|f - \lambda g\|\|y\| = \|f - \lambda g\|.$$

By the Hahn–Banach Theorem (Corollary 2.1.15) there is a continuous linear functional $\tilde{f}$ on X, $\tilde{f}|_{\operatorname{Ker} g} = f$, such that

$$\sup_{\substack{\langle g,y\rangle=0\\ \|y\|=1}} \langle f, y\rangle = \|\tilde{f}\|.$$

Since $\operatorname{Ker} g \subset \operatorname{Ker}(f - \tilde{f})$, there exists $\lambda \in \mathbb{R}$ such that $f - \tilde{f} = \lambda g$ (see Proposition 1.1.19). Hence we obtain

$$\|\tilde{f}\| = \|f - \lambda g\|.$$ □

The above lemma immediately yields the following assertion.

Proposition 7.8.26. *If $\varphi \in C^1(X, \mathbb{R})$ and $u \in \mathscr{V} := \{v \in X : \psi(v) = 1\}$, then*

$$\|\varphi'(u)\|_* = \min_{\lambda\in\mathbb{R}} \|\varphi'(u) - \lambda\psi'(u)\|.$$

In particular, u is a critical point of $\varphi|_{\mathscr{V}}$ if and only if there exists $\lambda \in \mathbb{R}$ such that

$$\varphi'(u) = \lambda\psi'(u).[77]$$

Next we define a tangent pseudogradient vector field on $\mathscr{V}$.

Definition 7.8.27.[78] Let $\varphi \in C^1(\mathscr{V}, \mathbb{R})$. A *tangent pseudogradient vector field* for φ on

$$\mathcal{M} := \{u \in \mathscr{V} : \|\varphi'(u)\|_* \neq 0\}$$

is a locally Lipschitz continuous vector field $g\colon \mathcal{M} \to X$ such that $g(u) \in T_u\mathscr{V}$ and

$$\|g(u)\| \le 2\|\varphi'(u)\|_*, \qquad \langle\varphi'(u), g(u)\rangle \ge \|\varphi'(u)\|_* \qquad \text{for every} \quad u \in \mathcal{M}.$$

Lemma 7.8.28.[79] *Let $\varphi \in C^1(X, \mathbb{R})$. Then there exists a tangent pseudogradient vector field for φ on $\mathcal{M}$.*

Proof. For every $v \in \mathcal{M}$ there exists $x \in T_v\mathscr{V}$ such that

$$\|x\| = 1 \qquad \text{and} \qquad \langle\varphi'(v), x\rangle > \frac{2}{3}\|\varphi'(v)\|_*.$$

There is also $z \in X$ such that

$$\langle\psi'(v), z\rangle = 1.$$

[77] Cf. Theorem 7.8.2.
[78] Cf. Definition 7.4.16.
[79] Cf. Lemma 7.4.17.

Set $y := \frac{3}{2}x\|\varphi'(v)\|_*$ and for $u \in \mathscr{V}$ such that $\langle\psi'(u), z\rangle \neq 0$, set

$$g_v(u) := y - \frac{\langle\psi'(u), y\rangle}{\langle\psi'(u), z\rangle}z.$$

Since $\langle\psi'(v), y\rangle = 0$, we have $g_v(v) = y$ and

$$\|g_v(v)\| < 2\|\varphi'(v)\|_*, \qquad \langle\varphi'(v), g_v(v)\rangle > \|\varphi'(v)\|_*^2.$$

Since φ' and g_v are continuous, there exists an open neighborhood $\mathcal{N}(v)$ of v such that

$$\|g_v(u)\| < 2\|\varphi'(u)\|_*, \qquad \langle\varphi'(u), g_v(u)\rangle > \|\varphi'(u)\|_*^2 \qquad \text{for every} \quad u \in \mathcal{N}(v).$$

The family

$$\mathscr{N} := \{\mathcal{N}(v) : v \in \mathcal{M}\}$$

is an open covering of $\mathcal{M}$. Since $\mathcal{M}$ is a metric space, there exists a locally finite open covering

$$\mathscr{M} = \{\mathcal{M}_i : i \in \mathbb{N}\}$$

of $\mathcal{M}$ which is subordinate to $\mathcal{N}$, i.e., such that for every $i \in \mathbb{N}$ there exists $v \in \mathscr{V}$ satisfying

$$\overline{\mathcal{M}_i} \subset \mathcal{N}(v)$$

(cf. the proof of Lemma 7.4.17). For any $i \in \mathbb{N}$ choose one such $v := v_i$ and define

$$g_i(u) := \begin{cases} g_{v_i}(u), & u \in \mathcal{N}(v_i), \\ o, & u \notin \mathcal{N}(v_i), \end{cases}$$

and

$$\varrho_i(u) := \operatorname{dist}(u, X \setminus \mathcal{M}_i), \qquad g(u) := \sum_{i\in\mathbb{N}} \frac{\varrho_i(u)g_i(u)}{\sum\limits_{j\in\mathbb{N}} \varrho_j(u)}.$$ [80]

It is now straightforward to verify that g is a tangent pseudogradient vector field for φ on $\mathcal{M}$. (The interested reader is invited to check it in details and realize that the fact $\psi \in C^2(X, \mathbb{R})$ is used!) □

The proof of the following version of the Quantitative Deformation Lemma follows the lines of the proof of Lemma 7.4.18 (cf. Exercise 7.8.41).

Lemma 7.8.29 (Quantitative Deformation Lemma). *Let $\varphi \in C^1(X, \mathbb{R})$, $\mathcal{S} \subset \mathscr{V}$, $c \in \mathbb{R}$, $\varepsilon, \delta > 0$ be such that*

$$\|\varphi'(u)\|_* \geq \frac{8\varepsilon}{\delta} \qquad \textit{for any} \quad u \in \varphi_{-1}([c - 2\varepsilon, c + 2\varepsilon]) \cap \mathcal{S}_{2\delta} \cap \mathscr{V}.$$

Then there exists $\eta \in C([0,1] \times \mathscr{V}, \mathscr{V})$ such that

(i) *$\eta(t, u) = u$ if $t = 0$ or $u \notin \varphi_{-1}([c - 2\varepsilon, c + 2\varepsilon]) \cap \mathcal{S}_{2\delta} \cap \mathscr{V}$,*

(ii) *$\eta(1, \varphi^{c+\varepsilon} \cap \mathcal{S}) \subset \varphi^{c-\varepsilon}$,*

(iii) *$\varphi(\eta(\cdot, u))$ is decreasing for any $u \in \mathscr{V}$.*

[80] Note that the sums contain only a finite number of nonzero terms. Note also that separability of X can be dropped and substituted by paracompactness.

We are now ready to prove the *General Minimax Principle on the manifold* $\mathscr{V}$. We assume that $\varphi \in C^1(X, \mathbb{R})$ is bounded below on $\mathscr{V}$. For $j \geq 1$, $j \in \mathbb{N}$, we define

$$\mathscr{A}_j := \{\mathcal{A} \subset \mathscr{V} : \mathcal{A} \text{ is closed, } \operatorname{cat}_{\mathscr{V}}(\mathcal{A}) \geq j\}, \qquad c_j := \inf_{\mathcal{A} \in \mathscr{A}_j} \sup_{u \in \mathcal{A}} \varphi(u). \tag{7.8.14}$$

Theorem 7.8.30 (General Minimax Principle). *Assume that φ and c_j are as above. If*

$$c := c_k = c_{k+1} = \cdots = c_{k+m}, \tag{7.8.15}$$

then for every $\varepsilon > 0$, $\delta > 0$, $\mathcal{A} \in \mathscr{A}_{k+m}$ and $\mathcal{B} \subset \mathscr{V}$ closed such that

$$\sup_{u \in \mathcal{A}} \varphi(u) \leq c + \varepsilon, \qquad \operatorname{cat}_{\mathscr{V}}(\mathcal{B}) \leq m, \tag{7.8.16}$$

there exists $u_0 \in \mathscr{V}$ such that

(i) $c - 2\varepsilon \leq \varphi(u_0) \leq c + 2\varepsilon$,

(ii) $\operatorname{dist}(u_0, \mathcal{A} \setminus \operatorname{int} \mathcal{B}) \leq 2\delta$,

(iii) $\|\varphi'(u_0)\|_* \leq \frac{8\varepsilon}{\delta}$.

Proof. We assume by contradiction that there exist numbers $\varepsilon > 0$, $\delta > 0$, and closed sets $\mathcal{A} \in \mathscr{A}_{k+m}$, $\mathcal{B} \subset \mathscr{V}$ such that (7.8.16) holds and for all $u \in \mathscr{V}$ satisfying (i) and (ii)[81] the inequality (iii) is false. We apply Lemma 7.8.29 with

$$\mathcal{S} := \mathcal{A} \setminus \operatorname{int} \mathcal{B}.$$

We obtain in virtue of Lemma 7.8.29(ii) that

$$(\mathcal{A} \setminus \operatorname{int} \mathcal{B}) \prec \varphi^{c-\varepsilon}.$$

It follows from Lemma 7.8.21(ii), (iii) and from the definition of c_k that

$$k + m \leq \operatorname{cat}_{\mathscr{V}}(\mathcal{A}) \leq \operatorname{cat}_{\mathscr{V}}(\mathcal{A} \setminus \operatorname{int} \mathcal{B}) + \operatorname{cat}_{\mathscr{V}}(\mathcal{B}) \leq \operatorname{cat}_{\mathscr{V}}(\varphi^{c-\varepsilon}) + m \leq k - 1 + m,$$

a contradiction. ☐

Definition 7.8.31. A functional φ satisfies the *Palais–Smale condition* $(\mathrm{PS})_c$ *on* $\mathscr{V}$ if any sequence $\{u_n\}_{n=1}^{\infty} \subset \mathscr{V}$ such that

$$\varphi(u_n) \to c, \qquad \|\varphi'(u_n)\|_* \to 0,$$

has a convergent subsequence.

Theorem 7.8.32. *Let φ be bounded below on $\mathscr{V}$, satisfy* $(\mathrm{PS})_c$ *on $\mathscr{V}$, and let* (7.8.15) *hold. Let*

$$\mathcal{K}_c := \{u \in \mathscr{V} : \varphi(u) = c, \ \|\varphi'(u)\|_* = 0\}.$$

Then

$$\operatorname{cat}_{\mathscr{V}}(\mathcal{K}_c) \geq m + 1.$$

[81] Cf. Exercise 7.8.46.

Proof. Assume that $\operatorname{cat}_{\mathscr{V}}(\mathcal{K}_c) \le m$. Then Proposition 7.8.24 implies the existence of a closed neighborhood $\mathcal{B}$ of $\mathcal{K}_c$ in $\mathscr{V}$ such that

$$\operatorname{cat}_{\mathscr{V}}(\mathcal{B}) \le m.$$[82]

By Theorem 7.8.30 for $\mathcal{A} = \mathscr{V}$ there exists a sequence $\{u_n\}_{n=1}^{\infty} \subset \mathscr{V}$ satisfying

$$\varphi(u_n) \to c, \qquad \operatorname{dist}(u_n, \mathscr{V} \setminus \operatorname{int} \mathcal{B}) \to 0, \qquad \|\varphi'(u_n)\|_* \to 0.$$

It then follows from the $(\mathrm{PS})_c$ condition on $\mathscr{V}$ that $\mathcal{K}_c \cap (\mathscr{V} \setminus \operatorname{int} \mathcal{B}) \neq \emptyset$, a contradiction with the definition of $\mathcal{B}$. □

Theorem 7.8.33. *Let φ be bounded below on $\mathscr{V}$, $d \ge \inf_{u \in \mathscr{V}} \varphi(u)$ and φ satisfy $(\mathrm{PS})_c$ on $\mathscr{V}$ for any $c \in \left[\inf_{u \in \mathscr{V}} \varphi(u), d\right]$. Then $\varphi|_{\mathscr{V}}$ has a minimum and φ^d contains at least $\operatorname{cat}_{\mathscr{V}}(\varphi^d)$ critical points of $\varphi|_{\mathscr{V}}$.*

Proof. If $n := \operatorname{cat}_{\mathscr{V}}(\varphi^d)$, then

$$\inf_{u \in \mathscr{V}} \varphi(u) = c_1 \le c_2 \le c_3 \le \dots \le c_n \le d$$

where c_i, $i = 1, \dots, n$, are given by (7.8.14). The critical points corresponding to different critical levels are mutually different. If some levels coincide, we apply Theorem 7.8.32 to get the assertion. □

Remark 7.8.34. Note that

(i) $\operatorname{cat}_X(\overline{B(o;1)}) = 1$ for the closed ball $\overline{B(o;1)}$ in X where X is a Banach space (see Figure 7.8.5);

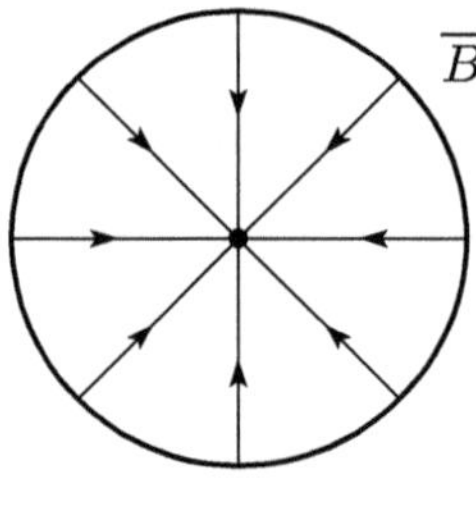

Figure 7.8.5.

(ii) $\operatorname{cat}_{\mathcal{S}^{N-1}}(\mathcal{S}^{N-1}) = 2$ for the unit sphere $\mathcal{S}^{N-1} = \partial B(o;1) \subset \mathbb{R}^N$, $N \ge 1$. Indeed, Figure 7.8.2 suggests that $\operatorname{cat}_{\mathcal{S}^{N-1}}(\mathcal{S}^{N-1}) \le 2$. On the other hand, it follows from Lemma 5.1.10 that $\operatorname{cat}_{\mathcal{S}^{N-1}}(\mathcal{S}^{N-1}) > 1$.

Definition 7.8.35. Let $\mathcal{S}^{N-1} \subset \mathbb{R}^N$ be the unit sphere. Then

$$\mathcal{P}^{N-1} = \{(u, -u) : u \in \mathcal{S}^{N-1}\}$$

is called an $(N-1)$-*dimensional projective space*.

[82]Note that the fact that $\mathscr{V}$ is an absolute neighborhood extensor is used here, cf. page 544.

The geometrical interpretation of $\mathcal{P}^{N-1}$ is the following: the $(N-1)$-dimensional projective space $\mathcal{P}^{N-1}$ results from $\mathcal{S}^{N-1}$, $N \geq 1$ by *identifying antipodal points* (see Figure 7.8.6).

The following identity is the key to the proof of existence of a sequence of eigenvalues of nonlinear problems:

$$\mathrm{cat}_{\mathcal{P}^{N-1}}(\mathcal{P}^{N-1}) = N. \tag{7.8.17}$$

To see that $\mathrm{cat}_{\mathcal{P}^{N-1}}(\mathcal{P}^{N-1}) \leq N$ we can proceed by induction as follows. $\mathcal{S}^1$ can be covered by two closed symmetric sets which are contractible to a point in $\mathcal{P}^1$ (see Figure 7.8.7).

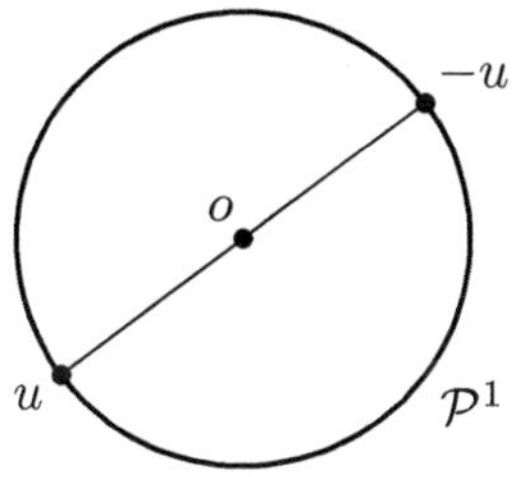

Figure 7.8.6.

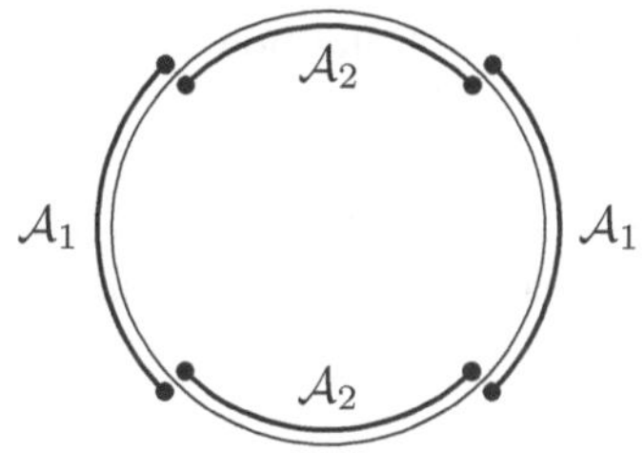

Figure 7.8.7.

The closed strip along the equator on $\mathcal{S}^2$ can be covered by two closed symmetric sets which are contractible to a point in $\mathcal{P}^2$ as well. If we add the closed north and south caps, we get a covering of $\mathcal{S}^2$ by three closed symmetric sets which are contractible to a point in $\mathcal{P}^2$, etc. (see Figure 7.8.8).

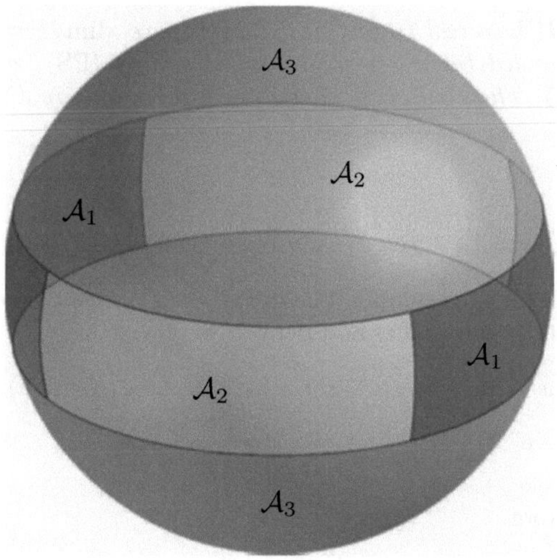

Figure 7.8.8.

To prove the reversed inequality we proceed by contradiction. Assume that $\operatorname{cat}_{\mathcal{P}^{N-1}}(\mathcal{P}^{N-1}) < N$. Then according to Exercise 7.8.42 there exist $M < N$ and closed symmetric sets $\mathcal{A}_i$, $i = 1, \dots, M$, such that

$$\mathcal{S}^{N-1} = \bigcup_{i=1}^{M} \mathcal{A}_i, \qquad \mathcal{A}_i = \tilde{\mathcal{A}}_i \cup (-\tilde{\mathcal{A}}_i), \qquad \tilde{\mathcal{A}}_i \cap (-\tilde{\mathcal{A}}_i) = \emptyset.$$

Then $\tilde{\mathcal{A}}_1, \dots, \tilde{\mathcal{A}}_M, (-\tilde{\mathcal{A}}_1) \cup \dots \cup (-\tilde{\mathcal{A}}_M)$ is a covering of $\mathcal{S}^{N-1}$ by $M+1$ closed sets and none of them contains antipodal points. This contradicts the covering result of Lusternik and Schnirelmann (if $M = N-1$, we can apply directly Exercise 5.7.40; if $M < N-1$, we complete the above covering by $N-1-M$ empty sets and apply again Exercise 5.7.40).

Similarly to Definition 7.8.35 we can define an infinite-dimensional projective space

$$P^{\infty} := \{(u, -u) : u \in \mathcal{S}\}$$

where $\mathcal{S} = \partial B(o;1)$ is the boundary of the unit ball $B(o;1)$ in an infinite-dimensional Banach space. Then (7.8.17) immediately yields that

$$\operatorname{cat}_{P^{\infty}}(P^{\infty}) = \infty.$$

Example 7.8.36. Let $f\colon \mathbb{R}^N \to \mathbb{R}$ be a continuously differentiable function. Since $\mathcal{S}^{N-1}$ is compact, it follows from the Extreme Value Theorem that there exists $d > \sup\limits_{u \in \mathcal{S}^{N-1}} f(u)$. It follows then from Theorem 7.8.33 that the number of critical points on $\mathcal{S}^{N-1}$ is greater than or equal to $\operatorname{cat}_{\mathbb{R}^N}(\mathcal{S}^{N-1}) = 2$. However, this result is trivial. On the other hand, if f is even, we can think of f as a continuous mapping from $\mathcal{P}^{N-1}$ into $\mathbb{R}$. Then by (7.8.17) and Theorem 7.8.33, f has at least N critical points in $\mathcal{P}^{N-1}$ to which N pairs $(-u, u)$ of critical points of f on $\mathcal{S}^{N-1}$ correspond. This is a nontrivial result. ○

If we combine this example and Theorem 7.8.33, we get the following assertion.

Theorem 7.8.37. *Let H be a real (separable) Hilbert space, $\dim H = \infty$, let the functional $\varphi \in C^1(H, \mathbb{R})$ be bounded below, even and let it satisfy $(\mathrm{PS})_c$ on $\partial B(o;1) \subset H$ for any $c \geq \inf\limits_{u \in \partial B(o;1)} \varphi(u)$. Then $\varphi|_{\partial B(o;1)}$ possesses infinitely many distinct pairs of critical points.*

Proof. Since

$$\partial B(o;1) := \{u \in H : \psi(u) = 1\} \qquad \text{where} \quad \psi(u) = (u, u)$$

is of the class $C^2(H, \mathbb{R})$, we can apply Theorem 7.8.33. Indeed, since φ and ψ are even, we can identify the antipodal points and define

$$X := \{x = (u, -u) : u \in H\}, \qquad \mathscr{V} := \{x \in X : \psi(u) = 1\}.$$

Since $\mathscr{V} = \mathcal{P}^{\infty}$, we have

$$\operatorname{cat}_{\mathscr{V}}(\mathscr{V}) = \infty.$$

This completes the proof. □

Now we illustrate the connection between the critical points of functionals on manifolds in Banach spaces and the nonlinear eigenvalue problems. We present this fact on a simple example.

Example 7.8.38. Set $X := W_0^{1,p}(0,1)$, $p \geq 2$,

$$\varphi(x) := \int_0^1 |\dot{x}(t)|^p \,\mathrm{d}t, \qquad \psi(x) := \int_0^1 |x(t)|^p \,\mathrm{d}t, \qquad x \in X.$$

Then φ and ψ satisfy all the above assumptions. The functional φ is bounded below on $\mathscr{V}$ by λ_1(see Example 7.8.5) and satisfies $(\mathrm{PS})_c$ on $\mathscr{V}$ for any level $c \geq \lambda_1$. Indeed, let

$$\varphi(x_n) \to c, \qquad \|\varphi'(x_n)\|_* \to 0 \qquad \text{for} \quad x_n \in \mathscr{V}. \tag{7.8.18}$$

The first convergence in (7.8.18) implies that $\{x_n\}_{n=1}^{\infty}$ is a bounded sequence in X. Then the reflexivity of X implies that without loss of generality we can assume $x_n \rightharpoonup x$ in X, and by the compact embedding $X = W_0^{1,p}(0,1) \hookrightarrow\hookrightarrow L^p(0,1)$ also $x_n \to x$ in $L^p(\Omega)$. But then $x \in \mathscr{V}$, i.e., $x \neq o$. It follows from (7.8.18) that

$$\left\langle \left(\frac{\varphi}{\psi}\right)'(x_n), w \right\rangle = \langle \varphi'(x_n), w\rangle - \varphi(x_n)\langle \psi'(x_n), w\rangle \to 0 \tag{7.8.19}$$

uniformly for all $w \in X$, $\|w\| \leq R$ (cf. Exercise 7.8.44). We can take $w := x_n - x$ in (7.8.19) (note that $\{x_n\}_{n=1}^{\infty}$ is bounded in X). Hence

$$\int_0^1 |\dot{x}_n(t)|^{p-2}\dot{x}_n(t)(\dot{x}_n(t) - \dot{x}(t)) \,\mathrm{d}t - \varphi(x_n) \int_0^1 |x_n(t)|^{p-2} x_n(t)(x_n(t) - x(t)) \,\mathrm{d}t \to 0.$$

Since also

$$\langle \varphi'(x), x_n - x\rangle = \int_0^1 |\dot{x}(t)|^{p-2}\dot{x}(t)(\dot{x}_n(t) - \dot{x}(t)) \,\mathrm{d}t \to 0$$

by the weak convergence $x_n \rightharpoonup x$ in X and

$$\int_0^1 |x_n(t)|^{p-2} x_n(t)(x_n(t) - x(t)) \,\mathrm{d}t \to 0$$

by the compact embedding $X = W_0^{1,p}(0,1) \hookrightarrow\hookrightarrow L^p(0,1)$, we obtain

$$\int_0^1 \left(|\dot{x}_n(t)|^{p-2}\dot{x}_n(t) - |\dot{x}(t)|^{p-2}\dot{x}(t)\right)(\dot{x}_n(t) - \dot{x}(t)) \,\mathrm{d}t \to 0.$$

However, for $p \geq 2$ we have

$$\int_0^1 \left(|\dot{x}_n(t)|^{p-2}\dot{x}_n(t) - |\dot{x}(t)|^{p-2}\dot{x}(t)\right)(\dot{x}_n(t) - \dot{x}(t)) \,\mathrm{d}t \geq \int_0^1 |\dot{x}_n(t) - \dot{x}(t)|^p \,\mathrm{d}t,$$

i.e., $x_n \to x$.

It follows from Theorem 7.8.37 that φ has infinitely many distinct pairs of critical points $(x_i, -x_i)$, $i = 1, 2, \dots$, $x_i \in \mathscr{V}$. It follows from Proposition 7.8.26 that there exist λ_i, $i = 1, 2, \dots$, such that

$$\varphi'(x_i) = \lambda_i \psi'(x_i).$$

But, since

$$\langle \varphi'(x_i), x_i\rangle = p\varphi(x_i), \qquad \langle \psi'(x_i), x_i\rangle = p\psi(x_i),$$

we have

$$\varphi(x_i) = \lambda_i, \qquad i = 1, 2, \dots,$$

i.e., the critical values of $\varphi|_{\mathscr{V}}$ are the eigenvalues of the problem

$$\int_0^1 |\dot{x}(t)|^{p-2}\dot{x}(t)\dot{y}(t)\,\mathrm{d}t = \lambda \int_0^1 |x(t)|^{p-2}x(t)y(t)\,\mathrm{d}t. \tag{7.8.20}$$

From the proof of Theorem 7.8.33 we have

$$\lambda_i = \inf_{\mathcal{A}\in\mathscr{A}_i} \sup_{x\in\mathcal{A}} \varphi(x) \tag{7.8.21}$$

$$\text{where} \quad \mathscr{A}_i := \{\mathcal{A} \subset \mathscr{V} : \mathcal{A} \text{ is closed, symmetric, } \mathrm{cat}_{\mathscr{V}}(\mathcal{A}) \geq i\}.$$

We prove that

$$\lim_{i\to\infty} \lambda_i = \infty. \tag{7.8.22}$$

For this purpose we have to exclude the following two cases:

Case 1. There exists $n \in \mathbb{N}$ such that $\lambda_m = \lambda_n$ for any $m \geq n$.

Case 2. Case 1 does not occur but there exists $\Lambda \in \mathbb{R}$ such that $\lambda_i \nearrow \Lambda$.

If Case 1 occurs, then necessarily $\mathrm{cat}_{\varphi^{\lambda_m+1}}(\mathcal{K}_{\lambda_m}) = \infty$ by Theorem 7.8.32. However, the $(\mathrm{PS})_{\lambda_m}$ condition implies that $\mathcal{K}_{\lambda_m}$ is a compact set and hence

$$\mathrm{cat}_{\varphi^{\lambda_m+1}}(\mathcal{K}_{\lambda_m}) < \infty$$

(see Exercise 7.8.43), i.e., Case 1 is excluded.

If Case 2 occurs, then we argue as follows. Let $\varepsilon > 0$ be specified later and denote

$$\mathcal{K} := \{x \in \mathscr{V} : \lambda_1 \leq \varphi(x) \leq \Lambda + \varepsilon,\ \|\varphi'(x)\|_* = 0\}.$$

By the $(\mathrm{PS})_c$ condition we know that $\mathcal{K}$ is compact, and hence $j := \mathrm{cat}_{\mathscr{V}}(\mathcal{K}) < \infty$. According to Proposition 7.8.24 there exists a closed neighborhood $\tilde{\mathcal{K}}$ of $\mathcal{K}$ in $\mathscr{V}$ such that $\mathrm{cat}_{\mathscr{V}}(\tilde{\mathcal{K}}) = j$. In particular, if we set

$$\mathcal{S} := \overline{\varphi^{\Lambda+\varepsilon} \setminus \tilde{\mathcal{K}}},$$

we can apply Lemma 7.8.29. Indeed, choose ε and δ such that the assumptions of Lemma 7.8.29 are satisfied with $c = \Lambda$, and let m be the smallest integer such that $\lambda_m > \Lambda - \varepsilon > \lambda_1$. Choose $\mathcal{A} \in \mathscr{A}_{j+m}$ such that $\sup\limits_{x\in\mathcal{A}} \varphi(x) \leq \Lambda + \varepsilon$ (this is possible due to the variational characterization of λ_{j+m}, see (7.8.21)) and set

$$\mathcal{B} := \overline{\mathcal{A} \setminus \tilde{\mathcal{K}}}$$

(the closure is taken in the topology of $\mathscr{V}$). Then according to Lemma 7.8.21(ii),

$$\mathrm{cat}_{\mathscr{V}}(\mathcal{B}) \geq m, \qquad \text{i.e.,} \qquad \mathcal{B} \in \mathscr{A}_m.$$

It follows from Lemma 7.8.21(iii) that

$$\mathrm{cat}_{\mathscr{V}}(\eta(1,\mathcal{B})) \geq \mathrm{cat}_{\mathscr{V}}(\mathcal{B}) \geq m, \qquad \text{i.e.,} \qquad \eta(1,\mathcal{B}) \in \mathscr{A}_m.$$

But then according to Lemma 7.8.29(ii),

$$\lambda_m \leq \sup_{x\in\mathcal{B}} \varphi(\eta(1,x)) \leq \Lambda - \varepsilon < \lambda_m,$$

a contradiction. Hence Case 2 is also excluded, and (7.8.22) is proved. ○

Remark 7.8.39. Using the technique of ordinary differential equations it is possible to prove that the set $\{\lambda_n\}_{n=1}^{\infty}$ represents all eigenvalues of (7.8.20), that every λ_n, $n = 1, 2, \dots$, is simple (see, e.g., Elbert [57], Došlý [46] and references therein). The same approach as above can be used to prove the existence of an infinite sequence of eigenvalues, approaching infinity, of the p-Laplacian in more dimensions. Contrary to the one-dimensional case it is not clear if such a sequence exhausts all the eigenvalues or not. This has been a long standing and challenging open problem of nonlinear analysis. Note that the assumption $p \geq 2$ can be relaxed to $p > 1$. However, $\mathscr{V}$ is not a manifold of the class C^2 for $1 < p < 2$ and so a more general approach must be employed (see, e.g., Ghoussoub [70]).

Remark 7.8.40. Similar and more general minimax arguments can be found in literature where instead of the (Lusternik–Schnirelmann) category a more general concept of the relative category is used.

One can develop an abstract index theory where the index of a set (an analogue of the category) satisfies certain axioms and similar results to those in this section can be proved. The reader can find various kinds of indices: Krasnoselski genus, S^1-index of Benci, cohomological index of Fadell–Rabinowitz, etc. (see, e.g., Zeidler [163] and references therein). The notion of a category is, in a certain sense, a maximal function satisfying the key properties of Lemma 7.8.21 (cf. Exercise 7.8.45).

Exercise 7.8.41. Give the proof of Lemma 7.8.29 in details. Watch carefully for the moment when the assumption $\psi \in C^2(X, \mathbb{R})$ is essentially used.

Exercise 7.8.42. Every closed set $\mathcal{A}^* \subset \mathcal{P}^{N-1}$ corresponds to a symmetric closed set $\mathcal{A} \subset \mathcal{S}^{N-1}$ and vice versa as follows:

$$x \in \mathcal{A} \qquad \text{if and only if} \qquad (x, -x) \in \mathcal{A}^*.$$

Prove that $\operatorname{cat}_{\mathcal{P}^{N-1}}(\mathcal{A}^*) = 1$ if and only if there exists $\tilde{\mathcal{A}}$ such that

$$\mathcal{A} = \tilde{\mathcal{A}} \cup (-\tilde{\mathcal{A}}) \qquad \text{and} \qquad \tilde{\mathcal{A}} \cap (-\tilde{\mathcal{A}}) = \emptyset.$$

Hint. $\operatorname{cat}_{\mathcal{P}^{N-1}}(\mathcal{A}^*) = 1$ if and only if there exist an odd continuous mapping $f\colon [0,1] \times \mathcal{S}^{N-1} \to \mathcal{S}^{N-1}$ and a point $a \in \mathcal{S}^{N-1}$ such that for $(x, -x) \in \mathcal{A}^*$ we have

$$(f(0,x), f(0,-x)) = (-x, x) \qquad \text{and} \qquad (f(1,x), f(1,-x)) = (a, -a);$$

take $\tilde{\mathcal{A}} = \{x \in \mathcal{S}^{N-1} : f(1,x) = a\}$.

Exercise 7.8.43. Prove that if $\mathcal{K}$ is a compact subset of a manifold $\mathscr{V}$ of the class C^2, then

$$\operatorname{cat}_{\mathscr{V}}(\mathcal{K}) < \infty.$$

Hint. For any $u \in \mathcal{K}$ there exists $B(u; R(u))$ such that $B(u; R(u)) \cap \mathscr{V}$ is contractible to a point in $\mathscr{V}$ (use Remark 7.8.34(i) and the fact that $\mathscr{V}$ is a manifold of the class C^2); $\bigcup\limits_{u \in \mathcal{K}} (B(u; R(u)) \cap \mathscr{V})$ is an open covering of $\mathcal{K}$, choose $\bigcup\limits_{i=1}^{n} B(u_i; R(u_i))$ a finite subcovering of $\mathcal{K}$, use Lemma 7.8.21(ii) to show $\operatorname{cat}_{\mathscr{V}}(\mathcal{K}) \leq n$.

Exercise 7.8.44. Prove (7.8.19).

Hint. Split $w := t_n x_n + y_n$, $t_n \in \mathbb{R}$, $\langle \psi'(x_n), y_n \rangle = 0$. Using the facts $\{x_n\}_{n=1}^{\infty}$ bounded in X and $x_n \to x$ in $L^2(\Omega)$ prove that for any $R > 0$ there exists $\hat{R} > 0$ such that $\|w\| \le R$ implies $\|y_n\| \le \hat{R}$ for all $n \in \mathbb{N}$. Now, take also into account $\psi(x_n) = 1$, $\langle \varphi'(x_n), x_n \rangle = p\varphi(x_n)$, $\langle \psi'(x_n), x_n \rangle = p$, in order to get

$$\left\langle \left(\frac{\varphi}{\psi}\right)'(x_n), w \right\rangle = \frac{\langle \varphi'(x_n), w\rangle \psi(x_n) - \varphi(x_n)\langle \psi'(x_n), w\rangle}{\psi^2(x_n)}$$
$$= pt_n\varphi(x_n) + \langle \varphi'(x_n), y_n\rangle - pt_n\varphi(x_n) - \varphi(x_n)\langle \psi'(x_n), y_n\rangle = \langle \varphi'(x_n), y_n\rangle \to 0$$

uniformly with respect to $w \in X$, $\|w\| \le R$.

Exercise 7.8.45. Prove the following assertion:

> *Let Φ_Y be a function defined on the class $\mathscr{A}$ of closed subsets $\mathcal{A}$ of Y. If Φ_Y possesses properties* (i)–(iii) *of Lemma* 7.8.21 *and $\Phi_Y(\mathcal{A}) = 1$ when $\mathcal{A}$ consists of a single point, then*
>
> $$\Phi_Y(\mathcal{A}) \le \operatorname{cat}_Y(\mathcal{A}) \qquad \textit{for all} \quad \mathcal{A} \in \mathscr{A}.$$

Hint. Let $\operatorname{cat}_Y(\mathcal{A}) = 1$. Since $\mathcal{A}$ is contractible to a point $u_0 \in Y$, hence, by the fact that Φ_Y satisfies (iii) of Lemma 7.8.21,

$$\Phi_Y(\mathcal{A}) \le \Phi_Y(u_0) = 1.$$

Now the assertion follows by using a covering and (ii) of Lemma 7.8.21.

Exercise 7.8.46. Prove that the set of all $x \in \mathscr{V}$ satisfying (i) and (ii) from Theorem 7.8.30 is not empty.

7.8B Krasnoselski Potential Bifurcation Theorem

Let us recall the definition of a potential operator.

Definition 7.8.47. Let $\mathcal{O}$ be an open subset of a real Hilbert space H, $f\colon \mathcal{O} \to H$. We say that f has a *potential* (in $\mathcal{O}$) if there exists a functional $F\colon \mathcal{O} \to \mathbb{R}$ which is Fréchet differentiable in $\mathcal{O}$, and for any $x \in \mathcal{O}$ we have

$$f(x) = F'(x). \tag{7.8.23}$$

Remark 7.8.48. Let us recall how to interpret the equality (7.8.23). The Fréchet derivative $F'(x)$ is a continuous linear operator from H into $\mathbb{R}$. It follows from the Riesz Representation Theorem (see Theorem 1.2.40) that there is a unique point $z := z(x) \in H$ such that

$$F'(x)y = (y, z), \qquad \|z\| = \|F'(x)\| \qquad \text{for any} \quad y \in H.$$

In what follows we will identify $F'(x)$ with $z(x) \in H$ and study bifurcation points of the equation

$$\lambda x - F'(x) = o. \tag{7.8.24}$$

The main objective of this appendix is to prove that (under the assumptions $F(o) = o$, $F'(o) = o$ and some assumptions concerning the smoothness of F)

every point (λ, o) *where* λ *is a nonzero eigenvalue of* $F''(o)\colon H \to H$ *is a bifurcation point of* (7.8.24).

Theorem 7.8.49 (Krasnoselski Potential Bifurcation Theorem). *Let* F *be a (nonlinear) functional on a Hilbert space* H*. Assume that*

$$F \text{ is twice differentiable in a certain neighborhood } \mathcal{U}(o) \text{ of } o \in H, \tag{7.8.25}$$

$$F' \text{ is compact on } \mathcal{U}(o), \tag{7.8.26}$$

$$F'' : \mathcal{U}(o) \to \mathcal{L}(H) \text{ is continuous at } o, \tag{7.8.27}$$

$$F(o) = o, \qquad F'(o) = o. \tag{7.8.28}$$

Then (λ_0, o) *where* $\lambda_0 \neq 0$ *is a bifurcation point of*

$$\lambda x - F'(x) = o \tag{7.8.29}$$

if and only if λ_0 *is an eigenvalue of the operator* $A := F''(o)$.

Remark 7.8.50. Note that the equation (7.8.29) is a special case of the equation

$$o = \lambda x - Ax + G(\lambda, x)$$

from Theorem 5.8.12. Indeed, the left-hand side of (7.8.29) can be written as

$$\lambda x - F''(o)x + [F''(o)x - F'(x)]$$

where $F''(o)$ is a compact linear operator (see Proposition 5.8.10), and

$$F''(o)x - F'(x) = o(\|x\|), \qquad \|x\| \to 0.$$

Note first that the implication

if $\lambda_0 \neq 0$ *and* (λ_0, o) *is a bifurcation point of* (7.8.29), *then* λ_0 *is an eigenvalue of* A,

follows from Exercise 5.8.14.

So we will concentrate on the proof of the reversed implication. Roughly speaking, we know that the "linearization of (7.8.29)", i.e., the equation

$$(\lambda I - F''(o))x = o$$

has a nontrivial solution, and we want to show that there is also a nontrivial solution of the "close" but nonlinear equation (7.8.29).

The basic idea of the proof consists in the fact that (7.8.29) is a necessary condition for x to be a critical point of F subject to the sphere

$$\partial B(o; r) := \left\{ x \in H : J(x) = \frac{1}{2} r^2 \right\} \qquad \text{where} \quad J(x) = \frac{1}{2}\|x\|^2.$$

Here we use the fact that identity is the differential of the functional J, and the Lagrange Multiplier Method. Later we will prove the existence of a sufficiently large number of critical points of F on $\partial B(o; r)$. If we restrict ourselves to spheres with sufficiently small radii ($B(o; r) \subset \mathcal{U}(o)$ at least), we get critical points converging to zero. The last part of the proof consists in showing that the corresponding Lagrange multipliers can be chosen close to λ_0.

Let us assume that $\lambda_0 \neq 0$ is an eigenvalue of the operator A. The assumption (7.8.27) guarantees that $F''(o)$ is a linear self-adjoint operator (see Proposition 3.2.28). We can assume, without loss of generality, that $\lambda_0 > 0$.

Let us start with a geometrical interpretation of the points $x \in \partial B(o;r)$ such that

$$\lambda x = F'(x). \tag{7.8.30}$$

In this case the differential $F'(x)$ is perpendicular (recall that $F'(x) \in H$ in our interpretation) to the sphere $\partial B(o;r)$ at x. Then x can be looked for as a limit of those points of the sphere $\partial B(o;r)$ at which the *tangent projections* (see (7.8.31) below and Figure 7.8.9) of $F'(x)$ converge to zero. More precisely, we have

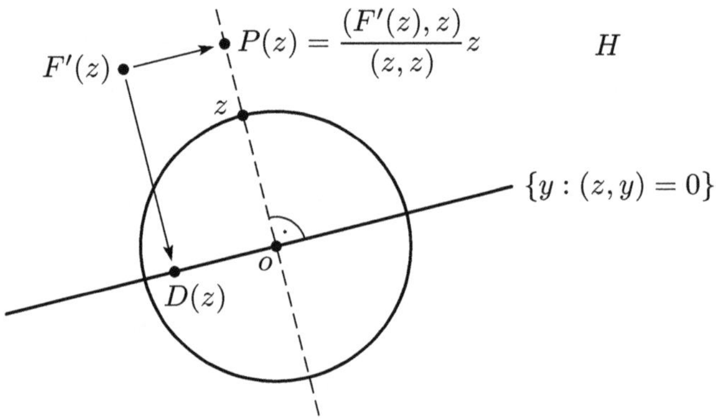

Figure 7.8.9.

Lemma 7.8.51. *For $z \in H$, $z \neq o$, set*

$$D(z) = F'(z) - \frac{(F'(z),z)}{(z,z)}z \tag{7.8.31}$$

($D(z)$ is the orthogonal projection of $F'(z)$ to the tangent space of $\partial B(o;\|z\|)$ at z[83]). Let $y_n \in \partial B(o;r)$, $y_n \rightharpoonup x_0$, and let F' be continuous, and

$$\lim_{n\to\infty} F'(y_n) = y \neq o, \qquad \lim_{n\to\infty} D(y_n) = o.^{84} \tag{7.8.32}$$

Then $y_n \to x_0$, $y = F'(x_0)$, $x_0 \neq o$, and

$$\lambda x_0 - F'(x_0) = o \qquad \textit{where} \quad \lambda = \frac{1}{r^2}(F'(x_0),x_0). \tag{7.8.33}$$

[83]This tangent space is equal to $\{x \in H : (x,z) = 0\}$ – see Remark 4.3.40.
[84]Both limits are considered with respect to the norm in H.

Proof. From the weak convergence $y_n \rightharpoonup x_0$ and from (7.8.32) we obtain

$$(F'(y_n), y_n) \to (y, x_0) \qquad \text{and hence} \qquad \frac{(F'(y_n), y_n)}{r^2} y_n \rightharpoonup \frac{(y, x_0)}{r^2} x_0.$$

At the same time, from the definition of $D(y_n)$ and (7.8.32) we have

$$\frac{(F'(y_n), y_n)}{r^2} y_n = F'(y_n) - D(y_n) \to y.$$

Hence

$$y = \frac{1}{r^2}(y, x_0)x_0.$$

Since $y \neq o$, we have $x_0 \neq o$ and also $(y, x_0) \neq 0$. The definition of $D(y_n)$ and the fact that $D(y_n) \to o$ yield

$$y_n = r^2 \frac{F'(y_n) - D(y_n)}{(F'(y_n), y_n)} \to r^2 \frac{y}{(y, x_0)} = x_0.$$

Continuity of F' at x_0 then implies

$$y = F'(x_0), \qquad \text{i.e.,} \qquad F'(x_0) = \frac{(y, x_0)}{r^2} x_0 = \frac{(F'(x_0), x_0)}{r^2} x_0. \qquad \square$$

We will look for a curve on the sphere $\partial B(o; r)$ which starts at a fixed point x, the values of F along this curve do not decrease, and after a finite time (even if large) we "almost" reach the critical point of F. In other words, we are looking for a curve $k = k(t, x)$, $t \in [0, \infty)$, $x \in \partial B(o; r)$ such that

$$k(0, x) = x, \tag{7.8.34}$$

and for all $t \in (0, \infty)$ we require

$$k(t, x) \in \partial B(o; r), \qquad \text{i.e.,} \qquad \|k(t, x)\|^2 = r^2.$$

The last relation implies

$$\frac{\mathrm{d}}{\mathrm{d}t} \|k(t, x)\|^2 = 0,$$

which is equivalent to

$$\left(\frac{\mathrm{d}}{\mathrm{d}t} k(t, x), k(t, x) \right) = 0 \qquad \text{for all} \quad t \in (0, \infty). \tag{7.8.35}$$

The equality (7.8.35) states that for all $t \in (0, \infty)$ the element $\frac{\mathrm{d}}{\mathrm{d}t} k(t, x)$ is perpendicular to $k(t, x)$. This will be satisfied if we look for a solution of the initial value problem

$$\begin{cases} \dfrac{\mathrm{d}}{\mathrm{d}t} k(t, x) = D(k(t, x)), & t \in (0, \infty), \\ k(0, x) = x. \end{cases} \tag{7.8.36}$$

The assumption (7.8.27) implies that F' is Lipschitz continuous in a neighborhood of o. Hence, for $r > 0$ sufficiently small, D is Lipschitz continuous. Then, by virtue of Corollary 3.1.6, there exists a unique solution of (7.8.36) which is defined on the whole interval $(0, \infty)$. It follows from Remark 3.1.7 that this solution depends continuously on the initial condition $x \in \partial B(o; r)$.

Let k be a solution of the initial value problem (7.8.36). Then it has the following important properties:

(i) For any $t \in (0, \infty)$ we have

$$\|k(t, x)\| = \|x\|.$$

(ii) For any $t \in (0, \infty)$ we have

$$\frac{\mathrm{d}}{\mathrm{d}t} F(k(t, x)) = (F'(k(t, x)), D(k(t, x))) = \|D(k(t, x))\|^2 \geq 0.$$

In other words, the values of the functional F increase along k regardless of the choice of $x \in \partial B(o; r)$.

(iii) For any $t \in (0, \infty)$ we have

$$F(k(t, x)) = F(x) + \int_0^t \|D(k(\tau, x))\|^2 \, \mathrm{d}\tau.$$

Since F is bounded on $\partial B(o; r)$ (by the Mean Value Theorem and (7.8.28)), there exists a sequence $\{t_i\}_{i=1}^{\infty} \subset (0, \infty)$ such that

$$\lim_{i \to \infty} D(k(t_i, x)) = o.$$

(iv) Since $\{k(t_i, x)\}_{i=1}^{\infty}$ is bounded, we can select a weakly convergent subsequence.[85]

Summarizing, we have

Lemma 7.8.52. *For any $x \in \partial B(o; r)$ there exist a sequence $\{t_i\}_{i=1}^{\infty} \subset (0, \infty)$ and $x_0 \in H$ such that*

$$k(t_i, x) \rightharpoonup x_0, \tag{7.8.37}$$

$$D(k(t_i, x)) \to o, \tag{7.8.38}$$

$$\{F(k(t_i, x))\}_{i=1}^{\infty} \text{ is an increasing sequence.} \tag{7.8.39}$$

It follows from (7.8.37) and (7.8.26) that $F'(k(t_i, x)) \to y$. If we prove that $y \neq o$, then the assumptions of Lemma 7.8.51 are verified with $y_n = k(t_n, x)$, and so the existence of a solution x_0 of (7.8.29) with λ described by (7.8.33) will be proved.

By an appropriate choice of the initial condition $x \in \partial B(o; r)$, we show that the above convergence takes place and that λ given by (7.8.33) is sufficiently close to λ_0.

Recall that $A = F''(o)$ is a compact linear self-adjoint operator in the Hilbert space H (see Proposition 5.8.10). Its spectrum consists of a countable set of real eigenvalues with one possible limit point $\lambda = 0$. We split the set of all eigenvalues into the parts $\lambda \geq \lambda_0$ and $\lambda < \lambda_0$, respectively. We denote by H_1 and H_2, respectively, the corresponding closed linear subspaces generated by the eigenvectors (see Theorem 2.2.16). Note that $\lambda_0 > 0$ implies that $\dim H_1 < \infty$. The eigenspace associated with λ_0 will be denoted by H_0. Let P_1, P_2 be the orthogonal projections of H onto H_1, H_2, respectively (see Figure 7.8.10).

Let us denote

$$\mathcal{S}_1 = \{x \in H_1 : \|x\| = r\}.$$

Lemma 7.8.53. *There exists $r_0 > 0$ such that $\partial B(o; r_0) \subset \mathcal{U}(o)$ (see (7.8.25)), and for all $0 < r < r_0$ we have*

---

[85] The reader is invited to justify (i)–(iv).

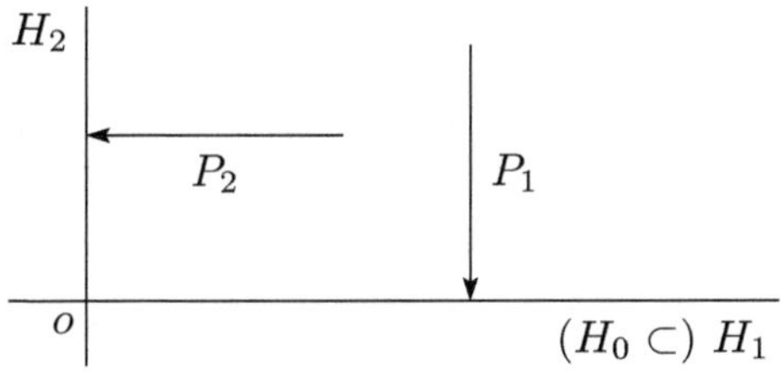

Figure 7.8.10.

(i) *there is no* $t \in [0, \infty)$ *for which the set* $k(t, \mathcal{S}_1)$ *is contractible to a point (see Definition* 5.1.8*) in*

$$R = \{x \in H : P_1 x \neq o\},$$

(ii) *for any* $t \in [0, \infty)$ *there exists* $x_t \in \mathcal{S}_1$ *such that*

$$P_1 k(t, x_t) \in H_0, \qquad i.e., \qquad k(t, x_t) \in H_0 \oplus H_2.$$

Proof. Lemma 5.1.11 and (i) imply (ii) (see Exercise 7.8.56). Hence we prove only (i). According to Lemma 5.1.9 it is sufficient to prove that for any t the set $\mathcal{S}_1$ is contractible into $k(t, \mathcal{S}_1)$ in R. Indeed, according to Lemma 5.1.10 the set $\mathcal{S}_1$ is not contractible to a point in R. Since k is a continuous function of both variables, it is sufficient to prove that it assumes only values from R: we want to prove that

$$P_1 k(t, x) \neq o \qquad \forall t \in [0, \infty), \quad x \in \mathcal{S}_1. \tag{7.8.40}$$

We have

$$F(k(0, x)) = F(x) \geq \frac{1}{2}(F''(o)x, x) - \varepsilon(\|x\|)\|x\|^2 \geq \left(\frac{1}{2}\lambda_0 - \varepsilon(\|x\|)\right)\|x\|^2$$

where $\varepsilon(r) \to 0$ as $r \to 0$ (see (7.8.28) and Proposition 3.2.27). Note that the last inequality holds due to $x \in H_1$. Since $F(k(t, x))$ is increasing in t, we conclude from here that

$$F(k(t, x)) \geq \left(\frac{1}{2}\lambda_0 - \varepsilon(r)\right) r^2. \tag{7.8.41}$$

On the other hand, we have an estimate from above (we write k instead of $k(t, x)$ for the sake of brevity):

$$\begin{aligned} F(k) &= \frac{1}{2}(F''(o)k, k) + \left[F(k) - \frac{1}{2}(F''(o)k, k)\right] \\ &\leq \frac{1}{2}(F''(o)P_1 k, P_1 k) + \frac{1}{2}(F''(o)P_2 k, P_2 k) + \varepsilon(\|k\|)\|k\|^2 \end{aligned}$$

(note that $(F''(o)P_1 k, P_2 k) = 0$ due to $H_1 \perp H_2$). Denote

$$\mu = \max\{\lambda : \lambda \in \sigma(F''(o))\}, \qquad \nu = \sup\{\lambda \in \sigma(F''(o)) : \lambda < \lambda_0\}.$$

Then

$$F(k) \le \frac{\mu}{2}\|P_1k\|^2 + \frac{\nu}{2}\|P_2k\|^2 + \varepsilon(\|k\|)\|k\|^2 = \frac{\nu}{2}\|k\|^2 + \frac{\mu-\nu}{2}\|P_1k\|^2 + \varepsilon(\|k\|)\|k\|^2.$$ [86]

Hence, due to the fact that $\|k\| = r$, we have

$$F(k) \le \frac{\nu}{2}r^2 + \frac{\mu-\nu}{2}\|P_1k\|^2 + \varepsilon(r)r^2. \tag{7.8.42}$$

It follows from (7.8.41) and (7.8.42) that

$$\|P_1k(t,x)\|^2 \ge \frac{\lambda_0 - \nu}{\mu-\nu}r^2 - \frac{4}{\mu-\nu}\varepsilon(r)r^2.$$

This implies the existence of r_0 such that

$$\|P_1k(t,x)\|^2 \ge ar^2 \qquad \text{for any} \quad r \le r_0 \qquad \text{where} \quad a = a(r_0) > 0. \tag{7.8.43}$$

This completes the proof of Lemma 7.8.53. □

Proof of Theorem 7.8.49.

Step 1. Let $t_n \to \infty$ be an arbitrary sequence of positive numbers. Let x_n be a point from $\mathcal{S}_1$ for which

$$P_1k(t_n, x_n) \in H_0$$

(its existence follows from (ii) of Lemma 7.8.53). Since $\mathcal{S}_1$ is compact, we can select a strongly convergent subsequence (denoted again by $\{x_n\}_{n=1}^\infty$) such that

$$\lim_{n\to\infty} x_n = \tilde{x}. \tag{7.8.44}$$

Step 2. It follows from Lemma 7.8.52 that there is a sequence $\{\tau_i\}_{i=1}^\infty$ such that

$$k(\tau_i, \tilde{x}) = y_i \rightharpoonup x_0 \qquad \text{in} \quad H,$$

and at the same time also

$$D(y_i) \to o.$$

Step 3. The compactness of F' implies that (passing again to a subsequence if necessary) there exists $y \in H$ such that

$$\lim_{i\to\infty} F'(y_i) = y.$$

We show that $y \ne o$. Indeed, we have

$$(F'(y_i), P_1y_i) \to (y, P_1x_0).$$

Also, for all $i \in \mathbb{N}$, we have the estimate

$$\begin{aligned}(F'(y_i), P_1y_i) &= (F''(o)y_i, P_1y_i) + (F'(y_i) - F''(o)y_i, P_1y_i)\\ &\ge \lambda_0\|P_1y_i\|^2 - \varepsilon(\|y_i\|)\|y_i\|^2 \ge \frac{1}{2}\lambda_0 ar^2\end{aligned}$$

for all r small enough due to (7.8.43). This immediately implies

$$(y, P_1x_0) \ne 0, \qquad \text{and so} \quad y \ne o,\ x_0 \ne o.$$

[86] We use the identity $\|P_1k\|^2 + \|P_2k\|^2 = \|k\|^2$.

Step 4. We have just verified the assumptions of Lemma 7.8.51. Hence $y_i \to x_0$ in H, and x_0 solves (7.8.29) with λ given by (7.8.33):

$$\lambda x_0 - F'(x_0) = o, \qquad \lambda = \frac{1}{r^2}(F'(x_0), x_0).$$

Step 5. The last step consists in proving the fact that for $r > 0$ small enough λ is arbitrarily close to λ_0. Let us estimate

$$\begin{aligned}|\lambda - \lambda_0| &= \frac{1}{r^2}|(F'(x_0), x_0) - \lambda_0(x_0, x_0)| \\ &\le \frac{1}{r^2}\left[|(F'(x_0) - F''(o)x_0, x_0)| + 2\left|\left(\frac{1}{2}F''(o)x_0, x_0\right) - F(x_0)\right| + |2F(x_0) - \lambda_0(x_0, x_0)|\right] \\ &= \frac{1}{r^2}|2F(x_0) - \lambda_0(x_0, x_0)| + \varepsilon(r).\end{aligned}$$

Since $\varepsilon(r) \to 0$ as $r \to 0$, it suffices to estimate

$$\frac{1}{r^2}|2F(x_0) - \lambda_0(x_0, x_0)|.$$

The continuity of F implies

$$F(x_0) = \lim_{i\to\infty} F(y_i). \tag{7.8.45}$$

Since F is increasing along k, we also have

$$F(y_i) = F(k(\tau_i, \tilde{x})) \ge F(\tilde{x}). \tag{7.8.46}$$

Since $\tilde{x} \in \mathcal{S}_1$,

$$\left(\frac{1}{2}F''(o)\tilde{x}, \tilde{x}\right) \ge \frac{\lambda_0}{2}r^2. \tag{7.8.47}$$

Then (7.8.45)–(7.8.47) imply an estimate from below:

$$F(x_0) \ge \left(\frac{\lambda_0}{2} - \varepsilon(r)\right) r^2. \tag{7.8.48}$$

Now we derive an estimate from above for $F(x_0)$. Since $x_n \to \tilde{x}$ and $k(\tau_i, \cdot)$ is continuous with respect to the second variable, for fixed $i \in \mathbb{N}$ we have

$$k(\tau_i, x_n) \to k(\tau_i, \tilde{x}) = y_i.$$

The continuity of F implies that for fixed $i \in \mathbb{N}$ and $r > 0$ there exists $n_0 \in \mathbb{N}$ such that for all $n \ge n_0$ we have

$$F(y_i) \le F(k(\tau_i, x_n)) + r^3. \tag{7.8.49}$$

However, for any fixed $i \in \mathbb{N}$ we find $n_i \ge n_0$ such that $t_{n_i} > \tau_i$, and the monotonicity of F along k then implies

$$F(k(\tau_i, x_{n_i})) \le F(k(t_{n_i}, x_{n_i})). \tag{7.8.50}$$

The choice of x_n from Step 1 guarantees that $k(t_{n_i}, x_{n_i}) \in H_0 \oplus H_2$, and so (writing k_i instead of $k(t_{n_i}, x_{n_i})$) we have the estimate

$$F(k_i) \le \frac{1}{2}(F''(o)k_i, k_i) + \varepsilon(\|k_i\|)\|k_i\|^2 \le \left(\frac{\lambda_0}{2} + \varepsilon(r)\right) r^2. \tag{7.8.51}$$

However, (7.8.45), (7.8.49) and (7.8.50) reduce (7.8.51) to

$$F(x_0) \le \left(\frac{\lambda_0}{2} + \varepsilon(r)\right) r^2. \tag{7.8.52}$$

Both the estimates (7.8.48) and (7.8.52) yield that

$$\frac{1}{r^2}|2F(x_0) - \lambda_0(x_0, x_0)| \to 0 \qquad \text{as} \quad r \to 0.$$

This completes the proof of Theorem 7.8.49. □

Remark 7.8.54. It follows from the Krasnoselski Potential Bifurcation Theorem that every point (λ_0, o) where λ_0 is a nonzero eigenvalue of the operator A is a bifurcation point. But there is no warranty that there is a curve (or continuum) of nontrivial solutions which departs from (λ_0, o). In fact, there are counterexamples even in the finite dimension which prove that such a curve need not exist.

Böhme [16] gave an example of a real function of two independent real variables, $F \in C^\infty(\mathbb{R}^2)$, for which $(\lambda_0, (0,0))$ is a bifurcation point of

$$f(z,\lambda) = \lambda z - F'(z) = o, \qquad z = (x,y) \in \mathbb{R} \times \mathbb{R}, \qquad \lambda \in \mathbb{R} \tag{7.8.53}$$

and there is no continuous curve of nontrivial solutions of (7.8.53) which contains the point $(\lambda_0, (0,0))$.

Example 7.8.55 (Application of the Krasnoselski Potential Bifurcation Theorem). We will consider a periodic problem similar to that studied in Example 4.3.25:

$$\begin{cases} \ddot{x}(t) + \lambda x(t) + g(\lambda, t, x(t)) = 0, & t \in (0, 2\pi), \\ x(0) = x(2\pi), \ \dot{x}(0) = \dot{x}(2\pi). \end{cases} \tag{7.8.54}$$

The difference between (7.8.54) and (4.3.12) consists in the fact that now we do not allow g to depend on $\dot{x}$. The reason for this restriction consists in the fact that the boundary value problem (4.3.12) cannot be written in the form (7.8.60) if g depends on $\dot{x}$.

We simplify the situation even more and write g in the form

$$g(\lambda, t, s) = (\lambda + 1)\tilde{g}(t, s).$$

Set

$$\tilde{G}(t,s) = \int_0^s \tilde{g}(t,\tau)\,\mathrm{d}\tau = \int_0^1 \tilde{g}(t, s\sigma)s\,\mathrm{d}\sigma,$$

i.e., $\tilde{G}$ is the primitive of $\tilde{g}$ with respect to the second variable s.

Put

$$F(x) = \int_0^{2\pi} \left[\frac{1}{2}|x(t)|^2 + \tilde{G}(t, x(t))\right] \mathrm{d}t. \tag{7.8.55}$$

We work in the Hilbert space

$$H := \{x \in W^{1,2}(0, 2\pi) : x(0) = x(2\pi)\} \tag{7.8.56}$$

with the scalar product on H given by

$$(x,y) = \int_0^{2\pi} [\dot{x}(t)\dot{y}(t) + x(t)y(t)]\,\mathrm{d}t, \qquad x, y \in H.$$

Then

$$(F'(x), y) = \int_0^{2\pi} [x(t)y(t) + \tilde{g}(t, x(t))y(t)]\,\mathrm{d}t \qquad \text{for any} \quad x, y \in H. \tag{7.8.57}$$

A *weak solution* of the periodic problem is a function $x \in H$ which satisfies the integral identity

$$\int_0^{2\pi} [\dot{x}(t)\dot{y}(t) - \lambda x(t)y(t) - (\lambda+1)\tilde{g}(t, x(t))y(t)]\,\mathrm{d}t = 0 \tag{7.8.58}$$

for any $y \in H$. The last equality (7.8.58) can be written as

$$\int_0^{2\pi} [\dot{x}(t)\dot{y}(t) + x(t)y(t) - (\lambda+1)x(t)y(t) - (\lambda+1)\tilde{g}(t, x(t))y(t)]\,\mathrm{d}t = 0. \tag{7.8.59}$$

The integral identity (7.8.59) can be written for $\lambda \neq -1$ as the operator equation

$$\mu x - F'(x) = o \qquad \text{where} \quad \mu = \frac{1}{\lambda+1}. \tag{7.8.60}$$

Let us define an operator $B\colon H \to H$ by

$$(B(x), y)_H = \int_0^{2\pi} x(y)y(t)\,\mathrm{d}t.$$

It follows easily that B is a bounded linear operator and the compact embedding

$$H \hookrightarrow\hookrightarrow Y := \{x \in C[0, 2\pi] : x(0) = x(2\pi)\}$$

(see Theorem 1.2.28) yields that B is compact. Since n^2 is an eigenvalue of

$$\begin{cases} \ddot{x}(t) + \lambda x(t) = 0, & t \in (0, 2\pi), \\ x(0) = x(2\pi), \ \dot{x}(0) = \dot{x}(2\pi), \end{cases}$$

then $\mu = \frac{1}{n^2+1}$ is an eigenvalue of B. We make the following assumptions:

$$\tilde{g}, \frac{\partial \tilde{g}}{\partial s} : \mathbb{R} \times \mathbb{R} \to \mathbb{R} \text{ are continuous functions,} \tag{7.8.61}$$

$$\tilde{g}(t, 0) = 0, \quad \frac{\partial \tilde{g}}{\partial s}(t, 0) = 0 \qquad \text{for all} \quad t \in \mathbb{R}. \tag{7.8.62}$$

Now we prove that F verifies the assumptions of Theorem 7.8.49: Note that F can be written as

$$F(x) = \frac{1}{2}\int_0^{2\pi} |x(t)|^2\,\mathrm{d}t + \int_0^{2\pi} \left[\int_0^1 \tilde{g}(t, sx(t))x(t)\,\mathrm{d}s\right]\,\mathrm{d}t. \tag{7.8.63}$$

(i) $F(o) = 0$ is an immediate consequence of (7.8.62).

(ii) *Differentiability of* F follows directly from (7.8.63).

(iii) *Compactness* of $F'(x)$. This is a consequence of the compactness of the embedding $H \hookrightarrow\hookrightarrow Y$ (cf. Exercise 7.8.58).

(iv) $F'(o) = o$ is a consequence of (7.8.62).

(v) $F''(o) = B$ and F'' is continuous at o (cf. Exercise 7.8.58).

Theorem 7.8.49 now implies that every point $\left(\frac{1}{n^2+1}, o\right)$ is a bifurcation point of the equation

$$\mu x - F'(x) = o.$$

In other words, for any $n = 0, 1, \dots$ we have the following assertion:

Under the assumptions (7.8.61) and (7.8.62), for an arbitrarily small neighborhood $\mathcal{U}$ of the point $(n^2, o) \in \mathbb{R} \times H$ there exists $(\lambda, x) \in \mathcal{U}$ such that $x \neq o$ is a weak solution of the periodic problem

$$\begin{cases} \ddot{x}(t) + \lambda x(t) + (\lambda + 1)\tilde{g}(t, x(t)) = 0, & t \in (0, 2\pi), \\ x(0) = x(2\pi), \ \dot{x}(0) = \dot{x}(2\pi). \end{cases}$$

Note that the continuity of $\tilde{g}$ and the regularity argument imply that every such nontrivial solution satisfies $x \in C^2[0, 2\pi]$ and $\dot{x}(0) = \dot{x}(2\pi)$. ○

Exercise 7.8.56. Prove that Lemma 5.1.11 and Lemma 7.8.53(i) imply the statement of Lemma 7.8.53(ii).

Hint. Argue by contradiction.

Exercise 7.8.57. Prove that H defined in Example 7.8.55 by (7.8.56) is a closed subspace of $W^{1,2}(0, 2\pi)$, i.e., H is a Hilbert space.

Exercise 7.8.58. Prove that F from Example 7.8.55 is twice Fréchet differentiable, F' compact and that F'' is continuous at o.

Exercise 7.8.59. Apply Theorem 7.8.49 to the Dirichlet and the Neumann boundary value problem.

Chapter 8

Some Applications to Partial Differential Equations

8.1 Linear Evolution Equations and Semigroups

We remind the reader that e^{tA} was constructed in Exercise 3.1.28(ii) provided A was a densely defined self-adjoint operator with a compact resolvent and with the spectrum bounded from above. In this section we generalize e^{tA} for a rather general class of linear operators $A : X \to X$ on a Banach space X.

Definition 8.1.1. Let $A : \operatorname{Dom} A \subset X \to X$ be a linear operator. Then A is said to be

(1) *dissipative*, provided

$$\|\lambda x - Ax\|_X \geq \lambda \|x\|_X \text{ for all } x \in \operatorname{Dom} A \text{ and } \lambda > 0,$$

(2) *m-dissipative*, provided it is dissipative and $\lambda I - A$ is surjective for all $\lambda > 0$.

Remark 8.1.2.

(i) If X is a Hilbert space then a linear operator A is dissipative if and only if

$$\operatorname{Re}(Ax, x)_X \leq 0$$

(see Exercise 8.1.10). For example, if A is as in Exercise 3.1.28 and $u(\cdot)$ is a solution of the initial value problem (3.1.17) then

$$\frac{1}{2}\frac{d}{dt}\|u(t)\|_X^2 = (Au(t), u(t))_X$$

and the dissipativity means that "the energy" $\|u(\cdot)\|^2$ is non-increasing.

(ii) If A is dissipative and there exists $\lambda_0 > 0$ such that $\operatorname{Im}(\lambda_0 I - A) = X$ then A is actually m-dissipative (see Exercise 8.1.11).

(iii) If A is a densely defined dissipative operator then A is closable and its closure $\overline{A}$ is dissipative.

(iv) If A is an m-dissipative operator then it is closed.

Example 8.1.3. We will introduce two different functional settings in which the Laplacian Δ is an m-dissipative operator.

(i) We set $X = L^2(\Omega)$ where Ω is an open set in $\mathbb{R}^N$ (bounded or unbounded) and define

$$\operatorname{Dom} A := \{u \in W_0^{1,2}(\Omega) : \Delta u \in X\}.$$

Here we understand Δ in the distributional sense, i.e., $\Delta u = w \in X$ if and only if

$$\int_\Omega w(x)\varphi(x)dx = \int_\Omega u(x)\Delta\varphi(x)dx$$

for all $\varphi \in \mathcal{D}(\Omega)$.
We put

$$Au = \Delta u \quad \text{for} \quad u \in \operatorname{Dom} A.$$

Since $\mathcal{D}(\Omega) \subset \operatorname{Dom} A$ then A is densely defined in X and it is dissipative by Remark 8.1.2(i). Indeed, for $u \in \operatorname{Dom} A$, $\varphi \in \mathcal{D}(\Omega)$ we have

$$\int_\Omega \Delta u(x)\varphi(x)dx = -\int_\Omega \nabla u(x)\nabla\varphi(x)dx.$$

By the density argument this identity holds also for all $\varphi \in W_0^{1,2}(\Omega)$. In particular, for $\varphi = u$ we get $(Au, u) \le 0$.

In order to show that A is m-dissipative, we apply Remark 8.1.2(ii) and show that the equation

$$u - Au = f$$

has a solution for any $f \in X$. For this purpose we define a bilinear form

$$a(u,v) := \int_\Omega (\nabla u(x)\nabla v(x) + u(x)v(x))\, dx,$$

$u, v \in Y := W_0^{1,2}(\Omega)$. This form is continuous on Y and we have

$$a(u,u) = \|u\|_Y^2.$$

Let $f \in X$. Then the linear form

$$v \mapsto (f, v)_X$$

is continuous on Y by the embedding $Y \hookrightarrow X$. The Riesz Representation Theorem (Theorem 1.2.40) yields the existence of $F \in Y$ such that

$$(f, v)_X = (F, v)_Y \quad \text{for all} \quad v \in Y.$$

By the Lax-Milgram Theorem (Proposition 1.2.41) there is a continuously invertible $B \in \mathcal{L}(Y)$ with $\|B^{-1}\|_{\mathcal{L}(Y)} \leq 1$ and $u \in Y$ such that

$$a(u, v) = (Bu, v)_Y = (F, v)_Y = (f, v)_X$$

for all $v \in Y$, in particular for $v \in \mathcal{D}(\Omega)$. This means that in the distributional sense we have

$$\Delta u = u - f \in X, \text{i.e.,} u \in \operatorname{Dom} \Delta \quad \text{and}$$

$$u - Au = f.$$

Since $f \in X$ was arbitrary, we have shown what we wanted.

The choice $Y = W_0^{1,2}(\Omega)$ corresponds to the Dirichlet boundary conditions (cf. the Trace Theorem, Theorem 5.5.1). We can also prove that A is a non-negative self-adjoint operator on X. Moreover, A has a compact resolvent if Ω is bounded and its boundary is locally Lipschitz continuous (cf. Theorem 1.2.28).

We note that for bounded Ω with C^2-boundary it can be proved that $\operatorname{Dom} A = W^{2,2}(\Omega) \cap W_0^{1,2}(\Omega)$ (see Brézis [20, Theoréme IX.25]).

(ii) Let Ω be a bounded open set in $\mathbb{R}^N$ with a Lipschitz continuous boundary $\partial\Omega$. Set $Z := C_0(\Omega) = \{u \in C(\overline{\Omega}) : u|_{\partial\Omega} = 0\}$, $\operatorname{Dom} \widetilde{A} := \{u \in Z \cap W_0^{1,2}(\Omega) : \Delta u \in C(\Omega)\}$ and $\widetilde{A}u = \Delta u$, $u \in \operatorname{Dom} \widetilde{A}$. Since $\mathcal{D}(\Omega)$ is a dense subset of Z, the operator $\widetilde{A}$ is densely defined. In order to show that $\tilde{A}$ is m-dissipative we need an estimate of a solution of the equation

$$u - Au = f$$

for a given $f \in Z$. We have just seen that this equation has a solution $u \in \operatorname{Dom} A$. Since there is no embedding $W^{1,2}(\Omega) \hookrightarrow C(\overline{\Omega})$ in higher dimensions, we first prove that $u \in L^\infty(\Omega)$. Let $k := \|f\|_Z$ and define $v := (u - k)^+$. It follows from Gilbarg & Trudinger [71, Section 7.4] (cf. also proof of Theorem 7.5.1) that $v \in W^{1,2}(\Omega)$ and

$$\nabla v(x) = \begin{cases} \nabla u(x) & \text{if} \quad u(x) \geq k, \\ 0 & \text{if} \quad u(x) < k, \end{cases}$$

a.e. in Ω. Since u is a solution in $\operatorname{Dom} A$, we have

$$\int_\Omega \nabla u(x) \nabla v(x) dx + \int_\Omega u(x) v(x) dx = \int_\Omega f(x) v(x) dx,$$

i.e., also

$$\int_\Omega \nabla u(x)\nabla v(x)dx + \int_\Omega (u(x)-k)\,v(x)dx = \int_\Omega (f(x)-k)\,v(x)dx.$$

Since the first integral on the left-hand side is non-negative and the integral on the right-hand side is non-positive (by the choice of k), we deduce

$$\int_\Omega v^2(x)dx \le 0,$$

i.e.,

$$u(x) \le \|f\|_Z \quad \text{a.e. in } \Omega.$$

Similarly, we derive

$$u(x) \ge -\|f\|_Z \text{ a.e. in } \Omega, \text{ i.e., } u \in L^\infty(\Omega) \text{ and } \|u\|_\infty \le \|f\|_Z.$$

Now, we use a nontrivial fact that $L^\infty(\Omega)\cap W_0^{1,2}(\Omega)$ is continuously embedded into $C_0(\Omega)$ if Ω has a Lipschitz boundary (see, e.g., Gilbarg & Trudinger [71, Theorem 8.30]. We thus arrive at an estimate

$$\|u\|_Z = \|u\|_{L^\infty(\Omega)} \le \|f\|_Z.$$ ○

Definition 8.1.4. A system $T(\cdot) = \{T(t)\}_{t\ge 0}$ of linear continuous operators on a Banach space X is called a C_0-*semigroup* if the following conditions are satisfied:

(S1) $T(0) = I$,

(S2) $T(t+s) = T(t)T(s) = T(s)T(t)$ for all $t, s \ge 0$,

(S3) $\lim\limits_{t\to 0_+} \|T(t)x - x\|_X = 0$ for all $x \in X$.

A semigroup satisfying $\|T(t)\|_{\mathcal{L}(X)} \le 1$ for all $t \ge 0$ is called contractive.

The operator $\Lambda : X \to X$ with domain

$$\operatorname{Dom}\Lambda := \left\{ x \in X : \exists \lim_{t\to 0+} \frac{T(t)x - x}{t} \in X \right\}$$

given by

$$\Lambda x := \lim_{t\to 0+} \frac{T(t)x - x}{t}$$

is called a *generator* of the semigroup $T(\cdot)$.

Below we summarize some basic facts about C_0-semigroups. For more details see Engel & Nagel [58] and Pazy [120].

Proposition 8.1.5. *Let $T(\cdot)$ be a C_0-semigroup on a Banach space X and let Λ be its generator.*

(i) $\exists \omega \in \mathbb{R}\, \exists c > 0\, \forall t \geq 0 : \|T(t)\| \leq ce^{\omega t}$.

(ii) *For any* $x \in X$ *function* $u(t) := T(t)x$ *belongs to* $C([0,\infty); X)$.

(iii) *If* $x \in \operatorname{Dom}\Lambda$ *then* $u \in C^1([0,\infty); X)$ *and* $u = u(t)$ *is a solution of the equation*

$$\dot{u}(t) = \Lambda u(t),\ u(0) = x;$$

actually $u(t) = T(t)x$ *is a unique solution of this equation.*

(iv) *If* $\operatorname{Re}\lambda > w$ *then* $\lambda \in \rho(\Lambda)$ *and*

$$(\lambda I - \Lambda)^{-1}x = \int_0^{+\infty} e^{-\lambda t}T(t)x\,dt.$$

(v) *The generator* Λ *is closed and densely defined.*

(vi) $\int_0^t T(s)x\,ds \in \operatorname{Dom}\Lambda$ *and* $\Lambda \int_0^t T(s)x\,ds = T(t)x - x$ *for any* $x \in X$ *and* $t \geq 0$.

Proof (a sketch).

(i) From the Banach-Steinhaus Theorem (Corollary 2.1.6) and the property $(S3)$ it follows that $\|T(t)\|$ is bounded uniformly on some interval $[0,\delta]$. The rest of the proof follows from the semigroup property $(S2)$.

(ii) The continuity from the right follows directly from $(S2)$ and $(S3)$. In order to prove the continuity from the left, we first write

$$T(t)x - T(t-h)x = T(t-h)\left[T(h)x - x\right]$$

and then use (i).

(iii) Notice that $T(t)\Lambda x = \Lambda T(t)x$ for $x \in \operatorname{Dom}\Lambda$. The proof is similar to that of (ii). For uniqueness see Exercise 8.1.12.

(iv) The idea of the proof is similar to the special case stated in Exercise 3.1.28.

(v) and (vi) The closedness of Λ follows from the fact that $\rho(\Lambda) \neq \phi$ (see property (iv)).

Let $y := \int_0^t T(s)x\,ds$. Then $\frac{T(\sigma)y - y}{\sigma} = \frac{1}{\sigma}\int_t^{t+\sigma} T(\xi)x d\xi - \frac{1}{\sigma}\int_0^{\sigma} T(\xi)x d\xi$. From (ii) it follows that $\frac{T(\sigma)y-y}{\sigma} \to T(t)x - x$ as $\sigma \to 0_+$, i.e., $y \in \operatorname{Dom}\Lambda$, $\Lambda y = T(t)x - x$. Moreover, $\frac{1}{t}y \to x$ for $t \to 0_+$, i.e., $\operatorname{Dom}\Lambda$ is dense in X. □

The next theorem states the basic existence result.

Theorem 8.1.6 (Lumer–Philips).

(i) *Let A be a densely defined m-dissipative operator on a Banach space X. Then there exists a C_0-semigroup $T(\cdot)$ on X with generator A. Moreover, T is a contractive semigroup.*

(ii) *If $T(\cdot)$ is a contractive semigroup on a Banach space X, then its generator is a densely defined m-dissipative operator.*

Proof.

(i) This proof is due to K. Yosida and has an approximative character. For $n \in \mathbb{N}$ we set

$$J_n x := n(nI - A)^{-1}x, \quad x \in X.$$

We have

$$J_n(x) \in \operatorname{Dom} A \text{ and by dissipativity also } \|J_n\| \le 1.$$

Note that for any $x \in X$, $\lim_{n\to\infty} \|J_n(x) - x\| = 0$.
Indeed, for $x \in \operatorname{Dom} A$ we can write

$$\begin{aligned} J_n x &= n(n\,I - A)^{-1}x = (n\,I - A)^{-1}(n\,I - A + A)x \\ &= (n\,I - A)^{-1}(n\,I - A)x + (n\,I - A)^{-1}Ax \end{aligned}$$

and

$$\|(n\,I - A)^{-1}\| \le \frac{1}{n}$$

by the dissipativity of A. We have

$$\|J_n(x) - x\| \le \|(n\,I - A)^{-1}\|\|Ax\| \to 0 \quad \text{as} \quad n \to \infty.$$

For $x \in X$ we prove $J_n x \to x$ as $n \to \infty$ using the above fact and the density of $\operatorname{Dom} A$ in X.

The continuous linear operator $A_n : X \to X$ defined by

$$A_n x := n(J_n x - x), \quad x \in X$$

is called *Yosida's approximation* of the operator A.

We have the following equivalent expression for A_n and $x \in \operatorname{Dom} A$,

$$\begin{aligned} A_n x &= n(J_n x - x) = n\left(n(nI - A)^{-1}x - x\right) \\ &= n(nI - A)^{-1}\left(nx - (nI - A)x\right) = n(nI - A)^{-1}Ax \\ &= J_n Ax. \end{aligned}$$

Then using above convergence $J_n x \to x$ for any $x \in X$ yields

$$A_n x \to Ax \quad \text{for} \quad x \in \operatorname{Dom} A.$$

For any $n \in \mathbb{N}$,

$$e^{tA_n} x = e^{-nt} \cdot e^{ntJ_n} x = e^{-nt} \sum_{k=0}^{\infty} \frac{n^k t^k}{k!} J_n^k x$$

forms a C_0-semigroup with generator A_n and the following estimate holds:

$$\|e^{tA_n} x\| \le e^{-nt} \cdot e^{nt} \|x\| = \|x\|.$$

The key idea of the proof is to show that the system $\{e^{tA_n}x\}_{n=1}^{\infty}$converges to the desired C_0-semigroup $T(t)x$. For this purpose we prove that for any fixed $x \in X$, $\left\{e^{tA_n}x\right\}_{n=1}^{\infty}$ is a Cauchy sequence. Indeed, for $n, m \in \mathbb{N}$ we put

$$w(s) = e^{sA_n} \cdot e^{(t-s)A_m} x.$$

Then we have

$$\begin{aligned}
\|e^{tA_n}x - e^{tA_m}x\| &= \|w(t) - w(0)\| = \|\int_0^t \dot{w}(s)ds\| \\
&\le \int_0^t \|\dot{w}(s)\|ds = \int_0^t \|e^{sA_n} e^{(t-s)A_m}(A_n - A_m)x\|ds \\
&\le t\|A_n x - A_m x\|.
\end{aligned}$$

Therefore $\{e^{tA_n}x\}_{n=1}^{\infty}$ converges locally uniformly in $[0, +\infty)$ for $x \in \operatorname{Dom} A$. The convergence for arbitrary $x \in X$ follows from the density of $\operatorname{Dom} A$ and the contractivity of e^{tA_n}. If we denote its limit as $T(t)x$, it is not difficult to prove that $T(\cdot)$ is a contractive C_0-semigroup.

Let us denote by Λ the generator of $T(\cdot)$. We prove that $\Lambda = A$. The idea of proving $\operatorname{Dom} A \subset \operatorname{Dom} \Lambda$ is to consider the equation

$$\dot{u}(t) = Au(t), \quad u(0) = x \in \operatorname{Dom} A$$

and to show that $T(t)x$ is its solution. Indeed, let $u_n(t) = e^{tA_n}x$. We get $\dot{u}_n(t) = A_n e^{tA_n} x$ and

$$\|\dot{u}_n(t) - T(t)Ax\| \le \| \left(e^{tA_n} - T(t)\right) Ax\| + \|e^{tA_n}(A_n x - Ax)\| \to 0$$

locally uniformly in $[0, +\infty)$. Hence

$$u_n(t) = x + \int_0^t \dot{u}_n(s)ds \to u(t) := T(t)x = x + \int_0^t T(s)Axds$$

and

$$\dot{u}(t) = T(t)Ax.$$

It follows that

$$\frac{d}{dt}T(t)x = T(t)Ax,$$

and, in particular, for $t = 0$ we get that

$$\Lambda x = T(0)Ax = Ax. \tag{8.1.1}$$

Hence, taking $x \in \operatorname{Dom} A$ we have proved that $x \in \operatorname{Dom} \Lambda$ and (8.1.1) holds.

Let now $x \in \operatorname{Dom} \Lambda$ be arbitrary. Set $y := \lambda x - \Lambda x$ and take λ so large that both equations

$$\lambda x - Ax = y \quad \text{and} \quad \lambda x - \Lambda x = y$$

have a unique solution (see the dissipativity of A and Proposition 8.1.5 (iv)). Then $x \in \operatorname{Dom} A$ follows, i.e., $\operatorname{Dom} \Lambda \subset \operatorname{Dom} A$ and $\Lambda x = Ax$, $x \in \operatorname{Dom} \Lambda$.

(ii) This statement is a consequence of Proposition 8.1.5 (iv) and (v). □

Let us consider the initial value problem for an *abstract differential equation*

$$\dot{u} = Au + f, \quad u(0) = x_0. \tag{8.1.2}$$

We assume that A is a generator of C_0-semigroup $T(\cdot)$ on a Banach space X, $f : [0, T) \to X$ is Bochner integrable function on an interval $(0, T)$ (we write $f \in L(0, T; X)$). As in the finite-dimensional case suitable candidate for a solution of (8.1.2) is given by the variation of constants formula (cf. Exercise 3.1.28(vi))

$$u(t) = T(t)x_0 + \int_0^t T(t-s)f(s)ds, \quad t \in [0, T). \tag{8.1.3}$$

The fundamental question here is:

"How do we define a solution of (8.1.2)?"

It follows from Proposition 8.1.5 (iii) that the first term on the right-hand side of (8.1.3) is a C^1-solution of the homogeneous equation $\dot{u} = Au$, $u(0) = x_0$, provided $x_0 \in \operatorname{Dom} A$. We are left with the question of how to interpret the second (convolution) term. Under our assumption on f, the function

$$s \mapsto T(t-s)f(s), \quad s \in (0, t)$$

is strongly measurable (see Definition 3.1.9) for any $t \in [0, T)$. By the Bochner Theorem (Proposition 3.1.11) the value of the integral term is in X, and, since $T(\cdot)$ is strongly continuous, it belongs to $C([0, T); X)$. This motivates the following definition.

Definition 8.1.7.

(1) A function $u \in C([0,T);X)$ given by (8.1.3) is called a *mild solution* of the problem (8.1.2).

(2) A function $u \in C^1([0,T);X)$ such that $u(0) = x_0$, $\dot{u}(t) \in X$, $u(t) \in \operatorname{Dom} A$ for all $t \in [0,T)$ and the equation (8.1.2) is satisfied for every $t \in [0,T)$ is called a *classical solution* of the problem (8.1.2).

The following two assertions show up the relation between the above notions.

Proposition 8.1.8. *A classical solution of* (8.1.2) *is unique and it is also a mild solution.*

Proof. Let u be a classical solution of (8.1.2) on $[0,T)$. Integrating the equation we get

$$u(t) - x_0 = \int_0^t Au(s)ds + \int_0^t f(s)ds = A\int_0^t u(s)ds + \int_0^t f(s)ds, \quad t \in [0,T).$$

Note that the second equality is due to Proposition 3.1.12 (ii). Set

$$w(s) := T(t-s)\int_0^s u(\sigma)d\sigma \quad \text{for} \quad 0 \le s \le t \le T.$$

Then

$$\begin{aligned}\int_0^t u(\sigma)d\sigma &= w(t) - w(0) = \int_0^t \dot{w}(s)ds \\ &= \int_0^t \left[-T(t-s)A\int_0^s u(\sigma)d\sigma + T(t-s)u(s)\right] ds \\ &= \int_0^t T(t-s)\left[x_0 + \int_0^s f(\sigma)d\sigma\right] ds = \int_0^t T(s)\left[x_0 + \int_0^{t-s} f(\sigma)d\sigma\right] ds.\end{aligned}$$

Taking the derivative with respect to t (Exercise 3.1.25), we get

$$u(t) = T(t)x_0 + \int_0^t T(s)f(t-s)ds.$$

In particular, a classical solution is uniquely determined. □

In order to build up a satisfactory theory we should give some sufficient condition for a mild solution to be a classical one. It is obvious that this concerns the convolution term.

Proposition 8.1.9. *If* $x \in \operatorname{Dom} A$ *and* $f \in W^{1,1}(0,T;X)$ *then the mild solution given by* (8.1.3) *is a classical solution of* (8.1.3).

Proof. According to Proposition 3.1.18 we have

$$f(t) = f(0) + \int_0^t f'(s)ds, \quad t \in [0,\tau].$$

Notice that for a uniformly continuous function f there exist limits at 0_+ and τ_-. We can put this expression into the convolution term to get

$$g(t) = \int_0^t T(t-s)f(s)ds = \int_0^t T(\sigma)f(0)d\sigma + \int_0^t T(t-s)\left(\int_0^s f'(\sigma)d\sigma\right)ds.$$

We denote the first summand on the right-hand side by $g_1(t)$ and the second one by $g_2(t)$. By Proposition 8.1.5 (vi), $g_1 \in \operatorname{Dom} A$ and

$$Ag_1(t) = T(t)f(0) - f(0), \quad t \in [0,\tau].$$

Moreover, $g_1'(t) = T(t)f(0)$. For the term g_2 we first use the Fubini Theorem:

$$g_2(t) = \int_0^t \left(\int_0^{t-\sigma} T(\xi)f'(\sigma)d\xi\right)d\sigma.$$

But $\int_0^{t-\sigma} T(\xi)f'(\sigma)d\xi \in \operatorname{Dom} A$ and, by Proposition 8.1.5(vi),

$$A\int_0^{t-\sigma} T(\xi)f'(\sigma)d\xi = T(t-\sigma)f'(\sigma) - f'(\sigma).$$

Since $T(t-\cdot)f'(\cdot) - f'(\cdot) \in L^1(0,t;X)$, Proposition 3.1.12(ii) yields

$$g_2(t) \in \operatorname{Dom} A \text{ and } Ag_2(t) = \int_0^t T(t-\sigma)f'(\sigma)d\sigma - f(t) + f(0).$$

Furthermore,

$$g_2'(t) = \frac{d}{dt}\int_0^t T(s)\left(\int_0^{t-s} f'(\sigma)d\sigma\right)ds = \int_0^t T(s)f'(t-s)ds$$

(see Exercise 3.1.25). Putting all partial results stated above together, we get

$$g'(t) = Ag(t) + f(t), \quad g(0) = 0. \qquad \square$$

Exercise 8.1.10. Let A be a linear operator on a Hilbert space H. Prove that A is dissipative if and only if

$$\operatorname{Re}(Ax, x)_H \le 0 \quad \text{for all} \quad x \in \operatorname{Dom} A.$$

We note that this is true in a general Banach space provided the scalar product is replaced by the duality mapping.

Exercise 8.1.11. Let A be a dissipative operator on a Hilbert space H. Prove that A is m-dissipative provided $\operatorname{Im}(\lambda_0 I - A) = H$ for some $\lambda_0 > 0$.

Hint. Notice that $\lambda_0 \in \varrho(A)$ and use the Contraction Principle to solve $\lambda x - Ax = y$ for $\lambda \in (0, 2\lambda_0)$.

Exercise 8.1.12. Let Λ be the generator of a C_0-semigroup $T(\cdot)$ on a Banach space X. Prove that for $x \in \operatorname{Dom} \Lambda$ there is a unique classical solution to the equation $\dot{u} = \Lambda u$ satisfying the initial condition $u(o) = x$.

Hint. The existence is shown in Proposition 8.1.5 (iii). To see the uniqueness assume that u_1, u_2 are two classical solutions and put $r(s) = T(t-s)[u_1(s) - u_2(s)]$, $s \in [0, t]$. Prove that $\dot{r} \equiv 0$.

8.2 Semilinear Evolution Equations

The notion of a mild solution is crucial for solving semilinear equation

$$\dot{u}(t) = Au(t) + f(t, u(t)), \quad u(0) = x_0. \tag{8.2.1}$$

Here A is the generator of a C_0-semigroup $T(\cdot)$ on a Banach space X. Suppose that this equation has a classical solution $\widetilde{u}$ on an interval $[0, T)$ and the function $g : t \mapsto f(t, \widetilde{u}(t))$ belongs to $L^1(0, T; X)$. Then $\widetilde{u}$ is also a mild solution of the problem

$$\dot{u}(t) = Au(t) + g(t), \quad u(0) = x_0,$$

i.e.,

$$F(\widetilde{u})(t) := T(t)x_0 + \int_0^t T(t-s) f(s, \widetilde{u}(s))\, ds = \widetilde{u}(t), \tag{8.2.2}$$

$t \in [0, T)$. In other words, any classical solution to (8.2.1) is a fixed point of the mapping F. This motivates the following definition.

Definition 8.2.1. A function $\widetilde{u} \in C([0, T); X)$ which satisfies the identity (8.2.2) on the interval $[0, T)$ is called a *mild solution of semilinear problem* (8.2.1).

In order to prove the existence of a mild solution to (8.2.1), we have to find conditions on the Nemytski operator

$$u \mapsto f(\cdot, u(\cdot))$$

which guarantee the existence of a fixed point of $F : C([0,T];X) \to C([0,T];X)$.

Let $f : [0,T] \times X \to X$ be continuous. Then the convolution term in (8.2.2) belongs to $C([0,T];X)$ for any $u \in C([0,T];X)$ and hence F is well-defined. Our plan is to find conditions on f which guarantee that F is a contraction. For this purpose assume that f satisfies the Lipschitz condition with respect to the second variable:

$\forall \tau \in (0,T)\, \exists\, L_\tau > 0\, \forall\, t \in [0,\tau]\, \forall\, y,z \in X :$

$$\|f(t,y) - f(t,z)\| \le L_\tau \|y - z\|. \tag{8.2.3}$$

Proposition 8.2.2. *Let $f : [0,T) \times X \to X$ be continuous and satisfy the Lipschitz condition* (8.2.3). *Then the equation* (8.2.1) *has a unique mild solution on the interval* $[0,T)$.

Proof. First we show the uniqueness. Let u, v be mild solutions on an interval $[0,\tau]$. Then for $t \in [0,\tau]$ we have

$$\begin{aligned} \|u(t) - v(t)\| &= \left\| \int_0^t T(t-s)\,[f(s,u(s)) - f(s,v(s))]\,ds \right\| \\ &\le L_\tau \sup_{\sigma \in [0,\tau]} \|T(\sigma)\| \int_0^t \|u(s) - v(s)\| ds. \end{aligned}$$

It follows from the Gronwall Lemma (Proposition 2.3.9) that $\|u(t) - v(t)\| = 0$ for $t \in [0,\tau]$ and the uniqueness follows.

Let c and ω be the constants from the estimate of $\|T(\cdot)\|$ given in Proposition 8.1.5 (i). Let us choose some $\tau_1 \in (0,T)$ and set $L_1 := L_{\tau_1}$.

Let $q \in (0,1)$ be fixed. Then there exists $\tau \in (0,\tau_1]$ such that

$$\frac{c}{\omega}\left(e^{\omega\tau} - 1\right) L_1 = q. \tag{8.2.4}$$

For $\rho > 0$ denote

$$M_\rho^\tau := \left\{ u \in C([0,\tau];X) : \sup_{t \in [0,\tau]} \|u(t) - x_0\| \le \rho \right\}.$$

Since M_ρ^τ is a closed subset of the space $C([0,\tau];X)$, it is a complete metric space equipped with the "sup" metric. Our aim is to find $\rho > 0$ such that

$$F\left(M_\rho^\tau\right) \subset M_\rho^\tau \tag{8.2.5}$$

and, moreover, F is a contraction on M_ρ^τ.

In order to prove the former fact, we take $u \in M_\rho^\tau$ and perform the following estimates with constants k_τ and $\widetilde{k}_\tau$ which depend only on τ and on the initial condition x_0:

$$\sup_{t\in[0,\tau]} \|F(u)(t) - x_0\| \le \sup_{t\in[0,\tau]} \|T(t)x_0 - x_0\| + \sup_{t\in[0,\tau]} \|\int_0^t T(t-s)f(s,u(s))ds\|$$

$$\le k_\tau + \sup_{t\in[0,\tau]} \int_0^t \|T(t-s)\| \left[\|f(s,u(s)) - f(s,x)\| + \|f(s,x)\|\right] ds$$

$$\le k_\tau + \sup_{t\in[0,\tau]} \int_0^t ce^{\omega(t-s)} \left[L_1\rho + \|f(s,x)\|\right] ds$$

$$\le k_\tau + \widetilde{k}_\tau + \frac{c}{\omega}\left[e^{\omega\tau} - 1\right] L_1\rho = k_\tau + \widetilde{k}_\tau + q\rho \le \rho.$$

The last inequality holds for sufficiently large $\rho > 0$. Hence (8.2.5) holds.

In order to prove that F is a contraction on M_ρ^τ, we take $u, v \in M_\rho^\tau$ and estimate

$$\sup_{t\in[0,\tau]} \|F(u)(t) - F(v)(t)\| = \sup_{t\in[0,\tau]} \|\int_0^t T(t-s)\left[f(s,u(s)) - f(s,v(s))\right] ds\|$$

$$\le \sup_{t\in[0,\tau]} \int_0^t ce^{\omega(t-s)} L_1 \|u(s) - v(s)\| ds \le \frac{c}{\omega}\left(e^{\omega\tau} - 1\right) L_1 \|u - v\|_{C([0,\tau];X)}$$

$$= q\|u - v\|_{C([0,\tau];X)}.$$

The Contraction Principle (Theorem 2.3.1) yields that F has a unique fixed point in M_ρ^τ and it is also a unique mild solution $\widetilde{u}$ of (8.2.1) on the interval $[0,\tau)$.

The uniqueness and the local existence of a mild solution allow us to construct a *maximal solution.* It remains to show that this maximal solution is actually defined on the entire interval $[0,T)$.

Denote

$$\widetilde{\tau} := \sup\{t : \exists \,\text{mild solution on}\, [0,t]\}.$$

We show that $\widetilde{\tau} = T$. Assume that $\widetilde{\tau} < T$. Then there exist $\tau_1 \in (\widetilde{\tau}, T)$ and corresponding $\tau \in (0,\tau_1]$ satisfying (8.2.4). Choose $t \in (0,\widetilde{\tau})$ such that $\widetilde{\tau} - t < \tau$ and let $\widetilde{u}$ be a mild solution defined on the interval $[0,t]$. We have already proved that the equation

$$\dot{v}(s) = Av(s) + f(s+t, v(s)), \quad v(0) = \widetilde{u}(t)$$

has a mild solution $\widetilde{v}$ on the interval $[0,\tau]$. But then the function

$$w(s) = \begin{cases} \widetilde{u}(s), & 0 \le s \le t, \\ \widetilde{v}(s-t), & t \le s \le t+\tau \end{cases}$$

is a mild solution on the interval $[0, t+\tau]$ which is strictly larger than $[0,\widetilde{\tau}]$, a contradiction. □

Remark 8.2.3. The global Lipschitz condition (8.2.3) is rather strong. From the proof of Proposition 8.2.2 one can see that for the existence and also the uniqueness of a maximal solution the following weaker form of the *Lipschitz condition* is sufficient:

$$\begin{aligned} &\forall \tau \in (0,T)\ \forall r > 0\ \exists L > 0\ \forall t \in [0,\tau]\ \forall y, z \in X : \\ &\|y\| \le r,\ \|z\| \le r \Rightarrow \|f(t,y) - f(t,z)\| \le L\|y-z\| \end{aligned} \tag{8.2.6}$$

(cf. Theorem 2.3.4). But this condition does not guarantee the global existence of solutions, i.e., solutions defined on the entire interval [0,T) (cf. Remark 2.3.6).

Remark 8.2.4. The fixed point method can be adapted to prove certain regularity of solutions to (8.2.1) provided that the semigroup T has some smoothing effect. In order to avoid technical details we assume that the generator A is self-adjoint in $X = L^2(\Omega)$ and has compact resolvent. For such A there is an orthonormal basis $\{e_n\}_{n=1}^{\infty}$ of X which consists of the eigenfunctions of A. Suppose that $Ae_n = \lambda_n e_n$, where the sequence of eigenvalues $\{\lambda_n\}_{n=1}^{\infty}$ is decreasing, $\lambda_1 < 0$, $\lim_{n\to\infty} \lambda_n = -\infty$. For $\alpha \in \mathbb{R}$ we define spaces X_α as follows:

$$X_\alpha = \left\{ x = \{\xi_n\}_{n=1}^{\infty} : \sum_{n=1}^{\infty} \xi_n^2 (-\lambda_n)^{2\alpha} := \|x\|_\alpha^2 < \infty \right\}.$$

If $\alpha \ge 0$ we can interpret $x \in X_\alpha$ as a point of X with $\xi_n = (x, e_n)_X$. The space $X_{-\alpha}$ is then the dual space to X_α. It is possible to extend the semigroup T to spaces X_α as follows:

$$T(t)x = \left\{ e^{t\lambda_n} \xi_n \right\}_{n=1}^{\infty}.$$

For $\alpha, \beta \in \mathbb{R}$, $\beta \le \alpha$, $t > 0$ we have

$$\begin{aligned} \|T(t)x\|_\alpha^2 &= \sum_{n=1}^{\infty} e^{2t\lambda_n} \xi_n^2 (-\lambda_n)^{2\alpha} = \sum_{n=1}^{\infty} \frac{e^{2t\lambda_n} t^{2(\beta-\alpha)}}{(-\lambda_n t)^{2(\beta-\alpha)}} (-\lambda_n)^{2\beta} \xi_n^2 \\ &\le c^2 t^{2(\beta-\alpha)} \|x\|_\beta^2. \end{aligned} \tag{8.2.7}$$

This estimate allows us to prove that the convolution $T*F \in X_\alpha$ provided $\alpha < \beta+1$ and the Nemytski operator $F(u)(\cdot) = f(\cdot, u(\cdot)) \in L^2(0,T;X_\beta)$ for $u \in X_\alpha$. For example, let the eigenfunctions e_n be continuous and uniformly bounded and let $\alpha \in (0,1)$ be such that the series $\Sigma \frac{1}{(-\lambda_n)^{2\alpha}}$ is convergent. Then $X_\alpha \subsetneq BC(\Omega)$ and for $F(u) = u^k$, $k \in \mathbb{N}$, we get $F(u) \in L^2(0,T;X)$. Moreover (8.2.6) is satisfied and, therefore, the equation (8.2.1) has a unique maximal mild solution in the space $C((0,\tau);X_\alpha)$ for any x_0. We notice that the crucial estimate (8.2.7) holds for more general semigroups, namely for analytic semigroups. For details and more information see, e.g., Henry [79], Pazy [120] or Lunardi [106] and references given there.

In the proof of Proposition 8.2.2 the Lipschitz condition (8.2.3) yields actually an *a priori estimate* of solutions – to see this fact use the Gronwall Lemma. Various a priori estimates can be used to prove the convergence of suitable approximations. We describe this idea in a simple case for the most popular Galerkin approximations.

Let V be a separable Banach space which is continuously embedded into a Hilbert space H. We can assume that V is dense in H. For $y \in H$, $\varphi_y(x) = (x,y)_H$ is a continuous linear form on V. Since V is dense in H, the mapping $y \to \varphi_y$ is injective, i.e., it is an embedding of H into V^* and it can be shown that H is dense in V^*. We will consider an "energy" functional $E : V \to \mathbb{R}$ and assume that $E \in C^1(V)$. Denote

$$\mathrm{Dom}(\nabla_H E) = \{v \in V : \exists x \in H\ \forall\, w \in V,\ E'(v)w = (w,x)_H\}.$$

Since x is uniquelly determined by v we put $x = \nabla_H E(v)$ and call $\nabla_H E$ the *H-gradient* of E.

Many differential equations can be written in the form

$$\dot{u}(t) + \nabla_H E\,(u(t)) = f(t), \tag{8.2.8}$$

where $f \in L^2(0,T;H)$. A function $u \in W^{1,2}(0,T;H) \cap L^\infty(0,T;V)$ is said to be a solution of the *gradient equation* (8.2.8) if for almost all $t \in (0,T)$ the equation (8.2.8) is satisfied. It follows from the definition of $W^{1,2}$ that $u \in W^{1,2}(0,T;H)$ can be changed on a set of measure zero such that u is continuous on $[0,T]$ (cf. Proposition 3.1.18), i.e.,

$$u(t) - u(s) = \int_s^t \dot{u}(\sigma)d\sigma \quad \text{for all} \quad 0 \le s \le t \le T.$$

We will always suppose that this change has been done. The continuity of solution allows us to prescribe an initial condition $u(0) = u_0 \in H$.

Theorem 8.2.5. *Let V be a separable and reflexive Banach space which is densely embedded into a Hilbert space. Let $E \in C^1(V,\mathbb{R})$ be convex, i.e.,*

$$E(\lambda u + (1-\lambda)v) \le \lambda E(u) + (1-\lambda)E(v), \quad \lambda \in (0,1), u, v \in V$$

and let E' maps bounded sets in V into bounded sets in V^* and let $\lim\limits_{\|u\|_V \to \infty} E(u) = \infty$. If $f \in C([0,T];H)$[1] *and $u_0 \in V$ then there exists a unique solution u to* (8.2.8) *which satisfies the initial condition $u(0) = u_0$.*

[1]See comments on this assumption below the proof.

Proof. Proof is done in several steps and contains a good deal of functional analysis. It is somewhat typical, even if it is rather simple in comparison with techniques used, e.g., in Navier-Stokes equations (see, e.g., Temam [151], Lions [105] or Feireisl [61]).

(i) If $E \in C^1(V)$ is convex on V, then

$$\langle E'(u) - E'(v),\ u - v\rangle_{V^* \times V} \geq 0 \tag{8.2.9}$$

for every $u, v \in V$. Indeed, from the convexity we have

$$\frac{E(v + \lambda(u - v)) - E(v)}{\lambda} \leq E(u) - E(v), \quad \lambda > 0$$

and for $\lambda \to 0+$,

$$E'(v)(u - v) \leq E(u) - E(v),$$

and similarly

$$E'(u)(v - u) \leq E(v) - E(u).$$

Summing up these inequalities we get (8.2.9).

(ii) *Uniqueness of a solution.* Let $u, \tilde{u}$ be two solutions to (8.2.8) satisfying the same initial condition. By the inequality (8.2.9), for $w = u - \tilde{u}$ we obtain

$$\frac{1}{2}\frac{d}{dt}\|w(t)\|_H^2 = (\dot{w}(t), w(t))_H = -\langle E'(u) - E'(\tilde{u}), u - \tilde{u}\rangle_{V^* \times V} \leq 0,$$

for a.a. $t \in (0, T)$. Since w is continuous and $w(t) = 0$, w must vanish everywhere in $[0, T]$.

(iii) *Galerkin approximations* . Since V is separable there exists a sequence $\{v_n\} \subset V$ such that for $V_n = \operatorname{Lin}\{v_1, \dots, v_n\}$ the union $\bigcup\limits_{n=1}^{\infty} V_n$ is a dense subspace of V. Let $u_{0,n} \in V_n$ be such that $u_{0,n} \to u_0$ in V. Consider the following system of ordinary differential equations for $w \in C^1([0, T]; V_n)$:

$$(\dot{w}(t), v_k)_H + \langle E'(w(t)), v_k\rangle_{V^* \times V} = (f(t), v_k)_H\,, \ k = 1, \dots, n,$$
$$w(0) = u_{0,n}. \tag{8.2.10}$$

We have proved in Example 2.3.8 that this system has a solution w_n. Similarly as in step (ii) this solution is unique.

(iv) *A priori estimates.* By integration, we get from (8.2.10)

$$\int_0^t \|\dot{w}_n(s)\|_H^2 ds + \int_0^t \frac{d}{ds} E(w_n(s))\, ds$$
$$\leq \frac{1}{2}\int_0^T \|f(s)\|_H^2 ds + \frac{1}{2}\int_0^t \|\dot{w}_n(s)\|_H^2 ds,\ 0 \leq t \leq T.$$

As E is continuous, $E(u_{n,0}) \to E(u_0)$ and therefore the sequence $\{E(w_n(\cdot))\}$ is uniformly bounded on the interval $[0,T]$. Using the assumption

$$\lim_{\|u\|\to\infty} E(u) = \infty$$

we obtain that

$$\sup_{n\in\mathbb{N}} \|w_n\|_{L^\infty(0,T;V)} < \infty. \tag{8.2.11}$$

By this and the a priori estimate, $\{\dot{w}_n\}$ is a bounded sequence in $L^2(0,T;H)$. Since E' maps bounded subsets of V into bounded subsets of V^* we also deduce from (8.2.11) that $\{E'(w_n)\}_{n=1}^\infty$ is a bounded sequence in $L^\infty(0,T;V^*)$.

(v) The *compactness argument* relies on Theorem 2.1.25 and Remark 2.1.26. Since V is a reflexive space, $L^\infty(0,T;V) = \left[L^1(0,T;V^*)\right]^*$ [2] and thus there exists a subsequence of $\{w_n\}_{n=1}^\infty$ (we will denote all such subsequences by $\{w_n\}$) such that

$$w_n \overset{*}{\rightharpoonup} w \in L^\infty(0,T;V).$$

It follows from boundedness of $\{w_n\}$ in the Hilbert space $W^{1,2}(0,T;H)$ that

$$w_n \rightharpoonup u \in W^{1,2}(0,T;H)$$

and

$$w_n \rightharpoonup u \in L^2(0,T;H).$$

By continuous embedding $L^\infty(0,T;V) \subset L^2(0,T;H)$, $w = u$.

Since $W^{1,2}(0,T;H)$ is continuously embedded into $C([0,T];H)$, the point evaluation $u \in W^{1,2} \mapsto u(t)$ is continuous and therefore (a continuous linear operator sends weakly convergent sequences into weakly convergent ones) $w_n(0) = u_{0,n} \rightharpoonup u_0$ in H. But $u_{0,n} \to u_0$ in V so that $u(0) = u_0$.

Further,

$$w_n(t) = w_n(0) + \int_0^t \dot{w}_n(s)ds \rightharpoonup u_0 + \int_0^t v(s)ds, \ t \in [0,T],$$

i.e.,

$$u(t) = u_0 + \int_0^t v(s)ds, \quad \text{and} \quad \dot{u} = v.$$

The last weakly converging subsequence we obtain from the boundedness of $\{E'(w_n)\}_{n=1}^\infty$ and the fact that $L^\infty(0,T;V^*) = \left[L^1(0,T;V)\right]^*$, i.e.,

$$E'(w_n) \overset{*}{\rightharpoonup} \kappa \in L^\infty(0,T;V^*).$$

[2] It is not generally true that $[L^p(0,T;X)]^*$ is $L^{p'}(0,T;X^*)$ $\left(\frac{1}{p} + \frac{1}{p'} = 1,\ 1 \le p < \infty\right)$, but reflexivity of X is sufficient in order for this duality to hold. The interested reader can consult (Diestel, J., Uhl, J. J. [43]). We also need that $L^p(0,T;X)$ is separable provided X is separable and $1 \le p < \infty$ (Exercise 8.2.7). Moreover, if X^* is separable then X is separable, too.

(vi) *The limit u is a solution to (8.2.8).* From the definition (8.2.10) of w_n we obtain for every $\varphi \in L^2(0,T)$ and $w \in V_n$:

$$\int_0^T (\dot{w}_n, \varphi w)_H + \int_0^T \langle E'(w_n), \varphi w\rangle_{V^*\times V} = \int_0^T (f, \varphi w)_H.$$

By the limit

$$\int_0^T (\dot{u}, \varphi w)_H + \int_0^T \langle \kappa, \varphi w\rangle_{V^*\times V} = \int_0^T (f, \varphi w)_H.$$

Since $\bigcup_{n=1}^{\infty} V_n$ is dense in V, the set $\left\{\varphi w : w \in \bigcup_{n=1}^{\infty} V_n,\ \varphi \in L^2(0,T)\right\}$ is dense in $L^2(0,T;V)$ (Exercise 8.2.7) and, therefore, the last inequality yields

$$\int_0^T (\dot{u}, v)_H + \int_0^T \langle \kappa, v\rangle_{V^*\times V} = \int_0^T (f, v)_H \tag{8.2.12}$$

for every $v \in L^2(0,T;V)$. We need to show that $\kappa = E'(u)$. We notice that the limit in a nonlinear term is a crucial point of any existence proof by approximations. For this purpose we need two inequalities.

(1) Put in (8.2.10) $v_k = w_n$ and integrate to get

$$\begin{aligned}\int_0^T \langle E'(w_n), w_n\rangle_{V^*\times V} &= -\frac{1}{2}\int_0^T \frac{d}{dt}\|w_n\|_H^2 + \int_0^T (f, w_n)_H\\ &= \frac{1}{2}\|w_n(0)\|_H^2 - \frac{1}{2}\|w_n(T)\|_H^2 + \int_0^T (f, w_n)_H.\end{aligned}$$

We already know that $w_n(0) \to u_0$ and $w_n(T) \rightharpoonup u(T)$ in H. The norm is lower semicontinuous (Proposition 2.1.22(iii)), i.e., $\|u(T)\|_H \le \liminf_{n\to\infty} \|w_n(T)\|_H$. This yields

$$\begin{aligned}\limsup_{n\to\infty} \int_0^T \langle E'(w_n), w_n\rangle_{V^*\times V} &\le \frac{1}{2}\|u(0)\|_H^2 - \frac{1}{2}\|u(T)\|_H^2 + \int_0^T (f, u)_H\\ &= \int_0^T \langle \kappa, u\rangle_{V^*\times V}.\end{aligned}$$

For the last equality we have used (8.2.12) with $v = u$.

(2) To prove the converse inequality we use the convexity of E. By the first step (i) we have

$$\int_0^T \langle E'(w_n), w_n\rangle_{V^*\times V} \geq \int_0^T \langle E'(w_n), u\rangle_{V^*\times V} + \int_0^T \langle E'(u), w_n - u\rangle_{V^*\times V}.$$

The right-hand side tends to $\int_0^T \langle \kappa, u\rangle_{V^*\times V}$ as $n \to \infty$, i.e.,

$$\liminf_{n\to\infty} \int_0^T \langle E'(w_n), w_n\rangle_{V^*\times V} \geq \int_0^T \langle \kappa, u\rangle_{V^*\times V}.$$

It follows from (1) and (2) that

$$\lim_{n\to\infty} \int_0^T \langle E'(w_n), w_n\rangle_{V^*\times V} = \int_0^T \langle \kappa, u\rangle_{V^*\times V}.$$

Now we choose arbitrary $v \in L^\infty(0,T;V)$ and put $z_\lambda = (1-\lambda)u + \lambda v$ for $\lambda > 0$. Using convexity again we obtain

$$\begin{aligned}\int_0^T \langle E'(w_n), w_n\rangle_{V^*\times V} &- \int_0^T \langle E'(w_n), z_\lambda\rangle_{V^*\times V} \\ &\geq \int_0^T \langle E'(z_\lambda), w_n - z_\lambda\rangle_{V^*\times V},\end{aligned}$$

and passing to the limit for $n \to \infty$,

$$\int_0^T \langle \kappa, u\rangle_{V^*\times V} - \int_0^T \langle \kappa, z_\lambda\rangle_{V^*\times V} \geq \int_0^T \langle E'(z_\lambda), u - z_\lambda\rangle_{V^*\times V},$$

i.e.,

$$\int_0^T \langle \kappa, u - v\rangle_{V^*\times V} \geq \int_0^T \langle E'(z_\lambda), u - v\rangle_{V^*\times V}.$$

Finally, we send $\lambda \to 0+$ to obtain

$$\int_0^T \langle \kappa, u - v\rangle_{V^*\times V} \geq \int_0^T \langle E'(u), u - v\rangle_{V^*\times V}.$$

Since v has been arbitrary, we derive the equality $E'(u) = \kappa$, i.e.,

$$\int_0^T (\dot{u}, v)_H + \int_0^T \langle E'(u), v\rangle_{V^*\times V} = \int_0^T (f, v)_H$$

holds for every $v \in L^2(0,T;V)$. In particular, set $v = \varphi \cdot w$, where $w \in V$, $\varphi \in L^2(0,T)$. This choice gives

$$(\dot{u}(t), w)_H + \langle E'(u(t)), w\rangle_{V^*\times V} = (f(t), w)_H \quad \text{for} \quad \text{a.a. } t \in [0,T].$$

But this means that

$$f(t) - \dot{u}(t) \in \text{Dom}\,(\nabla_H E) \quad \text{for} \quad \text{a.a. } t, \text{ i.e., (8.2.8) is satisfied a.e.} \qquad \square$$

We recommend to the reader to go through the proof again to recall the main steps of an approximation proof:

1. A finite-dimensional system of ordinary differential equations has a global solution (the interval of existence should not depend on n).
2. Sufficiently many a priori estimates yield boundedness of linear and non-linear terms in various norms. Those allow us to choose weakly convergent subsequences.
3. The limits in nonlinear terms need an extra effort. Here we used the convexity of an "energy functional" E. Another possibility consists in compact embeddings (Theorem 1.2.28) which send weakly convergent sequences into the strongly (in norm) convergent ones.
4. The uniqueness of a solution was simple to prove here, but in other situations (e.g., for the Navier-Stokes equations) this can be a very difficult issue or an open problem.

The reader might have noticed that we used the assumption $f \in C$ only for the existence of a classical solution of (8.2.10). In order to prove the statement of Theorem 8.2.5 for the more natural case of $f \in L^2(0,T;H)$ we only need to extend the notion of solutions of ordinary differential equations from C^1 to $W^{1,2}$, i.e., to consider continuous solutions which satisfy an equation almost everywhere. Such solutions are often called *Caratheodory solutions.* It is not too difficult to extend Theorem 2.3.4 to such solutions.

The bibliography on the existence of solutions of evolution equations and their properties, e.g., asymptotic behavior, including various attractors, is enormous. In Theorem 8.2.5 we followed the lecture notes Chill & Fasangova [30]. A lot of information can be found in Cazenave & Haraux [26], Temam [150] or Sell & You [140] and references therein.

Exercise 8.2.6. Apply Remark 8.2.3 to the equations

$$\frac{\partial u}{\partial t} = \Delta u \pm u^3,$$

where Δ is either the Dirichlet or the Neumann Laplacian.

Exercise 8.2.7. Let X be a separable Banach space and $1 \le p < \infty$. Then $L^p(0,T;X)$ is separable.

Hint. Consider dense sequences $\{\varphi_k\} \subset L^p(0,T)$, $\{x_n\} \subset X$ and show that $\{\varphi_k x_n\}_{k,n}$ is dense in $L^p(0,T;X)$.

Exercise 8.2.8. Let a be a symetric bilinear form on $X \times X$. If the quadratic form $E(u) = a(u,u)$ is non-negative then E is convex on X. Moreover, $E \in C^1(X)$. Prove these statements and apply them to $a(u,v) = \int\limits_\Omega \nabla u \nabla v$ on $X = W_0^{1,2}(\Omega)$ (Dirichlet boundary conditions) and on $X = W^{1,2}(\Omega)$ (Neumann boundary conditions). Does E satisfy all assumptions of Theorem 8.2.5?
Answer. E is not coercive on $X = W^{1,2}(\Omega)$, i.e., there exists $\{u_n\}_{n=1}^\infty \subset W^{1,2}(\Omega)$ such that $\|u_n\|_X \to \infty$ and $\lim\limits_{n\to\infty} E(u_n) \ne \infty$.

Exercise 8.2.9. Prove that the equation

$$\frac{\partial u}{\partial t} = \Delta u - u^3 + f$$

with Dirichlet or Neumann boundary conditions satisfies all assumptions of Theorem 8.2.5 provided $f \in C\left([0,\infty); L^2(\Omega)\right)$.

8.3 Linearization of Quasilinear PDEs and Fixed Point Theorems

Let $\Omega \subset \mathbb{R}^N$ be a bounded domain, $\Omega \in C^{2,\alpha}$ with some $\alpha \in (0,1)$. Let a_{ij}, b_i, $c \in C^{0,\alpha}(\overline{\Omega})$, $a_{ij} = a_{ji}$, $i,j = 1,\dots,N$. We assume, moreover, that there exists a constant $\kappa > 0$ such that

$$\sum_{i,j=1}^N a_{ij}(x)\xi_i\xi_j \ge \kappa \sum_{i=1}^N \xi_i^2 \tag{8.3.1}$$

for all $x \in \Omega$ and $\xi_i \in \mathbb{R}$.

The linear operator $L : C^{2,\alpha}(\overline{\Omega}) \to C^{0,\alpha}(\overline{\Omega})$ defined by

$$Lu(x) := -\sum_{i,j=1}^N a_{ij}(x)\frac{\partial^2 u}{\partial x_i \partial x_j} + \sum_{i=1}^N b_i(x)\frac{\partial u}{\partial x_i} + c(x)u$$

is called a *uniformly elliptic (second-order) linear differential operator.*[3] The inequality (8.3.1) is called an *ellipticity condition* .

[3]Notice that L is well-defined due to the fact that $C^{0,\alpha}(\overline{\Omega})$ is an algebra, see Exercise 5.3.6.

We consider the Dirichlet problem

$$\begin{cases} Lu(x) = f(x) & \text{in} \quad \Omega, \\ u(x) = 0 & \text{on} \quad \partial\Omega \end{cases} \tag{8.3.2}$$

where $f \in C^{0,\alpha}(\overline{\Omega})$. We set

$$X := \left\{u \in C^{2,\alpha}(\overline{\Omega}) : u(x) = 0,\ x \in \partial\Omega\right\}, Y := C^{0,\alpha}(\overline{\Omega})$$

and write the Dirichlet problem (8.3.2) as an operator equation

$$Lu = f. \tag{8.3.3}$$

The operator $L : X \to Y$ has all the properties stated in Section 5.3. In particular, we have the following assertion (cf. Theorem 5.3.1).

Proposition 8.3.1. (cf. Gilbarg & Trudinger [71, Theorem 6.14]) *Let $c \geq 0$ in Ω, $f \in C^{0,\alpha}(\overline{\Omega})$. Then the Dirichlet problem* (8.3.2) *has a unique solution $u \in C^{2,\alpha}(\overline{\Omega})$ and there exists a constant $k > 0$ (independent of f) such that*

$$\|u\|_{C^{2,\alpha}(\overline{\Omega})} \leq k\|f\|_{C^{0,\alpha}(\overline{\Omega})}. \tag{8.3.4}$$

Remark 8.3.2. The inequality (8.3.4) is called the *Schauder estimate* for classical solution (cf. Theorem 5.3.1). Note that besides (8.3.4) the following estimate also holds:

$$\|u\|_{W^{2,p}(\Omega)} \leq k_p\|f\|_{L^p(\Omega)} \tag{8.3.5}$$

for $p \in (1, \infty)$ with $k_p > 0$ independent of f, cf. Gilbarg & Trudinger [71, Chapter 9]. The inequality (8.3.5) is called the L_p*-estimate* for a strong solution (cf. Remark 8.3.10 below).

In several mathematical models the coefficients a_{ij}, b_i and c, may depend on a solution. The equation (and the differential operator) is then called *quasilinear.* There are several methods for solving this type of equations depending on the assumptions about a_{ij}, b_i and c. The *method of linearization* is a typical one and it combines the results from the linear theory with suitable methods from the nonlinear theory. In order to avoid rather complicated technical assumptions, we present the method of linearization under the following special hypotheses on the coefficients: *there exist positive constants α, β and γ such that for any $i, j = 1, \dots, N$ the real functions $a_{i,j}, b_i, c \in C^1(\mathbb{R})$ satisfy*

$$\begin{cases} a_{ij}(s) = a_{ji}(s),\ \forall s \in \mathbb{R}, \\ \alpha\|\xi\|^2_{\mathbb{R}^N} \leq \sum\limits_{i,j=1}^{N} a_{ij}(s)\xi_i\xi_j \leq \beta\|\xi\|^2_{\mathbb{R}^N},\ \forall \xi \in \mathbb{R}^N\ \forall s \in \mathbb{R}, \\ |b_i(s)| \leq \gamma,\ |c(s)| \leq \gamma,\ \forall s \in \mathbb{R}. \end{cases} \tag{8.3.6}$$

In order to guarantee that some useful embeddings hold true, we need a technical assumption concerning the domain Ω.

Definition 8.3.3 (cf. Kufner, John & Fučík [96]). A domain $\Omega \subset \mathbb{R}^N$ is said to satisfy *condition* (S) if there exists a constant $M > 0$ with the following properties: For every pair of points $x, y \in \Omega$ there exists a finite sequence of points $x = z_0, z_1, \dots, z_n = y$ such that the segments with endpoints $z_i, z_{i+1} (i = 0, \dots, n-1)$ are subsets of Ω and

$$\sum_{i=1}^{n-1} \|z_i - z_{i-1}\|_{\mathbb{R}^N} \le M\|x - y\|_{\mathbb{R}^N}.$$

Here n may depend on x and y.

If Ω satisfies condition (S) then continuous embedding $C^{k+1,\alpha}(\overline{\Omega}) \subsetneq C^{k,\alpha}(\overline{\Omega})$ holds with $k = 0, 1, \dots, \alpha \in (0, 1]$. This fact combined with the compact embedding $C^{k,\alpha}(\overline{\Omega}) \subsetneq\subsetneq C^k(\overline{\Omega})$ (see [96, Theorem 1.5.10]) implies the compact embedding

$$C^{k+1,\alpha}(\overline{\Omega}) \subsetneq\subsetneq C^k(\overline{\Omega}), \tag{8.3.7}$$

$k = 0, 1, \dots,$ and $\alpha \in (0, 1]$.

In particular, it follows from C^1-smoothness of a_{ij}, b_i, c and Lemma 5.3.2 that for a given $u \in C^{2,\alpha}(\overline{\Omega})$, $a_{ij}(u(\cdot))$, $b_i(u(\cdot))$ and $c(u(\cdot))$ belong to $C^{0,\alpha}(\overline{\Omega})$ and hence the *quasilinear differential operator*

$$Q(u)(x) := -\sum_{i,j=1}^N a_{ij}(u(x))\frac{\partial^2 u}{\partial x_i \partial x_j} + \sum_{i=1}^N b_i(u(x))\frac{\partial u}{\partial x_i} + c(u(x))u$$

is a well-defined mapping from $C^{2,\alpha}(\overline{\Omega})$ into $C^{0,\alpha}(\overline{\Omega})$.

We consider the *quasilinear Dirichlet problem*

$$\begin{cases} Q(u)(x) = f(x) & \text{in} \quad \Omega, \\ u(x) = 0 & \text{on} \quad \partial\Omega \end{cases} \tag{8.3.8}$$

where $f \in C^{0,\alpha}(\overline{\Omega})$ is a given function. We apply the method of linearization to solve (8.3.8). The following observation is the starting point for this method. For a fixed $v \in C^{2,\alpha}(\overline{\Omega})$ set

$$L_v(u)(x) := -\sum_{i,j=1}^N a_{ij}(v(x))\frac{\partial^2 u}{\partial x_i \partial x_j} + \sum_{i=1}^N b_i(v(x))\frac{\partial u}{\partial x_i} + c(v(x))u.$$

We deduce from (8.3.6) that $L_v : C^{2,\alpha}(\overline{\Omega}) \to C^{0,\alpha}(\overline{\Omega})$ is a uniformly elliptic linear differential operator and observe that $L_u u = Q(u)$. It follows from Proposition 8.3.1 that for given $v \in C^{2,\alpha}(\overline{\Omega})$ and $f \in C^{0,\alpha}(\overline{\Omega})$ the *linear problem*

$$\begin{cases} L_v(u)(x) = f(x) & \text{in} \quad \Omega, \\ u(x) = 0 & \text{on} \quad \partial\Omega \end{cases} \tag{8.3.9}$$

has a unique solution $u = u(v, f)$ provided $c \geq 0$. If we keep f fixed and allow v to vary, we get a mapping $S : C^{2,\alpha}(\overline{\Omega}) \to C^{2,\alpha}(\overline{\Omega})$, $u = S(v)$, where u is a unique solution of (8.3.9). If S has a fixed point u, then it must be a solution of *quasilinear problem* (8.3.8).

Our next goal is to show that S satisfies the assumptions of the Schauder Fixed Point Theorem (Theorem 5.2.5). We need the following technical assertion.

Lemma 8.3.4. *Let $v_n \to v$ in $C^{0,\alpha}(\overline{\Omega})$. Then $a_{ij}(v_n) \to a_{ij}(v)$, $b_i(v_n) \to b_i(v)$ and $c(v_n) \to c(v)$ in $C^{0,\alpha}(\overline{\Omega})$.*

Proof. This assertion follows directly from Lemma 5.3.2. □

For a fixed $f \in C^{0,\alpha}(\overline{\Omega})$ we denote by $u(v)$ a unique solution of (8.3.9). Assume that $\{v_n\}_{n=1}^{\infty}$ is a bounded sequence in $C^{2,\alpha}(\overline{\Omega})$. Due to (8.3.7) there is a subsequence (denoted again by $\{v_n\}_{n=1}^{\infty}$) such that $v_n \to v$ in $C^{0,\alpha}(\overline{\Omega})$. It follows from Lemma 8.3.4 that for any fixed $w \in C^{2,\alpha}(\overline{\Omega})$ we have

$$(L_v - L_{v_n})w \to 0 \quad \text{in} \quad C^{0,\alpha}(\overline{\Omega}) \tag{8.3.10}$$

as $n \to \infty$. We also have

$$\begin{aligned}
\|u(v_n) - u(v)\|_{C^{2,\alpha}(\overline{\Omega})} &= \|L_{v_n}^{-1} f - L_v^{-1} f\|_{C^{2,\alpha}(\overline{\Omega})} \\
&= \|L_{v_n}^{-1}(L_v - L_{v_n})L_v^{-1} f\|_{C^{2,\alpha}(\overline{\Omega})} \\
&\leq \|L_{v_n}^{-1}\|_{\mathcal{L}(C^{0,\alpha}(\overline{\Omega}),C^{2,\alpha}(\overline{\Omega}))} \|(L_v - L_{v_n})L_v^{-1} f\|_{C^{0,\alpha}(\overline{\Omega})}.
\end{aligned} \tag{8.3.11}$$

It follows from uniform estimates (8.3.6) that Proposition 8.3.1 applies uniformly with respect to v_n. Hence there is a constant $\delta > 0$ independent of v_n such that

$$\|L_{v_n}^{-1}\|_{\mathcal{L}(C^{0,\alpha}(\overline{\Omega}),C^{2,\alpha}(\overline{\Omega}))} \leq \delta. \tag{8.3.12}$$

(Actually, δ depends only on α, β and γ.)

Hence, from (8.3.10) - (8.3.12) we deduce that

$$u(v_n) \to u(v) \quad \text{in } C^{2,\alpha}(\overline{\Omega}).$$

A similar argument yields continuity of S, i.e., the operator S is compact. Taking advantage of (8.3.12) once again, we also have

$$\|u(v)\|_{C^{2,\alpha}(\overline{\Omega})} \leq \delta \|f\|_{C^{0,\alpha}(\overline{\Omega})}$$

for any $v \in C^{2,\alpha}(\overline{\Omega})$. It means that there exists a positive number $R > 0$ such that

$$S(\overline{B(o;R)}) \subset \overline{B(o;R)},$$

where $B(o;R)$ is an open ball in $C^{0,\alpha}(\overline{\Omega})$ centered at the origin and with radius R. It then follows from Theorem 5.2.5 that there exists $u \in \overline{B(o;R)}$ such that

$$u = S(u).$$

We have thus proved the following assertion.

Theorem 8.3.5. *Let $\Omega \in C^{2,\alpha}$ satisfy condition (S). Under the assumptions on a_{ij}, b_i, c stated above, given $f \in C^{0,\alpha}(\overline{\Omega})$, quasilinear Dirichlet problem* (8.3.8) *has at least one classical solution $u \in C^{2,\alpha}(\overline{\Omega})$.*

Remark 8.3.6. The assertion similar to Theorem 8.3.5 holds also for a more general quasilinear Dirichlet problem, where the coefficients of the differential operator depend not only on $u(x)$ but also on x and $\nabla u(x)$:

$$\begin{aligned}\widetilde{Q}(u)(x) := &- \sum_{i,j=1}^{N} a_{ij}(x, u(x), \nabla u(x)) \frac{\partial^2 u}{\partial x_i \partial x_j} \\ &+ \sum_{i=1}^{N} b_i(x, u(x), \nabla u(x)) \frac{\partial u}{\partial x_i} \\ &+ c(x, u(x), \nabla u(x))u.\end{aligned}$$

However, in this case the smoothness of a_{ij}, b_i and c has to be replaced by more technical hypotheses and the results from Appell & Zabreiko [9] must be employed rather than Lemma 5.3.2.

Let $\Omega \subset \mathbb{R}^N$ be a bounded domain, $\Omega \in C^{0,1}$. Let a_{ij}, b_i, c, $i, j = 1, \dots, N$, be measurable functions in Ω and consider an operator of the form

$$\widehat{L}u(x) := - \sum_{i,j=1}^{N} \frac{\partial}{\partial x_i} \left(a_{ij}(x) \frac{\partial u}{\partial x_j} \right) + \sum_{i=1}^{N} b_i(x) \frac{\partial u}{\partial x_i} + c(x)u. \tag{8.3.13}$$

An operator $\widehat{L}$ takes the form of the operator L from the beginning of this section provided the principal coefficients a_{ij} are differentiable. Then the theory of classical solutions can be applied. However, the *divergence form of the operator* $\widehat{L}$ has the advantage that it may be defined for significantly broader classes of functions than $C^{2,\alpha}(\overline{\Omega})$ as well as more general domains may be considered than in the classical case.

If we assume that the function u belongs to $W^{1,1}_{loc}(\Omega)$ and that the functions $\sum\limits_{j=1}^{N} a_{ij} \frac{\partial u}{\partial x_j} + b_i \frac{\partial u}{\partial x_i} + cu$, $i = 1, \dots, N$, are locally integrable, then in a *weak sense*, u is said to satisfy $\widehat{L}u = 0$ according as

$$\mathcal{L}(u, v) := \int_{\Omega} \left\{ \sum_{i,j=1}^{N} a_{ij} \frac{\partial u}{\partial x_j} \frac{\partial v}{\partial x_i} + (b_i \frac{\partial u}{\partial x_i} + c)v \right\} dx = 0$$

for all nonnegative functions $v \in \mathcal{D}(\Omega)$.

Provided the coefficients a_{ij}, b_i, c of $\widehat{L}$ are locally integrable, it follows from the Green formula that a function $u \in C^2(\overline{\Omega})$ satisfying $\widehat{L}(u) = 0$ in the classical sense (pointwise in Ω) also satisfies this relation in the weak sense. Moreover, if the coefficients a_{ij} have locally integrable derivatives, then a weak solution $u \in C^2(\Omega)$ is also a classical solution.

Let f be locally integrable in Ω. Then a function u is called a weak solution of the nonhomogeneous equation

$$\widehat{L}u = f \tag{8.3.14}$$

if

$$\mathcal{L}(u,v) = \int_\Omega fv\,dx \tag{8.3.15}$$

for all $v \in \mathcal{D}(\Omega)$ (cf. Section 5.5). Similarly as above, the classical solution of (8.3.14) is also a weak solution and a weak solution which belongs to $C^2(\Omega)$ is also a classical solution when the coefficients of $\widehat{L}$ are sufficiently smooth.

We are going to study the weak solution of the Dirichlet problem associated with (8.3.14). It depends on the coefficients of $\widehat{L}$ whether or not this problem is naturally well-posed.

We shall assume that $\widehat{L}$ is uniformly elliptic operator, i.e., (8.3.1) is satisfied for a_{ij}. We also assume that $\widehat{L}$ has coefficients in $L^\infty(\Omega)$. That is, there exist constants $K > 0$ and $\nu > 0$ such that for a.a. $x \in \Omega$ we have

$$\sum_{i,j=1}^N |a_{ij}(x)| \le K, \ \sum_{i=1}^N |b_i(x)| + |c(x)| \le \nu. \tag{8.3.16}$$

It follows from (8.3.16) that there exists a constant $C > 0$ such that

$$\begin{aligned} |\mathcal{L}(u,v)| &\le \int_\Omega \left\{ \left| \sum_{i,j=1}^N a_{ij} \frac{\partial u}{\partial x_j} \frac{\partial v}{\partial x_i} \right| + \sum_{i=1}^N \left| b_i v \frac{\partial v}{\partial x_i} \right| + |cuv| \right\} dx \\ &\le C \|u\|_{W^{1,2}(\Omega)} \|v\|_{W^{1,2}(\Omega)} \end{aligned} \tag{8.3.17}$$

for any $u, v \in W^{1,2}(\Omega)$ by the Hölder inequality. Hence, for fixed $u \in W^{1,2}(\Omega)$, the mapping $v \mapsto \mathcal{L}(u,v)$ is a bounded linear functional on $W_0^{1,2}(\Omega)$ (cf. Section 5.6). Consequently, the validity of $\mathcal{L}(u,v) = 0$ for $v \in \mathcal{D}(\Omega)$ implies its validity for $v \in W_0^{1,2}(\Omega)$ as well.

The estimate (8.3.17) shows that for $u \in W^{1,2}(\Omega)$, $\widehat{L}u$ may be defined as an element of the dual space of $W_0^{1,2}(\Omega)$ by setting $\widehat{L}u(v) = \mathcal{L}(u,v)$, $v \in W_0^{1,2}(\Omega)$. By virtue of the Riesz Representation Theorem (Theorem 1.2.40), $W_0^{1,2}(\Omega)$ may be identified with its dual and consequently the operator $\widehat{L}$ can be interpreted as a map $W^{1,2}(\Omega) \to W_0^{1,2}(\Omega)$.

The weak solution $u \in W_0^{1,2}(\Omega)$ of the Dirichlet problem

$$\begin{cases} \widehat{L}u(x) = f(x) & \text{in} \quad \Omega, \\ u = 0 & \text{on} \quad \partial\Omega \end{cases} \tag{8.3.18}$$

where $f \in L^2(\Omega)$ is then a solution of the operator equation

$$\widehat{L}u = \widehat{f}, \tag{8.3.19}$$

where $\widehat{f} \in W_0^{1,2}(\Omega)$ is an element determined by

$$(\widehat{f}, v)_{W_0^{1,2}(\Omega)} = \int_\Omega f v \, dx$$

for all $v \in W_0^{1,2}(\Omega)$ and

$$\widehat{L} \colon W_0^{1,2}(\Omega) \to W_0^{1,2}(\Omega).$$

Note that $u = 0$ on $\partial\Omega$ is understood in the sense of traces (see Section 5.5). The following existence and uniqueness result holds.

Proposition 8.3.7. *Let $c \geq 0$ in Ω, $f \in L^2(\Omega)$. Then there exists unique weak solution $u \in W_0^{1,2}(\Omega)$ of the Dirichlet problem* (8.3.18). *Moreover, there exists a constant $k > 0$ (independent of f) such that*

$$\|u\|_{W_0^{1,2}(\Omega)} \leq k\|f\|_{L^2(\Omega)}.$$

Proof. Proof of this assertion is a direct consequence of the Lax-Milgram Theorem (Proposition 1.2.41). □

We assume now that the coefficients of $\widehat{L}$ depend not only on x but also on the solution u. More precisely, we assume that $a_{ij} = a_{ij}(x,s)$, $b_i = b_i(x,s)$ and $c = c(x,s)$ satisfy the following assumptions: $a_{ij}, b_i, c \in CAR(\Omega \times \mathbb{R})$ and there exists a constant $\kappa > 0$ such that

$$\sum_{i,j=1}^N a_{ij}(x,s)\, \xi_i \xi_j \geq \kappa \sum_{i=1}^N \xi_i^2 \tag{8.3.20}$$

for a.a. $x \in \Omega$ and $s, \xi_i \in \mathbb{R}$; there are constants $K > 0$ and $\nu > 0$ such that $c(x,s) \geq 0$, and

$$\sum_{i,j=1}^N |a_{ij}(x,s)| \leq K, \ \sum_{i=1}^N |b_i(x,s)| + |c(x,s)| \leq \nu \tag{8.3.21}$$

for a.a. $x \in \Omega$ and $s \in \mathbb{R}$. We shall consider quasilinear operator

$$\begin{aligned}\widehat{Q}(u)(x) := &- \sum_{i,j=1}^N \frac{\partial}{\partial x_i}\left(a_{ij}\left(x, u(x)\right) \frac{\partial u}{\partial x_j}\right) \\ &+ \sum_{i=1}^N b_i\left(x, u(x)\right) \frac{\partial u}{\partial x_i} + c\left(x, u(x)\right) u\end{aligned}$$

and look for a weak solution of the Dirichlet problem

$$\begin{cases} \widehat{Q}(u)(x) = f(x) & \text{in} \quad \Omega, \\ u = 0 & \text{on} \quad \partial\Omega \end{cases} \tag{8.3.22}$$

where $f \in L^2(\Omega)$ is a given function. Similarly as in the case of a classical solution, for a fixed $v \in W_0^{1,2}(\Omega)$ we define

$$\begin{aligned} \widehat{L}_v u(x) := &- \sum_{i,j=1}^{N} \frac{\partial}{\partial x_i} \left(a_{ij}\left(x, v(x)\right) \frac{\partial u}{\partial x_j} \right) \\ &+ \sum_{i=1}^{N} b_i\left(x, v(x)\right) \frac{\partial u}{\partial x_i} + c\left(x, v(x)\right) u. \end{aligned}$$

It follows from (8.3.20), (8.3.21) that $\widehat{L}_v$ is of the form (8.3.13) and its coefficients satisfy (8.3.1) and (8.3.16). We also observe that $\widehat{L}_u u = \widehat{Q}(u)$ for $u \in W_0^{1,2}(\Omega)$. The idea of solving (8.3.22) by the linearization method remains the same as in the case of a classical solution, i.e., for a fixed $v \in W_0^{1,2}(\Omega)$ we find unique weak solution $u \in W_0^{1,2}(\Omega)$ of the linear problem

$$\begin{cases} \widehat{L}_v u(x) = f(x) & \text{in} \quad \Omega, \\ u = 0 & \text{on} \quad \partial\Omega \end{cases} \tag{8.3.23}$$

and then apply the Schauder Fixed Point Theorem to get a solution of (8.3.22). Existence and uniqueness results for the linear problem (8.3.23) follow from Proposition 8.3.7. It remains to prove that the mapping $S : W_0^{1,2}(\Omega) \to W_0^{1,2}(\Omega)$, $u = S(v)$, has a fixed point. For this purpose we need the following technical assertion.

Lemma 8.3.8. *Let $v_n \to v$ in $L^2(\Omega)$. Then there exists a subsequence denoted again by $\{v_n\}_{n=1}^{\infty}$ such that for any $w \in W_0^{1,2}(\Omega)$ we have*

$$\| \left(\widehat{L}_v - \widehat{L}_{v_n} \right) w \|_{W_0^{1,2}(\Omega)} \to 0$$

as $n \to \infty$.

Proof. The assumption $v_n \to v$ in $L^2(\Omega)$ yields the existence of a subsequence, denoted again by $\{v_n\}_{n=1}^{\infty}$, such that $v_n \to v$ a.e. in Ω (see Remark 1.2.18). The Dual Characterization of the Norm (Corollary 2.1.16), the Hölder inequality and the Lebesgue Theorem (cf. (8.3.21) for suitable integrable majorant) imply

$$
\begin{aligned}
\|(\widehat{L}_v - \widehat{L}_{v_n})w\|_{W_0^{1,2}(\Omega)} &= \sup_{\|f\|_{W_0^{1,2}(\Omega)}\le 1} \left| \left((\widehat{L}_v - \widehat{L}_{v_n})w, \widehat{f} \right)_{W_0^{1,2}(\Omega)} \right| \\
&\le \sup_{\|\widehat{f}\|_{W_0^{1,2}(\Omega)}\le 1} \int_\Omega \left\{ \sum_{i,j=1}^N |a_{ij}(x,v(x)) - a_{ij}(x,v_n(x))| \left|\frac{\partial w}{\partial x_j}\right| \left|\frac{\partial \widehat{f}}{\partial x_i}\right| \right. \\
&\quad + \sum_{i=1}^N |b_i(x,v(x)) - b_i(x,v_n(x))| \left|\frac{\partial w}{\partial x_i}\right| \left|\widehat{f}\right| \\
&\quad \left. + |c(x,v(x)) - c(x,v_n(x))|\,|w|\left|\widehat{f}\right| \right\} dx \\
&\le k \left\{ \sum_{i,j=1}^N \|a_{ij}(\cdot,v(\cdot)) - a_{ij}(\cdot,v_n(\cdot))\|_{L^\infty(\Omega)} \right. \\
&\quad + \sum_{i=1}^N \|b_i(\cdot,v(\cdot)) - b_i(\cdot,v_n(\cdot))\|_{L^\infty(\Omega)} \\
&\quad \left. + \|c(\cdot,v(\cdot)) - c(\cdot,v_n(\cdot))\|_{L^\infty(\Omega)} \right\} \|w\|_{W_0^{1,2}(\Omega)} \to 0
\end{aligned}
$$

as $n \to \infty$ (with a constant k independent of n). □

For a fixed $f \in L^2(\Omega)$ we denote by $u(v) = \widehat{L}_v^{-1}\widehat{f} \in W_0^{1,2}(\Omega)$ a unique weak solution of (8.3.23). Assume that $\{v_n\}_{n=1}^\infty$ is a bounded sequence in $W_0^{1,2}(\Omega)$. Passing to subsequences, we may assume that there exists $v \in W_0^{1,2}(\Omega)$ such that $v_n \rightharpoonup v$ weakly in $W_0^{1,2}(\Omega)$ and hence $v_n \to v$ strongly in $L^2(\Omega)$ and $v_n \to v$ a.e. in Ω. Proposition 8.3.7 applied to (8.3.23) together with Lemma 8.3.8 yield

$$
\begin{aligned}
\|u(v_n) - u(v)\|_{W_0^{1,2}(\Omega)} &= \left\| \widehat{L}_{v_n}^{-1}\widehat{f} - \widehat{L}_v^{-1}\widehat{f} \right\|_{W_0^{1,2}(\Omega)} \\
&= \left\| \widehat{L}_{v_n}^{-1}(\widehat{L}_v - \widehat{L}_{v_n})\widehat{L}_v^{-1}\widehat{f} \right\|_{W_0^{1,2}(\Omega)} \\
&\le \left\| \widehat{L}_{v_n}^{-1} \right\|_{\mathcal{L}(W_0^{1,2}(\Omega))} \left\| (\widehat{L}_v - \widehat{L}_{v_n})\widehat{L}_v^{-1}\widehat{f} \right\|_{W_0^{1,2}(\Omega)} \to 0
\end{aligned}
$$

due to the fact that the uniform estimates (8.3.20) and (8.3.21) ensure that

$$
\left\| \widehat{L}_{v_n}^{-1} \right\|_{\mathcal{L}(W_0^{1,2}(\Omega))} \le k
$$

with a constant $k > 0$ independent of v_n. Hence $u(v_n) \to u(v)$ in $W_0^{1,2}(\Omega)$ which proves the compactness of the operator S (note that the continuity of S is proved similarly). Taking advantage of the uniform estimates once again, we also have

$$
\|u(v)\|_{W_0^{1,2}(\Omega)} \le k \left\| \widehat{f} \right\|_{W_0^{1,2}(\Omega)}
$$

for any $v \in W_0^{1,2}(\Omega)$. This implies that there exists a ball $B(o;R) \subset W_0^{1,2}(\Omega)$ such that

$$S(\overline{B(o;R)}) \subset \overline{B(o;R)}.$$

It then follows from Theorem 5.2.5 that there exists $u \in \overline{B(o;R)}$ such that $u = S(u)$. This proves the following assertion.

Theorem 8.3.9. *Let $\Omega \in C^{0,1}$ be a bounded domain. Under the assumptions on a_{ij}, b_i and c stated above, given $f \in L^2(\Omega)$, quasilinear Dirichlet problem in the divergence form* (8.3.22) *has at least one weak solution $u \in W_0^{1,2}(\Omega)$.*

Remark 8.3.10. Let us consider the linear differential operator L with coefficients a_{ij}, b_i and c satisfying (8.3.1), $a_{ij} \in C(\overline{\Omega})$, b_i, $c \in L^\infty(\Omega)$. Let $f \in L^p(\Omega)$. A function $u \in W^{2,p}(\Omega) \cap W_0^{1,p}(\Omega)$ is called a *strong solution* of the Dirichlet problem (8.3.2), provided the equation is satisfied almost everywhere. The concept of the strong solution represents an intermediate case between the classical and weak solution. Let $\Omega \in C^{1,1}$ and $c \geq 0$. Then the Dirichlet problem (8.3.2) has unique strong solution and the estimate (8.3.5) holds (cf. Gilbarg & Trudinger [71]).

Exercise 8.3.11. Formulate the assumptions on $a_{ij} = a_{ij}(x, u(x))$, $b_i = b_i(x, u(x))$, $c = c(x, u(x))$ which guarantee the existence of at least one strong solution of the quasilinear problem

$$\begin{cases} \overline{Q}(u)(x) = f(x) \text{ in } \Omega, \\ u = 0 \text{ on } \partial\Omega, \end{cases}$$

$f \in L^p(\Omega)$, where

$$\overline{Q}(u)(x) := -\sum_{i,j=1}^{N} a_{ij}(x, u(x)) \frac{\partial^2 u}{\partial x_i \partial x_j} + \sum_{i=1}^{N} b_i(x, u(x)) \frac{\partial u}{\partial x_i} + c(x, u(x))u.$$

Exercise 8.3.12. Provide the details of the proof of Proposition 8.3.7.

Exercise 8.3.13. Consider the quasilinear operator $\widetilde{Q}$ from Remark 8.3.6. Use the results from [9] about continuity of the Nemytski operator between Hölder spaces and formulate the existence result for the Dirichlet problem

$$\begin{cases} \widetilde{Q}(u)(x) = f(x) \text{ in } \Omega, \\ u = 0 \text{ on } \Omega, \end{cases} \tag{8.3.24}$$

where $f \in C^{0,\alpha}(\overline{\Omega})$.

Exercise 8.3.14. Use the linearization method to solve (8.3.24) but in contrast with Exercise 8.3.13 use the Contraction Principle (Theorem 2.3.1) instead of the Schauder Fixed Point Theorem.

8.4 Equations on Riemann Manifolds

In Appendix 4.3C we have defined a Riemann manifold as a differentiable manifold $\mathscr{M}$ with a smooth function (i.e., Riemann metric) $g : \mathscr{M} \to T\mathscr{M} \times T\mathscr{M}$ such that $g_x \in T_x\mathscr{M} \times T_x\mathscr{M}$ is a symmetric positive definite bilinear form for each $x \in \mathscr{M}$, i.e.,

$$g_x(v, w) =: (v, w)_g$$

is a scalar product on $T_x\mathscr{M}$.

In order to study global properties of a Riemann manifold we need to compare vectors in different tangent spaces. In particular, we look for a definition of derivative of vector fields in the direction of other vector fields (cf. the notion of Lie derivative). To this end we introduce additional structure on a manifold (not necessarily Riemann) which is called a (linear) connection.

Definition 8.4.1. Let a mapping $D : T\mathscr{M} \times \mathcal{V}(\mathscr{M}) \to T\mathscr{M}$ be smooth and let it satisfy the following conditions:

(i) For $v \in T_x\mathscr{M}$ and $w \in \mathcal{V}(\mathscr{M})$ the value

$$D_v w := D(v, w) \in T_x\mathscr{M}.$$

(ii) For any $x \in \mathscr{M}$ the restriction of D to $T_x\mathscr{M} \times \mathcal{V}(\mathscr{M})$ is a bilinear mapping.

(iii)
$$D_v(fw) = \mathcal{L}_v(f)w + fD_v w$$
holds for any $v \in T_x\mathscr{M}$, $f \in \mathcal{F}(\mathscr{M})$, $w \in \mathcal{V}(\mathscr{M})$. Here $\mathcal{L}_v f(x) = \mathrm{d}f(x)v$ is the Lie derivative of f in the direction v).

Then D is called a *connection* on $\mathscr{M}$ and $D_v w$ is said to be a *covariant derivative* of the vector field w in the direction v.

Remark 8.4.2. Suppose that $\frac{\partial}{\partial y_1}, \dots, \frac{\partial}{\partial y_M}$ form a basis of $T_x\mathscr{M}$. If $v = \sum\limits_{i=1}^{M} v^i \frac{\partial}{\partial y_i}$ and $w = \sum\limits_{j=1}^{M} w^j(x) \frac{\partial}{\partial y_j}$ then

$$D_v w = \sum_{i=1}^{M} v^i D_{\frac{\partial}{\partial y_i}} w(x) = \sum_{i,j=1}^{M} v^i \left[\frac{\partial w^j}{\partial y_i}(x) \frac{\partial}{\partial y_j} + w^j(x) D_{\frac{\partial}{\partial y_i}} \left(\frac{\partial}{\partial y_j} \right) \right].$$

Since $D_{\frac{\partial}{\partial y_i}} \left(\frac{\partial}{\partial y_j} \right) \in T_x\mathscr{M}$ it can be expressed in the form

$$D_{\frac{\partial}{\partial y_i}} \left(\frac{\partial}{\partial y_j} \right) = \sum_{k=1}^{M} \Gamma_{ij}^k \frac{\partial}{\partial y_k}. \tag{8.4.1}$$

These coordinates Γ_{ij}^k are called the *Christoffel symbols* of the connection D with respect to local coordinates $y_1, \dots y_M$ at the point $x \in \mathscr{M}$. These symbols are smooth functions on $\mathscr{M}$ and they determine D. For the sake of brevity we will write D_i instead of $D_{\frac{\partial}{\partial y_i}}$ and $D_i w := \sum\limits_{k=1}^{M} (D_i w)^k \frac{\partial}{\partial y_k}$,

i.e.,

$$(D_i w)^k = \frac{\partial w^k}{\partial y_i} + \sum_{j=1}^{M} w^j \Gamma_{ij}^k. \tag{8.4.2}$$

Definition 8.4.3. Let $\gamma : \mathcal{I} \to \mathscr{M}$ be a C^1- curve on a manifold $\mathscr{M}$. A vector field w defined on $\gamma(\mathcal{I})$ is said to be *parallel* along γ provided $D_{\dot{\gamma}(t)} w\,(\gamma(t)) = 0$ for all $t \in \mathcal{I}$. There is also another formulation: A vector field w is given by the *parallel transport* of $w(a) \in T_a\mathscr{M}$ $(\gamma(t_0) = a)$.

The tangent vectors to γ can be expressed in the form $\dot{\gamma}(t) = \sum\limits_{i=1}^{M} v^i(t) \frac{\partial}{\partial y_i}$. A vector field $w(t) = \sum\limits_{j=1}^{M} w^j\,(\gamma(t))\,\frac{\partial}{\partial y_j}$ is parallel along γ provided

$$D_{\dot{\gamma}} w = \sum_{i=1}^{M} v^i D_i w = \sum_{i,j=1}^{M} \left(v^i \frac{\partial w^j}{\partial y_i} + \sum_{k=1}^{M} \Gamma_{ik}^j v^i w^k \right) \frac{\partial}{\partial y_j} = 0.$$

Since $\frac{\mathrm{d}w^j}{\mathrm{d}t} = \sum\limits_{i=1}^{M} \frac{\partial w^j}{\partial y_i} v^i$, a parallel vector field w has to satisfy the following system of linear ordinary differential equations

$$\frac{dw^k}{dt}\,(\gamma(t)) + \sum_{i,j=1}^{M} \Gamma_{ij}^k\,(\gamma(t))\, v^i\,(\gamma(t))\, w^j\,(\gamma(t)) = 0, k = 1, \dots, M, t \in \mathcal{I}. \tag{8.4.3}$$

By the theorem on existence and uniqueness for linear systems, each vector $w(a) \in T_a\mathscr{M}$ can be parallely transported along C^1- curve γ which passes through $a \in \mathscr{M}$.

Now we can ask whether there exists a curve γ the tangent vectors of which are parallel to γ – such a curve will be called *geodesic* . If we write $\gamma(t)$ locally in coordinates, i.e., $\gamma(t) = (y_1(t), \dots, y_M(t))$, then it follows from (8.4.3) that γ is a geodesic provided

$$\frac{\mathrm{d}^2 y^k}{\mathrm{d}t^2} + \sum_{i,j=1}^{M} \Gamma_{ij}^k\,(\gamma(t))\,\frac{\mathrm{d}y^i}{\mathrm{d}t}\frac{\mathrm{d}y^j}{\mathrm{d}t} = 0,\ k = 1, \dots, M. \tag{8.4.4}$$

This system is no longer linear, so only local existence and uniqueness for given initial values $\gamma(0) = a \in \mathscr{M}$, $\dot{\gamma}(0) = v_0 \in T_a\mathscr{M}$, hold. From the continuous dependence of a solution on initial data it follows that there exists a neighborhood $\mathcal{V}$ of $a \in \mathscr{M}$ and $r > 0$ such that for any $x \in \mathcal{V}$ and $v \in T_x\mathscr{M}$, $\|v\| < r$, there is a geodesic γ, $\gamma(0) = x$, $\dot{\gamma}(0) = v$ and γ is defined at least on the interval $[0, 1]$.

We put $\exp_x v := \gamma(1)$. For fixed $x \in \mathcal{V}$, the mapping

$$v \mapsto \exp_x v$$

is called an *exponential mapping*. It is a diffeomorphism of a ball $\mathcal{B}(o; r) \subset T_x\mathcal{M}$ onto a neighborhood (the so-called *normal neighborhood*) $\mathcal{U}$ of $x \in \mathcal{M}$. This fact is a consequence of smooth dependence of a solution of (8.4.4) on data and the Inverse Function Theorem. If an orthonormal basis in $T_x\mathcal{M}$ is chosen then the coordinates of $v \in \mathcal{B}(o; r)$ can be taken as coordinates of the point $\exp_x v$. They are called *normal coordinates*. It follows from the geodesic equation (8.4.4) that $\gamma_{\alpha v}(t) = \gamma_v(\alpha t)$ for any geodesic γ and $\alpha \in [0, 1]$, $\|v\| < r$.

Let $x, y \in \mathcal{U}$ (where $\mathcal{U}$ is defined above). Then there is unique $v \in T_x\mathcal{M}$, $\|v\| < r$, such that $\exp_x v = y$. Put $\gamma(t) = \exp_x tv$, $t \in [0, 1]$. Then γ is a geodesic which joins x and y.

In order to study the global behavior of geodesics it is convenient to introduce the length of a curve. Let g be a Riemann metric on $\mathcal{M}$. Then the length of a curve $\gamma : [\alpha, \beta] \to \mathcal{M}$ is defined as

$$l(\gamma) := \int_\alpha^\beta |\dot{y}(t)|_g \,\mathrm{d}t \left(= \int_\alpha^\beta (\dot{y}(t), \dot{y}(t))_g^{1/2} \,\mathrm{d}t \right).$$

Since geodesics are determined by choice of a connection on $\mathcal{M}$, we want to choose a connection on a Riemann manifold $(\mathcal{M}, g)$ which corresponds to g.

Consider first a surface $\mathcal{M} \subset \mathbb{R}^3$. A natural definition of covariant derivative of a smooth vector field $w \in \mathcal{V}(\mathcal{M})$ along a smooth curve γ is the vector field $D_{\dot{\gamma}} w(t)$ which is equal to the orthogonal projection (in $\mathbb{R}^3$) of $\frac{\mathrm{d}}{\mathrm{d}t} w(\gamma(t))$ onto $T_{\gamma(t)}\mathcal{M} (\subset \mathbb{R}^3)$. It is matter of calculation to show that the corresponding connection D has the following two properties:

(i) Let $\gamma : \mathcal{I} \to \mathcal{M}$ be a smooth curve and let v, w be smooth vector fields on $\mathcal{M}$ which are parallel to γ. Then

$$(v(t), w(t))_{\mathbb{R}^3} = (v(t_0), w(t_0))_{\mathbb{R}^3}$$

for all $t_0, t \in \mathcal{I}$.

(ii) Let v, w be smooth vector fields on $\mathcal{M}$. Then

$$D(v, w) - D(w, v) = [v, w],$$

where $[v, w]$ is the Lie bracket of (v, w) (see p. 204).

Our aim is to generalize this "natural" connection to a Riemann manifold. We are starting with the second property.

Definition 8.4.4. Let D be a connection on a manifold $\mathcal{M}$. Then the mapping T_D given by the formula

$$T_D(v, w) = D_v w - D_w v - [v, w], \ v, w \in \mathcal{V}(\mathcal{M}), \tag{8.4.5}$$

is called the *torsion* of D.

The torsion T_D is given in local coordinates $y_1, \dots, y_M$ by the following expression

$$T_D(v,w) = \sum_{i,j,k=1}^{M} \left(\Gamma_{ij}^k - \Gamma_{ji}^k\right) v^i w^j \frac{\partial}{\partial y_k} \tag{8.4.6}$$

for

$$v = \sum_{i=1}^{M} v^i \frac{\partial}{\partial y_i}, \quad w = \sum_{j=1}^{M} w^j \frac{\partial}{\partial y_j}.$$

Proposition 8.4.5. *Let $(\mathscr{M}, g)$ be a Riemann manifold. Then there exists a unique connection D on $\mathscr{M}$ (it is called the Levi-Civita connection or the induced connection) with the vanishing torsion and such that the metric g is invariant with respect to parallel transports given by D.*

Proof. We omit calculations and give only formulas for the Christoffel symbols of D:

$$\Gamma_{ij}^k = \Gamma_{ji}^k \text{ (torsion free connection – see (8.4.6))}$$

$$\Gamma_{ij}^k = \frac{1}{2} \sum_{l=1}^{M} \left[\frac{\partial g_{lj}}{\partial y_i} + \frac{\partial g_{il}}{\partial y_j} - \frac{\partial g_{ji}}{\partial y_l} \right] g^{lk} \tag{8.4.7}$$

where (g^{lk}) is the inverse matrix to (g_{ij}), i.e.,

$$\sum_{j=1}^{M} g_{ij} g^{jk} = \delta_{ik}.$$

The uniqueness of the induced connection follows from (8.4.7). □

Let D be the induced connection of a Riemann manifold $(\mathscr{M}, g)$ and let $z_1, \dots, z_M$ be corresponding normal coordinates at a with respect to the g-orthonormal basis $\frac{\partial}{\partial z_1}, \dots, \frac{\partial}{\partial z_M}$ in $T_a\mathscr{M}$, i.e., $g_a\left(\frac{\partial}{\partial z_i}, \frac{\partial}{\partial z_j}\right) = \delta_{ij}$. Since the parallel transport is g-invariant, we get $\frac{\partial g_{ij}}{\partial z_k}(a) = 0$ for all i, j, k and, by (8.4.7), $\Gamma_{ij}^k(a) = 0$ in normal coordinates. This means that the covariant differentiation is a standard one (see (8.4.2)).

Let us recall a metric structure on a Riemann manifold given by the metric (distance function) 4.3.117 (iii).

It can be shown (see Exercise 8.4.17 below) that $\widetilde{\gamma}(t) = \exp_x(tv)$ is a unique geodesic of minimal length which joins x and $y = \widetilde{\gamma}(1)$. Now we consider global assertion.

Proposition 8.4.6. *Let $\widetilde{\gamma}$ be a C^1-curve joining points $x, y \in \mathscr{M}$ such that $\varrho(x,y) = l(\widetilde{\gamma})$. Then $\widetilde{\gamma}$ is a geodesic.*

Proof. We need to show that $\widetilde{\gamma}$ satisfies the equation (8.4.4) which should be the Euler equation (see Remark 7.1.12) for the length functional $l(\gamma)$. There is an obstacle consisting in the fact that we are not working on a linear space but on a manifold. Fortunately, this can be overcome by using normal coordinates. Divide $\widetilde{\gamma}$ in a finite number of parts $\widetilde{\gamma}_1, \dots, \widetilde{\gamma}_n$ in such a way that each part $\widetilde{\gamma}_i$ belongs to one normal neighborhood $\mathcal{U}_i$. If $\widetilde{\gamma}_i$ joins points x_{i-1}, x_i ($x_0 = x$, $x_n = y$) then obviously $\varrho(x_{i-1}, x_i) = l(\widetilde{\gamma}_i)$ and, therefore $\widetilde{\gamma}_i$, is a geodesic (see Exercise 8.4.17). □

Theorem 8.4.7 (Hopf-Rinow). *Let* $(\mathscr{M}, g)$ *be a connected Riemann manifold with the distance function* ϱ. *Then* (i) *and* (ii) *are equivalent:*

(i) *The metric space* $(\mathscr{M}, \varrho)$ *is complete.*

(ii) *Any geodesic can be extended on the entire* $\mathbb{R}$.

Moreover, (ii) *(or* (i)*) implies*

(iii) *Any two points* $x, y \in \mathscr{M}$ *can be joined by a geodesic* γ *and* $l(\gamma) = \varrho(x, y)$. [4]

Proof. We will assume that all curves are parametrized by arc length s, in particular, their tangent vectors are unit vectors.

(i) $\Rightarrow$ (ii): Let γ be a geodesic defined on an interval $[0, \tau)$, $\tau < \infty$. Choose an increasing sequence $\{s_n\} \subset [0, \tau)$, $\lim_{n \to \infty} s_n = \tau$. Put $y_n = \gamma(s_n)$. Since $\varrho(y_n, y_m) \le |s_n - s_m|$, the sequence $\{y_n\}$ is convergent to some $y \in \mathscr{M}$. Choose now a normal neighborhood $\mathcal{U}$ of y. Since $y_n \in \mathcal{U}$ for n large enough, there is a unique geodesic which joins y_n and y. This means that we can put $\gamma(\tau) = y$ and extend γ beyond τ, i.e., γ can be prolonged to the entire interval $[0, \infty)$ in this way. A similar argument applies for the interval $(-\infty, 0]$.

(ii) $\Rightarrow$ (iii): Let $x, y \in \mathscr{M}$ and denote $S_\delta(x) := \{\exp_x v : v \in T_x\mathscr{M}, |v|_g = \delta\}$. Assume that $S_\delta(x)$ is a subset of a normal neighborhood $\mathcal{U}$ of x. We can suppose that $y \notin \mathcal{U}$ for otherwise the existence of γ has been already proved, see Exercise 8.4.17 below. Since $S_\delta(x)$ is compact, there is $x_0 \in S_\delta(x)$, $x_0 = \exp_x v_0$, such that $\operatorname{dist}(y, S_\delta(x)) = \varrho(x_0, y)$. Notice that $\varrho(x, x_0) = |v_0|_g = \delta$. Let κ be a curve joining x and y. Then κ has to intersect $S_\delta(x)$, say at a point z. By the definition of ϱ, we have

$$l(\kappa) \ge \varrho(x, z) + \varrho(z, y) \ge \varrho(x, x_0) + \varrho(x_0, y).$$

Since κ is arbitrary,

$$\varrho(x, y) \ge \varrho(x, x_0) + \varrho(x_0, y).$$

But the opposite inequality holds too, i.e.,

$$\varrho(x, y) = \varrho(x, x_0) + \varrho(x_0, y).$$

By the assumption (ii), the curve $\gamma(s) = \exp_x\left(\frac{s v_0}{\delta}\right)$ is defined on $[0, \infty)$ and it is a geodesic. Denote $A := \{s \ge 0 : \varrho(y, \gamma(s)) + s = \varrho(x, y)\}$ and $\sigma = \sup A (\delta \in A)$.

[4]Notice that the geodesic γ solves the minimization problem for the length functional.

Since A is closed, $\sigma \in A$. It is sufficient to show that $\sigma = \varrho(x,y)$, i.e., $y = \gamma(\sigma)$. Assume that $\sigma < \varrho(x,y)$. We now use the procedure described above for $\widetilde{y} = \gamma(\sigma)$ and y instead of x and y, respectively, to obtain $\delta_1 > 0$ and $x_1 \in \mathscr{M}$ such that $\varrho(\widetilde{y}, x_1) = \delta_1$ and

$$\varrho(y,\widetilde{y}) = \varrho(y,x_1) + \varrho(x_1,\widetilde{y}).$$

Using the fact $\sigma \in A$ we have

$$\varrho(x,y) = \varrho(y,\widetilde{y}) + \sigma = \varrho(y,x_1) + \delta_1 + \sigma \le \varrho(x,x_1) + \varrho(x_1,y)$$

and the inequality

$$\varrho(x,x_1) \ge \delta_1 + \sigma$$

follows. But $\delta_1 + \sigma$ is the length of $\gamma|_{[0,\sigma]} \cup \gamma_1 = \widetilde{\gamma}$, where γ_1 is the geodesic joining $\widetilde{y}$ and x_1. Therefore $l(\widetilde{\gamma}) = \varrho(x,x_1)$ and, by Proposition 8.4.6, $\widetilde{\gamma}$ is a geodesic. By the uniqueness, $\widetilde{\gamma} \subset \gamma$. In particular, $x_1 = \gamma(\sigma+\delta_1)$ and $\sigma+\delta_1 \in A$, a contradiction with the assumption $\sigma < \varrho(x,y)$.

(ii) $\Rightarrow$ (i): It is sufficient to prove that any ϱ-bounded and closed subset $A \subset \mathscr{M}$ is compact. Let $\{y_n\} \subset A$. Choose $a \in \mathscr{M}$ and let γ_n be a geodesic joining a and y_n and $\varrho(a,y_n) = l(\gamma_n) = s_n$ (see (iii)), in particular, $y_n = \exp_a s_n v_n$ for $v_n \in T_a\mathscr{M}$, $|v_n|_g = 1$. There are subsequences $\{v_n\}$ and $\{s_n\}$ such that $v_n \to v_0 \in T_a\mathscr{M}$, $|v_0|_g = 1$, $s_n \to s_0$. By the assumption (ii), the geodesic $s \mapsto \exp_a s v_0$ is defined on $\mathbb{R}$. Denote $y_0 = \exp_a s_0 v_0$. By continuity of $\exp_a$, we have

$$y_n = \exp_a s_n v_n \to \exp_a s_0 v_0 = y_0,$$

i.e., A is compact. □

We have already defined the most important partial differential operator on a differentiable manifold, namely the Laplace–Beltrami operator, in Remark 4.3.104. Now we show how it can be defined on a Riemann manifold $(\mathscr{M}, g)$. Let D be the Levi-Civita connection on $\mathscr{M}$ and for a fixed choosen $v \in \mathcal{V}(\mathscr{M})$ consider the linear operator $A : u \to D_u v$ on $T_a\mathscr{M}$. Set

$$(\operatorname{div} v)(a) = \operatorname{tr} A$$

(the *divergence of a vector field* v), where $\operatorname{tr} A$ is the trace of A, see Exercise 1.1.44 (ii). If $\frac{\partial}{\partial y_1}, \dots, \frac{\partial}{\partial y_M}$ is a basis of $T_a\mathscr{M}$ then, by Exercise 1.1.44 (ii)

$$\operatorname{div} v = \sum_{i=1}^{M} \left(D_i v, \frac{\partial}{\partial y_i} \right)_g = \sum_{i,j=1}^{M} \left(\frac{\partial v^j}{\partial y_i} + \sum_{k=1}^{M} v_k \Gamma^j_{ik} \right) g_{ij},$$

where $g_{ij} = \left(\frac{\partial}{\partial y_i}, \frac{\partial}{\partial y_j} \right)_g$. In particular, if $g_{ij} = \delta_{ij}$ then

$$\operatorname{div} v = \sum_{i=1}^{M} \left(\frac{\partial v^i}{\partial y_i} + \sum_{k=1}^{M} v_k \Gamma^i_{ik} \right). \tag{8.4.8}$$

Let f be a smooth function on $(\mathscr{M}, g)$. Then $\mathrm{d}f(a) \in (T_a\mathscr{M})^*$ and, since g is a scalar product on $T_a\mathscr{M}$, there is unique $\nabla f(a) \in T_a\mathscr{M}$ (the so-called *g-gradient* of f) such that

$$\mathrm{d}f(a)(v) = (v, \nabla f(a))_g.$$

In local coordinates,

$$\nabla f = \sum_{j=1}^{M} \left(\sum_{i=1}^{M} \frac{\partial f}{\partial y_i} g^{ij} \right) \frac{\partial}{\partial y_j}, \tag{8.4.9}$$

where (g^{ij}) is the inverse matrix to (g_{ij}). If f is a smooth function on $\mathscr{M}$ then the *Laplace–Beltrami operator* Δf on a Riemann manifold $(\mathscr{M}, g)$ is defined by

$$\Delta f = \operatorname{div}(\nabla f). \tag{8.4.10}$$

For an equivalent expression see Exercise 8.4.18 (ii). If $(\mathscr{M}, g)$ is a Riemann manifold, which can be oriented by the volume form $\varpi = \sqrt{\det g} dy^1 \wedge \ldots \wedge dy^M$, it can be shown that the Laplacian Δ with $\operatorname{Dom} \Delta = \mathscr{D}(\mathscr{M})$ (C^∞-functions with compact supports) is symmetric and negative definite in $L^2(\mathscr{M}, \varpi)$. Its eigenfunctions form on an orthonormal basis in $L^2(\mathscr{M}, \varpi)$. Its spectrum, which is purely a point spectrum, is sometimes called the spectrum of $(\mathscr{M}, g)$. One can ask to what extent this spectrum determines geometric properties of $\mathscr{M}$. In other words "Can one hear the shape of a drum?", see Kac [83].

The role played by curvature of a curve in $\mathbb{R}^2$ and $\mathbb{R}^3$ is well known from elementary differential geometry. The notion of curvature can be extended also to a surface in $\mathbb{R}^3$ (or more generally in $\mathbb{R}^{N+1}$), in order to explain how a given surface is bent in various directions. Assume that $\mathscr{M}$ is a two-dimensional surface in $\mathbb{R}^3$. At any point $a \in \mathscr{M}$ there is the one-dimensional orthogonal complement (in $\mathbb{R}^3$) to $T_a\mathscr{M}$. A unit vector in this complement is called a normal vector. The choice of one of them, say n, yields orientation on $\mathscr{M}$ (cf. Example 4.3.91). The so-called Weingarten map $L_a(v) = -D_v n \in T_a\mathscr{M}$ is a self-adjoint map on $T_a\mathscr{M}$ (with respect to the scalar product inherited from $\mathbb{R}^3$, i.e., the induced metric g on the Riemann manifold $\mathscr{M}$). The classical geometry of surfaces is based on investigation of eigenvalues of L_a (they are called principal curvatures at a), e.g., their product is the Gauss curvature. The reader can consult, e.g., textbooks by Berger & Gostiaux [14] and Thorpe [152].

There is no reasonable notion of a normal to an abstract Riemann manifold $(\mathscr{M}, g)$. So the following definition of a special trilinear form is ambiguous.

Definition 8.4.8. Let D be a connection on $\mathscr{M}$. The mapping $R : \mathcal{V}(\mathscr{M}) \times \mathcal{V}(\mathscr{M}) \times \mathcal{V}(\mathscr{M}) \to \mathcal{V}(\mathscr{M})$ given by

$$R(v, w)u = D_v(D_w u) - D_w(D_v u) - D_{[v,w]}u$$

is called the *curvature* of D.

It is possible to express the curvature in local coordinates. If we denote the n-th coordinate of $R\left(\frac{\partial}{\partial y_i}, \frac{\partial}{\partial y_j}\right)\frac{\partial}{\partial y_k}$ by R^n_{kij} (notice the order of indices) we obtain after straightforward computation

$$R^n_{kij} = \frac{\partial \Gamma^n_{jk}}{\partial y_i} - \frac{\partial \Gamma^n_{ik}}{\partial y_j} + \sum_{l=1}^{M} \left(\Gamma^n_{il}\Gamma^l_{jk} - \Gamma^n_{jl}\Gamma^l_{ik}\right). \tag{8.4.11}$$

In particular,

$$\sum_{n=1}^{M} (D_i D_j u - D_j D_i u)^n \frac{\partial}{\partial y_n} = \sum_{k,n=1}^{M} R^n_{kij} u^k \frac{\partial}{\partial y_n}$$

represents a "measure of non-commutativity" of the second covariant derivatives $D_i D_j$ due to a bent manifold.

Definition 8.4.9. Let D be the induced connection of a Riemann manifold $(\mathscr{M}, g)$ and let R be the curvature of D. Let S be a 2-dimensional subspace of $T_a\mathscr{M}$. For a g-orthonormal basis v, w of S the value

$$K(S) := g\left(v, R(v,w)w\right)$$

is said to be the *sectional curvature* of S. If $K(S)$ is constant ($= c$) for all $S \subset T_a\mathscr{M}$ and each $a \in \mathscr{M}$ then $\mathscr{M}$ has the *constant curvature* c. If $K(S) > 0$ for all $S \subset T_a\mathscr{M}$ and each $a \in \mathscr{M}$ then $\mathscr{M}$ has a positive curvature. Manifolds of negative curvature are defined similarly.

Remark 8.4.10.

(i) It can be shown that the sectional curvature $K(S)$ does not depend on the choice of orthonormal basis of S.

(ii) If $\mathscr{M} \subset \mathbb{R}^3$ is 2-dimensional surface then the sectional curvature is the Gauss curvature.

Example 8.4.11.

(i) $\mathbb{R}^M$ as a Riemann manifold with $g(u,w) = (u,w)_{\mathbb{R}^M}$ has constant curvature $c = 0$.

(ii) The M-dimensional sphere of radius r with standard (i.e., induced) metric has constant curvature $c = \frac{1}{r^2}$.

(iii) The open unit ball with the metric

$$g_a(u,v) = 4\frac{(u,v)_{\mathbb{R}^2}}{1 - \|a\|^2}$$

(the so-called hyperbolic metric) has constant curvature $c = -1$. $\bigcirc$

For fixed vectors $v, w \in T_a\mathscr{M}$ consider the linear operator

$$P_{v,w} : u \to R(u,v)w$$

(the so-called Ricci operator). Its trace is given (see Exercise 1.1.44)

$$\mathrm{tr}(P_{v,w}) = \sum_{i=1}^{M} (P_{v,w}e_i, e_i)_g,$$

where $e_1, \dots, e_M$ is an orthonormal basis of $T_a\mathscr{M}$. As it was mentioned, the trace does not depend on the choice of basis and it is equal to the sum of eigenvalues. It is called the *Ricci curvature* and usually denoted by $\mathrm{Ric}_a(v,w)$. The *scalar curvature* at the point $a \in \mathscr{M}$ is defined by

$$s(a) := \sum_{j=1}^{M} \mathrm{Ric}_a(e_j, e_j),$$

where $e_1, \dots, e_M$ is any orthonormal basis of $T_a\mathscr{M}$. Choosing normal coordinates we get

$$s(a) = \sum_{i,j=1}^{M} R^i_{jij}, \tag{8.4.12}$$

where R^n_{kij} are given by (8.4.11).

It is not difficult to see that for an M-dimensional Riemann manifold $(\mathscr{M}, g)$ with constant curvature c, the equality

$$\mathrm{Ric}_a = c(M-1)g_a$$

holds for the Ricci curvature. More generally, the *Einstein manifold* is a Riemann manifold for which the Ricci curvature is proportional to the metric g.

In the sequel we will study transformations of Riemann metrics on a given manifold $\mathscr{M}$.

Definition 8.4.12. Let $(\mathscr{M}, g)$ and $(\mathscr{N}, h)$ be Riemann manifolds and let $F : \mathscr{M} \to \mathscr{N}$ be a diffeomorphism. Denote

$$(F^*g)_a(u,v) = h_{F(a)}(F_*u, F_*v)$$

for $u, v \in T_a\mathscr{M}$. Here F_* is the push-forward mapping $T_a\mathscr{M} \to T_{F(a)}\mathscr{N}$ (see p. 198). The mapping F is called

(i) an *isometry* if $(F^*g)_a = g_a$ for all $a \in \mathscr{M}$;

(ii) an *almost isometry* if there is $c > 0$ such that $(F^*g)_a = cg_a$ for each $a \in \mathscr{M}$;

(iii) a *conformal map* if there is a smooth function φ on $\mathscr{M}$ such that

$$(F^*g)_a = e^{\varphi(a)} g_a$$

for all $a \in \mathscr{M}$. In this case the metrics g, F^*g are said to be *conformal metrics*.

Remark 8.4.13.

(i) If F is a conformal mapping then F preserves angles between vectors, i.e.,

$$\cos \sphericalangle_g(u,v) := \frac{(u,v)_g}{\|u\|_g \|v\|_g} = \cos \sphericalangle_h(F_* u, F_* v).$$

(ii) If Riemann manifolds $(\mathscr{M}, g)$ and $(\mathscr{N}, h)$ have the same dimension and both have constant curvature (not necessarily the same) then there is a conformal map $F : \mathscr{M} \to \mathscr{N}$.

(iii) F. Gauss proved that (in modern language) any Riemann manifold of dimension 2 has a conformal metrics of constant curvature.

Let g and $\widetilde{g} = e^{\varphi} g$ be conformal metrics on $\mathscr{M}$ with scalar curvatures s and $\widetilde{s}$ given by (8.4.12). We are looking for a relation between these curvatures. After computation (with preference for normal coordinates) we obtain

$$\widetilde{s} = e^{-\varphi} \left[s + (M-1)\Delta\varphi - \frac{(M-1)(M-2)}{4} |\nabla \varphi|^2 \right].$$

This expression can be simplified for $M > 2$ by introducing a positive function f such that $f^{p-2} = e^{\varphi}$, where $p = \frac{2M}{M-2}$.[5] We have

$$\widetilde{s} f^{p-1} = [s + a\Delta] f =: \Box f. \tag{8.4.13}$$

Here $a = \frac{4(M-1)}{M-2}$ and $\Box$ is the so-called conformal Laplacian. Now we can formulate the *Yamabe problem* (Yamabe [160]):

Let $(\mathscr{M}, g)$ be a compact connected Riemann manifold of dimension M greater than 2. Is there a conformal metric $\widetilde{g}$ on $\mathscr{M}$ with a constant scalar curvature $\widetilde{s}$?

In other words: Does the nonlinear eigenvalue problem (8.4.13) have a positive solution? In accordance with the linear eigenvalue problem we could expect a positive eigenfunction for the first eigenvalue for which we have a variational definition (cf. Example 7.8.5). Denote $\overline{\omega} = \sqrt{\det g}\, dy^1 \wedge \dots \wedge dy^M$ the Riemann measure on $(\mathscr{M}, g)$ and set $\|f\|_p = \left(\int_{\mathscr{M}} |f|^p \overline{\omega} \right)^{1/p}$. Consider the functional

$$\Phi_p(f) = \frac{\int_{\mathscr{M}} [a|\operatorname{grad} f|^2 + s f^2] \overline{\omega}}{\|f\|_p^2}$$

on the space $W^{1,2}(\mathscr{M}, \overline{\omega})$. It is not difficult to see that $\widetilde{f}$ is a critical point of Φ_p if and only if $\widetilde{f}$ is a weak solution of (8.4.13) with $\widetilde{s} = \Phi_p(\widetilde{f}) \|\widetilde{f}\|_p$ (the equation (8.4.13) is the Euler-Lagrange equation for Φ). Since $W^{1,2} \subsetneq L^p \subsetneq L^2$ and s is continuous on $\mathscr{M}$, the functional Φ_p is bounded from below. Denote

[5] Notice that p is the critical exponent for compact embedding of $W^{1,2}$ into L^2 (Theorem 1.2.28).

$\Lambda(\mathscr{M}) = \inf\{\Phi_p(f);\ f$ is a smooth positive function on $\mathscr{M}\}$.[6] An idea for finding of an appropriate critical point consists in the following steps:

1) Prove the strong convergence of minimizing sequence to some $\widetilde{f}$. From the pointwise convergence a.e. get $\widetilde{f} \geq 0$.
2) Use some type of elliptic regularity (left-hand side in (8.4.13) is $\Lambda \widetilde{f}^{p-1}$) and the bootstrap argument to show that $\widetilde{f} \in C^\infty$.
3) With the help of the strong maximum principle prove that $\widetilde{f}$ is positive everywhere.

A little consideration shows that the first step is far from being simple, since from the boundedness in $W^{1,2}$ of a sequence $\{f_n\}$, $\Phi_p(f_n) \to \Lambda(\mathscr{M})$ (we can choose f_n such that $\|f_n\|_p = 1$) we get a subsequence weakly convergent to $\widetilde{f} \in W^{1,2}$. Without compactness of embedding of $W^{1,2}$ into L^p we cannot assure that $\|\widetilde{f}\|_p > 0$ and $\widetilde{f} \geq 0$ a.e. in $\mathscr{M}$. Instead of Φ_p we can consider Φ_q for $q < p$ and obtain nonnegative minimizers $\widetilde{f}_q$ of Φ_q. It can be proved (not easy!) that there is $r > p$ such that $\{\widetilde{f}_q\}$ is bounded in L^r provided $\Lambda(\mathscr{M}) < \Lambda(S^M)$ (S^M is the unit sphere in $\mathbb{R}^M$) and using the bootstrap argument that $\{\widetilde{f}_q\}$ is bounded also in C^2 and there exists a subsequence uniformly convergent to some $\widetilde{f}$, $\Phi_q(\widetilde{f}_q) \to \Phi_p(\widetilde{f}) = \Lambda(\mathscr{M})$. The last step in the proof that $\Lambda(\mathscr{M}) < \Lambda(S^M)$, whenever $(\mathscr{M}, g)$ is not conformal to S^M, requires a very complicated construction of function h such that $\Phi_p(h) < \Lambda(S^M)$. The interested reader can consult Aubin [12] and references therein.

There are similar problems, e.g., characterization of all scalar curvatures of conformal metrics. This problem for the sphere S^2 is known as the Nirenberg problem:

For which smooth g does there exist a solution on S^2 to the equation

$$\Delta\varphi + s = g e^{\varphi}?$$

Here $s = 8\pi$ is the scalar curvature of S^2 (with the standard metric).[7] See, e.g., Aubin [12], Chen & Xu [29] and references therein.

There are other important partial differential equations on a Riemann manifold, namely

$$\frac{\partial g(t)}{\partial t} = -2\,\mathrm{Ric}\,(g(t))\,, \tag{8.4.14}$$

where Ric $(g(t))$ is the Ricci curvature given by the metric $g(t)$ on $\mathscr{M}$. A one-parameter system of metrics g solving (8.4.14) is called a *Ricci flow* . Investigation of asymptotic behaviour of Ricci flows was started by R. Hamilton and it was continued by G. Perelman who has recently proved the Poincaré Conjecture: *"Every closed, orientable and simply connected* 3*-manifold is homeomorphic to* S^3*"*.

[6]It can be shown that $\Lambda(\mathscr{M})$ is invariant with respect to conformal transformations of $\mathscr{M}$.
[7]For $s < 8\pi$ the above described procedure works.

Details of this approach is out of the scope of this book. The interested reader can consult, e.g., Aubin [12], Chow, Lu & Ni [32] or Morgan & Tsz-Ho Fong [114].

Exercise 8.4.14. Let $S^2 \subset \mathbb{R}^3$ be the sphere $x_1^2 + x_2^2 + x_3^2 = 1$.

(i) Show that geodesics are parts of great circles $\gamma(t) = (\cos t)e_1 + (\sin t)e_2$ (where e_1, e_2 is an orthonormal basis in $\mathbb{R}^2$).

(ii) Is it true for any $x, y \in S^2$ that there exists a unique geodesic joining them?

Exercise 8.4.15. Let $\mathscr{M} = \{(x,y) \in \mathbb{R}^2; y > 0\}$ and let $g(u,v) = \frac{(u,v)_{\mathbb{R}^2}}{y^2}$ be the Riemann metric at the point (x,y). The manifold $(\mathscr{M}, g)$ is called the Poincaré half plane.

(i) Find all geodesics on $\mathscr{M}$.

(ii) Prove that $(\mathscr{M}, g)$ has constant curvature $c = -1$.

Exercise 8.4.16. Let $\mathscr{M}$ be a manifold with a connection which is given by Christoffel symbols Γ_{ij}^k in coordinates $y_1, \dots, y_M$. Show that the normal coordinates $z_1, \dots, z_M$ are given the transformation

$$y_k = z_k - \frac{1}{2} \sum_{i,j=1}^{M} \Gamma_{ij}^k z_i z_j.$$

Exercise 8.4.17. Let $(\mathscr{M}, g)$ be a Riemann manifold and let U be a normal neighborhood of $a \in \mathscr{M}$. Let $x = \exp_a v$. Show that $\gamma(t) = \exp_a(tv)$ which joins points a and x is a geodesic which has minimal length, namely

$$l(\gamma) = \varrho(a, x) = \|v\|_g$$

and it is a unique curve with this property.

Hint. Write $B_r = \bigcup_{\alpha \in [0,r)} S_\alpha$, where $S_\alpha = \{v \in T_x\mathscr{M} : g(v,v) = \alpha^2\}$, express g on S_α (in the polar coordinates) and show that γ is g-perpendicular to S_α. For computing $l(\gamma)$ use the arc-parametrization.

Exercise 8.4.18. Define a covariant derivative of a function f (i.e a 0-form) in the direction

$$v = \sum_{i=1}^{M} v^i \frac{\partial}{\partial y_i} \in T_a\mathscr{M} \text{ as } D_v f := \mathrm{d}f(a)v = \sum_{i=1}^{M} \frac{\partial f}{\partial y_i}(a) v^i$$

and a covariant derivative of a 1-form $\omega = \sum_{k=1}^{M} \omega_k \, \mathrm{d}y^k$ as a 1-form $D_v\omega$ such that the Leibniz rule is satisfied, i.e., $D_v(\omega(w)) = (D_v\omega)w + \omega(D_v w)$.

(i) Show that for $D_i\omega = \sum_{k=1}^{M} (D_i\omega)_k \, \mathrm{d}y^k$ we get

$$(D_i\omega)_k = \frac{\partial \omega_k}{\partial y_i} - \sum_{j=1}^{M} \Gamma^j_{ik}\omega_j.$$

(ii) Put $A : u \mapsto D_u\omega$ and define the so-called *co-differential* δ of a smooth 1-form ω as

$$\delta\omega := \operatorname{tr} A.$$

Express $\delta\omega$ in local coordinates and show for a smooth function f that

$$\Delta f = \delta(\,\mathrm{d}f),$$

where the Laplace-Beltrami operator Δ is defined by (8.4.10).

Exercise 8.4.19. Let L be the Weingarten map on a surface $\mathscr{M} \subset \mathbb{R}^3$. Show that the induced connection on $\mathscr{M}$ has its curvature R in the form

$$R(v, w)u = (Lw, u)Lv - (Lv, u)Lw.$$

Summary of Methods Presented in This Book

Fixed Point Methods

- Contraction Principle (Theorem 2.3.1)
- Browder Theorem for non-expansive mappings (Proposition 2.3.12)
- Brouwer Fixed Point Theorem (Theorem 5.1.3)
- Schauder Fixed Point Theorem (Theorem 5.2.5, Examples 5.8.3 and 5.8.4)
- Fixed Point Theorem for condensing mappings (Theorem 5.2.18)
- Linearization and Fixed Point Theorem (Theorems 8.3.5 and 8.3.9)

Local Differentiability Methods

- Local Inverse Function Theorem (Theorem 4.1.1)
- Implicit Function Theorem (Theorem 4.2.1)
- Crandall–Rabinowitz Local Bifurcation Theorem (Theorem 4.3.22)

Topological Methods

- Brouwer degree (Theorems 5.7.7 and 5.7.26)
- Leray–Schauder degree (Theorem 5.8.2)
- Topological degree of Browder and Skrypnik (Theorem 5.8.34)
- Krasnoselski Local Bifurcation Theorem (Theorem 5.8.12)
- Rabinowitz and Dancer Global Bifurcation Theorems (Theorems 5.8.22 and 5.8.26)

Monotonicity Methods

- Monotone operators theory in Hilbert space (Theorem 6.1.4)
- Monotone operators theory in Banach space (Theorem 6.1.21)
- Leray–Lions Theorem for operators which are monotone in the principal part (Theorem 6.1.22)

Methods in Ordered Spaces

- Monotone iterations, subsolutions and supersolutions (Theorem 6.3.16)
- Supersolutions, subsolutions and topological degree (Theorems 6.3.51 and 6.3.52)
- Supersolutions, subsolutions and global extrema (Theorem 7.2.33)

Variational Methods

- Necessary and sufficient conditions for local extrema (Propositions 7.1.2 and 7.1.4, Theorem 7.1.5)
- Global extrema (Theorems 7.2.4, 7.2.8 and 7.2.11)
- Relative extrema and Langrange Multiplier Method (Theorem 7.8.2)
- Krasnoselski Potential Bifurcation Theorem (Theorem 7.8.49)
- Mountain Pass Theorem (Theorem 7.4.5 in the Hilbert space setting, Theorem 7.4.25 in the Banach space setting)
- Minimax Principles (Theorems 7.8.12 and 7.8.14)
- Lusternik–Schnirelmann Method (Theorems 7.8.30, 7.8.33 and 7.8.37)
- Saddle Point Theorem (Theorem 7.6.3 in the Hilbert space setting, Theorem 7.6.12 in the Banach space setting)
- Linking Theorem (Theorem 7.6.13)

Approximative Methods

- Contraction Principle (Theorem 2.3.1)
- Newton Method (Appendix 3.2A)
- Ritz Method (Proposition 7.2.41)
- Finite Element Method (Theorem 7.2.45)
- Galerkin Method (Theorem 8.2.5)

Typical Applications

For the reader's convenience we stress here some typical problems and the methods illustrated by them.

Semilinear Problems — Ordinary Differential Equations

$$\boxed{\dot{x}(t) = f(t, x(t)), \qquad x(0) = x(1)}$$

- Lyapunov–Schmidt Reduction and Implicit Function Theorem (Example 4.3.15)
- Coincidence degree (Example 5.8.7)

$$\boxed{\ddot{x}(t) = f(t, x(t)), \qquad x(0) = x(1) = 0}$$

- Contraction Principle (Example 2.3.10)
- Schauder Fixed Point Theorem (Example 5.2.8)
- Supersolutions and subsolutions (Example 6.3.19)
- Supersolutions and subsolutions combined with global extrema (Appendix 7.2A)

$$\boxed{-\ddot{x}(t) + g(x(t)) = f(t), \qquad x(0) = x(1) = 0}$$

- Monotone operators (Example 6.1.10)
- Extreme Value Theorem (special case $g(s) = s^3$, Examples 7.2.6, 7.2.14 and 7.2.21)
- Saddle Point Theorem (Example 7.6.5)

$$\boxed{-\ddot{x}(t) + \lambda x(t) = |x(t)|^{p-2}x(t), \qquad x(0) = x(\pi) = 0}$$

- Mountain Pass Theorem (Example 7.4.7)

$$\boxed{-\ddot{x}(t) + a(t)x(t) = f(t, x(t)), \qquad x(0) = x(1) = 0}$$

- Linking Theorem (Example 7.6.14)

$$\boxed{\ddot{x}(t) + \lambda x(t) = 0, \qquad x(0) = x(1) = 0}$$

- Courant–Weinstein Variational Principle (Example 7.8.15)

$$\boxed{-(p(t)\dot{x}(t))^{\cdot} + q(t)x(t) = \lambda x(t), \qquad x(a) = x(b) = 0}$$

- Hilbert–Schmidt Theorem (Example 2.2.17)
- Krein–Rutman Theorem (Exercise 6.3.48)

$$\boxed{\dot{x}(t) = \mathbf{A}(t)x(t) + g(t, x(t)), \qquad x(0) = x(T)}$$

- Brouwer Fixed Point Theorem (Example 5.1.6)

$$\boxed{\ddot{x}(t) + \lambda \sin x(t) = 0, \qquad x(0) = x(2\pi), \quad \dot{x}(0) = \dot{x}(2\pi)}$$

- Lyapunov–Schmidt Reduction (Example 4.3.24)

$$\boxed{\ddot{x}(t) + \lambda x(t) + g(\lambda, t, x(t)) = 0, \qquad x(0) = x(\pi) = 0}$$

- Krasnoselski Local Bifurcation Theorem (Exercise 5.8.21)
- Dancer Global Bifurcation Theorem (Example 5.8.27)

$$\boxed{\ddot{x}(t) + \lambda x(t) + g(\lambda, t, x(t)), \qquad x(0) = x(2\pi), \quad \dot{x}(0) = \dot{x}(2\pi)}$$

- Krasnoselski Potential Bifurcation Theorem (Example 7.8.55)

$$\boxed{\ddot{x}(t) = f(t, x(t), \dot{x}(t)), \qquad x(0) = x(1) = 0}$$

- Leray–Schauder degree (Example 5.8.5)

$$\boxed{\ddot{x}(t) + \lambda x(t) + g(\lambda, t, x(t), \dot{x}(t)), \qquad x(0) = x(2\pi), \quad \dot{x}(0) = \dot{x}(2\pi)}$$

- Crandall–Rabinowitz Bifurcation Theorem (Example 4.3.25)

Semilinear Problems — Partial Differential Equations

$$\boxed{-\Delta u(x) = \mu u(x) \quad \text{in } \Omega, \qquad u = 0 \quad \text{on } \partial\Omega}$$

- Krein–Rutman Theorem (Example 6.3.42)

$$\boxed{-\Delta u(x) = g(u(x)) + f(x) \quad \text{in } \Omega, \qquad u = 0 \quad \text{on } \partial\Omega}$$

- Schauder Fixed Point Theorem — existence of a classical solution (Theorem 5.4.1)
- Global extrema — existence of a weak solution (Theorem 7.3.1)

$$\boxed{-\Delta u(x) = g(x, u(x)) \quad \text{in } \Omega, \qquad u = 0 \quad \text{on } \partial\Omega}$$

- Contraction Principle — existence of a weak solution (Theorem 5.6.1)
- Schauder Fixed Point Theorem — existence of a weak solution (Proposition 5.6.2, Theorems 5.6.3 and 5.6.6)
- Monotone operators — existence of a weak solution (Theorem 6.2.2)

$$\boxed{-\Delta u(x) = \lambda u(x) + h(x, u(x)) \quad \text{in } \Omega, \qquad u = 0 \quad \text{on } \partial\Omega}$$

- Leray–Schauder degree — existence of a weak solution (Theorems 5.9.2 and 5.9.3, Exercises 5.9.8, 5.9.9, 5.9.10 and 5.9.11)
- Saddle Point Theorem — existence of a weak solution (Theorem 7.7.1)

$$\boxed{-\Delta u(x) + \lambda u(x) = |u(x)|^{p-2}u(x) \quad \text{in } \Omega, \qquad u = 0 \quad \text{on } \partial\Omega}$$

- Mountain Pass Theorem — existence of a positive weak solution (Theorem 7.5.1)

Quasilinear Problems — Ordinary Differential Equations

$$(|\dot{x}(t)|^{p-2}\dot{x}(t))^{\cdot} + \lambda|x(t)|^{p-2}x(t) = 0, \qquad x(0) = x(1) = 0$$

- Lagrange Multiplier Method (Example 7.8.5)
- Lusternik–Schnirelmann Method (Example 7.8.38)

$$-(|\dot{x}(t)|^{p-2}\dot{x}(t))^{\cdot} + g(x(t)) = f(t), \qquad x(0) = x(1) = 0$$

- Topological degree of Browder and Skrypnik (Example 5.8.38)
- Browder Theorem (Example 6.1.23)

$$-(|\dot{x}(t)|^{p-2}\dot{x}(t))^{\cdot} = f(t, x(t)), \qquad x(0) = x(1) = 0$$

- Supersolutions and subsolutions combined with the topological degree (Appendix 6.3B)

$$-(|\dot{x}(t)|^{p-2}\dot{x}(t))^{\cdot} + \lambda x(t) = |x(t)|^{r-2}x(t), \qquad x(0) = x(1) = 0$$

- Moutain Pass Theorem (Exercise 7.4.27)

Quasilinear Problems — Partial Differential Equations

$$-\Delta_p u(x) + g(x, u(x), \nabla u(x)) = f(x) \quad \text{in } \Omega, \qquad u = 0 \quad \text{on } \partial\Omega$$

- Leray–Lions Theorem — existence of a weak solution (Appendix 6.2A)

$$-\Delta_p u(x) = \lambda_1 |u(x)|^{p-2}u(x) + f(x, u(x)) - h(x) \quad \text{in } \Omega, \qquad u = 0 \quad \text{on } \partial\Omega$$

- Saddle Point Theorem — existence of a weak solution (Appendix 7.7A)

$$-\Delta_p u(x) = \lambda|u(x)|^{p-2}u(x) + f(\lambda, x, u(x)) \quad \text{in } \Omega, \qquad u = 0 \quad \text{on } \partial\Omega$$

- Topological degree of Browder and Skrypnik — bifurcation result (Section 5.9A)

$$-\sum_{i,j=1}^{N} a_{ij}(u(x))\frac{\partial^2 u}{\partial x_i \partial x_j} + \sum_{i=1}^{N} b_i(u(x))\frac{\partial u}{\partial x_i} + c(u(x))u = f \text{ in } \Omega, \quad u = 0 \quad \text{on } \partial\Omega$$

- Linearization of Quasilinear PDEs and Fixed Point Theorem (Theorems 8.3.5 and 8.3.9)

Semilinear Evolution Equations — Partial Differential Equations

$$\boxed{\dot{u}(t) = Au(t) + f(t, u(t)), \qquad u(0) = u_0}$$

- Semigroup approach and Contraction Principle (Proposition 8.2.2)

$$\boxed{\dot{u}(t) + \nabla_H E\,(u(t)) = f(t), \qquad u(0) = u_0}$$

- Approximation proof — Galerkin Method (Theorem 8.2.5)

Comparison of Bifurcation Results Presented in This Book

Bifurcation results presented in this book are based on the following *three basic tools*:

- *Implicit Function Theorem* (Crandall–Rabinowitz Local Bifurcation Theorem — Theorem 4.3.22),
- *Degree Theory* (Krasnoselski Bifurcation Theorem — Theorem 5.8.12, Rabinowitz Global Bifurcation Theorem — Theorem 5.8.22, Dancer Global Bifurcation Theorem — Theorem 5.8.26),
- *Variational Principles* (Krasnoselski Potential Bifurcation Theorem — Theorem 7.8.49)

Below we present a brief discussion of these results and point out differences and links among them.

A bifurcation result based on the *Implicit Function Theorem* provides very precise information about the structure of the set of nontrivial solutions near the bifurcation point — it is expressed in terms of a differentiable curve. Moreover, the result is obtained for "non potential" equations. On the other hand, to verify the assumptions, relatively strong smoothness assumptions on the nonlinearity are required and the information about the set of all nontrivial solutions has only local character. Also, the assumption that the dimension of the kernel of the linear part must be equal to 1 represents a relatively strong restriction.

Bifurcation results based on the *Degree Theory* do not require smoothness of the nonlinearity at all. They allow one to treat "non potential" equations as well and provide also some information about the global structure of the set of nontrivial solutions. On the other hand, the multiplicity of an eigenvalue of the "linear part" must be odd, the set of nontrivial solutions need not be a curve (even in a small neighborhood of the bifurcation point) and its global structure may be unclear if there is no additional information about "higher order terms".

A bifurcation result based on the *Variational Principles* holds for any multiplicity of an eigenvalue of the "linear part". The price to be paid for that consists in the fact that the equation has to possess a potential. It provides only local information about nontrivial solutions and the structure of the set of all nontrivial solutions might be "very wild" in general.

We can summarize the above discussion in the following table.

Bifurcation result	*Basic tool to prove it*	*Multiplicity m of an eigenvalue of the linear part*	*Form of the equation*	*Character of the information*	*Smoothness required*
Crandall–Rabinowitz Theorem (Theorem 4.3.22)	Implicit Function Theorem	$m = 1$	non potential in general	local	twice continuously differentiable in a neighborhood
Krasnoselski Theorem (Theorem 5.8.12)	Degree Theory	$m \geq 1$ m odd	non potential in general	local	continuous
Rabinowitz Theorem (Theorem 5.8.22)	Degree Theory	$m \geq 1$ m odd	non potential in general	global	continuous
Dancer Theorem (Theorem 5.8.26)	Degree Theory	$m = 1$	non potential in general	global	continuous
Krasnoselski Theorem (Theorem 7.8.49)	Lagrange Multiplier Method	$m \geq 1$ arbitrary	potential only	local	twice continuously differentiable at the point

List of Symbols

Sets and spaces

$\mathcal{M} \times \mathcal{N}$	Cartesian product of sets $\mathcal{M}$ and $\mathcal{N}$, 2
$\overline{\mathcal{M}}$	closure of the set $\mathcal{M}$, 25
$\overline{\mathcal{M}}^{w}$	weak closure of the set $\mathcal{M}$, 78
$\partial\mathcal{M}$	boundary of the set $\mathcal{M}$, 26
$\operatorname{int}\mathcal{M}$	interior of the set $\mathcal{M}$, 22
$\exp\mathcal{M}$	set of all subsets of $\mathcal{M}$, 130
$\sup\mathcal{M}$	lowest upper bound (supremum) of the set $\mathcal{M}$, 3
$\inf\mathcal{M}$	greatest lower bound (infimum) of the set $\mathcal{M}$, 3
$\mathbb{N}$	set of all positive integers, 18: $\mathbb{N} = \{1, 2, \dots\}$
$\mathbb{Z}$	set of all integers, 49
$\mathbb{Q}$	set of all rational numbers, 4
$\mathbb{R}$	set of all real numbers, 1
$\mathbb{C}$	set of all complex numbers, 1
$\mathbb{R}^N$, $\mathbb{C}^N$	real, or complex space of dimension $N \in \mathbb{N}$, 1, 12
$\mathbb{\Lambda}$	set of scalars (in general), 1
$\mathcal{M}^{\perp}$, $\mathcal{N}_{\perp}$	nullsets, 12, 70
$\mathcal{M}^{\perp}$	orthogonal complement of the set $\mathcal{M}$, 47
$X^{\#}$	(algebraic) dual space of the linear space X, 10
X^*	dual (adjoint) space of the Banach space X, 55
$\mathcal{M} \perp \mathcal{N}$	orthogonality of the sets $\mathcal{M}$ and $\mathcal{N}$, 43
$[a, b]$	order interval in X, 391: $[a, b] = \{x \in X : a \le x \le b\}$
$\mathcal{U}(x)$, $\mathcal{V}(x)$, ...	neighborhood of the point x, 25
$\{\mathcal{U}_n\}$	covering, 26
$B(a; r)$	open ball centered at the point $a \in X$ with the radius $r > 0$, 25
$\partial B(a; r)$	sphere centered at the point $a \in X$ with the radius $r > 0$, 316
$\mathcal{S}^k$	k-dimensional sphere in $\mathbb{R}^N$, 193
$\mathcal{K}$, $\mathcal{K}^r$	order cone and $\mathcal{K}^r = \{x \in \mathcal{K} : \lVert x\rVert \le r\}$, 400
$X^{*,+}$	set of all positive functionals, 404
$\operatorname{Lin}\mathcal{M}$	span of the elements of the set $\mathcal{M}$, 2
$\operatorname{Co}\mathcal{M}$	convex hull of the elements of the set $\mathcal{M}$, 367
$\mathcal{M} + \mathcal{N}$	set of all $z = x + y$, $x \in \mathcal{M}$, $y \in \mathcal{N}$, 258
$X \oplus Y$	algebraic direct sum of the spaces X and Y, 5
$\dim X$	dimension of the linear space X, 4
$\operatorname{codim} X$	co-dimension of the linear subspace X, 9

$X\vert_Y$	factor space X over Y, 9
$\mathscr{M}$	manifold, 164
$T\mathscr{M}$	tangent bundle, 199
$T^*\mathscr{M}$	cotangent bundle, 199
$H^p(\mathscr{M})$	cohomology group of the manifold $\mathscr{M}$, 217
$H_1(\mathscr{M})$	fundamental group of the manifold $\mathscr{M}$, 218
$\{\alpha_n\}$	partition of unity, 221

Elements

0	zero number in $\mathbb{R}$ or $\mathbb{C}$, 2
o	zero element of a (topological, ..., Banach, Hilbert) space X including $\mathbb{R}^N$ and $\mathbb{C}^N$, 1

Special spaces and classes

l^p	space of all sequences $\{x_n\}_{n=1}^{\infty}$ with $\sum\limits_{n=1}^{\infty} \vert x_n\vert^p < \infty$, 37, 49
l^∞	space of all bounded sequences, 10
c_0	space of all sequences with zero limit and with the sup norm, 60
$L(X,Y)$	space of all linear operators from X into Y, 5
$\mathcal{L}(X,Y)$	space of all linear continuous operators from X into Y, 29
$\mathcal{B}_2(X,Y)$	space of all bilinear continuous operators from x into Y, 139
$C(X,Y)$, $C[a,b]$	space of all continuous maps from X into Y, and continuous functions on the closed interval $[a,b]$, 31, 4
$C^k(X,Y)$	space of all maps from X into Y with continuous derivatives up to the order k, 37
$C^k(\overline{\mathcal{M}})$	space of all functions from $C^k(\mathcal{M})$ for which continuous derivatives up to the order k are bounded in $\mathcal{M}$, 38
$C^{k,\gamma}(\overline{\mathcal{M}})$	space of all γ-Hölder continuous, bounded functions in $\mathcal{M}$ with continuous derivatives up to the order k, 38
$C_0^k(X,Y)$	space of all maps from X into Y with continuous derivatives up to the order k and with compact supports in X, 442
$\mathcal{D}(\mathcal{M})$	class of all infinite differentiable functions on $\mathcal{M}$ which have compact supports lying in $\mathcal{M}$, 36
$BC(X)$	space of all bounded, continuous maps on X, 31
$L^p(\mathcal{M})$	Lebesgue space on $\mathcal{M}$, 33
$L^\infty(\mathcal{M})$	space of all classes of essentially bounded functions, 34
$L^p_{\mathrm{loc}}(\mathcal{M})$	space of all classes of functions which are in L^p on every compact subset of $\mathcal{M}$, 38
$W^{k,p}(\mathcal{M})$, $W_0^{k,p}(\mathcal{M})$	Sobolev space on $\mathcal{M}$, 39, 52
$\mathcal{F}(\mathscr{M})$	set of all differentiable functions on a manifold $\mathscr{M}$, 198
$\mathcal{V}(\mathscr{M})$	set of all smooth vector fields on a manifold $\mathscr{M}$, 201
$T_a\mathscr{M}$	tangent space of the differentiable manifold $\mathscr{M}$ at the point a, 193
$(T_a\mathscr{M})^*$	cotangent space of the differentiable manifold $\mathscr{M}$ at the point a, 198
$\mathrm{CAR}(\mathcal{M})$	class of all Carathéodory functions on $\mathcal{M}$, 136
$\mathcal{H}(\mathcal{M})$	class of all holomorphic functions on a neighborhood of the closed set $\mathcal{M}$, 21
Γ_a	class of all smooth curves, 193
$\Lambda_n(x)$	class of all skew-symmetric n-linear forms, 207

$[f]$	class of mutual homotopic continuous maps containing f, 218
$C^{0,1}$	class of all domains with Lipschitz boundary, 474
$\mathscr{C}(X,Y)$	class of all compact operators from X into Y, 78
$\mathscr{C}_{\mathrm{f}}(X,Y)$	class of all finite-dimensional compact operators from X into Y, 253

Maps, functions and operators

$A^{\#}$	adjoint operator to A, 11
A^*	adjoint operator to A, 69
$\operatorname{Dom} f$	domain of the map f, 51
$\operatorname{Im} f$	range (image) of the map f, set of all values $f(x)$, $x \in \operatorname{Dom} f$, 7
$f\colon X \to Y$	f is a map from X into Y (does not automatically mean that either $\operatorname{Dom} f = X$ or $\operatorname{Im} f = Y$), 26
$\operatorname{Ker} f$	kernel of the linear operator f, null–space of f, 7
$f_{-1}(\mathcal{M})$	set of all preimages, 26: $f_{-1}(\mathcal{M}) = \{x \in \operatorname{Dom} f : f(x) \in \mathcal{M}\}$
f^{-1}	inverse map to the map f, 7
f^+, f^-	positive and negative part of f, respectively, 53: $f^+ = \max\{f, 0\}$, $f^- = \max\{-f, 0\}$, $f = f^+ - f^-$
$f \circ g$	composition of the maps f and g, 131: $f \circ g = f(g)$
O, $\boldsymbol{O}$	zero map and zero matrix, respectively, 18, 16: $Ox = o$, $x \in X$, $\boldsymbol{O}x = o$

Special functions and maps

$\boldsymbol{J}$	Jacobi matrix, 181
J_f	determinant of the Jacobi matrix of the map f (Jacobian), 127
$f'(a)$	Fréchet derivative of the map f at the point a, 131
f_1'	partial Fréchet derivative of f with respect to the first variable, 133
$\delta f(a; v)$, $\delta^k f(a; v_1, \dots, v_k)$	first (and k^{th}) Gâteaux derivative of f at the point a in the direction v (directions $v_1, \dots, v_k$), 126, 139
$Df(a)$, $D^k f(a)$	first (and k^{th}) Gâteaux differential of f at the point a, 126, 139
$D^\alpha f$	partial derivative of f, 37
$D^\alpha_{\mathrm{w}} f$	α-weak derivative of f, 38
∇f	gradient of f, 53
$\operatorname{curl} f$	curl of f, 213: $\operatorname{curl} f = \nabla \times f$
$\operatorname{div} f$	divergence of f, 238: $\operatorname{div} f = \sum\limits_{i=1}^{N} \frac{\partial f_i}{\partial x_i}$
Δf	Laplace operator of f (Laplacian), 156: $\Delta f = \sum\limits_{i=1}^{N} \frac{\partial^2 f}{\partial x_i^2}$, $\Delta f = \operatorname{div}(\nabla f)$
$\Delta_p f$	p-Laplacian of f, 353
$\Delta_{\mathscr{M}} f$	Laplace–Beltrami operator of f on the manifold $\mathscr{M}$, 239
$\mathcal{L}_g f$	directional (Lie) derivative of f (in the direction of the vector field g), 203
$[f, g]$	commutator (Lie bracket) of the vector fields f and g, 207
$\langle x, y\rangle$	duality between the Banach spaces X^* and X, 340, 461: $\langle x, y\rangle = x(y)$, $x \in X^*$, $y \in X$

(x, y)	scalar (inner) product in the linear space X or in $\mathbb{R}^N$, $\mathbb{C}^N$ (we often write $xy := (x, y)_{\mathbb{R}^N}$), 41, 42: $(x,y)_{\mathbb{R}^N} = \sum\limits_{i=1}^{N} x_i y_i$, $x, y \in \mathbb{R}^N$, $(x,y)_{\mathbb{C}^N} = \sum\limits_{i=1}^{N} x_i \overline{y}_i$, $x, y \in \mathbb{C}^N$, respectively
$\|x\|$, $\|x\|_{\mathbb{R}^N} \equiv \|x\|_2$	norm in the normed space X and Euclidean norm in $\mathbb{R}^N$ ($\lvert x \rvert := \|x\|_{\mathbb{R}^N}$ in Chapter 5), respectively, 29: $\|x\|^2_{\mathbb{R}^N} = \sum\limits_{i=1}^{N} x_i^2$, $x \in \mathbb{R}^N$
$x \wedge y$	exterior product in the linear space X, 208
$x \times y$	cross (vector) product in $\mathbb{R}^3$, 211
$X \hookrightarrow Y$, $X \hookrightarrow\hookrightarrow Y$	the space X is continuously (and compactly) embedded into the space Y, 37, 40
$\operatorname{meas} \mathcal{M}$	Lebesgue measure of the set $\mathcal{M}$, 37
$\operatorname{dist}(x, \mathcal{M})$, $\operatorname{dist}(\mathcal{M}, \mathcal{N})$	distance of the point x from the set $\mathcal{M}$, and of the sets $\mathcal{M}$ and $\mathcal{N}$, respectively, 52, 288: $\operatorname{dist}(\mathcal{M}, \mathcal{N}) = \sup\limits_{x \in \mathcal{M}} \inf\limits_{y \in \mathcal{N}} \varrho(x, y)$
$\operatorname{diam} \mathcal{M}$	diameter of the set $\mathcal{M} \subset (X, \varrho)$, 257: $\operatorname{diam} \mathcal{M} = \sup\limits_{x, y \in \mathcal{M}} \varrho(x, y)$
$\chi(\mathcal{M})$	measure of noncompactness, 257
$\operatorname{Re} f$	real part of f, 42
$\operatorname{Re} f$	real part of f, 42
$\operatorname{Im} f$	imaginary part of f, 62
$f_*(v)$	push-forward; the map f pushes forward a tangent vector v, 198
$f^* g$	pull-back; the map f pulls back a differential form g, 200
$\deg(f, \mathcal{M}, a)$	degree of the map f at the point a with respect to the set $\mathcal{M}$, 296
$\operatorname{ind} f$	index of the Fredholm operator f, 71: $\operatorname{ind} f = \dim \operatorname{Ker} f - \dim \operatorname{Ker} f^*$
$\operatorname{Ind}_\gamma a$	index of the point a with respect to the curve γ, 295
$\operatorname{supp} f$	support of the map f, 36
$\sigma(A)$	spectrum (set of all eigenvalues) of the operator A, 15
$\varrho(A)$	resolvent set of the operator A, 56: $\varrho(A) = \mathbb{C} \setminus \sigma(A)$
$r(A)$	spectral radius of a linear operator A, 57
$\omega \rfloor f$	$(M-1)$-form of the vector field f and of the M-form ω, 238

Others

$x := y$, $y =: x$	let x be y; x is defined by y; denote y as x, 4, 144
a.a.	almost all (in the sense of the Lebesgue measure), 71
a.e.	almost everywhere, almost every (in the sense of the Lebesgue measure), 33
$x_n \to x$	strong convergence of the sequence $\{x_n\}_{n=1}^\infty$ to the element x, 26
$x_n \rightharpoonup x$	weak convergence of the sequence $\{x_n\}_{n=1}^\infty$ to the element x, 66
$x_n \overset{*}{\rightharpoonup} x$	weak star convergence of the sequence $\{x_n\}_{n=1}^\infty$ to the element x, 68
$x_n \rightrightarrows x$	uniform convergence of the sequence $\{x_n\}_{n=1}^\infty$ to the element x, 38
$p' = \frac{p}{p-1}$	exponent conjugate to p, 34
$p^* = \frac{Np}{N-kp}$	critical Sobolev exponent, 40
$x > o$, $x \geq o$, $x \gg o$	ordering in the Banach space X, $o, x \in X$, 391

Index

Bibliography

[1] Adams, J. F., *Lectures on Lie Groups*, W. A. Benjamin, Inc., New York, 1969.

[2] Adams, R. A., *Sobolev spaces*, Academic Press, New York, 1975.

[3] Alexander, J. C., A primer on connectivity, pp. 455–483, in: *Fixed Point Theory* (Fadell, E. & Fournier, G., eds.), Lecture Notes in Math. № 886, Springer Verlag, Berlin–Heidelberg–New York, 1981.

[4] Amann, H., *Ordinary Differential Equations. An Introduction to Nonlinear Analysis*, de Gruyter Stud. in Math. № 13, de Gruyter, Berlin–New York, 1990.

[5] Amann, H. & Weiss, S., "*On the uniqueness of the topological degree*", Math. Z. **130** (1973), 39–54.

[6] Ambrosetti, A. & Prodi, G., *A Primer of Nonlinear Analysis*, Cambridge Univ. Press, Cambridge, 1993.

[7] Ambrosetti, A. & Rabinowitz, P., "*Dual variational methods in critical point theory and applications*", J. Funct. Anal. **14** (1973), 349–381.

[8] Anane, A., "*Simplicité et isolation de la premiére valeur propre du p-laplacien avec poids*", C. R. Math. Acad. Sci. Paris **305** (1987), 725–728.

[9] Appell, J. & Zabreiko, P. P., *Nonlinear Superposition Operators*, Cambridge Univ. Press, Cambridge, 1990.

[10] Arcoya, D. & Orsina, L., "*Landesman–Lazer conditions and quasilinear elliptic equations*", Nonlinear Anal. **28** (1997), 1623–1632.

[11] Aubin, J. P. & Ekeland, I., *Applied Nonlinear Analysis*, Wiley, New York, 1984.

[12] Aubin, T., *Some Nonlinear Problems in Riemann Geometry*, Springer Verlag, New York, 1988.

[13] Aubin, T., *A Course in Differential Geometry*, Amer. Math. Soc., Providence, RI, 2000.

[14] Berger, M. & Gostiaux, B., *Differential Geometry: Manifolds, Curves and Surfaces*, Springer Verlag, New York, 1988.

[15] Berkovitz, L. D., *Optimal Control Theory*, Springer Verlag, Berlin–Heidelberg–New York, 1974.

[16] Böhme, R., "*Die Lösung der Verzweigungsproblemen für Nichtlineare Eigenwertprobleme*", Math. Z. **127** (1972), 105–126.

[17] Bourbaki, N., *Éléments de Mathématique*, vol. Livre VI, series *Intégration*, Hermann et Cie, Paris, 1952.

[18] Bourbaki, N., *Groupes et Algèbres de Lie*, Hermann et Cie, Paris, 1975.

[19] Brenner, S. C. & Scott, L. R., *The Mathematical Theory of Finite Element Methods*, 2nd edition, Springer Verlag, Berlin–Heidelberg–New York, 2002.

[20] Brézis, H., *Analyse Functionelle*, Paris, 1983.

[21] Bröcker, T. & Dieck, T., *Representations of Compact Lie Groups*, Springer Verlag, Berlin–Heidelberg–New York, 1985.

[22] Browder, F. E., *Problémes Nonlinéaires*, Université de Montreal, Montreal, 1966.

[23] Browder, F. E., "*Nonlinear elliptic boundary value problems and the generalized topological degree*", Bull. Amer. Math. Soc. **76** (1970), 999–1005.

[24] Cañada, A., Drábek, P., & Fonda, A., (eds.), *Handbook on Differential Equations, Ordinary Differential Equations*, vol. 1, Elsevier, North-Holland, Amsterdam, 2004.

[25] Cartan, H., *Calcul différentiel, Formes différentielles*, Hermann et Cie, Paris, 1967.

[26] Cazenave, T. & Haraux, A., *An Introduction to Semilinear Evolution Equations*, Oxford Science Publ., 1998.

[27] Cesari, L., Hale, J. K., & La Salle, J., (eds.), *Dynamical Systems*, vol. I, Academic Press, New York, 1976.

[28] Chavel, I., *Eigenvalues in Riemann Geometry*, Academic Press, New York, 1984.

[29] Chen, X. & Xu, X., *The scalar curvature flow on S^n-perturbation theorem revisited*, D0I 10. 1007/s 00222-011-0335-6, Invent. Math., 2011.

[30] Chill, R., Fašangová E., *Gradient Systems*, Matfyzpress, Prague, 2010.

[31] Chillingworth, D. R. J., *Differential Topology with View to Applications*, Pitman Publ., Harlow, 1976.

[32] Chow, B., Lu, P., & Ni, L., *Hamilton's Ricci Flow*, Graduate Studies in Mathematics, Amer. Math. Soc., Providence, 2006.

[33] Chow, S.-N., Li, C., & Wang, D., *Normal Forms and Bifurcation of Planar Vector Fields*, Cambridge Univ. Press, Cambridge, 1994.

[34] Citlanadze, E. S., "*Existence theorems for minimax points in Banach spaces*", Tr. Mosk. Mat. Obs. **3** (1953), 235–274, in Russian.

[35] Coddington, E. A. & Levinson, N., *Theory of Ordinary Differential Equations*, McGraw–Hill, Inc., New York–Toronto–London, 1955.

[36] Conway, J. B., *A Course in Functional Analysis*, Springer Verlag, Berlin–Heidelberg–New York, 1990.

[37] Crandall, M. & Rabinowitz, P. H., "*Bifurcation from simple eigenvalues*", J. Funct. Anal. **8** (1971), 321–340.

[38] Dancer, E. N., "*On the structure of solutions of nonlinear eigenvalue problems*", Indiana Univ. Math. J. **23** (1974), 1069–1076.

[39] Davies, B. & Safarov, Y., (eds.), *Spectral Theory and Geometry*, Cambridge Univ. Press, Cambridge, 1999.

[40] de Boor, C., *A Practical Guide to Splines*, Appl. Math. Sci. № 207, Springer Verlag, Berlin–Heidelberg–New York, 2001.

[41] De Coster, C. & Habets, P., The lower and upper solutions method for boundary value problems, in Cañada et al. [24], pp. 69–160.

[42] Deimling, K., *Nonlinear Functional Analysis*, Springer Verlag, Berlin–Heidelberg–New York, 1985.

[43] Diestel, J. & Uhl, J. J., *Vector Measures*, Amer. Math. Soc., Providence, 1977.

[44] Dieudonné, J., *Foundations of Modern Analysis*, Academic Press, New York–London, 1960.

[45] Dold, A., *Lectures on Algebraic Topology*, Springer Verlag, Berlin–Heidelberg–New York, 1982.

[46] Došlý, O., Halflinear differential equations, in Cañada et al. [24], pp. 161–357.

[47] Drábek, P., "*Continuity of Nemyckij's operator in Hölder spaces*", Comment. Math. Univ. Carolin. **16** (1975), № 1, 37–57.

[48] Drábek, P., *Solvability and Bifurcations of Nonlinear Equations*, Pitman Research Notes Math. Ser. № 264, Longman Scientific & Technical, Harlow, 1992.

[49] Drábek, P., Girg, P., & Manásevich, R., "*Generic Fredholm alternative-type results for one dimensional p-Laplacian*", NoDEA Nonlinear Differential Equations Appl. **8** (2001), 285–298.

[50] Drábek, P., Krejčí, P., & Takáč, P., *Nonlinear Differential Equations*, CRC Research Notes Math. № 404, Chapman & Hall/CRC, Boca Raton, FL–London–New York–Washington, DC, 1999.

[51] Drábek, P., Kufner, A., & Nicolosi, F., *Quasilinear Elliptic Equations with Degenerations and Singularities*, de Gruyter Ser. Nonlinear Anal. Appl., de Gruyter, Berlin–New York, 1997.

[52] Dugundji, J., *Topology*, Brown Publ., Dubuque, IA, 1989.

[53] Dunford, N. & Schwartz, J. T., *Linear Operators*, vol. I. General Theory, Intersci. Publ., New York–London–Sydney–Toronto, 1958.

[54] Dunford, N. & Schwartz, J. T., *Linear Operators*, vol. II, Intersci. Publ., New York–London–Sydney–Toronto, 1963.

[55] Edmunds, D. E. & Evans, W. D., *Spectral Theory and Differential Operators*, Oxford Science Publications, Calderon Press, Oxford, 1987.

[56] Ekeland, I., "*Nonconvex minimization problems*", Bull. Amer. Math. Soc. **1** (1979), 443–474.

[57] Elbert, Á., A half-linear second order differential equation, pp. 153–180, in: *Qualitative Theory of Differential Equations*, vol. I and II (Szeged, 1979), Colloq. Math. Soc. János Bolyai, № 30, North-Holland, Amsterdam, 1981.

[58] Engel, K.-J. & Nagel, R., *One-Parameter Semigroups for Linear Evolution Equations*, Springer Verlag, New York–Berlin–Heidelberg, 2000.

[59] Evans, L. C., *Partial Differential Equations*, Amer. Math. Soc., Providence, RI, 1998.

[60] Fabian, M., Habala, P., Hájek, P., Montesinos, V., Pelant, J., & Zizler, V., *Functional Analysis and Infinite-Dimensional Geometry*, CMS Books Math. / Ouvrages Math. SMC № 8, Springer Verlag, Berlin–Heidelberg–New York, 2001.

[61] Feireisl, E., *Dynamics of Viscous Compressible Fluids*, Oxford Univ. Press, Oxford, 2004.

[62] Fitzpatrick, P. M., "*Homotopy, linearization, and bifurcation*", Nonlinear Anal. **12** (1988), 171–184.

[63] Flucher, M., *Variational Problems with Concentration*, Birkhäuser, Basel–Boston–Berlin, 1999.

[64] Folland, G., *A Course in Abstract Harmonic Analysis*, Chapman & Hall/CRC, Boca Raton, FL, 1995.

[65] Fučík, S., *Solvability of Nonlinear Equations and Boundary Value Problems*, D. Reidel Publ., Dordrecht, 1980.

[66] Fučík, S. & Kufner, A., *Nonlinear Differential Equations*, Elsevier, Amsterdam–Oxford–New York, 1980.

[67] Fučík, S. & Nečas, J., "*Ljusternik–Schnirelmann theorem and nonlinear eigenvalue problems*", Math. Nachr. **53** (1972), 277–289.

[68] Fučík, S., Nečas, J., Souček, J., & Souček, V., *Spectral Analysis of Nonlinear Operators*, Lecture Notes in Math. № 346, Springer Verlag, Berlin–Heidelberg–New York, 1973.

[69] Gaines, R. E. & Mawhin, J., *Coincide Degree and Nonlinear Differential Equations*, Lecture Notes in Math. № 568, Springer Verlag, Berlin–Heidelberg–New York, 1977.

[70] Ghoussoub, N., *Duality and Perturbation Methods in Critical Point Theory*, Cambridge Univ. Press, Cambridge, 1993.

[71] Gilbarg, D. & Trudinger, N. S., *Elliptic Partial Differential Equations of Second Order*, Springer Verlag, Berlin–Heidelberg–New York, 2001.

[72] Goebel, K., "*An elementary proof of the fixed-point theorem of Browder and Kirk*", Michigan Math. J. **16** (1969), 381–383.

[73] Greenberg, M., *Lectures in Algebraic Topology*, W. A. Benjamin, Inc., New York, 1967.

[74] Gripenberg, C., Londen, S. O., & Staffans, O., *Volterra Integral and Functional Equations*, Cambridge Univ. Press, Cambridge, 1990.

[75] Hale, J., *Ordinary Differential Equations*, Wiley, New York–London–Sydney–Toronto, 1969.

[76] Halmos, P., *Finite-Dimensional Vector Spaces*, Van Nostrand, Princeton, NJ, 1960.

[77] Hamilton, R. S., "*The inverse function theorem of Nash and Moser*", Bull. Amer. Math. Soc. **7** (1982), 65–222.

[78] Helgason, S., *Differential Geometry, Lie Groups, and Symmetric Spaces*, Academic Press, New York, 1978.

[79] Henry, D., *Geometric Theory of Semilinear Parabolic Equations*, Lecture Notes in Math. № 840, Springer Verlag, Berlin–Heidelberg–New York, 1981.

[80] Hirsch, M. W., *Differential Topology*, Springer Verlag, Berlin–Heidelberg–New York, 1976.

[81] Hlaváček, I. & Nečas, J., *Mathematical Theory of Elastic and Elastoplastic Bodies: An Introduction*, Elsevier, Amsterdam, 1981.

[82] Izé, J. A., *Bifurcation Theory for Fredholm Operators*, Mem. Amer. Math. Soc. № 174, Amer. Math. Soc., Providence, RI, 1971.

[83] Kac, M., "*Can one hear the shape of a drum?*", Amer. Math. Monthly **73** (1966), 1–23.

[84] Kaczmarz, S. & Steinhaus, H., *Theorie der Orthogonalreihen*, Monografje Matematyczne, Pańštwo Wydawnistwo Naukowe, Warszawa–Lwow, 1935.

[85] Kakutani, S., "*Some characterization of Eucledian space*", Japan. J. Math. **16** (1939), 93–97.

[86] Kantorovich, "*Functional analysis and applied mathematics*", Uspekhi Mat. Nauk **3** (1948), № 3, 89–185 (in Russian).

[87] Kato, T., *Perturbation Theory for Linear Operators*, Springer Verlag, Berlin–Heidelberg–New York, 1966.

[88] Katok, A. & Hasselblatt, B., *Introduction to the Modern Theory of Dynamical Systems*, Cambridge Univ. Press, 1995.

[89] Kelley, J. L., *General Topology*, Van Nostrand, Princeton, NJ, 1957.

[90] Kittel, Ch., Knight, W. D., & Ruderman, M. A., *Mechanics, Berkeley Physics Course*, vol. I, McGraw–Hill, Inc., New York–San Francisco–Toronto–London, 1965.

[91] Kosniowski, C., *A First Course in Algebraic Topology*, Cambridge Univ. Press, Cambridge, 1980.

[92] Krasnoselski, M. A., *Topological Methods in the Theory of Nonlinear Integral Equations*, Pergamon, Oxford, 1964.

[93] Krasnoselski, M. A. & Zabreiko, P. P., *Geometric Methods of Nonlinear Analysis*, Springer Verlag, Berlin–Heidelberg–New York, 1984.

[94] Krawcewicz, W. & Wu, J., *Theory of Degrees with Applications to Bifurcations and Differential Equations*, Wiley, New York, 1997.

[95] Křížek, M. & Neitaanmäki, P., *Finite Element Approximation of Variational Problems and Applications*, Pitman Mon. Surv. Pure Appl. Math. № 50, Longman Scientific & Technical, Harlow, 1990.

[96] Kufner, A., John, O., & Fučík, S., *Function Spaces*, Academia and Noordhoff, Prague and Leyden, 1975.

[97] Kuratowski, K., *Topology II*, Academic Press, New York, 1968.

[98] Landesman, E. N. & Lazer, A. C., "*Nonlinear pertubations of linear elliptic boundary value problems at resonance*", J. Math. Mech. **19** (1970), 609–623.

[99] Leinfelder, H. & Simader, C. G., "*Schroedinger operators with singular magnetic vector fields*", Math. Z. **176** (1981), 1–19.

[100] Leray, J. & Lions, J.-L., "*Quelques résultats de Višik sur les problémes elliptiques nonlinéaires par les méthodes de Minty–Browder*", Bull. Soc. Math. France **93** (1965), 97–107.

[101] Lindqvist, P., "*On the equation* $\operatorname{div}\big(|\nabla u|^{p-2}\nabla u\big) + \lambda|u|^{p-2}u = 0$", Proc. Amer. Math. Soc. **109** (1990), 157–164.

[102] Lions, P. L., "*The concentration–compactness principle in the calculus of variations. The locally compact case I and II*", Ann. Inst. H. Poincaré Anal. Non Linéaire **1** (1984), 109–145 and 223–283.

[103] Lions, P. L., "*The concentration–compactness principle in the calculus of variations. The limit case I*", Rev. Mat. Iberoamericana **1** (1985), 145–201.

[104] Lions, P. L., "*The concentration–compactness principle in the calculus of variations. The limit case II*", Rev. Mat. Iberoamericana **2** (1985), 45–121.

[105] Lions, P. L., *Mathematical Topics in Fluid Dynamics I*, Oxford Science Publications, Oxford, 1996.

[106] Lunardi, A., *Analytic Semigroups and Optimal Regularity in Parabolic Problems*, Birkhäuser, Basel, 1995.

[107] Lusternik, L. & Sobolev, V., *Elements of Functional Analysis*, "Nauka", Moscow, 1965 (in Russian; English translation published in Wiley, New York 1974).

[108] Mawhin, J., *Topological Degree Methods in Nonlinear Boundary Value Problems*, Regional Conference Series in Mathematics № 40, Amer. Math. Soc., Providence, RI, 1979.

[109] Mawhin, J., "*A simple approach to Brouwer degree based on differential forms*", Adv. Nonlinear Stud. **4** (2004), 535–548.

[110] Maz'ja, V. G., *Sobolev spaces*, Springer Verlag, Berlin–Heidelberg–New York, 1985.

[111] Michlin, S. G., *Variationsmethoden in der Mathematischen Physik*, Deutsche Verlag der Wissenschaften, Berlin, 1960.

[112] Milnor, J., *Topology from the Differentiable Viewpoint*, Univ. of Virginia Press, Charlottesville, 1965.

[113] Milota, J. & Petzeltová, H., "*An existence theorem for semilinear functional parabolic equations*", Časopis Pěst. Mat. **110** (1985), 274–288.

[114] Morgan, J. & Tsz-Ho Fong, F., *Ricci Flow and Geometrization of 3-Manifolds*, University Lecture Series 53, Amer. Math. Soc., Providence, 2010.

[115] Moser, J., "*A rapidly convergent iteration method and nonlinear partial differential equations I, II*", Ann. Scuola Norm. Sup. Pisa Cl. Sci. **20** (1966), 265–315, 499–535.

[116] Nash, J., "*The imbedding problem for Riemann manifolds*", Ann. of Math. **63** (1956), 20–63.

[117] Nečas, J., *Les méthodes directes en théorie des équations elliptiques*, Masson et Cie, Paris, 1967.

[118] Nirenberg, L., *Topics in Nonlinear Functional Analysis*, New York Univ., Courant Inst. Math. Sci., New York, 1974.

[119] Palis, J. & De Melo, W., *Geometry Theory of Dynamical Systems*, Springer Verlag, Berlin–Heidelberg–New York, 1982.

[120] Pazy, A., *Semigroups of Linear Operators and Applications to Partial Differential Equations*, Springer Verlag, Berlin–Heidelberg–New York, 1983.

[121] Protter, M. H. & Weinberger, H. F., *Maximum Principle in Differential Equations*, Prentice Hall, Englewood Cliffs, NJ, 1967.

[122] Rabinowitz, P. H., A global theorem for nonlinear eigenvalue problems and applicatons, pp. 11–36, in: *Contribution in Nonlinear Functional Analysis* (Zarantonello, E. H., ed.), Academic Press, New York, 1971.

[123] Rabinowitz, P. H., "*Some global results for nonlinear eigenvalue problems*", J. Funct. Anal. **7** (1971), 487–513.

[124] Rabinowitz, P. H., *Minimax Methods in Critical Point Theory with Applications to Differential Equations*, Amer. Math. Soc., Providence, RI, 1986.

[125] Reed, M. & Simon, B., *Methods of Modern Mathematical Physics*, vol. IV, Academic Press, New York, 1978.

[126] Rektorys, K., *Variational Methods in Mathematics, Science and Engineering*, 2nd edition, D. Reidel Publ., Dordrecht, 1990.

[127] Robinson, D. W., *Elliptic Operators and Lie Groups*, Oxford Univ. Press, Oxford, 1991.

[128] Rockefellar, R. T., *Convex Analysis*, Princeton Univ. Press, Princeton, NJ, 1970.

[129] Rosenberg, S., *The Laplacian on a Riemann Manifold*, London Math. Soc., London, 1991.

[130] Rothe, E. H., *Introduction to Various Aspects of Degree Theory in Banach Spaces*, Amer. Math. Soc., Providence, RI, 1986.

[131] Rudin, W., *Functional Analysis*, McGraw–Hill, Inc., New York, 1973.

[132] Rudin, W., *Real and Complex Analysis*, McGraw–Hill, Inc., New York, 1974.

[133] Ruelle, D., *Elements of Differentiable Dynamics and Bifurcation Theory*, Academic Press, New York, 1989.

[134] Runst, T. & Sickel, W., *Sobolev Spaces of Fractional Order, Nemytskij Operators and Nonlinear Partial Differential Equations*, De Gruyter Series in Nonlinear Analysis and Applications № 3, de Gruyter, Berlin–New York, 1996.

[135] Saaty, T. L., *Modern Nonlinear Equations*, McGraw–Hill, Inc., New York–Toronto–London–Sydney, 1967.

[136] Sard, A., "*The measure of the critical set values of differential mappings*", Bull. Amer. Math. Soc. **48** (1942), 883–890.

[137] Schatten, R., *Norm Ideals of Completely Continuous Operators*, Springer Verlag, Heidelberg, 1960.

[138] Schwartz, J. T., *Nonlinear Functional Analysis*, Gordon & Breach, New York–London–Paris, 1969.

[139] Sehgal, V. M., "*A fixed point theorem for mapping with a contractive iterate*", Proc. Amer. Math. Soc. **23** (1969), 631–634.

[140] Sell, G., You Y., *Dynamics of Evolutionary Equations*, Springer Verlag, Berlin–Heidelberg–New York, 2002.

[141] Singer, I., *Best Approximation in Normed Linear Spaces by Elements of Linear Subspaces*, Springer Verlag, Berlin–Heidelberg–New York, 1970.

[142] Skrypnik, I. V., *Nonlinear Elliptic Boundary Value Problems*, Teubner, Leipzig, 1986.

[143] Sokal, A.D., "*A really simple elementary proof of the uniform boundedness theorem*", Amer. Math. Monthly **118** (2011), 450–452.

[144] Spanier, E., *Algebraic Topology*, McGraw–Hill, Inc., New York, 1966.

[145] Stein, E. M., *Singular Integrals and Differentiability Properties of Functions*, Princeton Univ. Press, Princeton, NJ, 1970.

[146] Sternberg, S., *Lectures on Differential Geometry*, Prentice Hall, Englewood Cliffs, NJ, 1964.

[147] Stoer, J. & Bulirsch, R., *Introduction to Numerical Analysis*, 3rd edition, Texts in Applied Mathematics № 12, Springer Verlag, Berlin–Heidelberg–New York, 2002.

[148] Takáč, P., "*A short elementary proof of the Krein–Rutman theorem*", Houston J. Math. **20** (1994), № 1, 93–98.

[149] Taylor, M., *Partial Differential Equations*, vol. I, series *Basic Theory*, Springer Verlag, Berlin–Heidelberg–New York, 1996.

[150] Temam, R., *Infinite Dimensional Dynamical Systems in Mechanics and Physics*, Springer Verlag, Berlin–Heidelberg–New York, 1988.

[151] Temam, R., *Navier-Stokes Equations and Nonlinear Functional Analysis*, SIAM, Philadelphia, 1995.

[152] Thorpe, J. A., *Elementary Topics in Differential Geometry*, Springer Verlag, New York, 1979.

[153] Triebel, H., *Theory of Function Spaces*, vol. I, Birkhäuser, Basel, 1983.

[154] Triebel, H., *Theory of Function Spaces*, vol. II, Birkhäuser, Basel, 1992.

[155] Vejvoda, O. et al., *Periodic Solutions of Partial Differential Equations: Time Periodic Solutions*, Sijthoff Noordhoff, The Netherlands, 1981.

[156] Walter, W., *Ordinary Differential Equations*, Graduate Texts in Mathematics, Springer Verlag, Berlin–Heidelberg–New York, 1998.

[157] Whitney, H., "*A function not constant on a connected set of critical points*", Duke Math. J. **1** (1935), 514–517.

[158] Whitney, H., *Geometric Integration Theory*, Princeton Univ. Press, Princeton, NJ, 1957.

[159] Willem, M., *Minimax Theorems*, Birkhäuser, Boston–Basel–Berlin, 1996.

[160] Yamabe, H., "*On deformation of Riemann structures on compact manifolds*", Osaka J. Math. **12** (1960), 21–37.

[161] Yosida, K., *Functional Analysis*, Springer Verlag, Berlin–Heidelberg–New York, 1965.

[162] Zeidler, E., "*Zur Methode der monotonen Minoranten von Krasnoselskii*", Math. Nachr. **49** (1971), 69–83.

[163] Zeidler, E., *Nonlinear Functional Analysis and Its Applications*, vol. I, II/A, II/B, III and IV, Springer Verlag, Berlin–Heidelberg–New York, 1986.

[164] Ziemer, W. P., *Weakly Differentiable Functions: Sobolev Spaces and Functions of Bounded Variation*, Springer Verlag, Berlin–Heidelberg–New York, 1989.

MIX
Papier aus verantwortungsvollen Quellen
Paper from responsible sources
FSC® C105338
FSC
www.fsc.org

If you have any concerns about our products,
you can contact us on
ProductSafety@springernature.com

In case Publisher is established outside the EU,
the EU authorized representative is:
Springer Nature Customer Service Center GmbH
Europaplatz 3, 69115 Heidelberg, Germany

Printed by Libri Plureos GmbH
in Hamburg, Germany